An Introduction to Genetics

DAVID J. MERRELL

THE UNIVERSITY OF MINNESOTA

An Introduction to Genetics

W · W · NORTON & COMPANY · INC · NEW YORK

Library of Congress Cataloging in Publication Data
Merrell, David J
   An introduction to genetics.
   Includes bibliographies and index.
   1. Genetics. I. Title.
QH430.M47   1975   575.1   75-1076
ISBN 0-393-09247-X
Published simultaneously in Canada
by George J. McLeod Limited, Toronto
This book was designed by Robert Freese.
The typefaces are Caledonia and Americana Bold,
set by Fuller Typesetting of Lancaster.
Printed in the United States of America

5   6   7   8   9   0

To Jessie

Contents

Preface

Since innumerable textbooks of genetics have already been written, just about every conceivable title has already been preempted. Faced with a choice between a unique but obscure title or a more informative one, I finally settled on *An Introduction to Genetics,* because that is, in fact, the purpose of this book: to introduce students to the major genetic principles and concepts and to some of the implications and applications of these concepts to man and other species.

Every teacher has his own preferences about the way he presents his subject. In genetics the two major approaches appear to be the molecular and the Mendelian. There is a logical beauty in the molecular approach, which involves presenting the nature of the DNA molecule first and then showing how it is the basis for mutation, replication, recombination, and gene action. Despite the appeal of this approach, I have chosen the more traditional Mendelian approach to the subject. One reason is that this is the way genetics actually developed. Another is that the diversity of backgrounds among students taking an introductory genetics course is usually so great that to throw them at once into an alien world of molecules, bacteria, and viruses can be rather traumatic. Still another is to give them a chance to appreciate the power and sophistication of the genetic methodology developed prior to the advent of molecular genetics. This methodology, culminating, for example, in the work of H. J. Muller, is still basic to much of the work in genetics.

The historical approach has its pitfalls. Science students are notoriously uninterested in the history of who did what, when, and where. However, once they realize that these discoveries did not just happen, but were made by some quite fascinating people, many of whom are still living, a new dimension is added to their interest in genetics. It is somewhat of a puzzle that no one would think of studying English or philosophy without learning something about the major authors whose works they ponder, yet they willingly remain biological illiterates, ignorant of the leaders in a great intellectual breakthrough. The history is included, not as a source of exam questions with which to plague students, but as a necessary part of the record. Undoubtedly some persons whose work deserves recognition have been omitted—still another pitfall in a historical treatment.

In some ways learning genetics is like learning a foreign language. A whole new vocabulary must be mastered in order to speak the language of genetics. Every effort has been made to introduce new terms gradually, defining them when first used and using them in context, but a glossary is also available for quick reference. In other ways, learning genetics resembles learning a mathematical game, with one difference. The rules of a mathematical game can be readily changed, but the rules of genetics, the basic principles of heredity, are not so easily modified. We must learn to play the game according to the rules laid down in nature; we cannot change them to suit our fancy. For this reason laboratory experience in making crosses and observing progeny is invaluable in the study of genetics. In this way students best develop an appreciation of both the power and the limits of genetic methods, and realize that genetics is not a game to be played on a checkerboard.

The levels of achievement reached by genetics students include a general understanding of basic principles, the application of these principles to the solution of problems, the execution of experiments under direction, and the ability to design and carry on original research. The questions at the ends of the chapters give students a chance to determine whether they have reached the second level of achievement. Further development of the budding geneticist usually comes in the laboratory, since genetics is very much an experimental discipline.

Because the interpretation of experimental data depends on statistical analysis, statistics is an essential part of any introduction to genetics. The problem is how much statistics is necessary or appropriate. The usual solution is a cookbook approach, giving only the bare minimum of formulas needed to calculate probabilities or tests of significance. This approach is often unsatisfying or frustrating to the student, and frequently leads to a subsequent antipathy toward statistics. However, limitations of time and the fact that statistics is peripheral to the main objectives of a genetics course ordinarily lead to a very perfunctory treatment of statistics. Nonetheless, both genetics and statistics deal with variability, and statistical methods are essential in treating genetic variability. Concepts such as the mean, standard deviation, standard error, variance, correlation, regression, and heritability crop up regularly in the genetics literature so the student needs some handy source of reference. For this reason the statistical appendix has been included. It includes more material than can feasibly be covered in a genetics course, but an attempt is made to develop the material in such a way that it can be used not only for reference but also by the more inquisitive student to satisfy his desire for an understanding of the fundamentals of statistics.

The first three parts of the text, on classical genetics, the nature of the gene, and the genetics of populations, constitute, more or less, the usual subject matter for a beginning course in genetics and can stand alone. The last part, on genetics and man, is designed to show how the principles developed earlier are being applied to problems ranging from disease resistance to pollution and the education of children. This material is included because I regard genetics not as just a biological discipline but as an essential part of a liberal education. The chapter on the origin and evolution of domesticated animals and plants, for example, is not designed primarily for students in agriculture, although they may find it useful, but for students in the liberal arts, most of whom are ignorant of how recent and how tenuous is the dependence of our society on agriculture. Similarly, the material on human genetics is not designed primarily for premedical students, even though the training of medical students in the United States is generally deficient in genetics. It is hoped that this material will lead students to realize the increasing importance of genetics to medicine. The more basic purpose is to give each student some appreciation of his biological nature. The social, cultural, and environmental aspects of human society receive so much emphasis in undergraduate education that a biological counterbalance is needed. Although I hesitate to use the overworked, abused, and misused word "relevance," the last part of the book is an attempt to point up the relevance of genetics to some significant aspects of human affairs. Although some of the subjects are controversial, that hardly seems sufficient reason to ignore them, since genetics now has reached the point where it can contribute to the understanding and the solution of many of our problems.

January 1975   D. J. M.

Acknowledgments

I should like to acknowledge the indispensable help of my wife, Jessie Clark Merrell, to whom this book is dedicated. She patiently typed and retyped the various drafts and helped with the seemingly endless details of tying together the loose ends of the manuscript. Her constant support and encouragement were equally important to the completion of the project.

The efforts of Mrs. Marilyn L. Steere, the artist, were greatly appreciated. Although carrying an extremely heavy schedule, she was always cheerful and unflappable in meeting the many demands on her time. David C. Dapkus read the entire manuscript and made numerous perceptive and useful comments. Portions of the manuscript were read by David B. Gray, Donald B. Lawrence, Robert G. McKinnell, Charles F. Rodell, and James C. Underhill. Their thoughtful suggestions were most helpful, but I, of course, am responsible for any errors of commission or omission. It has been a pleasure to work with the editors at W. W. Norton & Co., Inc. for they have been unfailingly patient and understanding during the book's long gestation period.

I am indebted to the Literary Executor of the late Sir Ronald A. Fisher, F.R.S., to Dr. Frank Yates, F.R.S., and to Longman Group Ltd., London, for permission to reprint Tables 4, 5 and 7 from their book *Statistical Tables for Biological, Agricultural and Medical Research*. William D. Schmid very kindly reviewed the material on statistics.

Finally, I wish to acknowledge the influence of the hundreds of students to whom, over the years, I have attempted to convey some of the principles and fascination of genetics. Teaching is a two-way process, and I am sure that their influence is reflected in many ways throughout the book.

PART ONE

CLASSICAL GENETICS

CHAPTER ONE

Heredity and Variation

That human infants are recognizably human never seems to surprise their parents. Nor do we find it remarkable that dogs have puppies or even that bulldogs produce bull pups and Scotties have Scotty pups. The tendency for offspring to resemble their parents is such a well-established fact that such expressions as "Like father, like son," "It runs in the family," "Blood will tell," and many others are in common usage. Recognition of the validity of this observation can be seen in the prices often paid for pedigreed but untested horses and cattle. The similarity between parents and offspring is thus not generally regarded as at all remarkable. However, once the tenuous nature of the physical bridge between generations (Figs. 1.1 and 1.2) is realized, this fact may seem more noteworthy. A primary problem for the student of heredity has always been to find an explanation for the similarities between parents and offspring, to explain how the traits of the parents are transmitted through the microscopic sperm and and egg cells to their progeny.

Ordinarily the children of a particular set of parents show some traits from one parent and some from the other. They are never exact duplicates of either parent, or of each other (although identical twins, with identical sets of hereditary factors, come close). Often a child will show traits not seen in either parent. For instance, it may come as quite a surprise to a brown-haired, brown-eyed couple to find that they have produced a

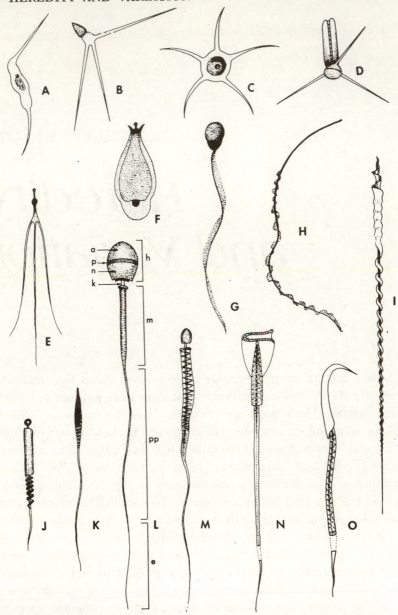

Fig. 1.1. The physical bridge between generations—types of animal sperm. A.–E. Crustacea. F. Nematode. G. Fish. H. Amphibian. I., J. Birds. K. Snake. L.–O. Mammals. A badger sperm is labeled: a, acrosome; p, perforatorium; n, nucleus; h, head; k, neck; m, middle piece; pp, principal piece; e, endpiece; m + pp + e = sperm tail. (Brown, W. V., and E. M. Bertke, 1969. *Textbook of Cytology.* C. V. Mosby Co., St. Louis, Mo., p. 530; redrawn from various 19th-century zoologists.)

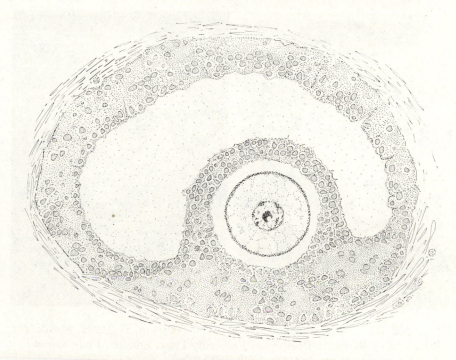

Fig. 1.2. The physical bridge between generations—an ovum within a Graafian follicle in the ovary of a cat.

blue-eyed, red-haired baby. Therefore, another problem for the geneticist is to explain the recombination of parental traits and also to account for the origin of new characteristics not seen in the parents. Thus, the science of heredity or genetics can be regarded as an investigation of the causes for the similarities and differences between parents and offspring.

1.1. The Origin of Variation

Variation had certainly long been observed, but never understood, when Francis Bacon, in his *Novum Organum*, urged the importance of the study of the nature and cause of biological variation. Centuries passed, but the explanation remained obscure. Charles Darwin based his theory of natural selection on the differential survival and reproduction of individuals varying to greater or lesser degree from one another. He recognized that a major drawback to his theory was the lack of an adequate

Fig. 1.3. (left). Gregor Mendel, 1822–1884. (Bateson, W., 1902. *Mendel's Principles of Heredity*. Cambridge University Press, London, Frontispiece.) Fig. 1.4. (right). Hugo de Vries, 1848–1935.

theory of heredity, which he tried to supply with his erroneous theory of pangenesis. This Lamarckian type of theory proposed that pangenes were assembled from all parts of the body in the gametes (the sperm and egg cells), but evidence favoring this theory has never been found.

Although Maupertius had proposed a particulate theory of heredity in the eighteenth century, the basic principles of heredity were discovered by Gregor Mendel (Fig. 1.3). In 1865 he presented the results of his hybridization experiments with the garden pea to the Natural History Society in Brünn, Czechoslovakia, which published them the following year. Darwin, who might have appreciated Mendel's work, apparently never became aware of it, and those who did failed either to understand or to appreciate its significance. Mendel postulated the existence of particulate factors (later called genes) in the gametes to account for the behavior of the traits he studied in crosses. It was not until 1900, when three botanists, Correns in Germany, de Vries in the Netherlands, and von Tschermak in Austria, independently confirmed Mendel's results and unearthed his paper, that the importance of Mendel's contribution was finally recognized. Genetics as a modern science thus had its beginnings in 1900 with the rediscovery of Mendel's paper.

Even before 1900, Bateson in England had become interested in the study of variation, and after 1900 he became a leading exponent of Mendelism. As a consequence of the work of the early Mendelians, it soon became clear that Mendelian recombination was responsible for much of the biological variation that was observed. In less than 20 years Morgan and his students showed that the Mendelian factors were located on the chromosomes in the nucleus of the cell and thus established the physical basis of heredity.

de Vries (Fig. 1.4), however, found some hereditary variations that did not seem to be explicable by genetic recombination, and as a result in 1901 proposed his mutation theory. This theory postulated that the appearance of new hereditary variations is the result of changes in the hereditary material itself, rather than merely a consequence of recombination.

A third major cause of variation is the effect of the environment. Lamarck, in 1809, suggested that evolutionary changes occurred because environmental effects on an organism could become hereditary. In this way the physiological adaptations of an individual could be transmitted to his descendants. This theory of the inheritance of acquired characteristics, as it is called, seems the simplest, most obvious explanation for the remarkable variety and precision of the adaptations of living things to their environments. Unfortunately, perhaps, convincing evidence for the theory is extremely difficult to find. In some parts of the world women have been piercing their ears for centuries, but the process still must be repeated each generation. The tails of some breeds of sheep and dogs are routinely docked, but the acquired short tail has never become hereditary. Weismann, in the late nineteenth century, did a controlled experiment to test the Lamarckian hypothesis, but even though (like the farmer's wife) he cut off the tails of mice for 22 generations, each new generation continued to produce a typical long tail. As mentioned above, Darwin's theory of pangenesis had Lamarckian implications, but no Lamarckian theory has ever been substantiated experimentally. Nevertheless, environmental influences may be a major cause of variation even though these variations are not hereditary.

1.2. Genotype and Phenotype

The Danish botanist, Johannsen, helped to clarify the relationship between hereditary and environmental variation. It was he who coined the terms *genotype* and *phenotype* and made clear the distinction between

Fig. 1.5. Identical twins = identical genotypes. Tweedledum and Tweedle-
dee from Lewis Carroll's *Through the Looking Glass.*

them. The genotype of an individual is the sum total of all of the hereditary material or genes he has received from his parents and is capable
of transmitting to his progeny. The phenotype is the sum total of all
of the traits of the individual, from biochemical to behavioral, resulting

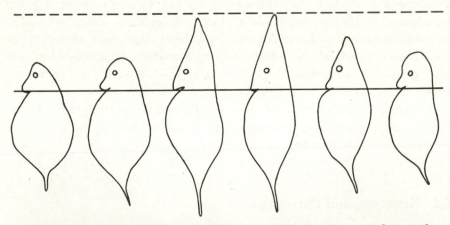

Fig. 1.6. Seasonal environmental effects on similar genotypes produce strikingly different phenotypes. Morphological cycle of head height and shell
length in *Hyalodaphnia.* Collection dates of specimens, left to right: June 3,
June 28, July 30, Sept. 15, Oct. 18, Jan. 3. (After Woltereck.)

from the interaction between the genotype and the environment. Thus, a person with one gene for red hair and one for nonred will have phenotypically nonred hair. However, since his genotype contains a factor or gene for red hair as well as one for nonred hair, he can transmit either gene to his offspring with equal frequency. When genotypes are alike, phenotypes will also be alike if the environmental conditions are the same. The extreme case of identical genotypes is that of identical twins, whose close resemblance to one another is often noted (Fig. 1.5).

If the environment is not constant, individuals identical in heredity may not be alike in phenotype and may in reality differ from one another to a marked degree (Fig. 1.6). This fact should make clear that traits as such are not inherited.

The sum total of the genes in the fertilized egg or zygote carries coded information that determines the course of development for that individual. The genotype determines whether the zygote becomes a man, a mouse, a moose, or a tomato plant. It sets limits within which the course of development flows. The exact course that is followed and the ultimate adult phenotype produced will be influenced and modified by the environmental conditions prevailing during development. Thus a single genotype may give rise to a wide range of phenotypes. If, for instance, a single plant is propagated asexually by cuttings, and each shoot is reared under different conditions, the plants, although all derived from one genotype, may appear markedly different (Fig. 1.7). However, they will still be recognizable as members of the same species; the range of possible reactions is not unlimited.

Furthermore, the phenotypic modifications that occur usually represent an adaptive response by the organism to the environmental conditions under which it develops. A lowland plant, for example, grown in the mountains under alpine conditions, generally assumes the growth form of a native alpine species if it survives at all.

1.3. The Meaning of Genetics

The essence of genetics, then, is the study of heredity and variation. Although Mendelian principles have been shown to apply in a wide array of plants, animals, and microorganisms, the bulk of the work has been carried out with relatively few species. Before you have mastered the principles of genetics you may well become satiated with discussions centering on peas and corn, fruit flies (*Drosophila*), and the common red

Fig. 1.7. Environmental effects on identical genotypes. Dandelions (*Tarax-acum*) derived from a single plant grown in a lowland (left) and an alpine (right) garden. (After Bonnier.)

bread mold (*Neurospora*). However, many of the major advances in genetics were made with these few organisms, to which must be added, more recently, bacteria and the bacterial viruses or bacteriophages.

When you discover that genetics deals with such seemingly innocuous subjects as fruit flies and corn, you may wonder that during the 1930's both genetics and geneticists came under virulent attack in the Soviet Union for promoting "the Mendelian-Morganian capitalistic science." Only recently has genetics started to emerge from the eclipse it underwent in Russia following these attacks. Genetics was apparently banned because the principles of genetics apply to man as well as to other species, and the findings of the geneticists seemed to contradict dialectical materialism, the governing philosophy in the Soviet Union. In essence, the Michurinism promoted as an alternative to Mendelian genetics was a Russsian version of Lamarckianism and was felt to be more compatible with the dominant political philosophy. The attack thus was political rather than scientific, and was made, not because fatal flaws were found

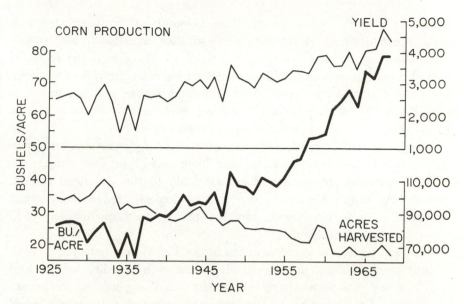

Fig. 1.8. Corn production in the United States. (Data to 1957 from *Historical Statistics of the United States. Colonial Times to 1957,* U.S. Dept. of Commerce, Bureau of the Census. Subsequent data from *Agricultural Statistics, 1969,* U.S. Dept. of Agriculture.)

in genetic principles, but because they did not meet the test of conformity with Marxism. This tragic error has had sad consequences not only for the Russian scientists involved but also for the Russian people, because of the ensuing adverse effects on Soviet agriculture.

The benefits to agriculture from the application of genetic principles can be demonstrated most readily by the development of hybrid corn for commercial use in the United States. The basic discoveries came as the result of the work of G. H. Shull, E. M. East, and D. F. Jones, of Princeton, Harvard, and Yale Universities, respectively. In the 1920s about 100 million acres were planted to corn each year, with an average yield of about 25 bushels per acre for a total production of some 2½ billion bushels per year. Hybrid corn was first grown commercially in the 1930s, and the percentage of the field-corn acreage planted to hybrid seed has gradually increased to about 99%. During this time the acreage planted has actually declined to about 65 million acres, total production has increased to some 4½ billion bushels, and the yield per acre is more than 75 bushels per acre (Fig. 1.8). Therefore, the total yield has nearly doubled while the acreage planted has decreased by one-third and the yield per acre has tripled. It has been estimated that the value of the increase in the yield of corn for a single year is greater than the total cost of federal research in plant improvement since 1900. Some of this increase is attributable to improved agricultural practices, but much of the gain is the result of controlled genetic crosses to obtain maximum hybrid vigor. (As a footnote, Henry A. Wallace, the former Vice President whose political image is generally that of an impractical visionary, made a fortune as a pioneer in the commercial development of hybrid corn.)

Although hybrid corn has been one of the more spectacular advances, equally impressive improvements have been made in the varieties of cereals, vegetables, fruits, and fibers available to the American farmer. Furthermore, improvements in the various domesticated mammals and birds, and even in honeybees, have led to improved quality and yield in animal products as well.

The recent history of genetics in the Soviet Union may help you to realize that genetics is not just a science dealing with heredity in corn and fruit flies, but has far-reaching implications and applications in human affairs. The scientist, unlike the historian or philosopher, is frequently asked the practical value of his research, since the success of a science is often measured by this standard. If so, the contributions of genetics to human welfare already loom large, and the future seems likely to hold even greater practical benefits.

Equally as significant, or perhaps even more so, however, are the contributions of genetics to man's understanding of himself and of the living world of which he is a part. A characteristic of all living things is their power of self-duplication. Advances in genetics in the last two decades have led to a notable increase in our knowledge of this process.

It is more than a century since Darwin's *On the Origin of Species* led to widespread acceptance of the concept of evolution. Today genetics, and specifically population genetics, provides an understanding of the mechanism of evolution, Thus, simply because all living things reproduce and are the products of the evolutionary process, genetics is of great scientific, theoretical, and even philosophical interest, entirely aside from its practical aspects. Man is an animal species, a mammal, and a primate, and Mendel's laws apply to him as well as to other species. He has evolved from some more primitive primate and is still continuing to evolve. Man may be studied in various ways: in relation to his environment, in relation to the development of culture and civilization, or as a biological species. Genetics, of course, contributes primarily to our understanding of man's biological nature. However, unless we first gain some insight into man's biological nature, we can hardly hope to appreciate human limitations or potentialities in the other dimensions of his existence.

Additional Reading

Bateson, W., 1894. *Materials for the Study of Variation.* Macmillan, London.

Carlson, E. A., 1966. *The Gene: A Critical History.* Saunders, Philadelphia.

Dunn, L. C., ed., 1951. *Genetics in the 20th Century.* Macmillan, New York.

Dunn, L. C., 1965. *A Short History of Genetics.* McGraw-Hill, New York.

Gabriel, M. L., and S. Fogel, eds., 1955. *Great Experiments in Biology.* Prentice-Hall, Englewood Cliffs, N.J.

Huxley, J., 1949. *Soviet Genetics and World Science.* Chatto and Windus, London.

Iltis, H., 1932. *Life of Mendel,* E. and C. Paul, trs. Norton, New York.

Medvedev, Z. A., 1969. *The Rise and Fall of T. D. Lysenko,* I. M. Lerner, tr. Columbia University Press, New York.

Peters, J. A., ed., 1959. *Classic Papers in Genetics.* Prentice-Hall, Englewood Cliffs, N.J.

Sirks, M. J., and C. Zirkle, 1964. *The Evolution of Biology.* Ronald, New York.

Stern, C., ed., 1950. "The Birth of Genetics," *Genetics* **35** (Suppl.): 1–47.

Stern, C., and E. R. Sherwood, eds., 1965. *The Origin of Genetics.* Freeman, San Francisco.

Sturtevant, A. H., 1965. *A History of Genetics.* Harper & Row, New York.

Questions

1.1. What are the contributions of each of the following to genetic thought?
  a. Lamarck
  b. Mendel
  c. Morgan
  d. de Vries
  e. Johannsen

1.2. Distinguish between genotype and phenotype.

1.3. What are the causes of variation?

1.4. Populations of the wild pansy, *Viola tricolor,* growing on wind-swept coastal dunes differ in appearance from inland populations. How would you determine experimentally the basis for these differences?

1.5. What is an acquired characteristic? What type of evidence would be needed to demonstrate the inheritance of acquired characteristics?

CHAPTER TWO

Mendel's Laws

In the preceding chapter genetic recombination was cited as one of the primary causes of variation. Mendel's great contribution to the study of heredity was his discovery of the orderly processes governing the recombination of genes where so many before him had failed. Notable among his predecessors were Kölreuter and Gärtner. Both men observed that the first generation (F_1, or first filial) hybrids were usually intermediate with respect to the parents and that the second (F_2) generation was more variable than the F_1. However, they did not deal with separate traits but rather with the entire organism. Since the parents differed in a number of traits, the data were complex and a description of the progeny was about all that was possible.

Darwin, the great synthesizer, gathered together a great deal of information about hybridization from breeding experiments of all kinds, both plant and animal. However, this approach, so fruitful with respect to evolution and natural selection, did not lead him to significant generalizations concerning heredity. He also conducted his own breeding experiments with pigeons and with plants, but made no notable advances.

Mendel's success was not just a happy accident. His experiments were designed to get at the essence of the problem of heredity, they were conducted with care and skill, and the results were subjected to a more refined analysis than had previously been attempted.

At first Mendel studied just one trait at a time rather than following all of the character differences in a cross as Kölreuter and Gärtner had tried to do. Only after this analysis had progressed did he attempt to follow two characters segregating in the same cross. The traits that he chose for study were well defined so that it was simple to classify them correctly. Because his peas were green or yellow, or round or wrinkled, or the flowers were red or white, with no intermediates, correct scoring of parents and progeny was never difficult. Thus by using well-defined unit characters, he reduced the problem to its simplest terms.

Perhaps his greatest innovation, which today is done routinely, was to count the numbers of each type of progeny to emerge from a given cross. He thereby reduced the phenomena of inheritance to a measurable basis that could be analyzed quantitatively.

His care in keeping meticulous pedigree records gave him an exact history of the ancestry of any given plant. Another factor in his success was his choice of experimental material. The common garden pea was well suited for the experiments he planned. It is normally self-fertilizing so that contamination of the crosses by stray foreign pollen was not a problem. Furthermore, many varieties with different well-defined characters were available for study, and because of regular self-fertilization, each variety bred true for the traits it carried. The different varieties were interfertile when crossed so that sterility and low viability did not complicate the results.

Finally, it must be added that Mendel was lucky. Altogether he worked with seven characters in peas. The garden pea has seven pairs of chromosomes, and it was later found that each of the factor pairs he studied was on a different chromosome pair. In 1865 chromosome behavior and the role of the chromosomes in heredity were not understood; in fact the word chromosome itself had not been coined. If he had happened to choose two factor pairs on the same chromosome pair, his results would have differed from those in all of the other crosses, and this lack of uniformity might have clouded the issue to the point where he could not see through to the generalizations we now call Mendel's laws. The chances of choosing seven traits on seven different chromosome pairs simply by chance can be estimated to be only about 6 in 1000. Thus, beyond question, he was lucky. However, it was a combination of luck and good judgment that produced Mendel's results.

2.1. Segregation

One of the first facts to emerge from Mendel's experiments was that when a cross was made between parents differing in a single trait, the hybrid offspring all resembled one parent but not the other. For example, if a plant pure for yellow seeds was crossed with one pure for green, all of the F_1 seeds turned out to be yellow and indistinguishable from those of the pure yellow strain. Since green reappeared in later generations, it obviously had not been eliminated, but must have been present, concealed, in the hybrid. Therefore, Mendel called yellow, which was expressed in the hybrid, *dominant*, and green, which was not expressed, *recessive*.

When the F_1 plants were self-fertilized, the F_2 seeds appeared in the ratio of 3 yellow to 1 green. Repetition of this experiment by seven independent investigators gave the following totals:

Yellow	134,707	75.09%
Green	44,692	24.91%
	179,399	

Mendel found similar 3-to-1 ratios for the other traits he studied (seed form, flower and seed coat color, pod form, unripe pod color, flower position, and stem length), and this type of result has been reported in many other species of animals and plants for a variety of traits. Not only does the recessive trait reappear in the F_2, but it appears in a definite proportion of the individuals, one-fourth of the total.

If only small numbers are involved, the ratio may deviate quite far from 3 to 1, but as the numbers of progeny get larger, the ratio usually comes closer and closer to an exact 3-to-1 ratio. The reason is that the ratio is dependent on the random union of gametes. In a similar manner, due to the vagaries of chance, the ratio obtained from 10 tosses of a coin may deviate rather far from the expected 1-to-1 ratio of heads to tails, but 1000 tosses will give a ratio that more nearly approximates 1 to 1.

These results formed the basis for Mendel's first law, the *law of segregation*. The facts of segregation may be simply stated as follows: When individuals differing in a single trait are crossed, each alternative behaves as a unit, passes through individuals of the first filial (F_1) generation where it may or may not be visibly expressed, and emerges unchanged in the second generation. Here, in the F_2, one-fourth of the individuals resemble in *appearance* and *breeding behavior* one of the original pure types (yellow), one-fourth resemble the other pure type (green), and one-half resemble the first-generation hybrids. More briefly, in a monohybrid cross when the hybrid reproduces, it transmits with equal frequency either the dominant character of one parent or the recessive character of the other, but not both.

From these observations, Mendel drew certain inferences. He knew that a variety pure for yellow seeds must carry some sort of factor for yellow, and that a variety pure for green seeds must carry only a factor for green. He assumed that the hybrid, even though the seeds were yellow like those of one parent, carried a factor for yellow and also a factor for green. Then, when a hybrid is self-fertilized (or when two such hybrids are crossed), the fact that true-breeding yellows are produced among the progeny indicates that both gametes giving rise to such plants must carry only a factor for yellow, and the true-breeding greens must carry only factors for green. Therefore Mendel inferred that a hybrid individual, carrying factors for both yellow and green, produces pure gametes of two different kinds in equal numbers, half carrying yellow and half carrying green.

The implication is, then, that every gamete is pure, never hybrid like the individual, and carries only one factor of the pair. In the formation of gametes there must be a reduction in the amount of hereditary material, each gamete carrying only half of what is found in the body or somatic cells of the individual. The union of gametes bearing different factors at fertilization brings the factors together in the hybrid where they coexist, but do not lose their identity or blend, as Darwin thought; instead, they emerge unchanged and are segregated from one another during gamete formation.

The 3-to-1 phenotypic ratio, or the 1:2:1 genotypic ratio, is explained by one further assumption: The different kinds of gametes unite at random. In other words, there is no greater likelihood that yellow will unite with yellow than with green, or vice versa.

The discussion of genetics is simplified by the use of symbols. A sort of shorthand notation and vocabulary has been developed to facilitate an understanding of genetics. The following are useful in dealing with segregation:

G = symbol for the Mendelian factor or gene for yellow

g = symbol for the Mendelian factor or gene for green

Thus, the capital letter is used as the symbol for the dominant gene; the same letter in lowercase is used as the symbol for the recessive gene for the same trait. G and g are said to be *alleles*.

GG = the genotype of an individual pure or *homozygous* for yellow

gg = the genotype of an individual pure or *homozygous* for green

Gg = the genotype of the hybrid or *heterozygous* individual,
       phenotypically yellow

$♀$ = symbol for female

$♂$ = symbol for male

The above cross can then be outlined as shown in Fig. 2.1. The genotypes and phenotypes of the F_2 generation can be generated from a

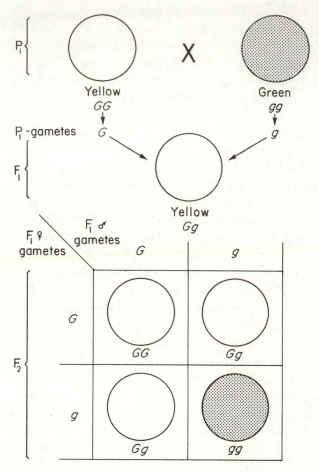

Fig. 2.1. Monohybrid cross in the garden pea (*Pisum sativum*). **G** = allele for yellow, g = allele for green, **P**₁ = parental generation, **F**₁ = first filial generation, **F**₂ = second filial generation.

checkerboard with the gametes along the sides, and their possible combinations in the squares, as shown in the figure. Thus, in a very brief but complete form, the cross and its results can be illustrated.

2.2. Independent Assortment

The next question becomes, "What are the results of a cross involving two segregating traits?" Mendel's experiments with these so-called dihybrid crosses led to his second major discovery, the ***principle of independent assortment***. It is simplest to state the law first and then to describe the experiments on which it is based. Mendel found that the segregation of one factor pair occurred independently of the segregation of a second factor pair.

For example, he crossed a variety of the garden pea with yellow round seeds to one with green wrinkled seeds. The results were as follows:

$$P_1 \quad \text{round yellow} \times \text{wrinkled green}$$
$$F_1 \qquad\qquad \text{round yellow}$$

Yellow and round, therefore, were dominant. Selfing or self-fertilization produced the F_2 shown below.

		n
F_2	round yellow	315
	round green	108
	wrinkled yellow	101
	wrinkled green	32
	total	556

Thus, in the F_2 both parental combinations were recovered, plus two new combinations not previously observed, green round and yellow wrinkled. His interpretation of the ratio led Mendel to the law of independent assortment.

If each trait is treated separately, it will be seen that both show quite good 3-to-1 ratios.

round	$315 + 108 = 423$	76.08%
wrinkled	$101 + 32 = 133$	23.92%
	556	

yellow	$315 + 101 = 416$	74.82%
green	$108 + 32 = 140$	25.18%
	556	

Thus, when the data are treated as two separate monohybrid crosses, they conform to Mendel's first law of segregation. If the two 3:1 segregations are occurring independently of one another, then three-fourths of the round plants should be yellow and one-fourth should be green; of the wrinkled plants, three-fourths should also be yellow and one-fourth should be green. Thus, the expected proportions would be as follows:

round yellow	$\tfrac{3}{4} \times \tfrac{3}{4} = \tfrac{9}{16}$
round green	$\tfrac{3}{4} \times \tfrac{1}{4} = \tfrac{3}{16}$
wrinkled yellow	$\tfrac{1}{4} \times \tfrac{3}{4} = \tfrac{3}{16}$
wrinkled green	$\tfrac{1}{4} \times \tfrac{1}{4} = \tfrac{1}{16}$

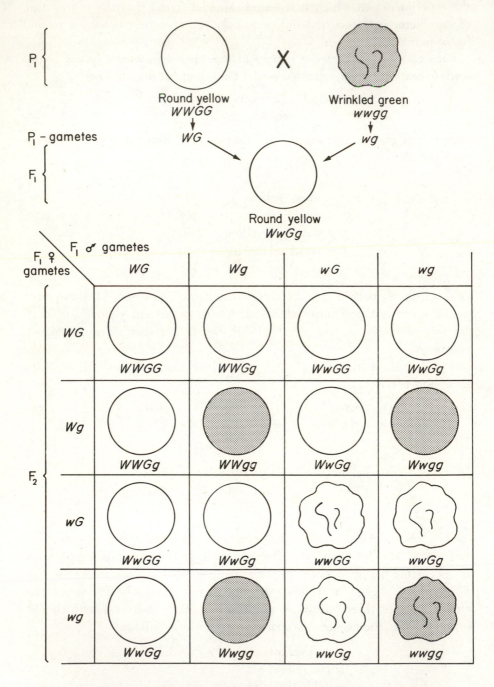

Fig. 2.2. Dihybrid cross in the garden pea (*Pisum sativum*). W = allele for round, w = allele for wrinkled, G = allele for yellow, g = allele for green.

To get the expected numbers of each type among a progeny of 556, one multiplies 556 by each fraction. Comparison of the observed with the expected numbers shows remarkably good agreement:

	Expected	Observed
round yellow	312.75	315
round green	104.25	108
wrinkled yellow	104.25	101
wrinkled green	34.75	32
	556.00	556

In fact, Mendel's results generally were so close to the exact ratios he expected that it has recently become fashionable to question his scientific objectivity (i.e., to wonder whether he "fudged" his data a bit.) However, it was obviously clear to him that in a dihybrid cross the expected ratio was 9:3:3:1, with $9/16$ of the F_2 being doubly dominant, and only $1/16$ the double recessive type.

Furthermore, it is even clearer that assortment is truly independent when it is known that the F_2 ratio is unchanged no matter how the traits enter the cross. In other words, the same F_2 ratio is obtained from a cross between round green and wrinkled yellow as from round yellow and wrinkled green.

Mendel's results can thus be interpreted as shown in Fig. 2.2. In the dihybrid cross the F_1 produces four different kinds of gametes in equal numbers, each gamete carrying only one factor of each type. The gametes are WG, Wg, wG, and wg, all the possible combinations and each kind equally frequent. The critical point for producing the 9:3:3:1 F_2 ratio is in F_1 gamete formation. Half will carry the gene for yellow, half for green. Pure chance determines the combination of the unrelated factors in the gametes. Hence, half of the gametes with the yellow factor will carry round, and the other half will carry wrinkled. The F_2 can be generated from a checkerboard comparable to that for the monohybrid cross, but with 16 rather than 4 squares.

Inspection of the checkerboard reveals several points of interest. There are 16 possible combinations of gametes, 9 of which have different genotypes; but with dominance there are only 4 different phenotypes, Twenty-five percent of the F_2 breeds true, and the four homozygous types lie on the diagonal from upper left to lower right. Twenty-five percent of the F_2 is like the F_1 in being heterozygous for both factor pairs, the dihybrids lying on the diagonal from upper right to lower left. The four phenotypes are composed of the following genotypes:

9 round yellow   1 *WWGG*; 2 *WWGg*; 2 *WwGG*; 4 *WwGg*
3 round green   1 *WWgg*; 2 *Wwgg*
3 wrinkled yellow   1 *wwGG*; 2 *wwGg*
1 wrinkled green   1 *wwgg*

The checkerboard above is known as a gametic checkerboard. Other, less cumbersome forms of the checkerboard may be used when appropriate. For instance, the same cross may be shown with a zygotic checkerboard, in this case with dominance:

W/w ⟍ G/g	¾ W–*	¼ ww
¾ G–	⁹⁄₁₆ W–G–	³⁄₁₆ wwG–
¼ gg	³⁄₁₆ W–gg	¹⁄₁₆ wwgg

* The dash indicates that the dominant allele or the recessive allele may be present.

If genotypic frequencies are of interest or there is no dominance, the following form of checkerboard is very useful.

B/b ⟍ A/a	¼ AA	½ Aa	¼ aa
¼ BB	¹⁄₁₆ AABB	²⁄₁₆ AaBB	¹⁄₁₆ aaBB
½ Bb	²⁄₁₆ AABb	⁴⁄₁₆ AaBb	²⁄₁₆ aaBb
¼ bb	¹⁄₁₆ AAbb	²⁄₁₆ Aabb	¹⁄₁₆ aabb

The genotypic ratio is thus 4:2:2:2:2:1:1:1:1. The following table gives the relation between the number of gene pairs segregating in a cross and the number of different F_2 genotypes and phenotypes.

Thus, with these seemingly simple experiments with peas, Mendel worked out the basic principles of heredity, showing that segregation or separation of the parental characters occurs in definite proportions in the progeny of hybrids and that the segregation of one gene pair is independent of the segregation of other factor pairs. These results are still the cornerstone for our understanding of heredity.

If Mendel's laws were applicable only to garden peas, they would be interesting, but not of singular importance. Their significance stems from the fact that they were soon shown to hold not just for peas but for all plants, for animals including man, and even for many microorganisms such as yeasts. Their simplicity and their generality meant that at a single

Gene pairs	Different F_1 gametes	Possible combinations of F_1 gametes	F_2 genotypes	F_2 phenotypes (complete dominance)
1	2	4	3	2
2	4	16	9	4
3	8	64	27	8
4	16	256	81	16
.	.	.	.	.
.	.	.	.	.
.	.	.	.	.
n	2^n	4^n	3^n	2^n

stroke Mendel had opened the door to the experimental study of heredity in a vast array of plant and animal species. The power of this new tool can be illustrated by its use to solve a taxonomic problem in frogs.

Among the thousands of leopard frogs received by F. J. Burns & Co., produce dealers of Chicago, were two types so distinctively different that they were tentatively described as new species. The unspotted type (Plate I) was named *Rana burnsi* in honor of F. J. Burns and J. J. Burns; the mottled type (Plate II) was called *Rana kandiyohi*, the Indian name of the county in Minnesota from which both types had been shipped. (Color plates follow page 26.)

Because the techniques for making crosses in frogs were slow to develop, decades passed before it was shown that these frogs were not separate species but that the unspotted pattern (Burnsi) and the mottled pattern (Kandiyohi) were due to two different dominant genes in the common spotted leopard frog, *Rana pipiens* (Plate III). The range of the leopard frog covers most of North America, but the Burnsi and Kandiyohi genes are found primarily in an area of some 100,000 square miles in Minnesota and adjacent states. The frequency of the Burnsi and Kandiyohi phenotypes is rather low, occasionally approaching a maximum of 10% in some wild populations. Thus, rather than three separate species coexisting in Minnesota, there is only a single species, *Rana pipiens*, in which two dominant pattern genes occur.

That the Burnsi condition was dominant was shown by crossing unspotted frogs to the usual wild-type spotted leopard frogs with the following results:

P₁   Burnsi × wild type
                                   ↓
F₁   50% Burnsi:50% wild type

Therefore, the unspotted frogs used in the crosses must have been heterozygous for a dominant gene. If the unspotted pattern were due to a homozygous recessive condition, all of the F_1 would have been the spotted wild type. Thus, the cross was as follows:

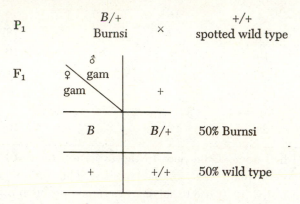

$$P_1 \qquad \begin{array}{c} B/+ \\ \text{Burnsi} \end{array} \quad \times \quad \begin{array}{c} +/+ \\ \text{spotted wild type} \end{array}$$

Here, the + symbol is used for the wild-type allele, which, when homozygous, gives rise to the spotted phenotype.

Similarly, the Kandiyohi type was shown to be dominant, because

$$P_1 \qquad \text{Kandiyohi} \times \text{wild type}$$
$$\downarrow$$
$$F_1 \qquad 50\% \text{ Kandiyohi} : 50\% \text{ wild type}$$

and the genotypes must have been

$$P_1 \qquad \begin{array}{c} K/+ \\ \text{Kandiyohi} \end{array} \quad \times \quad \begin{array}{c} +/+ \\ \text{wild type} \end{array}$$

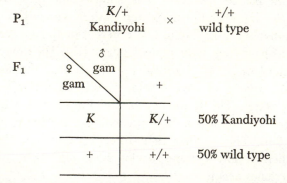

Here, the + symbol is again used to symbolize the wild-type allele, this time at the "K" locus.

Although these crosses show that both the gene for the unspotted pattern and the gene for the mottled pattern are dominant to the wild type, they do not reveal the relationship between the Burnsi and Kandiyohi genes. Although the symbols used suggest that these are different genes, nevertheless the possibility that they are alleles has not been ruled out by the crosses shown thus far.

The crucial cross to answer this question would be between a frog carrying both dominant genes and a common spotted wild-type frog. If the dominant pattern mutants were alleles, the cross of a double mutant with the wild type would give the results shown below. (The symbol B^K is used here for Kandiyohi because convention requires that alleles be symbolized by variations on the same symbol.)

P_1 B/B^K × $+/+$

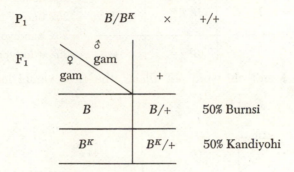

	$B/+$	50% Burnsi
	$B^K/+$	50% Kandiyohi

F_1 ♀ gam / ♂ gam $+$

| B | $B/+$ | 50% Burnsi |
| B^K | $B^K/+$ | 50% Kandiyohi |

If the genes are not alleles, the following results are expected:

P_1 $B/+; K/+$ × $+/+; +/+$

F_1 ♀ gam / ♂ gam $++$

BK	$B/+; K/+$	25% double mutant
B+	$B/+; +/+$	25% Burnsi
+K	$+/+; K/+$	25% Kandiyohi
++	$+/+; +/+$	25% spotted wild type

Because the two dominant genes are rather infrequent, the frequency of frogs in the wild populations of Minnesota carrying both the Burnsi and the Kandiyohi genes is roughly 1 in 10,000. Since Burnsi and Kandiyohi frogs are much more easily obtained, it may be asked why a cross between Burnsi and Kandiyohi heterozygotes would not serve equally well to answer the question of whether these two pattern genes are alleles. The reason can be shown by outlining the cross. If Burnsi and Kandiyohi are alleles, a cross of the heterozygotes would give the following:

P_1 $B/+$ × $B^K/+$

F_1

♀ \ ♂ gam	B^K	$+$	
B	B/B^K	$B/+$	25% double mutant
			25% Burnsi
			25% Kandiyohi
$+$	$B^K/+$	$+/+$	25% wild type

If Burnsi and Kandiyohi were not alleles, the cross would look like this:

P_1 $B/+; +/+$ × $+/+; K/+$

F_1

♀ \ ♂ gam	$+K$	$++$	
$B+$	$B/+; K/+$	$B/+; +/+$	25% double mutant
			25% Burnsi
			25% Kandiyohi
$++$	$+/+; K/+$	$+/+; +/+$	25% wild type

Thus, in either case a 1:1:1:1 ratio would be expected, and this cross cannot be used to determine whether Burnsi and Kandiyohi are alleles or not.

Oddly enough, because it was realized that a double mutant frog must be rare and because Mendel's discovery of dominance is not a law comparable to segregation and independent assortment, on which predictions can be based, no one was exactly sure what a double mutant would look like. (Dominance is dependent not only on the interactions between alleles, but also on the interactions of these genes with the rest of the genotype and with the environment. Therefore, the discussion of dominance has been deferred until Chapter 4, where gene interactions are discussed.) It was not known, and could not be predicted, whether the Burnsi gene would suppress the expression of the Kandiyohi gene, or whether Kandiyohi would suppress Burnsi, or whether frogs carrying both genes would have a phenotype more or less intermediate between the two types or quite different from either.

It might be supposed that the simplest way to obtain a double mutant

Plate I. Phenotype of leopard frog (*Rana pipiens*) with the dominant Burnsi gene.

Plate II. Phenotype of leopard frog (*Rana pipiens*) with the dominant Kandiyohi gene.

Plate III. The wild type of the common spotted leopard frog (*Rana pipiens*).

Plate IV. Phenotype of leopard frog (*Rana pipiens*) with both the dominant Burnsi gene and the dominant Kandiyohi gene.

Plate V. Phenotype of leopard frog (*Rana pipiens*) homozygous for the recessive gene for albinism.

frog would be to cross a Burnsi with a Kandiyohi and then rear the progeny to sexual maturity, for under either hypothesis, as shown above, one-fourth of the progeny of a cross of the two heterozygotes would be expected to carry both dominants. However, *Rana pipiens* tadpoles require nearly two years to reach sexual maturity when reared in the laboratory, and in the past rearing was so seldom successful that *Rana pipiens* genetics was for a long time limited to the study of the parents and their newly metamorphosed F_1 progeny. The question was first resolved, therefore, when a frog from the wild, intermediate in phenotype between Burnsi and Kandiyohi (Plate IV), was obtained and crossed with a spotted wild-type leopard frog. When the progeny turned out to be 25 percent spotted wild type, 25 percent Burnsi, 25 percent Kandiyohi, and 25 percent like the unusual parent, it was established that this unusual frog was, in fact, the double heterozygote for the dominant Burnsi and Kandiyohi genes and that these genes were not allelic but rather segregated independently of one another. The genotype of the unusual frog was, therefore, $B/+$; $K/+$.

The distribution and frequency of the Burnsi and Kandiyohi dominant genes suggest that they may have an adaptive role in the populations in which they are found. Albinism in *Rana pipiens* (Plate V), on the other hand, is so rare that it must be regarded as an aberrant form normally held in check by natural selection. When an albino frog was crossed to the spotted wild type, the following results were obtained:

$$P_1 \qquad \text{albino} \times \text{spotted}$$
$$\downarrow$$
$$F_1 \qquad \text{all spotted}$$

Therefore, the gene for wild-type spotting is dominant to the gene for albinism. When the spotted F_1 frogs were backcrossed to the albino parent, the progeny were 50 percent spotted and 50 percent albino. When the spotted F_1 frogs were crossed with one another, a 3 spotted to 1 albino ratio was obtained. Therefore, albinism in the leopard frog is due to a recessive gene just as it is in most other vertebrates. The color of a normally pigmented frog is due to the combined effects of three types of pigment cells, melanophores, iridophores, and xanthophores. The gene for albinism affects the production of black melanin pigment by the melanophores, but not the production of white pigments by the iridophores or yellow and orange pigments by the xanthophores, nor does it affect the spotting pattern. The result, then, is a cream-colored frog with yellowish-orange spots and red eyes, as seen in Plate V.

The crosses can be outlined as shown below:

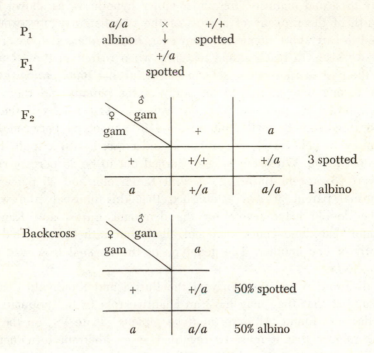

Here again the + symbol is used to designate the wild-type allele at the "A" locus. The use of this symbol rather than the capital and small letters used to describe Mendel's crosses with peas has two distinct advantages. For one, if all of the common wild-type alleles at all loci are designated by a +, the unusual genes can be picked out much more readily from the genotype than is the case if the genotype is written as a mixture of capital and small letters. Perhaps even more important, mutant genes may be dominant to the most common or wild-type allele, as is the case with Burnsi and Kandiyohi, or they may be recessive like albinism, and the use of + for the wild type, capital letters for dominants to the wild type, and lowercase letters for recessives to the wild type is the most effective way to bring out these relationships.

Thus, the application of Mendel's methods and his principles of segregation and independent assortment to the study of variation and heredity in a wide array of plant and animal species, including man, represented a major advance. Problems hitherto inaccessible to study could now be tackled, ranging from taxonomic problems in frogs to metabolic pathways in molds and men.

Additional Reading

Anderson, S. C., and E. P. Volpe, 1958. "*Burnsi* and *Kandiyohi* Genes in the Leopard Frog, *Rana pipiens*," *Science* **127**:1048–1050.

Auerbach, C., 1961. *The Science of Genetics*. Harper & Row, New York.

Bateson, W., 1909. *Mendel's Principles of Heredity*. Cambridge University Press, London.

Browder, L. W., 1972. "Genetic and Embryological Studies of Albinism in *Rana pipiens*," *J. Exp. Zool.* **180**:149–156.

Goldschmidt, R., 1952. *Understanding Heredity*. Wiley, New York.

Mendel, G., 1866. "Experiments on Plant Hybrids," E. R. Sherwood, tr. In *The Origin of Genetics*, C. Stern and E. R. Sherwood, eds., 1966. Freeman, San Francisco. Other translations in E. W. Sinnott, L. C. Dunn, and T. Dobzhansky, 1958, *Principles of Genetics*, 5th ed., McGraw-Hill, New York; and in W. Bateson, 1909, *Mendel's Principles of Heredity*, Cambridge University Press, London. Original German available in *J. Heredity* **42**:1–47.

Merrell, D. J., 1972. "Laboratory Studies Bearing on Pigment Pattern Polymorphisms in Wild Populations of *Rana pipiens*," *Genetics* **70**:141–161.

Moore, J. A., 1942. "An Embryological and Genetical Study of *Rana burnsi* Weed," *Genetics* **27**:408–416.

Olby, R. C., 1966. *Origins of Mendelism*. Constable, London.

Roberts, H. F., 1929. *Plant Hybridization Before Mendel*. Princeton University Press, Princeton, N.J.

Tschermak-Seysenegg, E. von, 1951. "The Rediscovery of Mendel's Work," *J. Heredity* **42**:163–171.

Volpe, E. P., 1955. "A Taxo-genetic Analysis of the Status of *Rana kandiyohi* Weed," *Syst. Zool.* **4**:75–82.

Questions

2.1 Which of the following F_2 ratios would indicate that the two traits in the original cross are due to allelic genes?

    a. 9:6:1

    b. 15:1

    c. 9:3:4

    d. 1:2:1

    e. 9:7

2.2. Schilder's disease, which causes a progressive degeneration of the central nervous system leading to death at about two years of age, behaves as a simple autosomal recessive trait. If a couple loses their first two children due to Schilder's disease, what is the chance that their third child will develop the disease?

2.3. If a trait shows up in both sexes with equal frequency and in half the children of an affected person, how must it be inherited?

2.4. Red and white coat colors in Shorthorn cattle are due to a pair of alleles, which when heterozygous produce a roan coat color.
What proportion of the progeny from the cross roan × roan will be expected to be roan also?

2.5. Two black female mice were bred with the same brown male. In three litters female A produced 9 black mice and 7 brown ones while female B produced 19 blacks. What conclusions can you draw about the genotypes of the parents?

2.6. In four o'clocks, red flower color (R) is only partially dominant over white (r), the heterozygotes being pink. How would you produce seed all of which would produce pink-flowered plants?

2.7. My cat has a short tail, as did his father, but his mother and some of his sibs have normal tails. What would you expect the tails of his progeny to be if he mated with a long-tailed female?

2.8. A mouse fancier discovers a yellow mouse and crosses it with a black one. Half the progeny are yellow and half are black. A cross of two of the yellow mice results in two-thirds yellow mice and one-third black. How do you account for these results?

2.9. In radishes, long and round are alleles, as are red and white. In a cross between a long, red variety and a round, white variety the F_1 is oval and purple. How many different phenotypes would you expect to find in the F_2? What would be the phenotypic ratio in the F_2?

2.10. In the F_2 of a dihybrid cross, what proportion of the individuals will be heterozygous at both loci? Homozygous at both loci?

2.11. In an ordinary F_2 involving three pairs of factors, what proportion of the progeny will show all three dominant traits?

2.12. In the F_2 from an ordinary trihybrid cross, what proportion of the progeny will show just two of the dominant characters?

2.13. In tomatoes, two-loculed fruit is dominant over many-loculed, red fruit over yellow, and tall vine over dwarf. A breeder has pure lines of two-loculed, red, dwarf plants and of many-loculed, yellow, tall plants. If he crosses the two lines and rears the F_1 and F_2 generations, what proportion of the F_2 will be phenotypically many-loculed, red, tall plants?

2.14. In pigeons, checkered is dominant to plain and red is dominant to brown. A checkered brown female produced two checkered red and three plain red offspring. Give the probable appearance and genotype of the father.

2.15. In guinea pigs rough coat (R) is dominant over smooth coat (r), and black coat (B) is dominant over white (b). A rough, black guinea pig bred to a smooth, white one gave 28 rough, black; 31 rough, white; 31 smooth, black; 29 smooth, white. What were the genotypes of the parents?

2.16. In poultry, feathered legs (F) are dominant over clean legs (f) and pea comb (P) over single comb (p), and the traits are independently inherited. A feather-legged, single-combed hen mated to a feather-legged, pea-combed rooster produced all together the following progeny: 15 feather-legged, pea-combed; 16 feather-legged, single-combed; 5 clean-legged, pea-combed; 6 clean-legged, single-combed. What were the genotypes of the parents?

2.17. "Goofus" ($o\ o$) is a recessive color mutant in mink with a pattern like a Siamese cat. "Green-eyed pastel" ($bg\ bg$) is another recessive color mutant with a chocolate brown color. In the F_2 from a cross of these two types, what proportion would you expect to be "green-eyed goofuses"?

2.18. Ebony body color and vestigial wings are independently inherited autosomal recessive traits in *Drosophila melanogaster*. An ebony male whose mother was vestigial was mated to a vestigial female whose mother was ebony. What are the genotypes of the parents and what are the expected genotypic and phenotypic ratios in their progeny?

2.19. In the garden pea, round seeds (W) are dominant to wrinkled (w) and tall (D) is dominant to dwarf (d). Since these traits show independent assortment, what phenotypic and genotypic ratios would be expected from the cross (a) $WwDd \times Wwdd$? From the cross (b) $Wwdd \times wwDd$?

2.20. In cattle, a recessive gene (u) results, when homozygous, in an udder abnormality, and another recessive, "parrot-beak" (pb), is lethal in the homozygous condition. If you wished to make sure that a bull that otherwise seemed suitable for use as a sire carried neither of these deleterious recessives, what sort of crosses would be most useful to determine his genotype with respect to these traits?

2.21. In corn, three dominant genes are necessary for aleurone color, the genotype $A/- C/- R/-$ being colored. Homozygosis for any one of the recessives results in the absence of color. An unknown color-less stock crossed with each of the three basic colorless stocks gave the following:

Unknown × a/a C/C R/R = colorless
Unknown × A/A c/c R/R = 50% colored
Unknown × A/A C/C r/r = colored

What was the genotype of the unknown stock?

2.22.   In the F_2 of a dihybrid cross, what proportion of the individuals is heterozygous for one locus or the other, but not both?

CHAPTER THREE

The Physical Basis of Heredity

An understanding of the process of fertilization, the union of sperm and egg, was not possible until the microscope was developed. In the late seventeenth century the pioneer microscopist, Leeuwenhoek, saw animal spermatozoa under a microscope and postulated that one sperm was enough to fertilize an egg. However, he did not observe fertilization directly, and his concept was not proven for another two centuries. Knowledge about fertilization in plants preceded comparable advances in animals. The growth of the pollen tube and fertilization in plants were observed in the midnineteenth century. Mendel himself believed that the union of one pollen cell with one egg cell was sufficient for the formation of a new organism, and apparently confirmed this belief experimentally. However, he never published this work and the point was not generally agreed upon.

Although reasonably exact descriptions of fertilization in animals had been reported in the 1850s, the observations of Hertwig, beginning in 1875, on fertilization in sea urchin eggs established beyond doubt the details of the process and the role of the nucleus in fertilization and cell division. It was soon shown that in the fertilization of the eggs of various animal species only one sperm normally enters, and thus, after two centuries, Leeuwenhoek's hypothesis was finally confirmed.

The last three decades of the nineteenth century marked the period during which cytology came of age. Within a decade after the start of Hertwig's work the chromosomes were recognized, and the regularity of their behavior in both mitosis and meiosis was described.

These discoveries about the orderly behavior of chromosomes led Roux in 1883 to postulate that the chromosomes bore the units of heredity. Furthermore, because of the linear structure of the chromosomes, he also suggested a linear arrangement for these hereditary units. Weismann, who in that same year published his germ line theory, stressing the distinctness between the germ cells and the somatic cells, elaborated Roux' ideas into a formidable theory of heredity and development. Although wrong in some details, it served as a stimulus to bring together the developing concepts of heredity and evolution on the one hand and the cell theory on the other.

Therefore, within a relatively short time after Mendel published his results, remarkable advances were made in understanding the nature of the physical bridge between generations. What the distinguished embryologist, von Baer, had regarded as parasites in 1827, and therefore named spermatozoa (spermato- = semen; zoa = animals), were now known to be the male's contribution to the heredity of his offspring, just as the ovum or egg was the physical link between mother and child. Furthermore, observation and reason had led to the conclusion, for which decisive evidence came much later, that the nucleus, and specifically the chromosomes within the nucleus, were the vehicles of heredity. For this reason, an understanding of chromosome behavior is essential to an understanding of genetics. The cytologists identified two forms of chromosome behavior, *mitosis* and *meiosis*. Since meiosis in some ways seems to be a modified form of mitosis, we shall consider mitosis first.

Mitosis is usually equated with cell division, particularly of the somatic cells. However, cell division consists of *karyokinesis*, or nuclear division, and *cytokinesis*, or division of the cytoplasm. These distinctions are necessary because at times (in so-called coenocytic or multinucleate tissues) the nucleus divides without a corresponding division of the cytoplasm. Cases are also known, for instance in insect embryos, where cytokinesis occurs without nuclear division. Mitosis, therefore, properly refers to the regular distribution of the chromosomes during nuclear division.

3.1. Mitosis

Mitosis is divided, for descriptive purposes, into a number of stages (Figs. 3.1 and 3.2). In the "resting" (or interphase or metabolic) nucleus, the chromosomes usually cannot be distinguished individually although chromatin material can be identified. Dark staining bodies called *nucleoli* are frequently present.

In early *prophase* the chromosomes start to become more distinct. The long, thin, threadlike structures appear to shorten somewhat and to become thicker. By mid-prophase, the chromosomes show a greater degree of condensation and are recognizable as discrete structures while the nucleoli appear to have diminished in size. In late prophase the chromosomes are even shorter and broader and show greater condensation. It may be possible to see that they are now duplicate structures, consisting of two strands or *chromatids*. The nucleoli and the nuclear membrane both have disappeared.

During the stage often called *prometaphase*, the chromosomes, which have been lying more or less free in the cytoplasm, move toward the central region of the cell. The spindle fiber apparatus appears, and the chromosomes, their *centromeres* (or *kinetochores*) having become attached to the spindle fibers, move to the equator of the cell.

In *metaphase* proper the chromosomes lie on the equator of the spindle. In reality, the centromeres are oriented along the equatorial plane, but the chromosome arms appear to extend freely in any direction. The chromosomes move independently of one another during mitosis, and are independently oriented on the equator at metaphase. The chromosome has its shortest and thickest appearance at metaphase. At this point it should be pointed out that the gradual shortening and thickening that appears to take place in the chromosomes as prophase progresses is in fact due primarily to a coiling cycle. The "condensation" results from the coiling up of the chromosome into a form similar to that of a spring. There is little actual change in the length of the strands, but their gross appearance under the microscope changes markedly when these strands are thrown into coils. The coiling cycle could more properly be described as an uncoiling cycle, since this is actually what seems to happen.

During metaphase the centromere of each chromosome divides so that each sister chromatid has its own centromere, and at this stage each chromatid can usually be easily seen.

Anaphase is the stage during which the sister chromatids separate from

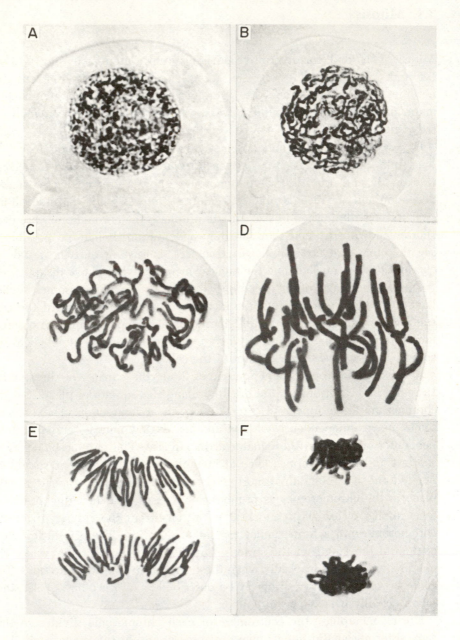

Fig. 3.1. Mitosis in root tip cells of *Lilium regale*. A. Interphase. B. Early prophase. C. Late prophase. D. Metaphase. E. Anaphase. F. Telophase. (McLeish, J., and B. Snoad, 1958. *Looking at Chromosomes.* St. Martin's Press, Inc., Macmillan & Co., Ltd., pp. 15, 17, 19, 20, 24, 27.)

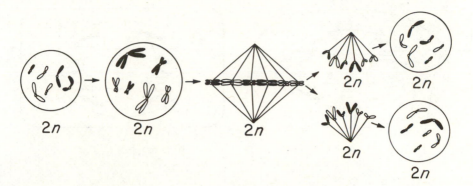

Fig. 3.2. Diagram of the stages of mitosis. Left to right: Prophase → prophase after chromatids become visible → metaphase → anaphase → telophase.

one another, and move to opposite poles of the spindle. The centromere is the mitotically active part of the chromatid, moving first toward the pole with the arms trailing behind. Exceptional chromosomes without centromeres do not orient on the equator at metaphase, nor do they move to the poles during anaphase. Anaphase ends when the two groups of sister chromatids, now more properly considered daughter chromosomes, arrive at opposite poles of the cell.

During *telophase* many of the nuclear events of prophase happen in reverse. The chromosomes uncoil, increasing their apparent length, and are less easily distinguished individually because their greater length and reduced width (and stainability) create the appearance of a tangled web. The nucleoli are reconstituted during telophase, and nuclear membranes form around each group of daughter chromosomes.

The differences between cell division in plants and animals are surprisingly few in number (Fig. 3.3). Animal mitosis usually involves a centriole that undergoes a regular division cycle. One of the two centrioles migrates around the nucleus, elements of the spindle begin to form between the separating centrioles, and they ultimately become the poles of the spindle apparatus. Centriolar division normally occurs prior to nuclear mitotic division so that it is the first significant event in the sequence leading to cell division. Centrioles have not been demonstrated in higher plants.

Cytokinesis in animal cells takes place by means of a cleavage furrow. The cell looks like a balloon with a string being tightened around its middle, eventually pinching it in two to form the daughter cells. In plant cells, cytokinesis begins at the center of the cell, with the formation of the cell plate gradually extending outward in the plane of the equator of the spindle

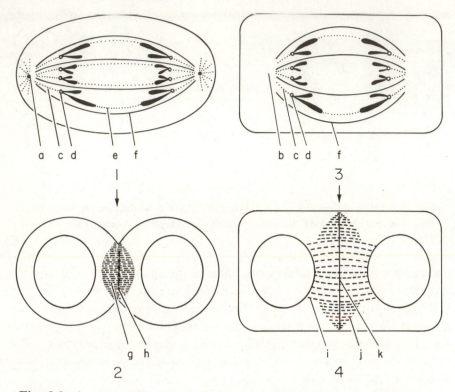

Fig. 3.3. A comparison of cell division in animal and plant cells. 1, 2. Animal cytokinesis by formation of a cleavage furrow. 3, 4. Plant cytokinesis by means of a cell plate. a, centriole; b, polar center; c, chromosomal fiber; d, centromere (kinetochore); e, spindle fiber; f, fiber or interzonal connection; g, spindle of cytokinesis; h, remains of the spindle; i, phragmoplast; j, intermediate bodies; k, cell plate. 1, 3. Anaphase. 2, 4. Telophase. (After De Robertis, Nowinski, and Saez.)

in the dividing cell. The cell plate thus separates the cytoplasm of the daughter cells and contributes to the formation of the middle lamella, which in turn gives rise to the new cell walls.

Although mitosis is described in stages, it is in reality a continuous process, typically requiring from 1 to 3 hr. Prophase and telophase are generally the longest stages, with anaphase the briefest (10–12 min or less) and metaphase slightly longer than anaphase.

Mitosis results in an equational division of the nucleus and the chromosomes, both quantitatively and qualitatively. Not only is the original number of chromosomes retained in each new daughter cell, but a representative of each of the chromosomes in the mother cell goes to each daughter cell. In other words, the chromosome composition (and the genetic constitution) of every cell derived from mitotic cell division is essentially the same.

3.2. Meiosis

Sexual reproduction involves the union at fertilization of gametes, usually a sperm and an egg. Each gamete has its own set or complement of chromosomes, and the total number of chromosomes in the fertilized egg or zygote is therefore twice that in either gamete. If the chromosome number is not to double each generation, there must be a reduction in chromosome number at some point in the life cycle. The process by which reduction is effected is *meiosis*. This reduction in number is brought about by two successive meiotic nuclear divisions during which the chromosomes divide only once. The resulting four daughter cells are *haploid*, having only half the number of chromosomes in the parent *diploid* cell. A diploid cell thus has two similar sets of chromosomes, one set from each parent. In other words, the chromosomes occur in pairs, each chromosome having a *homologous* partner much like itself. Meiosis produces not just a simple change from the diploid to the haploid condition, but a separation or segregation of the homologous chromosomes of each pair from one another.

Although there are fundamental differences between meiosis and mitosis, there are also many similarities, and meiosis is most easily studied as a modified form of mitosis. Let us now analyze more closely the events in meiosis (Figs. 3.4 and 3.5).

The first meiotic prophase (prophase I) is more complex and of longer duration than mitotic prophase. In fact, the most striking differences between meiosis and mitosis occur in prophase I, and for this reason several substages have been described.

Leptotene, the earliest prophase I stage, is marked by an increase in nuclear volume and by the extreme length and uncoiled state of the chromosomes, which cannot be identified individually. The time of chromosome duplication into sister chromatids varies in different species. In many species it seems probable that replication is complete by the end of leptotene, and in most species it is certainly complete by the end of the next stage, called *zygotene*. Leptotene, incidentally, derives its name from the threadlike condition of the chromosomes.

Zygotene is named because of the most significant event in this stage, the pairing or *synapsis* of homologous chromosomes. The pairing ordinarily starts at one point and proceeds in more or less zipperlike fashion. A most

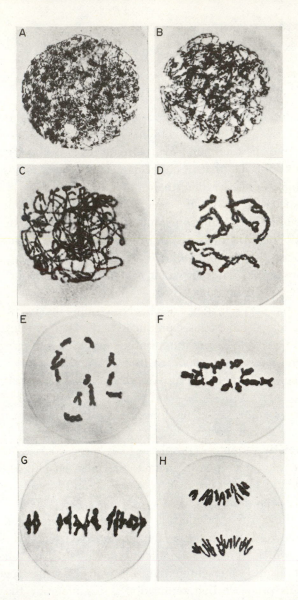

Fig. 3.4. Meiosis and pollen formation in *Lilium regale*. A. Leptotene. B. Zygotene. C. Pachytene. D. Diplotene. E. Diakinesis. F. Metaphase I. G. Early anaphase I. H. Later anaphase I. I. Telophase I. J. Interphase. K. Prophase II. L. Metaphase II. M. Anaphase II. N. Telophase II. O. The tetrad. P. Young pollen grains. (McLeish, J., and B. Snoad, 1958. *Looking at Chromosomes*. St. Martin's Press, Inc., Macmillan & Co., Ltd., pp. 48–55, 58, 60–62, 64, 66, 68, 69.)

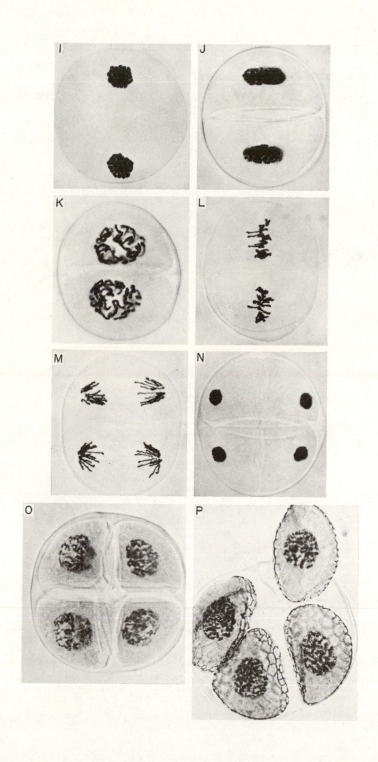

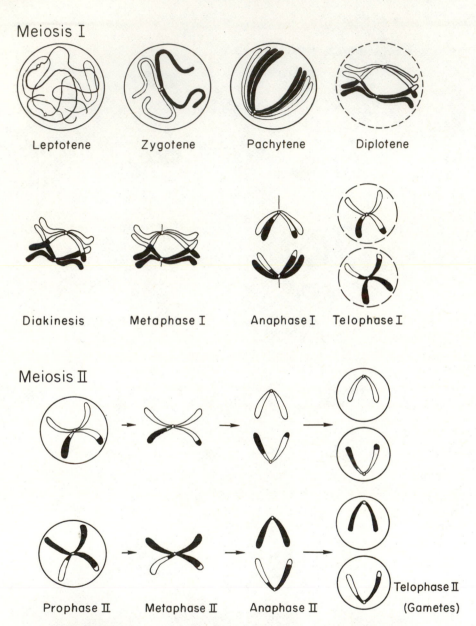

Meiosis I

Leptotene   Zygotene   Pachytene   Diplotene

Diakinesis   Metaphase I   Anaphase I   Telophase I

Meiosis II

Prophase II   Metaphase II   Anaphase II   Telophase II (Gametes)

Fig. 3.5. Diagram of the stages of meiosis. Two chiasmata are shown.

remarkable aspect of synapsis is the exactitude with which homologous chromosome regions pair. There is literally a point-by-point correspondence between the paired homologues at each level. If each chromosome already consists of two sister chromatids, each pair of homologous chromosomes contains four chromatids, and is called a *tetrad*. Just as mitotic prophase is marked by a gradual condensation of the chromosomes, due in part to coiling, so also in zygotene, the coiling and condensation of the chromosomes begins.

The completion of pairing marks the beginning of *pachytene*. The name refers to the shortening and thickening of the chromosomes during this stage (and uses the same Greek root found in pachyderm). In *diplotene*, individual pairs of homologous chromosomes can be readily identified, and the tetrad, because of its double appearance (whence the name diplotene), is also referred to as a *bivalent*. Whereas in zygotene synapsis occurs throughout the length of the pair, by diplotene sister chromatids continue to be closely paired, but the maternal and paternal homologues appear to repel each other except at points called *chiasmata*. A chiasma is formed by an actual physical exchange of material between nonsister maternal and paternal chromatids, and is the physical basis for genetic crossing over. Since crossing over will be pursued in more detail later (Chap. 8), it will suffice to say at this point that chiasma formation permits genetic recombination to occur between maternal and paternal hereditary material within a homologous chromosome pair.

The chromosomes continue to shorten and thicken during diplotene and *diakinesis*, the last stage of prophase I. The repulsion between homologous chromosomes leads to a tendency toward *terminalization* of the chiasmata. The configuration of a bivalent during diakinesis and metaphase I is determined by the number and position of its chiasmata. They seem to be all that prevents the complete separation of the homologues during late prophase.

Diakinesis is so named because during this time the bivalents migrate to the periphery of the nucleus and are maximally separated from one another. By the end of diakinesis the nucleoli have disappeared, and the nuclear membrane is dispersed. The spindle forms, and the axis of orientation of the cell is thereby established.

In *metaphase I* a prometaphase stage exists during which the bivalents move to the equatorial region of the spindle. The repulsion between the two centromeres of a bivalent leads to orientation of the bivalent on the equator in such a way that the centromeres lie toward the poles and the arms toward the equator. This orientation is different from that in mitosis.

The position of the two chromosomes in a bivalent with respect to the poles is random. Furthermore, the orientation of maternal and paternal homologues in one bivalent is independent of the orientation of the homologues in the other bivalents. This behavior of the chromosomes in meiosis forms the basis for Mendelian genetics.

Although the chromosomes divided into sister chromatids in early prophase, the centromere of each chromosome remains functionally single in the first meiotic division. Therefore, in *anaphase I* the two chromosomes of a bivalent move to opposite poles. The centromeres lead the way, the chiasmata being completely terminalized just before the arms of the chromosomes separate. Because the centromere has not divided, each chromosome still consists of two chromatids. These chromatids now appear to repel each other and are separate except at the centromere, giving a characteristic appearance to the chromosomes in anaphase I and prophase II different from the appearance of the chromosomes in comparable stages of mitosis. The result of the movement to the poles at anaphase is a reduction in the number of chromosomes to be observed cytologically from the diploid to the haploid number.

Telophase I is similar to mitotic telophase. The nucleoli reappear, nuclear membranes reform around each group of chromosomes, and the chromosomes elongate. The duration and extent of change in telophase I varies in different species. Some species skip telophase I and interphase and go directly into prophase II; in others the interphase may be very transient, or else it may be of some duration. In some cases no cytokinesis occurs after the first meiotic division.

Prophase II is simpler than prophase I. The chromosomes resume their more compact form, but as mentioned above, the arms of the chromatids remain widely separated as in anaphase I and not coiled about one another. The end of prophase II is marked by the breakdown of the nuclear membrane and the formation of the spindle.

In *metaphase II* the chromosomes are oriented on the equator as they are in mitotic metaphase, with the centromeres on the equator and the arms extending away from the central plane. At this stage each centromere becomes functionally double so that each chromatid has its own centromere, and the two chromatids are thus able to move to opposite poles during *anaphase II*.

In *telophase II* the daughter chromosomes (former chromatids) uncoil, the nucleoli reappear, the nuclear membranes form, and cytokinesis separates the four cells produced by meiosis. Therefore the two meiotic divisions produce a quartet of cells, each with the haploid number of chromosomes. The first division is disjunctional in that the centromeres of the homologues separate from one another without division. The second division is equational in a cytological sense. However, genetic reduction is not complete until both meiotic divisions have occurred. This last statement requires amplification.

3.3. Chiasmata and Crossing Over

If it were not for the formation of the chiasmata during prophase I, it would be proper to call the first meiotic division reductional and the second meiotic division equational. However, the chiasmata represent actual exchanges of segments between two nonsister chromatids of maternal and paternal origin. Any one chiasma involves an exchange between only two of the four chromatids in the tetrad. Therefore, as can be seen in Fig. 3.6, if a single chiasma were formed in prophase I, anaphase I would produce chromosomes of exclusively maternal or paternal origin up to the point of exchange. Therefore, this region of the chromosome is both cytologically and genetically haploid, and meiosis I is reductional for this region of the chromosome pair. However, beyond the point of exchange each chromo-

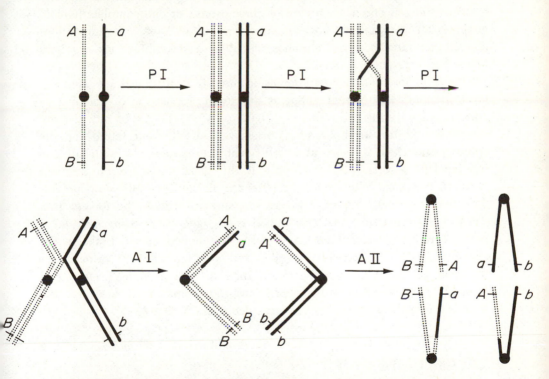

Fig. 3.6. Diagram of the formation of a chiasma. The exchange occurs in the four-strand stage between two nonsister chromatids. Note that genetic reduction is not complete until after the second meiotic division. P I = prophase I, A I = anaphase I, A II = anaphase II.

some in anaphase I carries chromatid segments of both maternal and paternal origin, and is thus still genetically diploid. Reduction to a truly haploid condition for this region is complete only after the second meiotic division. Only then does each chromatid consist of a strand of single origin throughout its length. A single chromatid (now a daughter chromosome) may, of course, contain segments of both maternal and paternal origin, but at any given position or locus on the chromosome only one kind of gene or allele will be present. Thus, each meiotic division is both reductional and equational for different regions of the chromosomes, depending on the number and placement of the chiasmata.

3.4. Genetic Implications of Meiosis

At the completion of meiosis, the chromosomes are present in the haploid number; there is one of each kind of chromosome, or one complete haploid genome. The haploid set, however, may contain a variety of combinations of maternal and paternal chromosomes. These recombinations are both intra- and interchromosomal.

The behavior of the chromosomes in meiosis is the basis for Mendel's laws. Mendel's law of segregation is due to the separation of homologous chromosomes during the reduction divisions. Mendel's law of independent assortment is due to the random orientation of different tetrads on the spindle, and their subsequent independent segregation. As we shall see, independent assortment occurs only if the gene pairs are on different chromosome pairs. When the gene pairs are on the same chromosome pair, they tend to remain together in crosses, and are said to be *linked*. If a chiasma is formed between two linked genes, genetic *crossing over* takes place, leading to recombination between the linked genes. Linkage and crossing over were discovered in the early twentieth century, primarily as the result of the research of Thomas Hunt Morgan and his colleagues, and marked the next major advance in our understanding of heredity. We shall consider these topics later in more detail in Chapter 8.

3.5. Life Cycles

In sexually reproducing species of animals and plants, four important phases of the life cycle can be identified: fertilization, mitosis, meiosis, and gamete formation. The sequence of the four phases varies, however, in different species. In the red bread mold, *Neurospora crassa,* meiosis immediately follows fertilization, so that the zygote is ordinarily the only diploid cell in the entire life cycle. In man, on the other hand, meiosis immediately precedes fertilization so that the sperm and egg cells are the only haploid cells in the life cycle. Between these extremes a wide range of possibilities exists, and a variety of sequences has been observed. In this alternation of haploid and diploid generations there has been a tendency in higher plants and animals toward a predominance of the diploid phase of the life cycle, and a reduction in the duration of the haploid phase. At this point it seems desirable to review the life cycles of corn, *Zea mays,* and the fruit fly, *Drosophila melanogaster,* two species much used in genetic research, in order to relate meiosis and mitosis to the life cycles of these species, representative of a higher plant and a higher animal.

Life Cycle of Drosophila melanogaster

The life cycle of the fruit fly consists of four stages: fertilized egg, larva, pupa, and adult or imago (Fig. 3.7). The rate of development is dependent on temperature, since the flies are cold-blooded invertebrates. Development from fertilization to emergence from the pupa case takes from 12 to 15 days at room temperature. The adults then may live for 60 days or more.

The fertilized egg is diploid (i.e., $2n = 8$, where n is the haploid number of chromosomes). It hatches within less than 24 hr after it is laid, the exact time depending on how long it is retained in the reproductive tract of the female after fertilization. The larva that hatches from the egg burrows into the food medium and feeds actively, typically on yeast. The larval period is a time of active growth and the larva molts twice, with each successive larval stage or *instar* being larger than the one before. The larvae crawl from the moist food to a drier environment before pupating. The pupa case is formed from the larval skin of the third larval instar as it becomes thick, hard, and dark. Within this pupa the wormlike larva metamorphoses into an adult fly. Differentiation and growth throughout the period of development are made possible by mitotic cell divisions.

Meiosis takes place in the gonads of the adult flies, giving rise in the males to haploid ($1n$) spermatids that change into functional spermatozoa without further cell division. In the female fly, meiosis gives rise to four haploid nuclei, only one of which becomes the nucleus of the functional ovum or egg. The egg is a large cell with a considerable amount of stored food substances in the cytoplasm, but the other three, known as polar bodies, have a minimum amount of cytoplasm associated with the nucleus. After the male inseminates the female, the spermatozoa are stored in the repro-

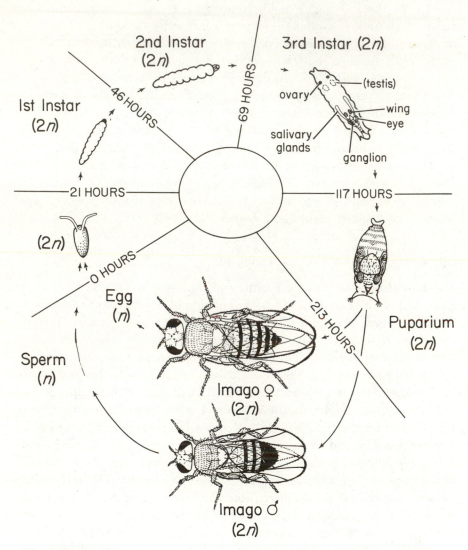

Fig. 3.7. The life cycle of the fruit fly, *Drosophila melanogaster*. The times are only approximate and are longer the lower the temperature.

ductive tract of the female in structures called the spermathecae and the ventral receptacle. These sperm then fertilize the eggs, produced at the rate of 50 to 100 a day for a number of days, and the cycle is then complete. The sequence is thus fertilization followed by mitosis, with meiosis and gametogenesis more or less coinciding.

In higher animals, spermatogenesis occurs in the germinal epithelium of the testis (Fig. 3.8). Diploid spermatogonial cells proliferate mitotically,

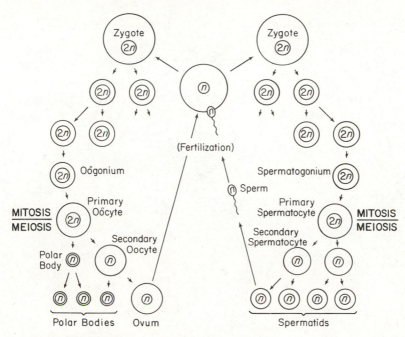

Fig. 3.8. Gametogenesis in animals. The formation of sperm and egg cells.

and some of the spermatogonia enlarge to form primary spermatocytes. The first meiotic division then results in the formation of two secondary spermatocytes from one primary spermatocyte. Each secondary spermatocyte undergoes the second meiotic division, and a quartet of haploid spermatids is produced. The spermatids undergo a form of metamorphosis to become mature functional spermatozoa.

Oögenesis occurs in the ovary, where the oögonia enlarge greatly to form diploid primary oöcytes. The first meiotic division produces a secondary oöcyte and a much smaller first polar body, the division of the cytoplasm being very unequal. The second meiotic division gives rise to the haploid ovum and a second polar body. If the first polar body also divides, a quartet of haploid cells is produced by oögenesis, genetically equivalent to the quartet of spermatids, but only one of these cells is functional.

Life Cycle of Zea mays

Among higher plants, the alternation of generations has two distinct stages, a diploid sporophyte and a haploid gametophyte. In mosses the green plant is the gametophyte, but in an angiosperm such as corn the gametophyte consists of just a few cells, and the green plant is the sporophyte generation (Fig. 3.9).

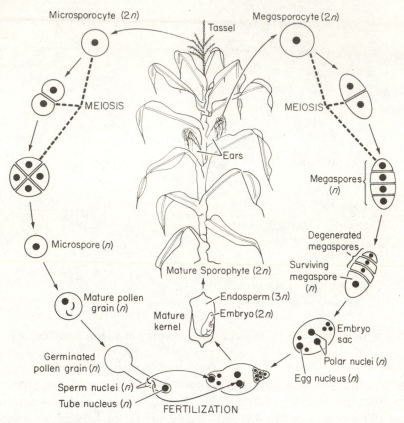

Fig. 3.9. The life cycle of Indian corn (*Zea mays*).

Meiosis occurs during the formation of haploid spores by the sporophyte. The spores, by a series of mitotic divisions, produce the haploid male and female gametophytes, which in turn produce haploid gametes. Union of the gametes forms a zygote that then develops into a diploid sporophyte.

In corn, a member of the grass family (Gramineae), the plant or sporophyte is monoecious, bearing separate male and female flowers on the same plant. The microspores or male spores are formed in the anthers of the terminal staminate (male) flowers or tassels. In the anther, microspore mother cells or microsporocytes enlarge and undergo two meiotic divisions to form a tetrad of microspores. Each haploid unicellular microspore divides once mitotically to form a tube nucleus and a generative nucleus. The generative nucleus divides again mitotically to form two sperm nuclei. This trinucleate structure, the pollen grain, is the male gametophyte, or microgametophyte.

The megaspores or female spores develop in the ovules of the pistillate (female) flowers (which later form the ears), in the axils of the upper leaves. The female spores form from megaspore mother cells or mega-sporocytes. Each ovule contains a megasporocyte that divides meiotically to form a row of four cells. Three of these cells degenerate, but the fourth enlarges to form a functional haploid megaspore. This haploid nucleus divides mitotically to form a two-, four-, and finally an eight-nucleate *embryo sac,* with four nuclei at each end. A nucleus migrates from each end to the center of the embryo sac to form the two *polar nuclei.* Of the three nuclei at one end, one becomes the *egg,* the other two the *synergids.* The three nuclei at the other end, the *antipodals,* divide mitotically to form a group of about 25 cells. The mature embryo sac at this stage is the female gametophyte or megagametophyte.

The pollen grain germinates after landing on the silk (the style), and a pollen tube grows down through the tissues of the style toward the ovule. When the pollen tube enters the embryo sac, the tube nucleus disin-tegrates, but both sperm nuclei participate in a double fertilization. One sperm nucleus fertilizes the egg to form the diploid zygote; the other unites with the two polar nuclei at the center of the embryo sac to form the triploid endosperm, a tissue for food storage. The zygote then develops into the new diploid sporophyte generation.

Additional Reading

Brachet, J., and A. E. Mirsky, eds., 1959–1964. *The Cell: Biochemistry, Physiology, Morphology,* 6 vols. Academic, New York.

Cohn, N. S., 1969. *Elements of Cytology,* 2nd ed. Harcourt Brace Jovanovich, New York.

Darlington, C. D., 1937. *Recent Advances in Cytology,* 2nd ed. Blakiston, Philadelphia.

Darlington, C. D., 1958. *The Evolution of Genetic Systems,* 2nd ed. Oliver and Boyd, Edinburgh.

Darlington, C. D., 1965. *Cytology.* J. and A. Churchill, London.

Darlington, C. D., and E. K. Janaki-Ammal, 1945. *Chromosome Atlas of Cultivated Plants.* Allen & Unwin, London.

Darlington, C. D., and L. F. LaCour, 1969. *The Handling of Chromosomes,* 5th ed. Allen & Unwin, London.

Darlington, C. D., and A. P. Wylie, 1955. *Chromosome Atlas of Flowering Plants,* 2nd ed. Allen & Unwin, London.

DuPraw, E. J., 1968. *Cell and Molecular Biology.* Academic, New York.

DuPraw, E. J., 1970. *DNA and Chromosomes.* Holt, Rinehart and Winston, New York.

Henderson, S. A., 1970. "The Time and Place of Meiotic Crossing-over," *Ann. Rev. Genetics* 4:295–324.

McLeish, J., and B. Snoad, 1958. *Looking at Chromosomes.* St. Martin's, New York.

Makino. S., 1951. *An Atlas of the Chromosome Numbers in Animals,* 2nd ed. Iowa State University Press, Ames.

Morgan, T. H., 1919. *The Physical Basis of Heredity.* Lippincott, Philadelphia.

Morgan, T. H., 1928. *The Theory of the Gene,* 2nd ed. Yale University Press, New Haven, Conn.

Morgan, T. H., A. H. Sturtevant, H. J. Muller, and C. B. Bridges, 1923. *The Mechanism of Mendelian Heredity,* 2nd ed. Holt, Rinehart and Winston, New York.

Moses, M. J., 1968. "Synaptinemal Complex," *Ann. Rev. Genetics* 2:363–412.

Ris, H., and D. F. Kubai, 1970. "Chromosome Structure," *Ann. Rev. Genetics* 4:263–294.

deRobertis, E. D. P., W. W. Nowinski, and F. A. Saez, 1971. *Cell Biology,* 5th ed. Saunders, Philadelphia.

Schrader, F., 1953. *Mitosis,* 2nd ed. Columbia University Press, New York.

Sharp, L. W., 1943. *Fundamentals of Cytology.* McGraw-Hill, New York.

Sutton, W. S., 1903. "The Chromosome in Heredity," *Biol. Bull.* 4:231–251.

Swanson, C. P., 1957. *Cytology and Cytogenetics.* Prentice-Hall, Englewood Cliffs, N.J.

Swanson, C. P., 1969. *The Cell,* 3rd ed. Prentice-Hall, Englewood Cliffs, N.J.

Swanson, C. P., T. Merz, and W. J. Young, 1967. *Cytogenetics.* Prentice-Hall, Englewood Cliffs, N.J.

White, M. J. D., 1961. *The Chromosomes.* Methuen, London.

White, M. J. D., 1973. *Animal Cytology and Evolution,* 3rd ed. Cambridge University Press, London.

Wilson, E. B., 1925. *The Cell in Development and Heredity,* 3rd ed. Macmillan, New York.

Questions

3.1. In what ways does meiosis resemble mitosis? How does meiosis differ from mitosis?

3.2. How many different kinds of gametes could be produced by an individual:

   a. With the genotype *Aa Bb Cc?*

   b. With the genotype *Aa Bc CC?*

   c. With the genotype *Aa Bb Cc Dd Ee?*

3.3. In an animal with the diploid number of chromosomes equal to eight, in what proportion of the gametes will all of the centromeres be of paternal origin if assortment is independent?

3.4. a. From how many secondary spermatocytes are 40 spermatids derived?

   b. How many secondary oöcytes are required to produce 40 functional ova?

3.5. The common red fox has 34 chromosomes in its somatic cells and the arctic fox has 52. How many chromosomes would a hybrid between these two species be expected to have in its somatic cells?

3.6. Which of the following have nuclei that are genetically completely diploid?

a. Human secondary spermatocytes

b. Endosperm in maize

c. *Drosophila* primary oöcyte

d. Onion root tips

e. Grasshopper spermatogonia

f. Mouse spermatids

g. Pollen mother cell in maize

h. Embryo sac in maize

3.7. An animal has 16 chromosomes in its somatic cells. Seventeen different loci have been identified. Six of the gene pairs fall into one linkage group, four other pairs belong to another linkage group, and three others to still another group. The remaining gene pairs do not fall into any of these linkage groups and are not linked to each other. How many chromosome pairs are there for which there are no marker genes?

CHAPTER FOUR

Gene Interactions

Thus far, the traits discussed have been treated as if they were the product of the action of a single pair of genes. Actually, each character results from the combined action of many different pairs of genes. In this chapter we shall consider some of the ways that genes can interact to produce the phenotype of the organism. The first description of an interaction between two different genes was Mendel's discovery of dominance. In this case the interaction was between genes at the same relative position on homologous chromosomes, that is, between alleles at the same locus. The term *dominance* refers only to the interactions between alleles. Interactions between genes at different loci are called *epistatic*. Such nonallelic interactions may occur between genes on homologous or nonhomologous chromosome pairs. In many respects dominance and epistasis are rather similar phenomena.

4.1. Dominance

Mendel's description of dominance helped to clarify some puzzling aspects of heredity, but because it lacks universality, dominance does not merit the status of a law or principle of heredity, comparable to segregation and independent assortment. Although Mendel reported clear-cut dominance

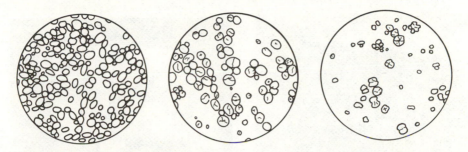

Fig. 4.1. Starch grains in round seeds (WW, left, Ww, center) and wrinkled seeds (ww, right) of the garden pea (*Pisum sativum*). The starch grains of Ww seeds are intermediate between those of WW and ww seeds, indicating incomplete dominance for this effect of the gene. However, the gene is dominant in its effect on seed shape, because the round WW and Ww seeds are indistinguishable by shape. (After Darbishire.)

for each of the traits he studied (that is, Aa was phenotypically indistinguishable from AA), not all traits behave in this way. Even the trait, round seeds, listed as dominant to wrinkled by Mendel, is not a complete dominant. In outward appearance both Ww and WW seeds are round, but microscopic examination of the hybrid Ww seeds reveals starch grains intermediate in appearance compared to those in seeds in the WW and ww homozygotes (Fig. 4.1). Thus, in this case the degree of dominance reported depends on how carefully the phenotype is scrutinized.

In some cases dominance appears to be absent. A cross between red and white zinnias, for instance, gives a pink F_1, and the F_2 segregates into one red, two pink, and one white, the genotype being shown directly by the phenotype. A similar result can be seen in Shorthorn cattle, where a cross between an individual with a red coat and one with a white coat gives roan offspring with coats consisting of a mixture of red and white hairs; the F_2 is one red, two roan, and one white (Fig. 4.2).

In the snapdragon, the phenotype of the F_1 hybrids from a cross between a plant with red and one with ivory-colored flowers depends on the environmental conditions. If the hybrids are reared in bright light at a relatively low temperature, their flowers will be red. If shaded and reared at a warmer temperature, the flowers will be ivory (Fig. 4.3). In this situation any attempt to ascribe dominance must first specify the environmental conditions.

A rather comparable example in animals is the inheritance and control of the appearance of yellow or white body fat in the rabbit. On its usual vegetarian diet, a rabbit consumes xanthophylls (a group of yellow compounds in green plants). In most rabbits, a liver enzyme breaks down the

Fig. 4.2. Intermediate roan heterozygote in Shorthorn cattle (left) together with red homozygote (above left) and white homozygote (above right). (From Hutt, F. B., 1964. *Animal Genetics*. Ronald Press Co., N.Y., p. 40, Fig. 3.1. Reprinted by permission of American Shorthorn Association, J. R. Henderson, Secy., P.O. Box 577, Hudson, Iowa 50643, and the author.)

xanthophylls to colorless derivatives. If a rabbit lacks this enzyme, the xanthophylls are stored intact in the body fat, which is then yellow. The presence or absence of the enzyme thus ordinarily determines the color of the body fat, and this in turn is controlled by a gene pair, with the dominant Y causing the enzyme to be formed. Therefore, YY and Yy rabbits will have the enzyme and white body fat, whereas yy rabbits will usually have yellow body fat. However, if such a yy rabbit is placed on a xanthophyll-free diet, it too will have white body fat. Thus here too, the expression of the gene is modified by the environmental conditions. Genes do not determine traits directly, but rather determine the range of possibilities that development may follow.

Not only external environmental conditions but the internal environment as well may modify dominance. In sheep H^1/H^1 is horned, and H/H hornless or polled, but the heterozygous H^1/H will be horned if male and hornless if female (Fig. 4.4). Such characters are called *sex-influenced* because dominance is influenced by the sex hormone differences between the males and females.

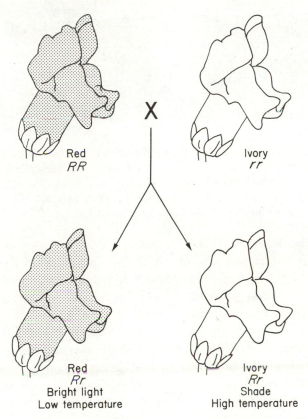

Fig. 4.3. Environmental effects on dominance in snapdragons.

The genetic environment in which a gene is placed may also modify its expression. Wild mice normally have long tails, but a mutation known as brachury produces short tails. The brachury gene behaves as a dominant in the European house mouse, *Mus musculus* (that is, heterozygotes are short tailed), but in the Asiatic house mouse, *Mus bactrianus*, the brachury gene acts as a recessive, for the heterozygotes have long tails.

Dominance, therefore, depends on the entire genotype and on the environment in which this genotype develops. It is not just the independent property of a single pair of factors. Dominance may or may not be complete, and the degree of dominance may vary under different conditions. About the only generalization that can be made about dominance is that if the parents belong to true-breeding strains, the F$_1$ hybrids will be uniform for the trait, provided, of course, that they are all raised under uniform conditions.

Most "wild-type" genes, the genes most commonly found in a population,

Fig. 4.4. Horns in sheep, a sex-influenced trait. Left, Dorset Horn ram. Center, F_1 ram from cross of Dorset Horn × polled breed. Right, polled Dorset ewe. F_1 ewes are also polled. (From Hutt, F. B., 1964. *Animal Genetics.* Ronald Press Co., N.Y., p. 162, Fig. 8.3; p. 163, Fig. 8.4. Reprinted by permission of Continental Dorset Club and the author.)

turn out to be at least partially dominant to the deleterious recessive mutants in the population. The origin of this dominance of the wild type has been a subject of considerable discussion. However, the implication that there exists one wild-type gene at each locus and that all of the other alleles are mutants and deleterious is an oversimplified picture of the actual situation. When a careful analysis is made, there turn out to be different alleles at a locus, all of which give the wild phenotype when homozygous. Nevertheless, they can be shown to be different because of significantly different phenotypic effects when heterozygous with a recessive mutant, or else they may differ in their physiological effects on such traits as temperature tolerance, longevity, or fecundity. Genes such as these are known as *iso-alleles*. Thus, although the concept of the "normal" or "wild-type" gene is a useful one, it is a considerable oversimplification to extend it to the idea of a "normal" individual homozygous at every locus for the "wild-type" gene at that locus. One has only to view the human species to realize that while we may have a few wild types, there is no one "wild type" in man.

4.2. Epistasis

The realization that the level of dominance itself may depend on the rest of the genotype leads us to a consideration of some of the simpler aspects of epistasis. In the dihybrid crosses studied earlier, the two factor pairs involved affected two different traits, such as color and seed coat. What happens when two pairs of genes affect the same trait? A classical example of this sort was reported by Bateson and Punnett on the inheritance of comb shape in fowl.

walnut   rose   pea   single

Fig. 4.5. Comb shape in fowl. (After Morgan.)

A type of comb called "rose" is found in the Wyandotte breed; "pea" is characteristic of the Brahma fowl; and the "single" blade is typical of such breeds as the Leghorn (Fig. 4.5). When Bateson and Punnett crossed these different types, the following results were obtained:

$$1. \quad P_1 \quad \text{rose} \times \text{single}$$
$$\downarrow$$
$$F_1 \quad \text{rose}$$
$$\downarrow$$
$$F_2 \quad 3 \text{ rose} : 1 \text{ single}$$

$$2. \quad P_1 \quad \text{pea} \times \text{single}$$
$$\downarrow$$
$$F_1 \quad \text{pea}$$
$$\downarrow$$
$$F_2 \quad 3 \text{ pea} : 1 \text{ single}$$

These crosses are perfectly good monohybrid crosses, with single recessive to both rose and pea. However, a cross between rose and pea had the following outcome:

$$3. \quad P_1 \quad \text{rose} \times \text{pea}$$
$$\downarrow$$
$$F_1 \quad \text{walnut}$$
$$\downarrow$$
$$F_2 \quad 9 \text{ walnut} : 3 \text{ rose} : 3 \text{ pea} : 1 \text{ single}$$

This cross has some unusual aspects. The F_1 has combs rather like the nut meat of an English walnut and unlike those of either parent. In the F_2 both rose and pea reappear, as well as walnut and a second new type in this cross, single. However, the F_2 ratio contains the clue to the interpretation of these results. The 9:3:3:1 ratio is typical of a dihybrid cross with independent assortment. The unusual results are due to the fact that two gene pairs on two different chromosome pairs both affect the same trait. With these facts in mind, the cross can be outlined as follows:

P₁	rose	×	pea
	RRpp	↓	*rrPP*
F₁		walnut	
		RrPp	
		↓	

		Genotype
F₂	9 walnut	*R–P–*
	3 rose	*R–pp*
	3 pea	*rrP–*
	1 single	*rrpp*

The F₂ may be derived from the gametic checkerboard,

♀ \ ♂	RP	Rp	rP	rp
RP	RRPP	RRPp	RrPP	RrPp
Rp	RRPp	RRpp	RrPp	Rrpp
rP	RrPP	RrPp	rrPP	rrPp
rp	RrPp	Rrpp	rrPp	rrpp

a zygotic checkerboard,

	¾ R–	¼ rr
¾ P–	⁹⁄₁₆ R–P–	³⁄₁₆ rrP–
¼ pp	³⁄₁₆ R–pp	¹⁄₁₆ rrpp

or a zygotic checkerboard with all genotypes shown,

	¼ RR	²⁄₄ Rr	¼ rr
¼ PP	¹⁄₁₆ RRPP	²⁄₁₆ RrPP	¹⁄₁₆ rrPP
²⁄₄ Pp	²⁄₁₆ RRPp	⁴⁄₁₆ RrPp	²⁄₁₆ rrPp
¼ pp	¹⁄₁₆ RRpp	²⁄₁₆ Rrpp	¹⁄₁₆ rrpp

Thus, R is a dominant factor for rose, and P is a dominant at a different locus for pea. The two dominants can be thought of as interacting to produce walnut; the two recessives, when homozygous, produce a single comb. A rose comb results from the interaction between dominant R and homozygous recessive p; pea results from that between P and rr. The cross has the usual 9:3:3:1 dihybrid ratio, but differs in having an F_1 with a trait not seen in either parent and another new type of comb appearing in the F_2.

Next let us consider an example of additive factors in the summer squash, which has the euphonious name, *Cucurbita pepo*. Squashes come in several shapes, among them a disk, a sphere, and an elongated form. In crosses sphere is recessive to disk, but in some crosses between different varieties with a spherical shape, rather unexpected results are observed. When crossed, the different spherical varieties give a disk F_1, and a 9 disk:6 sphere:1 elongate F_2 ratio. The explanation can be seen in the genotypes shown:

P_1	sphere	×	sphere
	AAbb	↓	*aaBB*
F_1		disk	
		AaBb	
		↓	

		Genotype
F_2	9 disk	9 *A–B–*
	6 sphere	3 *A–bb*
		3 *aaB–*
	1 elongate	1 *aabb*

The unusual aspect in this case is that two different genotypes (*A–bb* and *aaB–*) give identical phenotypes. The double recessive (*aabb*) has an elongate form. The addition of one dominant, either A or B, compresses this shape into a sphere. The addition of a dominant at the other locus flattens the squash still more into a disk.

The ancestor of all cultivated sweet peas (*Lathyrus odoratus*) is the wild sweet pea of Sicily, which has a purple flower with reddish wings. Purple acts as a dominant to white. A cross between two different white varieties sometimes produces an F_1 with flowers like the Sicilian ancestor, a reversion to the ancestral condition. In the F_2, nine purple and seven white are produced. The explanation for these results follows.

P_1 white   ×   white

       $CCpp$ ↓ $ccPP$

F_1 purple

        $CcPp$

         ↓

F_2 9 purple $9\ C\text{--}P\text{--}$

                                  $\begin{cases} 3\ C\text{--}pp \\ 3\ ccP\text{--} \\ 1\ ccpp \end{cases}$

       7 white

	$\frac{3}{4}\ C\text{--}$	$\frac{1}{4}\ cc$
$\frac{3}{4}\ P\text{--}$	$\frac{9}{16}\ C\text{--}P\text{--}$	$\frac{3}{16}\ ccP\text{--}$
$\frac{1}{4}\ pp$	$\frac{3}{16}\ C\text{--}pp$	$\frac{1}{16}\ ccpp$

The purple pigment is produced only if both independent dominant factors are present. White results whenever either dominant or both are absent. *C* and *P* are sometimes called **complementary factors** because both are required for the production of purple pigment. That they should have identical phenotypic effects can be most readily understood if their probable effects are visualized in chemical terms. The purple pigment must be derived from some precursor through a synthetic sequence. If the precursor and the intermediate compounds are colorless, then interruption of the sequence at different points will have the same phenotypic effect, the accumulation of a colorless compound behind a metabolic block.

In the shepherd's purse, *Capsella bursa-pastoris,* a small plant common as a weed in lawns in many areas, the shape of the seed capsules takes two forms, triangular or ovoid. In certain strains, a cross of triangular with ovoid gives a triangular F_1 and an F_2 consisting of 15 triangular to just 1 plant with a shape like a top. This situation arises because of the presence of independent duplicate dominant factors, either of which is sufficient to cause the formation of a triangular capsule.

P_1 triangular   ×   ovoid

       $T_1T_1T_2T_2$ ↓ $t_1t_1t_2t_2$

F_1 triangular

        $T_1t_1T_2t_2$

         ↓

F_2 15 triangular $\begin{cases} 9\ T_1\text{--}T_2 \\ 3\ T_1\text{--}t_2t_2 \\ 3\ t_1t_1T_2\text{--} \end{cases}$

       1 ovoid $1\ t_1t_1t_2t_2$

	$\frac{3}{4}\,T_1-$	$\frac{1}{4}\,t_1t_1$
$\frac{3}{4}\,T_2-$	$\frac{9}{16}\,T_1-T_2-$	$\frac{3}{16}\,t_1t_1T_2-$
$\frac{1}{4}\,t_2t_2$	$\frac{3}{16}\,T_1-t_2t_2$	$\frac{1}{16}\,t_1t_1t_2t_2$

Only the double recessive is top shaped. The result is rather similar to the cross with squashes, but there is no additive effect.

The color of the ordinary house mouse (*Mus musculus*) is known as agouti. This color is due to two pigments, most of the hair being black with a narrow yellow band near the top. In the light underparts the black or gray is near the base, and the rest of the hair is cream colored. A number of color varieties have been found in domesticated mice. Among them the albino has no pigment at all, and the black lacks the yellow pigment. A cross of black and albino can produce an agouti F_1 and a 9 agouti:3 black: 4 albino F_2. The explanation of these results is particularly helpful in understanding the concept of epistasis.

P_1	black	$\times$	albino
	$CCaa$	$\downarrow$	$ccAA$
F_1		agouti	
		$Co\Lambda a$	
		$\downarrow$	
F_2	9 agouti		9 $C-A-$
	3 black		3 $C-aa$
	4 albino		$\begin{cases} 3\ ccA- \\ 1\ ccaa \end{cases}$

	$\frac{3}{4}\,C-$	$\frac{1}{4}\,cc$
$\frac{3}{4}\,A-$	$\frac{9}{16}\,C-A-$	$\frac{3}{16}\,ccA-$
$\frac{1}{4}\,aa$	$\frac{3}{16}\,C-aa$	$\frac{1}{16}\,ccaa$

The gene c is said to be epistatic to the A locus because it prevents the expression of either color allele, A or a, at that locus. Because c is itself recessive at its own locus, this case is one of **recessive epistasis**. A similarity between dominance and epistasis can be seen in this instance. A dominant gene prevents the recessive allele from being expressed. The epistatic cc condition prevents the expression of either allele at the A locus.

A case of **dominant epistasis** is known in the summer squash. Fruit colors of white, yellow, and green are inherited as follows:

P₁   white   ×   yellow   yellow   ×   green
                        ↓   ↓
F₁   white   yellow
P₁   white   ×   green
        WWYY   ↓   *wwyy*
F₁   white
                        WwYy
                        ↓

F₂   12 white   ⎧ 9 *W–Y–*
                            ⎨ 3 *W–yy*
        3 yellow   ⎩ 3 *wwY–*
        1 green   1 *wwyy*

	¾ *W–*	¼ *ww*
¾ *Y–*	9/16 *W–Y–*	3/16 *wwY–*
¼ *yy*	3/16 *W–yy*	1/16 *wwyy*

W, which is dominant at its own locus, is epistatic to both yellow (Y) and green (y).

A final example of interactions between two loci is drawn from a cross between two white breeds of fowl, the Leghorn and the Wyandotte. The F_1 is also white, but the F_2 consists of 13 white to every 3 colored birds. The cross follows:

P₁   white Leghorn   ×   white Wyandotte
        IICC   ↓   *iicc*
F₁   white
                                IiCc
                                ↓

F₂   ⎧ 9 *I–C–*
                13 white   ⎨ 3 *I–cc*
                                                ⎩ 1 *iicc*
                3 colored   3 *iiC–*

	¾ *I–*	¼ *ii*
¾ *C–*	9/16 *I–C–*	3/16 *iiC–*
¼ *cc*	3/16 *I–cc*	1/16 *iicc*

In this cross the Leghorn carries a dominant color-inhibiting factor, I; the Wyandotte, a recessive inhibitor, cc. Only those birds lacking both inhibitors are able to develop pigment. This situation has been referred to as dominant and recessive epistasis, because both I and c are epistatic to any color factors.

The examples cited have shown six different modified dihybrid ratios (9:6:1, 9:7, 15:1, 9:3:4, 12:3:1, and 13:3). This series of dihybrid crosses has been considered in order to give some idea of the variety of interactions possible when just two pairs of factors affect a single trait. Since most traits are influenced by many gene loci, the possibilities for variation in these traits extend far beyond what has been indicated here.

As a final note to this chapter, it should be mentioned that in certain contexts dominance is often used to refer to all types of interactions between alleles, and epistasis to all types of interactions between genes at different loci. This somewhat broader meaning to these terms is not ordinarily a source of confusion, because the context usually makes clear which meaning is intended.

Additional Reading

Babcock, E. B., and R. E. Clausen, 1927. *Genetics in Relation to Agriculture.* McGraw-Hill, New York.

Bateson, W., 1909. *Mendel's Principles of Heredity.* Cambridge University Press, London.

Punnett, R. C., 1922. *Mendelism,* 6th ed. Macmillan, New York.

Searle, A. G., 1968. *Comparative Genetics of Coat Colour in Mammals.* Academic, New York.

Wagner, R. P., and H. K. Mitchell, 1964. *Genetics and Metabolism,* 2nd ed. Wiley, New York.

Wright, S., 1963. "Genic Interaction." In *Methodology in Mammalian Genetics,* W. J. Burdette, ed. Holden-Day, San Francisco.

Questions

4.1.   Why is it not appropriate to call dominance one of Mendel's laws?

4.2.   Distinguish between sex-influenced, sex-limited, and sex-linked traits.

4.3.   In the butterfly, *Colias,* a locus affecting wing color produces the following genotypes and phenotypes:

Genotype	Males	Females
W/W	orange	white
W/w	orange	white
w/w	orange	orange

How do you account for these results?

4.4. If a trait in man is more common in men than in women, can show up in men when neither parent shows it, shows up in all the sons of an affected woman, and is ordinarily transmitted directly from a father to half of his sons, how is this trait inherited?

4.5. A female albino mouse with a black father and mother produced, by an unknown father, a litter that was half agouti and half black. What was the color and the genotype of the unknown father?

4.6. In a cross in mice, the following ratio was obtained:

   3 agouti:1 black:4 albino

   What were the genotypes of the parents?

4.7. In a cross between albino and black mice the following ratio was observed:

   1 agouti: 1 black: 2 albino

   What were the genotypes of the parents?

4.8. If a *Drosophila* with brown eyes (*bw bw*) is crossed with one having scarlet eyes (*st st*), there will be two new eye phenotypes in the F_2. What proportion of the F_2 flies will show these two new phenotypes? The alleles *bw* and *st* are on different chromosomes.

4.9. One variety of yellow-centered daisy was crossed with a different variety of yellow-centered daisy. All the F_1 had purple centers and the F_2 produced a ratio of nine purple to seven yellow. If the F_1 were backcrossed to a homozygous double recessive, what would the progeny be like?

4.10. In poultry the genes for rose comb (*R*) and for pea comb (*P*), when present in the same individual, produce a walnut comb. Individuals homozygous for the recessive alleles (*rrpp*) have single combs. A rose-combed fowl crossed with a walnut-combed bird produced off-spring that were three-eighths walnut, three-eighths rose, one-eighth pea, and one-eighth single. What were the genotypes of the parents?

4.11. In poultry, white plumage may result from the homozygous condition of either or both of the recessive factors *c* and *o*. Color depends on the presence of both dominants *C* and *O*. White males with the genotype *CCoo* were crossed to white females of genotype *ccOO*. The colored F_1 birds were crossed to double recessives *ccoo*. a. What phenotypic ratio would you expect in these progeny? b. If the F_1

birds were crossed with one another, what phenotypic ratio would you expect in the F_2?

4.12. In addition to the complementary dominant genes C and O necessary for color in poultry, a dominant color inhibitor, I, is also known, which is epistatic to both C and O. What proportion of colored birds would be expected from crosses between birds of the following genotypes: $CcOOIi \times CcooIi$?

4.13. In *Drosophila*, a dominant gene, *Hw*, produces hairy wings except when a fly is homozygous for a recessive suppressor gene, *su*. Two normal flies are crossed and yield all hairy-winged offspring. What are the genotypes of the parents and their offspring? What will be the phenotypic ratio of the F_2?

4.14. A mouse fancier crossed an albino mouse with a solid black one. Half of the progeny were black-and-white spotted, which did not surprise him, but he could not account for the other half, which were solid black. Knowing that recessive white spotting is common in mice, how would you explain the parents' genotypes to him?

CHAPTER FIVE

Multiple Alleles

We have already defined alleles as alternative states of a gene found at the same relative position or locus on homologous chromosomes. For the most part we have dealt thus far with situations involving two alleles, one of which is usually dominant and the other recessive (*A* and *a*, for example). The possible genotypes in diploids, whether plant or animal, are then either *AA, aa,* or *Aa.* The diploid somatic cells of a single individual thus can contain at most only two kinds of alleles, and the gametes only one, since a haploid genome contains only one *A* gene, and the diploid genome just two.

However, a population of individuals, each of whom has two *A* loci, may contain more than two different kinds of *A* genes; a number of different alleles may exist at the *A* locus in the population. A series of such alleles occurring at the same locus on homologous chromosomes are referred to as *multiple alleles.* Although the entire population may contain a number of different alleles, nevertheless a single individual can carry at most two kinds of alleles if heterozygous, and only one kind if homozygous.

Multiple alleles were first discovered by Cuénot in 1904 in the course of his work with coat color in mice. He found that the genes for normal gray, for black, and for yellow each behaved as an allele of the other two, with yellow dominant to both gray and black, and gray dominant to black. Thus, his results indicated three alleles at the *A* (or agouti) locus, *A* (agouti),

A^y (yellow), and a (black). With three alleles taken two at a time (since mice are diploid), the number of possible genotypes is six, three homozygotes and three heterozygotes. Since this discovery many other multiple allelic series have been reported.

The C locus has been identified in a number of mammalian species, such as the mouse, rabbit, dog, guinea pig, and man. One recessive allele at this locus is responsible, when homozygous, for albinism. In the guinea pig several intermediate alleles have been identified in addition to the "intense" or "wild-type" allele.

Symbol	Designation	Coat color	Eye color
C	Intense	Black and red	Black
c^k	Dark dilution	Dark sepia and yellow	Black
c^d	Light dilution	Medium sepia and yellow	Black
c^r	Red-eyed dilution	Dark sepia and white	Red
c^a	Albino	White	Pink

The C allele is completely dominant to the other four alleles. However, the remaining alleles are not dominant to one another, heterozygotes being more or less intermediate. The number of different genotypes possible with n different alleles equals $(n^2 + n)/2$. In this case there are 15 different genotypes, of which only 11 are phenotypically distinguishable since the four heterozygotes, Cc^k, Cc^d, Cc^r, and Cc^a, will resemble CC. Hence, multiple alleles markedly increase the potential for variation. Not only does the number of possible genotypes increase rapidly as the number of alleles increases, but the possibilities for recombination among different loci are markedly enhanced. In red clover, a series of more than 200 different self-sterility alleles has been revealed, a mechanism that ensures a high level of heterozygosity in the population.

5.1. Blood Group Genetics

Because the blood groups in man are, for the most part, controlled by a series of multiple alleles and because of the inherent importance of the blood groups, we shall discuss them as examples of multiple alleles. Two terms basic to an understanding of blood grouping are antigen and antibody. An **antigen** has been defined as a foreign substance (typically a

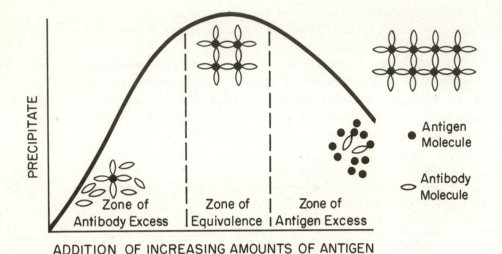

ADDITION OF INCREASING AMOUNTS OF ANTIGEN

Fig. 5.1. Diagrammatic representation of antigen-antibody reaction. (Figures 3–2, 3–3 of *Antibodies and Immunity* by G. J. V. Nossal, © 1969 by G. J. V. Nossal, published by Basic Books, Inc., Publishers, New York.)

protein) that, when introduced into the body of an animal, stimulates the production of a substance, or *antibody*, that reacts with the antigen (Fig. 5.1). The reaction seems to be protective in nature, for the antibody tends to neutralize or counteract the effects of the antigen. The various blood group antigens are to be found on the surface of the red blood cells. The antibodies are found in the globulin portion of the blood plasma or serum. (Plasma is the fluid part of blood; the serum is the fluid remaining after whole blood has clotted and the clot has been removed.) The antigen-antibody reaction may take place within the body, but the reaction will also occur if antigen and antibody are mixed in a test tube or on a microscope slide. A common type of antigen-antibody reaction is the aggregation of red cells into groups or clumps, a phenomenon called *agglutination*. In this case, the antigen is often referred to as an *agglutinogen* and the antibody as an *agglutinin*. Agglutination can be easily seen under the microscope if some red cells, say of blood type A, are mixed on a slide with an *antiserum* (serum containing antibodies) containing anti-A antibodies.

Some antibodies are produced by an organism only in response to the presence of a foreign antigenic substance. These induced antibodies are known as *immune* antibodies. Other antibodies are normally present and do not have to be induced. These may be referred to as *natural* antibodies.

The ABO System

The ABO blood group was discovered by Dr. Karl Landsteiner of Vienna in 1900, the same year that Mendel's laws were rediscovered. By testing the red cells from one person against the serum of another, he found that agglutination occurred in some cases but not in others. This finding indicated that normal antigens and antibodies must be present. The people tested could be classified into several groups on the basis of the kinds of antigens and antibodies present. The early results were explicable on the basis of the presence or absence of two antigens and two corresponding antibodies. The antigens are known as A and B. A person may have either, neither, or both, so that there are four possibilities, known as the ABO blood types. Everyone belongs to one of the four blood types, which, it should be noted, are dependent on the type of antigen present in the red cells. The ABO blood types and the types of antigens and antibodies present in each are shown in Table 5.1, together with the reactions they produce.

The ABO blood types were soon shown to be hereditary, and the genetic explanation most in accord with the facts was that a series of three multiple alleles was involved. Various symbols have been employed for this locus, but we shall use the following symbols for the three alleles:

$$A \ = \text{gene for antigen A}$$
$$A^B = \text{gene for antigen B}$$
$$a \ \ = \text{gene for neither}$$

Table 5.1

Blood type	Antigens in red cells	Antibodies in serum	Serum clumps cells of type	Red cells clumped by serum of type
O	—	Anti-A, Anti-B	A, B, AB	—
A	A	Anti-B	B, AB	O, B
B	B	Anti-A	A, AB	O, A
AB	A and B	—	—	O, A, B

A and A^B both act as dominants to a. However, in the heterozygote, AA^B, each allele produces its own antigen independently of the other. Thus, there is no dominance, and the alleles are sometimes referred to as *codominant*. The frequency of these alleles has been estimated in different populations as given below.

Allele	Frequency (%)			
	Norway	England	Japan	India
A	31.6	25.7	25.5	14.9
A^B	6.2	6.0	16.5	29.1
a	62.2	68.3	54.0	56.0

It is worth noting that mature human red blood cells lack nuclei, and therefore genes, but the gene products, the antigens, still remain. Furthermore, the presence of antigen A indicates the absence of anti-A antibody. Otherwise, a person would clump his own blood. Thus, some sort of control exists over the formation of both antigen and antibody.

The relatively simple and straightforward pattern outlined above has been somewhat complicated by further discoveries. One was the finding that there are two common serologically distinct A types, called A_1 and A_2, and that several more rare, weak A antigens exist. Consequently, several additional A alleles have been postulated. A few rare cases of weak B antigen have also been reported. Type O has been characterized as lacking both antigens A and B. This is correct, but more recently individuals of type O have been found to have an "H substance" as an antigen in their blood. Despite considerable investigation, a number of questions remain about the nature of these antigens and antibodies and their relation to the ABO system.

Attempts at blood transfusion had been made for centuries prior to 1900. Blood from animals as well as human blood had been tried, but the efforts usually failed—severe reactions, even death, frequently occurred. This discovery of the blood groups opened the way to an understanding of the reasons for the failure of interspecies and intraspecies transfusions, and for the first time made it possible to give massive blood transfusions to seriously ill patients with a high probability of success. Medical research on the blood groups has also led to applications of this new knowledge to human genetics, anthropology, and medico-legal problems.

The medico-legal applications of blood grouping most often involve cases of disputed paternity or mistakenly identified infants in hospitals. The blood groups and their corresponding genotypes are shown below:

Blood group	Genotype
O	aa
A	AA or Aa
B	A^BA^B or A^Ba
AB	AA^B

If the blood type of an infant and its mother are known, for example, the father's possible blood types can be determined, and any man not having one of these types can be excluded as the father of the child. For example, in a rather famous case the infant was type B, the mother type A; their genotypes must then have been the following:

	Blood type	Genotype
Mother	A	Aa
Infant	B	A^Ba

The baby obviously received her a gene from her mother, and must have received her A^B gene from her father, whose genotype thus could have been A^BA^B, A^Ba, or AA^B. The man brought to trial was type O with genotype aa, but in spite of the fact that he could not have been the father, he was found guilty. Such an outcome would be less likely today, but it points up the need to make the results of scientific discoveries intelligible to the public at large.

Blood tests of this sort can only *exclude* an individual as the father. Any individual with one of the appropriate blood types could not use the blood tests to demonstrate his innocence. However, a number of additional human blood group antigens has been discovered, and more are being reported periodically. The greater the number of antigens, the greater becomes the likelihood of proving innocence by exclusion. Conversely, the greater the number of antigens tested, the greater becomes the probability of identifying the actual father. This point has not yet been reached in man, where at least 14 blood group systems have now been established and several more are under investigation. With this battery of tests available, the usefulness of the method has been greatly enhanced. In cattle, some 40 different blood groups have been identified. It is thus possible to draw a blood group "portrait" of a particular animal that is as distinctive as a fingerprint. The chances that two individuals will be identical for all 40 antigens (unless they are identical twins) are extremely small.

Secretors

The ABO antigens have been found in an alcohol-soluble form not only in the red blood cells but also in many other tissues of the body. The A alleles are present in all the cells of the body, so this wide distribution of the antigens should not be surprising. Moreover, water-soluble forms of the ABO antigens have been found in body secretions, especially saliva. Not everyone, however, secretes the antigens so that two groups, "secretors" and "nonsecretors," can be identified. The secretor trait is controlled by a dominant gene, with *Se Se* and *Se se* being secretors and *se se*, the homozygous recessive nonsecretors. Apparently the only blood group antigens secreted are the ABO and the Lewis antigens (see below). The frequency of the secretor type has been estimated to be about 80% in the United States and varies in different races. The secretor locus, like the blood group antigens, has been useful in human genetics for studies both of families and of populations.

The MN System

In 1927, Landsteiner and Levine reported the second human blood group to be discovered, the MN system. The M and N antigens are inherited independently of the ABO system. The MN system differs from ABO in that natural antibodies are not ordinarily found in humans, and immune antibodies are not usually induced by transfusion. Thus, the MN group is generally not clinically significant; its main practical value is in cases of disputed paternity. Antibodies against these antigens are obtained by injecting rabbits with either M or N and inducing the formation of anti-M or anti-N antibodies. Human blood is then typed by testing against rabbit antiserum. Such tests revealed three phenotypes; individuals had one or the other antigen or both. There seemed to be nothing comparable to type O of the ABO system. Two alleles without dominance were postulated to account for the data, as shown below. (The symbol L used for this locus honors Landsteiner.)

Blood type	Genotype
M	$L^M L^M$
MN	$L^M L^N$
N	$L^N L^N$

Twenty years later, in 1947, Walsh and Montgomery discovered a new blood group antigen, S, that was associated with the MN system, and shortly thereafter a second antigen, s, was found. Family studies showed that, as with MN, persons with only S or s were homozygous, while the heterozygotes produced both kinds of antigens. Furthermore, these studies made it clear that the MN and Ss antigens were being inherited as if two antigens were controlled either by a single locus or by two extremely closely linked loci. In other words, the four types could be symbolized as being controlled by a series of four different multiple alleles, L^{MS}, L^{Ms},

L^{NS}, and L^{Ns}. The possible blood types and the corresponding genotypes, then, are as follows:

Phenotype	Genotype
MS	$L^{MS}L^{MS}$
Ms	$L^{Ms}L^{Ms}$
MSs	$L^{MS}L^{Ms}$
MNS	$L^{MS}L^{NS}$
MNSs	$L^{MS}L^{Ns}$ or $L^{Ms}L^{NS}$
MNs	$L^{Ms}L^{Ns}$
NS	$L^{NS}L^{NS}$
Ns	$L^{Ns}L^{Ns}$
NSs	$L^{NS}L^{Ns}$

The discovery of the S and s antigens permitted greater refinement of the blood typing possible at this locus. As with the ABO system, several rare alleles have been reported. The frequency of the common alleles in England has been estimated as follows:

Allele	Frequency (%)
L^{MS}	24.7
L^{Ms}	28.3
L^{NS}	8.0
L^{Ns}	39.0

Other Blood Groups

The discovery of the ABO and MN blood groups marked only the beginning of such discoveries; a number of additional blood group systems have now been found. The groups now fairly well understood and their year of discovery are shown below together with their allelic designations.

Blood group	Year	Major alleles
ABO	1900	A_1, A_2, A^B, a
MNSs	1927	L^{MS}, L^{Ms}, L^{NS}, L^{Ns}
P	1927	P^1, P^2, p
Rh	1939–1940	(See below)
Lutheran	1945	Lu^a, Lu^b
Kell	1946	K, k, k^a, K^o
Lewis	1946	Le, le
Duffy	1950	Fy^a, Fy^b
Kidd	1951	Jk^a, Jk^b

In addition to these blood groups, several others have been reported. The Diego and Sutter antigens are unusual in that both are extremely rare in Caucasians, but Diego is frequent in Mongols (e.g., Chinese, Japanese, and American Indians), and Sutter is apparently confined to Negroes. Other antigens are rare in all races, and may be confined to a single family. A number of these so-called "private" antigens have been reported, but thus far they have been of limited usefulness in human genetics. On the other hand, other antigens are now known that are so common that nearly everyone has them. These "public" antigens are difficult to identify because they are so common, and can be singled out only if a person is found who lacks them. Otherwise, they may be regarded as an antigen characteristic of the species. Some of these various kinds of antigens may be of clinical importance, while others are significant only as additional inherited human traits serving as chromosome markers and helpful in drawing individual blood group portraits.

The Rh System

The discovery of the Rh blood groups was undoubtedly the most significant advance in serology since the discovery of the ABO system some 40 years earlier. In 1939 Levine and Stetson reported that the mother of a stillborn infant had a severe hemolytic reaction to a transfusion of her husband's blood despite the fact that they were ABO compatible. Furthermore, her serum agglutinated her husband's red cells and those of 80 of 104 other ABO-compatible individuals. The antigen involved was shown to be independent of the groups known at the time, namely ABO, MN, and P. They suggested that the mother had been immunized against the new antigen by her fetus, which had inherited it from the father. Then, when the transfusion was made, the mother's immune antibodies reacted with her husband's red cell antigens. This seemingly complex interpretation was soon confirmed. The name of the system, however, came from the work on the rhesus macaque monkey reported in 1940 by Landsteiner and Wiener. They had injected blood from rhesus monkeys into rabbits in order to induce immune antibodies against the rhesus antigen. To their surprise, the antibodies not only agglutinated the monkey red cells, but also reacted with about 85% of the human red cells that were tested. Therefore, about 85% of the white New Yorkers tested had the rhesus antigen, and were called Rh positive, and the 15% without the rhesus antigen were called Rh negative. It was soon demonstrated that the antigen found by Levine and Stetson was the same as the rhesus antigen, but despite their priority of discovery, it is called the "Rhesus factor" or the Rh blood group system since they had not named it.

An investigation of the mode of inheritance of the antigen soon revealed that its presence is determined by a dominant gene, *Rh*, so that Rh positive people are *Rh Rh* or *Rh rh*, and Rh negative people are *rh rh*.

This blood group turned out to be of considerable clinical importance, because Rh incompatibility was found to cause not only problems in transfusion but also hemolytic disease of the newborn or *erythroblastosis foetalis*.

The Rh system differs from the ABO system in that natural anti-Rh antibodies are not present in the serum of Rh negative individuals, whereas an O individual has both anti-A and anti-B antibodies. It also differs from the MN system in that transfusion of Rh positive blood into an Rh negative person will induce antibody formation. The first such transfusion will ordinarily have no ill effects, but the introduced red cells stimulate antibody formation. Then, if a second transfusion from Rh positive to Rh negative occurs, a severe reaction will occur through the agglutination and destruction of the Rh positive red cells. The effects of this reaction may be so severe that death ensues.

A similar sort of reaction between mother and fetus leads to hemolytic disease in the newborn. Such cases may develop when an Rh negative mother is pregnant with an Rh positive child (Fig. 5.2). Although maternal and fetal circulatory systems are separate and independent of one another, they come into close contact in the placenta where exchanges between mother's and infant's capillary systems occur. Occasionally, however, Rh^+ antigens must get from the infant into the mother's circulatory system, perhaps because of capillary breaks or else at delivery. These antigens sensitize the mother, causing her to develop antibodies against the Rh^+ antigen. Usually the first pregnancy causes no difficulty, but if the mother becomes sensitized, she may have difficulty with subsequent pregnancies if the infant is again Rh^+. The antibodies do not affect the mother's red cells since, being Rh^-, they lack the antigen. However, if they cross the placental barrier, they may have severe effects because they destroy the baby's red cells, causing severe anemia and even attacking such tissues as those of the liver and spleen. As a result, the fetus or newborn child may die, or else may be seriously affected with symptoms of anemia and jaundice. The disease entity or syndrome is called *erythroblastosis foetalis* because erythroblasts, or immature nucleated red blood cells, ordinarily found only in the blood-forming tissues such as bone marrow, are found in the bloodstream. Today the Rhogam treatment of an Rh negative mother with anti-Rh^+ antibody within 72 hr after delivery of an Rh positive baby prevents her from becoming sensitized.

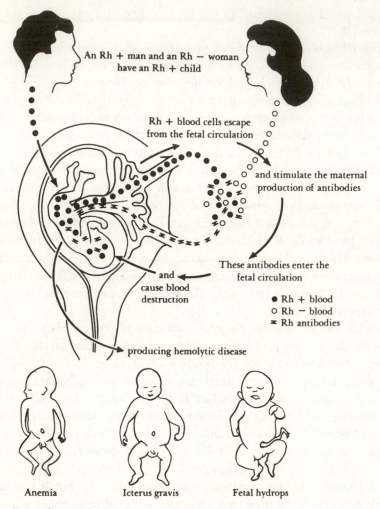

Fig. 5.2. The etiology of congenital hemolytic disease (*erythroblastosis foetalis*). (From *Rh* by Potter, E. L. Copyright © 1947 by Year Book Medical Publishers, Inc., Chicago. Used by permission.)

Although it is now known that other blood group incompatibilities (ABO, for example) may result in hemolytic disease of the newborn, more than 90% of the cases reported can be attributed to Rh incompatibility. This incompatibility between mother and child must have been a significant cause of infant mortality since the evolution of man, but it has been recognized for only about three decades. One may properly ask why such an incompatibility has been able to persist, and has not been eliminated during the course of human evolution. One possible answer is that in the long view of human evolution, viviparity (giving birth to live young) is a relatively recent innovation. Our reptilian ancestors reproduced by means of a shelled egg, and it may be that man has not yet completely adjusted to this relatively new form of reproduction.

Since about 15% of all women are Rh negative, and they may marry an Rh^+ Rh^+ or an Rh^+ Rh^- male, it can be estimated that one-tenth of all pregnancies are of the type Rh negative mother and Rh positive child. Despite this frequency, the incompatibility causes difficulty in only about one in 250 births rather than 1 in 10; in other words, only 1 in 25 such pregnancies actually results in symptoms of hemolytic disease in the newborn. Apparently, several factors are involved. The first pregnancy of an Rh negative woman usually causes no trouble, because the antibodies do not form rapidly enough. If the husband is heterozygous, only half the children will be involved either in sensitizing the mother or reacting to her antibodies. Furthermore, the penetration of the placenta may be variable, since some Rh negative women are known to have produced a number of Rh positive children without becoming immunized. Another possibility is that women differ in their sensitivity to the Rh^+ antigen. Some seem to form antibodies easily while others do not. Another major factor in the low incidence of Rh incompatibility is the interaction between Rh and ABO incompatibility. It has been discovered that an ABO incompatibility between an Rh negative mother and an Rh positive child actually reduces the probability of hemolytic disease markedly. The probable explanation is that if the mother is O and the father AB, for example, any of the child's red cells (which will be either type A or type B) that get into the mother's circulation will be destroyed by her natural anti-A and anti-B antibodies before they can induce the formation of anti-Rh^+ antibodies in her blood.

You may wonder why Rh incompatibility does not develop when an Rh positive mother carries an Rh negative child. The child's antibody-forming system does not mature until some 6 months after birth. Therefore, the fetus cannot produce antibodies against the mother's red cell antigens even if they do manage to cross the placental barrier.

The Rh blood group system was soon found to be considerably more complex than the simple Rh positive-Rh negative scheme outlined above. The situation has been still further complicated by differences in the interpretation of research findings and in the symbols used in describing results. Consequently, the facts are becoming increasingly clear, but the student learning about this system for the first time may find the various systems of nomenclature rather confusing. The basis for this difference is a fundamental one. Fisher, Race, and Sanger assume that the Rh system is controlled by a series of very closely linked genes, each controlling the formation of a different antigen, which they call the C, D, and E antigens. Wiener and his adherents believe that only a single locus is involved, that a series of multiple alleles is present, each of which can produce as many as three different antigens. The Wiener hypothesis may very well be correct, but its greater simplicity has led to wide use of the CDE system.

A number of different antigens have now been identified by the use of five different antisera. The so-called Rh antigens are known as C, D, and E; the Hr antigens are c, d, and e. As yet no anti-d antiserum has been found. Under the Wiener system each allele at the *Rh* locus controls the production of these three kinds of antigens, and eight alleles can be identified as follows:

| Alleles | | Antigens | Frequency in |
Fisher	Wiener	present	Caucasians (%)
Rh^{dce}	r	dce	38.9
Rh^{dCe}	r^1	dCe	0.6
Rh^{dcE}	r^{11}	dcE	0.5
Rh^{dCE}	r^y	dCE	0.01
Rh^{Dce}	R^0	Dce	2.6
Rh^{DCe}	R^1	DCe	40.8
Rh^{DcE}	R^2	DcE	14.1
Rh^{DCE}	R^z	DCE	0.2

Several points should be noted about this table. The total of allelic frequencies does not add up to 100 percent because other rare antigens have been identified (e.g., C^w, D^u, and E^u.) Since the D antigen is the one primarily responsible for hemolytic disease of the newborn, Wiener's system used a capital R to designate D positive individuals. However, the Rh negative genotype is r/r (Wiener) or Rh^{dce}/Rh^{dce} (Fisher and Race). In reality, the Fisher-Race designation would be *dce/dce*. In the absence of convincing evidence of crossing over, the modified system used here has been adopted as a compromise. It retains the simplicity of the Fisher-Race system, but is intended to carry the implication that a single complex *Rh* locus is involved rather than three separate, independent loci. The sequence *DCE* is used because blood has been found that lacks all the C and E antigens. Such a situation can be interpreted as being due to a single deletion involving both C and E (*D--*). If the order were *CDE*, two deletions would be required (*-D-*). Oddly enough, one instance has been reported in which no Rh substances at all could be detected.

Recently it was estimated that about 20 Rh antibodies and antigens have been identified and about 30 gene complexes or alleles have been distinguished. Obviously, the system is very complex, and more remains to be learned. However, the advances already made have led to considerable insight, and the knowledge can be applied clinically with some confidence.

Additional Reading

Giblett, E. R., 1969. *Genetic Markers in Human Blood.* Oxford University Press, London.
Hildemann, W. H., 1970. *Immunogenetics.* Holden-Day, San Francisco.
Moody, P. A., 1967. *Genetics of Man.* Norton, New York.
Race, R. R., and R. Sanger, 1968. *Blood Groups in Man,* 5th ed. Davis, Philadelphia.

Questions

5.1.   Four alleles have been identified at the same chromosomal locus. If dominance is lacking, how many different phenotypes are possible?

5.2.   At the brown locus in *Drosophila melanogaster,* three alleles are known: the wild type (bw^+); brown dominant (bw^D), dominant to the wild type; and brown recessive (bw), recessive to the wild type. (bw/bw) and $(bw^D/-)$ flies are virtually indistinguishable phenotypically. How many different phenotypes are possible with these three alleles?

5.3.   In rabbits, full color, Himalayan, and albino are multiple alleles at the C locus with decreasing dominance in the order given. In a dispute over the ownership of a pet white rabbit one boy, Joe, claims the rabbit even though his friend Jake claims that it is one member of a litter that consists of two Himalayan rabbits and four full-colored rabbits in addition to the white one. The parents of the litter are an albino and a full-colored rabbit. To whom would you award the white rabbit?

5.4.   What is the probability that a child will have either blood type A or B if his parents belong to groups AB and O?

5.5.   What are the chances that a child will have type AB blood if his parents are of types A and B?

5.6.   If one parent has type AB blood and the other has type O, what is the probability that their first child will be a type A girl?

5.7.   A woman with blood group A had a child of blood group O. What are the possible genotypes for the father?

5.8.   If you are of blood type O and neither of your parents is, what is the probability that your brother is also of type O?

5.9.   The first child of a normally pigmented woman of blood group AB belongs to group B and is albino. The father has normal pigmentation and type A blood. What is the chance that the next child born to this couple will be an albino boy with group A blood?

5.10. What is the probability that identical twins from MN parents will also both belong to type MN?

5.11. If there are two alleles, M and N, responsible for the MN blood type and three alleles A, A^B, and a for the ABO blood type, how many different genotypes are possible in a population?

5.12. How would you determine experimentally whether or not two recessive color mutants in *Drosophila* were allelic?

CHAPTER SIX

Probability and the Chi-Square Test

Experimental results obtained in crosses such as those discussed in Chapters 2 and 4 seldom conform to the exact ratios expected. The F_2 from the cross between round yellow and wrinkled green peas agreed quite well with a 9:3:3:1 ratio, so it is easy to accept these data as a reasonable approximation of such a ratio. However, data sometimes deviate markedly from expected ratios, and it becomes difficult to decide whether in fact the results conform to the theory on which the expectations were based. Under such circumstances the geneticist needs help in the decision-making process.

Experimental data can be presented and interpreted most clearly with the aid of statistics. Many of the statistical methods most commonly used today were first developed by such pioneers in the study of heredity as Francis Galton, Karl Pearson, and R. A. Fisher. For the most part, their methods were devised for use in the study of the inheritance of quantitative traits such as height or weight, but the techniques have been found applicable in many other ways as well. Statistics is now an integral part of the training of students in public health, psychology, sociology, economics, and just about all the physical and biological sciences. Statistics can be used not only to interpret scientific data, but also to assess the odds in games of chance and in a variety of other ways. Despite its usefulness, statistics is not a substitute for common sense. For instance, no one will be able to progress very far in statistics if he reasons like a certain Protestant woman in Northern Ireland who, having had three children, flatly refused to have a fourth because she had heard that every fourth child born there was Catholic.

In this chapter some very basic information necessary in the interpretation of genetic data is presented. A somewhat broader understanding of statistics is needed for the study of the inheritance of quantitative traits (Chapter 19 and the Appendix).

6.1. Probability

The concept of probability underlies all of the more elaborate statistical methods, and therefore we shall start with a consideration of this subject. If an event is certain to occur, it has a probability of 1. If it is certain not to occur, it has a probability of 0. If there is an element of uncertainty as to whether or not it will occur, the probability of its occurrence lies somewhere between 0 and 1.

It is simplest, perhaps, to discuss probability in terms of coin tossing. If a coin is tossed, and both the coin and the tosser are unbiased and honest, the coin can be expected to land heads half the time and tails half the time. The probability that the coin will come up heads is therefore ½. Any probability may be expressed as the number of times, m, that the event is expected to occur in n trials, or m/n. Thus in 10 tosses, the probability of heads is $5/10$. The expectation that a coin will fall heads half the time is based on the physical nature of the coin and on experience. This probability does not mean that we can expect to get exactly 5 heads in every series of 10 tosses, but rather that the *average* number of heads would turn out to be 5 in an infinitely large number of sets of 10 tosses. This is an empirical definition of probability; it is obvious that no probability can ever be known exactly under this definition, because no one can toss a coin an infinite number of times. However, confidence in the value obtained from a number of tosses increases as the number of tosses continues to increase.

The Probability of Independent Events

If two or more events are possible, the probabilities of the various combinations of events can be obtained from their separate probabilities. If two coins are tossed simultaneously, the way one falls has no influence on the fall of the other; they are *independent* events. There are four possible combinations of heads and tails. If, for example, a nickel and a dime are tossed, the combinations are as follows:

	Nickel	*Dime*
1.	Heads	Heads
2.	Heads	Tails
3.	Tails	Heads
4.	Tails	Tails

Each of the four combinations is equally likely. The first rule of probability deals with the probability of independent events such as these. The rule is that the probability of two or more independent events, occurring together, is the *product* of their separate probabilities. Thus

the probability of the nickel falling heads = $\frac{1}{2}$
the probability of the dime falling heads = $\frac{1}{2}$
the probability of both falling heads = $\frac{1}{2} \times \frac{1}{2} = \frac{1}{4}$
the probability of the nickel falling heads and the dime tails = $\frac{1}{2} \times \frac{1}{2} = \frac{1}{4}$
the probability of the nickel falling tails and the dime heads = $\frac{1}{2} \times \frac{1}{2} = \frac{1}{4}$
the probability of both falling tails = $\frac{1}{2} \times \frac{1}{2} = \frac{1}{4}$

If two unmarked nickels are used, tossing both at once gives a probability of $\frac{1}{4}HH:\frac{1}{2}HT:\frac{1}{4}TT$. The resemblance to the 1:2:1 monohybrid F_2 Mendelian genotype ratio is clear. The reason is that the Mendelian ratio is due to the random or independent combination of the two kinds of gametes at fertilization.

Suppose that three coins are tossed simultaneously. What are the chances of three heads? The probability for each is $\frac{1}{2}$; for all three it is $\frac{1}{2} \times \frac{1}{2} \times \frac{1}{2} = \frac{1}{8}$. On the average, three heads can be expected to come up once in eight trials. The same probability holds for three tails and for each of the other possible combinations, of which there is now a total of eight. Summation will show the following proportions of heads and tails:

$$\frac{1}{8}(3H):\frac{3}{8}(2H; 1T):\frac{3}{8}(1H; 2T):\frac{1}{8}(3T)$$

This method is rather cumbersome, especially as the number of coins tossed becomes larger. Therefore the **binomial expansion** has been found very useful in the calculation of probabilities. A binomial takes the form $(p+q)^n$. The rules for expansion are relatively simple:

1. There are $n+1$ terms. Thus $(p+q)^3$ has four terms.
2. The first term is p^n; the last is q^n. (Actually, the first is $p^n q^0 = p^n \cdot 1 = p^n$.)
3. The *exponent* of p in any term after the first is one less than in the preceding term; the exponent of q is one more than in the preceding term. The sum of the exponents of p and q for any one term equals n. Thus $(p+q)^3 = p^3 + -p^2q + -pq^2 + q^3$.
4. The *coefficient* of any term after the first is obtained from the preceding term. The product of the coefficient of the preceding term and the exponent of p is divided by the rank of the preceding term. Thus,

$$
\begin{array}{llllll}
 & \text{1st term} & \text{2nd} & \text{3rd} & \text{4th} & \text{5th} \\
(p+q)^3 = & p^3 & +3p^2q & +3pq^2 & +q^3 & \\
(p+q)^4 = & p^4 & +4p^3q & +6p^2q^2 & +4pq^3 & +p^4
\end{array}
$$

If the chance of heads is $p = \frac{1}{2}$, and of tails is $q = \frac{1}{2}$, the binomial expansion will give the exact probability for each combination of heads and tails. For example,

$$(p+q)^4 = (\tfrac{1}{2})^4 + 4(\tfrac{1}{2})^3(\tfrac{1}{2}) + 6(\tfrac{1}{2})^2(\tfrac{1}{2})^2 + 4(\tfrac{1}{2})(\tfrac{1}{2})^3 + (\tfrac{1}{2})^4$$

$$
\begin{array}{ccccc}
\text{4H} & \text{3H:1T} & \text{2H:2T} & \text{1H:3T} & \text{4T} \\
\tfrac{1}{16} & \tfrac{4}{16} & \tfrac{6}{16} & \tfrac{4}{16} & \tfrac{1}{16}
\end{array}
$$

In this case the probability of four heads is only 1 in 16, but the probability of three heads and one tails is 4 in 16. One method of looking at this relationship may aid in understanding it. If the four coins are a penny, a nickel, a dime, and a quarter, all four must come up heads to get four heads. However, three heads and one tail can occur four different ways, depending on whether the penny, the nickel, the dime, or the quarter comes up tails.

The probabilities need not be confined to coin tossing. They may also be applied to such problems as the inheritance of sex in man. For instance, the expansion of $(p+q)^5$ will give the expected frequency distribution of the various possible combinations of boys and girls in families of five children.

Let $p = \frac{1}{2} =$ the chance of a boy, and $q = \frac{1}{2} =$ the chance of a girl. Then

$$(p+q)^5 = p^5 + 5p^4q + 10p^3q^2 + 10p^2q^3 + 5pq^4 + q^5$$

$$
\begin{array}{cccccc}
\text{5B} & \text{4B:1G} & \text{3B:2G} & \text{2B:3G} & \text{1B:4G} & \text{5G} \\
\tfrac{1}{32} & \tfrac{5}{32} & \tfrac{10}{32} & \tfrac{10}{32} & \tfrac{5}{32} & \tfrac{1}{32}
\end{array}
$$

Thus, among 32 families of five children, only one would be expected to be all boys, and only one in 32 would be expected to be all girls. The most common types of family to be expected would be either three boys and two girls or two boys and three girls, which make up 20/32 of the total.

In some cases, only one particular probability is of interest. The complete expansion of the binomial to get one term is therefore laborious and unnecessary. To find, for example, the expected frequency of families with two boys and three girls among families of five children, the following formula is useful:

$$\text{The } p^r \text{ term} = \frac{n!}{r!\,(n-r)!} p^r q^{n-r}$$

r = the exponent of p in the desired term
n = the power to which the binomial is being expanded
$n!$ = n factorial (e.g., if $n = 5$, then $n! = 5 \cdot 4 \cdot 3 \cdot 2 \cdot 1$

Therefore, the p^2 term in $(p+q)^5$ is

$$\frac{5 \cdot 4 \cdot 3 \cdot 2 \cdot 1}{(2 \cdot 1)(3 \cdot 2 \cdot 1)} p^2 q^3 = 10 p^2 q^3$$

and the p^3 term in $(p+q)^6$ is

$$\frac{6 \cdot 5 \cdot 4 \cdot 3 \cdot 2 \cdot 1}{(3 \cdot 2 \cdot 1)\,(3 \cdot 2 \cdot 1)} p^3 q^3 = 20 p^3 q^3$$

One final word is needed to identify two distinct types of problem that are sometimes confused. One is the type just dealt with—the expected frequency of families with five boys, which is only 1 in 32 families of five children. The other deals with the probability that, in a family already having four boys, the next child will also be a boy. In this case the probability that the fifth child will be a boy is ½. There is neither a greater likelihood that the next youngster will be a boy nor a higher probability that it will be a girl. When the events are truly independent, what has gone before has no influence on the probability of the next event.

The Probability of Dependent Events

When a coin is tossed and comes up heads, it cannot come up tails. Similarly, if one die from a pair of dice is rolled, only one face at a time can be up. If the face with six dots comes up, none of the other five faces can come up on the same roll. Any such set of events, in which the occurrence of one event in a single trial excludes the possibility of the others in the same trial, is said to be a set of **mutually exclusive** or **dependent** events.

The probability that any particular face, for instance, the face with one dot, will come up is ⅙. The probability that the face with two dots will come up is also ⅙. The probability that on a single roll either the face with one dot or the face with two dots will come up is equal to the *sum* of their separate probabilities or $\frac{1}{6} + \frac{1}{6} = \frac{2}{6}$. The rule for the calculation of the probabilities of mutually exclusive events is as follows: *The probability that one or another of a set of mutually exclusive events will occur in a single trial is the sum of the probabilities of the separate events.*

When both dice are rolled, of course, independent as well as dependent probabilities may be involved, depending on the probabilities under consideration. Before any probabilities can be estimated, a decision must be reached as to whether the events are independent, dependent, or a combination of both. If the events are independent, the probabilities are multiplied; if dependent, the separate probabilities are summed. If there is a combination of the two types of probability, care must be exercised in dealing with the probabilities. For example, to the question of the chances of rolling either a 2 or a 3 on both of the dice, the answer is that

the probability of a 2 or a 3 on the first die = $\frac{1}{6} + \frac{1}{6} = \frac{1}{3}$
the probability of a 2 or a 3 on the second die = $\frac{1}{6} + \frac{1}{6} = \frac{1}{3}$
the probability of a 2 or a 3 on both dice = $\frac{1}{3} \times \frac{1}{3} = \frac{1}{9}$

Let us consider the probabilities related to a Mendelian ratio. Albinism in man is due to a single recessive gene in the homozygous condition. Suppose that a husband and wife are both heterozygous (Aa) for the gene for albinism. If they have three children, what are the chances that all three will be normally pigmented? In this case, let

$$p = \tfrac{3}{4} = \text{probability of normal pigmentation}$$
$$q = \tfrac{1}{4} = \text{probability of albinism}$$

Then $(p + q)^3 =$

$$p^3 = (\tfrac{3}{4})^3 = \tfrac{27}{64} \quad \text{(all normally pigmented)}$$
$$3p^2q = 3(\tfrac{3}{4})^2(\tfrac{1}{4}) = \tfrac{27}{64} \quad \text{(2 normal:1 albino)}$$
$$3pq^2 = 3(\tfrac{3}{4})(\tfrac{1}{4})^2 = \tfrac{9}{64} \quad \text{(1 normal:2 albino)}$$
$$q^3 = (\tfrac{1}{4})^3 = \tfrac{1}{64} \quad \text{(3 albino)}$$

Thus there are 27 chances in 64 that none will be albino and only 1 chance in 64 that all three children will be albino.

What is the probability that parents heterozygous for albinism would have a family of five boys, all albino?

$$\text{chance of albino boy} = \tfrac{1}{4} \times \tfrac{1}{2} = \tfrac{1}{8}$$
$$\text{chance of family of five albino boys} = (\tfrac{1}{8})^5$$

These examples will give some notion of the simpler aspects of probability. Experience with actual probability problems is the best way to develop proficiency in dealing with the laws of probability.

6.2. Tests of Significance—Chi-Square

Sometimes the deviation of the data from a Mendelian ratio is large, and we suspect that perhaps something more than chance is involved. It then becomes necessary to determine whether or not the observed data are in agreement with expectation, and how large the discrepancies may be and still be explained on the basis of random sampling error or chance deviations. For this purpose, various tests of significance have been devised. These tests do not tell us whether or not a hypothesis is right, but simply what the probability is that results as bad or worse than those obtained could occur by chance if the hypothesis is correct. One of the most useful tests of significance in genetics is the chi-square (χ^2) test.

Let us take an example to illustrate the use of χ^2. In swine, hydrocephalus, a lethal condition marked by an excessive amount of cerebrospinal fluid in the cranial cavity, was observed among certain litters from normal parents. Data from a number of such litters gave the following numbers of piglets:

normal	136
hydrocephalic	42
total	178

This looks like a 3:1 ratio, with hydrocephalus a recessive lethal. The calculated 3:1 ratio for a sample of 178 pigs would be 133.5:44.5. It is possible to find the exact probability of getting a ratio of 136:42, or deviations from the 3:1 ratio greater than this, by expansion of the binomial $(\frac{3}{4}+\frac{1}{4})^{178}$ and selection of the appropriate terms. However, such an approach would obviously be extremely laborious. Therefore, methods have been devised by statisticians to simplify this chore, among them the χ^2 test, for which the distribution was first worked out by Karl Pearson in 1900. The formula for χ^2 is

$$\chi^2 = \sum \frac{(O-E)^2}{E}$$

O = observed

E = expected

Σ = the sum of (capital sigma)

The calculation of χ^2 is shown below:

Class	Observed O	Expected (3:1) E	Deviation $O-E$	$(O-E)^2/E$
normal	136	133.5	2.5	6.25/133.5 = 0.047
hydrocephalic	42	44.5	–2.5	6.25/44.5 = 0.140
	178	178.0		$\chi^2 = 0.187$

The expected values are calculated on the basis of some hypothesis, in this instance that the trait is inherited as a simple Mendelian recessive. Thus, among 178 offspring, $178/4 = 44.5$, the expected number of hydrocephalic pigs from heterozygous parents, and $44.5 \times 3 = 133.5$, the expected number of normal offspring, both homozygous and heterozygous. Note that in the calculation of χ^2, the E value in the denominator is the E for that particular class. Furthermore, the actual numbers are used in the calculation of χ^2; the χ^2 test is not appropriate for use with percentages or ratios.

The larger the deviation of the observed value from expected, the larger is $(O - E)$, and the larger the value of χ^2. The larger χ^2 becomes, the less likely that the deviations are simply due to chance. To determine the meaning of any particular value of χ^2, we must use prepared tables of χ^2 values. (Table 6.1).

Table 6.1   The Cumulative χ^2 Distribution [a]

	Probability of a larger value of χ^2								
df.	0.99	0.95	0.90	0.50	0.20	0.10	0.05	0.01	0.001
1	0.000	0.004	0.016	0.455	1.642	2.706	3.841	6.635	10.827
2	0.020	0.103	0.211	1.386	3.219	4.605	5.991	9.210	13.815
3	0.115	0.352	0.534	2.366	4.642	6.251	7.815	11.341	16.268
4	0.297	0.711	1.064	3.357	5.989	7.779	9.488	13.277	18.465
5	0.554	1.145	1.610	4.351	7.289	9.236	11.070	15.086	20.517
6	0.872	1.635	2.204	5.348	8.558	10.645	12.592	16.812	22.457
7	1.239	2.167	2.833	6.346	9.803	12.017	14.067	18.475	24.322
8	1.646	2.733	3.490	7.344	11.030	13.362	15.507	20.090	26.125
9	2.088	3.325	4.168	8.343	12.242	14.684	16.919	21.666	27.877
10	2.558	3.940	4.865	9.342	13.442	15.987	18.307	23.209	29.588
11	3.053	4.575	5.578	10.341	14.631	17.275	19.675	24.725	31.264
12	3.571	5.226	6.304	11.340	15.812	18.549	21.026	26.217	32.909
13	4.107	5.892	7.042	12.340	16.985	19.812	22.362	27.688	34.528
14	4.660	6.571	7.790	13.339	18.151	21.064	23.685	29.141	36.123
15	5.229	7.261	8.547	14.339	19.311	22.307	24.996	30.578	37.697
16	5.812	7.962	9.312	15.338	20.465	23.542	26.296	32.000	39.252
17	6.408	8.672	10.085	16.338	21.615	24.769	27.587	33.409	40.790
18	7.015	9.390	10.865	17.338	22.760	25.989	28.869	34.805	42.312
19	7.633	10.117	11.651	18.338	23.900	27.204	30.144	36.191	43.820
20	8.260	10.851	12.443	19.337	25.038	28.412	31.410	37.566	45.315

Table 6.1.   The Cumulative χ^2 Distribution [a] (*Continued*)

			Probability of a larger value of χ^2						
df.	0.99	0.95	0.90	0.50	0.20	0.10	0.05	0.01	0.001
21	8.897	11.591	13.240	20.337	26.171	29.615	32.671	38.932	46.797
22	9.542	12.338	14.041	21.337	27.301	30.813	33.924	40.289	48.268
23	10.196	13.091	14.848	22.337	28.429	32.007	35.172	41.638	49.728
24	10.856	13.848	15.659	23.337	29.553	33.196	36.415	42.980	51.179
25	11.524	14.611	16.473	24.337	30.675	34.382	37.652	44.314	52.620
26	12.198	15.379	17.292	25.336	31.795	35.563	38.885	45.642	54.052
27	12.879	16.151	18.114	26.336	32.912	36.741	40.113	46.963	55.476
28	13.565	16.928	18.939	27.336	34.027	37.916	41.337	48.278	56.893
29	14.256	17.708	19.768	28.336	35.139	39.087	42.557	49.588	58.302
30	14.953	18.493	20.599	29.336	36.250	40.256	43.773	50.892	59.703

[a] The table gives the probabilities most frequently used for significance tests based on χ^2. For example, the probability of observing a χ^2 with three degrees of freedom *greater* in value than 7.815 is 0.05. The probability of a χ^2 with five degrees of freedom *less* in value than 1.145 is $1 - 0.95 = 0.05$.

Table 6.1 is taken from Table IV of Fisher and Yates: *Statistical Tables for Biological, Agricultural and Medical Research,* published by Longman Group Ltd., London. (previously published by Oliver & Boyd, Edinburgh), and by permission of the authors and publishers.

In order to use the table, one must know the number of degrees of freedom (df). At this point, it is sufficient to say that where n is the number of classes of data, $df = n - 1$. In this case there are two classes of data (normal and hydrocephalic), so df = 1. Entering the table on the row with df = 1, we find that a χ^2 of 0.187 falls between values of χ^2 of 0.016, with a probability (p) of 0.90, and 0.455, with a probability of 0.50. What do these probabilities tell us? They mean that deviations from the expected numbers (133.5:44.5) as large as the ones observed (136:42) can be expected to occur by chance alone more than 50 times in 100 but less than 90 times in 100. In other words, if we collected data from 100 more groups of 178 piglets from parents both heterozygous for hydrocephalus, we would expect more than 50 of these groups to deviate further from the expected 3:1 ratio than in this case merely as the result of chance. Therefore, the agreement with expectation is excellent.

Hydrocephalus is also found in the mouse (Fig. 6.1). By analogy, one would expect that it might be inherited in a similar fashion in both species since both are mammals. From a number of litters the following data were obtained.

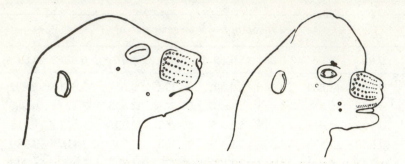

Fig. 6.1. Congenital hydrocephalus in the mouse (18-day embryos). Left, normal, right, hydrocephalic, showing abnormal head shape and failure of eyelids to close. (Hadorn, E., 1961. *Developmental Genetics and Lethal Factors.* Methuen & Co., Ltd., London.)

Class	O	E (3:1)	$(O-E)$	$(O-E)^2/E$
normal	2069	1855.5	213.5	45,582.25/1855.5 = 24.566
hydrocephalic	405	618.5	−213.5	45,582.25/618.5 = 73.698
total	2474	2474.0		χ^2 = 98.264

This large a value of χ^2 has a probability much lower than 1 in 100 (or even 1 in 1000) of being due to chance, as can be seen in the χ^2 table. Thus, the data do not appear to fit the hypothesis of a simple recessive trait, since there is a deficiency of hydrocephalic mice. However, in mice the mothers have often been observed to eat their defective young. When 43 litters from the same type of cross were watched carefully to prevent the mothers from consuming the young before they could be scored, the following results were obtained:

Class	O	E (3:1)	$(O-E)$	$(O-E)^2/E$
normal	280	291.75	−11.75	$(11.75)^2/291.75 = 138.0625/291.75 = 0.473$
hydro-cephalic	109	97.25	11.75	$(11.75)^2/97.25 = 138.0625/97.25 = 1.420$
Total	389	389.00		χ^2 = 1.893

In this case $0.20 > p > 0.10$. Thus, under these circumstances somewhat more hydrocephalic mice were observed than expected, but the chances are between 10 and 20 in 100 that these deviations are due to chance. It appears, therefore, that hydrocephalus is inherited as a recessive lethal

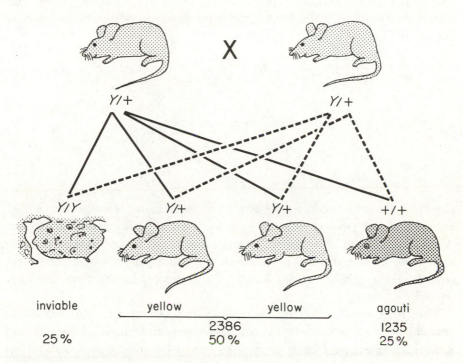

Fig. 6.2. The inheritance of yellow lethal in the mouse. (After Hadorn.)

trait in the mouse after all. These results suggest the usefulness of tests of significance. If the results of the first test had been accepted blindly, the 3:1 hypothesis would have been rejected out of hand. However, the test pointed up the likelihood that some unknown factor was influencing the data. The resulting search for this unknown bias led to its discovery and elimination so that a more meaningful test of the hypothesis could be carried out.

Next let us consider a hereditary trait in the mouse that at first did not seem to fit any Mendelian pattern. A mutant with yellow coat color could not be bred as a pure stock because whenever yellow mice were mated together, agouti (wild-type) mice appeared among the offspring as well as yellow. From a large number of such matings, 2386 yellow and 1235 agouti mice were obtained (Fig. 6.2). This result obviously does not fit a 3:1 ratio but rather fits a 2:1 ratio ($\chi^2_{1\ df} = 0.975$; $0.50 > p > 0.30$.) The difficulty with this result is that the magic number in Mendelian genetics is 2. The very nature of meiosis and fertilization leads to F_2 ratios whose sum is divisible by 2 (e.g. 3:1, 9:3:3:1, etc.) In this case the sum is divisible by 3 rather than 2. Crosses of yellow with agouti gave 177 yellow:178 agouti or a 1:1 ratio, suggesting that the yellow mice were heterozygous at the A locus. The cross of yellow and agouti then must be, with A^Y the dominant gene for yellow, and $+$ used as the symbol for the wild-type agouti,

$$A^Y/+ \ \times \ +/+$$
$$\text{yellow} \downarrow \text{agouti}$$
$$A^Y/+ \ : \ +/+$$

The cross of yellow with yellow, then, would be

$$A^Y/+ \quad \times \quad A^Y/+$$
$$\downarrow$$
$$A^Y/A^Y \quad : \quad 2A^Y/+ \quad : \quad +/+$$

However, no true-breeding yellows were ever found, so that the A^Y/A^Y group is absent from among the F_2 mice. Further study revealed that the A^Y/A^Y zygotes die at an early stage *in utero*. Thus the A^Y acts as a dominant color gene, but a recessive lethal, and the 2 yellow:1 agouti ratio results from the absence of one-fourth of the zygotes by the time the F_2 is scored.

This brief examination of the use of the χ^2 test has also served to introduce the subject of lethal genes. Differences in viability of the different genotypes emerging from a cross are perhaps the most common cause for deviations from expected Mendelian ratios. The most extreme effect on viability is lethality, but even among the lethals there is variation in the time of action of the gene that causes the death of the embryo. Thus the homozygous A^y/A^y mice die at a much earlier stage of development than those with hydrocephalus. With other mutants of less severe effect, only a fraction of the progeny may die or the rate of development of the mutant type may be slower than that of the non-mutant. The data from such crosses may be even more difficult to interpret than data involving lethals where every individual with a lethal genotype dies. Once one realizes that whole classes of data may be missing because of lethality, the treatment of such data poses few problems.

A χ^2 test with more degrees of freedom can be illustrated by still another type of inheritance. Cultivated flax, *Linus usitatissimum*, is infected by the flax rust *Melampsora lini*. (A field of flax in full bloom, with its blue flowers matching the blue Dakota sky, is a rare and beautiful sight.) There are a number of varieties of cultivated flax and a number of different "races" of the flax rust. The varieties of flax differ in their susceptibility to rust, and the different "races" of flax rust differ in their ability to attack the flax. These differences are genetically controlled by relatively few loci in both the host and the pathogen. Table 6.2 outlines the results of

Table 6.2   Segregation of F_2 Plants of Ottawa 770B × Bombay for Reaction to Races 22 and 24 of *Melampsora lini*

Reaction (S = susceptible; I = immune) and genotype of

	Parent varieties		F_2 plants			
	Ottawa 770B	Bombay				
Race and pathogenic genotype	*LLnn*	*llNN*	*L–N–*	*L–nn*	*llN–*	*llnn*
Race 22						
$a_L a_L A_N A_N$	S	I	I	S	I	S
Race 24						
$A_L A_L a_N a_N$	I	S	I	I	S	S
Number of plants observed			110	32	43	9
Number of plants expected (9:3:3:1)			109	36	36	12
$O - E$			1	–4	7	–3
$(O - E)^2/E$			1/109	16/36	49/36	9/12
			(0.009)	(0.444)	(1.361)	(0.750)

$\chi^2_{3df} = 2.564$ $0.5 > p > 0.3$

From Flor, H. H., 1956. "The Complementary Genic Systems in Flax and Flax Rust," *Adv. Genetics* 8:29–54, p. 42, Table 3.

crossing two varieties of flax and testing both parents and F_2 progeny against two "races" of flax rust. In this case there are four classes of F_2 data. Therefore, the number of degrees of freedom equals $(n-1)$ or $(4-1)$ or 3. The χ^2 table is entered on the row with three degrees of freedom and the probabilities are determined from the relation between the value obtained (2.564) and those shown in that row.

Not only the results but the postulated genotypes of the flax and of the rust have been shown in Table 6.2. In essence, the Ottawa variety is susceptible to Race 22 but immune to Race 24, while Bombay is immune to Race 22 and susceptible to Race 24. The F_2 shows immunity in this cross to be controlled by two different dominant genes at independently segregating loci, since 9/16 of the progeny are immune to both races, 3/16 are immune to Race 22, 3/16 are immune to Race 24, and only 1/16 are susceptible to both races.

The results of the cross between Races 22 and 24 of the flax rust are outlined in Table 6.3. Again in the F_2 the four classes of data give three

Table 6.3 Segregation of F_2 Cultures of Race 22 and Race 24 of *Melampsora lini* for Reaction to Ottawa 770B and Bombay

Reaction: I = immune (avirulent); S = susceptible (virulent) of variety to

Variety and reaction genotype	Parent Race		F_2 genotype			
	22	24				
	$a_La_LA_NA_N$	$A_LA_La_Na_Ñ$	A_L-A_N-	$a_La_LA_N-$	$A_L-a_Na_N$	$a_La_La_Na_N$
Ottawa 770B						
LLnn	S	I	I	S	I	S
Bombay						
llNN	I	S	I	I	S	S
Number of cultures observed			78	27	23	5
Number of cultures expected (9:3:3:1)			75	25	25	8
O–E			3	2	–2	–3
(O–E)²/E			9/75	4/25	4/25	9/8
			(0.20)	(0.160)	(0.160)	(1.125)
$\chi^2_{3df} = 1.565$			$0.70 > p > 0.50$			

From Flor, H. H., 1956. "The Complementary Genic Systems in Flax and Flax Rust," *Adv. Genetics* 8:29–54, p. 43, Table 4.

degrees of freedom. The data fit the hypothesis that the pathogenicity of the rust in relation to these two varieties of flax is determined by two independently segregating loci. All host-parasite combinations of genes other than A_L-L and A_N-N result in susceptibility.

It cannot be overemphasized how strikingly this cross shows that whether or not the disease develops depends on both the genotype of the host and the genotype of the parasite. Both the susceptibiliy of the host and the pathogenicity of the parasite are strongly influenced by their genotypes.

6.3. Levels of Significance

In the examples of χ^2 given thus far, the probabilities have been either quite large so that it has been reasonable to accept the hypothesis being tested or else, as in the first set of data on hydrocephalic mice, χ^2 was so large that the probability that the data at hand fitted the hypothesis was extremely remote. However, there are cases in which the deviations from expectation are large enough to give a sizable χ^2 and a fairly low probability that these deviations are due to chance. The question then becomes, how large must χ^2 be and how low the probability value that will lead us to reject the hypothesis under test. A decision is required, and the levels of significance used (usually indicated by the Greek letter a) are more or less governed by custom. One level of significance frequently used is that a be no larger than 0.05. A p value of 0.05 is conventionally referred to as "statistically significant." This means that if you use this level of significance to reject a hypothesis, there is one chance in 20 that the deviations are due merely to chance and that you will be wrong in rejecting it. Put another way, among 20 replications of an experiment in which the data fit the hypothesis, you would expect on the average to get one, by chance, with a χ^2 value giving a p of 0.05 or less.

The more serious the consequences of wrongly rejecting a hypothesis, the smaller a should be. However, if a is made too small, one runs the risk of accepting a hypothesis that is, in fact, wrong. An a of 0.01 is conventionally referred to as "statistically highly significant." In this case there is only one chance in 100 that you would be wrong if you rejected the hypothesis. In other words, you are betting against the one in 100 likelihood that you would get chance deviations this large on the very first experiment in a series of 100 experiments. Nevertheless, that probability does exist. Thus, the selection of levels of significance is arbitrary, and the interpretation of the data, in the final analysis, depends on judgment. However, the statistical tests provide a much firmer basis for judgment than is otherwise available, because the results can be stated with precision in terms of probabilities. If doubt remains, it may be necessary to gather further data, which ordinarily will tip the scales one way or the other.

Additional Reading

Bailey, N. T. J., 1967. *The Mathematical Approach to Biology and Medicine.* Wiley, New York.

Dixon, W. J., and F. J. Massey, 1969. *Introduction to Statistical Analysis,* 3rd ed. McGraw-Hill, New York.

Fisher, R. A., 1958. *Statistical Methods for Research Workers,* 13th ed. Hafner, New York.

Flor, H. H., 1956. "The Complementary Genic Systems in Flax and Flax Rust," *Adv. Genetics* 8:29–54.

Mather, K., 1964. *Statistical Analysis in Biology,* 5th ed. Methuen, London.

Simpson, G. G., A. Roe, and R. C. Lewontin, 1960. *Quantitative Zoology.* Harcourt Brace Jovanovich, New York.

Weaver, W., 1963. *Lady Luck. The Theory of Probability.* Doubleday, Garden City, N.Y.

Woolf, C. M., 1968. *Principles of Biometry.* Van Nostrand, New York.

Questions

6.1.   When tossing pennies, what is the probability that you will toss two heads, two tails, and then two heads in that order?

6.2.   What is the probability of throwing a 7 on one roll of a pair of dice?

6.3.   What is the probability that fraternal twins from MN parents will both also have type MN blood?

6.4.   If 3% of all normal men and women are heterozygous for the recessive gene for albinism, what is the probability that the husband of a heterozygous woman is also heterozygous?

6.5.   If 13% of all men and women are Rh negative, what is the probability that a husband and wife will both be negative?

6.6.   Huntington's chorea is a rare dominant disease that develops at about age 45. a. What are the chances that a grandchild of a person affected with Huntington's chorea will later develop the disease? b. What proportion of the grandchildren can be expected to be afflicted by the disease?

6.7.   "Blaze" or white forelock in man is inherited as a simple dominant. What is the probability that a grandniece of a man with "blaze" will also have a white forelock?

6.8.   An albino man marries a normal woman. Before you could say (with no more than one chance in 100 of being wrong) that the woman is not heterozygous for the recessive gene for albinism, how many normal children must the couple produce?

6.9.   a. What is the probability that an albino son, a normal son, and an albino daughter will be born, in that order, to parents heterozygous for albinism? b. What is the probability that these three types of children would be born to these parents in any order?

6.10. If you collect data on all three-child families in your county in which there is at least one boy, what proportion of the collected families would contain three boys?

6.11. Some dogs bark when they are trailing an animal; others, such as the Basenji, are silent. The barking trait is inherited as a dominant. A barker, with a Basenji mother, sired a litter of three puppies by a Basenji dam. What is the probability that one will bark when trailing and two will not?

6.12. In poultry, the genes for black and for splashed white are alleles, but neither is dominant, the heterozygous condition resulting in blue (the Blue Andalusian). A pair of Blue Andalusians are mated, and the hen lays three eggs. What are the chances that the chicks from these eggs will be one black, one splashed white, and one Blue Andalusian?

6.13. Shortly after her marriage, a woman with a very determined character announced that she was planning her family to be two boys and a girl, in that order. What are the chances that she will be able to fulfill her ambition?

6.14. A married couple has three children, two boys and a girl. After their experiences with these three, they decide they would like to have two more children if they could be sure both would be girls. What are the chances that they will realize their hope for a family of three girls and two boys?

6.15. Among families of five children, what is the probability that there will be two or more boys?

6.16. The inability to taste phenyl-thio-carbamide (PTC) is inherited as a recessive trait and so is red hair. What are the chances that the first child of parents heterozygous for these genes will be a red-haired girl who cannot taste PTC?

6.17. In man, polydactyly (which causes the formation of extra digits) is due to a dominant gene. a. When one parent is polydactylous but heterozygous, and the other parent is normal, what proportion of three-child families would you expect to consist of three normal children? b. What proportion would you expect to consist of one polydactylous and two normal children?

6.18. In man, an abnormality of the large intestine called intestinal polyposis is dependent on a dominant gene A, and a neurological disorder, Huntington's chorea, is determined by a dominant gene H. A man carrying the gene A (genotype $Aahh$) married a woman carrying the gene H (genotype $aaHh$). Assume that A and H are not linked. If they have two children, what is the probability that both children will be normal?

6.19. In an experiment with the shepherd's purse (*Capsella bursa pastoris*), a ratio of 15 plants with triangular seed capsules to 1 with ovoid capsules was expected. The observed numbers were 141 to 19. What is the value of χ^2 and what is its p value? How do you interpret the results of this experiment?

6.20. In an experiment with sweet peas, a 9:7 ratio of colored to white flowers was expected. The observed numbers were 80 colored:80 white. What is the value of χ^2 and what is its p value? How do you interpret the results of this experiment?

6.21. In an experiment with coat color in mice, a ratio of 9 agouti:3 black: 4 albino was expected. The observed numbers were 95 agouti:35 black:30 albino. What is the value of χ^2 and what is its p value? How do you interpret the results of this experiment?

6.22. In an experiment with comb shape in fowl, a ratio of 9 walnut: 3 rose:3 pea:1 single was expected. The observed numbers were 87 walnut:37 rose:25 pea:11 single. What is the value of χ^2 and what is its p value? How do you interpret the results of this experiment?

CHAPTER SEVEN

Sex Linkage

After Mendel, the next major advance in genetics was the development of the chromosome theory of heredity. Mendel's laws emerged primarily from the solitary efforts of one man, but the chromosome theory developed gradually as the result of the efforts of many workers over a number of decades. Mendel, of course, formulated his laws in ignorance of the nature and behavior of chromosomes, because the study of cytology really did not begin to make significant progress until the decade following 1865. Cell structures that we now realize are chromosomes had been described in 1848, but the word "chromosome" itself was not even coined until 1888. The concept of the continuity of the germ plasm, which drew attention to the nucleus and its chromosomes, was suggested in 1858 by Virchow and later elaborated by Roux and Weismann, but proof of the role of chromosomes in heredity was still far in the future.

After 1900 a number of workers discussed the relation between the segregation and independent assortment of Mendelian factors on one hand and the behavior of the chromosomes at meiosis on the other. W. S. Sutton published the first detailed, correct statement of this relationship in 1903. He postulated that the separation of paired homologous maternal and paternal chromosomes during meiosis could account for Mendelian segregation, and that the random orientation of the different pairs of homologues on the meiotic spindle could account for the independent assortment of different pairs of genes. The proof of this theory was not immediately forthcoming. Nevertheless, it brought together the rapidly advancing but separate fields of cytology and genetics, and henceforth the two became inseparably intertwined.

7.1. Sex Determination

Since the time of Aristotle, there have been a number of theories about the determination of sex. Practically all these theories were environmental, suggesting that phase of the moon, age of the parents, direction of the wind, and so on, at the time of conception or later were responsible for the sex of the offspring. Two centuries ago Drelincourt listed 262 such groundless hypotheses, to which must be added his own and many others proposed subsequently. Since the chances of being right were about 50% in every case, any theory could have considerable viability. It was not until the turn of this century that Cuénot in animals and Strasburger in plants reported that sex was determined by some internal mechanism. The nature of this internal mechanism was first indicated in 1901 by McClung, who suggested that an accessory chromosome, called the X chromosome, determined sex. In a few years the details of sex determination were established, and it was shown by Wilson and others that in the insects they studied the female had two X chromosomes while the male had only one X and a different kind of sex chromosome called the Y. The

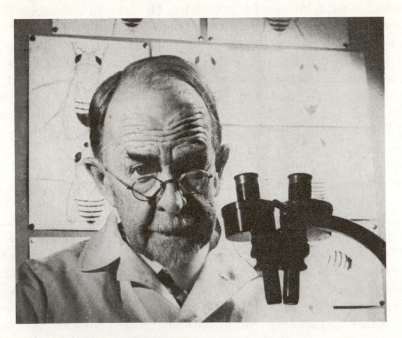

Fig. 7.1. Thomas Hunt Morgan, 1866–1945. Nobel laureate, 1934. (From *Portraits of Geneticists, 1967. Genetics.***)**

XY male produced two kinds of gametes, one bearing the X and the other the Y chromosome, and was therefore referred to as the *heterogametic* sex. The XX female was called *homogametic* because all of the eggs were alike in carrying a single X chromosome.

7.2. Sex Linkage

In the early 1900s Columbia University was an extremely active center of biological research. E. B. Wilson in cytology and T. H. Morgan in genetics (Fig. 7.1) were preeminent in their fields. Among their students were many who made major scientific contributions, including Sutton, Sturtevant, Bridges, and Muller. Many of these advances were the result of work with the fruit fly, first used in a genetic study on the effects of inbreeding by W. E. Castle at Harvard. Morgan learned of the advantages of *Drosophila* as an experimental animal from Castle at Wood's Hole, Massachusetts, where biologists from all over the United States still congregate in the summer.

· The discovery of sex linkage resulted from the appearance of a single white-eyed male fly in one of the culture bottles. This male was mated to the usual red-eyed or wild-type females with the following outcome:

$$P_1 \qquad \text{red} \; ♀ \quad × \quad \text{white} \; ♂$$
$$\downarrow$$
$$F_1 \qquad \text{all red}$$
$$\downarrow$$
$$F_2 \qquad \text{3 red:1 white}$$

These results looked like simple Mendelian segregation except for one fact: All the F_2 white-eyed flies were males. In some way white was associated with sex. Morgan hypothesized that these results could be explained if he assumed that the factors responsible for this trait were on the X chromosomes, that the Y carried no factor for eye color, and that the male *Drosophila* was the heterogametic sex. He then verified the hypothesis by four sets of experiments; in each case the predictions based on the hypothesis were borne out by the data.

In *Drosophila*, man, and most other species of animals, the male is the heterogametic sex. The male produces two types of gametes, one bearing an X sex chromosome plus a haploid set of the other chromosomes or autosomes, the other bearing the Y sex chromosome plus a haploid set of autosomes. Sex, then, is determined as follows:

```
        female  XX                    male  XY
                  ↓                          ↓
eggs          X or X          sperm      X or Y
                  |                          |
zygotes       XX ←                        → XY
            female                        male
```

In this way also the equality of the sexes was explained.

However, in birds and in moths and butterflies (Lepidoptera) the female turned out to be the heterogametic sex. By convention, the female is designated as ZW, the male as ZZ. In these species sex determination is as shown below:

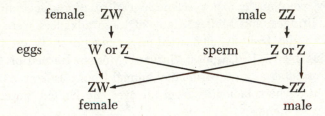

```
        female  ZW                    male  ZZ
                  ↓                          ↓
eggs          W or Z          sperm      Z or Z
                   \                         |
            ZW ←                          → ZZ
          female                          male
```

As we shall see subsequently, still other mechanisms of sex determination are known, but the heterogametic male system and the heterogametic female system, in which one sex is heterozygous for the sex chromosomes, are by far the most common in animals.

The cross between a white-eyed male fruit fly and a red-eyed female is diagrammed below, incorporating Morgan's assumptions that the factors for red or white are on the X, the male is the heterogametic sex, and the Y carries no factor for eye color.

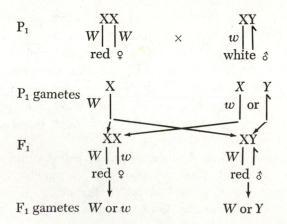

P_1 XX   ×   XY
    $W| |W$ $w| |↑$
    red ♀   white ♂

P_1 gametes W | X w | X or ↑ Y

F_1 XX   XY
    $W| |w$ $W| ↑$
    red ♀   red ♂

F_1 gametes W or w W or Y

	δ gam		
$\female$ gam		W	Y
F$_2$ W		WW red $\female$	WY red δ
w		Ww red $\female$	wY white δ

Thus Morgan's hypothesis clearly accounted for the results, for under this scheme one-fourth of the F$_2$ flies would be expected to be white-eyed males. The trait in some way is associated or linked with the sex chromosome and the phenomenon has come to be called sex linkage.

If Morgan's assumptions were right, it should be possible to predict the results of the reciprocal cross between a white-eyed female and a red-eyed male. The cross is shown below.

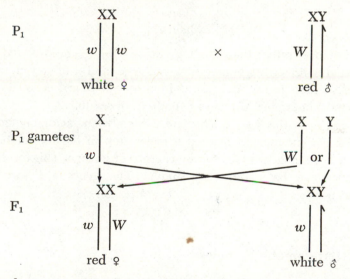

This result is most unusual compared to ordinary Mendelian segregation. The parents were white-eyed females and red-eyed males. Rather than the F$_1$ being all alike, the usual outcome of crosses thus far, both eye colors appear in the F$_1$ but in the opposite sex from the parents, for now the females are red-eyed and the males white-eyed. Although the trait is still associated with sex, it has switched to the opposite sex, a result at one time called criss-cross inheritance. The reason, of course, that the white-eyed trait appears in the F$_1$ males is that they have only one X chromosome (inherited from their homozygous white-eyed mother) and thus all carry and express a single white gene. Because the males have but one X, they cannot be heterozygous or homozygous for sex-linked genes, but are customarily spoken of as being *hemizygous*.

Continuing the cross, we see:

F_1	Ww red ♀	wY white ♂
	↓	↓
F_1 gametes	W or w	w or Y

♀ \ ♂	w	Y
W	Ww red ♀	WY red ♂
w	ww white ♀	wY white ♂

In this instance, rather than a 3:1 F_2 ratio, a 1:1 F_2 ratio of red to white is expected with equal numbers of red-eyed and white-eyed males and females. Needless to say, the results from the reciprocal cross did turn out as expected, confirming Morgan's hypothesis of sex linkage.

This case was the first to associate a Mendelian factor or gene with a particular chromosome. It marked a major step in the formulation of the next major principle of heredity, the concept of linkage. This third principle of genetics can be stated very simply: The genes are borne on the chromosomes.

7.3. Heterogametic Females

When the female is the heterogametic sex, as in birds and butterflies, the same principles apply but in reverse. Since this shift may at first be confusing, a case of inheritance in canaries will be considered. Some birds are green with black eyes (C); others are cinnamon with red eyes (c). The traits are sex-linked, with cinnamon recessive to green in the males. Therefore, a cinnamon female by green male cross gives the following results:

P₁ cW ♀   × CC ♂
cinnamon   ↓   green

F₁ CW ♀ Cc ♂
green   green

♀ \ ♂	C	c
C	CC green ♂	Cc green ♂
W	CW green ♀	cW cinnamon ♀

The reciprocal cross is as follows:

P₁ CW ♀   × cc ♂
green   ↓   cinnamon

F₁ cW ♀ Cc ♂
cinnamon   green

♀ \ ♂	C	c
c	Cc green ♂	cc cinnamon ♂
W	CW green ♀	cW cinnamon ♀

This bit of knowledge about sex linkage in birds has been put to use by commercial poultry breeders. It is important to them to know the sex of their chicks at an early age, but determining the sex of young chicks is by no means easy. However, if they cross hens with barred feathers (a dominant trait) with nonbarred roosters (i.e., BW ♀ ♀ × bb ♂ ♂), they will get all of the male chicks barred and all of the female chicks nonbarred in the F₁. This procedure has never been widely used, however, because a fast, accurate method of determining the sex of newly hatched chicks by cloacal examination was discovered shortly thereafter.

7.4. Other Forms of Sex Linkage

In man, faulty tooth enamel is inherited as a dominant trait. In some families it is an autosomal dominant, in others it is sex-linked. In order to distinguish between the two types of inheritance, the crucial mating is that between an affected man and a normal woman. If it is a sex-linked dominant, all of the man's daughters receive his X chromosome and will be affected; none of his sons receive an X from their father and therefore they will have teeth with normal enamel. If it is autosomal, both the sons and daughters may inherit the gene from their father.

In the case of a sex-linked dominant, the male transmits the dominant to all of his daughters but none of his sons. There is another, rather rare type of sex linkage in which the father transmits a trait to all of his sons and none of his daughters. This situation can exist only for traits controlled by genes on the Y chromosome. Even in *Drosophila*, very few such genes have been identified. Although we have used the expression, sex linkage, to refer to genes on the X chromosome, it would be more proper to distinguish between X-linked and Y-linked types of sex linkage. However, Y linkage or *holandric* inheritance is so rare and exceptional that confusion is unlikely, and we shall continue to use the term sex-linked to refer to X-linked genes.

The one-to-one sex ratio is very readily explained by the chromosome mechanism of sex determination. However, once in a while a sex ratio of two females to every male is found, for example in *Drosophila*. Here again, as in the case of yellow mice, the results suggest that some progeny are missing. The explanation of the unusual sex ratio is the presence of a sex-linked recessive lethal in the heterozygous condition in the female parent. The cross can be outlined as follows:

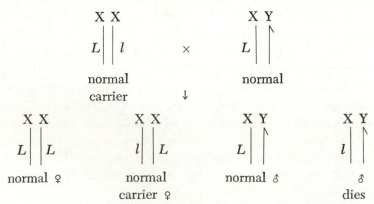

In species with heterogametic females, the sex ratio when a sex-linked lethal is present in a carrier male will be two males to one female. Similarly, the critical mating for the detection of a sex-linked dominant would be that between an affected female and a normal male. Finally, if a gene were present on the W chromosome, inheritance would be hologynic, from the mother to all of her daughters and none of her sons. As mentioned earlier, sex-linked heredity in birds and butterflies is just the reverse of the usual type of sex linkage.

Additional Reading

Bacci, G., 1965. *Sex Determination*. Pergamon, Oxford.

McKusick, V. A., 1964. *On the X-Chromosome of Man*. AIBS, Washington, D.C.

Mittwoch, U., 1967. *Sex Chromosomes*. Academic, New York.

Montague, A., 1959. *Human Heredity*. Mentor, New York.

Morgan, T. H., 1910. "Sex Limited Inheritance in *Drosophila*," *Science* **32**:120–122. (Reprinted in Peters. See references, Chapter 1.)

Morgan, T. H., 1925. *The Theory of the Gene*. Yale University Press, New Haven, Conn.

Morgan, T. H., and C. B. Bridges, 1916. "Sex-Linked Inheritance in *Drosophila*," Carnegie Inst. of Washington Pub. No. 237, 1–88.

Morgan, T. II., A. H. Sturtevant, H. J. Muller, and C. B. Bridges, 1922. *The Mechanism of Mendelian Heredity*, 2nd ed. Holt, Rinehart and Winston, New York.

Questions

7.1. a. In *Drosophila*, in which sex will sex-linked recessive traits be expressed more frequently?

  b. In turkeys, in which sex will sex-linked recessive traits be expressed more frequently?

7.2. a. In *Drosophila*, in which sex will sex-linked dominant traits be expressed more frequently?

  b. In bluebirds, in which sex will sex-linked dominant traits be expressed more frequently?

7.3. Barring in chickens is due to a sex-linked dominant gene (*B*). The sex of chicks at hatching is difficult to determine, but barred chicks can be distinguished fron nonbarred at that time. To use this trait so that all chicks of one sex are differently colored from those of the opposite sex, what cross would you make?

7.4. a. In cats, black is unlucky and is due to the sex-linked recessive gene (b); the dominant allele (B) causes yellow. The Bb heterozygote is tortoise shell. What kinds of offspring would be expected from the cross of a yellow male and a black female?

b. What color will the sons of a tortoise-shell female always be, no matter whether the father is black or yellow?

7.5. Hemophilia is inherited in man as a sex-linked recessive. A woman whose maternal grandfather suffered from hemophilia has parents who appear to be normal. She and her husband also seem normal. What are the chances that her first son will be normal?

7.6. An albino nonhemophilic man marries a normally pigmented, non-hemophilic woman whose father was hemophilic and whose mother was an albino. What is the expected ratio of genotypes and phenotypes among their sons and daughters?

7.7. In canaries, the green variety with black eyes is dependent on a dominant sex-linked factor (B), and the cinnamon variety with red eyes on its recessive allele (b). What is the expected appearance of the F_1 progeny from a cross between a cinnamon male and a green female?

7.8. In a flock of turkeys, if you found that in some matings you got a ratio of two males to one female, what explanation seems most likely?

CHAPTER EIGHT

Autosomal Linkage and Chromosome Mapping

8.1. Linkage

The first experimental evidence indicating that the Mendelian factors or genes are borne on the chromosomes was the work done by Morgan with the white eye trait in *Drosophila melanogaster*. Since the pattern of transmission of the white eye gene paralleled the transmission of the X chromosome, he inferred that the gene must be on the X chromosome. This inference seemed highly probable but was, nonethless, an inference.

The next major piece of evidence for the chromosome theory was reported in 1916 by Calvin Bridges, one of Morgan's students. In crosses between white-eyed females and red-eyed males, Bridges sometimes obtained white-eyed daughters and red-eyed sons in the F_1 instead of the expected red-eyed daughters and white-eyed sons. When checking revealed that this result could be repeated, an explanation became essential.

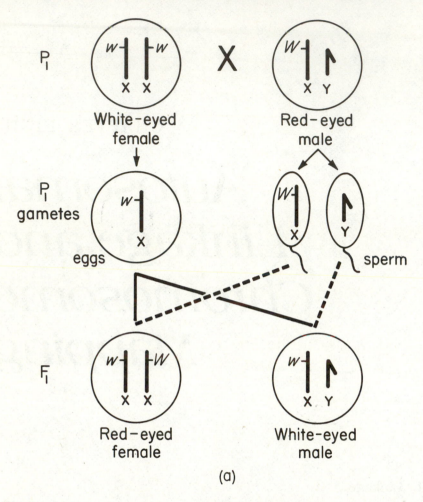

(a)

Fig. 8.1. (a) Normal segregation and recombination of the sex chromosomes and the sex-linked genes at the white locus in *Drosophila melanogaster*. (b) The effects of nondisjunction of the X chromosomes in the female. (See text.)

Bridges proposed that this result could be obtained if, in these exceptional females, their two X chromosomes failed to separate during gametogenesis. In this event, half the eggs would have two X chromosomes, the other half none. Figure 8.1 outlines the usual and the exceptional crosses so that the differences between them can be compared.

If Bridge's theory of nondisjunction of the X chromosomes in the female were correct, it meant that the normal sex-determining mechanism was completely out of kilter, but it also made possible predictions about the chromosomal makeup (or karyotype) of the exceptional flies. The ex-

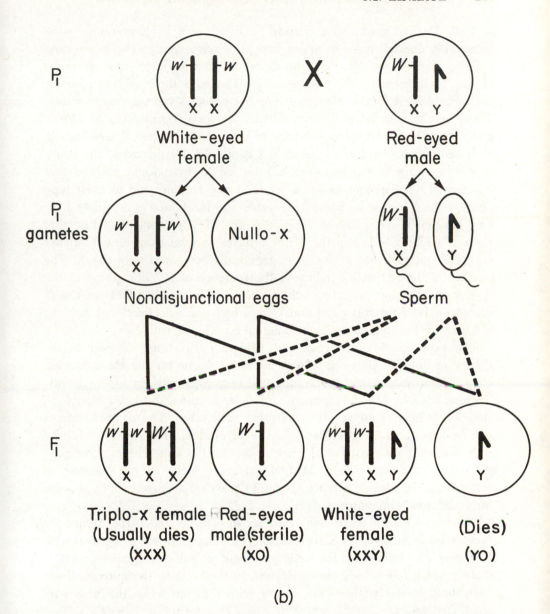

P_I

White-eyed female   X   Red-eyed male

P_I gametes

Nondisjunctional eggs   Sperm

F_I

Triplo-x female (Usually dies) (xxx)   Red-eyed male (sterile) (xo)   White-eyed female (xxy)   (Dies) (yo)

(b)

ceptional white-eyed females should have two X's and a Y chromosome, and the red-eyed males should have an X but no Y. Cytological examination of their chromosomes confirmed the prediction in both instances. This experiment was rather impressive evidence for the theory of linkage, since it demonstrated that abnormal chromosome behavior was accompanied by abnormal genetic behavior.

In addition to its bearing on linkage, this cross had other implications as well. The red-eyed males, despite their lack of a Y chromosome, were reasonably normal males in appearance, but were sterile. The white-eyed females were female-like in appearance and fertile despite the presence of the Y chromosome. The red-eyed females with three X chromosomes had accentuated female characteristics but reduced viability and fertility. These results stimulated additional work on sex determination in *Drosophila* that ultimately led to Bridge's balance theory of sex determination, to be considered in greater detail in Chapter 16. Furthermore, the death in the egg stage of the flies with a Y but no X chromosome showed that for normal development to occur, at least one chromosome of each type (except the Y) must be present. In other words, at least a complete haploid genome is necessary to support normal development. The fact that the red-eyed males lacked the Y but survived seemed to confirm Morgan's earlier hypothesis that it was inert genetically compared with the X. The sterility of the XO males indicated the presence of fertility genes on the Y, but even today very few other loci have been identified on the Y chromosome. Thus this exceptional cross had seminal effects not only on the study of linkage but on developmental genetics as well.

The experiments of Morgan and Bridges gave strong support to the theory of linkage. However, if the genes are borne on the chromosomes, they obviously will tend to be inherited in groups and not independently as demanded by Mendel's principle of independent assortment. Therefore, the two principles apparently contradict each other. The contradiction is resolved by the fact that Mendel's law of random assortment holds true only when the different factor pairs are not on the same chromosome pair. As we pointed out before, Mendel was very lucky in his selection of traits when he picked seven segregating traits, each governed by a locus on a different chromosome pair.

Thus far we have dealt only with sex linkage, in which the Mendelian factors are located on the sex chromosomes. What happens if we deal with two gene loci located on the same autosome, or with an organism such as maize, which has no sex chromosomes? In these cases, the outcome from reciprocal crosses is the same, rather than different as is the case with sex-linked genes.

In maize, for instance, colorless and shrunken are kernel characteristics controlled by recessive genes (Fig. 8.2). The gene loci involved are found on the same chromosome pair so that the genes are said to be linked. If both mutants are on the same chromosome, they are said to be linked in *coupling* phase. The genes for colored and full are also in coupling.

Let c = recessive gene for colorless, s = recessive gene for shrunken, and let the symbol + stand for the corresponding "normal" or "wild-type" gene in both cases. Then the cross can be outlined as follows (C.O. = crossover):

$$P_1 \qquad \frac{++}{++} \qquad \times \qquad \frac{cs}{cs}$$

$$\downarrow$$

$$F_1 \qquad\qquad \frac{++}{cs} \quad \text{backcross to} \quad \frac{cs}{cs}$$

F_2 phenotypes	F_2 genotypes	n	Type of F_1 gamete
colored full	$\dfrac{++}{cs}$	4032	non-C.O.
colored shrunken	$\dfrac{+s}{cs}$	149	C.O.
colorless full	$\dfrac{c+}{cs}$	152	C.O.
colorless shrunken	$\dfrac{cs}{cs}$	4035	non-C.O.
	total	8368	

Data from Hutchison, 1922.

Several points should be observed about this cross. First, it does not matter whether cs/cs or $++/++$ serves as the female parent and the other type as

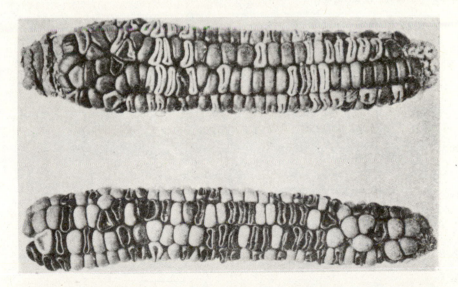

Fig. 8.2. Kernel characters in maize. The cob above, from the backcross of $++/cs \times cs/cs$ has a preponderance of colored full and colorless shrunken kernels. The cob below, from the backcross of $+s/c+ \times cs/cs$, has a preponderance of colored shrunken and colorless full kernels. (Hutchison, C. B., "The Linkage of Certain Aleurone and Endosperm Factors in Maize, and Their Relation to Other Linkage Groups," from *Mem. Cornell Univ. Agr. Expt. Stat.* 60:1425–1473. 1922.)

the male parent; the results of the reciprocal crosses will be similar. Second, the most effective way to determine the amount of linkage and crossing over is to test cross the heterozygous F_1 with the homozygous recessive type. In this way the recombinants or crossover types can be distinguished immediately from the noncrossover or parental types. It should be noted that if these genes were not linked, independent assortment would lead to the production of approximately equal numbers of each of the four F_2 phenotypes. Instead, there is a great preponderance of parental types (colored full and colorless shrunken.) Nevertheless, some recombination or crossing over has occurred, so that genes formerly located on the same chromosome are now found in new combinations. To calculate the amount of crossing over, the following simple equation is used:

$$\% \text{ C.O.} = \frac{\text{no. of C.O. individuals in } F_2}{\text{total No. of individuals in } F_2} \times 100$$

In this case,

$$\% \text{ C.O.} = \frac{149 + 152}{4032 + 149 + 152 + 4035} \times 100 = \frac{301}{8368} \times 100 = 3.6\%$$

Suppose that the recessive genes are in repulsion; in other words, suppose that the cross is as follows:

$$P \qquad \frac{c+}{c+} \quad \times \quad \frac{+s}{+s}$$

$$\downarrow$$

$$F_1 \qquad \frac{c+}{+s} \quad \text{test cross with} \quad \frac{cs}{cs}$$

F_2 phenotypes	F_2 genotypes	n	Type of F_1 gamete
colored full	$\dfrac{++}{cs}$	638	C.O.
colored shrunken	$\dfrac{+s}{cs}$	21,379	non-C.O.
colorless full	$\dfrac{c+}{cs}$	21,906	non-C.O.
colorless shrunken	$\dfrac{cs}{cs}$	672	C.O.
		total 44,595	

Data from Hutchison, 1922.

In this case, the percentage of crossing over is calculated as follows:

$$\% \text{ C.O.} = \frac{1,310}{44,595} \times 100 = 2.9\%$$

Thus, in the above crosses, not only are the results from reciprocal crosses similar, but the same crossover percentage (about 3%) is obtained no matter whether the genes are linked in coupling or in repulsion.

At this point it may be added that it is possible to calculate linkages from the usual F_2 population obtained by inbreeding the F_1. Because of dominance, however, the F_2 genotypes often cannot be determined directly from the phenotypes. For this reason, crossover percentages are not ordinarily calculated from F_2 data; but if the need arises, formulas for making such calculations are available (Immer, 1930).

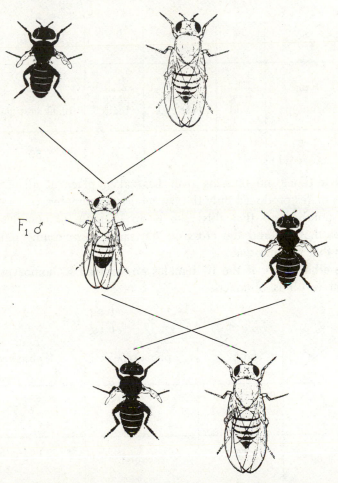

Fig. 8.3. A male homozygous for the linked autosomal recessive genes for vestigial wings and black body crossed to a homozygous wild-type female gives a wild-type F_1. The backcross of an F_1 male to a black, vestigial female gives only black, vestigial, and wild-type progeny, in equal numbers. In this cross black and vestigial are completely linked. (Morgan, T. H., 1928. *The Theory of the Gene*, 2nd ed. Yale University Press, New Haven, Conn., p. 13.)

8.2 Linkage in *Drosophila*

Next let us consider a cross in *Drosophila melanogaster* involving the linked autosomal recessives, black (b) and vestigial (vg) (Fig. 8.3).

$$\frac{b\ vg}{b\ vg} \times \frac{++}{++}$$

F$_1$ $\dfrac{++}{b\ vg}$ ♂ ♂   crossed to $\dfrac{b\ vg}{b\ vg}$ ♀ ♀

♀ \ ♂	++	+ vg	b +	b vg
b vg phenotype	$\dfrac{++}{b\ vg}$ wild	$\dfrac{+vg}{b\ vg}$ vestigial	$\dfrac{b+}{b\ vg}$ black	$\dfrac{b\ vg}{b\ vg}$ black vestigial
% observed	50	0	0	50

In this instance, no crossing over has taken place at all. The heterozygous F$_1$ males produced only the parental-type gametes, $(++)$ or $(b\ vg)$, in equal numbers so that this case is an example of complete linkage. The genes that entered the cross on the same chromosome remained together on that chromosome.

On the other hand, if the F$_1$ females are crossed to homozygous males, a different result is obtained:

$$\frac{++}{b\ vg}\ ♀\ ♀ \qquad \times \qquad \frac{b\ vg}{b\ vg}\ ♂\ ♂$$

♀ \ ♂	b vg	phenotype	% observed
++	$\dfrac{++}{b\ vg}$	wild	41.5
+ vg	$\dfrac{+vg}{b\ vg}$	vestigial	8.5
b +	$\dfrac{b+}{b\ vg}$	black	8.5
b vg	$\dfrac{b\ vg}{b\ vg}$	black vestigial	41.5

Recombinant types do appear from this cross, and the crossover percentage is 17%.

This particular cross has been purposely chosen to illustrate the fact that there is no crossing over in the *Drosophila* male. This failure of crossing over in the male applies to all chromosome pairs and is apparently true for males in all species of the genus. However, crossovers do occur in male spermatogenesis as well as in female oögenesis in all of the other species commonly studied, including man. Because *Drosophila* has been so widely used in genetic research, this peculiarity of its genetic system is worth emphasizing.

The few examples of linkage given thus far have produced crossing-over percentages of 0, 3, and 17%. Even these few cases reveal that crossover values may vary. In fact, they may range from 0 to 50%. Your immediate reaction may be to wonder why there could not be 100% crossing over between two linked genes. To this point we shall devote some attention, in the hope that the discussion will lead to further insight into the phenomena of linkage and crossing over.

A heterozygote, *AaBb*, from *AABB* and *aabb* parents, can produce four kinds of gametes, *AB*, *Ab*, *aB*, and *ab*. If the genes are on different chromosome pairs, the gametes are produced in equal numbers, and the new combinations, *Ab* and *aB*, will constitute only 50% of the total. Thus, even with random assortment, 100% recombination is not obtained. If the genes were on the same chromosome pair, but the *Ab* and *aB* gametes equalled 50% of the gametes formed, then it would not be possible to distinguish between linkage and independent assortment. Thus, random assortment looks the same as 50% crossing over. However, more than one crossover may occur, which can be expected to lead to still further recombination. Why, then, is it possible to say that the upper limit for the crossover percentage between any two linked genes is 50%? One might expect that as the number of crossovers increases, the number of recombinant gametes would increase to the point where the upper limit for recombinants would be 100% rather than 50%, but this is not the case.

Let us review the mechanism of crossing over during meiosis. The homologous chromosomes, each consisting of two sister chromatids, pair with one another. Chiasmata may form between any two nonsister chromatids, leading to a reciprocal exchange of chromatin material. As may be seen in Fig. 8.4, a single chiasma at once leads to 50% recombination. Of the four gametes from this tetrad, two will show recombination between the *A* and *B* loci. If this is true, then what is the effect of a second chiasma?

If the four chromatids are identified individually by number as shown in Fig. 8.4, and the first crossover occurs between 2 and 3, then a second crossover between the *A* and *B* loci may occur with equal probability

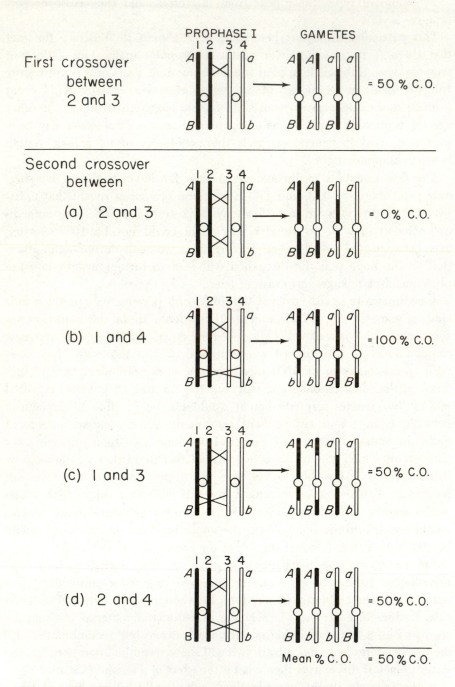

Fig. 8.4. The effect of a second crossover on the percentage of crossing over between two loci. (See text.)

between 2 and 3, or 1 and 4, or 1 and 3, or 2 and 4. The effect of these chiasmata on recombination between A and B is shown diagrammatically.

Hence, the diagram shows that when a second crossover occurs between A and B, it has an equal chance of restoring the parental combinations, and thus will not raise the detectable crossovers between A and B above the 50% mark. An increase in the number of crossovers between two loci will not lead to an increase in detectable recombination, approaching 100% as an upper limit, for the simple reason that any even number of crossovers involving the same two nonsister chromatids will restore the parental combination of genes and thus become undetectable. Only the presence of additional segregating loci between A and B will make possible the detection of these double crossovers.

8.3. Linear Order of the Genes

The work of Mendel established two major principles of heredity, segregation and independent assortment. The work of Morgan and his co-workers in developing the chromosome theory of heredity led to two additional fundamental principles. The first was linkage: The genes are borne on the chromosomes. The second was the linear order of linked genes: Not only are some genes linked, but their geometrical configuration is linear.

If the genes are in fact linearly arranged, it should be possible to determine both the sequence or linear order of the genes and also the distances between the different genes. A number of conditions may influence the frequency of crossing over—for example, temperature, the age of the organism, and genic modifiers. Furthermore, crossovers are much more frequent in some chromosome regions than in others. For instance, they are relatively rare near the centromere in many species. The presence of an intrachromosomal rearrangement such as an inversion in one chromosome pair, which in effect causes reduced crossing over in that pair, may lead to increased crossover frequencies in other chromosome pairs. Nevertheless, despite these possible influences, crossover frequencies are remarkably constant and can therefore be used to map the order of the genes and the distances between them. As was shown above, the crossover frequencies are obtained by counting the numbers of recombinant progeny in a cross and converting the proportion to a percentage. These percentages are now going to be used to obtain information about the way the very small entities, the genes, are arranged on the chromosome. Obviously the information so obtained is rather indirect. It is somewhat like trying to visualize a trip by bus through unknown territory simply by an examination of the bus schedule. You will have no difficulty in determining from the schedule the sequence of places through which the bus passes, but your guesses as to the distances between them may be considerably in error if based simply on times of arrival and departure, If some parts of the route are on freeways or toll roads while others are through city traffic at much lower rates of speed, then equal periods of time may not represent equal distances traveled. Similarly, equal percentages of crossing over may not necessarily represent equal physical distances along the chromosome, but the gene sequence based on crossover percentages should be correct.

Let us examine a cross involving three gene loci in *Drosophila melanogaster*. The three recessives at these loci are yellow (y), causing a light yellow body color, bifid (bi), affecting the wings, and white eye (w). The crossover percentage between y and w is 1.5%; between w and bi, it is 5.5%. This information tells us something about relative distances, but nothing about the sequence of the genes. The order could be either

(a)

$$\begin{array}{ccc} y & w & bi \\ \vdash\!\!\!-\!\!\!-\!\!\!\dashv & - - - - - - \dashv \\ 1.5 & 5.5 \end{array}$$

or

(b)

$$\begin{array}{ccc} bi & - - - & y & w \\ & & 1.5 \\ & 5.5 \end{array}$$

Only when the percentage of crossing over between y and bi is known to be 7.0% can the problem be resolved. The order must be

$$\begin{array}{ccc} y & w & bi \\ \vdash\!\!\!-\!\!\!-\!\!\!\dashv\!\!\!-\!\!\!-\!\!\!-\!\!\!\dashv \\ 1.5 & 5.5 \\ 7.0 \end{array}$$

or, since the first point is arbitrarily chosen, bi w y.

When a chromosome map is constructed, the genes are placed in a linear order determined by their crossover relations. The locus of each is indicated by a number. The difference between the numbers of two adjacent genes gives the percentage of crossing over that has been found between these two genes. Thus, a map of y w bi can be drawn as follows:

$$\begin{array}{ccc} y & w & bi \\ \vdash\!\!\!-\!\!\!-\!\!\!+\!\!\!-\!\!\!-\!\!\!-\!\!\!-\!\!\!-\!\!\!-\!\!\!\dashv \\ 0 & 1.5 & 7.0 \end{array}$$

When, as in the example above, the crossover percentages and map distances are small, crossing over between the extremes (*y* and *bi*) turns out to equal the sum of the two intervening crossover values. When the distances are greater, matters become somewhat more complicated. Another cross, involving ruby eye color (*rb*), cut wing (*ct*), and vermilion eye color (*v*), will illustrate the problems involved. Between *rb* and *ct* the % C.O. equals 16.7; between *ct* and *v* it is 14.5. If crossover values were always additive, and the order is *rb ct v*, the % C.O. between *rb* and *v* should be 16.7 + 14.5 = 31.2. However, when just *rb* and *v* are tested without *ct* segregating in the cross (i.e., *rb v/++* × *rb v*), the % C.O. between them turns out to be 29.4 rather than 31.2. This difference is not due to errors in counting or scoring or to chance fluctuations. Repeated tests between *rb* and *ct* and between *ct* and *v* on the one hand, and between just *rb* and *v* on the other, have revealed a significant difference between the two figures obtained. The biological basis for this discrepancy is the phenomenon of double crossing over. To demonstrate this effect, let us consider first the kinds of gametes produced by the heterozygote.

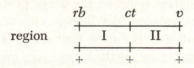

	Type of gamete	Origin
1.	+ + +	
2.	rb ct v	non-C.O.
3.	+ ct v	
4.	rb + +	single C.O. region I
5.	+ + v	
6.	rb ct +	single C.O. region II
7.	+ ct +	
8.	rb + v	double-C.O. regions I and II

The important point to note is that if segregation were not occurring at the cut locus in the double C.O. group to give a third point of reference, this group would be included with the non-C.O. because they would have the parental combinations, *rb v* or + +. In reality, of course, each individual represents two separate crossovers, one in region I, the other in region II. Instead of not being counted at all, they should each be counted twice because each gamete carries two C.O. In one set of data from the cross of the triple heterozygote with the corresponding recessive type, there were 15 individual flies among 1622 progeny that were of the double C.O. type (i.e., phenotypically either $+ ct +$ or $rb + v$). Thus,

$$\frac{15}{1622} = 0.9\%$$

but since each of the 15 represents two crossovers, the % C.O. is

$$2 \times 0.9\% = 1.8\%$$

Adding the 1.8% to the 29.4% C.O. value obtained when C.O. was determined between rb and v alone, we get 31.2%, which equals the sum of 16.7% + 14.5%. Thus the apparent discrepancy between the 31.2% expected C.O. between rb and v and the 29.4% observed is accounted for by the presence of undetected double crossovers.

Let us now map from scratch the relations among three loci, starting from a set of data obtained by crossing triple heterozygotes from the cross $+++/+++ \times cn\,vg\,b/cn\,vg\,b$ to triple recessive individuals. Black (b) is a body color recessive, cinnabar (cn) an eye color mutant, and vestigial (vg) a recessive mutant reducing the wings to straplike vestiges. There are various ways for working out linkage and map relations, all of which, properly used, will give the same final result. After some practice, you will probably adopt the method that you find simplest and quickest for your own use. The method used below is adopted here because it is relatively foolproof, and the student is not likely to be misled if by chance the gene sequence written down in a preliminary fashion when the data are recorded happens to be incorrect.

The data are as follows:

	Phenotype	n	%
1.	$+ + +$	332	
2.	$cn\,vg\,b$	326	82.25
3.	$cn + b$	36	
4.	$+ vg +$	34	8.75
5.	$+ + b$	35	
6.	$cn\,vg +$	31	8.25
7.	$+ vg\,b$	2	
8.	$cn + +$	4	0.75
		800	100.00

The ($+++$) and ($cn\,vg\,b$) are known to be the parental types, and it can be seen that these two types are equal in number and considerably more frequent than any of the other classes. [If $(cn\, vg\, b)$, for example, were not equal to $(+++)$ in number, the most likely explanation is that the two groups differ in viability.] Note that reciprocal types of gametes are paired.

The first step is to calculate the percentage of the total for each reciprocal pair of gametes. The next step is to determine the percentage recombination between each pair of loci in turn. This is done by including in the numerator any arrangement of these two gene pairs other than that seen in the parents.

$$\% \text{ C.O. } (cn \text{ and } vg) = 8.75\% + 0.75\% = 9.5\%$$

Similarly,

$$\% \text{ C.O. } (cn \text{ and } b) = 8.25\% + 0.75\% = 9.0\%$$

and

$$\% \text{ C.O. } (vg \text{ and } b) = 8.75\% + 8.25\% = 17.0\%$$

From these percentages, it is clear that b and vg are farthest apart and that cn must lie between them.

The order given with the data therefore turns out to have been incorrect. In this case the double C.O. classes were $(+ vg\, b)$ and $(cn ++)$, which when written in this way, look like the result of a single C.O. between cn and vg. However, double crossovers will be rarer than single crossovers, and this fact alone can often help in the identification of the double crossover group. Here the double C.O. classes can be recognized by inspection.

Now that the correct order has been established, the next problem is to determine the map distances between the loci. The map is constructed, as before, by arbitrarily setting one of the outside loci equal to zero and adding the map distances between adjacent loci. A map of these three loci then would be

Once again, a difference between the sum of the crossover percentages between b and cn and between cn and vg (18.5%) and the crossover percentage between b and vg calculated separately (17.0%) can be seen. However, the double crossover class was omitted from the numerator entirely for the b–vg calculation because the gene order was originally written incorrectly. When this adjustment is made (% C.O. between b and vg = $8.75\% + 8.25\% + 2 \times 0.75\% = 18.5\%$), the difference disappears. Of course, if the correct gene order can be determined by inspection of the data, this type of problem will not arise.

When an entire chromosome is mapped, double crossing over will ordinarily cause no trouble because the map is drawn on the basis of the addition of crossover percentages between adjacent loci. The map will be constructed by finding the linkage relations between closely linked genes in this general fashion, $A\,B\,C$, $C\,D\,E$, $E\,F\,G$, and so on. In this way only the shortest distances are involved. The units of map distance are derived from crossover percentages; 1% crossing over is equivalent to one unit of map distance. This unit is sometimes called a *morgan*.

8.4. Interference

If a crossover in one region of a chromosome has no influence on the frequency of crossing over in another region of the chromosome (i.e., if the crossovers are independent events), then the expected frequency of double crossovers should equal the product of the frequencies or probabilities of the respective single crossovers. In the $b\,cn\,vg$ example above, the expected frequency of double C.O. individuals is therefore equal to $9.0\% \times 9.5\%$ or 0.86%. However, the observed frequency is 6/800 or 0.75%. For the $rb\,ct\,v$ example, the expected frequency on the assumption of independence is $14.5\% \times 16.7\%$ or 2.4%, but the observed frequency is 15/1622 or 0.9%, a value considerably lower than expected. These and many similar observations support the conclusion that the occurrence of one crossover tends to reduce the likelihood of another. This phenomenon has been referred to as interference. An estimate of the amount or strength of the interference is obtained from the coefficient of coincidence, which is the ratio of the observed to the expected frequency of double crossover individuals based on the assumption of independence. Thus, for $b\,cn\,vg$,

$$\text{coefficient of coincidence} = \frac{0.75}{0.86} = 0.87$$

In the $rb\,ct\,v$ cross the amount of interference appears much greater, for the coefficient is considerably lower.

$$\text{coefficient of coincidence} = \frac{0.9}{2.4} = 0.37$$

It has been found that the amount of interference varies according to the distance between the loci being studied. There is a certain minimal distance (about 10 C.O. units in *Drosophila*) below which double crossing over does not occur: in other words, interference is complete. Beyond this point, the amount of interference decreases in direct proportion to the distance between the loci. Beyond about 45 map units there is little or no interference. The actual mechanism of interference is still unknown. The analogy to a spring is often used in explaining interference. The chromosome is likened to a spring under tension. A crossover "break" then relieves this tension not only at the point of the "break," but in adjacent regions so that a second "break" is less probable in these regions. Only in the more remote regions of the chromosome will the tension remain sufficiently great to lead to additional crossover "breaks." While useful as an analogy, this physical explanation of interference probably has little relation to reality.

A most important aspect of interference is the insight it gives into what is actually exchanged at crossing over. Complete interference means that blocks of genes, rather than single genes, are always exchanged during crossing over. The integrity of the chromosome is not completely destroyed by crossing over because linear segments or blocks of genes are exchanged. Favorable gene combinations within a chromosome thus tend to be preserved, while at the same time the recombination permitted by crossing over leads to the formation of new and possibly better adapted gene combinations.

With the methods outlined above, chromosome maps have been constructed for a number of species, a few of which are illustrated in Fig. 8.5. It can be seen that the map length of some chromosomes not only exceeds 50 units but even 100 map units in some cases. Since the point was made earlier that the percentage of crossing over between any two linked genes cannot exceed 50%, a discrepancy may seem to exist. However, map distances are obtained by adding together the C.O. percentages between adjacent genes. As long as the genes are close together, the map distances and crossover percentages will be in good agreement. For longer intervals, they may not agree at all. For example, in *Drosophila* the map distance between roughoid and minute on the third chromosome is 106.2, but the observed crossover percentage between these two mutants alone is 45.8%. The explanation, as you should now realize, is that the map distances were determined by the use of all available information on linkage relations between genes on the third chromosome, but the crossover percentage between roughoid and minute is much lower than the map distance because of undetected double crossovers.

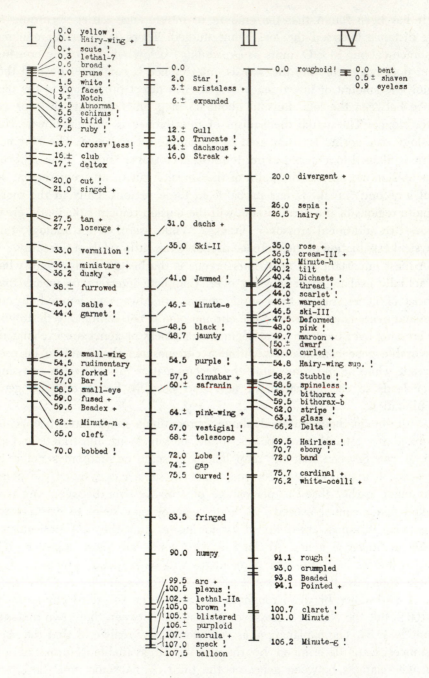

Fig. 8.5(a). Linkage map of *Drosophila melanogaster*. (Morgan, T. H., 1928. *The Theory of the Gene*, 2nd ed. Yale University Press, New Haven, p. 23.)

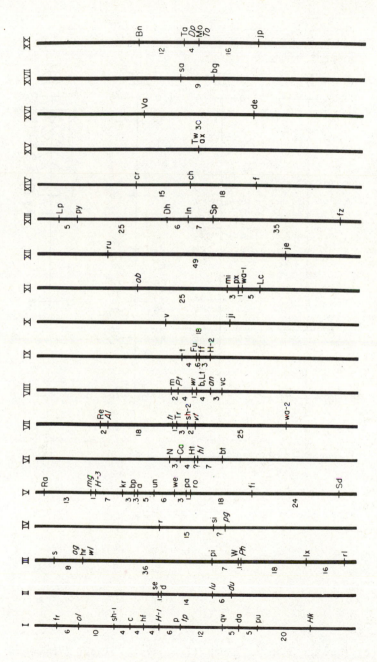

Fig. 8.5(b). Linkage map of the house mouse (*Mus musculus*). (Green, M. C., and M. M. Dickie, "Linkage Map of the Mouse." Reprinted with permission from *The Journal of Heredity* 50:2–5, 1959. Copyright 1959, by the American Genetic Association.)

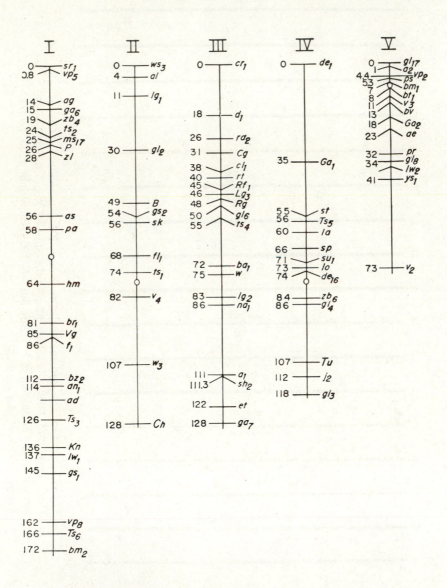

Fig. 8.5(c). Linkage map of maize (*Zea mays*). (From *Elementary Genetics* by W. Ralph Singleton, © 1967. Reprinted by permission of D. Van Nostrand Company.)

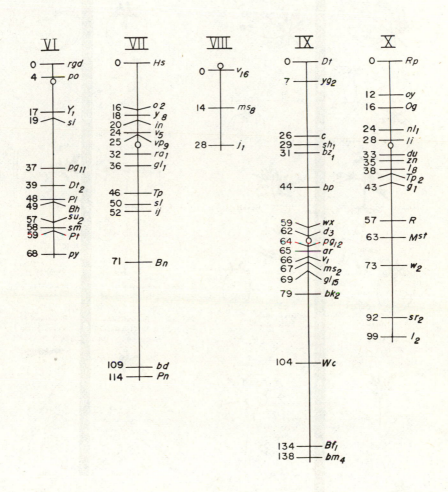

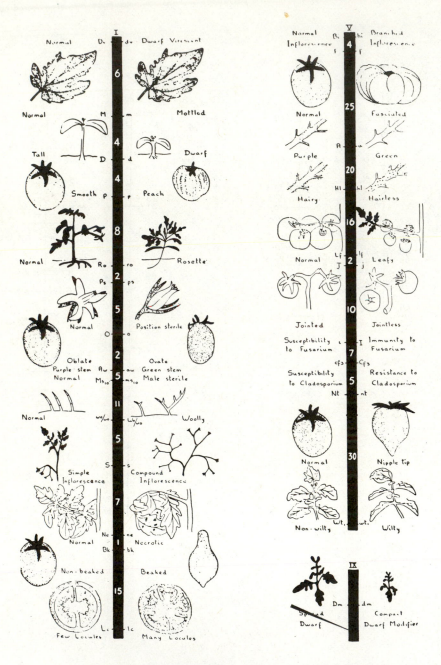

Fig. 8.5(d). Linkage map of the tomato (*Lycopersicon esculentum*). (Butler, Dr. Leonard, Department of Zoology, University of Toronto.)

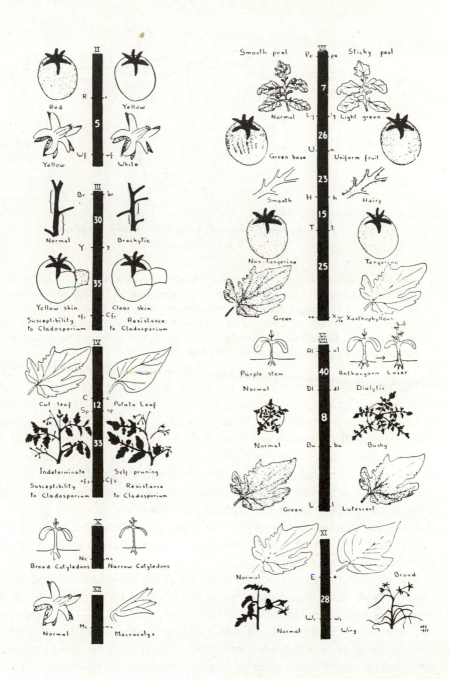

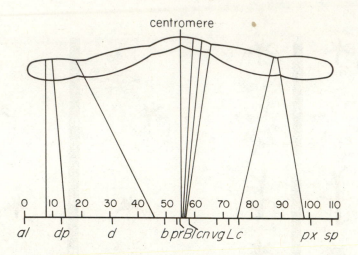

Fig. 8.6. A comparison of the genetic linkage map, based on crossing over, with the cytological map of the second chromosome of *Drosophila melanogaster*. The cytological map is determined by the use of overlapping deficiencies discussed in Chapter 9. (After Dobzhansky, as it appears in Sturtevant, A. H. and G. W. Beadle, 1939. *An Introduction to Genetics*. W. B. Saunders Co., Philadelphia, p. 187, Fig. 78.)

The mapping of chromosomes by use of the different proportions of the various types of test-cross progeny is, of course, an indirect way to determine chromosome structure. The next question to consider is the relationship between the chromosome map and the actual living structure of the chromosome. When such comparisons have been made, the linear order of the genes turns out to be the same, but equal map distances in many cases do not represent equal physical distances along the chromosomes. This is especially true in *Drosophila* near the centromere (Fig. 8.6). In this region the low frequency of chiasma formation and crossing over means that a crossover unit represents a greater physical distance than in other chromosome regions.

The discovery that the Mendelian factors or genes are located on the chromosomes and are arranged in a linear order of such constancy that the genes can be mapped was a major advance in the development of the science of genetics. It means that each gene (in the absence of chromosomal rearrangements) occupies a fixed position on a particular chromosome and that its alleles occupy corresponding positions on homologous chromosomes. Despite the bewildering amount of variability generated by the hereditary mechanism, the mechanism itself is an orderly and relatively stable one. If the genes are borne on the chromosomes, the number of linkage groups can be no greater than the number of pairs of chromosomes. This is actually the case in all organisms carefully studied, as in the few examples shown below:

Species	Haploid chromosome number	Number of linkage groups
D. melanogaster	4	4
D. pseudoobscura	5	5
D. virilis	6	6
Neurospora crassa	7	7
Pisum sativum	7	7
Z. mays	10	10
M. musculus	20	20

That the hereditary units tend to be inherited in a rather small number of groups is important not only to theoretical genetics but also to evolution. The possible amount of genetic recombination is greatly restricted by linkage. While multiple alleles, segregation, and independent assortment all tend to increase the amount of recombination, linkage tends to decrease it. Furthermore, the task of the plant and animal breeder in getting desirable new combinations of genes is rendered far more difficult by linkage. The practical problems of combining the best traits from two distinct parental types are greatly complicated by the fact that the favored genes may be closely linked to undesirable genes.

Additional Reading

Barratt, R. W., D. Newmeyer, D. D. Perkins, and L. Garnjobst, 1954. "Map Construction in *Neurospora crassa,*" *Adv. Genetics* 6:1–93.

Emerson, R. A., G. W. Beadle, and A. C. Fraser, 1935. "A Summary of Linkage Studies in Maize," Cornell Univ. Agr. Sta. Mem. *180.*

Hutchison, C. B., 1922. The linkage of certain aleurone and endosperm factors in maize, and their relation to other linkage groups. Cornell Agr. Exp. Sta. Mem. *60.*

Immer, F. R., 1930. "Formulae and Tables for Calculating Linkage Intensities," *Genetics* 15:81–98.

Janssens, F. A., 1909. "La théorie de la chiasmatypie," *La Cellule* 25:389–411.

Lindsley, D. L., and E. H. Grell, 1968. "Genetic Variations of *Drosophila melanogaster,*" Carnegie Inst. Washington Pub. No. 627, pp. 1–471.

Mather, K., 1951. *The Measurement of Linkage in Heredity*. Methuen, London.

Morgan, T. H., A. H. Sturtevant, H. J. Muller, and C. B. Bridges, 1923. *The Mechanism of Mendelian Heredity*, 2nd ed. Holt, Rinehart and Winston, New York.

Sturtevant, A. H., 1913. "The Linear Arrangement of 6 Sex-Linked Factors in *Drosophila* as Shown by Their Mode of Association," *J. Exp. Zool.* 14:43–59.

Sturtevant, A. H., and G. W. Beadle, 1939. *An Introduction to Genetics*. Saunders, Philadelphia. (Also Dover, New York, 1962.)

Questions

8.1. Why are the F_2 progeny from a cross not ordinarily used in linkage studies?

8.2. What is the expected number of linkage groups in man?

8.3. How many linkage groups would you expect to find in a triploid?

8.4. Calculate the crossover value between the autosomal recessive genes for black body and arc wings in *Drosophila melanogaster* from the following data. Black females were crossed with arc males and the resulting F_1 females were crossed with black, arc males. The progeny were 279 normal, 333 arc, 338 black, and 241 black arc. What is the crossover value?

8.5. Black body and purple eyes are recessive traits 6 crossover units apart on the second chromosome of *Drosophila melanogaster*. From a cross between black-bodied stock males and purple-eyed stock females, what proportion of the F_2 flies would have both black bodies and purple eyes?

8.6. There is 19.8% crossing over between yellow and cut, and 12.7% between cut and raspberry in *Drosophila*. Cut is located between yellow and raspberry. If there were no chromosome interference, what would be the expected percentage of double crossovers?

8.7. In *Drosophila melanogaster*, Beaded (*Bd*) causes a dominant wing abnormality but is homozygous lethal. Another homozygous lethal (*l*) is located on the homologous chromosome, and a crossover suppressor prevents crossing over between *Bd* and *l*. What results would be expected from a cross between two flies with $Bd\,l^+/Bd^+\,l$ genotypes?

8.8. In a diploid species, three pairs of alleles are linked on a single chromosome pair. One parent is homozygous for the three dominant alleles and the other for the three recessive alleles. In a test cross of the F_1 to the triple recessive strain, there were 14 *aBC/abc* and

16 *Abc/abc* progeny among a total of 1000. All the other six possible genotypes were much more frequent. What is the correct order of these three genes on the chromosome?

8.9. Black, cinnabar, and vestigial are linked autosomal recessive genes in *Drosophila melanogaster*. An F_1 male from a cross between a homozygous wild-type male and a female homozygous for these three recessives will produce how many different kinds of gametes with respect to these three genes?

8.10. In sweet peas, blue flowers are the result of the dominant *B,* and red of its recessive allele *b*. Long pollen is due to a dominant factor *R,* and round pollen to its recessive allele *r*. A plant homozygous for blue flowers and round pollen was crossed with one homozygous for red flowers and long pollen. The F_1 results were blue long, 23; blue round, 153; red long, 155; red round, 21. What is the crossover percentage between these loci?

8.11. In tomatoes, tall vine (*T*) and spherical fruit (*S*) are dominant alleles of dwarf vine (*t*) and pear-shaped fruit (*s*). Two homozygous plants were crossed and the F_1 crossed to the double recessive yielded the following offspring: 406 *Ttss;* 394 *ttSs;* 93 *TtSs;* 107 *ttss*.
a. What is the relationship between these loci?
b. What were the phenotypes of the original parents?

8.12. In rabbits, short velvetlike fur (rex) is produced by the homozygous state of either or both of the recessive mutations r_1 and r_2. These two varieties of rex rabbits were crossed and the F_1 progeny were all normal. The F_1 was crossed to double recessive rex rabbits and produced the following offspring: rex, 45; normal, 5. What is the percentage of crossing over between r_1 and r_2?

8.13. In rabbits, black (*B*) is dominant over brown (*b*), and recessive albinism (*c*) is epistatic to both. Albino rabbits of the genotype (*ccBB*) were crossed to brown rabbits (*CCbb*) and the F_1 backcrossed to the double recessive. The progeny were 35 brown, 49 albino, and 16 black. What do these results indicate?

8.14. A female heterozygous for the sex-linked recessive mutants, yellow (*y*), white (*w*), and cut (*ct*) was backcrossed to a male carrying all three recessives. The progeny were, phenotypically:

y w ct	399		*y w +*	89
+ + +	404		+ + *ct*	93
+ *w ct*	5		+ *w* +	2
y + +	7		*y* + *ct*	1

If yellow is at 0.0, what are the map locations of white and cut?

8.15. Normal maize was crossed to a stock homozygous for three recessive mutants: brown midrib ((bm_1)); red aleurone (pr); and virescent (v_2). The F_1 plants were backcrossed to the recessive parental type and the following progeny were obtained:

+ + +	232	pr bm +	194
pr + v	84	+ bm +	77
+ + v	201	pr bm v	235
pr + +	40	+ bm v	46

What is the linear order of these three gene loci? Construct a map showing their relationship.

8.16. In a mating between a triple heterozygote (ABC/abc) and a triple recessive (abc/abc), there were 1000 offspring with the following phenotypes:

ABC	428	ABc	44
abc	420	abC	38
Abc	34	aBc	1
aBC	33	AbC	2

If locus A is at 0.0 on the chromosome map, where will locus C be?

CHAPTER NINE

Chromosomal Variation

Thus far, we have dealt with the genetics of normal diploid organisms, in which the somatic cells are diploid and the gametes haploid. Of the two sets of chromosomes in diploid cells, one is maternal in origin, the other paternal. During synapsis in the first meiotic prophase, precise gene-by-gene pairing occurs between homologous chromosomes throughout their length, since the genes are in the same linear order in both members of a pair.

However, chromosome number and structure do not always conform to the usual scheme. Therefore, it is necessary to learn something about the possible variations in chromosome number and structure in order to understand their genetic consequences. The cytological knowledge of this sort has been acquired primarily from work with relatively few species, among which should be mentioned *Drosophila*, corn (*Zea mays*), spiderwort (*Tradescantia*), and the Jimson weed, *Datura*. The types of changes can be broadly categorized into three classes: those involving chromosome fragments, those involving whole chromosomes, and those involving whole sets of chromosomes.

9.1. Rearrangements Involving Fragments of Chromosomes

In rearrangements involving chromosome fragments, a chromosome segment is shifted in location within the chromosome complement. In order for a segment to form, a transverse break must occur in a chromosome, and the

139

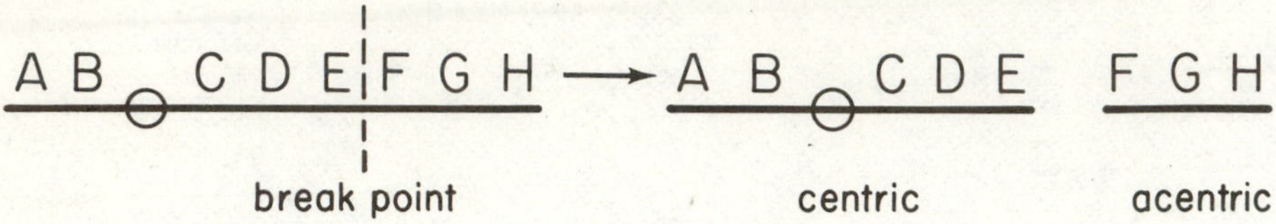

A B ₒ C D E¦F G H ⟶ A B ₒ C D E   F G H

break point   centric   acentric

Fig. 9.1. Terminal deletion or deficiency. A single break produces a **centric** fragment and a mitotically inactive acentric fragment.

broken ends must unite in some new manner; alternatively, loss of the fragment may occur. "Spontaneous" chromosome breakage does occur, but it is rare, and much of the work has been done with induced chromosome breaks. Rearrangements have been induced experimentally in several ways: by irradiation, especially by ionizing radiations such as X-rays, by chemical agents such as the mustard gases, and also by temperature shock (i.e., exposure of the cells to extreme high or low temperatures for relatively short periods). It is usually thought that the breakage due to irradiation, for example, results from the ionizations caused by the treatment. After a chromosome is broken, it may undergo restitution, the broken ends rejoining in the original form, or the broken ends may rejoin in some different fashion to form a rearrangement. The restituted breaks are usually impossible to detect cytologically, but the various rearrangements can often be observed in suitable chromosome preparations.

Deletions

A chromosome deletion or deficiency occurs when a chromosome fragment is lost from the complete set of chromosomes. Terminal deletions result from a single chromosome break and the loss of the terminal fragment. In cytological preparations a centric fragment that is mitotically active and an acentric fragment that is mitotically inactive can be observed (Fig. 9.1). The acentric fragment is usually quickly lost from the nucleus because of its lack of a centromere.

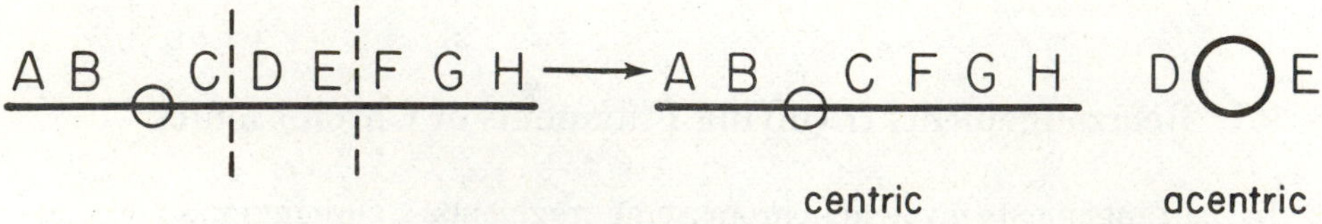

A B ₒ C¦D E¦F G H ⟶ A B ₒ C F G H   D◯E

centric   acentric

Fig. 9.2. Intercalary deletion or deficiency. Two breaks lead to the formation of a centric fragment and a mitotically inactive ring fragment.

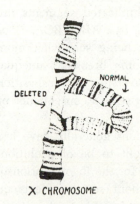

NORMAL

DELETED

X CHROMOSOME

Fig. 9.3. Chromosome pairing in a deletion heterozygote. Somatic pairing of normal and deleted X chromosomes in the salivary gland of Drosophila melanogaster. (Painter, T. S., "Salivary Chromosomes and the Attack on the Gene." Reprinted with permission from *The Journal of Heredity* 25:464–476, 1934.)

Intercalary deletions occur when two breaks in a single chromosome lead to the formation of a centric fragment and an acentric ring fragment (Fig. 9.2). The acentric fragment, as before, is quickly lost.

The precision of the synaptic mechanism can be readily observed in cells heterozygous for a normal chromosome and a chromosome with a deletion. The homologous regions pair in a "gene-by-gene" or band-by-band manner, but the region of the normal chromosome corresponding to the deletion is unpaired and thus forms an easily observed loop (Fig. 9.3). Homozygosity for a deletion is often lethal. In cases where the deficiency is quite large, it may even be lethal in heterozygotes, acting therefore as a dominant lethal. As shown in Fig. 9.11, Sect. 9.2, deletions, especially overlapping deletions, have been very useful in relating particular genes to specific regions of the chromosomes.

Duplications

The repetition of a given chromosome segment within the genome is called a *duplication*. The duplication may be interchromosomal, with the dupli-

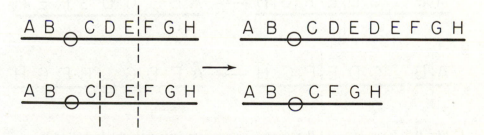

Fig. 9.4. Duplication. A tandem repeat.

cated segment incorporated into a nonhomologous chromosome. More commonly, it is intrachromosomal, with the duplicated segments contained within a single chromosome (Fig. 9.4). If the duplicated segments are adjacent, they may be simple tandem repeats (e.g., . . . ABCABC. . .) or reverse repeats (. . . ABCCBA. . .), the latter being somewhat more stable. Duplications may arise following chromosome breakage, unequal crossing over, or as the result of crossing over in inversion or translocation heterozygotes. Duplications and deficiencies often arise as a consequence of a single rearrangement.

Whereas deficiencies tend to be lethal, duplications generally have less drastic effects on viability. Duplications are thought to be of evolutionary significance because they provide a mechanism for the incorporation of additional genetic material into the genome without serious effects on viability. However, duplications sometimes have significant phenotypic effects, for example, the Bar-eye duplication in *Drosophila* (see Figs. 14.4 and 14.5).

Inversions

An inversion results if, after a chromosome is broken in two places, the fragment between the breaks swaps ends (or is rotated through 180°) before reunion of the broken ends occurs. After an inversion occurs, the genes on the chromosome continue to be linked, but the linkage relations are changed. Genes formerly far apart may become very closely linked while others, more closely linked, may show weaker linkage as a result of the inversion. Another genetic effect sometimes observed is a "position effect." Placing a gene in a new relationship with its fellow genes may cause the expression of the gene to be different in its new location. That this change

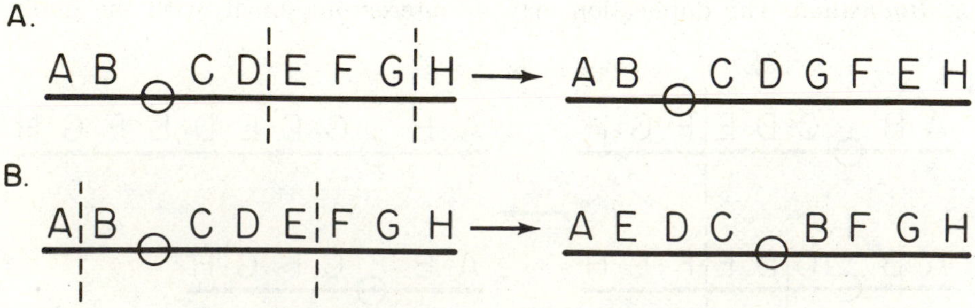

Fig. 9.5. Inversions. (a) Paracentric or one-arm inversion not including the centromere. (b) Pericentric or two-arm inversion including the centromere.

differs from mutation is shown when the gene is returned to its original position, for it then assumes its original expression. This discovery of position effect has had far-reaching consequences for the gene concept and will enter our discussion again later in Chapter 14.

Two types of inversions have been identified, the paracentric or one-arm inversion, and the pericentric or two-arm inversion (Fig. 9.5). The paracentric inversion is "beside" the centromere, while the pericentric inversion includes the centromere. When synapsis occurs in an inversion heterozygote, a loop is formed in order to satisfy the demands of gene-by-gene pairing. However, in this case, unlike that of the heterozygote for a deletion, both chromosomes are involved in the loop (Fig. 9.6). One reason for making the distinction between para- and pericentric inversions is that the consequences of crossing over within the inversion in heterozygotes are somewhat different. As a result of the formation during meiosis of dicentrics (chromosomes with two centromeres) and acentric fragments (paracentric), and of duplications and deficiencies (pericentric), the presence of an inversion in a population frequently results in a reduction in fertility.

Translocations

A translocation is the transfer of a chromosome segment to another part of the same chromosome, or to a homologous chromosome, or most often, to a different chromosome. Most translocations are reciprocal, resulting from breaks in each of two chromosomes with the terminal fragments being exchanged (Fig. 9.7). As a consequence, some genes that were formerly linked now assort independently while others that were formerly in differ-

SYNAPSIS OF A NORMAL
WITH A DELTA 49
X CHROMOSOME

NORMAL

Fig. 9.6. Pairing in an inversion heterozygote for a normal X chromosome and an X chromosome with the "delta 49" inversion in the salivary gland of a female larva of *Drosophila melanogaster*. (Painter, T. S., "Salivary Chromosomes and the Attack on the Gene." Reprinted with permission from *The Journal of Heredity* 25:464–476, 1934.)

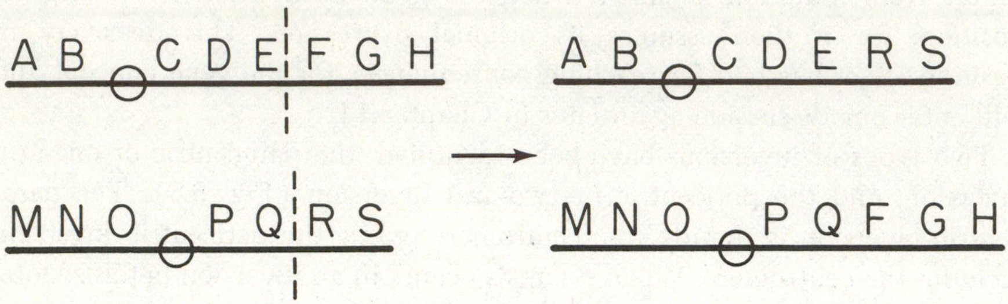

A B C D E｜F G H   A B C D E R S

⟶

M N O P Q｜R S   M N O P Q F G H

Fig. 9.7. Reciprocal translocations.

ent linkage groups now show linkage. As shown in Fig. 9.7, the broken ends may unite in different ways. If a dicentric chromosome and an acentric fragment are formed, both are likely to be soon lost from the nucleus, the acentric fragment because it is mitotically inert, the dicentric because of the formation of a dicentric bridge during cell division. On the other hand, if the result is a set of chromosomes, each with a single centromere, they will behave normally in mitosis, but in meiosis in translocation heterozygotes, sets of four chromosomes, or even more, may form complex pairing configurations (Fig. 9.8). As a result of crossing over and chromosome segregation during meiosis in these heterozygotes, gametes with abnormal chromosome complements are often formed, and a high degree of sterility may be observed.

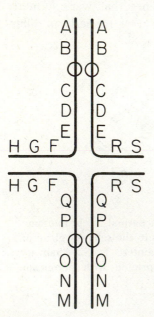

Fig. 9.8. Pairing in a translocation heterozygote. A tetravalent configuration involving four chromosomes is formed.

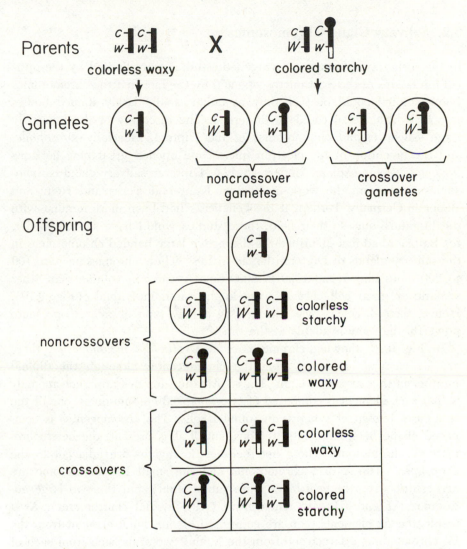

Fig. 9.9. Heteromorphic homologues as proof of the chromosome theory of heredity. Diagrammatic illustration of the work of Creighton and McClintock with maize showing that cytological crossing over parallels genetic crossing over.

The decisive proof of the chromosome theory of heredity involved the use of translocations in *Drosophila* by Stern and in corn by Creighton and McClintock (Figs. 9.9 and 14.2).

A somewhat different form of translocation, often called an insertion, requires three chromosome breaks for its formation. In this case a fragment with two broken ends is inserted between the broken ends of the third break.

9.2. Salivary Gland Chromosomes

In the early days of genetics, cytogenetic studies were limited by the optical limitations of the light microscope and by the fact that the chromosomes being studied could be seen only as rather small, darkly stained bodies in which little detail could be discerned. The discovery of the unusually large salivary gland chromosomes of the Diptera, especially *Drosophila*, and the invention of the electron microscope opened great new horizons for cytologists to explore. Modern study of Dipteran salivary gland chromosomes dates from the work in 1933 of Painter at Texas and Heitz and Bauer in Germany. Perhaps if the geneticists had been more familiar with the literature outside their field, these studies would have started sooner, for Balbiani had first described the distinctive large banded chromosomes in the salivary glands of Dipteran larvae in 1881. These chromosomes are 100 to 200 times longer and 1000 to 2000 times greater in volume than other somatic or germ cell chromosomes in the same individuals (Fig. 9.10). Hence, their discovery was like finding a new type of microscope more powerful than any available at the time.

In Fig. 9.10, showing chromosomes of *Drosophila melanogaster*, there are six elements radiating from the chromocenter although the diploid number in this species is eight. The explanation for this apparent anomaly is that each element is composed of two synapsed chromosomes, one of the few cases known of synapsis in somatic cells. The chromocenter is composed of the heterochromatic regions, and if it is present, the heterochromatic Y chromosome. These heterochromatic regions are adjacent to the centromere. The X or I chromosome and the small IV chromosome are acrocentric (i.e., the centromeres are subterminal), but the two large autosomes (II and III) are metacentric (with medial centromeres). As a result, the six elements seen are comprised of one small element from the IV chromosome, a larger one from the X, and two arms each from each of the two large autosomes for a total of six.

Not only are the salivary gland chromosomes larger, but far more detail can be observed in them because of the unique banding patterns in each of the chromosomes. Because of these patterns, the individual chromosomes and regions within the chromosomes can be identified.

The large size of the chromosomes has been explained by the fact that each is composed of a large number of chromosome strands or chromonemata. The numerous strands result from chromosome duplication without cell division. These polytene chromosomes, as they are called, have been estimated to have 512, 1024, or even more strands.

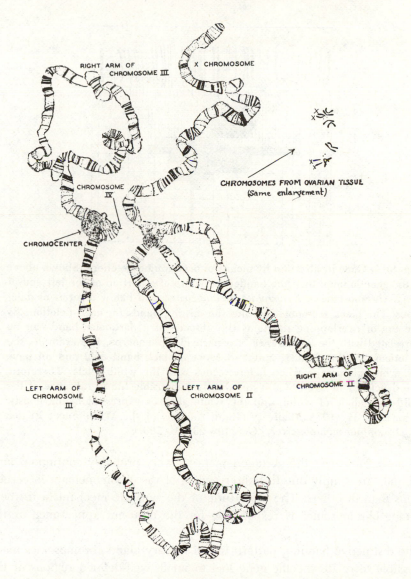

RIGHT ARM OF
CHROMOSOME III

X CHROMOSOME

CHROMOSOME
IV

CHROMOSOMES FROM OVARIAN TISSUE
(Same enlargement)

CHROMOCENTER

RIGHT ARM OF
CHROMOSOME II

LEFT ARM OF
CHROMOSOME
III

LEFT ARM OF
CHROMOSOME II

9.10. A *camera lucida* **drawing of the salivary gland chromosomes of a female larva of** *Drosophila melanogaster.* **All of the chromosomes have undergone somatic pairing except a portion of the right arm of chromosome II.
(Painter, T. S., "Salivary Chromosomes and the Attack on the Gene." Reprinted with permission from** *The Journal of Heredity* **25:464–476, 1934.)**

The banding pattern is best seen in stained material. Each band is composed of the numerous chromomeres of the individual chromonemata arranged in a plane perpendicular to the axis of the chromosome. The inter-
band regions are so lightly stained that they appear like spaces between

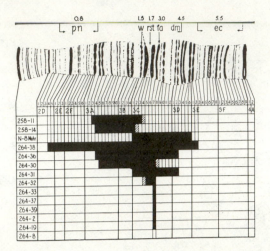

Fig. 9.11. Gene localization by means of overlapping deletions. Shown above is the genetic map, then the banded structure of a portion of the left end of the X chromosome of *Drosophila melanogaster*, and below a series of deletions. The black segments indicate the missing bands for each deletion. By the use of overlapping deletions the absence of a particular band can be correlated with the appearance of a particular phenotype. For example, the deletions 258-14, 264-31, and N-8 Mohr all lack band 3C2 and all have the white phenotype when heterozygous with the white allele. Therefore, the deficiency of band 3C2 is apparently responsible for the absence of the wild-type allele of white, and the white allele is expressed hemizygously. (Slizynska, H., 1938. "Salivary Gland Analysis of the White-Facet Region of *Drosophila melanogaster*," *Genetics* 23:291–299.)

the bands. However, the chromonemata are very probably continuous, and the bands are simply due to lighter coiling of the chromonemata in certain regions than in others. The exactitude of the synaptic mechanism in these polytene chromosomes is responsible for the clear-cut appearance of the bands.

The distinctive banding pattern in the salivary gland chromosomes made it possible to relate specific gene loci to small, well-defined regions of the chromosomes by means of overlapping deletions. Some conception of the method can be gained from Fig. 9.11. In favorable cases the absence of a specific band can be associated with the loss of a particular gene.

9.3. Aneuploidy

The second major type of chromosome variation is that involving whole chromosomes. A nucleus with some number of chromosomes other than an exact multiple of the haploid number is aneuploid. (The term euploid is used for cells with an exact multiple of the haploid number of chromosomes.)

The simplest types of aneuploidy are the simple trisomic, with $2n + 1$ chromosomes, and the simple monosomic, with $2n - 1$ chromosomes. In the $2n + 1$ trisomic, all of the chromosomes are present in pairs except one, which is represented three times. In the $2n - 1$ monosomic, all the chromosomes have a partner but one, which is without a homologue. One major cause for the formation of aneuploids is nondisjunction, as shown in Fig. 8.1.

The most common types of aneuploids observed are undoubtedly the simple trisomics, which are frequently viable but usually have distinctive effects on the phenotype as compared with the normal diploid. Monosomics are seen less often, in part because they are frequently nonviable. Other types of aneuploids can be conceived ($2n + 2$, $2n + 1 + 1$, etc.), and are sometimes observed, but they are less frequent.

For a long time aneuploidy was only of academic interest, primarily to plant cytologists such as Blakeslee, who made a classic study of all the possible trisomics in *Datura* (Fig. 9.12). The discovery that Mongolian idiocy (Down's syndrome) in man is the consequence of a simple trisomy initiated a revolution in human cytogenetics, the effects of which are still being felt (see Figs. 26.1 and 26.2).

It should be noted that duplications, inversions, or even translocations have been incorporated into the normal genetic system of various different species, and thus cannot, at least in those species, be regarded as chromosome aberrations. The maintenance of aneuploidy is, however, apparently so difficult, and the effects so marked, that aneuploids are truly aberrant.

9.4. Polyploidy

The final major type of chromosome variation is polyploidy. Polyploid cells or organisms have more than two haploid sets of chromosomes in their nuclei. There are two distinct types of polyploids with different modes of origin. The *autopolyploids* are derived from parents with very similar genomes, and an autopolyploid may therefore have three, four, or more haploid sets of chromosomes of the same type. If a haploid chromosome set is represented by A, an autotriploid will be AAA, an autotetraploid will be AAAA, an autopentaploid AAAAA, and so on. An autotetraploid may arise from the doubling of the diploid number of chromosomes in somatic cells, as sometimes occurs in callus formation in plants, or from the union of two

Fig. 9.12. The phenotypes of the seed pods of the normal diploid and of the 12 different trisomics of the Jimson weed, *Datura*. (Blakeslee, A. F., "New Jimson weeds from old chromosomes." Reprinted with permission from *The Journal of Heredity* 25:80–108, 1934.)

diploid gametes in which reduction has failed to occur. An autotriploid may arise from a cross between an autotetraploid with diploid gametes and a diploid with haploid gametes. Autopolyploidy is rare in natural populations

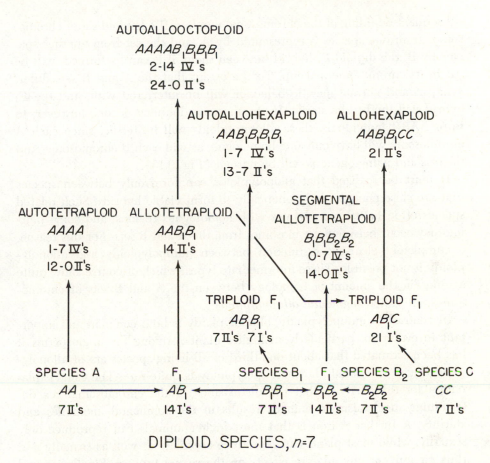

AUTOALLOOCTOPLOID

$AAAAB_1B_1B_1$

2-14 IV's

24-0 II's

AUTOALLOHEXAPLOID   ALLOHEXAPLOID

$AAB_1B_1B_1$ AAB_1B_1CC

1-7 IV's   21 II's

13-7 II's

AUTOTETRAPLOID   ALLOTETRAPLOID   SEGMENTAL

$AAAA$ AAB_1B_1 ALLOTETRAPLOID

1-7 IV's   14 II's $B_1B_1B_2B_2$

12-0 II's   0-7 IV's

14-0 II's

TRIPLOID F_1 TRIPLOID F_1

AB_1B_1 AB_1C

7 II's 7 I's   21 I's

SPECIES A F_1 SPECIES B_1 F_1 SPECIES B_2 SPECIES C

AA ———→ AB_1 ←——— B_1B_1 ——→ B_1B_2 ←— B_2B_2 CC

7 II's   14 I's   7 II's   14 I's 7 II's   7 II's

DIPLOID SPECIES, $n=7$

Fig. 9.13. Diagram showing interrelationships, genomes, and mode of origin of typical autopolyploids, allopolyploids, segmental allopolyploids, and autoallopolyploids. (After Stebbins, G. L., *Variation and Evolution in Plants*, p. 315, Fig. 34. Reprinted by permission of Columbia University Press.)

and has apparently been of little evolutionary importance. One reason is the rather high degree of sterility in autopolyploids, which in large measure is due to abnormal synaptic configurations during meiosis when three or more homologues are involved in pairing. However, colchicine, derived from the autumn crocus, and other drugs have been used to induce polyploidy experimentally. This research tool has made available such horticultural wonders as the Tetrasnap, an autotetraploid snapdragon, of more impressive dimensions and intense hues than the more pedestrian diploids.

An *allopolyploid* organism has more than two haploid sets of chromosomes derived from two (or more) species following hybridization and subsequent doubling of the chromosome number. If a haploid set of chromosomes from one species is represented by A, and the set from another species by B, the diploid F_1 hybrid between them, if it can be formed, will be AB in its chromosome makeup. To the extent that the A and B sets differ, synapsis and normal meiotic behavior will be interfered with, and the F_1 hybrid will tend to be sterile. If chromosome doubling occurs, however, to form $AABB$ individuals, these allotetraploids will be fertile, since each A chromosome will have only one homologue, as will each B chromosome, and meiosis and gametogenesis will be normal (Fig. 9.13).

It must be realized that allopolyploids can form only between species that are close enough phylogenetically to form viable hybrids. Such related species often have chromosomes with homologous regions, and some pairing may occur between chromosomes from the A and B sets. For this reason, a completely clear-cut distinction between autopolyploidy and allopolyploidy is not possible, These intermediate types, which of course show quite a range in the amount of homology between the A and B sets of chromosomes, are called *segmental allopolyploids*.

In contrast to autopolyploidy, allopolyploidy is both common and important in evolution, particularly in higher plants. Among the angiosperms it has been estimated that about one-third of all living species are of allopolyploid origin. In animals, however, polyploidy is quite rare. The usual explanation for this rarity is that the disturbance of the chromosomal sex determining mechanism in animals results in developmental anomalies and sterility. A further reason is that most higher animals can reproduce only sexually, while most plants can reproduce asexually as well as sexually and thus circumvent any adverse effects on the sexual process. Finally, animal development is generally a more complex process than development in plants, and thus is more apt to be influenced by the disharmonies induced by polyploidy. In view of the above, it is of interest to note that the few polyploids that have been observed in animals have been found primarily in species that are either hermaphroditic or parthenogenetic.

Like many autopolyploids, many of the allopolyploids have favorable traits compared to the corresponding diploids, and a number of cultivated plants are allopolyploid in origin (wheat, tobacco, cotton, and many others). In some cases, it has been possible to resynthesize allopolyploids from their putative ancestors (e.g., tobacco, *Nicotiana tabacum*, from *N. sylvestris* and *N. tomentosiformis*). In others, new, potentially useful allopolyploids have been created. The classical example is *Raphanobrassica*, an allotetraploid created by Karpechenko from the hybrid between the radish (*Raphanus sativus*) and the cabbage (*Brassica oleracea*) (Fig. 9.14). By any criterion, this experiment represents the creation of a new species, fully viable and

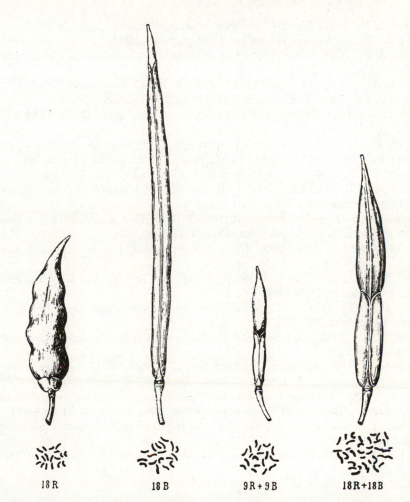

Fig. 9.14. Seed pods of, from left to right, diploid radish (*Raphanus*), diploid cabbage (*Brassica*), the F$_1$ sterile hybrid between the radish and the cabbage, and the fertile allotetraploid *Raphanobrassica*. (After Karpechenko.)

fertile, but reproductively isolated from its parental species. From a practical standpoint, however, the experiment was less successful—the plants had roots like cabbages and heads like radishes.

Additional Reading

Avery, A. G., S. Satina, and J. Rietsema, 1959. *Blakeslee: The Genus Datura*. Ronald, New York.

Benirschke, K., and T. C. Hsu, eds., 1971. *Chromosome Atlas: Fish, Amphibians, Reptiles and Birds*. Springer-Verlag, New York.

Blakeslee, A. F., 1931. "Extra Chromosomes a Source of Variations in the Jimson Weed." In *Smithsonian Report for 1930*, pp. 431–450.

Blakeslee, A. F., 1934. "New Jimson Weeds from Old Chromosomes," *J. Heredity* 25:80–108.

Bridges, C. B., 1936. "The Bar "Gene" a Duplication," *Science* 83:210–211.

Burnham, C. P., 1962. *Discussions in Cytogenetics*. Burgess, Minneapolis.

Cleland, R. E., 1962. "The Cytogenetics of *Oenothera*," *Adv. Genetics* 11:147–237.

Creighton, H. S., and B. McClintock, 1931. "A Correlation of Cytological and Genetical Crossing Over in *Zea mays*," *Proc. Nat. Acad. Sci.* 17:492–497.

Darlington, C. D., 1956. *Chromosome Botany*. Allen and Unwin, London.

Hsu, T. C., and K. Benirschke, eds., 1971. *An Atlas of Mammalian Chromosomes*. Springer-Verlag, New York.

Karpechenko, G. D., 1928. "Polyploid Hybrids of *Raphanus sativus* (L.) × *Brassica oleracea* (L.)," *Z. Ind. Abst. Vererb.* 48:1–85.

Lewis, K. R., and B. John, 1963. *Chromosome Marker*. J & A. Churchill, London.

Rhoades, M. M., 1955. "The Cytogenetics of Maize." In *Corn and Corn Improvement*. Academic, New York.

Sears, E. R., 1948. "The Cytology and Genetics of Wheat and Their Relatives," *Adv. Genetics* 2:239–270.

Stebbins, G. L., 1950. *Variation and Evolution in Plants*. Columbia University Press, New York.

Stebbins, G. L., 1971. *Chromosomal Evolution in Higher Plants*. Addison-Wesley, Reading, Mass.

Stern, C., 1931. "Zytologisch-genetische Untersuchungen als Beweise für die Morgansche Theorie des Faktorenaustauchs," *Biol. Zentralbl.* 51:547–587.

Sturtevant, A. H., and G. W. Beadle, 1939. *An Introduction to Genetics*. Saunders, Philadelphia.

Yunis, J. J., ed., 1965. *Human Chromosome Methodology*. Academic, New York.

Questions

9.1. The diploid number of chromosomes in *Drosophila melanogaster* is 8, yet salivary gland nuclei show what appear to be only six chromosomes. What is the explanation for this apparent discrepancy?

9.2. When a cytological map of *Drosophila* salivary gland chromosomes is compared with the genetic map, the various genes lie in the same linear order on both maps. However, there are discrepancies in the relative distances between genes on the two maps. How are these two types of maps constructed? Why are there discrepancies between them?

9.3. What is the interpretation of the puffs observed in the salivary gland chromosomes of the *Diptera*?

9.4.   How would you distinguish between an inversion loop and a deletion loop in salivary gland chromosomes of *Drosophila*?

9.5.   What are the consequences of a single chiasma within the inversion in a heterozygote for a paracentric inversion?

9.6.   What are the consequences of a single chiasma within the inversion in a heterozygote for a pericentric inversion?

9.7.   How can it be determined whether crossing over occurs in the two-strand stage before the parental chromosomes divide or after they divide to form four chromatids?

9.8.   The occurrence of a ring of four chromosomes during meiosis I in a species that otherwise forms only ordinary chromosome pairs is reliable evidence for what kind of heterozygosity?

9.9.   What is the probable origin of most cases of simple trisomy?

9.10.   Two genes are linked with 9% crossing over. If you found a strain in which the crossover percentage between these genes was 23%, what has probably occurred? If you found another strain in which recombination between these same two genes was 50%, what would you suspect?

9.11.   If examination of a slide prepared from the somatic cells of a normal vertebrate female revealed a total of 27 chromosomes in each of many cells, to what class of vertebrates would this female probably belong?

9.12.   What significant differences are there between the Bar-eye type of position effect and the white-apricot type of position effect?

9.13.   Yellow body color in *Drosophila melanogaster* is a sex-linked recessive. Usually yellow females crossed to wild-type males produce an F_1 consisting of normal daughters and yellow sons. Occasionally, an exceptional yellow female, crossed to a normal, wild-type male, produces only yellow daughters and normal sons. What is the most reasonable explanation for these results?

9.14.   The diploid chromosome number of the most primitive species of wheat is 14. Ordinary bread wheat (*Triticum vulgare*) has 42 chromosomes in its somatic tissue cells. What level of ploidy is found in bread wheat?

9.15.   How many linkage groups would you expect to find in an autotraploid African violet with a somatic chromosome number of 60?

9.16.   In plant species A, $n = 11$; in species B, $n = 19$.

   a. What is the chromosome number in the allotetraploid produced from A and B?

   b. How many linkage groups are there?

9.17.   Why are F_1 hybrids from crosses between two distinct species so often sterile?

9.18. Plant species A has $n = 7$, species B has $n = 9$, and species C has $n = 10$. What would be the chromosome number of an allohexaploid plant synthesized from these species?

9.19. American cotton, thought to be the result of hybridization and subsequent polyploidy, has 26 pairs of chromosomes: 13 large pairs and 13 small pairs. Asiatic cotton has 13 large pairs of chromosomes and is believed to be one of the ancestral species. The other is as yet unknown. When it is identified, how many and what kind of chromosomes would you expect it to have?

PART TWO

THE NATURE
OF THE GENE

CHAPTER TEN

Gene Structure

10.1. The Gene Concept

By 1925 the theory of the gene had been developed to the point that it furnished a very satisfactory explanation for the available facts about heredity. Mendel had studied the transmission of alternative factors for specific traits from one generation to the next in the garden pea; from this information, he developed the laws of segregation and independent assortment. Mendel assumed that the differences among individuals were transmitted as if they were controlled by the segregation and recombination of unitary factors that behaved in a predictable manner in crosses. To Mendel these factors were only hypothetical units, postulated as an aid in the interpretation of his data. They had no known physical counterparts.

The great contribution of Morgan and his collaborators was their demonstration of the parallelism between the behavior of the Mendelian factors and the behavior of the dark staining nuclear bodies, the chromosomes. Morgan showed that the genes must be located within the cell in the chromosomes of the nucleus and extended the knowledge of the transmission of the genes by showing that they were inherited in linkage groups and that the genes within a given linkage group were arranged in a linear order. The alternative traits were shown to be controlled by specific regions of the chromosomes. Chromosome deficiencies and duplications, inversions and translocations were found to affect not only the chromosome structure but the mechanism of heredity as well.

The genes themselves were thought to have several distinctive characteristics. First, the gene was a hereditary unit, transmitted from one generation to the next. It was one of a number of alternative states of a hereditary element, occurring in an individual only if he had inherited it from
one of his parents. It was incapable of being resolved into separately transmitted subentities and would be transmitted asexually to all offspring, but
sexually to only a proportion (usually 50%). In other words, the gene was
thought of as a very small chromosome region within which no crossing over
or chromosome breakage had been observed.

The gene was also a functional unit; it produced a phenotypic effect
that modified the expression of one or more traits. The effect of a particular
gene was constant within a given genetic and environmental background.
Despite changes in its expression in intervening generations, during which
it might exist in a variety of genetic and environmental conditions, the gene
would (barring mutation) show its original expression when returned to the
original conditions.

The other functional aspect of the gene was self-duplication. The gene
was able to replicate independently of other genes, the cell, the organism,
and the environment. The wild type ($+$) gene at the vestigial (vg) locus
was subjected to repeated backcrosses for 300 generations by the cross
($+/vg \times vg/vg$), but despite its extended exposure to the possible influence
of the vg allele, it retained its identity without change or modification of
its expression. Even after crosses between different species, a given gene
could be recovered unchanged. Thus the gene formed an exact duplicate
of itself and was considered to be a remarkably stable unit.

Despite this stability the gene occasionally underwent an abrupt change
or mutation to an alternative state. This new type of mutant allele then
persisted as a new sort of self-duplicating unit at this locus. By 1925, therefore, the gene was thought to be a stable, self-replicating hereditary unit
with a recognizable phenotypic effect and capable of an occasional rare
mutation.

Classical genetics arrived at the gene concept by indirect methods.
Crosses were made and the progeny counted, and inferences were drawn
about the mechanism of heredity from the numbers of different types of
progeny. This approach was used by both Mendel and Morgan. When the
results of Morgan and his collaborators led to the conclusion that the hereditary factors were on the chromosomes, cytology and genetics made common cause in pursuit of the nature of the genetic material. Since the cytologists of the time relied for their data primarily on observations of the
chromosomes under the light microscope, their approach to the gene was

also indirect. A major problem was to relate their cytological observations to the genes as delimited by the genetic tests. The method of overlapping deletions sometimes made it possible to relate a given gene, identified by its behavior in crosses, to a particular, and in some cases quite limited, region of the chromosome. Nevertheless, the exact limits of the gene within this region or even whether it had exact limits, remained conjectural.

10.2. Identification of the Genetic Material

Direct biochemical analysis of the genetic material originated with the work of Miescher published in 1871. He described a material that he called nuclein, derived from the nuclei of pus cells and fish sperm. In 1889, Altmann showed that nuclein could be split into protein and nucleic acid, a complex organic acid rich in phosphorus. A few years later the cytologist E. B. Wilson suggested that the nucleic acid component was the hereditary material, but proof for this hypothesis did not become available for almost half a century.

Soon two different types of nucleic acid, called "thymus nucleic acid" and "yeast nucleic acid," were identified. Thymus nucleic acid is now known as deoxyribose nucleic acid (DNA), and yeast nucleic acid as ribose nucleic acid (RNA). Just after 1900, it was shown that both DNA and RNA contain the bases, adenine, cytosine, and guanine, but that the thymine of DNA is replaced by uracil in RNA. Furthermore, it was shown that the five-carbon sugar (pentose) present in both was deoxyribose in DNA and ribose in RNA.

The early analyses seemed to indicate that each of the four bases was present in equivalent or equimolar amounts. These findings led to the "tetranucleotide" theory, that the fundamental unit of the nucleic acid molecule is composed of one of each of the four bases. However, in 1950 Chargaff and his colleagues showed that in DNA the bases are not present in equimolar amounts but that, instead, the amount of cytosine equals the amount of guanine, and the amount of adenine equals that of thymine.

With the development of the Feulgen stain, specific for DNA, and of enzymatic methods for distinguishing DNA from RNA, the distribution of these compounds in the cell could be studied. While DNA was found almost exclusively in the nucleus, RNA could be identified in both nucleus and cytoplasm.

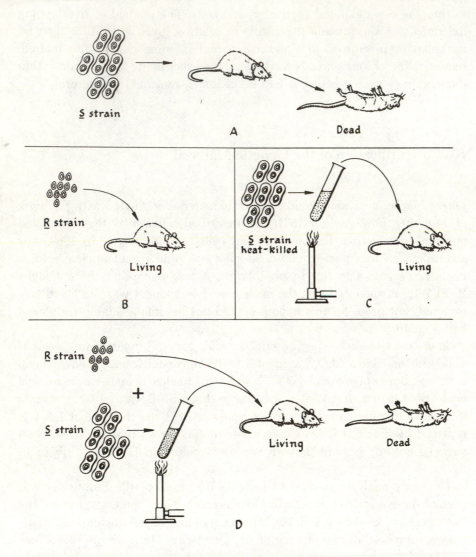

Fig. 10.1. Diagram of Griffith's experiments on transformation in *Pneumococcus*. A. Mice die when injected with a pathogenic, encapsulated S strain of *Pneumococcus*. B. Mice survive when injected with a nonpathogenic, nonencapsulated R mutant strain. C. Mice survive when injected with the heat-killed S strain. D. Mice die when injected with a mixture of a living culture of the R strain and a heat-killed culture of the S strain. The living bacteria recovered from the mice had a capsule like that of the S strain. (Sager, R., and F. J. Ryan, 1961. *Cell Heredity.* John Wiley & Sons, Inc., New York, p. 14, Fig. 1.4.)

Bacterial Transformation

One of the first steps leading to the identification of DNA as the genetic material was the discovery of bacterial transformation in the pneumococcus bacteria (*Diplococcus pneumoniae*) by Griffith in 1928. (Fig. 10.1) Earlier workers had described a number of serologically distinct types of pneumococcus. The type specificity is determined by the polysaccharide capsule that encloses the cell. Grown in culture (*in vitro*), the encapsulated cells form a "smooth" colony; grown in a susceptible mammalian host (*in vivo*), the encapsulated cells are virulent and usually cause a severe illness. Occasionally, mutants lacking a capsule arise in laboratory cultures. Such mutants give rise to "rough" colonies in the cultures and have lost not only their type specificity but also their virulence.

Griffith used two strains of bacteria, a rough avirulent strain derived from type II and a smooth virulent type III strain. Mice injected with live cells from the avirulent strain or with heat-killed cells from the type III strain remained healthy. However, mice injected with both the avirulent live cells and the heat-killed virulent cells developed a severe septicemia that was quickly fatal. From these mice it was possible to recover virulent type III strains of pneumococcus. One possible explanation is that somehow the heat-killed type III cells had been brought back to life, but the true explanation was later shown to be a transformation of the avirulent cells derived from the type II strain to virulent cells with a type III capsule. When the transformation was carried out *in vitro*, it became clear that the mouse was not an essential part of the transformation and that somehow genetic information was being transferred from dead to living cells.

The next major step in the study of transformation was the effort to identify the active transforming principle among the various fractions obtained from killed encapsulated cells. After more than a decade of work, Avery, MacLeod, and McCarty announced in 1944 that DNA was the active transforming principle. They showed that the transforming activity of the extract was not destroyed by depolymerizing or by denaturing the proteins, or by enzymes that attacked the RNA, or by antibodies that bound the pneumococcal proteins or the polysaccharides. On the other hand, the transforming activity was rapidly destroyed by DNAase, the enzyme that depolymerizes DNA. Tests of the purified active fraction of the extract were negative for protein but contained hydrogen, carbon, nitrogen, and phosphorus in the proportions characteristic of DNA. Thus, even though the change was detected by the appearance of a polysaccharide capsule, it was mediated by DNA.

Since the initial discovery of bacterial transformation in pneumococci, transformation by DNA has been demonstrated for a variety of traits in a number of different species of bacteria.

These experiments clearly pointed toward DNA as the genetic material, but the possibility remained that DNA was not the hereditary material but was somehow able to act upon the genes of the recipient cells to produce directed mutations. In other words, the DNA from the donor cells might be able to change the genes of the recipient cells to resemble the genes of the donor from which the DNA was derived. However, further experiments to be described below have even more clearly implicated DNA as the genetic material.

Bacterial Viruses

The bacteriophages or bacterial viruses can multiply only within bacterial cells. The simple phage life cycle consists of infection of a bacterium by a phage particle, followed about 20 min later by the bursting of the bacterium with the release of a few hundred new infective phage particles like the original one. The T2 phage, which infects the colon bacillus, *Escherichia coli,* was used in the classic experiments of Hershey and Chase that led to the identification of DNA as the genetic material in phage.

The T2 phage particles consist of about 50% protein and 50% DNA. They have a hexagonal head and a protruding tail ending in six fibers. The DNA is found within the head of the virus, while the protein forms the outer coat. In 1952 Hershey and Chase labeled some phages with radioactive phosphorus (P^{32}) and others with radioactive sulfur (S^{35}). (Fig. 10.2.) Phosphorus is present only in the viral DNA, sulfur only in the viral protein. Hershey and Chase mixed suspensions of labeled virus with suspensions of unlabeled *E. coli.* After only a few minutes of contact, the mixture was agitated in a Waring blender to break up the phage-bacterial complex. The bacteria were then separated from the phage by centrifugation and analyzed for radioactivity. The bulk of the viral P^{32} was now found to be associated with the bacteria, but most of the S^{35} was associated with the remains of the viruses. Furthermore, if the phage life cycle was then completed, the labeled phages could transmit P^{32} to some of their progeny, but the S^{35} label was not transmitted to the progeny.

When the phage attacks a bacterial cell, it becomes attached by its tail to the surface of the bacterium. The phage DNA passes into the interior of the cell, but the protein coat remains as a "ghost" on the outer surface. Since only the DNA passes into the bacterium where the formation of hundreds of new phage particles is initiated, the DNA rather than the protein

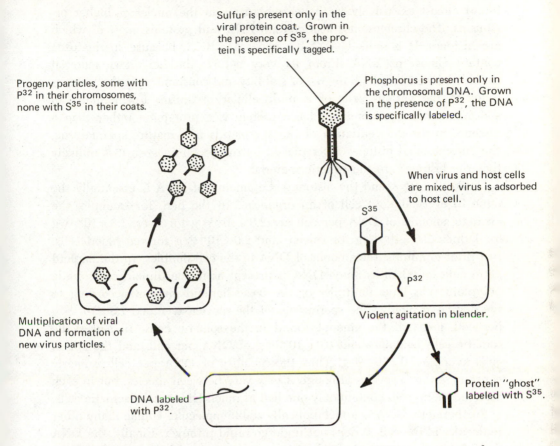

Sulfur is present only in the viral protein coat. Grown in the presence of S³⁵, the protein is specifically tagged.

Progeny particles, some with P³² in their chromosomes, none with S³⁵ in their coats.

Phosphorus is present only in the chromosomal DNA. Grown in the presence of P³², the DNA is specifically labeled.

When virus and host cells are mixed, virus is adsorbed to host cell.

S^{35}

P^{32}

Violent agitation in blender.

Multiplication of viral DNA and formation of new virus particles.

Protein "ghost" labeled with S³⁵.

DNA labeled with P³².

10.2. The Hershey-Chase experiment, showing that the DNA component of T2 phage carries the genetic information and the protein acts as a protective coat. (After Watson.)

must be the genetic material. Thus, the viral protein serves as a protective coat for the virus and mediates its attachment to the host bacterial cell, but the DNA, as the chemical basis of heredity in these viruses, controls not only its own replication, but the synthesis of the proteins necessary for the formation of the new phage particles as well.

DNA Constancy

The role of DNA in bacterial transformation and in viral reproduction implicated DNA as the physical basis of heredity, but the generality of this discovery for other organisms, especially higher organisms, remained problematical. However, there was circumstantial evidence that DNA is the hereditary material in species other than bacteria and viruses. DNA is found almost exclusively in the chromosomes in the nuclei of higher organisms. The chromosomes also contain RNA and proteins, most of which are histones. It is unlikely that the genes are RNA, because spermatozoa contain almost no RNA. From its very nature, the hereditary material should show a fairly high degree of stability and constancy; yet the histones differ in the various cells of a multicellular organism. Furthermore, in spermatozoa the predominant basic protein is a protamine rather than a histone. In the differentiation of the spermatids into mature spermatozoa, the chromosomal histones are replaced by the protamines, so it is unlikely that the histones are the genetic material.

Unlike the RNA and the histones, the amount of DNA is essentially the same in every diploid cell of an organism. In the hen, for example, the average amounts of DNA per cell are 2.6×10^{-12} g for liver, 2.3×10^{-12} g for kidney, 2.6×10^{-12} g for spleen, and 2.6×10^{-12} g for red blood cells. Just prior to mitosis, the amount of DNA in the cell doubles. In the haploid germ cells only half as much DNA is present as in the diploid cells, while polyploid cells have multiples of the basic haploid amount according to the degree of ploidy. The sperm cells of the rooster averaged 1.3×10^{-12} g per cell, just half the amount found in the somatic cells. In cattle, the somatic cells contain about 6.0×10^{-12} g of DNA per cell, and the sperm cells average 3.0×10^{-12} g. Thus the quantity of DNA per cell is nearly constant within a species although it may vary between species, but in each case the sperm cells contain only one half as much DNA as the somatic cells.

Furthermore, DNA is a metabolically stable molecule. Unlike many other molecules in the cell, it does not undergo rapid turnover. Finally, the DNA from any cell from any individual in a given species always seems to have the same relative proportions of the four kinds of nucleotides. Hence the composition of the DNA is similar in cells sharing a common ancestry.

Thus, of the chemical components of the chromosomes of higher organisms, DNA is the only one that fulfils the stability and constancy requirements of the hereditary material.

The Effect of Ultraviolet Light on DNA

Ultraviolet light is similar to visible light, but the wavelengths are too short to be detected by the human eye. Although ultraviolet light has less energy than X rays, it can be a very effective mutagenic agent. The most effective wavelength for inducing mutations is about 2600 Å. The absorption of energy by DNA also reaches a maximum at this wavelength, but proteins absorb little energy at 2600 Å. The fact that maximum mutagenicity by ultraviolet and maximum energy absorption by DNA coincide has been

Pyrimidines

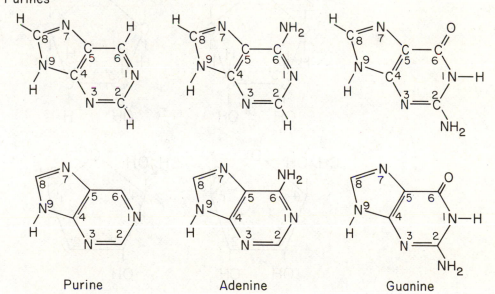

10.3. Structural formulas of the common purine and pyrimidine bases in nucleic acids.

interpreted to mean that the ultraviolet acts directly on the DNA to produce mutations and that the DNA is the hereditary material.

Thus many lines of evidence converge to point to DNA as the physical basis of heredity. It is rather ironic that, having reached this conclusion, we must qualify it by pointing out that in certain viruses, such as the tobacco mosaic virus (TMV), RNA rather than DNA serves as the genetic material. However, except for these viruses, the evidence has become increasingly clear that in many viruses and in bacteria and the higher organisms, both plant and animal, DNA is the "stuff" of heredity.

10.3. The Structure of DNA

DNA occurs as very long, unbranched molecules made up of many similar units called nucleotides, a polymer composed of a number of nucleotide monomers. The DNA chains generally contain just four kinds of nucleotides, characterized by the four organic bases, adenine, thymine, guanine, and

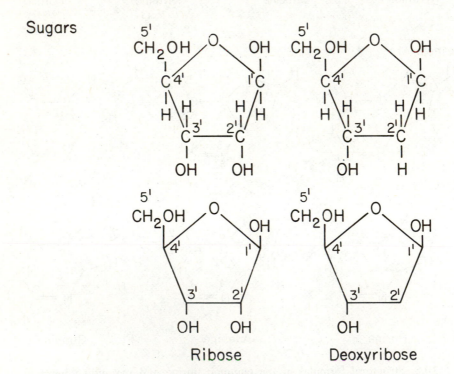

Fig. 10.4. The pentoses, ribose and deoxyribose.

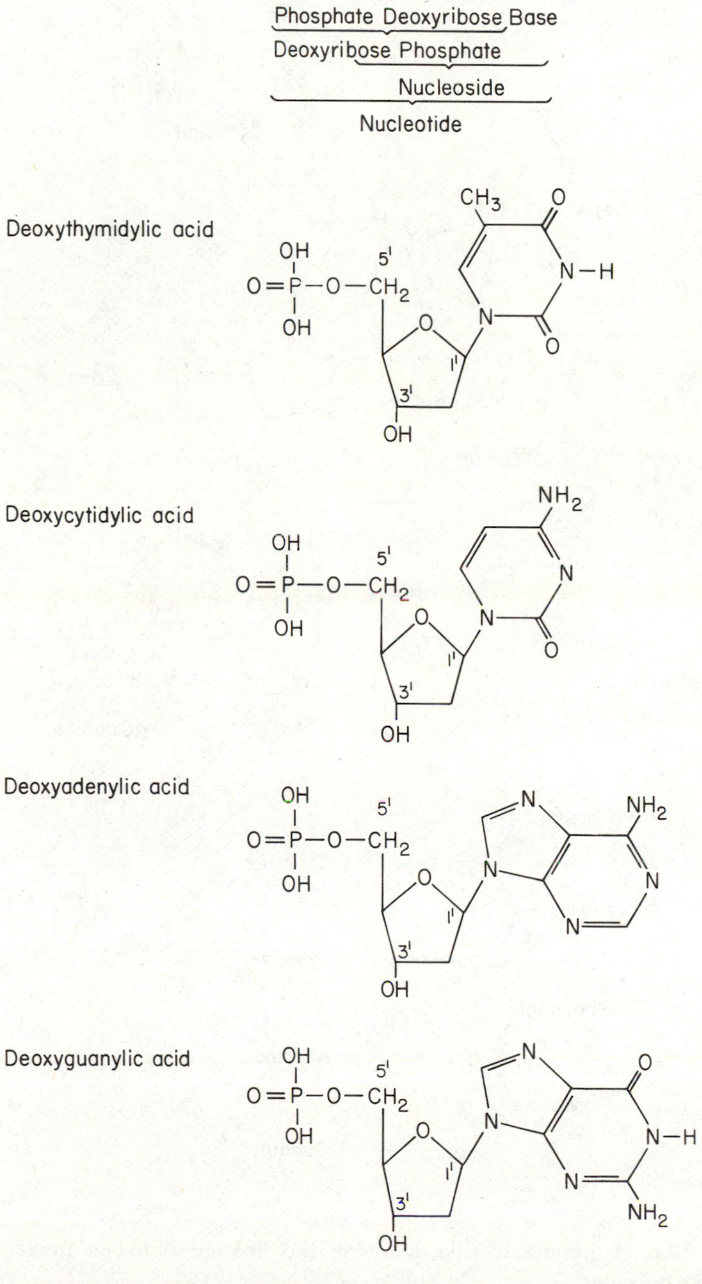

Fig. 10.5. The four common nucleotides of DNA.

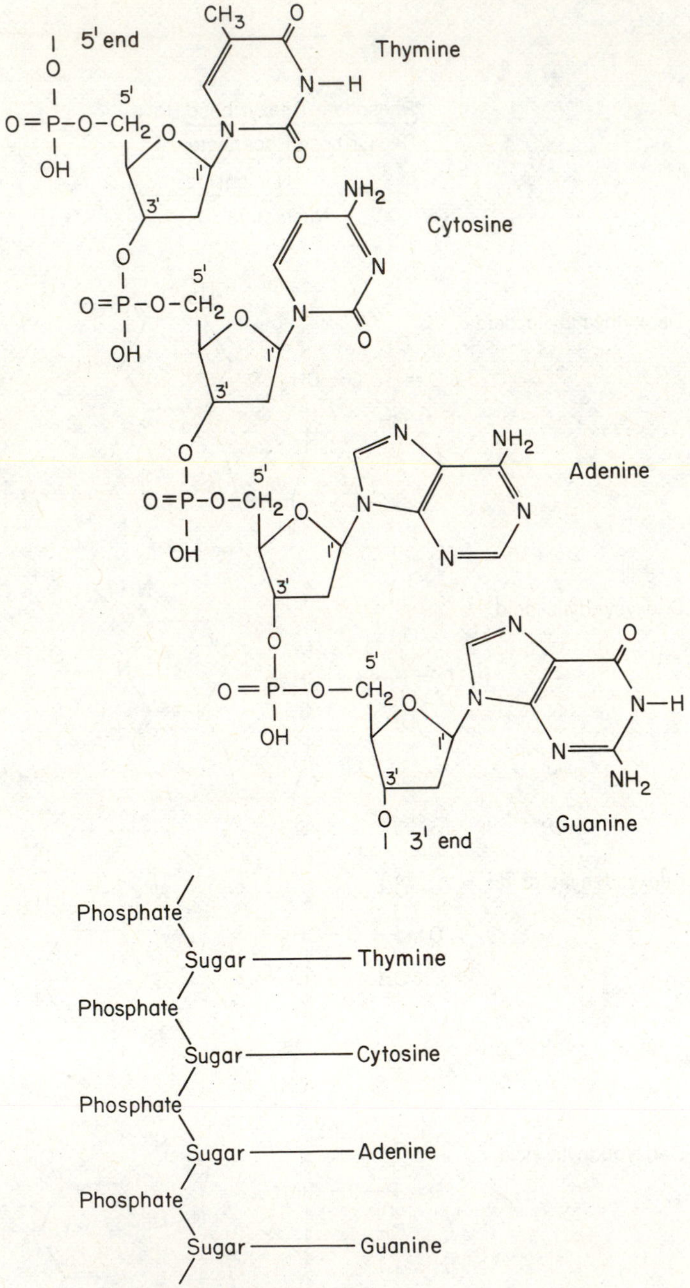

Fig. 10.6. A polynucleotide, showing the linkage between nucleotides to form a polymer.

cytosine. In addition to a base, each nucleotide contains a five-carbon sugar (pentose) called deoxyribose, and a phosphate (PO_4) group.

The organic bases are of two kinds, purines and pyrimidines, with the structure shown in Fig. 10.3. As a memory aid, you might note that the smaller molecules, the pyrimidines, have the longer name. Two versions of the structural formulae are shown, with the simpler version using the conventions adopted by chemists to bring out the salient features of the molecules. The same bases are present in RNA except for thymine, which is replaced by uracil.

The pentoses, ribose and deoxyribose, characteristic of RNA and DNA respectively, are shown in Fig. 10.4.

A purine or pyrimidine base linked with a pentose forms a nucleoside. The union of a nucleoside with a phosphate group forms a nucleotide. The four common nucleotides of DNA are shown in Fig. 10.5. The phosphoric acid and deoxyribose are joined by an ester linkage at either the 3' or 5' carbon of the sugar. In Fig. 10.5 the linkage is shown at the 5' carbon. The nitrogenous bases are linked to the sugar at its 1' carbon atom. These four nucleotide molecules are the main building blocks for DNA. The polynucleotides are formed by the formation of a phosphate "bridge" from the 5' carbon atom of the deoxyribose in one nucleoside to the 3' carbon atom of the deoxyribose in the next, as shown in Fig. 10.6. The polynucleotide molecule, then, is a long chain with alternating residues of sugar and phosphate as the backbone and the nitrogenous bases as side groups attached to the 1' carbons of the sugars. In Fig. 10.6 each of the bases is arbitrarily shown once, but in the long DNA molecules, the bases are found in a variety of sequences.

10.4. The Watson-Crick Double Helix

Even after the evidence had become fairly conclusive that DNA was the hereditary material and its chemical composition was known in a general way, the exact structure of DNA remained an open question. In 1953 James D. Watson and Francis H. C. Crick proposed a model for the structure of DNA that accounted for the available facts about the substance and also made some exceedingly useful predictions about how it would behave. Their proposal was that the DNA molecules consist of two long polynucleotide strands coiled about one another to form a double helix. In each strand the nitrogenous base of the nucleotides is oriented toward the other strand and the phosphate group away from the other strand. The two chains

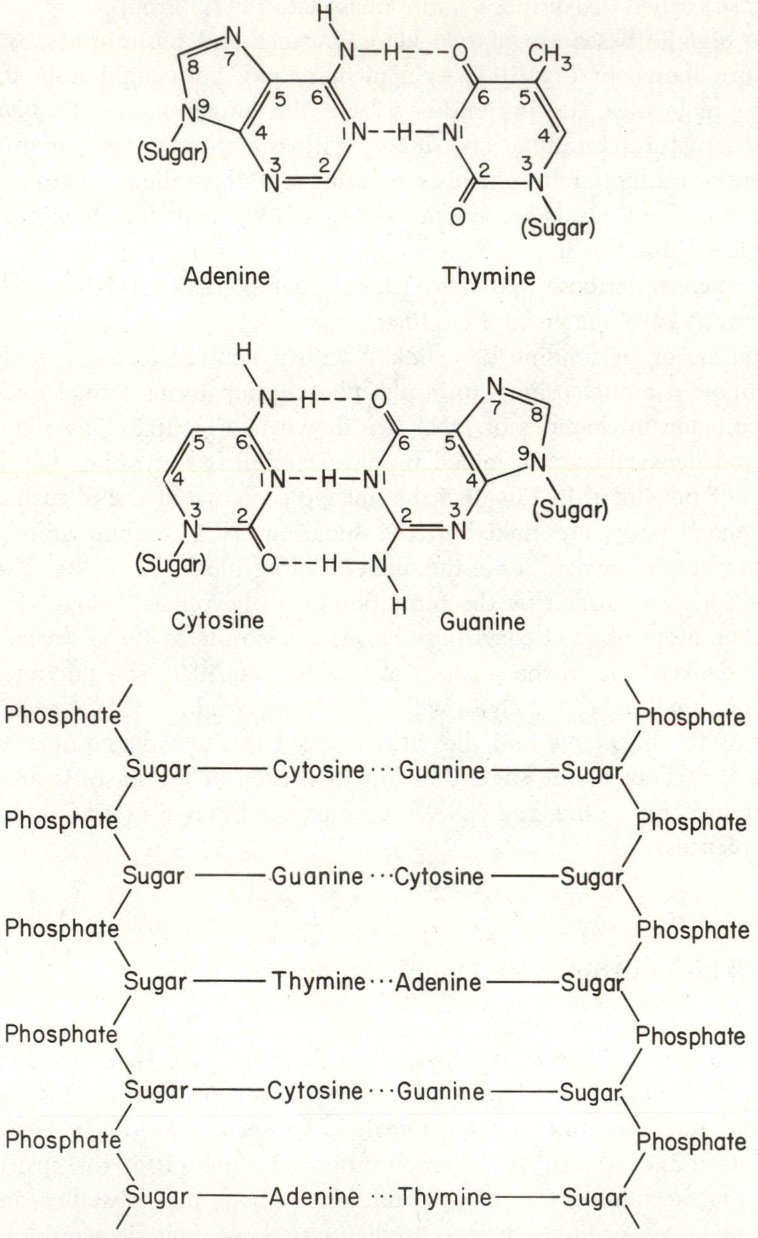

Fig. 10.7. Hydrogen bonding between bases in the DNA double helix. Adenine is always paired with thymine and cytosine with guanine. This association of a purine with a pyrimidine base permits the sharing of hydrogens by the two bases in the manner shown.

are held together by hydrogen bonds formed between the adjacent bases on the two chains. In other words, the DNA takes the form of a turning ladder, with the sugar-phosphate chains forming the uprights and the base pairs the steps of the ladder. In each complete turn of the double helix there are 10 nucleotide pairs.

A few words about chemical bonds may be appropriate at this point. The bonds governed by the usual rules of valence are known as covalent bonds and result from the sharing of electrons by two atoms. Such bonds are rather strong, the atoms are close together, and considerable energy is required to break the bond. Hydrogen bonds, on the other hand, are rather weak, the atoms are not held so close together, and less energy is needed to break them. A hydrogen bond arises when a single hydrogen atom is shared between two other atoms. In such a case, one atom covalently bound to a hydrogen atom with some positive charge serves as a hydrogen donor and another atom with some negative charge acts as a hydrogen acceptor. The biologically important donors involve hydrogen atoms covalently bound to oxygen or to nitrogen. The negative acceptor atoms are also usually oxygen or nitrogen. Each of the purine and pyrimidine bases in DNA contains both donor and acceptor atoms. However, Watson and Crick discovered that, because of the restrictions imposed by the orientation and spacing of the bases in the nucleotides, hydrogen bonding was possible only between certain base pairs. Adenine on one DNA strand would form hydrogen bonds only with thymine on the other strand, and guanine would form hydrogen bonds only with cytosine (Fig 10.7). Therefore, the strands are complementary, and if the two molecules in the double helix are held together by hydrogen bonds throughout their length, wherever adenine is found on one strand, thymine will be found as its partner on the other and vice versa. Similarly, the occurrence of guanine on one strand will be matched by cytosine on the other. If the base sequence on one strand is –T–A–G–C–C–A–, the complementary sequence on the other strand will be –A–T–C–G–G–T–. The obvious implication of this finding is that in DNA from any source the number of adenine molecules will equal the number of thymine molecules (A = T) and the molar amounts of guanine and cytosine will also be equal (G = C). (In some DNA, part of the cytosine is in the form of 5-methyl or 5-hydroxymethyl cytosine, but this does not affect the points being made here.) As pointed out earlier, Chargaff and his colleagues had already demonstrated the equivalence of A and T and of G and C. There was no restriction, however, in Watson and Crick's model on the relative numbers of AT pairs and GC pairs, which may vary widely. The base composition of DNA from several different species is shown in Table 10.1.

Table 10.1 Base Composition of DNA from Different Species (%)

Species	Adenine	Thymine	Guanine	Cytosine
Man (sperm)	31.0	31.5	19.1	18.4
Salmon	29.7	29.1	20.8	20.4
Sea urchin	32.8	31.1	17.7	17.7
Tuberculosis bacillus	15.1	14.6	34.9	35.4
Escherichia coli	26.1	23.9	24.9	25.1
Vaccinia virus	29.5	29.9	20.6	20.3

From Herskowitz, I. H., 1967. *Basic Principles of Molecular Genetics*. Little, Brown and Co., Boston, p. 20, Fig. 2.11. Courtesy of the author.

In addition to the equivalence of A to T and G to C, Watson and Crick had a number of other facts available that had to be incorporated into their model. It was known that the DNA molecules were very large, apparently the largest of the biological macromolecules, with molecular weights then estimated to be from 10^6 to 10^7. Those estimates were based on DNA that had probably been fragmented during purification, because estimates of the molecular weight of DNA in *E. coli* now range as high as 2.5×10^8. Such weights are clearly greater than those of the largest protein molecules, which seldom get as large as 10^6.

DNA solutions in water are very viscous, suggesting that the DNA molecules are in the form of long fibers. From the manner in which DNA was broken down by acid treatment or by the enzyme DNAase to form first nucleotides, then nucleosides by the splitting off of phosphate, and finally separate bases and sugars, the polynucleotide structure in Fig. 10.6 was inferred.

Because the fibrous DNA molecules tend to aggregate in parallel bundles at high concentrations, they take on some of the characteristics of a crystal. The structure of the DNA molecule could thus be studied by X-ray diffraction techniques comparable to those used with inorganic crystals. Studies of the X-ray diffraction patterns of DNA by Maurice Wilkins and his colleagues at the University of London revealed certain regularities in the structure of the DNA molecule, with periodic spacings being observed at 34, 20, and 3.4 Å.

With all of this information as their starting point, Watson and Crick constructed their model, which was consistent with the available facts. The double helix was long and thin, consistent with the fibrous nature of DNA. The two-stranded molecule held together by the specific AT and GC hydrogen bonding accounted for the equivalence between A and T and between G and C. The sugar-phosphate backbones on the outside of the molecular

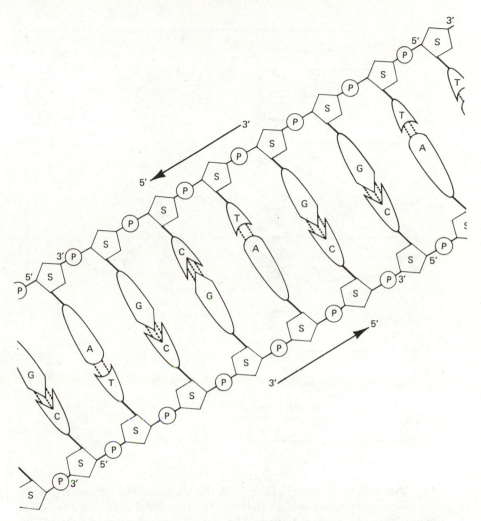

Fig. 10.8. Diagram of the DNA double helix. The two strands are held together by hydrogen bonds and are complementary rather than identical. The strands are of opposite polarity; the 3′–5′–deoxyribose–phosphodiester linkages run in opposite directions in the two strands.

helix and the bases in the interior held together by hydrogen bonds gave an overall configuration of considerable stability. The periodic spacing at 3.4 Å in the X-ray diffraction pattern marked the distance between successive nucleotides in each strand, while that at 34 Å corresponded to the distance between successive turns in the helix. The periodic spacing of 20 Å marked the diameter of the double helix (Figs. 10.8 and 10.9).

One further aspect of the Watson-Crick model has not yet been touched upon. It can be seen in Fig. 10.6 that a single DNA strand has polarity,

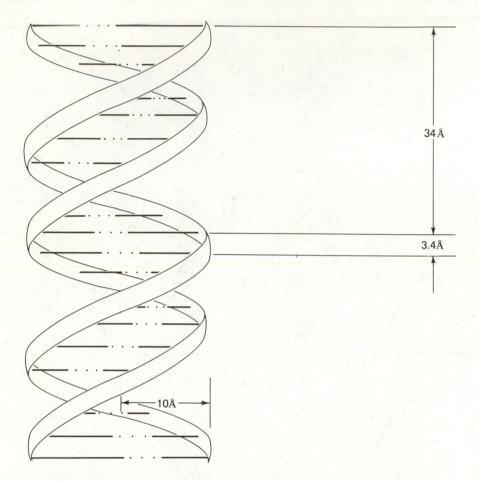

34 Å

3.4 Å

10 Å

Fig. 10.9. Schematic model of the DNA double helix. The ribbons represent the deoxyribose–phosphodiester chains; the horizontal rods, the base pairs held together by hydrogen bonds. The vertical line merely marks the axis of the molecule.

that one end can be distinguished from the other because the sugar residues are not placed symmetrically. Each sugar is linked by its 3′ carbon to the phosphate group toward one end of the molecule and by its 5′ carbon in the other direction, so that a 3′ end and a 5′ end of the strand can be identified. In the double helix (Figs. 10.8 and 10.9) the two strands can be seen to run in opposite directions; the complementary strands are of opposite polarity or antiparallel. The two strands are complementary in a very real sense, for the sequence of bases along one strand specifies the sequence of bases along the other strand.

The implications of the Watson-Crick model of DNA structure were of considerable significance. The specificity characteristic of the genetic material had to reside in the sequence of the bases along the DNA strands. Mutation could then be explained by the substitution of one base for another or by the addition or deletion of bases from the DNA chain. Although there were restrictions in the base pairing between strands, there appeared to be none on the sequence of bases or on the total length of the DNA molecule. The number of DNA nucleotides per chromosome set has, in fact, been shown to be quite different in different species, with the organisms higher on the evolutionary scale (or of greater complexity in structure and function) having greater amounts of DNA than other species, as shown in Table 10.2. The complementary nature of the DNA strands in the double helix immediately suggested a method by which duplication of the DNA could occur. The Watson-Crick model stimulated a burst of research that has led to new understanding and insight into the nature and function of the genetic material. The Nobel Prize was awarded in 1962 to Watson, Crick, and Wilkins for their contribution to our understanding of heredity.

Table 10.2 DNA Nucleotides per Chromosome Set

Species	Number of nucleotides
Man, mouse, corn	$1–1.4 \times 10^{10}$
Drosophila	1.6×10^{8}
Escherichia coli	2×10^{7}
Bacteriophage T4	4×10^{5}
Polyoma virus	1×10^{4}

From Herskowitz, I. H. 1967. *Basic Principles of Molecular Genetics.* Little, Brown and Co., Boston, p. 20, Fig. 2.12. Courtesy of the author.

Additional Reading

Adelberg, E. A., ed., 1966. *Papers on Bacterial Genetics,* 2nd ed. Little, Brown, Boston.

Avery, O. T., C. M. MacLeod, and M. McCarty, 1944. "Studies on the Chemical Nature of the Substance Inducing Transformation of Pneumococcal Types. Induction of Transformation by a Desoxyribonucleic Acid Fraction Isolated from Pneumococcus Type III," *J. Exptl. Med.* **79**:137–158.

Caspari, E. W., and A. W. Ravin, eds., 1969. *Genetic Organization,* Vol. 1. Academic, New York.

Chargaff, E., 1950. "Chemical Specificity of Nucleic Acids and Mechanism of Their Enzymatic Degradation," *Experientia* 6:201–209.

Griffith, F., 1928. "The Significance of Pneumococcal Types," *J. Hygiene* 27:113–159.

Hall, R. G., 1971. *The Modified Nucleosides in Nucleic Acids*. Columbia University Press, New York.

Hershey, A. D., and M. Chase, 1952. "Independent Functions of Viral Protein and Nucleic Acid in Growth of Bacteriophage," *J. Gen. Physiol.* 36:39–56.

Herskowitz, I. H., 1967. *Basic Principles of Molecular Genetics*. Little, Brown, Boston.

Stent, G. S., ed., 1965. *Papers on Bacterial Viruses*, 2nd ed. Little, Brown, Boston.

Taylor, J. H., ed., 1965. *Selected Papers on Molecular Genetics*. Academic, New York.

Watson, J. D., 1970. *Molecular Biology of the Gene*, 2nd ed. Benjamin, Menlo Park, Calif.

Watson, J. D., and F. H. C. Crick, 1953. "Molecular Structure of Nucleic Acids. A Structure for Deoxyribose Nucleic Acid," *Nature* 171:737–738.

Watson, J. D., and F. H. C. Crick, 1953. "Genetical Implications of the Structure of Deoxyribonucleic Acid," *Nature* 17:964–967.

Watson, J. D., and F. H. C. Crick, 1953. "The Structure of DNA," *Cold Spring Harbor Symp. Quant. Biol.* 18:123–131.

Wilkins, M. H. F., 1956. "Physical Studies of the Molecular Structure of Deoxyribose Nucleic Acid and Nucleoprotein," *Cold Spring Harbor Symp. Quant. Biol.* 21:75–90.

Questions

10.1.   Compare the nature of the gene as conceived by Mendel in 1865, Morgan in 1925, and by modern geneticists.

10.2.   Among the carbohydrates, fats, proteins, and other organic molecules found in living cells, the proteins, because of their complex structure, were long suspected to be the genetic material. What evidence led to the conclusion that the nucleic acids, specifically deoxyribonucleic acid, was the genetic material?

10.3.   What knowledge and techniques had to be available before Hershey and Chase could perform their elegant experiments demonstrating the nature of the genetic material in T2 phage?

10.4.   What are the differences between DNA and RNA? How might you distinguish between them experimentally?

10.5.   Given the composition and structure of DNA, where does it seem most logical that the specificity known to be a property of the genes must reside?

10.6. Most vertebrates have been found to have a genome size of 10^9 or more nucleotide pairs per cell.

    a. What is the length of this amount of DNA in centimeters?

    b. How many turns of the DNA double helix will occur in 10^9 nucleotide pairs?

    c. What are the implications of your answer in (b) for the process of DNA replication?

10.7. Explain the probable reason that a tetranucleotide unit was first postulated as an essential part of DNA structure. What, in fact, is the regularity in the relationships among the nucleotides?

10.8. a. What will be the base sequence in the complementary DNA strand of a single strand of DNA with the base sequence . . . AGCTTAGGCGATC. . . ?

    b. What will be the base sequence in a strand of messenger RNA complementary to this strand of DNA?

10.9. What observations about DNA did the Watson-Crick double helix model account for?

10.10. What implications did the model have with respect to mutation, gene replication, and gene specificity?

CHAPTER ELEVEN

Replication

DNA is a unique substance, self-duplicating, mutable, and controlling cell function. The Watson-Crick model of DNA structure made it possible to consider all of these characteristics at the molecular level in quite specific chemical terms. No longer were the genes rather obscure, inaccessible entities that could only be studied indirectly by means of breeding experiments. The double-helix DNA model had exciting implications for the replication, function, and mutation of genes. Each of these aspects of the behavior of the genes will be considered in turn in the next three chapters; first, we shall discuss replication.

When they proposed the double-helix model, Watson and Crick suggested that DNA replication could occur by breakage of the hydrogen bonds, separation of the two DNA strands, and formation of a new complementary chain along each strand (Fig. 11.1). Thus, they postulated that each of the existing strands serves as a template for the addition of a new complementary strand. The net result is the formation of two identical double helices where only one existed before. Furthermore, the gene-copying mechanism is direct; there is no need to postulate the transfer of gene specificity from DNA to protein and then back to DNA. Earlier theories of self-duplication had usually postulated a major role for protein, but the Watson-Crick theory made it possible to think in terms of DNA alone. This mode of replication, in which each new DNA molecule consists of one old nucleotide strand and one new one, has been called semiconservative. Several experiments have now established that DNA replication is, in fact, semiconservative (Fig. 11.2).

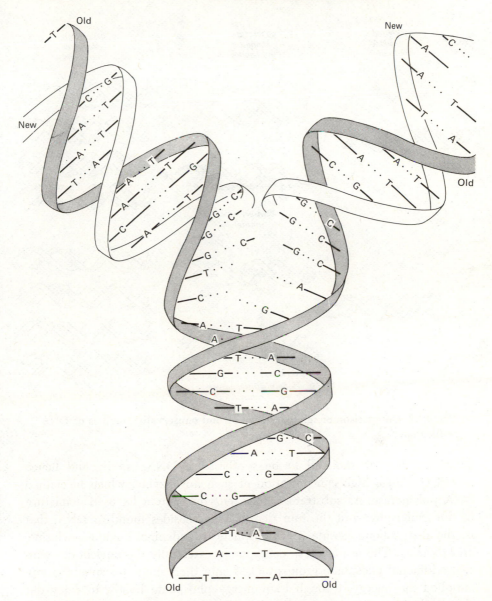

Fig. 11.1. The Watson-Crick model of DNA replication.

11.1. DNA Synthesis *in Vitro*

In one of these experiments Kornberg and his associates showed that a purified enzyme from the bacterium *Escherichia coli* catalyzed the synthesis of DNA *in vitro*, a discovery that brought him the Nobel Prize in 1959. He chose *E. coli* as his enzyme source because he reasoned that bacterial cells

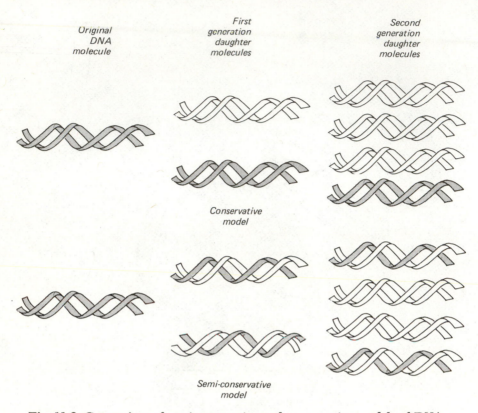

Fig. 11.2. Comparison of semiconservative and conservative models of DNA replication.

dividing every 20 min had to be synthesizing DNA rapidly and hence would contain a high concentration of such an enzyme, which he named DNA polymerase. As substrate for the enzyme system he used a mixture of the triphosphates of the four deoxyribonucleosides found in DNA, that is, the deoxyribonucleoside 5′-triphosphates of adenine, guanine, cytosine, and thymine. The use of these compounds, essentially the nucleotides with two additional phosphate groups tacked onto their single phosphate group, supplied the energy, through high-energy phosphate bonds, to carry out the synthesis in the presence of the enzyme. In addition to the enzyme and the four nucleoside triphosphates, magnesium ions were also necessary, but most interesting of all was the requirement for a polymerized DNA primer to make the reaction go (Fig. 11.3). The DNA primer could come from a variety of sources and did not need to be related to the source of the enzyme system. Thus, primer DNA from other bacterial species, from T2 phage, and from yeast and cattle was used with the *E. coli* polymerase.

Kornberg's group achieved a synthesis *in vitro* of more than 20 times the

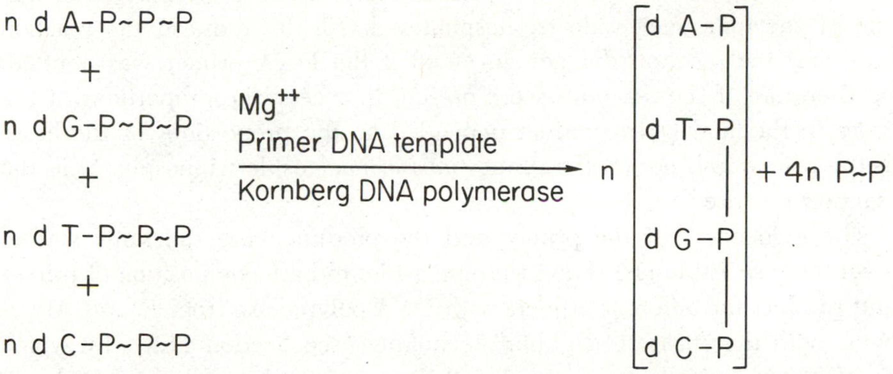

n d A-P~P~P

$+$

n d G-P~P~P   Mg^{++}

$+$ $\dfrac{\text{Primer DNA template}}{\text{Kornberg DNA polymerase}}$ n

n d T-P~P~P

$+$

n d C-P~P~P

$\begin{bmatrix} d\ A-P \\ | \\ d\ T-P \\ | \\ d\ G-P \\ | \\ d\ C-P \end{bmatrix}_n$ $+\ 4n\ P{\sim}P$

Fig. 11.3. The reaction system used by Kornberg for the *in vitro* synthesis of DNA.

initial amount of primer DNA. Even more significant was the finding that the newly synthesized DNA was identical in length, base composition, and sequence to the primer DNA and not to the DNA in *E. coli* from which the DNA polymerase came. The amount of diphosphate liberated was

Table 11.1 Chemical Composition of Enzymatically Synthesized DNA, Synthesized with Different Primers [a]

DNA	A	T	G	C	$\dfrac{A+G}{T+C}$	$\dfrac{A+T}{G+C}$
Mycobacterium phlei						
Primer	0.65	0.66	1.35	1.34	1.01	0.49
Product	0.66	0.65	1.34	1.37	0.99	0.48
Escherichia coli						
Primer	1.00	0.97	0.98	1.05	0.98	0.97
Product	1.04	1.00	0.97	0.98	1.01	1.02
Calf thymus						
Primer	1.14	1.05	0.90	0.85	1.05	1.25
Product	1.12	1.08	0.85	0.85	1.02	1.29
Bacteriophage T2						
Primer	1.31	1.32	0.67	0.70	0.98	1.92
Product	1.33	1.29	0.69	0.70	1.02	1.90

[a] A = adenine, T = thymine, G = guanine, C = cytosine.
From Kornberg, A., 1960. "Biological Synthesis of Deoxyribonucleic Acid," *Science* 131:1503–1508, p. 1507, Table 1.

equivalent to the amount of nucleotide incorporated. The omission of any one of the four nucleoside triphosphates drastically reduced the reaction rate, and the reaction did not go at all if the DNA primer was omitted. Furthermore, if the reactants were present in excess, the proportions of the bases in the final DNA product depended on the proportions of the bases in the primer and not on the proportions of nucleoside triphosphates in the reaction mixture.

The evidence that the primer and the product were the same was of several kinds. Table 11.1 shows the similarities in base composition of primer and product for different primers with DNA polymerase from *E. coli*. Moreover, with the "nearest-neighbor" technique (see Section 11.3), by which the DNA is partially fragmented and the terminal deoxyribonucleotides in the fragments are identified, the sequence of bases in primer DNA and product DNA was compared and found to be very much alike. The new DNA was also found to have the same molecular weight and viscosity as the primer. All of the evidence, therefore, indicated that the DNA polymerase had the capacity to catalyze the self-duplication of the primer DNA by a template mechanism of the sort proposed by Watson and Crick.

11.2. DNA Replication in *E. coli*

That DNA replication is semiconservative was demonstrated by the elegant experiments of Meselson and Stahl in 1958. They grew *E. coli* for a number of generations in a culture medium in which the only nitrogen source was the heavy isotope of nitrogen, N^{15}. The bacterial DNA was thereby thoroughly labeled with the heavy isotope. The bacteria were then placed in a medium containing nitrogen in its usual N^{14} form. At intervals thereafter, samples of cells were removed, the DNA extracted, and the density of the DNA molecules measured. The method used is known as equilibrium density gradient centrifugation. With this technique the DNA in a concentrated salt solution (in this case, cesium chloride) is centrifuged in an analytical ultracentrifuge at some 45,000 rpm for 20 hr. The CsCl, being somewhat denser than water, tends to concentrate toward the bottom of the whirling tube. However, the process of sedimentation is opposed by diffusion so that, in time, an equilibrium is reached in which the concentration of CsCl, and hence the density of the solution, gradually increases toward the bottom of the tube. The DNA molecules become concentrated in the region of this density gradient that corresponds to their

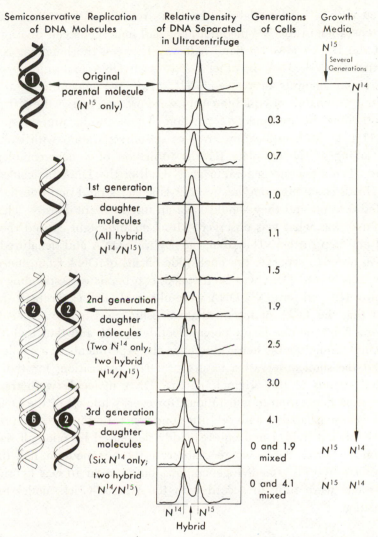

Fig. 11.4. The Meselson-Stahl experiment demonstrating semiconservative DNA replication in *E. coli*. The central tracings from photographs show the banding of the DNA in the centrifuge cells. The interpretation is based on the nature of the DNA double helix after various generations. (DeBusk, A. G., *Molecular Genetics*, copyright © 1968 by A. G. Debusk. Reprinted with permission of Macmillan Publishing Co., Inc.)

own density. The position of the DNA bands can then be determined by ultraviolet absorption photographs.

The results of the experiments are shown in Fig. 11.4. The DNA labeled with N^{15} was found to have a measurably higher density than the DNA labeled with N^{14}. When labeled bacteria from the N^{15} medium had spent one generation in the N^{14} medium (i.e., had undergone cell division once), all of the DNA showed a band half-way between the band for N^{15} DNA and that for N^{14} DNA. In other words, each DNA molecule appeared to have equal amounts of N^{15} and N^{14}. After two generations in the N^{14} medium, two bands of equal intensity were present, one at the N^{14} position, the other at the intermediate position. These results are interpreted in Fig. 11.4. If DNA replication involves strand separation with each strand then acting as the template for the formation of a new complementary strand, then after one generation in N^{14}, all of the DNA molecules should be hybrids containing one heavy (N^{15}-labeled) strand and one light (N^{14}-labeled) strand and they should have an intermediate density. This result, of course, was what was observed. Meselson and Stahl carried the experiment one step further. They took the "hybrid" DNA and denatured it with heat in order to separate the nucleotide chains of DNA from one another. When this heated DNA was centrifuged, two bands corresponding to those for pure N^{14} and pure N^{15} DNA were obtained. The experiment thus indicated that the DNA of intermediate density consisted of molecules with just one light and one heavy polynucleotide strand, and thus took the form of a single, long, double helix. Moreover, the experiments as a whole demonstrated the semiconservative nature of DNA replication, for the isotope-labeled strands of the original "parent" DNA molecule were transmitted intact from generation to generation. However, while all of the DNA molecules were half-labeled with N^{15} after one generation in N^{14}, the proportion of half-labeled hybrid molecules steadily decreased by one-half each generation thereafter. After two generations in the N^{14} medium, half the molecules were hybrid; after five generations, only one in 16 was hybrid. Thus, the results were fully in accord with the Watson-Crick model for DNA replication.

11.3. Polarity in DNA

The Watson-Crick model postulated that the complementary strands of the DNA double helix are of opposite polarity. To test this concept, Kornberg and his collaborators used their "nearest-neighbor" technique. The essence of their method was to synthesize the DNA molecule using 5′ nucleotides (i.e., with the phosphate group attached to the 5′ carbon of deoxyribose) and to break it down again, but into the 3′ nucleotides. This step is possible because of an enzyme that cleaves only the bond between the 5′ carbon and the phosphate, thus producing the 3′ nucleotides. They used radioactive phosphorus (P^{32}) to label the 5′ nucleotides used for synthesis. When the 3′ compounds were recovered, the P^{32} had been transferred from one nucleotide to its neighbor. Sufficient nucleotides of all four types were provided to permit extensive DNA synthesis, but in any one experiment only one of the four was labeled. Each of the four, however, was labeled and used in separate experiments.

The interpretation of the results was based on the equivalence of base pairing between adenine and thymine, and between guanine and cytosine. If both strands had the same polarity, the frequency of transfer of P^{32} from G to A, for example, should equal the frequency from C to T. If the strands had opposite polarity, the frequency of transfer of P^{32} from G to A should equal the frequency from T to C. All the experiments showed the latter type of result and thus supported the concept of opposite polarity proposed by Watson and Crick. In other words, the two strands of the double helix are antiparallel, the 5′ end of one strand being paired with the 3′ end of the other.

11.4. Chromosome Structure in Prokaryotes

Cell organization is much simpler in the viruses, bacteria, and blue-green algae (the prokaryotes) than in the eukaryotes (the higher plants and animals). In the eukaryotes, the genetic material is organized into chromosomes and enclosed by a nuclear membrane in a nucleus, which undergoes division by mitosis and meiosis. In contrast, the genetic material of prokaryotes consists essentially of a single long, thin, naked DNA molecule.

The structurally simple viruses consist of a DNA core and a protein sheath, from which the DNA can be isolated as a single long macromolecule. In the T4 bacteriophage, for example, the DNA has a diameter of 20 Å (the diameter of a Watson-Crick double helix) and a length of about 60 μ. One of the puzzles about the organization of the DNA within the virus is the "packing" problem. The length of the T4 DNA is roughly 1000 times greater than the internal diameter of the phage head that contains it. The puzzle lies in the way the DNA leaves the phage to enter the bacterium on infection and how the newly synthesized viral DNA is "packed" into the protein sheath after viral multiplication has occurred within the bacterial cell.

Different viruses vary markedly in size, shape, and certain other features. Most, however, have chromosomes in the form of naked molecules of DNA.

The small lambda (λ) bacteriophage has a chromosome 20 Å wide and 17 μ long. Its molecular weight is equal to 3.2×10^7. Since there are 10 nucleotide pairs per 34 Å in length of DNA, this chromosome has been estimated to have some 50,000 nucleotide pairs. The number of nucleotide pairs per gene cannot yet be specified, but if there were a thousand nucleotide pairs per gene, the number of genes in the lambda phage would be of the order of 50. Presumably, the greater the length of DNA, the greater the number of genes.

Another variation among phages is that in some, such as T4, the DNA chromosome appears to be linear in physical form but behaves in crosses as if it were a single continuous, circular structure. The λ phage, on the other hand, shows a circular structure in electron micrographs, but is genetically linear.

The fact that in some viruses RNA rather than DNA seems to be the genetic material has already been mentioned. The genetic nature of the RNA was demonstrated by testing the protein and RNA fractions of the tobacco mosaic virus (TMV) separately. Only the RNA fraction was able to infect tobacco plants and produce a new crop of TMV particles complete with protein sheath. The isolated viral RNA thus carried the genetic information needed both for its own replication and for the synthesis of the protein coat, and was clearly the hereditary material.

Another point of difference among viruses is that in some the nucleic acid is double stranded while in others there is only a single polynucleotide strand. Thus, most of the best-known DNA viruses, such as the T2, T4, and T6 group of bacterial viruses, form the Watson-Crick double helix, but the RNA of TMV, influenza virus, and poliomyelitis virus is single-stranded as is the DNA of the small ϕx174 virus. The single-strandedness of ϕx174 is proved both by the X-ray diffraction patterns and the nonequivalence of A and T and of G and C. Double-stranded RNA viruses (the Reo viruses) have also been found so that both DNA and RNA viruses may be either single- or double-stranded.

The replication of the single-stranded nucleic acids poses no special problem, for as soon as this strand (the plus strand) enters its host cell, it serves as a template for the formation of its complementary strand (the minus strand). This double helix then acts as the template for the formation of the new plus strands, which are incorporated into the mature progeny viruses.

In bacteria such as *E. coli*, the chromosome is also a single naked DNA molecule as in the viruses. The bacterial chromosomes, however, are considerably larger. The λ phage, for instance, has a single DNA molecule with a molecular weight of 3.2×10^7, and the T2 phage molecule weighs 1.2×10^8, but the single DNA double helix that constitutes the chromosome of *E. coli* has a molecular weight of about 2×10^9. The λ phage is about 17 μ in length and the T2 and T4 phages about 50 μ. The *E. coli* chromosome, however, is about 1000 μ in length, that is, nearly a millimeter long, and is ordinarily found in the form of a ring of double-stranded DNA. Since viral and bacterial chromosomes both have a diameter of 20 Å, the difference in weight is obviously due to the greater length of the bacterial chromosome. This greater length allows for the greater number of genes necessary to carry on the more complex synthetic and metabolic activities of the bacterial cell.

In prokaryotes, each gene or cistron is represented only once; that is, the DNA sequences are unique. The DNA cistrons that code for ribosomal RNA are exceptions in that some five or six copies of these genes are present.

11.5. Replication in Prokaryotes

We have already seen that replication in prokaryotes is semiconservative, that the DNA, in the form of a long slender macromolecule often joined at the ends to form a ring, serves as a template for its own replication, and that DNA polymerase mediates the process.

Further studies have shown that the chromosome in prokaryotes serves as a single replicating unit or replicon with a rate of DNA synthesis of some 30 μ/min. It is now believed that replication is initiated only when a break in at least one of the strands permits the two parental strands to unravel from one another and thus serve as templates for the formation of their complementary strands. The two strands do not completely separate before synthesis of new strands begins; instead, strand separation and replication are concurrent processes. Replication normally begins with the enzymatic breaking of the circular DNA double helix by an endonuclease at a unique starting point. These points have been mapped in both phage and *E. coli*.

From the starting point, replication continues in one direction along the chromosome and can be observed in suitable preparations as a single replicating Y fork in each chromosome, giving rise to a θ-shaped configuration (Fig. 11.5). The replicating Y fork moves around the chromosome, and replication is normally terminated when the two double helices are completed.

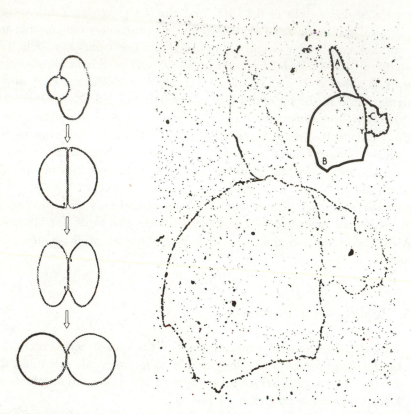

Fig. 11.5. Replication in the circular DNA molecule of *E. coli*. Replication begins at a single point and proceeds around the circle at a rate of 20 to 30 μ/min. The *E. coli* chromosome is about 1100 μ in diameter. In the autoradiograph at right, the daughter segments A and C always equal one another in length while the length of the unreplicated segment B plus either A or C equals the circumference of one circle. (From DuPraw, E. J., *The Biosciences: Cell and Molecular Biology*, CMB Council, Stanford, California, 1972.)

There are two puzzles associated with the replication process for which no completely satisfactory explanation is yet available. Although replication is in a single physical direction, nevertheless since both nucleotide strands are involved in replication and they are antiparallel, synthesis must proceed in opposite chemical directions, from the 5′ to the 3′ end in one strand and the 3′ to the 5′ end in the other. However, the Kornberg *E. coli* DNA polymerase catalyzes DNA synthesis only in the 5′ to 3′ direction *in vitro*, and no other replicating enzyme has yet been identified. Therefore, the mechanism of replication in the 3′ to 5′ direction as well as the 5′ to 3′ direction *in vivo* presents a problem. One suggested solution is that DNA

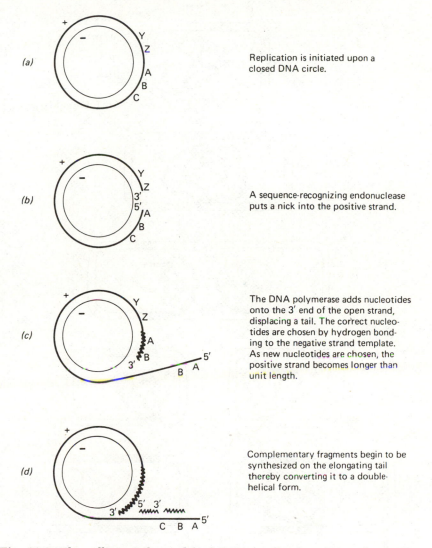

Replication is initiated upon a closed DNA circle.

A sequence-recognizing endonuclease puts a nick into the positive strand.

The DNA polymerase adds nucleotides onto the 3' end of the open strand, displacing a tail. The correct nucleotides are chosen by hydrogen bonding to the negative strand template. As new nucleotides are chosen, the positive strand becomes longer than unit length.

Complementary fragments begin to be synthesized on the elongating tail thereby converting it to a double-helical form.

Fig. 11.6. The rolling circle model of DNA replication. (From James D. Watson, *Molecular Biology of the Gene*, Second Edition, copyright © 1970 by J. D. Watson; W. A. Benjamin, Inc., Menlo Park, California.)

replication at the Y fork proceeds by assembly of short polynucleotide fragments.

An even greater problem is the mechanism of separation of the two nucleotide chains, or the unraveling problem. Even the DNA of the T2 virus, only some 50 μ in length, has 15,000 turns in its double helix. Since strand separation precedes replication, and the molecule replicates a number of times in the 30 min between infection of the bacterial host and re-

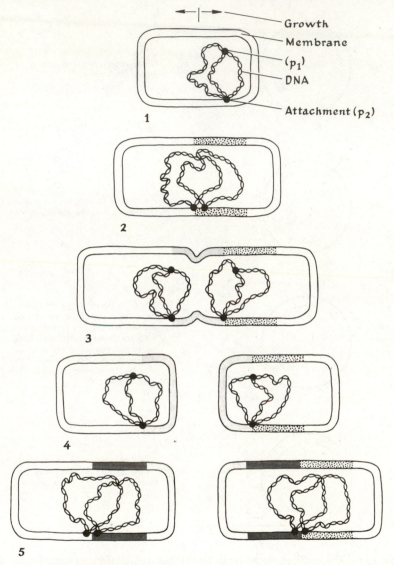

Fig. 11.7. The relation between DNA replication and the cell membrane in *E. coli*. Replication started at p_1 in the parental cell. The partially replicated DNA molecule is attached to the bacterial membrane at the site of replication p_2. Growth of the bacterial membrane, shown stippled or gray, leads to separation of the points of attachment. 1. Daughter bacterium just formed. 2. Replication of DNA molecule now complete. 3. Bacterium near second division, each chromosome has started a new cycle of replication. 4. Division of bacterium into two cells complete. 5. Replication of the circular chromosomes in the two daughter cells is now complete. (From *Molecular Genetics* by Gunther S. Stent. W. H. Freeman and Company. Copyright © 1971.)

lease of the virus progeny, the molecules would have to spin like whirling dervishes to unwind the requisite number of times during this period. It has been estimated that the single DNA molecule of *E. coli,* 1 mm long and containing 3×10^6 nucleotide pairs or 300,000 turns of the double helix, would have to spin 10,000 times per minute to replicate once in the normal 30-min generation interval. Obviously, some questions remain to be answered concerning the mode of separation and unraveling of the two strands during replication.

There is now evidence that new DNA strands may also form by addition to the free ends of existing strands. This finding has led to the development of rolling circle models for DNA replication (Fig. 11.6). Moreover, several types of repair synthesis have been discovered that repair DNA damaged by X-ray or ultraviolet radiation. In addition, the cell membrane has been implicated in DNA replication in *E. coli,* for the replicating DNA molecule has been found to be attached to the bacterial cell membrane (Fig. 11.7). It now appears that several enzymes rather than just DNA polymerase are involved in replication and that some of them are associated with the bacterial membrane. All of these discoveries have implications for the mechanism of replication in prokaryotes. Thus, it appears that even though progress is being made, much remains to be learned.

11.6. Chromosome Structure in Eukaryotes

The chromosomes of higher organisms are considerably more complex in structure than those of the prokaryotes. There seem to be no intermediate forms between the prokaryotes and the eukaryotes, except perhaps the dinoflagellates. The eukaryotes are distinguished by the presence of a nucleus containing a number of chromosomes enclosed within a nuclear membrane and by the mitotic apparatus controlling nuclear division.

Eukaryotic chromosomes are large enough to be easily visible under the light microscope. They show a considerable range in size, varying both among species and within a species in different tissues or different stages of the mitotic cycle. At mitotic metaphase, when the chromosomes are most contracted due to coiling, they may range in size from about 0.2 to 2 μ in diameter and from about 0.2 to more than 30 μ in length. Human chromosomes are of the order of 4 to 6 μ long, and the length of mitotic metaphase chromosomes of *Drosophila* and maize averages about 3.5 μ and 8–10 μ, respectively. The larger salivary gland chromosomes of *Drosophila* and meiotic pachytene chromosomes of maize are, however, the ones usually studied cytologically.

Chemical analysis of chromosomes has shown them to contain DNA, basic protein, RNA, and the so-called residual proteins. The basic proteins are histones and protamines, rich in the basic amino acids, lysine, arginine, and histidine, while the residual proteins are acidic due to the presence of relatively large amounts of the amino acids tryptophane and tyrosine. The structural and functional relationships of these four kinds of macromolecules in the chromosomes are still a matter of investigation and speculation. The DNA-histone complex seems to be the fundamental structural unit in the chromosomes, because RNA is not an integral part of the structure and the function of the residual protein is still unknown.

The lampbrush chromosomes of amphibian oöcytes and the salivary gland chromosomes of Dipteran larvae are so large that their structures have been intensively studied. The salivary gland chromosomes of *Drosophila melanogaster* are more than 100 times the length of the somatic metaphase chromosomes. The X chromosome, for example, is less than 2 μ in length at somatic metaphase, but in the salivary glands it is more than 400 μ in length. A corresponding increase in both length and diameter is observed in all the other chromosomes. The estimates of the total length for all these chromosomes have ranged as high as 2000 μ or 2 mm. The salivary gland chromosomes are unusual, not only because of their size but also because they show somatic pairing of homologous chromosomes comparable to the synapsis of homologues seen during the first meiotic prophase. Furthermore, they have banding patterns so distinctive that individual regions and bands can be identified, and synapsis can be shown to be a specific pairing of identical bands. The increase in diameter is due to the continued duplication of the chromosomes without any separation of the products of division so that a multistranded, or polytene, chromosome is built up that may have as many as 1024 strands. The chromosomes appear to have transverse bands because of the side-by-side alignment of the chromomeres of each strand of the polytene chromosome.

The estimates for the length of the amphibian lampbrush chromosomes (Fig. 11.8) are considerably greater than those of *Drosophila* salivaries. In frogs of the genus *Rana* ($n = 13$), each chromosome at maximum extension is from 800 to 1000 μ in length. The term "lampbrush" is derived from the hundreds of lateral loops that give the chromosomes their fuzzy, brushlike appearance. Each loop extends laterally for 10 to 15 μ or more. In the newt, *Triturus viridescens* ($n = 11$), the chromosomes range in length from 350 to 800 μ, with the total length for all being about 5900 μ. However, because of the presence of the loops, estimates of the actual total length of the chromosomes are much greater. The length of the loops plus the interchromomeric regions in a haploid set of *Rana pipiens* chromosomes has been

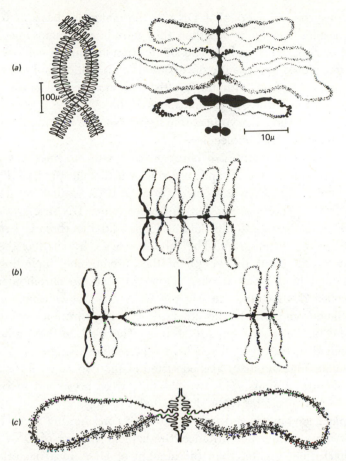

Fig. 11.8. Diagrammatic view of lampbrush chromosomes from amphibian oöcytes. (a) Homologous chromosomes (left) joined together by two chiasmata during meiosis. Identity of lateral loops at a given point (right) suggests that each chromosome is already split into two chromatids. (b) Stretching of a chromosome reveals that the loop axis and central axis are continuous. (c) A pair of loops showing RNA molecules being synthesized on the DNA axis. (Joseph G. Gall, "On the submicroscopic structure of chromosomes," *Brookhaven Symposia in Biology*, Number 8. Brookhaven National Laboratory, June 1955.)

estimated to be 20 cm, and in *Triturus* the minimal length has been estimated to be 50 cm. Because the chromomeres seem to represent regions of coiling, only about one-twentieth of the DNA is extended at any one time. Therefore, the total length of the DNA in a haploid set of *Triturus* chromosomes has been estimated to be 10 m. Such a length poses a real packing problem.

Ever since the discovery that DNA is the molecular basis of heredity, a major question has been whether the germ-line chromosome is single-stranded or multiple-stranded. The single-stranded model has been favored by most geneticists because the observations on mutation and recombination are difficult to explain with the multistrand model, but many cytologists have favored the latter. At present most of the evidence is consonant with the concept that the eukaryotic chromosome is single-stranded, forming one long, continuous DNA molecule.

The treatment of lampbrush chromosomes with enzymes has supported this interpretation: pepsin and trypsin, which split peptide linkages in protein, and RNAase, which breaks down RNA, failed to destroy the integrity of the chromosome. On the other hand, DNAase, which breaks down DNA, caused both the lateral loops and the threads between the chromomeres to disintegrate. The inference was drawn that the DNA is continuous throughout the length of the chromosome, including the axis of the lateral loops. Gall further suggested that the chromomeres represented coiled DNA strands so that the DNA axis was continuous from one end of the chromosome to the other through the chromomeres, the lateral loops, and the interchromomeric strands. The data further indicated that in the paired lateral loops, each loop represented a single chromatid with two subunits, and the main axis consisted of the two sister chromatids with a total of four subunits. The two subunits were presumed to be the two nucleotide chains of the DNA double helix.

A haploid genome in eukaryotes contains considerably more DNA than a prokaryote genome. For instance, the haploid amount of DNA in man is about 1000 times greater than the amount of DNA in a bacterial genome (Table 11.2). Such a finding was anticipated on the grounds that more genes would be needed to control the differentiation, development, and function of the higher plants and animals, which are considerably more complex than the viruses and bacteria. What was unexpected, however, was the wide range in DNA content within a single group such as the vertebrates. Man and the other mammals are rather similar in DNA content, but man's conception of himself as the pinnacle of evolution must have been shaken by the discovery that the lowly leopard frog *Rana pipiens* has twice as much DNA per diploid nucleus as he, while the salamander *Necturus maculosus* has 27 times as much and a primitive lungfish, *Lepidosiren paradoxa*, has 35 times as much (Table 11.3). In all, a 175-fold difference was found within the vertebrate subphylum. (Moreover, even within a single genus, striking differences in DNA content were sometimes found; Table 11.4). If we assume that about the same amount of genetic information is required to make any vertebrate, these differences in DNA content present a paradox.

Table 11.2   DNA per Haploid Genome

Organism	DNA (pg) [a]	Length (μ) [b]
Viruses		
ϕx 174	2.6×10^{-6}	2
λ	50×10^{-6}	17
T2	208×10^{-6}	55
Fowlpox	382×10^{-6}	91
Bacteria		
Mycoplasma	840×10^{-6}	265
Escherichia coli	$4,000 \times 10^{-6}$	1200
Fungi		
Yeast	$22,000 \times 10^{-6}$	
Protozoa		
Astasia longa	1.5	
Trypanosoma evansi	0.2	
Plasmodium berghei	0.06	
Sponges		
Tube sponge	0.06	
Coelenterate		
Cassiopeia	0.33	
Echinoderms		
Sea urchin (Lytechinus)	0.9	
Molluscs		
Snail (Tectarius)	0.7	
Crustacea		
Crab (Plagusia)	1.5	
Insects		
Drosophila	0.2	
Chironomus	0.2	
Chordates		
Amphioxus	0.6	
Fishes		
Lamprey (Petromyzon)	2.5	
Carp	1.7	
Shad	1.0	

[a] 1 pg = 1 picogram = 1×10^{-12} g.   [b] 1 μ = 1×10^{-3} mm.

From Ris, H., and D. F. Kubai, 1970. "Chromosome Structure," *Ann. Rev. Genetics* 4:263–294, p. 265, Table 1.

Table 11.2   DNA per Haploid Genome (*cont.*)

Organism	DNA (pg) [a]
Amphibians	
Toad, *Bufo bufo*	7.3
Amphiuma	84.0
Reptiles	
Snapping turtle	2.5
Birds	
Domestic fowl	1.2
Mammals	
Man	3.2
Plants	
Euglena gracilis	3
Aquilegia sp.	0.6
Gladiolus sp.	3
Lilium longiflorum	53
Tradescantia	58

[a] 1 pg = 1 picogram = 1×10^{-12} g.

This seeming paradox was resolved, at least in part, by the discovery by Britten of DNA redundancy. In contrast to the prokaryotes, in which nearly all genes are represented only once, some of the genes in the higher forms are represented a number of times. Instead of a DNA double helix consisting of a series of unique nucleotide sequences coding for all the necessary types of proteins and RNA molecules, the chromosomes in eukaryotes contain repetitive DNA sequences. The degree of repetition varies for different genes because some DNA sequences are present in only a single copy, while others have been found to have a hundred, a thousand, or even up to a hundred thousand or a million copies. These copies have been found to contain similar but not identical DNA sequences.

Three obvious possibilities for the form of the DNA redundancy are polyploidy, polyteny, or tandem linear repetition of the genes. Polyploidy is clearly excluded, and there is little evidence for lateral redundancy in cell types other than the salivary glands in Diptera, but there is evidence to suggest that gene multiplicity in other species takes the form of linear gene repetition (Fig. 11.9).

Table 11.3 Relative DNA Contents from Feulgen Photometry (1.00 = 6×10^{-12} g of DNA per diploid nucleus)

T2 bacteriophage	0.0007
Escherichia coli	0.0033
Sponge (Dysidea crawshagi)	0.02
Coelenterate (Cassiopeia)	0.11
Echinoderm (Lytechinus pictus)	0.30
Flatfish (Pleuronichthys verticalis)	0.20
Trout (Salmo gairdneri)	0.80
Mouse (Mus musculus)	1.00
Frog (Rana pipiens)	2.17
Newt (Triturus viridescens)	13.00
Salamander (Necturus maculosus)	27.80
Lungfish (Lepidosiren paradoxa)	35.40

From Markert, C. L., 1968. "Panel Discussion: Present Status and Perspectives in the Study of Cytodifferentiation at the Molecular Level," *J. Cell. Physiol.* 72 (Suppl. 1):213–230, p. 214, Table 1.

Many questions remain to be answered about the role of redundant DNA sequences in the life of the eukaryotic cell. One possibility is that multiple copies of a single gene would provide a means for an increased rate of synthesis of proteins or other molecules. One clear-cut case of a gene present in tandem duplicate form is the region of the chromosome known as the nucleolar organizer, which codes for ribosomal RNA. In all the eukaryotes studied, from 100 to 1000 copies of these rRNA genes have been found in one or more nucleoli. In amphibian oöcytes the nucleolus organizer DNA, containing hundreds of copies of the rRNA sequence, is multiplied about a thousand times to produce about a thousand extrachromosomal nucleoli. These nucleoli then provide the large numbers of ribosomes needed by the developing embryo. The best guess at present for the function of repetitive DNA sequences in higher forms is that they are involved in differential gene expression and in the regulation of gene action, a topic to be pursued in more detail in Chapter 15. This appears to be true for the lateral redundancy of the polytene salivary gland chromosomes of Diptera as well as the linear repetitive DNA sequences in other species.

Because the redundant DNA sequences are similar but not necessarily identical, Callan proposed his master-slave model for the chromosomes. In essence, he suggested that one copy of the gene serves as a master template against which the other sequences (the slaves) are compared and corrected, if necessary. In this way the gene is transferred intact from one generation to the next (unless changes occur in the master copy itself), and

Table 11.4 Variation in Chromosome Number and DNA Content within Groups of Related Organisms

Organism	Chromosome number (2n)	DNA content (2n)
Animals		in picograms:
Amphibians		
(Urodeles) *Triturus viridescens*	22	89
Necturus maculosus	24	48
Amphiuma means	24	168
(Anurans) *Bufo regularis* (toad)	22	8.9
Bufo viridis	22	11.4
Bufo bufo	22	14.6
Rana sp.	26	15
		relative values:
Bufo calamita	22	100
Bufo viridis	22	107
Bufo bufo	22	149
Rana temporaria (frog)	26	100
Rana arvalis	24	128
Rana esculenta	26	262
Insects *Liturgousa cursor* (mantid)	33	100
Liturgousa maya	17	106
Liturgousa actuosa	23	152
Liturgousa sp. n.	21	160
Pseudaulacaspis pentagona	16	100
Chrysomphalus ficus	8	200
Thyanta calceata (Hemiptera)	28	100
Thyanta pseudocasta	14	100
Murgantia	14	193
Euschistus obscurus	14	257
Acrosternum	14	332

Table 11.4 Variation in Chromosome Number and DNA Content within Groups of Related Organisms (*cont.*)

Organism		Chromosome number (2n)	DNA content (2n)
Plants			*in picograms:*
Ranunculaceae	*Aquilegia alpina*	14	1.3
	Pulsatilla occidentalis	16	12
	Anemone virginiana	16	21
	Anemone fasciculata	14	52
			relative values:
Juncaceae	*Luzula pilosa*	66	100
	Luzula sudetica	48	150
	Luzula luzuloides	12	200
	Luzula multiflora	36	400
	Luzula purpurea	6	600

From Ris, H., and D. F. Kubai, 1970. "Chromosome Structure," *Ann. Rev. Genetics* 4:263–294, p. 269, Table 4.

the effects of independent mutation or recombination in the slaves are eliminated.

In contrast to the naked DNA molecule of the prokaryotic chromosome, the characteristic chromosome structure in eukaryotes is a fiber about 100 Å thick formed by the association of DNA and histones in approximately equal amounts. This nucleohistone fiber seems designed to permit the packing of a large amount of DNA into a manageable unit and the control of the complex functions of eukaryotic cells. Five major types of histones have been identified in a wide variety of plants and animals and in a variety of cell types at all stages of mitosis, development, and gene activity. An exception is found in the mature sperm of some animals, where the histones are replaced by smaller protein molecules, the protamines. The basic nucleohistone fiber contains a single DNA double helix, which is highly compacted or folded.

The eukaryotic chromosome manifests a variety of forms. At mitotic metaphase it is a highly compressed body, autocatalytically and heterocatalytically inactive, in a form that can be readily moved to the site of the

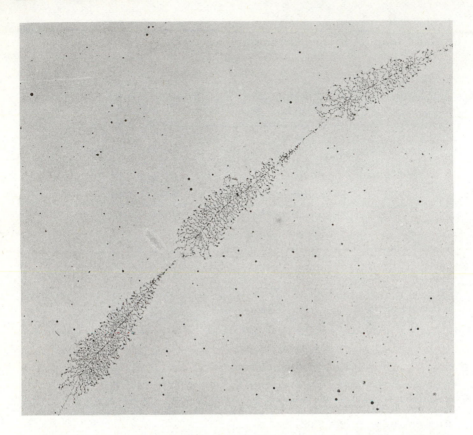

Fig. 11.9. A linearly arranged series of nucleolar rRNA genes from *Triturus viridescens* lampbrush chromosomes showing growing RNA chains along their DNA template. The black dots at the RNA attachment sites are RNA polymerase molecules. (Watson, J. D., 1970. *Molecular Biology of the Gene*, 2nd ed. W. A. Benjamin, Menlo Park, Calif., p. 542, Fig. 16.16. Photograph courtesy of Oscar L. Miller, Jr. and Barbara R. Beatty.)

new daughter nucleus. At interphase, it is usually no longer visible under the light microscope, but nonetheless retains its identity, having unfolded from its tight metaphase package into long, metabolically active nucleohistone fibers. These changes in state parallel the life cycle in phages, with the metaphase chromosome analogous to the infective mature phage particle and the interphase chromosome analogous to the vegetative phage DNA molecule, which directs events within the bacterial host cell.

Chromomeres and euchromatin and heterochromatin have been identified as specialized chromosome regions, and metabolically active and inactive chromosome segments have been recognized as well, but the basic nucleohistone thread seems to run through all of these regions. The modifications observed seem related to the functional control of gene activity in a way not yet fully understood. At present it is believed that the polypeptide chains of the histones lie in the two spiral grooves left by the unequal spacing of the two nucleotide chains in the DNA double helix, but in the nucleoprotamines from sperm cells, the polypeptide chain occupies only the smaller groove, thus permitting tighter packing in the sperm head.

11.7. DNA Replication in Eukaryotes

The experiments of J. H. Taylor and his colleagues in 1957 revealed that DNA replication in the chromosomes of higher organisms is semiconservative just as it is in the prokaryotes. They worked with a plant with the alliterative name, the British Broad Bean (*Vicia faba*). DNA synthesis goes on rather rapidly in the growing root tips of Broad Bean seedlings. In these experiments the root tips were immersed for a period in a solution containing radioactive thymidine, the nucleoside that is incorporated only into DNA. The thymidine was labeled with tritium (radioactive hydrogen —H^3). Tritium was used because the β electrons that it emits on decay are of such low energy that they penetrate only about 1 μ in a photographic emulsion. Therefore, when film is placed in contact with a root-tip preparation on a microscope slide, a picture can later be developed that reveals the form of the structures emitting the particles. Since the labeled DNA, by the emission of β particles, causes the deposition of silver grains in the film and thus takes its own picture, the technique is called *autoradiography*. When the slide preparations were made, the root tips were fixed and stained with Feulgen's method. In this way the same chromosomes could be examined both under the light microscope and in the autoradiographs, and direct comparisons were possible between observations made by the two methods.

In *Vicia faba* the mitotic cycle takes about 24 hr at room temperature, but the divisions are not synchronized and cells in all stages of mitosis are found at any one time. DNA replication requires about 8 hr and is completed some 8 hr before the cell goes into metaphase. Therefore, the seedlings were placed in tritiated thymidine for 8 hr, then thoroughly washed with water and transferred to a nonlabeling solution containing colchicine. The colchicine inhibits the formation of the mitotic spindle but not chromosome replication. In this way cell divison is prevented, and the number of chromosomes in a cell is doubled at each mitosis. Hence, a chromosome count will reveal the number of mitoses that have occurred since the roots were placed in colchicine. A further advantage of the method is that the chromatids of each chromosome tend to lie free of one another, being held together only at the centromere. It is thus possible to observe the pattern of labeling in the chromatids as well as the chromosomes. Some of the roots were kept in the colchicine for 10 hr, others for 34 hr before the root tips were fixed and stained.

The cells fixed after 10 hr in colchicine showed the normal diploid number of 12 at metaphase, and both chromatids of each chromosome were equally labeled with tritium. The cells fixed after 34 hr showed either 24 or 48 metaphase chromosomes, and had thus undergone one or two mitoses. In the nuclei with 24 chromosomes, only one of the two chromatids in each chromosome was radioactive. However, in the cells with 48 chromosomes, half of the chromosomes were completely unlabeled and the other half had just one of the two chromatids labeled. The amount of label in the labeled chromatids appeared to be about the same in all three cell types, diploid, tetraploid, and octoploid (Figure 11.10). From these results Taylor concluded that before duplication the chromosomes are composed of two subunits and that at duplication these subunits are conserved. He further concluded that the subunits separate at duplication and that a new subunit is laid down along each old one. Thus, chromosome replication is semiconservative.

Even though chromosome duplication is semiconservative, it does not necessarily follow that DNA replication in the chromosomes is also semiconservative. However, the discovery of rare exchanges between labeled and unlabeled sister chromatids led to the conclusion that the two strands of a chromosome or chromatid are structurally different in a way that prevents reunion of unlike strands. Apparently, reunion was possible only between strands having the same polarity. These results are in full accord with the Watson-Crick concept of two complementary polynucleotide chains of opposite polarity and suggest that DNA replication as well as chromosome replication is semiconservative in higher organisms.

In the prokaryotes such as T2 phage and its host *E. coli*, the chromosome, some 50 μ long in T2 and about 1000 μ long in *E. coli*, constitutes a single replicating unit or replicon. The eukaryote chromosome, however, in which the DNA is many centimeters in length, appears to consist of a linear series of simultaneously replicating segments, each of the order of 50 to 100 μ in length. The chromosomes show individual patterns of early and late replicating regions, indicating that replication must be initiated simultaneously at a number of different sites along the chromosome. The rate of replication is less than 2.5 μ/min, considerably slower than the 30 μ/min rate in the prokaryotes. These observations have led to the con-

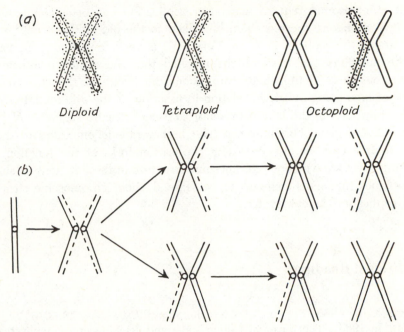

Fig. 11.10. Diagram of Taylor's autoradiograph experiments with *Vicia faba* root-tip chromosomes to demonstrate semiconservative chromosome replication. (a) Drawings represent chromosomes at mitotic metaphase with the dots representing silver grains in the photographic emulsion and indicating the position of tritiated thymidine incorporation into DNA. The root tips were grown in tritiated thymidine for 8 hr and then removed. (b) The interpretation of the observations, with the broken lines representing subunits containing tritium and the solid lines subunits without tritium. For each set of four lines representing a colchicine metaphase chromosome, the outer lines represent the old subunits and the inner lines the new subunits. See text. Whitehouse, H. L. K., 1969. *Toward an Understanding of the Mechanism of Heredity*, 2nd ed. Edward Arnold, London, p. 184, Fig. 12.2.)

cept that the chromosome consists of a series of hundreds of replicons linked together in some unspecified manner. It has also been suggested that the chromomere corresponds to the replicon.

This replicon theory is helpful in the solution of the unraveling problem during replication of the eukaryotic chromosome. An average human chromosome has been estimated to require about 400 hr to unwind during replication if it were one continuous DNA double helix. Autoradiography has shown, however, that the unwinding takes only about 6 hr rather than 400. If the chromosome consists of a series of replicons with free rotation of the strands at the replicon junction, the unwinding problem is considerably simplified. However, Taylor's data showing semiconservative replication of plant chromosomes require that, after replication has been completed in adjacent replicons, the old chain be linked to the old, and the new to the new across the replicon junction. Very little is yet known about the nature of the junctions, but it should be recalled that the evidence indicates a continuous DNA double helix throughout the length of the chromosome. Another interesting puzzle in relation to junctions is the differential rates of replication seen in the heterochromatic and euchromatic segments of polytene salivary gland chromosomes. The higher rate in euchromatin should lead to free ends or else persistent replication forks at the junction, but nothing is yet known about chromosome structure in these regions. It should be obvious that much remains to be learned about chromosome structure and replication in eukaryotes.

Additional Reading

Bridges, C. B., 1935. "Salivary Chromosome Maps," *J. Heredity* **26**:60–64.

Cairns, J., 1963. "The Bacterial Chromosome and Its Manner of Replication as Seen by Autoradiography," *J. Mol. Biol.* **6**:208–213.

Cairns, J., 1963. "The Chromosome of *Escherichia coli*," *Cold Spring Harbor Symp. Quant. Biol.* **28**:43–46.

Callan, H. G., 1963. "The Nature of Lampbrush Chromosomes," *Intern. Rev. Cytol.* **15**:1–34.

Callan, H. G., 1967. "On the Organization of Genetic Units in Chromosomes," *J. Cell Sci.* **2**:1–7.

Callan, H. G., and L. Lloyd, 1960. "Lampbrush Chromosomes of Crested Newts *Triturus cristatus* (Laurenti)," *Phil. Trans. Roy Soc. B* **243**:135–219.

Gall, J. G., 1956. "On the Submicroscopic Structure of Chromosomes," *Brookhaven Symp. Biol.* **8**:17–32.

Jacob, F., and S. Brenner, 1963. "Sur la régulation de la synthèse du DNA chez les bactéries: l'hypothèse du réplicon," *C.R. Acad. Sci.* (Paris) **256**:298–300.

Jacob, F., S. Brenner, and F. Cuzin, 1963. "On the Regulation of DNA Replication in Bacteria," *Cold Spring Harbor Symp. Quant. Biol.* **28**:329–347.

Jacob, F., and J. Monod, 1963. "Elements of Regulatory Circuits in Bacteria. In *Biological Organization*, R. J. C. Harris, ed. Academic, New York.

Josse, J., A. D. Kaiser, and A. Kornberg, 1961. "Enzymatic Synthesis of Deoxyribonucleic Acid. VIII. Frequencies of Nearest Neighbor Base Sequences in Deoxyribonucleic Acid," *J. Biol. Chem.* **236**:864–875.

Kornberg, A., 1960. "Biologic Synthesis of Deoxyribonucleic Acid," *Science* **131**:1503–1508.

Kornberg, A., I. R. Lehman, M. J. Bessman, and E. S. Simms, 1956. "Enzymic Synthesis of Deoxyribonucleic Acid," *Biochim. Biophys. Acta* **21**:197–198.

Meselson, M., and F. W. Stahl, 1958. "The Replication of DNA in *Escherichia coli*," *Proc. Nat. Acad. Sci.* **44**:671–682.

Pelling, C., 1966. "A Replicative and Synthetic Chromosomal Unit—The Modern Concept of the Chromosome," *Proc. Roy. Soc. B* **164**:279–289.

Ris, H., and D. F. Kubai, 1970. "Chromosome Structure," *Ann. Rev. Genetics* 4:263–294.

Sueoka, N., 1967. "Mechanisms of Replication and Repair of Nucleic Acid. In *Molecular Genetics*, Part II, J. H. Taylor, ed. Academic, New York.

Taylor, J. H., 1959. "The Organization and Duplication of Genetic Material," *Proc. 10th Intern. Congr. Genetics* 1:63–78.

Taylor, J. H., 1963. "The Replication and Organization of DNA in Chromosomes." In *Molecular Genetics*, Part I, J. H. Taylor, ed. Academic, New York.

Taylor, J. H., P. S. Woods, and W. L. Hughes, 1957. "The Organization and Duplication of Chromosomes as Revealed by Autoradiographic Studies Using Tritium-Labeled Thymidine," *Proc. Nat. Acad. Sci.* **48**:122–128.

Questions

11.1.   What is the difference between the "semiconservative" and the "conservative" models of replication?

11.2.   What is the evidence that replication is semiconservative in prokaryotes?

11.3.   What is the evidence for semiconservative replication in eukaryotes?

11.4.   Demonstrate how the "nearest-neighbor" technique used by Kornberg proved that the two strands of the DNA double helix must be of opposite polarity.

11.5.   DNA usually exists in the form of a double helix, but single-stranded DNA may also be formed. If you had a preparation of DNA and wished to determine whether it was in the single-stranded or double-stranded condition, what differences might you expect to find?

11.6.   A bacterial chromosome contains 4 million nucleotide pairs and replicates once in 25 min, the length of a generation.

   a. What is the rate of synthesis of nucleotide pairs per minute? Per second?

   b. If the double helix must unwind for replication to occur, how many revolutions per minute will there be? Per second?

11.7.   If 1800 bases of a double-stranded nucleic acid are uracil, and this RNA consists of 10,000 nucleotides, what are the percentages of the four types of bases present?

11.8.   What are the differences between the genetic material in prokaryotes and in eukaryotes?

11.9.   What evidence gave rise to the concept of the replicon as the replicating unit in eukaryotes?

11.10. What is the explanation for the wide variations in DNA content sometimes observed within related groups of organisms?

CHAPTER TWELVE

Gene Action

The hereditary material has two capabilities, self-duplication and control of the phenotype. In this chapter we shall consider how genes produce their phenotypic effects. Mendel postulated the existence of genes to account for the segregation of such phenotypic traits as flower color and seed shape. However, he had no knowledge of the physical nature of genes and no understanding of how they produced their effects.

12.1. Levels of Gene Action

The phenotypic effects of genes can be studied at various levels of organization. A goat, for instance, may be considered as a constellation of genes organized into a genotype that gives rise to the phenotype that we recognize as a goat by its capriform and capricious qualities. If this herbivorous constellation of genes is introduced into an oceanic island, gene action can be studied at the ecological level. On Pitcairn Island in the Pacific, the grazing of the goats has prevented the growth of trees to such an extent that the islanders, descendants of the *Bounty* mutineers, must go to other islands for lumber. The introduction of a semidomesticated array of carnivorous genes in the form of a house cat has had a comparable effect on the native fauna in the Galapagos Islands. Thus, the introduction of a single new kind of genotype may dramatically upset the ecological balance of an area.

Gene action may also be studied in terms of its effect on behavior. A stock bottle of *Drosophila melanogaster* carrying the cut wing mutation, for example, can be recognized even at a distance, because the flies tend to rest quietly on the food medium rather than flying about or crawling up the sides of the bottle as other strains do. The cut flies show a low activity level even though they are capable of flight and movement. Similarly, poor water escape performance in mice has been shown to be associated with albinism in segregating F_2 populations.

The most familiar type of phenotypic effect of genes is on morphology. Most of the mutations in *Drosophila* have been identified by their effects on the external morphology of the flies. Mutations of this sort are often referred to as "visible" mutants, for they are easy to identify and to score in crosses. In general, these effects result from a change in form or in the production of some substance such as a pigment. A large number of mutations are known that affect the shape or size of the wings or legs, for example. Dumpy, miniature, and vestigial wings are three such mutants, each affecting wing size and shape to a greater degree than the previous one.

The white eye trait in *Drosophila* is due to the absence of the normal eye pigments that cause the bright red eye color of the wild type. Yellow body color is due to a reduction in the amount of melanin in the exoskeleton. On the other hand, flies with the ebony trait have bodies darker than wild type because more melanin has been produced.

At the morphological level, the gene is often thought to have one phenotypic effect, and is equated with the character it affects most conspicuously. However, it is quite clear that at this level the relationship between gene and character is not one to one. In addition to its effect on eye color, the white eye gene also affects the shape and color of the spermatheca (the sperm receptacle of the female), and reduces viability and male mating success. A minute mutant has been shown to reduce bristle size as well as wing size, to roughen the surface of the eyes, to lower female fertility, and to prolong larval development. In situations such as these, in which a number of phenotypic effects can be traced to a single gene, the effect of the gene is said to be *pleiotropic*.

In the examples above, there seems to be no relation among the different effects of the pleiotropic gene. In other cases of pleiotropy, however, a number of quite diverse effects of a single gene can be traced to a single primary defect, as shown for a lethal gene in the rat (Fig. 12.1).

Just as one gene has been found to affect a number of different traits, so many genes at different loci have been discovered all of which affect a single characteristic. In *Drosophila melanogaster*, mutants affecting eye

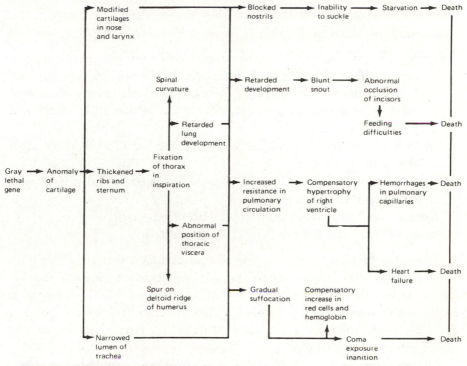

Fig. 12.1. Pleiotropic effects of a lethal gene in the rat, all stemming from an anomaly of the cartilage. (After Grüneberg.)

color have been found at more than 40 different gene loci. This means that the normal wild-type eye color is the result of the combined action of the wild-type alleles at these 40 loci and probably others at which mutants have not as yet been identified. Similarly, wing size is affected by at least 34 different loci and wing shape by 43 more loci, so that more than 75 genes are known to be involved in the production of normal wings. From the fact that one gene affects many characters and many genes affect a single trait, it is clear that at this level there is a complex network of interrelationships between genes and characters.

The observations on pleiotropy raise two fundamental questions about gene action. Just what is the primary action of the gene, and can a single gene have more than one primary effect? When gene action was studied at the morphological level, reliable answers to these questions were not forthcoming.

Gene action has also been studied at the physiological level. Such traits as fertility, female fecundity, viability, and DDT or disease resistance fall into this category. Viability especially has been frequently studied. Although authors vary somewhat in the limits they use to define each group, the classes may be listed as follows:

Lethals—all (or nearly all) of the individuals with this genotype die. Nonetheless, lethals are a heterogeneous group. Some are cell lethals, others block development at one particular stage, while with others, the developing organisms seem to run down like clocks and die at various stages.

Semilethals—less than 50% of the expected number of individuals with this genotype survive.

Subvitals—from 50% to 90% of the expected number of individuals with this genotype survive.

"Normal"—viability ranges from 90% to 110% of expected.

Supervital—viability is more than 110% of expected.

One of the problems with this type of classification is that viabilities must be compared with that of some "normal" or standard type. A suitable choice of a standard genotype is not always as simple a problem as it might seem. Nevertheless, the use of lethals, because their effects are clear cut, has been common ever since Muller's pioneering study on the mutagenic effects of X rays.

You may wonder why there should be a problem in finding a standard genotype against which to compare the viability of mutants, and why quotation marks have been placed around such expressions as "wild-type" and "normal." The explanation is that there is no one genotype that can be regarded as the standard against which all others are to be compared. In *Drosophila* many different genotypes produce the wild phenotype. No single genotype is common enough to represent the "wild type." The very concept of the "wild type" has been undermined by the discovery of *iso-alleles*. Iso-alleles were first demonstrated when it was shown that three "wild-type" alleles of independent origin, all of which gave the normal phenotype when homozygous, nevertheless differed from one another in subtle, cryptic ways that could be demonstrated only under special conditions.

Genes have sometimes been classified in terms of their effects when different doses of the gene have been inserted in an otherwise constant genotype. The effects of increasing the dosage of a given gene from a single dose to two, three, or even more doses may give some insight into the way the gene is acting. An *amorph* shows no effects of added doses of the gene. A *hypomorph* has an effect similar to the "wild-type" gene but is less efficient; added doses produce individuals more like the wild type. The *hypermorph* exceeds the wild type. An ***antimorph*** is opposite to the standard allele in its effects, while a ***neomorph*** has an activity apparently different from and unrelated to that of the "wild type." (The standard type then can be considered an amorph of the neomorph!)

The physiological studies of gene effects have not contributed greatly to our understanding of the fundamental nature of gene action. As with the morphological traits, the chain of events between the primary action of the gene and the trait under study is too long to extrapolate back to the action of the gene itself.

12.2. Biochemical Genetics

The studies that ultimately led to a real understanding of the nature of gene action were made at the biochemical level. Sir Archibald Garrod, in his 1908 Croonian Lecture dealing with four hereditary conditions in man, first made the suggestion that later became the keystone in the development of biochemical genetics. The title of his lecture was "Inborn Errors of Metabolism." Its essence was that in such human traits as albinism and alkaptonuria, the basic defect was a block at some point in a sequence of metabolic reactions and that the block was due to the lack of the specific enzyme needed to carry out that step. There seems to be a tendency to discount Garrod's contribution because it had little influence on the development of genetics in the early 1900s. However, like Mendel, he seems to have been ahead of his time, and perhaps deserves all the more credit for his insight at a time when enzyme biochemistry was in its infancy.

Another early effort in biochemical genetics was made by a group of English workers in a study of the genetics of the flower pigments, anthocyanin and anthoxanthin. They found that a particular gene substitution led to a specific color change for which the basis was a specific chemical substitution. Thus, different genes were found to control such chemical processes as hydroxylation, methylation, methoxylation, and glycosidation, and the observed color changes were due to the addition or removal of hydroxyl or methyl or methoxy or sugar groups from the pigment molecules. Each gene seemed to control a single type of reaction in the modification of the pigment.

Another biochemical approach to gene action was through the study of eye pigmentation in insects, particularly *Drosophila melanogaster* and the flour moth, *Ephestia kühniella*. These studies during the 1930s were fruitful and suggestive, but even though the genetic and experimental results were clear enough, the biochemistry was still obscure, largely because an insect is a complex organism with a complex metabolic system.

For this reason, Beadle and Tatum, who had been working with *Drosophila*, sought another, simpler system to work with. They turned to the common red bread mold, *Neurospora crassa*, because they reasoned that the chain of events between primary gene action and the phenotypic effects observed in the cells would be shortest in microorganisms. This decision, to initiate studies in microbial genetics, was paralleled by similar decisions by a number of others, and it seems safe to say that the rapid breakthroughs

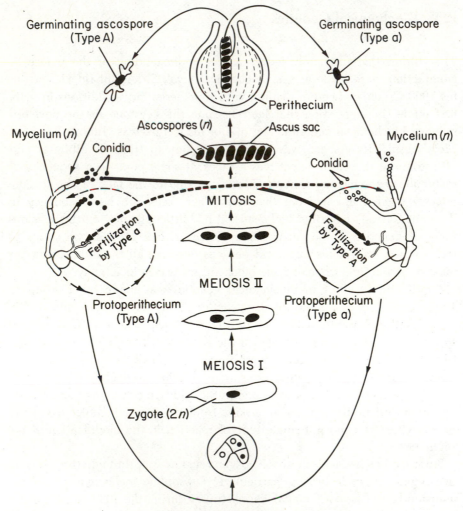

Fig. 12.2. Life cycle of *Neurospora crassa*. (See text.)

that came in the area now referred to as molecular genetics were made possible to a large extent by the switch from the genetics of corn, *Drosophila,* and other higher organisms to the genetics of microorganisms. In addition to *Neurospora,* genetic studies were initiated with various species of bacteria, notably *Escherichia coli,* with *Paramecium,* with yeast, and with bacteriophages and other kinds of viruses.

The microorganisms offered a number of advantages over the more complex organisms previously used in genetic research. They have short life cycles, measured in minutes or days rather than in weeks or months or even years. Progeny numbering in the millions can be easily reared in the laboratory, a great boon, for example, to research on mutations, which is particularly cumbersome and difficult with higher organisms. Furthermore, microorganisms are typically haploid so that problems related to dominance do not arise, and the genotype is directly reflected by the phenotype. Since they can be easily propagated asexually, large numbers with a single genotype are easily reared, yet a refined genetic analysis is usually possible because of the nature of the sexual processes. In addition, pure cultures of microorganisms are easily maintained on chemically defined media, a distinct advantage for research in biochemical genetics. Because they are unicellular, the developmental complexities and cell interactions observed in multicellular species do not complicate matters. The actions of the genes can be studied within the confines of a single cell. Finally, Beadle and Tatum reversed the procedure they had used in the work on *Drosophila* eye color. Rather than attempting to learn the biochemistry of genetically known traits, they deliberately set out to study the genetics of known biochemical traits.

12.3. Life Cycle of *Neurospora*

The patch of mold sometimes seen growing on bread is composed of a number of long, branched filaments or **hyphae** that together form a tangled mass known as a **mycelium** (Fig. 12.2). In many fungi, the hyphae consist of chains of individual cells, but in *Neurospora* the cell crosswalls are incomplete, so that the cytoplasm is continuous throughout the hyphae. Each hyphal segment or cell is normally multinucleate, with all of the nuclei being haploid. Thus, the vegetative structure of *Neurospora* is haploid, and growth of the hyphae is due to mitotic divisions of the haploid nuclei.

Asexual reproduction occurs in two ways in *Neurospora,* by hyphal growth or by asexual spores. A strain can be propagated either by transfer of a piece of the mycelium or by means of the asexual haploid spores called *conidia*. The conidia are budded off from aerial hyphae and are normally carried by air currents to new environments, where they germinate to form new mycelia.

Sexual reproduction occurs in Neurospora when strains of opposite mating types come in contact. There are just two mating types, called *A* and *a*, which are determined by a single pair of alleles. Under suitable conditions fruiting bodies, known as *proto-perithecia*, develop with receptive filamentous extensions called *trichogynes*. If a conidium of the opposite mating type reaches the trichogyne, a sequence of events is initiated that leads ultimately to fusion of two haploid nuclei of opposite mating type to form a diploid zygote nucleus. This nucleus then undergoes two meiotic divisions followed by a single mitotic division in the process of forming an *ascus* with a linear array of eight haploid *ascospores*. The mature fruiting body or *perithecium* contains a large number of asci. When released from the ascus, the haploid ascospores germinate to form new mycelia.

A process useful in genetic studies that in some ways is similar to sexual reproduction has been observed in *Neurospora*. Under some conditions hyphae from different genetic strains of the same mating type will fuse when grown together. The nuclei, however, do not fuse, so that two genetically different nuclei are found in a common cytoplasm. Since a true diploid heterozygote is not formed, the process is referred to as *heterokaryon* formation.

12.4. Biochemical Mutants

The nutritional requirements of wild-type *Neurospora* are quite simple. It will grow on a "minimal" medium containing a sugar such as sucrose as a carbon source, inorganic salts, water, an inorganic nitrogen source (nitrate), and a single vitamin, biotin. For their work, Beadle and Tatum also concocted a "complete" medium that contained, in addition to the ingredients in the minimal medium, a rich potpourri of vitamins, amino acids, and other organic compounds found in living cells.

Wild-type *Neurospora* are able to grow on both minimal and complete media. Hence, the wild type can synthesize, from the simple ingredients in the minimal medium, all of its amino acids, vitamins, carbohydrates, fats, purines, pyrimidines, and so on.

Beadle and Tatum's strategy was a deliberate attempt to induce, isolate, and identify biochemical mutants in *Neurospora*, strains that could not grow on the minimal medium but would grow if one specific substance such as a single vitamin or amino acid were added to the minimal medium. Their technique was to induce mutations with X-rays or ultraviolet light in the asexual spores or conidia. The treated spores were then isolated and cultured on the complete medium, and from these isolates, conidia were transferred to the minimal medium. A number of different strains were found that were unable to grow on the minimal medium. The next step was to test each isolate systematically in a series of cultures, each of which

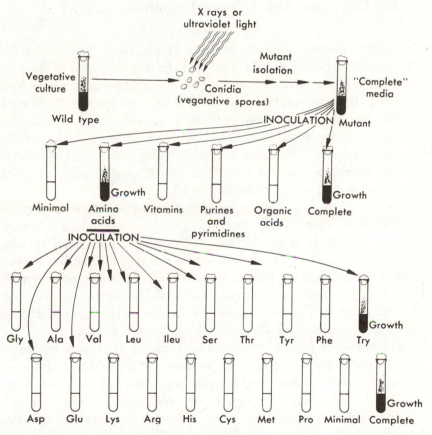

Fig. 12.3. Detection of biochemical mutants in *Neurospora crassa*. A biochemical mutant does not grow on a minimal medium but does grow on a complete medium. The type of requirement (amino acids, vitamins, etc.) is next identified. In this case, growth occurred only when the minimal medium was supplemented with amino acids. The particular amino acid required was identified when it was shown that growth occurs only when the minimal medium was supplemented with tryptophan. (DeBusk, A. G., *Molecular Genetics*, copyright © 1968 by A. G. DeBusk. Reprinted with permission of Macmillan Publishing Co., Inc.)

contained the minimal medium plus one additional specific compound such as an amino acid or a vitamin (Fig. 12.3). In this way mutants were found that could grow on the minimal medium plus the amino acid arginine, or on the minimal medium plus the vitamin thiamine, and so on. With this technique a large number of nutritional mutations were isolated and identified, each deficient in its ability to synthesize a single known compound.

12.5. The One-Gene, One-Enzyme Hypothesis

Once isolated, the biochemically deficient strains were then subjected to genetic analysis by crosses to the wild type. In this way, most of the nutritional deficiencies were shown to be due to mutation at a single locus, because wild-type and mutant ascospores were recovered in 50:50 ratios. In addition to these tests of the individual mutants, crosses between different mutant strains were also carried out in order to determine whether these mutations were at the same locus or at different loci. Just as in higher organisms, recombination and linkage permitted genetic mapping of the numerous mutants that had been discovered.

The more interesting and informative crosses were those involving mutants both of which had the same nutritional requirement. For example, crosses were made between different independent mutants all of which required tryptophan for growth on the minimal medium. In some crosses of pairs of independent tryptophan mutations, all of the ascospores required tryptophan. These mutations therefore behaved like alleles, and the mutations were either at the same locus or else were very closely linked and acting as pseudoalleles (see Chapter 14).

In other crosses of pairs of tryptophan-requiring mutations, a sizable proportion of wild-type ascospores was recovered. The wild-type ascospores arose through recombination, and from the proportions of wild-type, doubly mutant, and singly mutant ascospores in such crosses, it was possible to determine whether the mutations were independently inherited or linked, and to measure the amount of crossing over. The tryptophan-requiring mutants fell into three independently inherited groups. Members of the same group behaved as alleles, but were inherited independently of members of the other groups. Thus three different loci were involved.

The three classes of tryptophan-requiring mutants were next studied biochemically. One group was able to use anthranilic acid or indole as well as tryptophan to satisfy the growth requirement. A second group grew well on the minimal medium plus indole or tryptophan but could not utilize

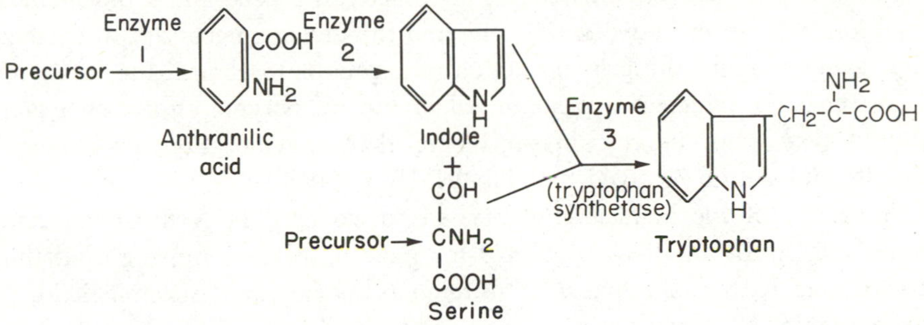

(a)

Class of Mutations	Substances satisfying nutritional requirement			Biochemical step blocked
	Anthranilic acid	Indole	Tryptophan	
1	+	+	+	1
2	−	+	+	2
3	−	−	+	3

(b)

Fig. 12.4. (a) The synthesis of tryptophan. (b) Growth requirements of tryptophan-requiring mutants. Plus indicates growth, minus indicates no growth.

anthranilic acid. In the third group, tryptophan could not be replaced by either anthranilic acid or by indole (Fig. 12.4). Furthermore, mutants of class 2 tended to accumulate anthranilic acid in the medium as if a metabolic block prevented the conversion of anthranilic acid to indole, and the anthranilic acid were piling up behind this blockade. Similarly, mutants of class 3 tended to accumulate indole.

Since it was already known that tryptophan biosynthesis involves a series of reactions, in which each step is catalyzed by a specific enzyme, it was logical to assume that the blocking of a particular step was due to lack of the necessary enzyme. This assumption received support from the finding that mutants of class 3 lacked tryptophan synthetase (enzyme 3), but that wild-type *Neurospora* and also mutants of classes 1 and 2 all had this enzyme, which couples indole and serine to form tryptophan.

The discovery that mutation at a given locus blocked a specific biosynthetic step due to lack of the functional enzyme needed to catalyze that step led Beadle and Tatum to formulate the one-gene, one-enzyme hypothesis. The association was not just between a gene and a biochemical reaction, but rather between the gene and the enzyme necessary to catalyze the reaction. Thus, the primary action of a gene is to control the formation of an enzyme, which is a protein. All of the phenotypic effects of a gene should flow from its role in producing that enzyme. For this concept, Beadle and Tatum received the Nobel Prize in 1958.

A variety of biochemical mutants was discovered in *Neurospora,* some requiring one or another amino acid for growth, others requiring a vitamin, and others with such diverse requirements as adenine, succinic acid, or sulfonamide. Furthermore, the one-to-one relationship between gene and enzyme was found to hold in various yeast species and also in bacteria such as *Escherichia coli.* In fact, the induction of biochemical mutants in *E. coli* made possible the discovery of genetic recombination in bacteria by Lederberg and Tatum in 1946. For his work on bacterial recombination Lederberg received the Nobel Prize along with Beadle and Tatum in 1958. Biochemical mutants are now known not only in yeasts, molds, and bacteria, but in a number of different species, including man. Thus, Garrod's interpretation of his findings on human inborn errors of metabolism has now been confirmed.

One consequence of the study of biochemical mutants in different species has been the realization that metabolic pathways are much the same in species ranging from microorganisms to mammals. A biochemical mechanism found in one group of organisms can be expected to hold in many different groups. A biochemical unity seems to underlie the diversity of form and function in different species. Another consequence of the studies of biochemical mutants has been the progress made in working out biosynthetic pathways. The accumulation of intermediate metabolites in quantity behind the genetic blocks has made it possible to isolate and identify these intermediates and thus to work out the sequence of intermediate steps along the metabolic pathways.

12.6. Protein Structure

One function of the genes is their autocatalytic role in self-duplication. Since enzymes are proteins, the one-gene, one-enzyme hypothesis led to the inference that the other, heterocatalytic function of the genes is the control of protein synthesis. If so, the problem becomes to explain how DNA, the genetic material, which consists of phosphate, a simple sugar, and four

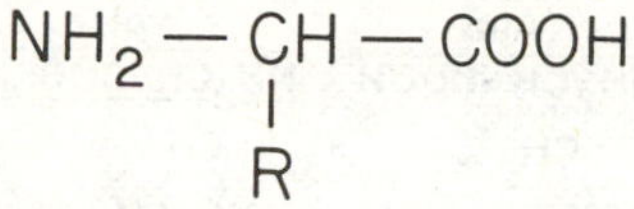

$$NH_2 - CH - COOH$$
$$|$$
$$R$$

Fig. 12.5. The general formula for an amino acid. (See text.)

organic bases made up into just four kinds of nucleotides, can control the synthesis of the great variety of proteins known to occur in living things. On the face of it, DNA might appear to be too simple chemically to carry out the complex job assigned to it. One of the notable advances in molecular biology has been the solution of this problem, the breaking of the genetic code.

We have already seen that DNA is a long, double-stranded polynucleotide molecule made up of the four common nucleotides, which contain adenine, guanine, thymine, or cytosine. Proteins, on the other hand, are formed from amino acids, of which there are 20 common types strung together in long polypeptide chains. All amino acids have in common an acidic carboxyl group and a basic amino group and a hydrogen atom attached to a carbon called the alpha carbon (Fig. 12.5). In Fig. 12.5 the R stands for the side chains, which differ in each amino acid. The 20 common amino acids and their structure are shown in Fig 12.6. The alpha carbon is asymmetric in all amino acids but glycine because it is joined by covalent bonds to four different groups. In glycine R is a hydrogen atom, so it is not asymmetric. The asymmetric amino acid molecule can exist in two different molecular configurations, known as the D and L isomers, which are mirror images of one another. For reasons not yet understood, virtually all of the proteins analyzed from living things contain just the 20 amino acids listed and no more, and the amino acids are practically always the L isomers.

In proteins the amino acids are linked together by a bond between the amino group of one amino acid and the carboxyl group of the next (Fig. 12.7). These covalent bonds, termed *peptide* bonds, form the backbone of the protein molecule and link the amino acids together to form a long *polypeptide* chain. As can be seen in the figure, one end of the polypeptide terminates in an amino group, the other in a carboxyl group.

The structure of proteins is complex because several factors come into play in its determination. The *primary* structure of a protein is established by the linear sequence of amino acids in the polypeptide chain. The *secondary* structure refers to the ordered regions of a polypeptide chain that form a coiled structure called the alpha (a) helix (Fig. 12.8). One turn of the helix encompasses 5.4 Å and includes 3.6 amino acid *residues*, as each amino acid in the chain is called. The helix is formed and stabilized by

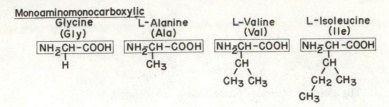

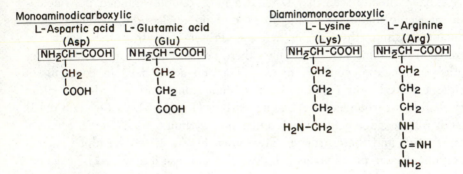

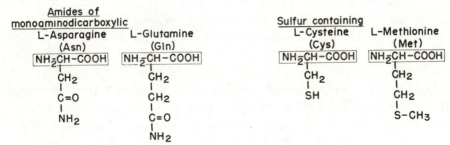

Fig. 12.6. The 20 most common amino acids in proteins.

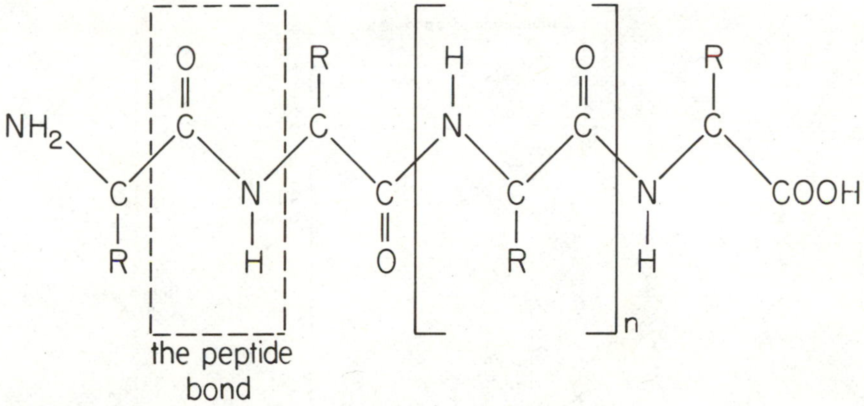

Fig. 12.7. The peptide bond links amino acids together into a polypeptide chain.

noncovalent chemical bonds between amino acids near one another in the chain. The hydrogen bonds in large measure responsible for the alpha helix in proteins are comparable to the hydrogen bonds responsible for the formation of the DNA double helix.

Many protein molecules are folded into a complex, three-dimensional pattern known as the *tertiary* structure. The tertiary structure is the result of linkages between reactive side groups not involved in the peptide linkage. Among these groups are the sulfhydryl (—SH) and hydroxyl (—OH) groups and the secondary amino (—NH$_2$) and carboxyl (—COOH) groups not tied up in the peptide backbone of the molecule.

Some proteins consist of several polypeptide subunits. The way in which the subunits are associated with one another is referred to as the *quaternary* structure.

At present the evidence indicates that the primary protein structure, the linear sequence of amino acid residues, also determines the secondary, tertiary, and even the quaternary structure of protein molecules. If true, then the genetic control of protein structure is reduced to the question of how the genes determine the amino acid sequence in proteins. In other words, the basic problem is to figure out how four kinds of nucleotides code for 20 different amino acids.

12.7. The Genetic Code

The first step toward the solution of the genetic code was the suggestion made independently by Dounce in 1952 and by Gamow in 1954 that the sequence of nucleotides in nucleic acids might determine the sequence of

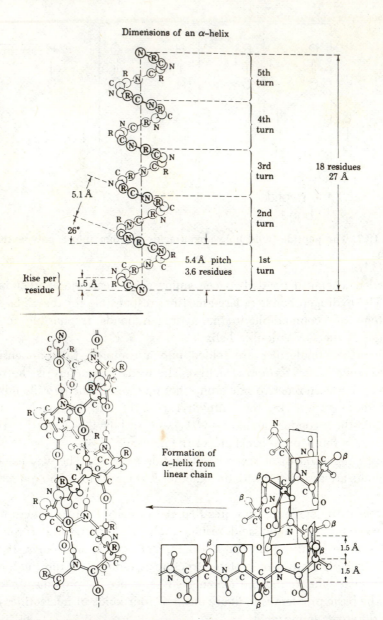

Dimensions of an α-helix

5th turn

4th turn

3rd turn

18 residues
27 Å

2nd turn

5.1 Å

26°

5.4 Å pitch
3.6 residues

1st turn

Rise per residue
1.5 Å

Formation of
α-helix from
linear chain

1.5 Å

1.5 Å

Fig. 12.8. The formation of an alpha (a) helix from an extended polypeptide chain. (Anfinsen, C. B., 1959. *The Molecular Basis of Evolution.* J. Wiley & Sons, New York, p. 101, Fig. 54.)

amino acids in polypeptides. If the nucleic acids form one language, with the four kinds of nucleotides as the letters, and the polypeptides form another language with the 20 amino acids as letters, the problem becomes one of translation from the nucleic acid language to that of the polypeptides.

A single nucleotide clearly cannot code for a single amino acid. If the four nucleotides containing adenine, guanine, cytosine, or thymine each coded for one amino acid, only four amino acids could be specified and 16 would be left over without a code designation. Therefore, combinations of nucleotides must code for each amino acid. Pairs of nucleotides would also form an inadequate code, since the maximum possible number of coding units or codons would be only 16 (4^2). Nucleotide triplets would permit 64 (4^3) different combinations of nucleotides, sequences of four would allow 256 codons (4^4), and so on. The most likely code is a triplet code, even though it has considerably more codons than the essential minimum of 20, one for each amino acid. Once the suggestion was made about the nature of the code, experimental efforts were begun to break the code and to learn in detail how it worked.

One of the first questions that arose was whether the code was overlapping or nonoverlapping. If the code letters are A, B, C, D, and the sequence were ABCDABCAC. . . , then, in a triplet nonoverlapping code, ABC would code for the first amino acid, DAB for the second, and CAC for the third. In a triplet overlapping code, the correspondence would be ABC = aa1, BCD = aa2, CDA = aa3, and so on (or it could be ABC = aa1, CDA = aa2, etc.) However, if the code were overlapping, a mutation affecting a single nucleotide would be expected to lead to changes in more than one amino acid. For example, if ABCDABCAC became ABADABCAC, at least two amino acids would be changed with an overlapping code, but only one with a nonoverlapping code. The available experimental evidence all indicates that the code is nonoverlapping, because a change in a single nucleotide leads to a change in just one amino acid. In tobacco mosaic virus a number of nitrous acid-induced mutations were analyzed by Wittman for their effects on the TMV proteins. The nitrous acid converts the $-NH_2$ groups of adenine, guanine, and cytosine to $-OH$, so that with brief treatment only one or a few of the RNA nucleotides in this RNA virus will react. Isolation and analysis of the virus proteins from the mutant strains showed that only one amino acid was affected in most cases. When more than one amino acid was different, they were not adjacent, and thus represented two separate mutations and not the effect of an overlapping code. Mutations affecting the primary structure of mammalian hemoglobin and of the bacterial enzyme, tryptophan synthetase, have also been found to change a single amino acid, and thus also suggest that the code is nonoverlapping.

Another problem was to determine whether the code was actually a triplet code as postulated by Dounce and Gamow. The study of the mutations induced by the acridine dye, proflavine, by Crick and his colleagues has supplied evidence that the genetic code is indeed a triplet code.

Proflavine, used as a chemical mutagen on the T4 virus of *E. coli,* was strongly mutagenic. Its mutagenic effects, however, were quite different from those of other chemical mutagens such as the base analogue, 5-bromouracil. 5-bromouracil appeared to be incorporated into the DNA in place of the base it resembled chemically, with the mutation resulting when, at replication, a base other than the one normally found at that site would be inserted in the DNA chain opposite the base analogue. Proflavine, however, does not appear to be substituted in this fashion in the DNA chain. Instead, it appears that the flat proflavine molecule is intercalated between the adjacent base pairs in the double-stranded DNA chain in such a way that the DNA molecule is distorted during replication or recombination. The resulting distortion almost doubles the distance between the nucleotides. As a consequence, the proflavine could lead to the insertion of an extra nucleotide in the DNA during replication or, by unequal crossing over during recombination, a base could be lost. Thus, the fundamentally different effect of proflavine, compared with 5-bromouracil, 2-aminopurine, nitrous acid, hydroxylamine, and other chemical mutagens, appeared to be due to its ability to induce "frameshift mutations," the addition or deletion of a single nucleotide rather than base substitutions.

Mutations induced by proflavine revert to wild type spontaneously, and proflavine itself increases the rate of reversion markedly. Proflavine does not, however, cause the mutations induced by base analogues to revert, nor do the base analogues enhance the rate of reversion of proflavine-induced mutations.

12.8. Fine-Structure Analysis in T4 Phage

The mutants used in much of this work on chemical mutagenesis were the so-called *r*II mutants of the T4 phage of *E. coli* (Fig. 12.9). The bacteria are cultured on plates containing a solid medium. When a bacterium is infected by a bacteriophage, lysis soon occurs and more phage particles are released, which in turn infect more bacteria. The spreading zone of lysis on the solid medium produces a plaque or clear area where all of the bacteria have been lysed. Several types of plaque morphology mutants have been found, among which the *r,* or *rapid* lysis, mutants are easy to identify because they form large, clear plaques with sharp edges. The *r*-type mutants have been found to occur at several locations in the phage genome. Benzer, who initiated the major work on fine-structure analysis in T4, concentrated his efforts on mutations in the *r*II region. The *r*II mutants differ from other

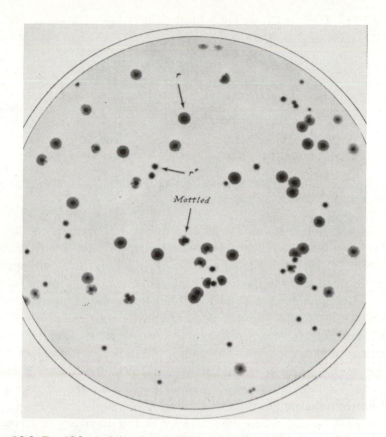

**Fig. 12.9. Rapid lysis (*r*) mutants in bacteriophage T2. *r*⁺ = wild type; r =
rapid lysis mutant, which forms a larger plaque in the same interval. The
mottled plaques owe their appearance to the growth of both r and r⁺
phages in the same site. (From *Molecular Biology of Bacterial Viruses* by
Gunther S. Stent. W. H. Freeman and Company. Copyright © 1963.)**

r mutants and from the "normal" or wild-type (*r*⁺) phage in being unable to
multiply on the K12 (λ) strain of *E. coli* although they grow perfectly well
on strain B (Table 12.1). Thus, the *r*II mutants affect not only plaque
morphology, but also the range of bacterial hosts that the virus can successfully infect. This latter trait was particularly useful, since it permitted
the selection and isolation of any wild-type revertants in the mutant phage
population.

Benzer isolated over 3000 independent *r*II mutants and showed that although they all seemed to be alike in their effect, in fact a number of
different *r*II mutants existed, all giving rise to the *r*II characteristic. This
finding suggested that the structure of the *r*II gene could be altered at a
number of different sites. The technique used by Benzer was mixed or

Table 12.1 Type of Growth of *r* Mutants of T-even Phages on the B and K Strains of *E. coli*.

Genotype	Plaques formed on host strain	
	B	K
*r*I or *r*III mutants	*r*	*r*
*r*II mutants	*r*	None
r⁺	+	+

From Herskowitz, I. J., 1965. *Genetics,* 2nd ed. Little, Brown and Co., Boston, p. 344, Fig. 26.5. Courtesy of the author.

double infection of K bacteria with two different *r*II mutant strains. The results showed that the *r*II mutants fell into two distinct groups, called A and B. None of the *r*II mutants alone could multiply in K, although the virus is adsorbed and the bacterial cell killed. Similarly, a mixed infection with two *r*IIA mutants or with two *r*IIB mutants gave no growth on K. In other words, all the *r*IIA mutants acted like alleles with one another, as did the *r*IIB mutants. However, mixed infection with an *r*IIA mutant and an *r*IIB mutant produced plaques on strain K of the host, with phages of both types being produced in normal numbers. Thus, the A and B mutants complement one another to permit normal growth, and the *r*II region must consist of two independent functional units, A and B.

The work with mixed infections also led to the discovery of recombination in bacteriophage by Delbrück and Bailey and by Hershey and Rotman. In other words, crosses could be made and recombinant progeny recovered in a manner analogous to that already familiar in higher organisms. The procedure used by Benzer was to infect strain B of *E. coli* with equal numbers of two different *r*II mutants so that there were several of each mutant type for each bacterial cell. After growth on strain B had occurred, the progeny virus particles were then tested for their ability to grow on strain K. Although neither *r*II parent could grow on K, nevertheless in nearly every cross, a few *r*⁺ progeny appeared capable of growth on strain K. Although rare, the wild-type phages could be shown to originate by recombination, for their frequency was still higher than that of the wild types arising by back mutation. This technique made it possible to screen millions of phage particles in order to recover the rare recombinant *r*⁺ types. Since only the recombinant type multiplies, it was possible to recover the *r*⁺ type when its frequency was as low as one in 10 million. The total population size was estimated from the growth of the progeny on strain B.

Benzer's work revealed that recombinants were formed not only in

rIIA × rIIB crosses, but also in A × A and in B × B crosses. Therefore, even though the independent rIIA mutants behaved as alleles—as parts of the same functional unit—nevertheless it was possible to get recombination between them. Thus, they could not be identical mutations, but represented changes at different sites within the functional unit. If we consider this functional unit to be the gene, then we are confronted by intragenic recombination.

When the many independent rII mutants were tested for their recombination frequency, a linear map of these mutants could be drawn. All of the A mutants mapped in one region, all of the B mutants in an immediately adjacent region with no overlap. Mapping some 3000 rII mutants by crossing them in all possible combinations would have been a tedious task despite the ease of working with the T4 phage-$E.$ $coli$ system. The discovery that some mutations in the rII region were deletions greatly simplified the mapping process.

The deletions were discovered when it was found that some rII mutants gave no recombination with any member of a series of rII mutants, all of which gave wild-type recombinants with one another. In addition, although most rII mutants are capable of reversion to wild type by back mutation, these nonrecombining mutants were incapable of back mutation. The most reasonable explanation for both the stability and the lack of recombination by these mutants is that a small segment has been deleted from the rII region, that the mutant covers a segment of the linkage map rather than a single point. About a fifth of the spontaneous rII mutants turned out to be deletion mutations, and a number of different deletion types were isolated and identified.

Crosses among these deletion mutants permitted the construction of a deletion map. If the deletions overlap, no r^+ recombinants can be formed; if the deletions do not overlap, r^+ recombinants will be recovered. The deletion map showed some deletions to be quite small while others were extensive. Furthermore, the deletion map turned out to be one-dimensional or linear in form, in agreement with the map obtained with the single-site or point mutations. With the use of a series of overlapping deletions, the mapping of point mutations was greatly facilitated (Figs. 12.10 and 12.11). A new rII mutant could be localized to the segment absent in a given deletion by the lack of r^+ recombinants when tested with that deletion. If the mutant lay outside that region, r^+ recombinants were formed. After the mutants had been sorted in this manner into regions, recombination studies with the mutants within each region permitted the construction of a more refined map. With these techniques Benzer mapped more than 3000 spontaneous and induced rII mutations. The point mutations were at more than 300 different sites, some 200 in the rIIA region and 108 in the rIIB region (Fig.

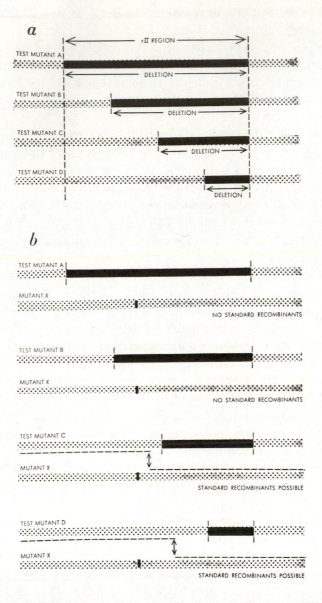

Fig. 12.10. Deletion mapping in the *r*II region of T4 phage. The unknown mutant is crossed with a group of reference mutants containing deletions of known length in the *r*II region. When the unknown falls within the limits of the reference deletions as in A and B, no standard recombinants can be formed because both DNA molecules are defective at the same place. With C and D, however, standard recombinants can be formed. (From Benzer, S., "The Fine Structure of the Gene." Copyright © January 1962 by *Scientific American*, Inc. All rights reserved.)

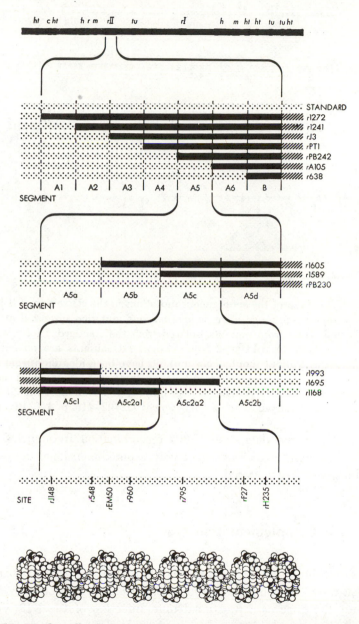

Fig. 12.11. Further illustration of mapping in the *r*II region of T4 phage, showing progressive refinement in the localization process. The seven mutational sites shown in the bottom row are tentative, but each site probably represents the smallest mutable unit in the DNA molecule, a single base pair. A DNA segment of about 40 base pairs is shown at the very bottom. (From Benzer, S., "The Fine Structure of the Gene." Copyright © January 1962 by *Scientific American*, Inc. All rights reserved.)

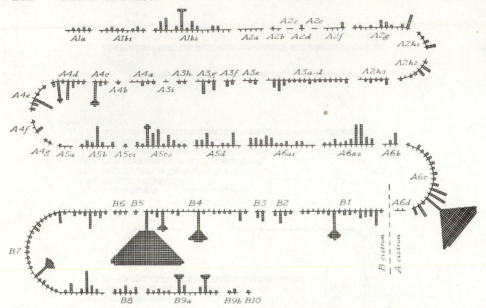

Fig. 12.12. Map of the spontaneous mutations in the *r*II region of T4 phage. Each square represents a single spontaneous mutation observed at the site indicated. Data from mutants isolated from the standard T4 r⁺ wild type and revertants to wild type from various *r*II mutants are pooled in this figure. Sites without spontaneous mutants have been identified from induced mutations. The arrangement of sites within each segment is arbitrary. (Benzer, S., 1961. "On The Topography of the Genetic Fine Structure," *Proceedings National Academy of Sciences* 47:403–416, p. 410, Fig. 6.)

12.12). All of the data, both from recombination frequencies and from deletion mapping, were consistent with a one-dimensional or linear model for the hereditary material.

12.9. The Complementation Test

In order to probe further into the nature of the *r*II region, Benzer adapted the "cis-trans" or complementation test developed earlier with *Drosophila* to study the gene as a functional unit. Infection of the K12(λ) strain of *E. coli* with a single *r*IIA mutant or a single *r*IIB mutant does not lead to viral multiplication. Similarly, the simultaneous infection of K12(λ) with two different *r*IIA mutants or two different *r*IIB mutants does not permit viral multiplication. However, if K12(λ) is simultaneously infected with an *r*IIA mutant and an *r*IIB mutant, viral multiplication occurs. It appears that the *r*IIA group of mutants affects one functional unit in the phage genome and the *r*IIB group of mutants affects a different functional unit, both of which must function normally for multiplication on K12(λ). Two *r*IIA mutants can only produce normal B-gene polypeptide; two *r*IIB mutants can only produce normal A-gene polypeptide. Therefore, no phage multiplication is possible in these mixed infections due to the lack of one or the other of the essential polypeptides. Mixed infection with an *r*IIA and an *r*IIB mutant results in normal phage multiplication because each mutant complements the other, producing the gene product that the other is unable to make.

As a result of these studies, Benzer suggested that the functional unit of the hereditary material, as defined by the *cis-trans* test for complementation, be called a **cistron**. In other words, a cistron would be a map segment within which no complementation occurred between mutants in the *trans* configuration. Thus, within the cistron the *trans* condition would result in the mutant phenotype. However, the *cis* condition would produce the wild phenotype. By this definition the *r*IIA segment would be one cistron and the *r*IIB segment a different cistron. The term *allele* was used to refer to mutations within a single cistron. In earlier conceptions of the gene it was thought that the unit of function, the unit of recombination, and the unit of mutation were one and the same. The discovery of mutation and recombination involving different sites within a single functional unit led to the realization that the old concept of the gene was an oversimplification. Therefore, for the smallest element undergoing recombination Benzer coined the term **recon**, and the smallest element capable of mutation was called a **muton**. These terms give recognition to the fact that the unit of function, the unit of recombination, and the unit of mutation are not identical. Benzer carried his fine-structure analysis a step further by relating the one-dimensional genetic map to the linear structure and the total content of the DNA in the T4 phage. With the assumption of the Watson-Crick structure for the DNA, he estimated that each *r*II cistron contained from several hundred to a few thousand nucleotide pairs. The muton and the recon, on the other hand, consisted of a very few nucleotide pairs at most. The most reasonable conclusion was that the smallest unit of recombination or of mutation was a single nucleotide pair.

12.10. The Triplet Code

After this digression into the ways of the T4 phage, let us return to a consideration of the reversion of proflavine-induced *r*II mutants to the wild type. Crick, Barnett, Brenner, and Watts-Tobin grew the *r*II mutant viruses on strain B of *E. coli* to build up their numbers. The viruses were then

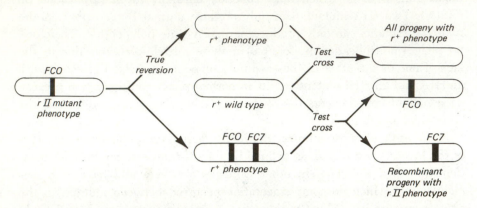

Fig. 12.13. Testing for suppressor mutations or true reversion to wild type with the FCO mutant at the *r*II locus in T4 phage.

sprayed on a culture plate covered with bacteria of strain K12(λ) to test whether reversion to the wild type had occurred. Since the *r*II mutants do not grow on *E. coli* strain K12(λ), only wild-type revertants will form plaques under these circumstances. Small numbers of plaques were found, from which the wild-type revertants were isolated. When these revertant phages were cultured on strain B, their plaque morphology differed slightly from that of the original wild type. Therefore, it seemed possible that they did not arise by an exact reversal of the original mutation, but rather by some sort of change that more or less compensated for that mutation. To test whether the revertant is truly like the original wild-type phage, double infection of a B bacterium by revertant phage and by the original wild-type T4 phage was produced (Fig. 12.13). If the two are the same, only wild-type phage capable of growth on both B and K should result. However, double infection of this sort gave rise to both mutant and wild-type phage. Therefore, the reversion to wild type was not due to true back mutation to the original nucleotide sequence. Recombination tests showed that the revertants had a second mutation close to the first in the same short segment (less than a fifth) of the *r*IIB gene. The second suppressor mutation, when separated from the original *r*II mutant by recombination, turned out to be an *r*II mutant itself even though these two *r*II mutants together produced the wild phenotype. Because the suppressor mutations were located within the same cistron, they functioned as intragenic or intracistronic suppressors. The interpretation offered for these results has furnished some of the best evidence that the genetic code is a triplet code, that is, that three nucleotides code for one amino acid, and that the code is read from a fixed starting point.

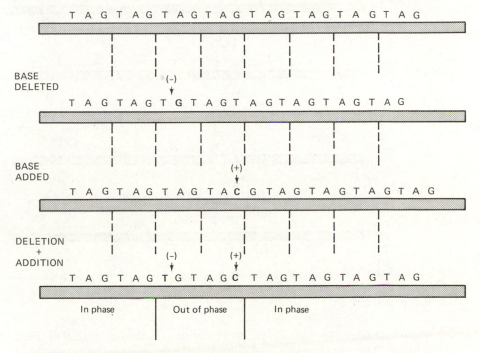

Fig. 12.14. The effect of mutations that add or remove a base from DNA. If the message is read from the left, the wild type reads TAG, TAG. . . . Adding the base C shifts the reading beyond that point to GTA, GTA. . . . Removing a base shifts the message to AGT, AGT. . . . Addition of a base and removal of a nearby base puts the message back in phase again.

To explain these results, Crick et al. arbitrarily assumed that the original proflavine-induced mutation was due to the addition of a single nucleotide pair at the site of the mutation. If the nucleotides in the gene are read in groups of three from a fixed starting point, the code would read properly from the starting point to the site of the "plus" mutant, but from there on, because of the addition of a nucleotide, the code would be misread, and abnormal amino acids would be incorporated in the polypeptide chain. As a consequence the mutant phenotype is produced. They further assumed that the suppressor mutations were due to the loss of a single nucleotide pair. The revertant types therefore had two mutations, the original "plus" mutant adding a nucleotide and a nearby "minus" mutant deleting a nucleotide (Fig. 12.14). The genetic code would be read correctly to the site of the first mutant, then incorrectly, because out of phase, to the site of the second mutant, and then correctly again to the

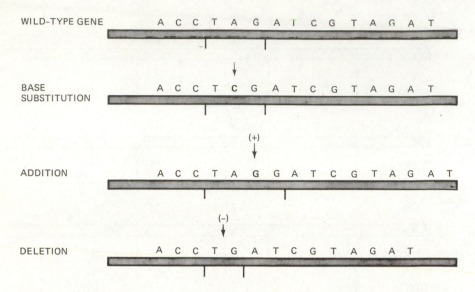

Fig. 12.15. Possible types of mutations in phage. In one group one base is replaced by another, for example, A by C. In the other group a base is added or deleted.

end of the gene since the "plus" and "minus" mutants canceled each other's effect. The phenotype of the suppressed type was not quite identical with the wild type because the small segment of the polypeptide chain coded between the two mutant sites was abnormal. The mutant sites were apparently always rather close together, because if they were too far apart the polypeptide chain would be so abnormal that the mutants would not grow on the strain K bacteria used for screening.

When the minus rII mutants that acted as suppressors to the original plus mutation were isolated from the plus mutant by recombination, they could be tested for their ability to revert to wild type. Reversions were found that again were due to a second mutation at a nearby site rather than to a true back mutation. These mutations were therefore acting as suppressors of the suppressors. Thus three types of mutants had been identified: the original rIIB single site plus mutant, the minus suppressors of the original mutant, and the suppressors of the suppressors. If the original hypothesis is correct, this last group of mutants should be plus like the original mutant. This inference turned out to be correct, for when two plus mutants were brought together in the same DNA strand by recombination, they still produced the mutant phenotype rather than reversion to wild type. Thus, in reality, the single site rIIB mutants fell into two groups, the plus mutants and the minus mutants (Fig. 12.15).

WILD-TYPE GENE

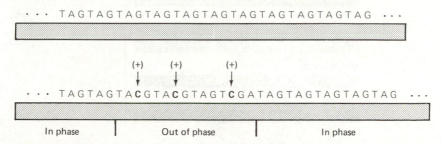

Fig. 12.16. When three bases are added close together, the genetic message is brought back in phase beyond the point where the third base has been added. The same effect is seen when three neighboring bases are deleted.

Recombination among the various single-site mutants was possible through genetic crosses. In nearly all cases, the combination of a plus and minus mutant resulted in the wild-type phenotype. The combination of two plus or of two minus mutants still formed the mutant phenotype. However, a combination of almost any three minus mutants or any three plus mutants formed the wild type rather than the mutant phenotype (Fig. 12.16). This finding led to the conclusion that the genetic code is indeed a triplet code, with three nucleotides coding for one amino acid.

The interpretation of these results is relatively straightforward. As noted above, the addition or deletion of a single nucleotide base would cause the genetic code to be misread from the mutant site on to the end of the gene. Nearby plus and minus mutants canceled out each other's effects except in the short segment between them. However, two plus or two minus mutants did not compensate for one another, and the reading frame of the triplet code remained out of kilter to the end. Since a combination of three plus mutants or three minus mutants within a short DNA segment gave a near wild-type phenotype, the presumption is that the reading frame, thrown off by one or two mutants of the same type, is restored to the correct sequence by the third mutation of the same kind. In other words, the genetic code is incorrect only over the short segment that includes the three mutants, and the original sequence exists from the third mutant site to the end of the cistron. The small segment disrupted by the three mutants was not large enough to prevent the formation of a functional gene product. Presumably, three plus mutants would lead to the formation of a polypeptide with one more amino acid than the wild type, whereas one less amino acid than usual would be found with three minus mutants.

Thus, the *coding ratio*, the ratio of nucleotide bases to amino acids,

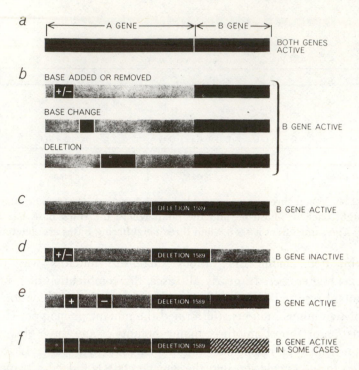

Fig. 12.17. The effect of the r1589 deletion. (a) The two wild-type genes, A and B, are read independently. (b) Mutations in the A gene do not affect the activity of the B gene. (c) Deletion r1589 inactivates the A gene, but the B gene remains active. (d, e, f) Mutations in the A gene, however, now will often inactivate the B gene also, suggesting that, as a result of the r1589 deletion, the two genes have been joined in such a way that they are read as a single unit. (From Crick, F. H. C., "The Genetic Code." Copyright © October 1962 by *Scientific American*, Inc. All rights reserved.)

appears to be three, that is, three bases code for one amino acid. The nucleotide triplet is referred to as a *codon*. The genetic code is read in only one direction from a fixed starting point.

Additional evidence on this last point was obtained from the use of a deletion (r1589) that covered parts of both the A and B cistrons in the rII region. This deletion covered the last portion of the A cistron, the point of union between A and B, and the first portion of the B cistron (Fig. 12.17). Although most such mutants lose both the A and the B function, this particular deletion lacked only A activity, but retained the B function. The explanation suggested for this result was that loss of this segment of the B cistron did not lead to its inactivation. The essential functional part of the molecule was not disturbed.

An rII mutant that reverts readily with proflavine and is therefore presumably due to the addition (or deletion) of a nucleotide pair will destroy the A function but leave the B activity unimpaired. However, the combination of such a mutant with the r1589 deletion caused the loss of function of both A and B. The most reasonable assumption seems to be that ordinarily the A and B cistrons are read independently with some sort of spacer between them. The deletion causes the loss of this spacer as well as portions of the A and B cistrons. As a result the remaining parts of the two cistrons code for a single polypeptide chain, initiated at the A starting point and containing a part of the A and a part of the B polypeptide. As noted above, the r1589 deletion left the B function unaffected. The addition mutation in the A region caused the loss of B function when combined with the r1589 deletion because it put the reading of the genetic code out of phase not only in the A region but also in the B region as well, since the two were now being read as a single code. This assumption was further supported by the finding that a suppressor of the A mutant restored the function of the B cistron.

Therefore, the evidence supports the concept that three nucleotides code for a single amino acid, that the code is linear and read from a fixed starting point, and that the code is nonoverlapping. Furthermore, although the data obtained from the r1589 deletion indicate some sort of spacer separating the rIIA and rIIB cistrons, there is no evidence for spacers or commas between adjacent nucleotides within the code for a single polypeptide so that the code is referred to as a commaless code.

12.11. The Role of RNA

We now have seen that DNA is the genetic material, that it is self-replicating, and that it is also heterocatalytic since it codes for the synthesis of proteins and thus controls the cell's metabolism. However, we still have not tackled the question of just how the genes exert their control over protein synthesis. Before we consider the way the genetic code was cracked so that the codons for the different amino acids can be specified, it is advisable to consider the mechanism by which DNA controls protein synthesis.

The simplest method for such control would be for the DNA to act directly as a template for the assembly of the amino acids into polypeptides. However, the evidence is clear that such direct control is not exerted by DNA. An early advance was the independent discovery by Brachet and by Caspersson that ribonucleic acid (RNA) is involved in protein synthesis. They found that whereas DNA is mostly confined to the nucleus,

RNA is present in both the nucleus and the cytoplasm of the cells of higher organisms. Furthermore, the amount of cytoplasmic RNA is directly correlated with the rate of protein synthesis in the cell. Cells not actively synthesizing protein have relatively little cytoplasmic RNA, but if the rate of protein synthesis is increased, the amount of RNA in the cytoplasm increases markedly. When Brachet treated cells with ribonuclease, an enzyme that causes the breakdown of RNA, protein synthesis stopped. In another experiment Brachet showed that protein synthesis continued in the cytoplasm of *Acetabularia*, a green alga, after the nucleus had been removed from the cell. From these and other experiments the idea grew that RNA played a vital role in protein synthesis. The concept was developed that the DNA code was *transcribed* to RNA, which then moved from the nucleus into the cytoplasm where *translation* from the base sequence in the RNA to the amino acid sequence in proteins took place.

Ribosomal RNA

The site of protein synthesis in the cytoplasm was identified as the result of work by Borsook and by Zamecnik and their colleagues. The essence of their experiments was to inject a radioactive amino acid into a small mammal and then to sacrifice the animal a short time later and analyze the various cell fractions for radioactivity. The experiments showed that proteins in the ribosomes had up to seven times more radioactivity than any other fraction, and thus the ribosomes, small cytoplasmic particles usually associated with the endoplasmic reticulum, appeared to be the site of protein synthesis in the cells of higher organisms. Other experiments with other organisms, including bacteria, also implicated the ribosomes.

The presumption was, then, that the RNA molecules moved through the 40-nm pores in the nuclear membrane to the ribosomes where they mediated protein synthesis. Since the ribosomes are about 60% RNA and 40% protein, it seemed reasonable to suppose that this RNA was directly involved in protein synthesis. If this were the case, there should be a direct relationship between the DNA in the nucleus and the ribosomal RNA. This relationship is based on the hydrogen bonding between the bases adenine and thymine, on the one hand, and guanine and cytosine on the other. When DNA replicates, the new strand that is formed is complementary to the existing strand. Thus, if the base sequence in the existing strand is A–C–A–T–G–C, the sequence in the new strand will be T–G–T–A–C–G. If the DNA serves as a template for RNA, presumably a similar process is involved, for the RNA molecule is very similar to the DNA molecule except that the sugar is ribose rather than deoxyribose and the base uracil (U) is present instead of thymine. Nevertheless, hydrogen bonds form just as well between A and U as between A and T. Therefore, if the DNA base sequence were A–C–A–T–G–C, the expected RNA sequence is U–G–U–A–C–G, and if the theory holds, the ratio $G + C/A + T$ in the DNA should equal the ratio $G + C/A + U$ in the RNA. The theory is most readily tested in bacteria, where rapid synthesis of many kinds of proteins occurs, rather than in the tissues of higher plants or animals where many genes are not active. When Belozersky tested these ratios in the total DNA and RNA from a number of different bacterial species, they turned out not to be the same. Therefore, the hypothesis seemed to fail. The bulk of the RNA in the cell is ribosomal RNA, but even when the ribosomal RNA was analyzed separately, its ratio still differed from that of DNA. In fact, the $G + C/A + U$ ratio was rather constant from one bacterial species to another, whereas the DNA composition varied widely.

The theory was rescued by the assumption that only a fraction of the ribosomal RNA served as a "messenger" carrying coded information from the genes in the nucleus to the ribosomes. This assumption was only confirmed after further advances in the study of the chemistry of RNA. These studies revealed that there are, in fact, three major kinds of RNA, *ribosomal* RNA (rRNA), *messenger* RNA (mRNA), and *transfer* RNA (tRNA). About 80% of the total RNA is ribosomal RNA, which is a stable component of the ribosomes. Ribosomal RNA consists of large molecules of high molecular weight and uniform size that remain intact as an integral part of the ribosomes during protein synthesis. Messenger RNA molecules are also large but rather heterogeneous in size. Furthermore, mRNA molecules tend to be less stable than ribosomal RNA and to break down during protein synthesis. When the ribosomes are placed in a solution with a reduced magnesium ion concentration, the mRNA molecules are released from the ribosomes, but the ribosomal RNA molecules are unaffected. The transfer RNA molecules are small, consisting of only about 80 ribonucleotides, and are thus rather soluble in aqueous solution. This soluble RNA is called transfer RNA or adaptor RNA because tRNA molecules carry amino acids to the ribosomes where they are incorporated into proteins. The unraveling of the function of these three kinds of RNA has contributed greatly to our understanding of the way that the genes control protein synthesis.

Messenger RNA

Messenger RNA was first identified experimentally by Volkin and Astrachan of Oak Ridge National Laboratory. They showed that when *E. coli* was infected with T2 phage, synthesis of bacterial DNA stopped at once, but synthesis of RNA and protein continued. The base composition of this newly synthesized RNA was determined by transferring recently infected bacteria to a medium containing labeled phosphorus (P^{32}), then hydrolysing the RNA after a few minutes, separating the four kinds of nucleotides, and measuring their radioactivity. They found that the $A + U/G + C$ base ratio of this RNA was similar to the $A + T/G + C$ base ratio of the virus DNA but different from that of the bacterial DNA and different from that of the total RNA, most of which is in the ribosomes. This finding indicated that the newly formed RNA was carrying information from the viral DNA to protein. Furthermore, this RNA showed a high turnover rate, suggesting that it is unstable. Further evidence that this RNA was carrying a transcription of the viral DNA code was obtained from experiments with DNA-RNA hybrid molecules by Hall and Spiegelman. In essence, they showed that the newly synthesized virus-specific RNA, when mixed with single-strand virus DNA, formed hybrid DNA-RNA molecules. A mixture of the T2 virus RNA with single-strand *E. coli* DNA or with DNA from other sources failed to form such hybrid molecules. Thus, the evidence suggests that the RNA formed after T2 infection was produced in association with the virus DNA. It was later shown that in normally growing *E. coli*, the bacterial DNA is constantly forming mRNA that becomes attached to the ribosomes. The amount of this mRNA is small compared to the ribosomal RNA, and the mRNA is continually being broken down and replaced by new mRNA molecules.

The enzyme catalyzing the formation of mRNA in *E. coli* has been isolated and is called RNA polymerase. This enzyme, incubated with the nucleotide diphosphates of RNA, will only catalyze mRNA formation if a small amount of DNA is present as a primer. The mRNA molecules formed carry a transcription of the code of this primer DNA. RNA polymerase has been found not only in bacteria, but in higher organisms as well, where the mRNA appears to be more stable and functions in conjunction with the ribosomes for longer periods than in bacteria.

Transfer RNA

Transfer RNA (also referred to as adaptor or soluble RNA) occurs in a number of different forms, each one specific for a particular amino acid. The different tRNA's are quite similar in general size and structure despite their different specificities. Each kind of tRNA molecule is probably synthesized in its own particular region of the DNA; in other words, the function of certain genes must be the production of tRNA. A given tRNA molecule combines with its amino acid and transports it to the mRNA on the ribosomes, where the amino acid is incorporated into the polypeptide and the tRNA is freed to continue with its transport function. A single tRNA molecule is relatively stable and may repeat the trip to the ribosomes many times.

In the tRNA the amounts of adenine and uracil and of guanine and cytosine are approximately equal. This finding suggests that there is complementary base pairing similar to that in DNA, but studies have shown that the tRNA molecule has only a single nucleotide strand with some 80 nucleotides. The evidence indicates that the molecule does have a double-stranded helical structure, and that the single strand is doubled back on itself like a hairpin, with the base sequence in one arm complementary to that in the other (Fig. 12.18). The indications are that there must be at least three unpaired bases at the folded end of the molecule, and these are thought to form hydrogen bonds with the coding region of the messenger RNA. The amino acid is attached at the opposite end of the tRNA molecule to a terminal sequence of three nucleotides that is the same in all tRNA molecules. Adenine is the terminal base and the next two are cytosine; thus, the sequence is –C–C–A. The specificity of the tRNA molecule therefore resides at the folded end, where it bonds with mRNA, and at the opposite end, where it pairs with its amino acid. Although the terminal sequence to which the amino acid is attached is always –C–C–A, the specificity for the amino acid apparently resides in the base sequence adjoining this terminal group and also in the sequence at the other end of the single strand. In other words, the specificity for mRNA lies at the loop end of the hairpin while the two ends of the hairpin bind the amino acid.

12.12. Genetic Control of Protein Synthesis

The unraveling of the functions of RNA has led to a far better understanding of the mechanism of protein synthesis. A number of different lines of evidence indicate that only one of the two strands in the DNA double helix serves as the template for messenger RNA. The mRNA, carrying a transcription of the information from this DNA strand, moves into the cytoplasm, where it becomes attached to the ribosomes (Figs. 12.19 and 12.20).

The ribosomes, containing about 60% RNA and 40% protein, are very similar in all organisms and appear to be a fundamental component of all cells. Bacterial ribosomes are found free throughout the cytoplasm, but the ribosomes in the cells of higher organisms are sometimes attached to the

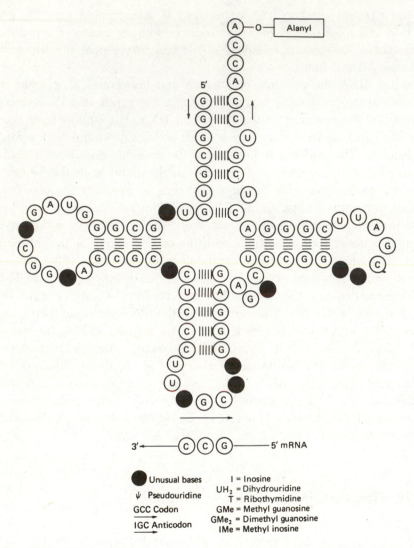

Fig. 12.18. The complete nucleotide sequence of a transfer RNA, alanine tRNA, with the unusual bases and the codon-anticodon position indicated. (From James D. Watson, *Molecular Biology of the Gene*, Second Edition, copyright © 1970 by J. D. Watson; W. A. Benjamin, Inc., Menlo Park, California.)

endoplasmic reticulum. The ribosomes are .approximately spherical in shape and are classified according to their sedimentation constants as determined by their rates of sedimentation in an ultracentrifuge. The ribosomes of *E. coli* have a sedimentation rate of about 70s (Svedberg units) but at low magnesium ion concentrations they dissociate into a 50s and a 30s

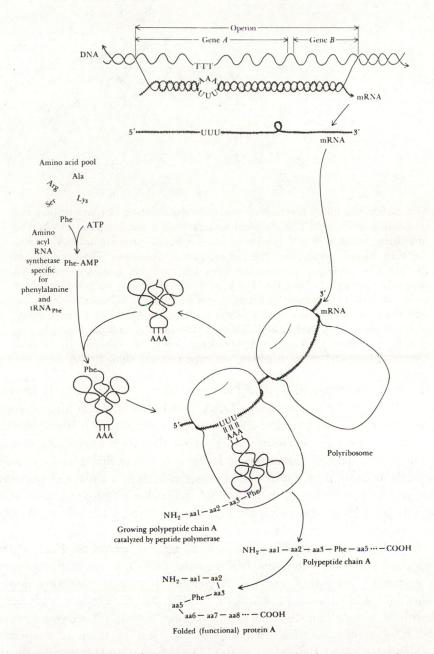

Fig. 12.19. Diagram showing the genetic control of protein synthesis. (Philip E. Hartman & Sigmund R. Suskind, *Gene Action*, Second Edition © 1969. Reprinted by permission of Prentice-Hall, Inc., Englewood Cliffs, New Jersey.)

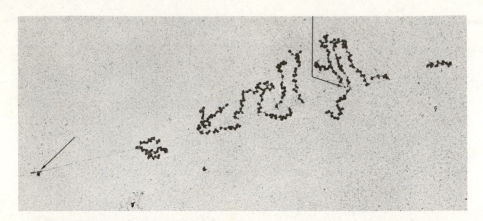

Fig. 12.20. The direct electron-microscopic visualization of transcription and translation in *E. coli*. The diagonal straight line is the bacterial DNA. The prominent black dots are ribosomes on the lateral-growing mRNA chains, the thin threads connecting the polyribosomal ribosomes. Transcription of the DNA is evidently progressing from left to right, because the mRNA molecules emerging from the DNA near the right are longer than those near the left. One arrow indicates a probable RNA polymerase initiation site, the other a structure visible at the juncture between DNA and mRNA, which is probably a molecule of RNA polymerase. (*E. coli* photograph; Fig. 2 in O. L. Miller, Jr., Barbara A. Hamkalo, and C. A. Thomas, Jr. "Visualization of bacterial genes in action," *Science* 169:392–395. 1970.)

component. The ribosomal RNA, of two classes (16s and 23s), is synthesized in specific regions of the DNA, and becomes stable after it combines with protein to form the 30s and 50s components. The 70s ribosomes are active in protein synthesis, during which they form aggregates known as polysomes. The ribosomes seem to have a nonspecific role in protein synthesis, because they can be used in the synthesis of a variety of proteins. The specificity is conferred by the mRNA molecules, which apparently hold the 70s particles of the polysome together by being adsorbed to the 30s components.

The initial step in protein synthesis is amino acid activation. Each amino acid combines with adenosine triphosphate (ATP), a reaction mediated by a specific "activating" enzyme. A complex is formed consisting of the amino acid, adenosine monophosphate (AMP), and the enzyme, with pyrophosphate being eliminated. There is a specific "activating" enzyme (amino acyl RNA synthetase) for each amino acid, just as there is a specific transfer RNA. Next the amino acid-AMP-enzyme complex reacts with its specific tRNA molecule, binding the amino acid to the tRNA and releasing the AMP and the enzyme. The linkage between the tRNA and the amino acid is between the ribose of the terminal adenine group of the tRNA and the carboxyl group of the amino acid. In the final step of protein synthesis, the amino acid-tRNA complex becomes associated with the mRNA on the ribosomes through hydrogen bonding between its coding nucleotides at the loop end of the tRNA and the appropriate complementary base sequence in the mRNA. In this way the amino acids are arranged in the order specified by the mRNA. The polypeptide chain is initiated at the N terminal or free amino end, the amino acids are added to the C terminal or carboxyl end, replacing the tRNA, and the free carboxyl group marks the end of the chain. This final step is mediated by a single enzyme (peptide polymerase) with a cofactor (guanosine triphosphate). When the polypeptide chain is completed, it is released from the polysomes and assumes its functional folded structure.

The rate of protein synthesis has been estimated to be remarkably high. In the case of hemoglobin, it was calculated that a new polypeptide chain was being synthesized every minute and a half. Since each chain contains about 150 amino acids, the rate of synthesis was approximately two amino acids per second.

12.13. Breaking the Genetic Code

The discoveries about the nature of the mechanism of protein synthesis in the cell paved the way for breaking the genetic code. The cell-free system by which mRNA can be synthesized in the presence of primer DNA and the enzyme, RNA polymerase, has already been mentioned. The discovery by Nirenberg and Matthaei in 1961 that polypeptide synthesis was still possible when artificially synthesized mRNA was substituted for natural mRNA in a cell-free system led to the initial breakthrough. They made a synthetic mRNA using the polynucleotide phosphorylase first isolated by Grunberg-Manago and Ochoa. The synthetic mRNA was then added to an *in vitro* protein-synthesizing system derived from *E. coli*. This system contained ribosomes, tRNA, ATP, and the necessary enzymes plus C^{14}-labeled amino acids.

The first synthetic mRNA used by Nirenberg and Matthaei contained only nucleotides with the base uracil, which formed polyuridylic acid or poly-U. Incubation of poly-U with the cell-free system revealed that, of all the amino acids, only phenylalanine was being incorporated into the polypeptide chains. Thus, the radioactive "protein" being synthesized was polyphenylalanine. Therefore, for the first time it became possible to deduce the DNA code for one of the amino acids.

We have already seen that the genetic code is apparently a triplet code, a fact reinforced by the finding that the number of nucleotide pairs in the DNA has been estimated for several genes and compared with the number of amino acid residues in the protein controlled by the gene. In each case, the coding ratio is close to three; in other words, there are three times as many nucleotides as amino acids. Therefore, the triplet –U–U–U– in mRNA must be formed *in vivo* in contact with the nuclear DNA triplet –A–A–A– under the control of RNA polymerase. It then moves to the ribosomes, where the tRNA's carrying phenylalanine are selectively directed to incorporate it into the growing polypeptide chain. After this breakthrough, a flood of experiments was directed toward resolving the genetic code meaning for each of the 64 possible triplets. Poly-A was shown to control lysine incorporation, and poly-C proline incorporation. Poly-G, however, did not work well in this system, but is now thought to code for glycine.

The next step was to synthesize artificial mRNA with mixtures of two different ribonucleotides in varied proportions. For example, in 1961 Lengyel, Speyer, and Ochoa used a mixture with the ratio 5U:1A. Since the ribonucleotides unite at random, it is possible to predict the relative frequency of the different types of triplets (UUU, UUA, UAA, and AAA) in the RNA polymer. The kinds of amino acids and the relative frequency with which they are incorporated into proteins can then also be determined. In theory, for every 125 triplets with the sequence UUU, there will be 25 each with UUA, UAU, and AUU, 5 each with UAA, AUA, and AAU, and 1 with AAA. The finding that the 5U:1A mixture led to the incorporation of phenylalanine and tyrosine in a ratio of about 4 to 1 led to the conclusion that the code for tyrosine contained two uracils and one adenine. That finding, however, does not tell us the sequence of the three bases, which could be UUA, UAU, or AUU.

The proper base sequences were revealed by experiments testing the ability of short polyribonucleotides to cause the binding *in vitro* of specific amino acyl-tRNA's to the ribosomes. For example, it was discovered that a nucleotide containing U and G leads to valine incorporation into protein at a frequency in relation to phenylalanine incorporation that suggests the codon for valine contains 2U and 1G in an unknown sequence. When the trinucleoside diphosphates GpUpU, UpGpU, and UpUpG were tested with C^{14}-labeled valine-tRNA, the binding of the radioactive valine to the ribosomes was directed only by GpUpU and not by UpGpU or UpUpG. Furthermore, the binding of val-tRNA is specific, because GpUpU had no effect on the amino acyl-tRNA's of a number of other amino acids. From these results it can be concluded that the codon for valine is GUU and that

First Position (5'end)	U	C	Second Position A	G	Third Position (3'end)
U	phe	ser	tyr	cys	U
	phe	ser	tyr	cys	C
	leu	ser	CT*	CT	A
	leu	ser	CT	try	G
C	leu	pro	his	arg	U
	leu	pro	his	arg	C
	leu	pro	gln	arg	A
	leu	pro	gln	arg	G
A	ile	thr	asn	ser	U
	ile	thr	asn	ser	C
	ile	thr	lys	arg	A
	met,fmet△	thr	lys	arg	G
G	val	ala	asp	gly	U
	val	ala	asp	gly	C
	val	ala	glu	gly	A
	val	ala	glu	gly	G

Fig. 12.21. The genetic code. Codon meanings have been established primarily by studies with *E. coli*, but appear to be universal. *CT = chain-terminating codons (formerly "nonsense" codons). Sometimes the CT codons are known as ochre (UAA), amber (UAG), and opal (UGA). △fmet is chain initiating in *E. coli*.

the valine-tRNA has the complementary anticodon CAA at the loop end of the molecule.

We cannot consider in detail all of the experiments that have led to our present knowledge of the code as shown in Fig. 12.21. However, the principles by which the work was carried on have been indicated above.

Examination of Fig. 12.21 shows that a number of different triplets may code for the same amino acid. Because of this multiple coding for a single amino acid, the genetic code has been called a *degenerate* code, a term that has some misleading implications. The degeneracy is primarily at the third position; for example, GUU, GUC, GUA, and GUG all code for valine. It may also be found at the first position, because AGA and CGA both code for arginine. For the most part, the triplets coding for a given amino acid differ from one another by only a single nucleotide.

When spontaneous mutations occur or when mutations are induced by mutagens that cause base-pair substitutions in the DNA, a high proportion of these mutations cause the substitution of just one amino acid in the protein for another. Such mutations are referred to as *missense* mutations, and the fact that they are so frequent suggests that most of the 64 possible triplets code for some amino acid. Since there are just 20 amino acids but 64 triplets, the genetic code is obviously degenerate.

In some cases, however, the mutations are *nonsense* mutations, because they produce codons that do not code for an amino acid. Again, this term may be misleading, for the nonsense codons are not entirely lacking in function. For example, the so-called "amber" mutations of the T4 virus of *E. coli* cause premature termination of polypeptide chain synthesis. Further study revealed that the RNA triplet coding for polypeptide chain termination was UAG. Similarly, in the "ochre" mutants of T4, which also terminate polypeptide chain synthesis, the codon indicating the end of the polypeptide was UAA. Altogether, three chain-terminating codons have been identified, UAA, UAG, and UGA ("opal"). All bacterial polypeptides have been found to start with N-formyl methionine, a modified methionine with a formyl group attached to its terminal amino group. The initiation codon is apparently AUG, the codon for methionine, but the process is not yet fully understood even in prokaryotes, because GUG (normally the codon for valine) has also been found capable of initiating polypeptide synthesis. Thus, in some cases at least, it appears that the codons may have functions other than simply coding for amino acids. The need to separate one cistron from another, to initiate and terminate polypeptide synthesis, and to regulate and control gene action all suggest that further study will reveal more about this aspect of the coding process.

Implied in our discussion thus far has been the idea that the sequence of nucleotide pairs in the DNA double helix specifies the sequence of amino acids in the polypeptide chain being synthesized. An elegant demonstration of the validity of this concept of *colinearity* between gene and protein emerged from the research of Yanofsky and his co-workers on tryptophane synthetase mutants in *E. coli* (Fig. 12.22). The A protein of this enzyme

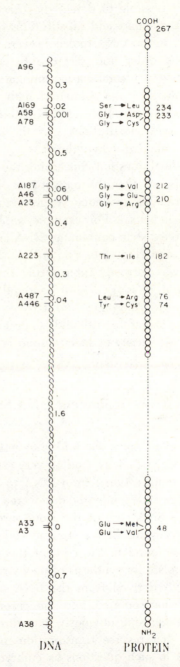

Fig. 12.22. The colinearity of the DNA genetic code and the corresponding protein as seen in the tryptophan synthetase A gene of *E. coli*. The positions of known mutations in the DNA (left) are exactly matched by the order of the amino acids altered in the enzyme (right). (From Du-Praw, E. J., *The Biosciences: Cell and Molecular Biology*, CMB Council, Stanford, California, 1972.)

contains 267 amino acids. The different mutants were shown to differ from the wild-type protein by a single amino acid substitution. Furthermore, the determination of the normal sequence of amino acids in the A protein made it possible to detect the exact location in the polypeptide chain at which the amino acid substitutions had occurred. The sites of the different mutants in the DNA molecule were established by recombination analysis, with the order and distance between the mutants determined by the frequency of recombination among them. When the genetic map based on recombination was compared with the map showing the location of the corresponding amino acid substitutions, the sequence of mutants and of amino acid substitutions was the same, thus demonstrating the validity of the colinearity hypothesis.

One final question concerning the genetic code is whether it is *universal* —that is, whether the meaning of the code is the same in all organisms. The available evidence suggests the tentative answer that the code is indeed universal. For example, rabbit hemoglobin can be synthesized in a cell-free system containing tRNA and amino acid-activating enzymes from *E. coli* and ribosomes and mRNA from the rabbit. Hence, the bacterial tRNA was correctly interpreting the rabbit mRNA. Furthermore, if synthetic poly-U is used with bacterial ribosomes and tRNA's or with rabbit reticulocyte ribosomes and tRNA's, both systems synthesize polyphenylalanine. Therefore, although some exceptions may occur, the genetic code for a wide variety of species appears to be the same.

12.14. RNA-Dependent DNA Synthesis

The story we have just told is sometimes referred to as the "central dogma" of molecular biology, and is, very simply, that the flow of genetic information is unidirectional from DNA to RNA to protein. Recently, Temin and Baltimore independently discovered that reverse transcription can occur, that the RNA core of the Rous sarcoma virus can make a DNA copy of itself by means of an enzyme called RNA-dependent DNA polymerase, which is found in the core of the virus. This DNA copy then serves as the template for synthesis of new virus particles. Temin also postulated that the DNA derived from the RNA viral template could act as a provirus and be incorporated into the chromosome of a virus-infected host cell. Such a concept obviously has direct implications for viral theories of cancer, but reverse transcription may also have broader implications in relation to such questions as embryonic differentiation, cellular regulation, and immunological memory. How widely distributed the process of reverse transcription is remains to be seen, but it is already clear that the central dogma should read that in the DNA-RNA-protein system, the only type of information transfer excluded is that initiated by protein.

Additional Reading

Adelberg, E. A., ed., 1963. *Papers on Bacterial Genetics,* 2nd ed. Little, Brown, Boston.

Ames, B. N., and P. E. Hartman, 1963. "The Histidine Operon," *Cold Spring Harbor Symp. Quant. Biol.* **28**:349–356.

Baltimore, D., 1970. "Viral RNA-Dependent DNA polymerase," *Nature* **226**: 1209–1211.

Beadle, G. W., 1946. "Genes and the Chemistry of the Organism," *Amer. Sci.* **34**:31–53.

Beadle, G. W., and B. Ephrussi, 1937. "Development of Eye Colors in *Drosophila:* Diffusable Substances and Their Interrelations," *Genetics* **22**:76–86.

Beadle, G. W., and E. L. Tatum, 1941. "Genetic Control of Biochemical Reactions in *Neurospora,*" *Proc. Nat. Acad. Sci.* **27**:499–506.

Belozersky, A. N., and A. S. Spirin, 1958. "A Correlation Between the Compositions of Deoxyribonucleic and Ribonucleic Acids," *Nature* **182**:111.

Benzer, S., 1957. "The Elementary Units of Heredity. In *The Chemical Basis of Heredity,* W. D. McElroy and B. Glass, eds. "Johns Hopkins University Press, Baltimore.

Benzer, S., 1962. "The Fine Structure of the Gene," *Sci. Amer.* **206** (Jan.):70–84.

Borsook, H., C. L. Deasy, A. J. Haagen-Smit, G. Keighley, and P. H. Lowy, 1950. "Metabolism of C^{14}-Labeled Glycine, L-Histidine, L-Leucine and L-Lysine," *J. Biol. Chem.* **187**:839–848.

Brachet, J., 1940. "La localisation de l'acide thymonucléique pendant l'oogénèse et la maturation chez les Amphibiens," *Archives de Biologie* **51**:151–165.

Brachet, J., 1942. "La localisation des acides pentose nucléiques dans les tissus animaux et les oeufs d'Amphibiens en voie de developpemont," *Archives de Biologie* **53**:207–257.

Brachet, J., 1955. "The Biological Role of the Pentose Nucleic Acids." In *The Nucleic Acids: Chemistry and Biology,* Vol. 2, pp. 475–519, E. Chargaff and J. E. Davidson, eds. Academic, New York.

Brenner, S., F. Jacob, and M. Meselson, 1961. "An Unstable Intermediate Carrying Information from Genes to Ribosomes for Protein Synthesis," *Nature* **190**:576–581.

Buttin, G., F. Jacob, and J. Monod, 1967. "The Operon. A Unit of Coordinated Gene Action." In *Heritage from Mendel,* R. A. Brink, ed. University of Wisconsin Press, Madison.

Cairns, J., G. S. Stent, and J. D. Watson, eds., 1966. *Phage and the Origins of Molecular Biology.* Cold Spring Harbor Lab. Quant. Biol.

Caspari, E., 1949. "Physiological Action of Eye Color Mutants in the Moths *Ephestia kühniella* and *Ptychopoda seriata,*" *Quart. Rev. Biol.* **24**:185–199.

Caspersson, T. O., 1941. "Studien über den Eiweissumsatz der Zelle," *Naturwiss.* **29**:33–43.

Caspersson, T. O., 1950. *Cell Growth and Cell Function.* Norton, New York.

Crick, F. H. C., 1962. "The Genetic Code," *Sci. Amer.* **207** (Oct.):66–74.

Crick, F. H. C., 1970. "Central Dogma of Molecular Biology," *Nature* **227**:561–563.

Crick, F. H. C., L. Barnett, S. Brenner, and J. R. Watts-Tobin, 1961. "General Nature of the Genetic Code for Proteins," *Nature* **192**:1227–1232.

Culliton, B. J., 1971. "Reverse Transcription: One Year Later," *Science* **172**:926–928.

De Busk, A. G., 1968. *Molecular Genetics*. Macmillan, New York.

Delbrück, M., and W. T. Bailey, Jr., 1946. "Induced Mutations in Bacterial Viruses," *Cold Spring Harbor Symp. Quant. Biol.* **11**:113–114.

Dounce, A. L., 1952. "Duplicating Mechanism for Peptide Chain and Nucleic Acid Synthesis," *Enzymologia* **15**:251–258.

Ephrussi, B., 1942. "Chemistry of 'Eye Color Hormones' of *Drosophila*," *Quart. Rev. Biol.* **127**:327–338.

Gamow, G., 1954. "Possible Relation Between Deoxyribonucleic Acid and Protein Structures," *Nature* **173**:318.

Gamow, G., and M. Ycas, 1955. "Statistical Correlation of Protein and Ribonucleic Acid Composition," *Proc. Nat. Acad. Sci.* **41**:1011–1019.

Garrod, A. E., 1909. *Inborn Errors of Metabolism*. Oxford University Press, London. (2nd ed., 1923.)

Goodman, H. M., and A. Rich, 1962. "Formation of a DNA-Soluble RNA Hybrid and Its Relation to the Origin, Evolution, and Degeneracy of Soluble RNA," *Proc. Nat. Acad. Sci.* **48**:2101–2109.

Grunberg-Manago, M., and S. Ochoa, 1955. "Enzymatic Synthesis and Breakdown of Polynucleotides; Polynucleotide Phosphorylase," *J. Amer. Chem. Soc.* **77**:3165–3166.

Hall, B. D., and S. Spiegelman, 1961. "Sequence Complementarity of T2-DNA and T2-Specific RNA," *Proc. Nat. Acad. Sci.* **47**:137.

Hartman, P. E., and S. R. Suskind, 1969. *Gene Action*, 2nd ed. Prentice-Hall, Englewood Cliffs, N.J.

Hayes, W., 1968. *The Genetics of Bacteria and Their Viruses*, 2nd ed. Wiley, New York.

Hershey, A. D., and R. Rotman, 1948. "Linkage among Genes Controlling Inhibition of Lysis in a Bacterial Virus," *Proc. Nat. Acad. Sci.* **34**:89–96.

Herskowitz, I. H., 1967. *Basic Principles of Molecular Genetics*. Little, Brown, Boston.

Hoagland, M. B., M. L. Stephenson, J. F. Scott, L. I. Hecht, and P. S. Zamecnik, 1958. "A Soluble Ribonucleic Acid Intermediate in Protein Synthesis," *J. Biol. Chem.* **231**:241–257.

Holley, R. W., J. Apgar, G. A. Everett, J. T. Madison, M. Marquiser, S. H. Merrill, J. R. Penswick, and A. Zamir, 1965. "Structure of a Ribonucleic Acid," *Science* **147**:1462–1465.

Lawrence, W. J. C., 1950. "Genetic Control of Biochemical Synthesis as Exemplified by Plant Genetics—Flower Colors," *Biochem. Soc. Symp.* **4**:3–9.

Lawrence, W. J. C., and J. R. Price, 1940. "The Genetics and Chemistry of Flower Colour Variation," *Biol. Rev.* **15**:35–58.

Lanni, F., 1964. "The Biological Coding Problem," *Adv. Genet.* **12**:1–141.

Lederberg, J., 1947. "Gene Recombination and Linked Segregations in *Escherichia coli*," *Genetics* **32**:505–525.

Lederberg, J., and E. L. Tatum, 1946. "Novel Genotypes in Mixed Cultures of Biochemical Mutants of Bacteria," *Cold Spring Harbor Symp. Quant. Biol.* **11**:113–114.

Lengyel, P., J. F. Speyer, and S. Ochoa, 1961. "Synthetic Polynucleotides and the Amino Acid Code," *Proc. Nat. Acad. Sci.* **47**:1936–1942.

Lewin, B. M., 1970. *The Molecular Basis of Gene Expression.* Wiley, New York.

Lindsley, D. L., and E. H., Grell, 1968. "Genetic Variations of *Drosophila melanogaster*," Carnegie Inst. Wash. Publ. No. 627.

Nirenberg, M. W., and J. H. Matthei, 1961. "The Dependence of Cell-Free Protein Synthesis in *E. coli* upon Naturally Occurring or Synthetic Polyribonucleotides," *Proc. Nat. Acad. Sci.* **47**:1588–1602.

Sadgopal, A., 1968. "The Genetic Code after the Excitement," *Adv. Genet.* **14**: 325–404.

Sager, R., and F. J. Ryan, 1961. *Cell Heredity.* Wiley, New York.

Srb, A. M., R. D. Owen, and R. S. Edgar, eds., 1970. *Facets of Genetics. Readings from Scientific American.* Freeman, San Francisco.

Stent, G. S., 1963. *Molecular Biology of Bacterial Viruses.* Freeman, San Francisco.

Stent, G. S., 1971. *Molecular Genetics.* Freeman, San Francisco.

Taylor, J. H., ed. *Molecular Genetics*, Part I, 1963; Part II, 1967. Academic, New York.

Taylor, J. H., ed., 1965. *Selected Papers on Molecular Genetics.* Academic, New York.

Temin, H. M., 1972. "RNA-Directed DNA Synthesis," *Sci. Amer.* **226** (Jan.): 24–33.

Temin, H. M., and S. Mizutani, 1970. "RNA-Dependent DNA-Polymerase in Virions of Rous Sarcoma Virus," *Nature* **226**:1211–1213.

Volkin, E., and L. Astrachan, 1956. "Intracellular Distribution of Labeled Ribonucleic Acid after Phage Infection of *Escherichia coli*," *Virology* **2**:433–437.

Wagner, R. P., and H. K. Mitchell, 1964. *Genetics and Metabolism*, 2nd ed. Wiley, New York.

Watson, J. D., 1970. *Molecular Biology of the Gene*, 2nd ed. Benjamin, Menlo Park, Calif.

Yanofsky, C., B. C. Carlton, J. R. Guest, D. R. Helsinki, and U. Henning, 1964. "On the Colinearity of Gene Structure and Protein Structure," *Proc. Nat. Acad. Sci.* **51**:266–272.

Yanofsky, C., and S. Spiegelman, 1962. "The Identification of the Ribosomal RNA Cistron by Sequence Complementarity. II. Saturation of and Competitive Interaction at the RNA Cistron," *Proc. Nat. Acad. Sci.* **48**:1466–1472.

Zamecnik, P. C., 1960. "Historical and Current Aspects of the Problem of Protein Synthesis," *Harvey Lectures* **54**:256–281.

Questions

12.1. Define a phenotypic character.

12.2. Tests of viability always measure relative viability by means of a comparison with some "standard" genotype.

a. What do you consider the most appropriate genotype to use as a standard?

    b. IIow would you perform experiments to measure the relative viability of different genotypes?

    c. What effect will your choice of a "standard" genotype have on the proportions of lethal, semilethal, subvital, normal, and super-vital genotypes that you observe?

12.3.   What was the rationale for the switch from higher plants and animals such as corn and *Drosophila* to microorganisms in order to study biochemical genetics?

12.4.   Compare the life cycle of *Neurospora* with that of Indian corn, *Zea mays,* and with that of the fruit fly, *Drosophila melanogaster.*

12.5.   What problems might you encounter if you tried to screen *Drosophila* for biochemical mutants using a technique comparable to that used with *Neurospora,* involving a minimal medium, a complete medium, and the minimal medium supplemented by one or another of the amino acids or vitamins?

12.6.   If the eight ascospores in the ascus of *Neurospora* were alternately *co* and +, what would you have to assume about when the reduction from diploid to the haploid condition occurred?

12.7.   If the formation of ascospores in a strain of *Neurospora* heterozygous for light and dark ascospores is compared with the formation of pollen grains in corn heterozygous for waxy and nonwaxy, what would you expect to find with respect to these traits?

12.8.   If you found independent biochemical mutants in *Neurospora* that required tryptophan for growth on minimal medium, what sort of progeny would you expect from crosses in which these mutants behaved as

    a. alleles

    b. linked genes at different loci

    c. unlinked genes at different loci

12.9.   a. If each of the four bases in DNA coded for one amino acid, how many amino acids could be handled by this code?

    b. If two of the purine and pyrimidine bases were necessary to code for an amino acid, how many amino acids could be handled by this code?

    c. If the coding unit or codon consisted of three bases, how many amino acids could be handled by this system if each codon coded for a different amino acid?

12.10. The genetic code has been found to be a 64-word triplet code based on all permutations of one, two, or three of the four purine and pyrimidine bases in DNA. If the genetic code were based only on different combinations of bases (i.e. if the order were immaterial), how many different code words would be possible?

12.11. How was it determined whether the genetic code is overlapping or nonoverlapping?

12.12. Give the essence of the experiments that demonstrated that the genetic code is a triplet code with a fixed starting point.

12.13. If an abnormal polypeptide is produced that lacks 27 amino acids found in the normal polypeptide and if this polypeptide results from a deletion, what must be the length of the deletion in terms of nucleotides?

12.14. If the base analogue 5-bromouracil were added to a bacterial culture medium, and all protein synthesis by the bacteria ceased in less than 5 min, what conclusion could you draw?

12.15. What are the distinguishing characteristics of the three classes of RNA molecules: ribosomal, transfer, and messenger?

12.16. The key that first unlocked the genetic code was the discovery that polyuridylic acid or poly-U coded for polyphenylalanine, suggesting that UUU was the code for phenylalanine. What was the rationale behind the next step, when mixtures of two kinds of ribonucleotides were used? Once the base composition for a particular codon was determined, how was the base sequence worked out?

12.17. Why is the genetic code called "degenerate"?

12.18. What is meant by "colinearity"? How has it been demonstrated?

12.19. Distinguish between missense mutations, nonsense mutations, and suppressor mutations.

12.20. What is the "central dogma" of molecular biology?

12.21. What are the mechanisms for chain initiation and chain termination in the genetic code?

12.22. What, if any, are the evolutionary implications of the universality of the genetic code?

12.23. In the light of the material in this chapter, how would you rephrase the "one-gene, one-enzyme" concept?

12.24. What is a gene?

CHAPTER THIRTEEN

Mutation

We have now considered the gene as a unit of heredity, as a self-replicating unit, and as a functional unit. In this chapter we shall discuss the gene as a mutating unit. In previous chapters we have discussed mutations in order to clarify certain other concepts, but now we shall deal with the development of the concept of mutation in a more orderly fashion.

13.1 The Mutation Theory of de Vries

Modern thought about mutation stems from the mutation theory of de Vries, first proposed in 1901, only a year after he, Correns, and von Tschermak rediscovered and independently confirmed Mendel's work. de Vries' theory was based primarily on his work with the evening primrose, *Oenothera lamarckiana*. This American species had escaped from cultivation near Amsterdam in Holland, and in this wild population de Vries discovered two new varieties that differed in a number of respects from the typical form. He brought all three types under cultivation and found that not only did the two variant forms breed true, but that the normal type continued to give rise to a number of mutant forms. Since many of these new types bred true and differed from the typical form in a number of ways, he felt that he had discovered a process, mutation, by which new species could arise in a single step. Thus mutation was offered as an alternative to Darwinian natural selection as the mechanism for the origin of species.

Oddly enough, most of de Vries' "mutations" turned out not to be what would now be called mutations, except perhaps in the very broadest sense of the word. *Oenothera lamarckiana* is now known to be a complex translocation heterozygote maintained as a balanced lethal system. [A balanced lethal system is one in which two nonallelic recessive lethal genes on homologous chromosomes appear to give rise to a true-breeding heterozygous stock. The homozygotes die because they are homozygous for one (25%) or the other (25%) of the recessive lethal genes, and only the heterozygotes (50%) survive.] The variants he observed were of several types, including tetraploids (*gigas* form), haploids (*nanella* form), and rare viable recombinants from the balanced lethal system. Only two were due to the change of a gene at a given locus to a new allele, the process that is usually called mutation today.

Furthermore, his idea that evolution occurs by "saltations" in a stepwise manner rather than by gradual changes due to selection is no longer generally accepted. Nevertheless, his theory is a landmark in the development of genetics, for it led to numerous studies of the mutation process, from which have come a better understanding of mutation, of the gene, and of the role that mutation plays in the origin of new hereditary variations on which natural selection acts.

13.2. The Presence-Absence Theory of Bateson and Punnett

In an early attempt to explain the nature of mutations, Bateson and Punnett suggested that a dominant trait is due to the presence of a gene and the recessive trait to its absence. However, this presence-absence hypothesis was soon shown to be untenable. The discovery of systems of multiple alleles, usually consisting of several genes recessive to the dominant wild type, was difficult to explain under this theory, as was the occurrence of true back mutations, which would require "nothing" to mutate to "something." Furthermore, the theory would seem to suggest that evolution has proceeded by means of a series of loss mutations, which seems extremely unlikely.

13.3. Definition of Mutation

Although the word mutation is still used occasionally in a broad sense to refer to any type of change in the hereditary material leading to an inherited change in phenotypic expression (a definition that would include chromosomal variation and position effect as well as gene mutation), it is now customary to restrict the meaning of mutation to changes at a given locus, the so-called "gene" or "point" mutations. As long as the classical theory of the gene remained intact, this definition was satisfactory. However, increased knowledge of the nature of the genetic material has led to difficulties with such constructs as "the gene" and "locus." Even this restricted definition of "mutation" must be regarded primarily as an operational definition. We shall continue to refer to hereditary changes that behave as if they have occurred at a specific "gene locus" as "gene" mutations, but must realize that our ability to distinguish such changes from other, more gross types of change may be more dependent on our sophistication in experimentation than on any clear-cut distinction between "gene" mutations and other types of change. An illustration may be helpful in clarifying this statement. When H. J. Muller demonstrated the mutagenic effects of X-rays in 1927, his success was due largely to the use of a specially designed stock of *Drosophila melanogaster,* the *ClB* stock. The *C* stood for a crossover suppressor, the *B* for bar eyes, both originally thought to be due to dominant sex-linked genes, and the *l* for a sex-linked recessive lethal. As more refined genetical and cytological analyses became possible, *C* turned out to be due to an inversion, and *B* to a duplication, both chromosomal rearrangements. Thus only *l* is still regarded as a "gene" mutation. However, since small deletions often act as recessive lethals, the possibility cannot be ignored that further study might move *l* out of the "gene" mutation category as well. Thus, the distinction between chromosomal changes and point mutations is not as clear cut as it may seem at first glance. The categories are to a large extent artificial. Nevertheless, it is useful to distinguish between the cytologically detectable changes and the "point" mutations so long as the operational nature of the distinction and the limitations of the definition are clearly understood.

The effects of genes on the phenotype of the organism are extremely diverse. Most research on mutations has been carried on by identifying new mutants by their phenotypic effects. One of the oddities of Mendelian genetics is that a gene can only be identified when it is known to exist in at least two alternative states, or as two different alleles. Only then is it possible to observe Mendelian segregation, a 3:1 or 1:2:1 ratio. If all individuals were homozygous for the same allele, the existence of that gene locus could not be discovered by the usual methods of Mendelian genetics, the study of progeny from segregating crosses.

13.4. The Site of Mutations

Apparently, mutations may occur in the cells of any type of tissue, somatic as well as germinal. Somatic mutations, of course, will not be inherited unless propagated asexually in some way, as for example, by cuttings in plants. Because the asexual process is mitotic, all of the cells descended from the mutant somatic cell will be similar to it.

Mutations affecting the germ cells are transmitted to subsequent generations. The dominant mutations will be expressed in the F_1 generation, but a recessive mutation may be carried through a number of generations without expression until it combines with a similar recessive to form a homozygote for the recessive gene.

13.5. Spontaneous Mutations

"Spontaneous" mutations are so named because their exact cause is unknown, and no mutagenic agent has been used to induce them. The natural or background radiation from radioactive minerals and cosmic rays is evidently insufficient to account for all "spontaneous" mutations. However, chemical and thermal effects plus the background radiation undoubtedly play a major role in the induction of naturally occurring or "spontaneous" mutations.

Most spontaneous point mutations are deleterious. A maximum of only 1 in 1000 has been estimated to be beneficial under existing conditions. The reason is that existing genes are the products of prior evolution, giving rise to well-adapted organisms. Since these genes have survived the selective sieve, any change in them is more likely to be harmful than beneficial. Because the genes' primary function seems to be control of the presence and specificity of proteins, particularly enzymes, new mutations, by disrupting the usual metabolic systems, are apt to be harmful. Available estimates of the ratio of detrimental to lethal induced mutations indicate that the ratio is about 4 to 1, but the proportion of detrimentals may actually be higher among spontaneous mutants.

Most new mutations in diploids are recessive, with less than 1 in 100 mutants completely dominant. Therefore, unless special genetic techniques are available with which to detect the mutants, most of them will go unobserved in the F_1 and subsequent generations. Only when the recessive mutants become homozygous will the traits be expressed. Therefore, even though the great majority of mutations are deleterious, an increased mutation rate does not ordinarily lead to an immediate increase in the proportion of defective progeny in the next generation.

13.6. Mutation Rates

Even though mutations are rare, they are recurrent rather than unique events. Because, given a large enough population, the same type of mutation can be observed repeatedly, it is possible to estimate mutation rates. The rate is usually expressed as the number of mutations per locus per gamete per generation although whole chromosome mutation rates are sometimes determined instead. Mutation rates have been found to vary at different loci, and even within a single locus the rates of mutation between different alleles may be quite different. Mutations from the wild-type allele (usually dominant) to other alleles (usually recessive) are generally the most frequent type, but reverse or back mutations have also been detected. When it has been possible to make comparable studies of mutation rates in different tissues, furthermore, the rates have been shown to differ significantly.

The spontaneous rates of mutation at specific loci can be most easily determined by crossing a homozygous multiple recessive stock with a homozygous wild-type stock and examining the F_1 for individuals showing the mutant phenotype for any of the recessive visible mutants. Such tests in *Drosophila* and the mouse have revealed spontaneous specific locus rates of the order of 1 in 250,000 to 1 in 1½ million, with the rates apparently somewhat higher in the mouse. The spontaneous mutation rates in a microorganism such as *Escherichia coli* are also of the same order of magnitude (Table 13.1). This fact poses an interesting question, for the generation length of *E. coli* is measured in minutes, that of *D. melanogaster* in days, that of the mouse in weeks, and that of man in years, yet they all have rather similar mutation rates per generation. If spontaneous muta-

Table 13.1 Spontaneous Mutation Rates at Specific Loci in Different Species

Organism	Trait	Estimated mutation rate
		Per cell per division
Bacteria: *Escherichia coli*	Lactose fermentation	
	$lac^- \rightarrow lac^+$	2×10^{-7}
	$lac^+ \rightarrow lac^-$	2×10^{-6}
	Histidine requirement	
	$his^- \rightarrow his^+$	4×10^{-8}
	$his^+ \rightarrow his^-$	2×10^{-6}
	Streptomycin sensitivity	
	$str\text{-}s \rightarrow str\text{-}d$	1×10^{-9}
	$str\text{-}d \rightarrow str\text{-}s$	1×10^{-8}
	Phage T1 resistance	
	$T1\text{-}s \rightarrow T1\text{-}r$	2×10^{-8}
	Radiation resistance	
	$rad\text{-}s \rightarrow rad\text{-}r$	1×10^{-5}
		Per gamete per generation
Flies: *Drosophila melanogaster*	Yellow body	
	$y^+ \rightarrow y$	12×10^{-5}
	Brown eyes	
	$bw^+ \rightarrow bw$	3×10^{-5}
	Ebony body	
	$e^+ \rightarrow e$	2×10^{-5}
	Eyeless	
	$ey^+ \rightarrow ey$	6×10^{-5}
	Eleven loci	
	$+ \rightarrow$ mutant	6×10^{-7}
Corn: *Zea mays*	Shrunken seeds	
	$sh^+ \rightarrow sh$	1×10^{-6}
	Colorless	
	$c^+ \rightarrow c$	2×10^{-6}
	Sugary	
	$su^+ \rightarrow su$	2×10^{-6}
	Purple	
	$pr^+ \rightarrow pr$	1×10^{-5}

Table 13.1   Spontaneous Mutation Rates at Specific Loci in Different Species (*cont.*)

Organism	Trait	Estimated mutation rate
	Color inhibitor	
	$i^+ \rightarrow i$	1×10^{-4}
Mouse: *Mus musculus*	Seven loci	
	$+ \rightarrow$ mutant	3×10^{-6}
Man: *Homo sapiens*	Hemophilia	
	$h^+ \rightarrow h$	2×10^{-5}
	Albinism	
	$a^+ \rightarrow a$	3×10^{-5}
	Achondroplasia	
	$A^+ \rightarrow A$	5×10^{-5}
	Aniridia	
	$An^+ \rightarrow An$	5×10^{-6}
	Epiloia	
	$Ep^+ \rightarrow Ep$	1×10^{-5}
	Retinoblastoma	
	$R^+ \rightarrow R$	2×10^{-5}
	Huntington's chorea	
	$Hu^+ \rightarrow Hu$	5×10^{-6}

Various sources.

tion rates were completely determined by external environmental factors, the mutation rates in these different species would be expected to be comparable on the basis of absolute time rather than generation length. If this were true, and man's gametes accumulated mutations at the bacterial rate on an absolute time scale, they would carry a far greater load of mutations than they do. Therefore, it seems probable that the mutation rates in these species with such different generation lengths are regulated in some way.

13.7. Mutator Genes

The mutation rate genes or *mutator* genes suggest a mechanism for the regulation of "spontaneous" mutation rates. The dominant Dotted (*Dt*) gene in corn exemplifies a mutator gene. When it is present, the recessive gene a_1 at the A_1 (aleurone color) locus, which is quite stable in the presence of the recessive dt gene, becomes highly mutable, reverting to the dominant A_1. Thus a gene at one locus has a specific effect on the mutability of a gene at an entirely different locus. Other mutator genes, such as hi in *Drosophila melanogaster*, appear to have a more generalized effect, raising mutation rates at a number of different loci. Natural selection, by favoring or eliminating the mutator genes in different species, could then regulate the mutation rates in these species.

13.8. Induced Mutations

In the early years of this century a number of attempts were made to induce mutations with a variety of physical and chemical treatments. The failure to produce clear-cut results was not so much due to failure to induce mutations as to the lack of adequate genetic techniques for detecting a significant increase in the frequency of mutations.

Nevertheless, genetic effects were detected in some cases. Plough, for example, reported in 1917 that crossover percentages in female *D. melanogaster* were higher at low and high temperatures than at intermediate temperatures. The first genetic effects of high-energy radiation were reported by Mavor (1923, 1924, 1925), who found that X-rays caused an increase both in the frequency of nondisjunction and of crossing over.

However, the first clear demonstration of the artificial induction of mutations was that of H. J. Muller, working with the effect of X-rays on *D. melanogaster*. L. J. Stadler, using the slower growing barley plant, reported similar results in 1928. This discovery marked a new era in genetics, since it now became possible to attack the nature of the gene through study of the mutation process.

13.9. The Detection of Mutations

Muller's success largely was due to the ingenious technique he developed for the detection of mutations. The methods used prior to Muller's *ClB* technique were so laborious that it was practically impossible to get adequate data to demonstrate significant differences between treated and control populations.

The *ClB* technique is so simple and the principles involved still so

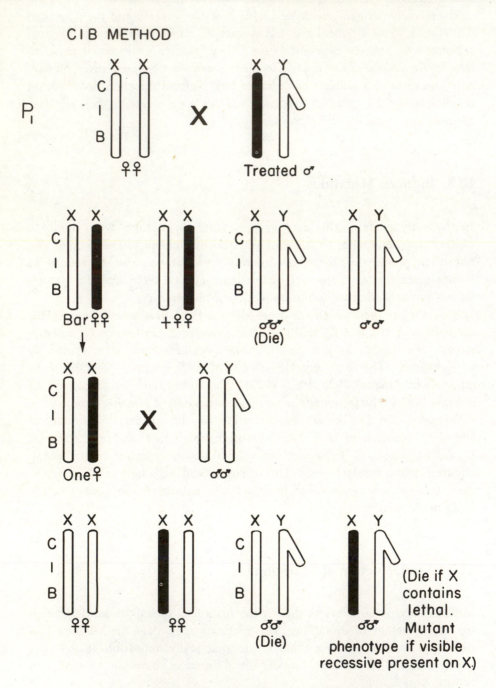

Fig. 13.1. Muller's *ClB* technique for detecting sex-linked lethals. (See text.)

ATTACHED-X METHOD

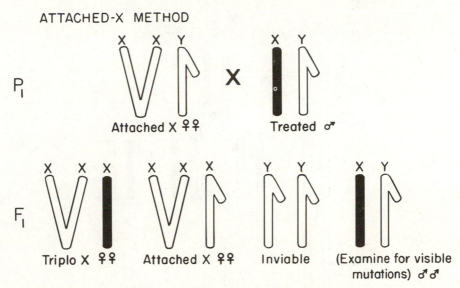

Fig. 13.2. The attached-X method for detecting sex-linked mutations.

widely used that it is worthwhile studying the method (Fig. 13.1). Furthermore, it is of interest to see the simplicity and elegance of the experiments that earned Muller a Nobel Prize in 1946.

In Muller's *ClB* stock, one X chromosome carried *C*, an inversion functioning as a crossover suppressor, *l*, a sex-linked recessive lethal, and *B*, the Bar eye duplication, which acts as a dominant marker for this chromosome. Males were irradiated with X-rays and then mated to females heterozygous for the *ClB* chromosome. The F_1 Bar-eyed females carried the *ClB* chromosome from their mothers and an irradiated X chromosome from their fathers. Each F_1 female was then mated individually, and her progeny reared in separate culture vials. In the F_2, half the sons of these Bar-eyed females died because of the lethal on the *ClB* chromosome. If a lethal was induced on the irradiated X chromosome, the rest of the males also died. Thus, it was possible to score for induced lethals by examining F_2 cultures for the presence or absence of males. No longer was it necessary to examine or count large numbers of flies under the microscope. The absence of males in an F_2 culture vial was evidence of an induced lethal, and possible biases or errors in scoring were minimized.

The method also permits the detection of recessive visible mutations. If a visible mutation is present, all surviving males in the F_2 will show the trait, again making detection easy. Another approach, the attached-X method, is even simpler to use for the detection of sex-linked genes with visible effects, because it requires one less generation (Fig. 13.2).

DETECTION OF AUTOSOMAL RECESSIVES

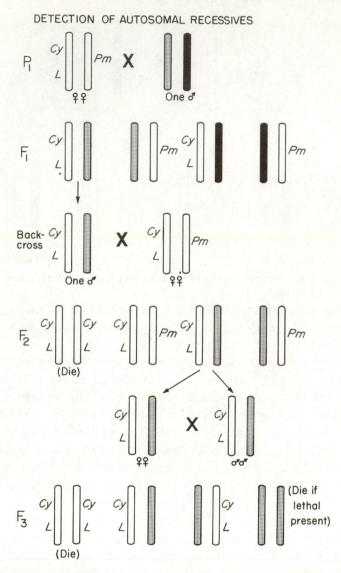

Fig. 13.3. Detection of autosomal recessive mutations on the second chromosome of *Drosophila melanogaster* by use of the *Cy L/Pm* technique. *Cy* = the dominant, Curly wings, associated with inversions. *L* = the dominant, Lobe eyes. *Pm* = Plum eye color, also dominant and also associated with inversions, now known to be allelic to brown (*bw*) and called brown-Variegated (*bw^{V1}*). The inversions, in effect, prevent crossing over, and there is no crossing over in *Drosophila* males so that intact chromosomes can be made homozygous in the F_3 and any recessive genes present anywhere along the length of the chromosome will be expressed.

Exposed males are mated to females whose X chromosomes are attached to one another. The F_1 sons of the cross all receive their single X chromosome from their father and will express any visible mutations induced.

The detection of autosomal recessive lethals is similar in principle to the procedure for detecting sex-linked lethals. It is more laborious, however, because an extra generation is required, and if only new lethal mutations are to be scored, the autosome must first be shown to be free of lethals at the outset of the experiment. The method puts a single autosome in the homozygous condition, thereby revealing any deleterious or lethal recessive mutants (Fig. 13.3).

In the F_3 a ratio of 2 Cy L:1 wild type is expected if no harmful mutations are present. Therefore, F_3 males chosen from a cross in which this ratio is realized will be free of deleterious genes on this chromosome. These males can then be irradiated and a number of the treated chromosomes recovered in the homozygous condition by putting them through the same sequence of crosses with the Cy L/Pm stock. In this way an estimate of the number of mutations resulting from the treatment can be obtained. It must be stressed that in all of these methods, it is essential to run comparable untreated controls in order to estimate the spontaneous mutation rate. Only when such controls are available is it possible to estimate the proportion of mutations actually resulting from the treatment.

13.10. The Biological Effects of Radiation

Following the discovery that X-rays were mutagenic, many other types of ionizing radiation were also shown to be mutagenic, including protons, neutrons, and electrons, and, among the emanations from radioactive substances, α (helium nuclei), β (fast electrons) and γ rays (gamma rays or "natural X-rays").

An atom consists of a positively charged nucleus containing one or more protons (and sometimes neutrons as well) surrounded by one or more "shells" of electrons orbiting around the nucleus. The atom as a whole is electrically neutral. Ionizing radiations passing through matter dissipate their energy by ejecting electrons from the outer electron shells of atoms. Loss of electrons leaves behind positively charged particles or ions that are then capable of chemical reaction. Thus, when the chromosomes are exposed to this type of radiation, the resulting ionizations may lead to both mutations and chromosome breaks. Mutation and chromosome breakage, therefore, appear to be due to chemical reactions requiring energy.

Radiation dosage is measured in roentgens or *r* units. One *r* unit is formally defined as the amount of radiation producing 2.08×10^9 ion pairs per cubic centimeter of air at $0°C$ and 760 mm pressure. A simpler, more easily remembered and more useful equivalent is to recall that one *r* of radiation causes approximately two ionizations (or the formation of two ion pairs) per cubic micron of water or tissue. In tissue the amount of ionization is some 800 times greater than in air.

X-rays and γ-rays are electromagnetic waves similar in character to visible light and ultraviolet light, but they have relatively shorter wavelengths and greater penetrating power in tissue than either visible or ultraviolet light. Ultraviolet light has shorter wavelengths than visible light and is also mutagenic, but it is not an ionizing radiation. Instead, the energy from ultraviolet light causes electrons to move from inner to outer shells in the atoms, leaving them in a state of *excitation* in which they are capable of chemical reaction. However, ultraviolet light has such a low penetrating power in tissue that most of the mutagenic work with ultraviolet has been done with microorganisms or with cells such as pollen grains that are easily exposed to effective doses of ultraviolet. Although ultraviolet does induce chromosome breakage as well as point mutations, it is relatively less effective in breaking chromosomes than the more energetic X-rays or ionizing particles such as protons.

It is of considerable interest that the most effective wavelength of ultraviolet for inducing mutations is about 2600 Å. This wavelength coincides with the peak of maximum energy absorption by DNA. As noted in Chapter 10, the coincidence between maximum mutagenic effect by ultraviolet and maximum energy absorption by DNA suggests that the ultraviolet light is acting directly on the DNA. However, the effects must be indirect in some instances, for the mutation rate in bacteria has been shown to be increased when bacterial cells are cultured on a medium exposed to ultraviolet light prior to the time the cells were plated on it. In this experiment the cells were never directly exposed to the radiation, yet the mutation rate increased.

The effects of radiation on biological systems are frequently discussed under the separate headings of genetic effects on the one hand and somatic or physiological effects on the other. There can be no doubt of the severity of the somatic effects. A dose of 500 *r* of whole-body radiation is lethal for most mammals. Lesser doses may cause skin cancer or leukemia or a breakdown in the intestinal mucosa or the hemopoietic (bloodforming) tissues. Even when no such gross effects have been detected, it has nonetheless been shown in mice that the average life span is significantly shortened by low level doses of radiation. Several of the effects noted above occur in dividing cells rather than resting cells, and cells undergoing mitosis seem to be more sensitive to radiation damage than others. This fact is the rationale for the use of radiation in the treatment of cancer. However, the underlying basis for these somatic effects may well be somatic mutations and chromosome breaks induced by radiation rather than some other type of physiological damage. If this is the case, then the distinction between physiological effects and genetic effects of radiation is rather artificial.

Muller's discovery of the mutagenic action of X-rays in *Drosophila* led to many careful quantitative studies of these effects. From these experiments, some generalizations of considerable interest have emerged.

1. The effect of X-rays on the lethal mutation rate is directly proportional

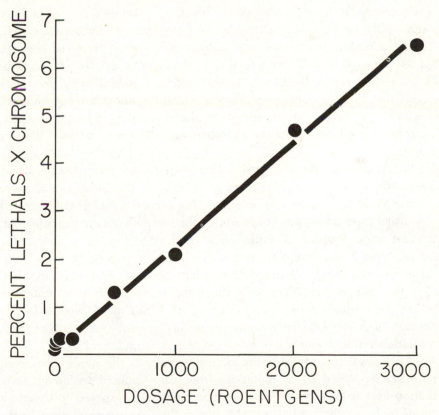

Fig. 13.4. The relation between X-ray dosage and mutation rate in *Drosophila melanogaster*. (Data from Spencer and Stern.)

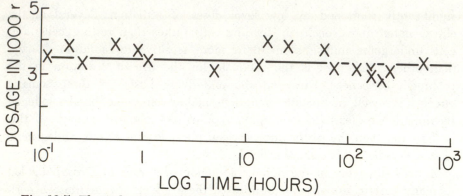

Fig. 13.5. The independence between intensity of dosage with X-rays and mutation frequency in *Drosophila melanogaster*. (Wagner, R. P., and H. K. Mitchell, 1964. *Genetics and Metabolism*, 2nd ed. John Wiley & Sons, New York, p. 189, Fig. 4.12.)

to the dose as measured in *r* units. Therefore, a doubling of the X-ray dosage would be expected to double the percentage of lethals recovered.

This relationship has been found to hold over a wide range of dosages, from as low as 5 and 25 *r* up to several thousand *r* in *Drosophila* (Fig. 13.4). The fact that a linear relationship is found even at very low doses suggests that there is no threshold below which mutations are not induced. Therefore, any increase in exposure to radiation can be expected to cause a corresponding increase in the mutation rate. There is no safe dose of radiation.

At higher dosages the observed data appear to fall away from a linear relationship. However, this result may be due to the induction of more than one lethal in a single chromosome. The second lethal is not detected by the usual experimental methods so that the total effect is underestimated.

2. The same amount of radiation at high intensity in a short time interval, or at a low intensity over a much longer period, or even given in intermittent doses, will induce equivalent numbers of mutations (Fig. 13.5). In other words, 500 *r* will cause the same number of mutations whether it is administered over a period of 5 min or 5 weeks. In the induction of mutations there seems to be no intensity effect. The effect of radiation in the production of mutations thus is cumulative.

W. L. Russell has reported an intensity effect in the mouse, however, for he has recovered fewer mutations from mice exposed to low-intensity radiation over a long period of time than from mice exposed to the same dose of high-intensity radiation. The reason for the dose-rate effect in the mouse is not yet known, but it could be due to a lower probability of

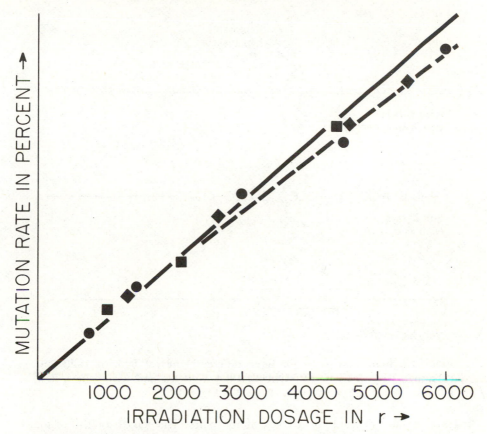

Fig. 13.6. The independence between mutation frequency and type of radiation used. ● = hard X-rays. ◆ = gamma rays. ■ = soft X-rays. (Data from Timoféeff-Ressovsky.)

recovering gametes carrying mutations in the chronically radiated mice rather than to an actual difference in induced mutation rate, or else to DNA repair at the mutant sites.

3. There is no wavelength effect with X-rays, both short and long wavelengths being equally effective at equal dosages (Fig. 13.6). The wavelengths vary from less than 1 to about 10 Å.

One conclusion to be drawn from these facts is that an induced mutation is a unitary event, often referred to as a "single-hit" event. However, this phrase should not be thought to suggest that induced mutations necessarily result from a single ionization. As the incident radiation passes through a cell, its energy is dissipated by the ionizations it causes as it knocks electrons free from the atoms. These electrons may in turn cause secondary ionizations. As a result, the path of the radiation through the cell is

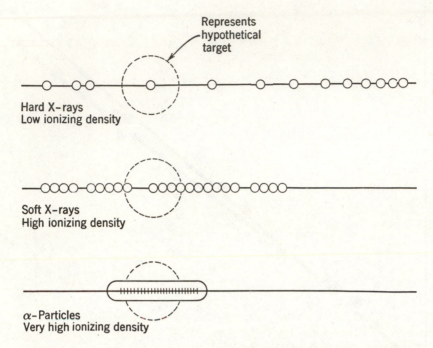

Fig. 13.7. Ionization paths for three different kinds of ionizing radiation. (Wagner, R. P., and H. K. Mitchell, 1964. *Genetics and Metabolism*, 2nd ed. John Wiley & Sons, New York, p. 192, Fig. 4.13.)

marked by an ion track, or a track of ion pairs. The length of the primary ion track, the number of side branches or secondary ionizations, and the density of the ion pairs along the track depend on the type and energy of the radiation (Fig. 13.7). Fast neutrons with high energy, for example, make a rather long track with a uniformly thick density of ion pairs. X-rays, on the other hand, have a relatively short track with few ions at first, but the track becomes moderately dense toward the end. Since the radiations typically produce clusters of ions, it cannot be stated with any degree of certainty that a mutation results from only one ionization. It does seem clear, however, from the linear relation between dosage and mutation frequency, that a single ion track is responsible for a single mutation, and it seems possible that at least some mutations may result from a single ionization.

To summarize the mutagenic effects of radiation, it can be stated that in general the mutation rate is linearly proportional to the dosage and independent of wavelength and intensity. There is no threshold, and the effects are cumulative.

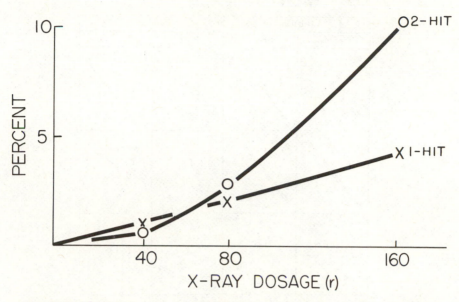

Fig. 13.8. The relation between X-ray dosage and the frequency of one-hit and two-hit chromatid rearrangements. One-hit rearrangements increase in direct proportion to the dosage, but two-hit rearrangements tend to increase as the square of the dosage at relatively high intensities. (Sax, K., 1948. "The Effects of X-rays on Chromosome Structure," *J. Cell. Comp. Physiol.* 35 (Suppl. 1):71–81, p. 74, Fig. 2.)

13.11. Induced Chromosome Breakage

There is also a linear relation between radiation dose in *r* units and the number of chromosome breaks produced (Fig. 13.8). This fact indicates that a single ion track or ion cluster is sufficient to break a chromosome. Some types of radiation are less efficient than others in breaking chromosomes. Fast neutrons, for example, produce fewer breaks per roentgen than X-rays because they produce fewer but larger ion clusters per *r* than the X-rays, and the larger ion clusters apparently exceed the size needed to break the chromosome.

The direct relation between radiation dosage and frequency of chromosome breaks is best demonstrated by the study of "one-hit" chromosome aberrations such as terminal deletions. However, many chromosomal rearrangements such as reciprocal translocations, inversions, intercalary deletions, and duplications require two breaks and therefore are referred to as "two-hit" rearrangements. In the case of the "two-hit" rearrangements there is an intensity effect (Fig. 13.8). The broken ends apparently remain

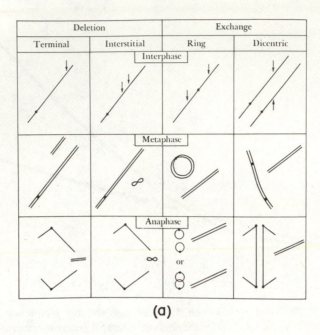

(a)

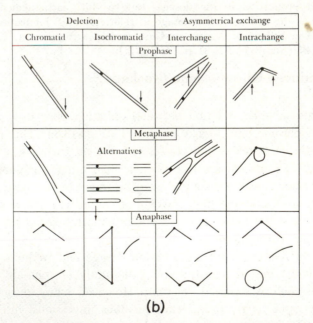

(b)

Fig. 13.9. Radiation-induced rearrangements. (a) Chromosome rearrangements. (b) Chromatid rearrangements. (© 1964 and 1969 by Harcourt Brace Jovanovich Inc., and reproduced with their permission from *Elements of Cytology* by Norman S. Cohn.)

open for only a limited period and then undergo restitution so that the break becomes undetectable. The "two-hit" aberrations induced by X-rays are intensity-dependent because both open breaks must be present simultaneously in the same nucleus before rearrangement is possible. At lower dosages the frequency of "two-break" rearrangements increases approximately as the square of the dosage. At higher doses the increase is still exponential, but at about the ¾ power of the dose, because many hits are wasted in the sense that they do not produce additional detectable rearrangements. In the case of fast neutrons with long, dense ion tracks, a single track may sometimes induce two chromosome breaks. Here too, the frequency of two-hit rearrangements will increase at less than the square of the dose and may approach linearity.

From the above discussion it should be obvious that a distinction must be made between *primary* breaks, the total number of breaks induced, and *recovered* breaks, the ones that are actually detected cytologically. The study of the effects of mutagens on chromosome breakage is of necessity limited to those breaks that lead to rearrangements, because the chromosome breaks that undergo restitution are not ordinarily detectable. A further complication lies in the fact that if the chromosomes have already divided into two chromatids at the time of irradiation, both or only one of the chromatids may be broken. The structural changes resulting from chromatid breaks are somewhat more complex than those from chromosome breaks (Fig. 13.9). The detection and interpretation of the induced rearrangements have been made possible by the skill and ingenuity of cytologists working with the cytogenetic effects of radiation.

The discovery that X-rays not only induced mutations but also broke chromosomes meant that the induced changes were a heterogeneous group that included point mutations and chromosomal rearrangements. Only by refined genetic and cytological analysis was it possible in many cases to distinguish between the two, and Stadler argued that in fact the induced "point mutations" were merely the residue of induced changes for which no cytological basis had been detected. His data on corn also seemed to indicate a difference between spontaneous and X-ray-induced mutations, with a much higher percentage of chromosomal changes among the induced mutations.

13.12. The Target Theory

Very soon after the discovery of the mutagenic effects of X-rays, scientists began to attempt to estimate the size and number of genes in organisms by means of the target theory. In essence the gene was regarded as a target to be hit by the incident radiation. Since it was possible to estimate the number of ionizations produced by the X-rays and to determine the number of mutations, the size of the gene could be calculated on physical grounds with the aid of certain assumptions. The concept of the target as identical with the gene itself was soon abandoned in favor of the concept of a "sensitive volume" as the target within which an ionization had a high probability of causing a mutation. This shift resulted from the realization that the radiation might induce mutations indirectly as the result, for example, of the formation of highly reactive free radicals that were mutagenic, rather than by the direct effect of the radiation on the genetic material itself.

However, the usefulness of the target theory for estimating gene size came into question because the assumptions on which it was based seemed dubious. For one thing, the efficiency of the radiation in inducing mutations was unknown. There was not necessarily a one-to-one relation between ionizations and mutations, for to an unknown extent restitution or simply no effect at all would modify the results. Furthermore, some types of induced mutations might not be detectable by the methods used, and finally, the relative importance of direct and indirect effects of the radiation was difficult to assess yet crucial to the target theory. For these reasons the target theory as a means of estimating gene size lost much of its appeal, and X-ray-induced mutations were thought to be due primarily to the indirect effects of free radicals. More recently, however, the direct action of X-rays on DNA has come to be regarded as the primary mutagenic mechanism. The diffusion radius of the free radicals is quite small, so that "sensitive volume" and gene size cannot be too different. Thus the target theory once again has won some measure of respectability. However, its significance probably lies more in the interest in biological problems it stimulated among physicists than in the actual estimates it produced. Among the physicists were Delbrück and Schrödinger, whose interest in applying physical principles to this problem led more or less directly to the development of molecular genetics. The flowering of molecular genetics in the last three decades has in turn led to a new and far more sophisticated understanding of the nature of mutation than was ever before possible.

13.13. Temperature Effects

Muller also demonstrated that mutation rates in *Drosophila* are dependent on temperature by rearing flies over the entire life cycle at different temperatures within the range at which they normally live. The mutation rates showed a fivefold increase for a 10°C rise in temperature when a correction was made for the shortened life span at the higher temperature. The temperature coefficient (Q_{10}) for most developmental processes or chemical reactions is of the order of 2 or 3, so the temperature coefficient for mutations is somewhat higher.

It has also been found that temperature shocks (exposing organisms to sublethal but abnormally high or low temperatures for limited periods) will raise mutation rates.

Although temperature has been shown to influence spontaneous mutation rates, it has been found that over a wide range of temperatures (10°C to 50°C in barley and 8°C to 34°C in *Drosophila*) there is no effect of temperature on X-ray-induced mutation rates. This fact, when considered in relation to the effect of temperature alone on mutation rates, suggests a difference in the way in which their effects are produced.

13.14. Chemical Mutagens

The induction of mutations by chemicals was attempted many times without success until, during World War II, Auerbach and Robson showed that mustard gas was mutagenic in *Drosophila melanogaster*. The study was initiated because of the similarity in the burns produced in mammals by mustard gas and by irradiation. Although the mutagenic effect was discovered in 1941, the information remained classified until the end of the war and was first published in 1946. Since that time the field of chemical mutagenesis has developed extensively, and a wide variety of compounds has been found to be mutagenic. Among them, the base analogues (e.g., 5-bromouracil, 2-aminopurine), acridine dyes (e.g., proflavine, acridine orange), alkylating agents (e.g., phenol, formaldehyde, ethyl methane sulfonate, and the mustard gases), nitrous acid, and inhibitors of the synthesis of nucleic acid precursors (e.g., caffeine and urethane) are of particular interest. Also worthy of note are hydroxylamine, hydrazine, peroxides, manganese, nitrosoguanidine, and such coal tar derivatives as dibenzanthracene and methylcholanthrene; the latter two are carcinogenic as well as mutagenic. In some cases the mode of action of the chemical mutagens is still not well understood, but study of the effects of chemical mutagens has shed considerable light on the nature of the mutation process.

13.15. The Effects of Combined Treatments

The combined effects of different treatments are of interest for the further light they shed on the mutation process. For example, infrared rays alone do not cause chromosome breakage. However, when used as either a pretreatment or a posttreatment in combination with X-rays, they increase the number of chromosome breaks recovered. It is not known whether this result is due to an increased number of primary breaks or to a reduction in the amount of restitution.

A similar sort of reinforcement or synergistic effect has been found when ultraviolet light and mustard gas are used to induce mutations in *Neurospora*. Even though the amount of mustard gas used is too small to have an appreciable mutagenic action by itself, it causes a significant increase in the number of mutations resulting from a given dose of ultraviolet light.

On the other hand, ordinary white light from the sun or a lamp counteracts the effects of ultraviolet light. If bacteria are spread on an agar plate and exposed to a heavy dose of ultraviolet light, they will all be killed. However, if the plate is subsequently exposed to strong white light, many of the bacterial cells will survive. This phenomenon is called *photoreactivation*. The white light seems to be capable of counteracting both the physiological and genetic effects of the ultraviolet. It is worth noting that white light has no apparent photoreactivating effect on X-ray-induced changes, suggesting that there is a fundamental difference between X-ray-induced mutations and the mutations induced by ultraviolet light.

The oxygen concentration at the time of irradiation with X-rays has been found to influence the number of structural changes induced. In the root-tip cells of *Vicia faba* (the British Broad Bean), the aberration frequency is about three times higher in air than in nitrogen. The crucial difference lies in the oxygen concentration because the frequency of structural changes increases as the oxygen concentration increases. Interestingly enough, although oxygen increases the number of breaks if present during irradiation, it increases the reunion of broken ends of chromosomes if present after irradiation. Apparently, reunion is dependent on active metabolism and protein synthesis. Both breakage and reunion seem to require energy, so these contrasting effects are not too surprising. However, it means that oxygen may have opposing effects on rearrangement frequency.

Concern about the possible harmful effects of exposure to radiation from atomic explosions or other sources has led to a search for ways to counteract radiation damage. The discovery that a lowered oxygen tension reduced the mutagenic effects of X-rays gave hope that radiation protection would be possible. As might be expected, reducing agents and respiratory inhibitors have been found effective, and such compounds as alcohol, BAL (British anti-Lewisite), and sulfhydryl compounds have also been found to reduce the amount of genetic change.

Thus, mutagenic radiations and chemical mutagens act in diverse and complex ways. In some cases the effects are directly upon the chromosomes and genes; in others, the effects are apparently indirect, the agent producing a chemical change that subsequently causes a mutation or rearrangement. The ability to induce such an array of genetic changes has given geneticists a weapon of considerable power in their efforts to study the nature and behavior of genes and chromosomes.

13.16. Mutations in Terms of DNA

With the recognition of DNA as the genetic material, the nature and kinds of mutations could be discussed with a precision not previously possible. Mutations could now be described in terms of changes in DNA rather than as "point mutations" or "chromosomal rearrangements." The

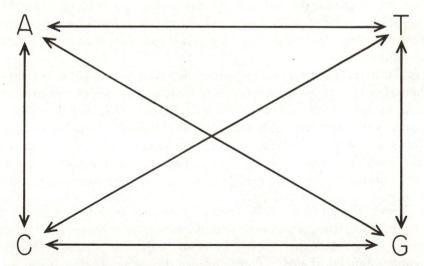

Fig. 13.10. Mutation by base-pair substitution: transitions and transversions. Transitions involve the substitution of a purine for a purine or of a pyrimidine for a pyrimidine. Transversions involve the substitution of a purine for a pyrimidine or vice versa. Diagonals = transitions. Horizontals and verticals = transversions.

possible types of change are *base-pair substitutions, frameshift mutations,* and *macrolesions.*

Base-pair substitutions result from the substitution of one purine or pyrimidine base for another. Freese has called the substitution of a purine for a purine ($A \rightleftharpoons G$) or a pyrimidine for a pyrimidine ($C \rightleftharpoons T$) a *transition*. He terms the substitution of a purine for a pyrimidine or vice versa (A or $G \rightleftharpoons C$ or T) a *transversion* (Fig. 13.10).

Frameshift mutations involve the addition or deletion of one or a few base pairs (see Chapter 12). Addition mutants are identified as (+) mutants and deletion mutants as (−) mutants. The reversion of a (+) mutant is accomplished by means of a (−) mutant and vice versa. There is no obvious gap between a "large" frameshift mutant and a "small" macrolesion, but it is convenient to restrict the term frameshift mutant to those that can be induced to revert by acridine dyes.

Macrolesions consist of deletions, duplications, or rearrangements (inversions, insertions, translocations, and the like) of segments of DNA. A variety of macrolesions has been studied in eukaryotes, but in the prokaryotes, deletions have been more extensively studied than duplications or rearrangements.

Transition mutations are routinely induced by base analogues (5-bromouracil and 2-aminopurine), hydroxylamine, nitrous acid, and alkylating agents. Although transversions occur spontaneously, they are not known to be specifically induced by any chemical mutagen. Frameshift mutations are induced in T4 phage and other systems by the acridine dyes. The mechanisms involved vary depending on the chemical mutagen (Fig. 13.11).

In their initial proposal of DNA structure, Watson and Crick suggested *tautomerism* as a possible mechanism of mutation. In our discussions thus far, only the most likely form for each of the bases in DNA has been shown, but each base can assume different forms as the result of tautomeric shifts of the hydrogen atoms within the molecules (Fig. 13.12). As a result, the usual pairing of adenine with thymine and guanine with cytosine may not occur. For example, the imino ($=NH$) tautomer of A can pair with C rather than T, and the enol ($=\overset{|}{C}-OH$) form of T can pair with G rather than A, and so on. Thus, when a purine or pyrimidine is in its rare tautomeric state, it is able to pair with the "wrong" base, and a "wrong" purine-pyrimidine base pair is formed. If such a pair is formed during replication, it can give rise to a mutation resulting from a base-pair transition. For example, when a "heteroduplex" AC pair replicates, the A will once again pair with T to form an AT pair like that in the original DNA from which it came. The

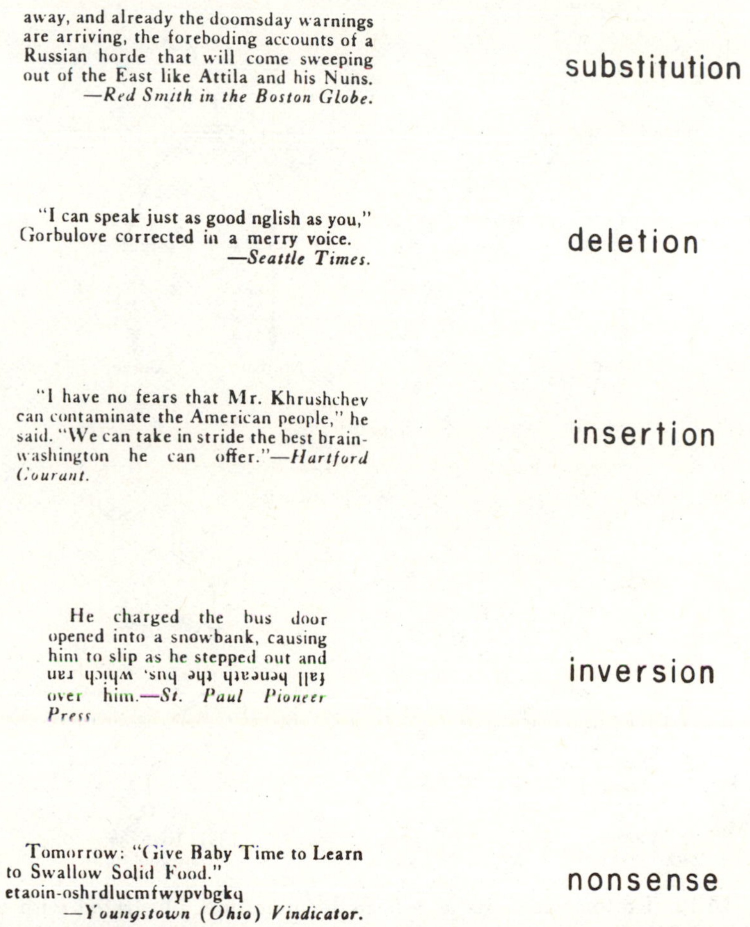

away, and already the doomsday warnings
are arriving, the foreboding accounts of a
Russian horde that will come sweeping
out of the East like Attila and his Nuns.
—*Red Smith in the Boston Globe.*

substitution

"I can speak just as good nglish as you,"
Gorbulove corrected in a merry voice.
—*Seattle Times.*

deletion

"I have no fears that Mr. Khrushchev
can contaminate the American people," he
said. "We can take in stride the best brain-
washington he can offer."—*Hartford
Courant.*

insertion

He charged the bus door
opened into a snowbank, causing
him to slip as he stepped out and
ran beneath the bus, which fell
over him.—*St. Paul Pioneer
Press*

inversion

Tomorrow: "Give Baby Time to Learn
to Swallow Solid Food."
etaoin-oshrdlucmfwypvbgkq
—*Youngstown (Ohio) Vindicator.*

nonsense

**Fig. 13.11. Various types of mutations illustrated by typographical errors.
(From Benzer, S., 1961. "Genetic Fine Structure," *Harvey Lectures*
56:1–21, p. 3, Fig. 1. [Courtesy of New Yorker Magazine.])**

C, on the other hand, will—with a high degree of probability—assume its
usual pairing relationship with G, and thus a GC base pair will now be
found where an AT pair originally existed in the DNA double helix. This
change in the genetic code may then be observed as a mutation.

Tautomeric shifts are thought to play a significant role in spontaneous
mutations. High-energy ionizing radiations are also believed to induce
transition mutations because ionization of a base at the time of duplication
may lead to "incorrect" base-pair formation. In these instances, the mutations can be thought of as originating as the result of *copy errors* during
DNA replication.

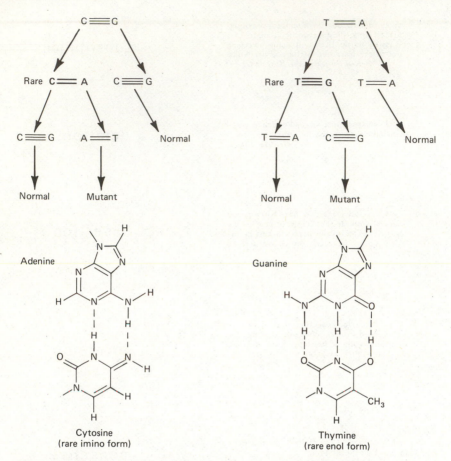

Fig. 13.12. Tautomeric shifts as a basis for mutation, illustrated with cytosine and thymine. Cytosine normally pairs with guanine and thymine with adenine. In their rare states, they pair abnormally and the result in both cases shown is a transition mutation.

Certain chemical mutagens are mutagenic only during DNA replication. The base analogues, 5-bromouracil (5-BU) and 2-aminopurine (2-AP) appear to be mutagens of this type. 5-BU, for instance, is an analogue of thymine and thus can be incorporated in its place during DNA synthesis. However, 5-BU occasionally pairs with guanine rather than adenine. Hence, as shown in Fig. 13.13, the pyrimidine, 5-BU, may give rise to purine transitions in either direction ($A \rightleftharpoons G$). Other base analogues seem to act in comparable ways. For the most part, the mutations induced by base analogues appear to be transition mutations.

One of the interesting discoveries in the work with chemical mutagens was that the mutations induced by base analogues such as 5-bromouracil and 2-aminopurine could be made to revert to wild type by the same

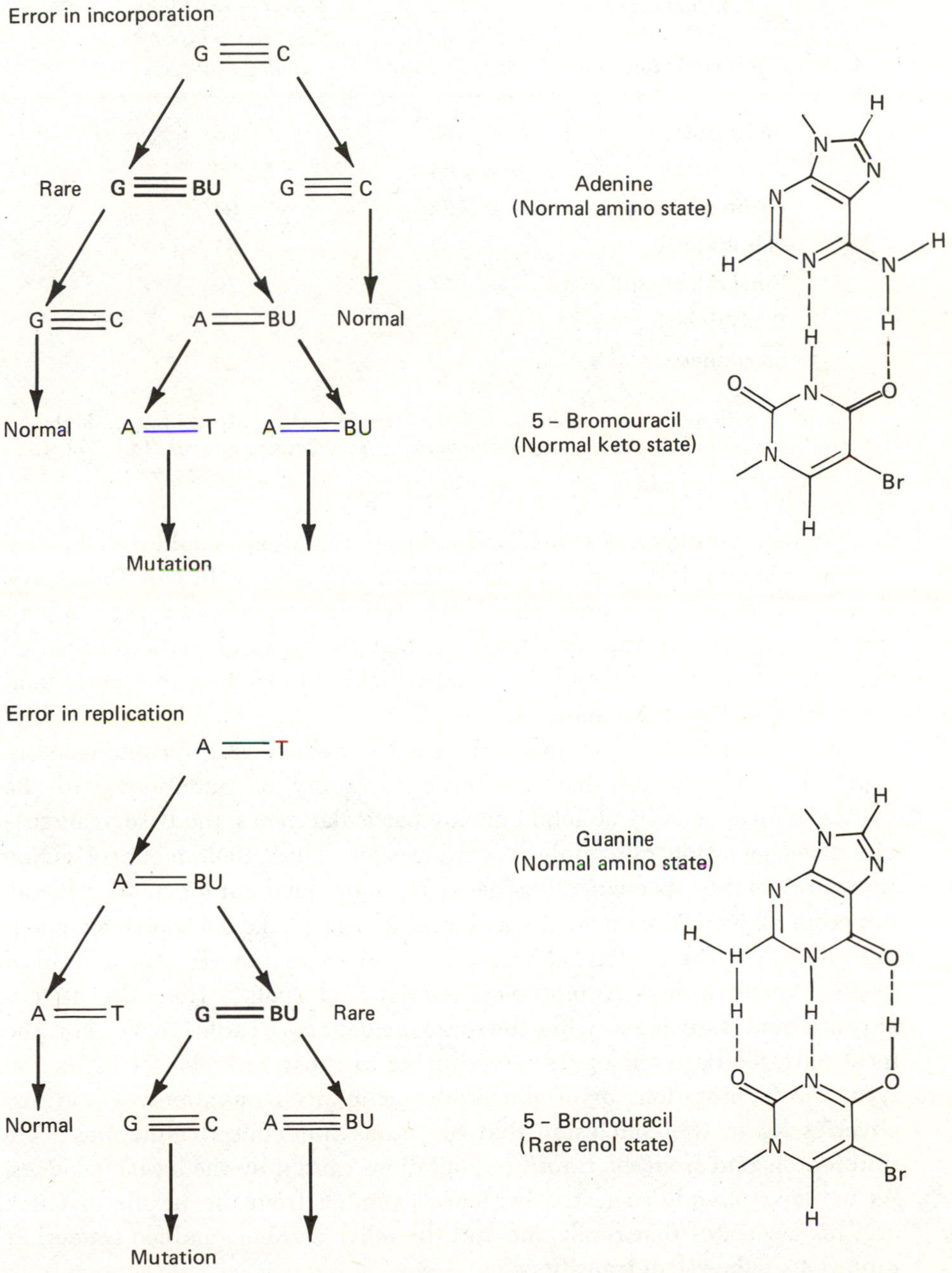

Fig. 13.13. The mode of mutagenic action of 5-bromouracil. See text.

Table 13.2   Induced Reverse Mutation of a Set of Spontaneously Reverting T4 *r*II Mutants

Mutagen used for induction of $r^+ \rightarrow r$II (forward mutation)	No. of rII mutants tested	Percent of rII mutants found inducible to revert to r^+ by base analog mutagen
Aminopurine	98	98
Bromouracil	64	95
Hydroxylamine	36	94
Nitrous acid	47	87
Ethyl ethane sulfonate	47	70
Proflavine	55	2
Spontaneous	110	14

Reprinted with permission from E. Freese, "The Molecular Mechanism of Mutations," *Proc. 5th Internat'l Cong. Biochemistry,* 1961, Pergamon Press Ltd. Vol. I, p. 220, Table 2.

compounds. Similarly, nitrous acid-induced mutations could be induced to revert to wild type by nitrous acid. Furthermore, it was discovered that these mutagens were capable of causing reversion of each other's induced mutants. It appears, therefore, that the types of mutations caused by the base analogues and by nitrous acid are similar in nature and are probably transition mutations.

A further discovery was that, whereas the earlier work on mutagenesis, especially with X-rays, had produced a variety of mutations with no evidence for specificity of action on any particular genes, the base analogues enhanced mutation rates only at certain sites. Thus, their mode of action appeared to be specific, affecting only some loci, rather than general, affecting all loci. Moreover, it was found that in phage T4 the spontaneous mutations and the proflavine-induced mutations in the *r*II region showed quite different groups of mutable sites (or "hot spots") from the pattern of mutations associated with the base analogues (Table 13.2). For the most part, the base analogues were unable to cause reversion of either the spontaneous mutations or of the proflavine-induced mutations. These discoveries led to the conclusion that the proflavine-induced mutations were different in kind from the transition mutations caused by the base analogues. As we have already seen, the evidence is strong from the results of Crick and his associates that proflavine and the other acridines induce frameshift mutations rather than transitions.

Proflavine induces mutations because the acridine dye is intercalated

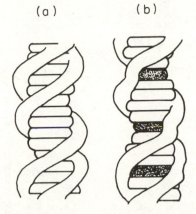

(a)   (b)

Fig. 13.14. Models for the second-ary structure of normal DNA (a) and DNA containing intercalated proflavine molecules (b). (Fishbein, L., W. G. Flamm, and H. L. Falk, 1970. *Chemical Mutagens.* Academic Press, New York, p. 30, Fig. 3.11.)

between adjacent base pairs (Fig. 13.14), which in turn leads to the insertion or deletion of one or more base pairs in the DNA molecule (hence, a shift in the reading frame). The most likely effect of the acridines is to cause errors in the DNA repair processes rather than errors in DNA replication because normal DNA replication is not required for frameshift mutations.

Reversion analysis is complicated by the fact that reversion to the wild type may be due to an intracistronic suppressor mutation close to the original mutation. This is true for both transition mutations and frameshift mutations. However, nearly all base-pair substitutions are only suppressed intracistronically by other base-pair substitutions, and nearly all frameshift mutants are only suppressed intracistronically by other frameshift mutations. The reversion of a plus frameshift mutation, for example, usually involves a nearby suppressor minus mutation, which results in the formation of a polypeptide similar to but not identical with the polypeptide produced by the original wild type. It differs in having one or a few substituted, added, or deleted amino acid residues.

Other evidence is available to show that both spontaneous and induced mutations may arise as changes in existing genes in the absence of DNA synthesis. For example, mutations of this sort have been detected in *Drosophila* sperm stored in the female's sperm receptacles and in static bacterial populations. In both cases, no DNA synthesis is occurring.

Nitrous acid (HNO_2) is a chemical mutagen that is thought to induce mutations by acting directly on existing genes. In this case, the nitrous acid removes amino groups from the bases, substituting keto groups ($=O$). Such deaminations cause base changes (e.g., adenine to hypoxanthine, cytosine to uracil, and guanine to xanthine), which in turn may lead to new base-pair formation and hence to transition mutations (Fig. 13.15).

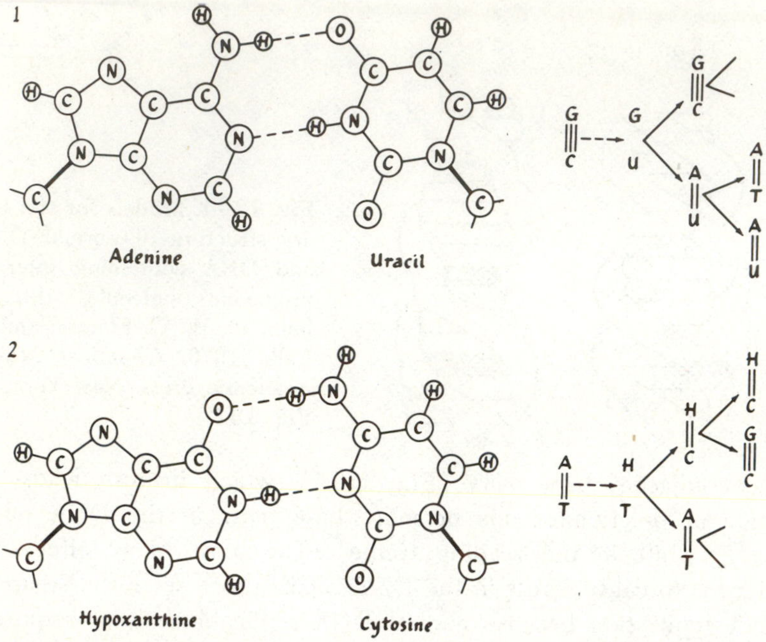

Fig. 13.15. The chemical basis for the mutagenic action of nitrous acid. 1. The 6-amino group of cytosine is deaminated to form uracil. At replication, the uracil pairs with adenine, thus leading to the replacement of the G-C pair by an A-T pair. 2. The 6-amino group of adenine is deaminated to form hypoxanthine, which then pairs with cytosine, and the A-T pair is ultimately replaced by a G-C pair. (Freese, E., 1963. "Moleculer Mechanism of Mutations." In J. H. Taylor, ed., *Molecular Biology, Part I*. Academic Press, New York, p. 226, Fig. 9.)

Careful analysis of bacteriophages treated with HNO_2 indicates that the treatment affects only one of the two DNA strands so that the progeny of a phage carrying an induced mutation will be half mutant and half nonmutant.

Alkylating agents such as nitrogen mustard and ethyl methane sulfonate are also mutagenic when acting on nonreplicating DNA. One of the unusual features of the effect of these agents is the production of delayed mutations. In phage and microorganisms several generations may pass before the mutation becomes manifest. In *Drosophila* the progeny from treated sperm may have mosaic phenotypes, again indicating a delay in the expression of the induced mutation. A completely satisfactory explanation for the phenomenon of delayed mutation has not yet emerged, but it now seems clear that some sort of replicating instability has been induced, which may periodically give rise to an observable mutation.

The process of photoreactivation, by which visible light counteracts the

(a)

(b)

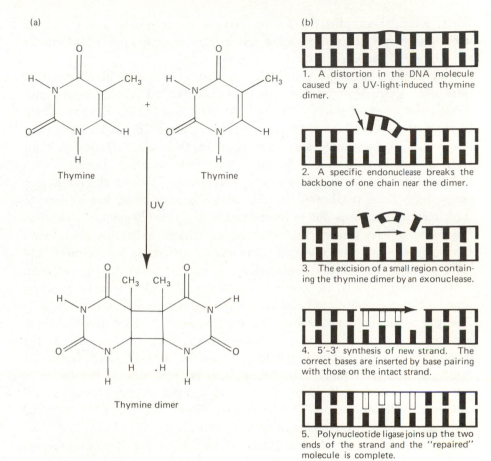

1. A distortion in the DNA molecule caused by a UV-light-induced thymine dimer.

2. A specific endonuclease breaks the backbone of one chain near the dimer.

3. The excision of a small region containing the thymine dimer by an exonuclease.

4. 5'-3' synthesis of new strand. The correct bases are inserted by base pairing with those on the intact strand.

5. Polynucleotide ligase joins up the two ends of the strand and the "repaired" molecule is complete.

Thymine   Thymine

UV

Thymine dimer

Fig. 13.16. The thymine-thymine dimer produced by ultraviolet radiation of DNA. (a) Chemical structure of the dimer. (b) Some enzymatic steps in the repair of DNA molecules containing thymine dimers. (From James D. Watson, *Molecular Biology of the Gene*, Second Edition, copyright © 1970 by J. D. Watson; W. A. Benjamin, Inc., Menlo Park, California.)

lethal effect of ultraviolet light on microorganisms, has already been mentioned. When first reported by Kelner in 1949, this phenomenon could not be satisfactorily explained because of the paucity of knowledge about DNA at that time. Now, however, it has been learned that the effect of ultraviolet is due to the formation of unusual chemical bonds between adjacent pyrimidine bases in one strand of the DNA molecule. Two pyrimidines bonded in this fashion are called dimers. Of the three possible types of pyrimidine dimer, the thymine dimer is the most common (Fig. 13.16). The formation of the pyrimidine dimers blocks the normal duplication of DNA and thus bacteria with even a few dimers are unable to grow. The energy from visible light activates an enzyme bound to the

DNA, which then splits the unusual bonds, returning the bases to their usual state. Thus photoreactivation activates an enzyme system that repairs the damage to the DNA caused by ultraviolet light.

Another type of repair mechanism has been identified in *E. coli* that does not require light and has therefore been called "dark reactivation." A radiation-resistant strain of *E. coli* was first isolated by Evelyn Witkin in 1946. More recently, radiation-sensitive mutants of this strain have been discovered. It has now been demonstrated that the radiation-resistant strain is able to remove the thymine dimers from its DNA, but the radiation-sensitive mutants lack this ability. Furthermore, at least three different genes have been implicated in the repair process. The mechanism of dark reactivation turns out to be different from that for photoreactivation. Experiments have shown that the radiation-resistant strains of bacteria have several enzymes that act sequentially to excise the dimers and replace them with bases complementary to the bases in the adjacent "normal" strand. Thus, the redundancy in the DNA double helix prevents any loss in specificity, and the repair mechanism produces DNA exactly like the original, intact DNA. The evidence for excision resulted from experiments with radioactive thymine. After repair had occurred, the ultraviolet-resistant strains were found to contain small DNA fragments with no more than three bases each. Such fragments were not found in the sensitive bacteria, because all the labeled thymine was associated with the intact DNA molecules. It was further shown that the DNA repaired by dimer excision was capable of normal replication.

DNA repair became of even greater significance when it was discovered that other types of defects in DNA could also be repaired, not only in *E. coli* but in other species as well. Nitrogen mustard apparently reacts primarily with guanine bases in DNA, producing guanine-guanine cross-links between the two strands of the helix. It has been found that the radiation-resistant strains were also able to repair the defects caused by nitrogen mustard and by such diverse mutagens as X-rays, nitrosoguanidine, and mitomycin C. Since the biological effects of these agents are different, it seems probable that the enzymatic DNA repair system does not recognize aberrant bases as such, but rather recognizes the distortions in the phosphodiester backbone of the DNA double helix. In this way a number of different types of defect in DNA can be repaired by a common mechanism.

The idea that the DNA has a built-in repair mechanism that helps to maintain the integrity of the genetic material has opened up a whole new vista in the study of mutation. Many problems still remain to be solved before all the questions related to mutation are answered. In most cases, the nature of the effect of the mutagen is still in doubt. The mechanism of delayed mutation, a fairly widespread phenomenon, is not yet understood. In some instances, such as the rII mutants, the way in which suppressor mutations act seems clear, but this is not the case for all types of suppressor mutations. Mutator genes that raise mutation rates at other loci, highly mutable loci in corn and other species, paramutation and gene conversion, in which mutationlike changes occur in heterozygotes, are all phenomena awaiting adequate explanations. Nevertheless, recent progress in understanding the nature of mutation makes it seem probable that new information will continue to accumulate rapidly.

In conclusion, it may be worthwhile to consider briefly the possible kinds of mutations that can occur. We now know that some of the DNA, the so-called structural DNA, is transcribed into messenger RNA, which codes for polypeptide synthesis. A mutation in this DNA may lead to an altered polypeptide. Chain-initiating and chain-terminating triplets are known; mutations at these sites may lead to aberrant chain-initiation or chain-termination during polypeptide synthesis. Some of the DNA codes for transfer RNA. Mutation in this DNA may lead to defective tRNA molecules. Similarly, mutation in the DNA that codes for ribosomal RNA may lead to aberrations in the ribosomes. The operon concept, involving regulator genes, operator genes, and promoter genes in bacteria, and the complexities of gene regulation in higher organisms, suggest that mutations may also be detected in the DNA responsible for these regulatory functions. Finally, since DNA is now known to occur in mitochondria and plastids, mutations may be expected in this DNA also. Thus, these kinds of mutations may provide tools for more sophisticated investigations into the nature of gene action.

Additional Reading

Altenburg, E., 1934. "The Artificial Production of Mutations by Ultraviolet Light," *Amer. Nat.* **68**:491–507.

Auerbach, C., 1962. *Mutation, an Introduction to Research on Mutagenesis.* Oliver & Boyd, Edinburgh.

Auerbach, C., and J. M. Robson, 1946. "Chemical Production of Mutations," *Nature* **157**:302.

Beadle, G. W., and E. L. Tatum, 1945. "*Neurospora.* II. Methods of Producing and Detecting Mutations Concerned with Nutritional Requirements," *Amer. J. Bot.* **32**:678–686.

Catcheside, D. G., 1948. "Genetic Effects of Radiations," *Adv. Genetics* **2**:271–358.

Davis, B. D., 1948. "Isolation of Biochemically Deficient Mutants of Bacteria by Penicillin," *J. Amer. Chem. Soc.* **70**:4267.

de Vries, H., 1901–1903. *Die Mutation Theorie*. Veit Leipzig. (*The Mutation Theory*, Tr. 1909. Open Court, Chicago.)

Drake, J. W., 1969. "Mutagenic Mechanisms," *Ann Rev. Genetics* **3**:247–268.

Drake, J. W., 1970. *The Molecular Basis of Mutation*. Holden-Day, San Francisco.

Freese, E., 1959. "On the Molecular Explanation of Spontaneous and Induced Mutations," *Brookhaven Symp. Biol.* **12**:63–73.

Freese, E., 1963. "Molecular Mechanism of Mutations." In *Molecular Genetics*, Part I, J. H. Taylor, ed. Academic, New York.

Glass, B., and R. K. Ritterhoff, 1956. "Spontaneous Mutation Rates at Specific Loci in *Drosophila* Males and Females," *Science* **124**:314–315.

Hanawalt, P. C., and R. H. Haynes, 1967. "The Repair of DNA," *Sci. Amer.* **216** (Feb.):36–43.

Hollaender, A., ed., 1954. *Radiation Biology*. McGraw-Hill, New York.

Kelner, A., 1949. "Photoreactivation of Ultraviolet Irradiated *E. coli* with Special Reference to the Dose Reduction Principle and to Ultraviolet-Induced Mutations," *J. Bact.* **58**:511–522.

Lea, D. E., 1955. *Actions of Radiations on Living Cells*, 2nd ed. Cambridge University Press, London.

Lindgren, D., 1972. "The Temperature Influence on the Spontaneous Mutation Rate. I. Literature Review," *Hereditas* **70**:165–178.

Mavor, J. W., 1923. "An Effect of X-rays on Crossingover in *Drosophila*," *Proc. Soc. Exp. Biol. Med.* **20**:335–338.

Mavor, J. W., 1924. "The Production of Nondisjunction by X-rays," *J. Exp. Zool.* **39**:381–432.

Mavor, J. W., 1925. "The Attack on the Gene," *Sci. Monthly* **21**:355–363.

Muller, H. J., 1927. "Artificial Transmutation of the Gene," *Science* **66**:84–87.

Muller, H. J., 1950. "Evidence of the Precision of Genetic Adaptation," *Harvey Lectures* **43**:165–229.

Muller, H. J., and I. I. Oster, 1963. "Some Mutational Techniques in *Drosophila*." In *Methodology in Basic Genetics*, W. J. Burdette, ed. Holden-Day, San Francisco.

Mutation Research, Vols. 1–. 1964–.

Plough, H. H., 1917. "The Effect of Temperature on Crossing-Over in *Drosophila*. *J. Exp. Zool.* **24**:147–208.

Punnett, R. C., 1911. *Mendelism*, 3rd ed. Macmillan, New York.

Rhoades, M. M., 1941. "The Genetic Control of Mutability in Maize," *Cold Spring Harbor Symp. Quant. Biol.* **9**:138–144.

Russell, W. L., 1963. "The Effect of Radiation Dose Rate and Fractionation on Mutation in Mice." In *Repair from Genetic Radiation Damage*, F. H. Sobels, ed. Pergamon, Oxford.

Russell, W. L., 1969. "Observed Mutation Frequency in Mice and the Chain of Processes Affecting It." In *Mutation as a Cellular Process*, G. E. W. Wolstenholme and M. O'Connor, eds. J. & A. Churchill, London.

Sax, K., 1941. "Types and Frequencies of Chromosomal Aberrations Induced by X-rays," *Cold Spring Harbor Symp. Quant. Biol.* **9**:93–101.

Setlow, R. B., and W. L. Carrier, 1964. "The Disappearance of Thymine Dimers from DNA: An Error-Correcting Mechanism," *Proc. Nat. Acad. Sci.* **51**:226–231.

Sobels, F. H., ed., 1963. *Repair from Genetic Radiation Damage.* Pergamon, Oxford.

Sparrow, A. H., 1951. "Radiation Sensitivity of Cells During Mitotic and Meiotic Cycles with Emphasis on Possible Cytochemical Changes," *Ann. N.Y. Acad. Sci.* **51**:1508–1540.

Stadler, L. J., 1928. "Mutations in Barley Induced by X-rays and Radium," *Science* **68**:186–187.

Stadler, L. J., 1954. "The Gene," *Science* **120**:811–819.

Stent, G. S., 1971. *Molecular Genetics.* Freeman, San Francisco.

Stern, C., and E. W. Schaeffer, 1943. "On Wild-type Iso-alleles in *Drosophila melanogaster*," *Proc. Nat. Acad. Sci.* **29**:361–367.

Stone, W. S., O. Wyss, and F. Haas, 1947. "The Production of Mutations in *Staphylococcus aureus* by Irradiation of the Substrate," *Proc. Nat. Acad. Sci.* **33**:59–66.

Swanson, C. P., 1957. *Cytology and Cytogenetics.* Prentice-Hall, Englewood Cliffs, N.J.

Swanson, C. P., and A. Hollaender, 1946. "The Frequency of X-ray Induced Chromatid Breaks in Tradescantia as Modified by Near Infrared Radiation," *Proc. Nat. Acad. Sci.* **32**:295–302.

Thoday, J. M., and J. M. Read, 1947. "Effect of Oxygen on the Frequency of Chromosome Aberrations Produced by X-rays," *Nature* **160**:608.

Timoféeff-Ressovsky, N. W., 1937. *Experimentelle Mutationsforschung in der Vererbungslehre. Beeinflussung der Erbanlagen durch Strahlung und andere Faktoren.* Steinkopff, Leipzig.

Timoféeff-Ressovsky, N. W., K. G. Zimmer, and M. Delbrück, 1935. "Über die Natur der Genmutation und der Genstruktur," *Nach. Ges. Wiss. Göttingen* **1**:189–245.

Witkin, E. M., 1966. "Radiation Induced Mutations and Their Repair," *Science* **152**:1345–1353.

Witkin, E. M., 1969. "Ultraviolet Mutation and DNA Repair," *Ann. Rev. Genetics* **3**:525–552.

Wolff, S., ed., 1963. *Radiation-Induced Chromosome Aberrations.* Columbia University Press, New York.

Wolff, S., 1967. "Radiation Genetics," *Ann. Rev. Genetics* **1**:221–244.

Zamenhof, P. J., 1966. "A Genetic Locus Responsible for Generalized High Mutability in *Escherichia coli*," *Proc. Nat. Acad. Sci.* **56**:845–852.

Zimmer, K. G., 1934. "Ein Beitrag zur Frage nach der Beziehung zwischen Röntgenstrahlendosis und dadurch ausgelöster Mutationsrate," *Strahlentherapie* **51**:179–184.

Questions

13.1. What is meant by an "operational" definition?

13.2. Is it still true that a gene can be identified only if two different alleles exist at that locus? If not, how would you identify a gene represented by only one allele?

13.3.   Why are most mutations deleterious?

13.4.   How are specific locus mutation rates determined in higher organisms such as *Drosophila*, mice, and corn?

13.5.   How are entire chromosomes tested for the presence of mutations? What are the differences between the tests for sex-linked recessive lethals and autosomal recessive lethals? How would you ensure that the lethals detected are new mutations not present at the start of your experiment?

13.6.   What classes of mutagenic agents have been discovered?

13.7.   After irradiation of *Drosophila* with X-rays, from which group would you expect to recover the greater total number of mutations?
   a. 100 males exposed to 50 r/min for 5 min.
   b. 100 males exposed to 2.5 r/min for 10 min on each of 10 successive days.

13.8.   If you examined *Tradescantia* cells for terminal deletions after radiation with X-rays, in which of the following would you expect to find the greatest total number of terminal deletions?
   a. 800 cells exposed to a total of 100 r during a 5-min period.
   b. 400 cells exposed to a total of 200 r during a 5-hr period.
   c. 100 cells exposed to a total of 800 r during a 5-day period.

13.9.   If you examined *Tradescantia* cells for reciprocal translocations and inversions after radiation with X rays, in which of the following would you expect to find the greatest number of translocations and inversions?
   a. 800 cells exposed to 100 r in 5 min.
   c. 800 cells exposed to a total of 100 r over a 5-day period.
   c. 800 cells exposed to a total of 100 r over a 5-day period.

13.10.   If the spontaneous mutation rate $his^+ \to his^-$ is 2×10^{-6} and *str-s* $\to$ *str-d* is 1×10^{-9} per cell per division in *E. coli*, how often would you expect to find spontaneous mutants for both his^- and *str-d* in the same bacterial cell?

13.11.   How would you distinguish in *Neurospora* between a back mutation from arg^- to arg^+ and a suppressor mutation at another locus?

13.12.   There is evidence that a strain of a *Drosophila* species from the tropics has a lower spontaneous mutation rate than a strain of the same species from the temperate region when both strains are reared under the same experimental conditions. How might this difference be explained?

13,13.   a. If an organism has 20,000 genes that mutate on the average at a rate of 1×10^{-6}, what proportion of the gametes will carry a new, probably deleterious, mutation?

b. If the average mutation rate is 1×10^{-5}, what proportion of the gametes will carry a new mutation?

13.14. In Question 13.13, do you think (a) or (b) is more nearly correct? Why? What difficulties do you envision in estimating the total number of genes in an organism? Why are estimates of the average mutation rate apt to be biased? Make other assumptions about gene number and average mutation rate to demonstrate how all-important the assumptions are to the conclusions about the effect of mutation.

13.15. If a single base substitution occurs, which amino acid, methionine or leucine, is more likely to be replaced by another amino acid?

13.16. Is DNA replication essential for mutation to occur? If not, is the mutation process then entirely independent of the process of replication?

13.17. Some base substitutions will not lead to the replacement of one amino acid in the polypeptide by another as a result of the degeneracy of the code. Does this fact necessarily mean that such mutations will be neutral in their effect on fitness? Explain.

13.18. How would you distinguish between a mutation and a phenocopy if a new trait appeared in a population of fruit flies? If it is inherited, how would you determine whether it is dominant or recessive, autosomal or sex-linked? How would you establish which chromosome it is located on? Having determined its chromosomal location, how would you proceed to map it in relation to the other genes on that chromosome?

13.19. How have mutations been classified? What do you consider the most useful classification? Why?

13.20. If transitions and transversions occur at random with respect to one another, what proportion of transversions to transitions would be expected?

13.21. How would you carry out experimental tests for the following in *Drosophila melanogaster?*
a. Sex-linked recessive lethals
b. Sex-linked recessive visibles
c. Recessive lethals on the II chromosome
d. Recessive visibles on the II chromosome
e. Sex-linked dominant lethals
f. Autosomal dominant lethals

13.22. What is your definition of mutation?

CHAPTER FOURTEEN

Recombination

The basis for genetic analysis is recombination. Despite its fundamental importance and the prolonged interest in the subject, recombination continues to be poorly understood. A number of recombination phenomena have been discovered, but the mechanisms underlying these phenomena have not yet been clarified. Mendel established the principles governing the recombination of alleles in his law of segregation, and of unlinked genes in his law of independent assortment. Morgan and his associates established the principles governing the recombination of linked genes in their studies of crossing over. Nevertheless, the forces controlling chromosome movement in mitosis and meiosis remain obscure, and the mechanisms involved in synapsis and crossing over are still not clear at the molecular level. The theories of genetic recombination were first developed to interpret studies of higher organisms, but the more recent work with microorganisms and viruses has added new dimensions to our knowledge. The question that has drawn most attention has been the mechanism by which crossing over occurs.

14.1. Theories of Recombination

The first major theory of recombination was Janssens' chiasmatype hypothesis, proposed in 1909 (Fig. 14.1). Janssens gave the name chiasma (or cross) to the points of contact seen at intervals along the length of the

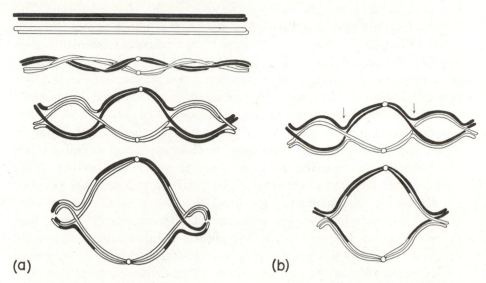

Fig. 14.1. (a) The chiasmatype or one-plane theory of crossing over. Note that the sister chromatids remain paired. (b) The classical or two-plane theory of crossing over. Note that at each chiasma pairing switches from a sister to a nonsister chromatid. See text. (Carl P. Swanson, *Cytology and Cytogenetics*, © 1957, p. 286. Reprinted by permission of Prentice-Hall, Inc., Englewood Cliffs, N. J.)

configuration of homologous chromosomes at diplotene and diakinesis in first meiotic prophase. He proposed that the paired homologous chromosomes each consisted of two sister chromatids and that the chiasmata resulted from the breakage of two nonsister chromatids and their subsequent reunion with an exchange of partners. In good preparations, it was possible to see at each chasma that two of the four strands crossed each other, but the other two did not. He further argued that in late meiotic prophase sister chromatids were always paired.

The so-called "classical" theory presented an alternative version of the breakage and reunion hypothesis. The essence of the classical theory was that at each chiasma there is a change in the chromatids' pairing partners (Fig. 14.1). On one side of the chiasma, sister chromatids would be paired, but on the other side, the nonsister strands would be paired. Thus, no prior breakage and reunion was required to produce the chiasmata, but instead subsequent breakage of the chromatids at the chiasmata and the reunion of homologues gave rise to genetic crossing over.

Both theories represented attempts to correlate the behavior of physical entities, the chromosomes, with the genetic data showing recombination in crosses. After a long and sometimes bitter controversy, the evidence appeared to favor the chiasmatype theory over the classical theory. The picture that emerged was that when homologous chromosomes are synapsed in meiosis, breakage and reunion of two nonsister chromatids gives rise to a chiasma. The two nonsister chromatids involved in each chiasma are random with respect to other chiasmata; that is, the fact that a given chromatid is involved in one exchange neither increases nor decreases its chances of participating in another exchange. In stages of first meiotic prophase subsequent to pachytene, the chromatids separate in the synaptic plane, with sister chromatids remaining in close apposition to one another. At this stage the homologous pairs seem to repel one another, being held together only at the chiasmata. The breakage and exchange between chromatids at the chiasmata are responsible for genetic crossing over, which must therefore involve segments of the chromatids rather than individual genes. Furthermore, at first meiotic anaphase, the maternal and paternal centromeres segregate from one another. Therefore, in meiosis alleles close to the centromere would show first-division segregation while alleles distal to a crossover would show second-division segregation. If chiasmata are in fact the physical basis for genetic crossing over, the frequency of chiasmata and of crossovers should be correlated. Such a relationship has been observed, and the intensity of interference has also been found similar for crossing over and chiasmata.

In 1911 Morgan interpreted his genetic results in terms of Janssens' hypothesis, and genetic evidence compatible with the theory continued to accumulate. There was no cytological evidence for breakage and reunion of chromatids during meiosis although Belling made the possibility somewhat more plausible by suggesting that it occurred during pachytene rather than early diplotene as Janssens had claimed. During the 1930s Darlington became the leading proponent of the chiasmatype theory. He suggested that the tension created by the coiling of the paired chromosomes led to breakage of two nonsister strands at exactly the same level and that reunion of the nonsister strands then occurred. This hypothesis explained interference as well as crossing over because the release of tension in the coiled chromosomes would reduce the probability of a nearby break. He also postulated that chiasma formation was essential for normal meiosis because chromosome segregation was abnormal if chiasmata were absent.

Darlington explained chromosome pairing in meiosis as resulting from a precocious entrance into first meiotic prophase before chromosome replication had occurred. He assumed that unduplicated homologues had an affinity for one another. After duplication, the sister chromatids satisfied

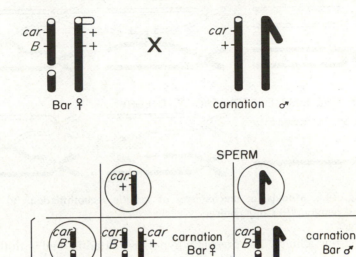

Fig. 14.2. Diagram of the chromosome system used by Stern to show that genetic crossing over involves an exchange of segments between homologous chromosomes. In the female parent, one X chromosome is broken into two pieces; the top segment, with carnation (*car*) and Bar (*B*), has its own centromere, the bottom piece has a centromere translocated from chromosome IV. The other X chromosome in the female parent, with the wild-type alleles of *car* and *B*, is distinguished because an arm derived from the Y chromosome has been added. The noncrossover and crossover chromatids and their genetic constitutions are shown below. When this heterozygous female was crossed to a *car* male with normal chromosomes, four types of female progeny were expected, each with a distinctive phenotype, genotype, and chromosome pattern. Invariably, Stern found that genetic crossing over was accompanied by an exchange of chromosome segments. (Carl P. Swanson, *Cytology and Cytogenetics*, © 1957, p. 107. Reprinted by permission of Prentice-Hall, Inc., Englewood Cliffs, N. J.)

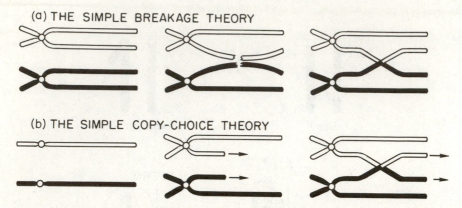

Fig. 14.3. Models for mechanisms of genetic recombination. (a) Breakage and reunion. (b) Copy choice.

this pairing affinity. The precocity hypothesis fitted in well with the chiasmatype theory that sister chromatids always remain paired in first meiotic prophase.

The classic experiments by Creighton and McClintock with corn and by Stern with *Drosophila* in 1931 using "heteromorphic homologues" demonstrated the association between a given gene and a particular chromosome segment (Figs. 9.10 and 14.2). The "heteromorphic homologues" were homologous chromosomes marked by cytologically detectable differences at both ends. They showed that when genetic recombination between marker genes occurred, it was accompanied by an exchange of segments between the homologous chromosomes. These experiments indicated that a given gene could be related to a particular chromosome segment, but they did not necessarily indicate that the recombinant chromosome had been produced by breakage and reunion.

An alternative explanation, now known as "copy choice," was proposed that same year, 1931, by Belling (Fig. 14.3). This theory related recombination to replication. In the original version, Belling visualized chromosome duplication as occurring in two stages: first, the duplication of the chromomeres, the beads on the chromosome thread; and second, the duplication of the thread itself, connecting together the newly formed chromomeres. If, in the second stage, when connections are being formed between chromomeres, the new chromatid being formed along the paternal chromosome switches its connections over to the new strand forming along the maternal chromosome, then the replica of the maternal chromosome, when it reaches the switch point, will be forced to copy the paternal chromosome from that point on. In this way two strands with reciprocal exchanges are formed. The most appealing facet of Belling's hypothesis was that it seemed to account for the precision of the reciprocal exchange between nonsister homologous chromatids. One of the difficulties of the breakage-reunion hypothesis is in finding a satisfactory mechanism to account for breakage of these nonsister chromatids at precisely the same point in each strand. The great weakness of Belling's hypothesis was that, if it were true, crossing over would involve only the two new chromatids, but three- and four-strand double crossovers were already known to occur.

Although both Darlington's breakage and reunion theory and Belling's copy-choice theory postulated that chromosome replication occurs after synapsis in meiosis, they were otherwise quite different. For some time Darlington's theory tended to dominate the thinking of geneticists and cytologists, but recent discoveries in microbial genetics, which have permitted a more molecular approach to the problem of recombination, have reopened the whole question. Some of the data clearly favor breakage and reunion of existing nucleotide chains, other data suggest a copy-choice mechanism, and some data seem to fit none of the theories. At present no one theory fits all the available data. One of the problems, in fact, is whether one theory can be expected to fit all the data because it is not known to what extent the behavior of the "chromosomes" of bacteria and viruses, which consist of a single DNA double helix, can be homologized with the much larger and more complex chromosomes of higher organisms. The assumption is generally made that these structures are fundamentally similar in nature, but more must be learned about chromosome structure before too much confidence can be placed in this assumption.

14.2. Recombination in Eukaryotes

Tests for Recombination

In order to establish that two genes are members of the same set of multiple alleles, it is necessary to study their behavior in crosses. The members of an allelic series will all affect the same trait. When two allelic mutants are crossed, the F_1 will be like one mutant or the other, or else more or less intermediate. In the F_2, either two classes of progeny appear in a 3:1 ratio or, if the F_1 was intermediate to the parents, in a 1:2:1 ratio. In either case, the more frequent F_2 class will resemble the F_1.

On the other hand, when two nonallelic recessive mutants are crossed, even though they may affect the same character, the F_1 will ordinarily have the normal wild-type phenotype because each strain carries the normal dominant allele of the other's mutation, and thus complements the other's deficiency. Furthermore, in the F_2, recombination will give rise to four classes of progeny, including some double-mutant and some wild-type individuals as well as single-mutant types like the original parents. If the mutants are not linked, a 9:3:3:1 ratio will be obtained. With linkage, the numbers of recombinants will depend on the amount of crossing over between the two loci.

Allelic genes do not recombine with one another in crosses, and they show the same crossover frequency with other loci on the same chromosome pair. Thus, they map at the same position in the linkage group. Nonallelic genes, on the other hand, recombine with one another and can be shown to occupy different loci either in the same or a different chromosome.

In this way a number of series of multiple alleles have been identified in many different species, both plant and animal. Classical examples are the A locus (agouti) and the C locus (albinism) in mammals such as the mouse and the rabbit, the A_1 locus (anthocyanin pigment) and the R locus (plant and aleurone color) in maize, and the white locus in *Drosophila melanogaster*. These series lent support to the concept of the gene as a single entity; the functional unit, the one undergoing recombination, and the unit of mutation were thought to be one and the same. An analogy was drawn between the genes on the chromosome and a series of beads on a string.

Position Effect

The first chink in the armor of the unitary gene concept was the discovery of position effect by Sturtevant in 1925. The trait called Bar eye in *Drosophila melanogaster* behaved as if it were due to a sex-linked gene with incomplete dominance. Male hemizygotes and female homozygotes have narrow eyes with fewer facets than the normal round eye while female heterozygotes have eyes shaped like lima beans (Fig. 14.4). The Bar gene appeared to mutate at an unusually high frequency, reverting to wild type at the rate of about 1 in 1600 gametes. It also mutated to a more extreme form of Bar, called Double-bar, at a lower rate, about 1 in 28,000 gametes. Double-bar caused very narrow, slitlike eyes, and seemed to have a similar high mutability.

However, these apparent mutations occurred only in the females and were soon shown to be related to crossing over. When the recessive mutants, forked bristles and fused wing veins, were present in the heterozygous condition in the females as outside marker genes close to and on

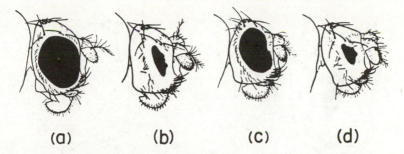

Fig. 14.4. Bar eye in *Drosophila melanogaster*. (a) Homozygous wild-type (+/+) female. (b) Homozygous Bar (*B/B*) female. (c) Heterozygous Bar (*B/+*) female. (d) Double-Bar (*BB/Y*) male. (Sturtevant, A. H., 1925. "The effects of unequal crossing over at the Bar locus in *Drosophila*," *Genetics* 10:117–147, p. 119, Plate 1.)

either side of the Bar locus, it was found that whenever a reversion from Bar to wild type occurred, there was also crossing over in the short distance (2.7 map units) between the marker genes, forked and fused. Crossing over was also associated with the origin of Double-bar. A complete explanation of this situation did not appear until 1933, when Painter and Heitz and Bauer rediscovered the salivary gland chromosomes, first described

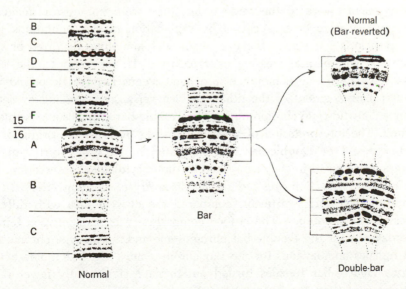

Fig. 14.5. The Bar region in the salivary gland chromosomes of *Drosophila melanogaster*. (Bridges, C. B., 1936. "The Bar 'Gene' a Duplication," *Science* 83:210–211, p. 210, Fig. 1.)

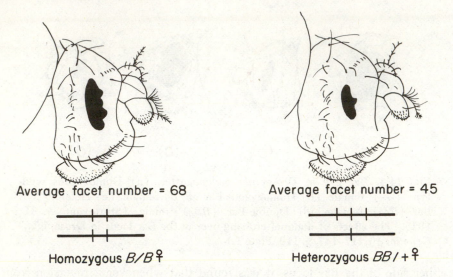

Fig. 14.6. Position effect at the Bar locus. Comparison of B/B and $BB/+$ females. See text.

by Balbiani in Diptera in 1881. Bar was shown to be, not a simple sex-linked dominant gene, but a duplication, and Double-bar a triplication, of a small chromosome segment. Thus, flies with normal eyes had one such segment, Bar flies had two, and Double-bar, three segments (Fig. 14.5). The segments were in the same order; they were not mirror images of one another or reverse repeats. The explanation of the "mutations" was that in meiosis synapsis in the Bar region was sometimes distorted or staggered so that unequal crossing over occurred. If, for example, the right-hand segment of one homologue in a homozygous Bar female paired with the left-hand segment of the other, or vice versa, and crossing over then occurred, a wild-type chromosome and a Double-bar chromosome would be formed. The interpretation of unequal crossing over for the origin of the wild-type and of Double-bar first suggested in 1923 by Sturtevant and Morgan was thus ultimately confirmed by direct cytological observation.

The work with Bar eyes led to the discovery of position effect. In the course of this work Sturtevant compared the eyes of flies with different kinds of Bar genotypes. The crucial comparison was that between females heterozygous for the Double-bar chromosome and a wild-type chromosome and females homozygous for the Bar chromosome (Fig. 14.6). The heterozygous Double-bar females turned out to have significantly fewer facets in their eyes than the homozygous Bar females. The importance of this observation lay in the fact that both types of females had four of the segments in the Bar region, the segments were genetically equivalent as indicated by their interconvertibility when crossing over occurred, and yet they differed in phenotype. Since there was no mutation and no gain or loss in the amount of genetic material, the only possible conclusion was that the phenotypic difference was due to the different arrangement of the hereditary material—a "position effect."

Such a conclusion undermined the unitary concept of the gene: It indicated that the gene was not a single entity operating independently of the rest of the genotype. Instead, the results suggested either that adjacent genes could influence one another's activity directly, or else that their immediate products interact to produce the observed phenotypic effects. This original case of position effect was complicated by the fact that chromosomal rearrangements were present. Numerous later examples of position effect have now been reported in which chromosome duplications and triplications are not involved.

Pseudoalleles

The first such case was reported in 1940 by Oliver, who was working with two recessive alleles, glossy (lz^g) and spectacle (lz^s), at the sex-linked lozenge eye locus in *Drosophila melanogaster*. As would be expected, heterozygous lz^g/lz^s females have a mutant phenotype. Among the progeny of these heterozygous females mated to mutant males of either type, however, Oliver found about 0.2% with normal, wild-type eyes. If the genes were truly allelic, only mutant progeny should emerge from the cross. Moreover, when marker genes bracketed the lozenge region, the reversions to wild type were always shown to be associated with crossing over in the lozenge region. Only one of the crossover types was recovered, however; the complementary crossover type was not found. Furthermore, in studies of the salivary gland chromosomes no indication of unequal crossing over was observed similar to that seen in Bar eye.

A similar case of crossing over between seeming alleles, involving the dominant Star and a recessive allele (later called asteroid), was soon reported by E. B. Lewis. In this case, both the wild-type reversion (++) and the double-mutant Star-asteroid (*S ast*) type were recovered. Since the two complementary crossover classes were recovered in this instance, normal crossing over appeared to be involved. What was abnormal was that it appeared to take place between alleles. In addition, it was learned that when the genes were in the "trans" configuration (*S +/+ ast*), the eyes were more abnormal than when the genes were in the "cis" configuration (*S ast/++*). Thus again a position effect was found. The phenomenon soon came to be called *position pseudoallelism*. Additional examples of pseudoallelism were soon found in *Drosophila*, in the fungi *Neurospora* and *Aspergillus*, in maize, and later in bacteria and bacteriophages. Such classical examples of multiple alleles as the white locus in *Drosophila* and the A and R loci in maize turned out to be pseudoallelic. In fact, it began to seem that whenever a refined genetic analysis at a locus was possible, pseudoallelism was revealed. Thus, the phenomenon was widespread rather than unique to *Drosophila*.

Two explanations for these results were suggested. Lewis hypothesized that the pseudoalleles represented neighboring genes concerned with some common function. He further suggested that these genes may have originated from a duplication followed by divergence in function due to mutation. Pontecorvo, on the other hand, regarded these alleles as belonging to a single functional unit or gene. He felt that they were changes at different sites within the functional units and that intragenic crossing over took place between these sites. He saw no need to invoke duplication to explain the origin of pseudoalleles.

Complementation

Study of linked mutants in both the cis configuration ($a\ b/{+}{+}$) and the trans configuration ($a{+}/{+}b$) gave a functional test whereby it could be determined whether the mutants belonged to the same or a different functional unit (Fig. 14.7). Benzer called the functional unit defined by the cis-trans test a *cistron*. If both the cis and the trans arrangements of two recessive mutants have a wild phenotype, the mutants complement one another and belong to two different functional units or cistrons. On the other hand, in the case of white and apricot at the white locus, the trans

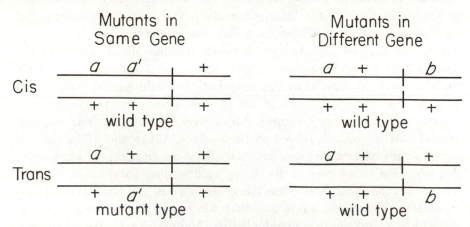

Fig. 14.7. The cis-trans complementation test for functional allelism. See text.

heterozygote ($w+/+apr$) has a mutant phenotype with light apricot eyes, but the cis heterozygote ($w\ apr/++$) has wild-type red eyes. Therefore, white and apricot behave as functional alleles and are members of the same cistron because they do not complement one another. However, they are not structural alleles, because recombination between them occurs. They must be very close together, because even with inversions on other chromosomes causing a sixfold increase in crossovers on the X chromosome, only 12/40,000 wild-type recombinants were recovered. The discovery of pseudoallelism caused still further difficulties for the simple gene concept as the unit of function, mutation, and recombination. Here, clearly, the functional unit is larger than the unit of recombination. Since this is so, a problem of definition arises. Lewis' solution was to use the word "gene" in relation to the unit of recombination. Pontecorvo defined the gene as the functional unit. Subsequent discoveries in microbial genetics led Benzer to coin the words, *muton, cistron*, and *recon* to describe the different entities being revealed by refined genetic analysis. As knowledge of the nature of the genetic material has increased, the semantic problems have tended to diminish. Mutation, recombination, and function can now be described in terms of DNA, and the idea that recombination can occur within the functional unit that controls synthesis of a polypeptide or that mutation may occur at any one of a number of sites within this functional unit is now well established.

The explanation for the position effect observed with pseudoalleles in cis and trans configuration appears in essence to be that in trans ($a+/+b$) neither DNA molecule can mediate transcription of normal messenger RNA, but in cis ($++/ab$) the ++ strand permits transcription of wild-type messenger RNA.

Just as the phenomenon of crossing over became the basis for genetic mapping, the phenomenon of complementation became the basis for complementation mapping based on a functional test. The genetic maps, of course, turned out to be linear. For the most part, complementation maps also turned out to be linear, but enough significant exceptions have been found to indicate that there is not strict colinearity between the genetic recombination map and the functional complementation map. The complexities of complementation studies were brought out in the work of Fincham with *Neurospora*, in which interallelic or intracistronic complementation was discovered, and of Carlson with the dumpy-wing series in *Drosophila*, in which there was no relation between complementation and the distance between the mutants on the genetic map. For this reason complementation mapping as an independent method of ordering the genetic material on a functional basis rather than by genetic recombina-

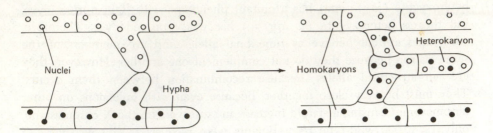

Fig. 14.8. The formation of heterokaryons in *Neurospora*. See text.

tion has not lived up to the hopes originally expressed for it. However, complementation studies may yet prove fruitful in making clear how gene products interact while carrying on their functions during development and differentiation.

In most species, complementation studies are conducted with diploid heterozygotes. In *Neurospora*, which has been used for a number of investigations of complementation, *heterokaryons* rather than heterozygotes have been studied (Fig. 14.8). Vegetative growth in *Neurospora* from a single ascospore gives rise to a mycelium composed of many filaments or hyphae. All the nuclei in such a mycelium will have identical genotypes. The cells in the mycelium of *Neurospora* are typically coenocytic or multinucleate so that a number of nuclei are found in a common cytoplasm. If mycelia of different genetic strains of *Neurospora* are grown in close proximity to one another, vegetative fusion may occur, with the result that nuclei of different genotypes come to reside within a common cytoplasm. Thus, in a heterokaryon, different alleles are located in separate nuclei but the same cytoplasm; in heterozygotes, the different alleles are contained within the same nucleus and cytoplasm.

Complementation tests with *Neurospora* heterokaryons have been conducted most often with nutritional mutants. A strain that requires *p*-aminobenzoic acid will not grow on minimal medium. Similarly, a strain that requires nicotinic acid will also not grow on minimal medium. Each type was shown by crosses to differ from the normal by a single gene. However, when such strains having the same mating type were grown together, cell fusion gave rise to a dikaryotic mycelium, and this heterokaryon grew normally on minimal medium. By isolating hyphal tips from the heterokaryon, it was shown that no nuclear fusion had occurred, for some of the progeny required *p*-aminobenzoic acid and others required nicotinic acid. Thus, in the heterokaryon, each kind of haploid nucleus complemented the deficiency of the other by supplying the metabolite the other was unable to synthesize.

The heterokaryon method has also been useful in tests for allelism. For instance, when a number of different arginine-requiring mutant strains were isolated, they were tested for complementation in heterokaryons. When this was done, some were found to be complementary, presumably because of nonallelic mutations at different loci, but others were not complementary, presumably because of allelic mutations at a single locus. These mutants were also tested through crosses, and the results from the heterokaryon tests and the crosses agreed. However, it should be noted that such phenomena as interallelic complementation, mentioned above, have led to some caution in the interpretation of the heterokaryon tests. Thus, complementation tests can be run either in heterozygotes or in heterokaryons. Only if the mutants are recessive can meaningful complementation tests be performed.

14.3. Recombination in Prokaryotes

Variation in Bacteria

The study of recombination received new impetus when genetic recombination was discovered in bacteria. The rise of the science of bacteriology in the latter part of the nineteenth century was marked by the isolation and description of bacterial species by such men as Pasteur, Koch, and Cohn. In general the species were thought to be monomorphic. From about 1900 to 1940, attention turned to the identification and description of bacterial variation in morphology, virulence, and biochemical and antigen properties. In the 1940s, efforts were begun to analyze heredity in bacteria in terms of classical genetic theory. It was soon discovered, however, that phenomena existed in the genetics of bacteria that had no counterparts in the genetics of higher organisms, and that could not be explained by classical genetic theory.

Among the types of bacterial variation, three kinds were most often used in the study of heredity in bacteria: resistance to viruses or drugs or antibiotics (e.g., streptomycin), ability to utilize various carbon sources (e.g., lactose), and ability to synthesize essential metabolites (e.g., tryptophan).

A major controversy developed over the origin of the hereditary variations in bacteria. One explanation was that the variations arose as a result of spontaneous mutations. Another was that they were adaptations induced by the changed environmental conditions, a belief that represented one of the last strongholds of Lamarckianism. The first convincing evidence that bacterial variation originated by spontaneous mutation was presented in 1943 in the "fluctuation test" of Luria and Delbrück. In their experiments they studied the development of resistance by the bacterium, *E. coli,* to a bacterial virus. If a number of similar culture tubes, each inoculated with a few bacteria, are all treated in the same way, the two theories lead to quite different expectations. If adaptation is the explanation, then each bacterial cell in every culture has the same chance of becoming resistant to the bacteriophage, and the numbers of adapted cells should be about the same in every culture. However, spontaneous mutations are rare, random events. If a mutation to phage resistance occurs early in the growth of the *E. coli* inoculum, there will be many resistant cells in the culture. If the mutation to resistance occurs late in the growth of the culture, there will be relatively few resistant bacteria in the culture. Instead of the homogeneity from culture to culture indicative of adaptation to the virus, Luria and Delbrück found wide fluctuations in the numbers of phage-resistant cells from one culture to another, strong evidence in support of the theory of spontaneous mutation.

This statistical approach to the problem was regarded with a somewhat jaundiced eye by some biologists, but the replica-plating experiments of the Lederbergs satisfied even the most hardened skeptics (Fig. 14.9). The essence of their technique was to introduce wild-type *E. coli* into a complete culture medium containing a broth rich in various organic molecules. These cells were then diluted and cultured on plates containing agar and the same complete medium. The dilution was such that the colonies on the agar plates arose from single cells. Members from each colony were then transferred to a different plate containing a minimal medium. The transfer was effected by means of a disk covered with velvet that was pressed first on the master plate and then on the replica plate with minimal medium. When the replica plate was incubated to permit the growth of visible colonies, these colonies formed a replica of the master plate in the position and identity of the colonies formed. Occasionally, however, one of the colonies on the master plate failed to grow on the minimal medium of the replica, and was therefore nutritionally deficient compared to the wild type. It was then possible to isolate cells from the colony on the master plate and identify the nature of the nutritional defect.

A modification of this technique involves the use of a selective medium. In this case the medium contains a particular virus, or antibiotic, or carbon source, so that only adapted individuals capable of growth in the presence of the selective agent will be able to form colonies. The replica-plating technique is then used to transfer cells from the colonies on the master plate to the plate with the selective medium, say, for example,

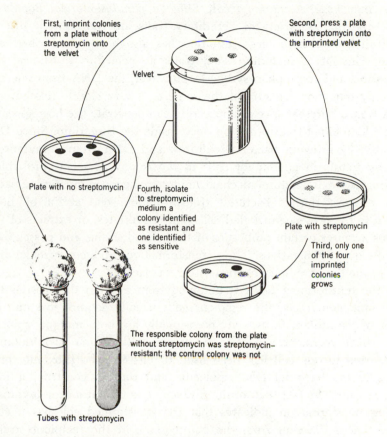

First, imprint colonies from a plate without streptomycin onto the velvet

Second, press a plate with streptomycin onto the imprinted velvet

Velvet

Plate with no streptomycin

Fourth, isolate to streptomycin medium a colony identified as resistant and one identified as sensitive

Plate with streptomycin

Third, only one of the four imprinted colonies grows

The responsible colony from the plate without streptomycin was streptomycin-resistant; the control colony was not

Tubes with streptomycin

Fig. 14.9. The replica-plating technique for the indirect selection of mutants. See text. (Sager, R., and F. J. Ryan, 1961. *Cell Heredity*, John Wiley & Sons, New York, p. 52, Fig. 2.7.)

streptomycin. It is then possible to return to the original master plate and remove cells from the parent colony of a streptomycin-resistant colony. From this colony a streptomycin-resistant strain of bacteria can be isolated that has never been exposed to streptomycin. Obviously, the resistance is not an adaptive response to the stress of exposure to streptomycin. This technique of indirect selection of mutations has been repeated for so many different cases that it seems reasonable to state as a general rule that heritable variations in bacteria ordinarily originate by mutation.

Transformation in Bacteria

The phenomenon of transformation in bacteria, discussed earlier, has no counterpart as yet in higher organisms. Experiments have shown that transformation occurs in bacterial species other than pneumococcus (e.g., *Hemo-*

*philus influenzae, Escherichia coli, Shigella paradysenteriae, Bacillus sub-
tilis*), and that a variety of hereditary traits are transmitted. Just about
any trait known to undergo mutation has also been transformed by the
appropriate DNA, including such traits as penicillin resistance, lactose
utilization, and tryptophan synthesis. However, the DNA from one species
can only transform bacteria of the same or a very closely related species.
DNA from a different species apparently can penetrate the host species, but
is unable to effect transformation once inside the bacterium. Since DNA is
liberated by growing bacterial cultures, and DNA from a mutant could
thereby transform nonmutant cells in the colony, it seems probable that
transformation is a natural mechanism for genetic recombination in bacteria.

In essence, during bacterial transformation, one strain of bacteria,
grown in the presence of killed cells, culture filtrates, or extracts of related
strains acquires certain properties of the related strains and passes them on
to subsequent generations. Thus, transformation produces specific, directed
changes in the hereditary characteristics of bacteria.

If the transforming DNA is obtained from a strain that carries two or
more mutant markers, the recipient cells usually acquire just one or the
other of the mutant traits; only occasionally are both mutants transmitted
to a single recipient cell. The frequency with which two mutants are
transferred to one cell has been used to measure the proximity of these
genes in the bacterial DNA molecule and thus to construct a form of
linkage map. The fact that, ordinarily, only one trait at a time is transferred
during transformation indicates that the transformed cell has a chromosome that is different from the chromosome of the recipient strain and
also from that of the donor strain. Therefore, the phenomenon is truly
one of genetic recombination and not the total substitution of one bacterial chromosome for another.

The mechanism of transformation is not yet completely understood.
However, in order to undergo transformation, the bacterial cells must be
in a state of physiological competence, a transient phase in growing
bacterial cultures. Competence has at least two components, one related
to the ability of the DNA fragment to penetrate the recipient cell, the
other related to the ability of the fragment to become integrated or incorporated into the chromosome of the host and thus complete the process.
The evidence is now reasonably clear that the double-stranded transforming
DNA, consisting of relatively small fragments of the DNA from the donor,
is taken up by the recipient cell and incorporated into its genome. There
is no evidence that the process involves replication of either donor or recipient DNA prior to or during transformation. The experiments of Griffith
and those of Avery, MacLeod, and McCarty were of fundamental significance

both in showing for the first time that DNA is the genetic material and also in revealing a new and unexpected method of genetic recombination.

Bacterial Conjugation

In 1946, shortly after transformation was reported, another mechanism of genetic recombination in bacteria was discovered by Lederberg and Tatum. They found that bacterial conjugation occurred in the K12 strain of *E. coli,* during which genetic material was transferred from one bacterial cell to another through a conjugation tube. Because, in order to study heredity in bacteria, it is essential to be able to effect crosses between different strains, sexual phenomena had long been sought in bacteria without success. Thus, Lederberg and Tatum's discovery opened the way for meaningful studies of the genetics of bacteria.

In designing their experiments, they assumed that recombination in bacteria is a rare event and that, in order to detect it, they needed some sort of selective process to winnow out the recombinants. Furthermore, it was essential that recombinants should not be confused with spontaneous mutants.

Wild-type *E. coli* strain K12 grows on a simple minimal synthetic medium and does not require any additional growth factors. The original experiments involved biochemically deficient mutants of *E. coli.* By successive mutagenic treatments with X-rays, two multiple mutant strains were obtained, one (A) having growth requirements for methionine and biotin, and the other (B) for threonine, leucine, and thiamine (vitamin B_1). If either strain A or strain B were grown on complete medium, and 10^8 washed cells of one or the other were then plated on minimal medium, no growth appeared. Spontaneous back mutations of each mutant do occur at a low frequency—less than 1 in 10^6—but the chances that two or three independent spontaneous mutations back to wild type would all occur in the same cell are so very low that no wild-type colonies could reasonably be expected.

However, if mixtures of cells from strains A and B were plated on minimal medium, a number of wild-type colonies or **prototrophs** were formed with a frequency of about 1 in 10^6 cells (Fig. 14.10). (The wild-type colonies are often referred to as **prototrophs**, and the nutritionally deficient colonies as **auxotrophs**.) In these experiments only one of the possible types of recombinants was recovered, the prototrophs or wild type for all five markers. It was soon shown, however, by the use of unselected markers in addition to those selected by the medium, that a variety of recombinations was possible and that they did not recombine at random, but

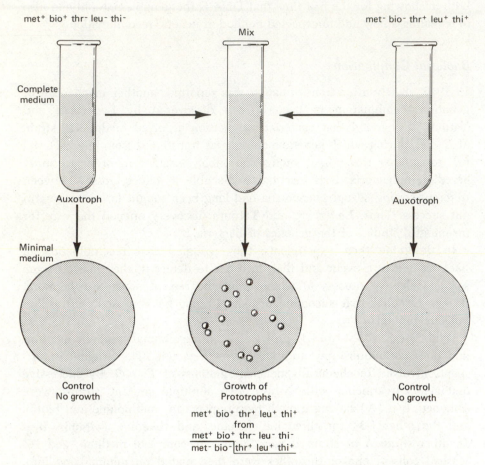

met⁺ bio⁺ thr⁻ leu⁻ thi⁻

Mix

met⁻ bio⁻ thr⁺ leu⁺ thi⁺

Complete medium

Auxotroph   Auxotroph

Minimal medium

Control
No growth

Growth of
Prototrophs

Control
No growth

met⁺ bio⁺ thr⁺ leu⁺ thi⁺
from
met⁺ bio⁺ thr⁻ leu⁻ thi⁻
met⁻ bio⁻ thr⁺ leu⁺ thi⁺

Fig. 14.10. Lederberg and Tatum's classic experiment demonstrating genetic recombination between two mutant strains of the bacterium *E. coli* K12.

rather that there was evidence of linkage, which could best be explained by assuming that all of the factors studied fitted on a single circular linkage map.

This phenomenon could be distinguished from transformation, in which only a single genetic trait is transferred to a transformed cell, because a number of traits regularly recombine during conjugation. It could also be shown that physical contact between the bacteria is necessary for conjugation to occur. If two conjugating strains of bacteria are placed in separate arms separated by a filter, no recombination occurs. The two cultures are growing in a common medium, which can be thoroughly mixed by alternating the pressure on the arms, and virus particles and

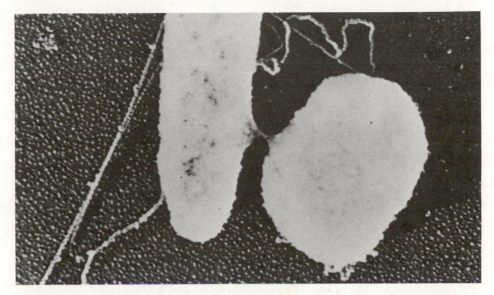

Fig. 14.11. Electron micrograph of conjugation between an elongated Hfr donor cell and a rounded F– recipient cell of *E. coli*. (Courtesy of T. F. Anderson, E. L. Wollman and F. Jacob [*Ann. Inst. Pasteur* 93:450–456, 1957].)

chemical molecules pass freely from one side to another, but the bacteria do not. This test helps to distinguish bacterial conjugation from transduction, to be described further in the next section. In transduction, recombination does take place through the filter, mediated by bacteriophage particles that transport small segments of the bacterial genome from one bacterium to another.

Studies with the electron microscope soon revealed conjugating pairs in intimate contact with one another through a conjugation tube (Fig. 14.11). Further details about the process of conjugation were rapidly discovered. The transfer of genetic material is unidirectional, from the donor or "male" cell to the recipient or "female" cell. Actively growing cells of *E. coli* contain from two to four bacterial chromosomes, all the replication products of a single two-stranded circular DNA molecule or chromosome. Thus, *E. coli* is genetically haploid, with a continuous linkage map and a circular chromosome. During conjugation, the DNA from one of the donor chromosomes passes through the conjugation tube into the recipient cell. Conjugation ceases when the tube ruptures, which may be before or after all of the DNA has been transferred. Hence, there may be a complete transfer or only a partial transfer of the DNA complement

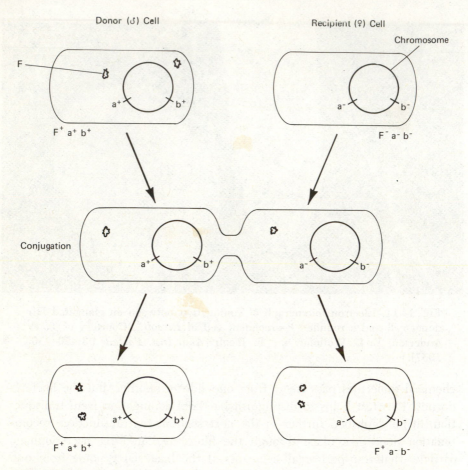

Fig. 14.12. Transfer of the autonomous F factor during conjugation in E. coli.

from the donor to the recipient cell. It is also possible to separate the conjugating pairs of cells at timed intervals by subjecting them to the shearing forces created by a Waring blender.

The donor or male cells differ from the recipient or female cells in having an autonomous F factor (or fertility factor or sex factor) extrachromosomally in the cell. The F factor is composed of DNA, but is only about one-fiftieth the size of the bacterial chromosome. F^+ cells are capable of acting as donors, apparently because the F factor modifies the cell surface so that coupling can occur. F^- cells, lacking the F factor, act as recipients during conjugation. When F^+ and F^- cell populations are mixed at densities exceeding 10^8 bacteria/ml, conjugating pairs are formed in less than 5 min. Each F^+ cell normally contains a number of F factors, and each F^-

cell receives at least one F factor during conjugation. Therefore, the efficiency of transfer of F during conjugation approaches 100%, and the F factor acts like an infectious agent in its ability to convert an F^- cell population into an F^+ population (Fig. 14.12).

However, even though nearly all F^- cells receive the sex factor during conjugation, only about one in 10^5 F^- cells receives any segment of the F^+ donor's chromosomal DNA. Thus, genetic recombination is a rare event when F^+ and F^- cell populations are mixed. Furthermore, it appears that chromosomal DNA is transferred only when the F factor combines with and becomes an integral part of the bacterial chromosome, where it may be integrated at any one of a number of different sites. Cells with the F factor integrated into the chromosome are called Hfr cells, for high frequency of recombination, because recombination involving chromosomal genes occurs at frequencies as high as 1 in 10 pairs rather than 1 in 10^5.

Isolation of Hfr cells from F^+ populations has permitted a much more efficient study of genetic recombination during conjugation. When conjugation occurs, the circular bacterial chromosome breaks at the point of attachment of the F factor. The linear chromosome then passes to the recipient F^- cell through the conjugation tube in an oriented way (Fig. 14.13). The broken end of the chromosome lacking F is transferred first and is called the origin, while the end to which F is attached brings up the rear and is the last bit of genetic information to be transferred. In different Hfr strains the bacterial genes first or last to enter the recipient cells may be different, but the linear order remains the same. Since the conjugation tube breaks spontaneously with a fairly high frequency, the entire chromosome is seldom transferred, and hence the efficiency of conversion of F^- to F^+ cells is considerably reduced when Hfr cells are used.

The F factor, which plays such a key role in bacterial conjugation, has been classed as an *episome* (see Chapter 17). Episomes are particles with genetic properties that may exist in the cell either in an autonomous state, multiplying independently of the chromosome, or in an integrated state on the chromosome, multiplying in synchrony with it (Fig. 14.14). Unlike bacterial viruses, they "infect" other cells only through the conjugation tube and lack a free, extracellular stage.

When conjugation between Hfr and F^- cells is permitted to go on without interruption, the frequency of recombination is not the same for all the genetic markers. Rather, every donor gene occurs with a characteristic frequency among the recombinants, but the frequency differs for different genes. The spontaneous disruption of the conjugation tube is apparently accompanied by breakage of the DNA molecule before the DNA

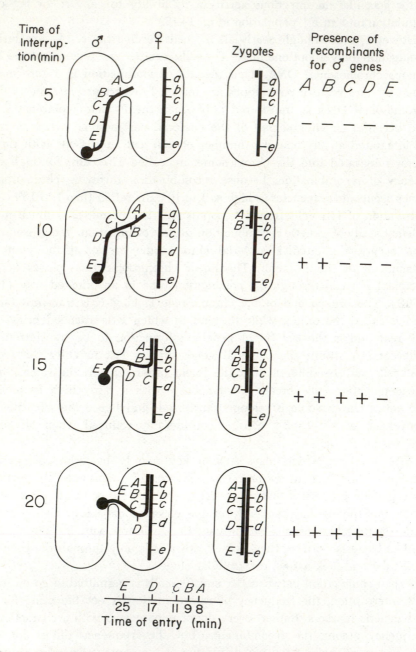

Fig. 14.13. Genetic recombination in an interrupted mating experiment in *E. coli.* (After Hayes.)

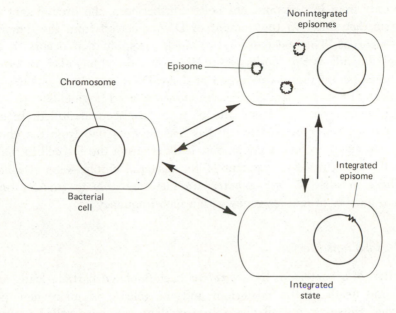

Fig. 14.14. The episome concept as illustrated by the F factor. The cell may be free of the episome or may carry it attached to the bacterial chromosome or free from the chromosome. Integrated with the chromosome, F replicates in synchrony with it; when not integrated, F replicates autonomously.

transfer is completed. Because of this fact, it is possible to construct a linear map based on the frequencies of recombination of the marker genes used. This map can then be compared with the results obtained from interrupted mating experiments, in which the conjugating cells are experimentally separated at various timed intervals from the start of conjugation up to 100 min (at 37°C). In this way the time of transfer of each of the donor's marker genes can be determined, and a time-sequence map of the genes constructed. This map, based on time of transfer, agreed with the map based on recombination frequencies.

The obvious conclusion is that the genes near the origin are transferred from donor to recipient at a high frequency, the genes further removed from the origin at lower frequencies, and those near the F factor at the terminal end of the linear DNA chromosome molecule with the lowest frequencies of all. In addition, since only part of the donor's genome is transferred ordinarily, the F⁻ exconjugants are only partial diploids or **heterogenotes**. This partial diploid may persist through several generations, because recombinant segregants continue to be recovered from the recipient F⁻ cells for as many as nine generations. The transfer is only in one direction, from donor to recipient, because recombinants are recovered only from F⁻ exconjugant cells. Furthermore, the exconjugant cells, because they differ in the amount of DNA received from the donor, are heterogeneous. The mechanism by which recombination occurs in these partial diploids is not understood. However, something akin to crossing over must permit the incorporation of donor DNA into the recipient's chromosomes with subsequent elimination of the replaced fragments.

Still another type of donor, the F′ type, has been identified. In an F′ cell, the sex factor in an Hfr cell has become detached from the chromosome and again become autonomous. It differs from the F⁺ cell in that the F factor now carries a fragment of the bacterial chromosome attached to it. These so-called F *merogenotes* are therefore able to transfer one or a small number of marker genes with a high frequency.

Bacterial Transduction

Infection of a bacterium by a **virulent** bacteriophage particle leads to the lysis and death of the bacterium and the release of many new phage particles. However, not all phages are virulent; some are called **temperate**

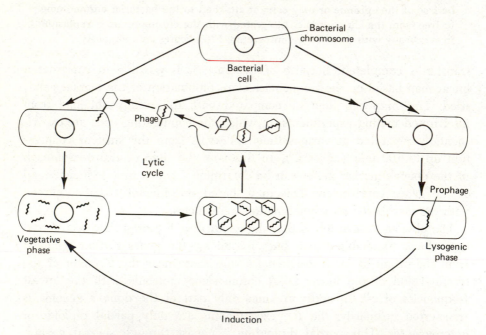

Fig. 14.15. The lytic cycle and the lysogenic phase of a temperate phage. When lysogeny occurs, the phage DNA is incorporated in the bacterial chromosome, where it is known as a prophage. Occasionally, the prophage is released, and the lytic cycle then ensues.

phages, because infection of a bacterium does not result in death. Instead, the bacterium and the bacteriophage form an association that permits the infected bacterium to survive and multiply with the phage DNA attached to the bacterial chromosome. Bacteria thus associated with a temperate phage are called *lysogenic*, because they can be induced by suitable treatment (chemical or ultraviolet) to undergo lysis and produce mature, active phage particles. When lysogenization occurs, the lysogenic bacteria not only acquire the capacity to produce phage as a hereditary trait, but they also become immune to further infection by the types of phage for which they are lysogenic (Fig. 14.15).

In lysogenic bacteria the phage DNA is incorporated into the bacterial genome and replicates in synchrony with it. In lytic phage infections, replication of the phage DNA is autonomous, and the vegetative phage particles control the production not only of phage DNA but also the phage proteins essential for the formation of mature phage particles. Thus, the phage DNA integrated into the bacterial chromosome in lysogenic bacteria is distinctly different from that of vegetative phage, and the bacteriophage in lysogenic bacteria is referred to as being in the *prophage* state.

In 1952 Zinder and Lederberg found that in *Salmonella typhimurium,* genes from a lysogenic strain could be transferred to a nonlysogenic strain and that the transfer was mediated by temperate bacteriophage particles. The phenomenon was called *transduction* (Fig. 14.16). Apparently, in transduction a few of the bacteria in a lysogenic population lyse spontaneously, releasing virus particles that contain a fragment of bacterial DNA. This DNA fragment is only about 1/100 of the total genome, and so ordinarily only one gene is transferred at a time. In the occasional cases in which genes are transduced together, they have also been found to be linked in conjugation experiments. Transduction can be distinguished from transformation by the fact that DNAase prevents transformation but not transduction. On the other hand, transduction can be distinguished from conjugation by the fact that conjugation requires direct cellular contact between donor and recipient cells, but such contact is not required in transduction. As noted above, transduction occurs through a filter separating cultures of donor and recipient bacteria in the arms of a U tube, but the filter effectively prevents conjugation.

Two types of transducing phages have been identified. The *generalized* transducing phages may carry a variety of bacterial genes, apparently because the phage may be attached as a prophage at any one of a number of attachment sites, and a given phage will carry only the bacterial gene or genes close to it during its period of attachment. Generalized transducing phages cannot be mapped because they have a number of attachment sites,

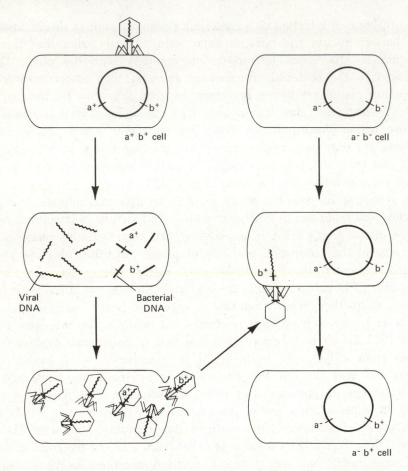

Fig. 14.16. Genetic recombination by the transduction of bacterial genes by viruses. Occasionally, when a bacterial virus infects a bacterium, a small segment of bacterial DNA is incorporated in the viral DNA and thus transferred with the viral DNA to a new host cell. If the viral DNA infects but does not kill the host cell, the bacterial genes thus transferred may be incorporated in the genome of the new host.

rather than just one as in the case of the *specific* transducing phages. Specific transducing phages become attached only to one specific site on the bacterial chromosome, and therefore can readily be mapped; furthermore, they will transduce only the restricted segment of the bacterial chromosome to which they are regularly attached as a prophage. Phage P22 of *Salmonella typhimurium* is a generalized transducer, while the λ phage of *E. coli* is a specialized transducer that only transfers a chromosome segment in the galactose region of the *E. coli* genome. The frequency of transduction for any particular gene by a generalized transducer is of the order 10^{-5} to

10^{-6} of the surviving infected cells. The frequency of transduction of the gal^+ gene by λ is of the same order of magnitude. Oddly enough, the transducing phage particles have been found to be defective; they have been shown to lack a complete viral genome, and in place of the missing viral segment a section of the bacterial genome has been substituted. Thus, the protein coat of the phage encloses DNA from both the virus and the bacterium.

When a bacterial cell is infected by such a defective transducing phage particle, the great majority of the infections result in *abortive transductions*. The transductions are aborted because the viral genome with its attached segment of bacterial genes is not integrated into the recipient's genome and does not replicate when the cell divides. Therefore, the fragment is passed on to just one of the two daughter cells at each cell division and, in an abortively transduced clone, only one bacterium ever carries the transduced genes. Since these genes are functional, it is sometimes possible to detect their presence. The most notable demonstration of an aborted transduction was the transfer of a gene controlling flagellum formation to a nonmotile strain of *Salmonella*. The recipient cell, after receiving this gene, formed a flagellum and started swimming about in the culture dish. At each division, a nonmotile cell and a motile cell were formed so that, as division occurred, the motile cell left in its wake a trail of nonmotile colonies (Fig. 14.17). Thus, it was possible to measure the number of divisions and the distance traveled between each division. More important, the cell is a partial diploid for the transduced segment of bacterial genes, and thus it is sometimes possible to test these genes for allelism and dominance.

Because the transducing phages are defective, the recipient bacteria become lysogenic only if they are infected simultaneously by a normal phage particle and by the transducing phage particle. Under these circumstances, both the defective phage and the normal phage multiply in a normal fashion, and the infected bacterial cell may either become lysogenic or enter the lytic mode of phage reproduction. If lysogenic, the cell may be lysogenic for both defective and normal phage. If lytic, the cell produces about equal numbers of the two kinds of phage particles.

In a small proportion of the infections, the DNA segment from the transducing phage becomes intimately associated with the DNA of the host. This association is apparently dependent on homologous DNA base sequences in the bacterial and phage chromosomes. Unlike the phage particles in abortive transduction, these integrated transducing fragments replicate in synchrony with the bacterial chromosome so that each daughter cell receives a copy of the original transducing DNA. There appears to be

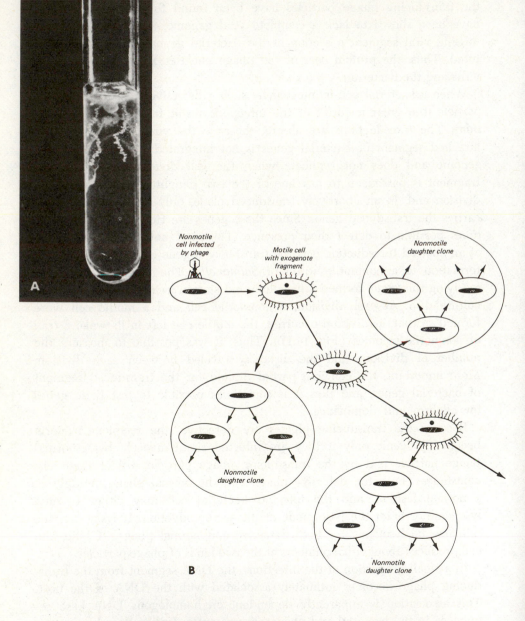

Fig. 14.17. (A) Photograph and (B) diagram of abortive transduction involving a motile and a nonmotile strain of *Salmonella*. (Lederberg, J., 1956. "Linear Inheritance in Transductional Clones," *Genetics* 41:845–871, p. 857, Fig. 4.)

a high probability that recombination will occur between the transducing phage DNA and the bacterial chromosome DNA leading to the insertion of the phage DNA into the bacterial chromosome. When recombination occurs, a free DNA fragment is formed that may persist in the cell for a time but is subsequently lost. In this way a donor's bacterial genes that have become attached to viral DNA of a defective transducing phage are transferred to a recipient bacterium and incorporated as functional units in the chromosome of the new host. Transduction, thus, is essentially a mode of genetic recombination in bacteria that is mediated by a virus. It resembles conjugation and transformation in being a one-way transfer of genetic material. Since only a small DNA fragment bearing one or a few genes is involved, transduction is more like transformation than conjugation.

Recombination in Phages

In the genetic study of bacteriophages, their small size, simple structure, and rapid multiplication have presented difficulties in the identification and counting of variant phenotypes. Nevertheless, several types of mutations have been discovered in phages, among which the plaque type and host range mutants have been particularly useful.

The number of phage particles in a given sample can be estimated by diluting the sample to such a concentration that, when an aliquot is added to a bacterial culture, the number of bacteria in the test tube far exceeds the number of virus particles. Under these conditions, each virus will infect a different bacterial host cell. If the contents of the tube are spread over an agar plate and incubated, the numerous bacteria form an opaque bacterial "lawn" on the surface of the plate. After the virus particle has completed its reproductive cycle within the bacterium, the bacterial cell is lysed and hundreds of daughter phage particles are released, which then infect the adjacent bacteria. After several such cycles, the growing zone of lysis becomes visible on the plate as a clear area or plaque in the "lawn" of bacteria. At these concentrations each plaque stems from a single phage particle, so it is possible to estimate the number of phages in the original sample.

The plaque-type mutants were recognized when it was observed that not all plaques are similar in appearance. One type, the so-called r mutants, produced a larger plaque with a sharper margin than that of the more or less arbitrarily chose "wild type." (See Fig. 12.9). The r (for *rapid* lysis) mutants caused a more rapid breakdown of the host cells than the normal r^+ type, which produce a smaller plaque with a turbid halo.

The host range (h) mutants were first recognized when it was discovered that different phages had different host specificity; that is, they differed from the normal h^+ type in their ability to grow on different strains of bacteria.

Genetic recombination in viruses was discovered in 1947 in the T-even phages (T2, T4, T6) of *E. coli* by Delbrück and Bailey and independently by Hershey and Rotman. A phage cross is made by culturing bacteria with two different types of phage in such numbers that there are about five phage particles of each type for each bacterium. At these concentrations nearly every bacterium is doubly infected with both types of phage.

A representative phage cross is that of h^+r and hr^+ strains of the phage T2. The h mutant phage will grow in both strain B and strain B/2 of *E. coli*, the host of the T2 phages. The wild-type h^+ phage will form a plaque on strain B but not on B/2. Thus, these two strains of bacteria can be used to distinguish between the h and h^+ genes, as shown below,

		E. coli strain	
		B	B/2
T2	h^+	+	−
phage			
type	h	+	+

where + indicates the ability to form a plaque. The technique is to plate the phage on a mixture of B and B/2 *E. coli* cells. The h mutant will give clear plaques because it lyses both B and B/2 cells; h^+ produces turbid plaques, which result from the presence of the unlysed resistant B/2 bacteria.

When the cross is made, h^+r and hr^+ phage are mixed and grown on *E. coli* strain B. The residual phages are removed, and the bacteria are diluted and allowed to lyse. Dilutions of the progeny phage in the lysate are then plated on a mixture of B and B/2 cells so that the plaques formed can be scored for markers from the parents. When this procedure was followed, four plaque types were obtained (Fig. 14.18):

Phage genotype	Plaque type
1. Large turbid	h^+r^+ recombinant
2. Small, clear	h^+r parental
3. Small, turbid	hr^+ parental
4. Large, clear	hr recombinant

Thus, quite clearly, a form of genetic recombination occurs in bacteriophages. The frequency of recombination observed in various crosses has

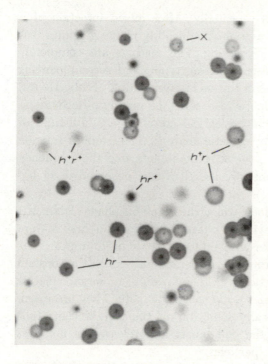

Fig. 14.18. The types of plaques produced by the progeny from a cross between h^+r and hr^+ parental strains of the T2 phage. (Hayes, W., 1968. *The Genetics of Bacteria and Their Viruses*, 2nd ed. John Wiley & Sons, New York, Plate 21 facing p. 488. [Photograph by Mrs. Maureen de Saxe and Miss Janet Mitchell.])

ranged from the order of 0.01% to as much as 40%. The reciprocal recombinant types are found in approximately equal numbers, and the same frequency of recombination is observed when the mutants are in coupling as in repulsion. Thus, the same frequency of recombinants is observed from the cross $h^+r^+ \times hr$ as from $h^+r \times hr^+$. These results permit genetic mapping similar to that in higher organisms. The original linkage maps for the T-even phages appeared to be linear, but later it was found that genes thought to be at opposite ends of the linear map were also close together, indicating that the genes must form a single, circular linkage map.

However, the recombination process does not appear to be comparable to the events in meiosis of higher organisms. Even though reciprocal recombinants are formed with equal frequency, they are not formed by a reciprocal event comparable to meiotic crossing over. Furthermore, Hershey and Rotman found that if a bacterium was infected with three different strains of T2 phage, some of the progeny had traits derived from all three parents. Hence, the process clearly differs from meiosis in diploids.

When the phage DNA first penetrates the sensitive bacterial cell, there is a latent period during which no infectious phage particles can be recovered from the cell. The phage is then said to be in the vegetative state. Recombination in phage apparently occurs only between phages in the vegetative condition when they consist only of DNA. During the latent period the phage DNA replicates repeatedly. After mixed infection of a bacterium, the phage DNA forms a pool of replicating phage genomes that participate in successive rounds of mating. The matings are completely random because there are no different mating types, and recombinant as well as parental types are involved in the successive mating events. Hence, a single chromosome may participate in a series of matings. Recombinations even occur between genetically identical chromosomes, although of course these cannot be detected by genetic analysis. Thus, only a fraction of the mating events gives rise to genetic recombinants, and the frequency of recombination in phages is generally less than that observed in higher organisms.

At the end of the latent period, when the pool of phage DNA has reached the equivalent of 50 or more phage particles, the first mature infective phage particles, complete with protein coat, start to appear. The phage genomes are withdrawn from the pool randomly and at a constant rate to form mature phage particles. Although replication and mating continue, the mature phage particles no longer participate. They are essentially physiologically and genetically inert during the rest of their time in the cell prior to lysis. Then the cell bursts and releases as many as several hundred infectious phage particles.

A discovery of fundamental importance in the study of the mechanism of recombination was the finding by Meselson and Weigle in 1961 that recombination in the virus λ of *E. coli* could occur independently of DNA replication. In other words, recombination took place through breakage and rejoining of existing DNA molecules. The essence of a later and even more decisive experiment by Meselson was the demonstration by density-gradient centrifugation that recombination between two different λ phages, both containing labeled DNA, had occurred. Replication had taken place in unlabeled medium, but recombinant genotypes were recovered from the fraction in the centrifuge tube that was completely labeled, and hence it indicated that no replication was necessary for recombination to occur.

A unique feature of recombination in phages is the formation of partial heterozygotes (hets) among the progeny of phage crosses, a discovery made by Hershey and Chase in 1951. Phages ordinarily behave genetically like haploids, but the hets act like partial diploids. If bacteria (*E. coli*) are infected with both r^+ (small plaque, turbid halo) and r (large, clear plaque) mutant T-even phages in a mixed infection, and then plated on indicator bacteria before lysis, mottled plaques are formed due to the release of a mixture of r^+ and r particles at lysis. The mottling is due to the different effects of the r^+ and r mutants on the bacterial "lawn" used as an indicator. However, when the progeny particles from an $r^+ \times r$ cross are diluted so that a single progeny particle can be plated and studied, about 2% of the plaques formed are also mottled. If particles from these mottled plaques are again plated, they produce r^+ and r plaques in equal proportions, and again 2% mottled plaques. Since a single progeny particle from the $r^+ \times r$ cross gives rise to both r^+ and r colonies as well as the mottled plaques, the most reasonable explanation is that such particles are unstable heterozygotes in the r region. They carry both r^+ and r, and by segregation during multiplication give rise to pure r^+- or r-type particles, but also continue to form a small percentage of heterozygotes.

It was later found that hets are not associated with any particular locus, but arise at about the same rate at many different loci.

With double heterozygotes carrying markers different distances apart, it was possible to estimate the length of the heterozygous region, which turned out to be quite small, of the order of a few cistrons. Furthermore, the association between het formation and recombination of outside markers was so close that it seems very probable that the origin of these heterozygotes is in some way related to the phenomenon of recombination.

An additional feature of recombination in bacteriophages, also observed in higher organisms, is negative interference. In contrast to the phenomenon of positive interference, discussed earlier, in which the presence of one crossover reduces the probability of a second crossover below the frequency expected if the two events are random, the probability of the second crossover is greater than expected on a random basis with negative interference.

Thus, genetic recombination does occur in bacteriophages. In several respects it differs from the recombination observed in other organisms. Studies of recombination in phages have led to better understanding of the process in some respects, but have also posed questions that require answers before a general theory of recombination can be formulated. One question is whether a single, all-encompassing theory of recombination is possible. Before discussing the theories of recombination further, we shall first consider some additional phenomena related to recombination.

14.4. Negative Interference

The clustering of several genetic exchanges within a short chromosome segment, called *localized negative interference*, is apparently a fairly general phenomenon. First studied in detail in the fungus *Aspergillus*, negative interference has also been detected in other fungi (e.g., yeast and *Neuro-*

spora), in bacteriophages, as noted above, in bacteria, and in *Drosophila*. The clustering of exchanges observed with negative interference differs from multiple crossovers over longer map distances in that, in a given cluster, only two parental homologues are involved rather than all four taken two at a time. On the assumption that the genetic markers used do not affect the pattern of recombination, it has been calculated for phages that the mean number of exchanges per cluster is two or three, and that the mean length of the region involved is about the size of a gene.

14.5. Gene Conversion

Another phenomenon of interest in relation to recombination is **gene conversion**. The term was first proposed in 1930 by Winkler in an attempt to formulate a theory of recombination. Stern's work, however, soon showed that crossing over actually involved exchanges between nonsister chromatids rather than gene conversion, and the concept was neglected until the term was revived by Lindegren in 1953 to explain some of his data for yeast. He found that, in tetrad analysis in yeast crosses, instead of getting the expected 2:2 segregation ratios, he sometimes obtained 3:1 ratios. He suggested that gene conversion was occurring, that at meiosis some sort of interaction between dominant and recessive alleles in heterozygotes resulted in the transformation of a dominant allele into a recessive or vice versa. In a sense, it appeared to be a "directed" mutation occurring at meiosis in heterozygotes as the result of an effect by one allele on another. Lindegren's finding was so unexpected that it was not widely accepted until similar results began to be reported by Mitchell in 1955 with *Neurospora crassa*.

However, conversion differs from mutation in that it occurs only in heterozygotes and not in homozygotes. If it were simply a mutation phenomenon, it should be equally frequent in both heterozygotes and homozygotes. Furthermore, it was distinguishable from negative interference by the apparent absence of the expected reciprocal crossover classes. Hence, in *Neurospora* and *Sordaria,* rather than the expected 4:4 segregation ratios from a single ascus, 5:3 and 6:2 ratios were found.

The discovery of 5:3 ratios as well as 6:2 ratios among the eight ascospores in a single ascus was significant because it indicated that postmeiotic segregation must occur during the mitotic division after meiosis. Furthermore, it seems probable that different phenomena are responsible for the 5:3 and the 6:2 segregation ratios.

A further observation in these experiments was that if marker genes were present on either side of the site of gene conversion, there was a high correlation between ordinary reciprocal recombination of the outside marker genes and the occurrence of gene conversion. The nature of the relationship between reciprocal crossing over and nonreciprocal gene conversion remains obscure.

An observation on 6:2 segregation ratios in still another fungus, *Ascobolus,* revealed a form of polarized recombination. In such cases the map position of the mutants involved determined which mutant type would undergo conversion. This discovery, though not typical of all cases of gene conversion, led Lissouba and Rizet in 1960 to propose the name *polaron* for these small chromosomal segments within which polarized recombination occurred.

Gene conversion is best studied in species in which tetrad analysis is possible, but conversionlike phenomena have been described in species other than fungi, such as corn, *Drosophila,* and *E. coli.* Thus, the phenomenon apparently can be regarded as general rather than restricted.

A phenomenon that superficially resembles gene conversion is paramutation in corn, studied most extensively by Brink. In paramutation, as in gene conversion, one allele produces an effect on the other in the heterozygote. However, conversion is a relatively rare event wheras, given the right combinations of heterozygous alleles, paramutation invariably occurs, and one allele is modified in a specific way by the presence of the other. Like a mutation, this change persists over more than one generation, but gradually it tends to revert toward its original phenotype. The nature of the phenomenon suggests that it more properly belongs in a discussion of the control of differentiation rather than of recombination or of mutation.

14.6. Somatic Crossing Over

One further facet of recombination is that crossing over, usually associated with meiosis, has also been discovered in somatic cells undergoing mitosis. This fact was first demonstrated by Stern in 1936 with *Drosophila* in which mosaic patches appeared on the body of flies heterozygous for the bristle mutation, singed (*sn*), and the body color mutation, yellow (*y*). The most common type was a twin spot with adjacent patches of singed and yellow on a wild-type background, the next most frequent was a small patch of yellow, and the least frequent was a single patch of singed tissue. The origin of these patches by somatic crossing over is shown in

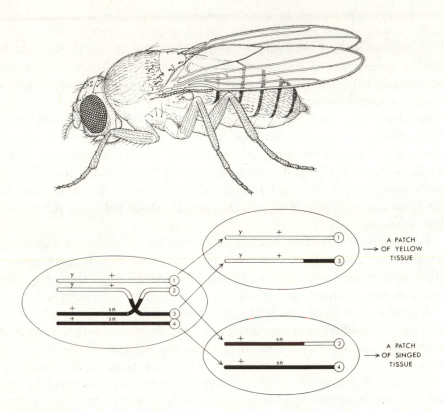

Fig. 14.19. Somatic crossing over in a female *Drosophila melanogaster* of the genotype *y* +/+ *sn* (where *y* = yellow body color and *sn* = singed bristles). Most of the body appears wild type, but there are adjacent patches of yellow (drawn to appear pale) and singed on the thorax and the abdomen. The mechanism for the origin of the twin spots by somatic crossing over is diagrammed below. (King, R. C., 1965. *Genetics*, 2nd ed. Oxford University Press, New York, p. 135, Fig. 9.8.)

Fig. 14.19. Just as in meiosis, mitotic crossing over occurs in the four-strand stage. Chromosome behavior continues to be typical of mitosis. The effect of mitotic crossing over is to produce, from a heterozygous nucleus, two daughter nuclei each of which is homozygous for one or the other chromosome segment distal to the crossover from the centromere. In this way recessives such as yellow and singed are brought to expression in the homozygous condition.

Somatic or mitotic crossing over has also been discovered in other organisms, particularly in yeast and fungi. The most careful studies have been made with the fungus, *Aspergillus nidulans*.

Aspergillus, like *Neurospora,* is able to form heterokaryons. Unlike *Neurospora,* however, it does not have separate mating types so that hyphal fusions between any two strains may occur or a strain may form fertile fruiting bodies by itself. The asexual vegetative spores formed by heterokaryons are uninucleate and ordinarily haploid, but occasionally, with suitable markers, heterozygous diploid spores can be derived from such heterokaryons. These heterozygous diploids are used in the detection of somatic crossovers.

An added finding in *Aspergillus* was the process of haploidization, by which heterozygous diploid cells give rise to haploid cells. If the diploids have marker genes on several chromosomes, the haploids derived from them have all the possible genotypes to be expected from recombination of intact (noncrossover) nonhomologous chromosomes. Since haploidization yields all possible recombinations among whole chromosomes, but virtually no intrachromosomal recombinants due to crossing over, this process permits quick determination of the proper linkage group for any unknown mutant.

Because mitotic crossing over is a rare process, occurring at the rate of only one exchange in several hundred nuclei in *Aspergillus,* some sort of selective process is necessary to permit identification of the rare crossover events. For this purpose selective markers, as far from the centromere as possible, are needed. The markers may be for color, or nutritional requirements, or resistance to harmful agents, but their function is to permit selection of the crossovers, which can then be analyzed for the rest of the genotype. With multiply marked chromosomes, it is possible to make a chromosome map similar to those obtained after meiotic crossing over, and it is also possible to map the centromere. This analysis is completed either by analyzing the haploid derivatives following haploidization, or else, when possible, by making crosses with other diploids to give a tetraploid zygote. Meiosis in this zygote ordinarily produces haploid segregants for analysis.

In *Aspergillus,* therefore, it is possible to have genetic recombination without the usual sexual processes involving meiosis. Pontecorvo, a leading worker on *Aspergillus,* refers to the sequence of diploid formation, mitotic crossing over, and haploidization as the parasexual cycle. In some fungi such as *Aspergillus niger* and *Penicillium chrysogenum,* in which no sexual cycle is known to occur, chromosome maps have been made by the use of mitotic recombination. In theory, it could also be used for chromosome mapping in man by the use of appropriate cells in tissue culture. This possibility has not yet been realized, but it is an intriguing suggestion for solving one of the more difficult aspects of human genetics.

14.7. Present Status of the Models

As the result of cytological studies in the 1930s, the breakage-reunion model of crossing over advanced by Darlington won wide acceptance. However, after Belling's copy-choice concept had been in eclipse for a number of years, Hershey and Rotman revived it in 1949 when they found that in some phage crosses the progeny from a single bacterium did not contain the recombinant types of phages in equal proportions, even though the sum total of the reciprocal recombinants for all progeny was approximately equal. This finding suggested to them that phage recombination is non-reciprocal rather than due to reciprocal breakage and reunion, and that partial replicas are formed by copy choice.

Copy choice was also invoked to explain the behavior of phage hetero-zygotes, and as the means by which chromosome fragments and episomes are integrated into the bacterial chromosome. Thus, the recombinations resulting from lysogenization, transduction, transformation, and conjugation in bacteria were all thought to involve copy choice.

However, despite the early popularity of the copy-choice theory in phage and bacterial genetics, an impressive array of evidence began to accumulate favoring breakage and reunion as the usual mechanism of recombination. One such experiment showed that phage recombination occurred while DNA synthesis was being inhibited by potassium cyanide. Under such conditions only existing DNA molecules could participate in any recombination events. Furthermore, in transformation, it was demonstrated that labeled donor DNA actually is physically incorporated into the recipient chromosome. Breakage and reunion seemed the most reasonable explanation for these findings. The most decisive experiments demonstrating breakage and reunion as the mechanism of recombination in phage were those of Meselson and Weigle with λ phage.

In higher forms, the copy-choice model was placed in difficulty by Taylor's demonstration that replication of both the DNA and the chromosomes is semiconservative rather than conservative. In other words, instead of a new chromatid being laid down beside the old, the DNA of each chromatid is half old and half new, and it is incorrect to designate parent and daughter chromatids. Chromosome duplication evidently occurs by the separation of the two strands of the existing DNA molecule, with a new complementary strand being formed adjacent to each of the existing strands.

The evidence favoring the breakage-reunion theory means that crossing over could occur at times other than that of DNA replication. In eukaryotes, in fact, it was found that DNA replication and histone synthesis precede the onset of meiosis. Moreover, the discovery that the effective doubling of the chromosomes into chromatids (as measured by X-ray breakage) also occurs prior to meiosis raised doubts about Darlington's "precocity" theory of synapsis and hence about his breakage reunion theory in its original form.

If recombination need not occur at the time of DNA replication, the question remains as to when it does occur. It had long been supposed that crossing over in higher forms occurs during the synapsis of homologous chromosomes in first meiotic prophase. The discovery of localized negative interference in bacteriophages, bacteria, various fungi, and in *Drosophila* led to a different concept. Pritchard's work on negative interference in *Aspergillus* led to the suggestion that effective pairing with respect to genetic recombination is discontinuous, with the paired segments quite short relative to the total length of the chromosome. This effective pairing was thought to be different from and prior to the cytological pairing seen at pachytene. Within the small paired region, the probability of exchanges is high, leading to a clustering of exchanges within the segment. Thus, rather than complete and continuous pairing with a low probability of genetic exchange at any one point, pairing was thought to be discontinuous with the probability of exchanges high within the small paired segments and absent elsewhere. With this theory, if the genes are closely linked within a single effectively paired region, the chances are better than random that two genetic exchanges will occur within that region (thus producing negative interference). On the other hand, for crossing over to occur between genes farther apart on the chromosome, two separate effectively paired regions must be formed, which will arise more or less at random. If, however, some sort of spacing between effectively paired regions is necessary, positive interference will occur.

The concept that crossing over occurs prior to meiotic prophase in discontinuous effective pairing segments makes the elaborate pairing behavior of the chromosomes during zygotene and pachytene seem superfluous. A number of experiments on a variety of plant and animal species have shown that crossing over can be influenced by physical or chemical agents before, during, or after premeiotic DNA synthesis is completed. The most clear-cut data indicate that the time of crossing over, or at least its completion, is not during the G_1, S, G_2 stages of interphase or during leptotene, but that it is a postleptotene event associated with the synaptinemal complex of chromosomes during zygotene and pachytene. This is not to say that earlier pretreatments may not modify the amount of crossing over or chiasmata formation observed, but rather that crossing over is a discrete event during meiotic prophase, probably during pachytene.

In mitosis, DNA replication occurs prior to prophase in a single discrete S period. In meiosis, on the other hand, a small amount of the chromosomal DNA replication is delayed, taking place not during the main S period, but later during zygotene and pachytene. Moreover, meiotic prophase also differs from mitotic prophase in having a marked delay in chromosomal histone replication. The discovery of this small amount of DNA synthesis during the presumed time of crossing over has led to the belief that it may in some way be involved in DNA repair or exchange during crossing over.

More recent attempts to provide a satisfactory explanation for crossing over are the hybrid DNA models, proposed independently by Whitehouse, Holliday, and Meselson. The essence of these models is that chromosome replication is completed during the S period of premeiotic interphase, but that recombination occurs during the zygotene or pachytene stages of meiosis while the homologous chromosomes are paired. One chromatid from each chromosome, selected at random, is involved in each crossover event. It is assumed that only one of the two strands of each of the DNA molecules involved is broken initially, and that reassociation in new ways then occurs. This association is not end-to-end following breakage, but rather a lateral association of complementary segments from homologous regions to produce hybrid DNA. End-to-end reunion would seem likely to generate duplications or deficiencies of nucleotides, whereas lateral association would provide a mechanism giving high precision to the recombination process, which, of course, is what is observed. Following the initial breakage and chain separation, synthesis of new segments of the DNA chain in the crossover regions occurs, and the bases in the complementary segments of the two molecules pair (see Fig. 14.20).

This concept takes into account a number of the facts observed in relation to crossing over, including the small amount of DNA synthesis observed during zygotene and pachytene. Moreover, it provides a reasonably plausible explanation for gene conversion and negative interference within the same framework as the mechanism for crossing over. The discovery of a mechanism for the excision and repair of small aberrant DNA segments also lends support to the idea. The hybrid DNA models, which will undoubtedly undergo further refinement, are more sophisticated versions of the breakage-reunion concept. Two problems under study are polarity in recombination in eukaryotes and whether this implies that crossovers are initiated only at certain fixed positions along the DNA molecules. At present, we seem to be about to emerge from a wilderness of facts about recombination into a clearing where it will be possible to get our bearings and a clearer view of the phenomenon.

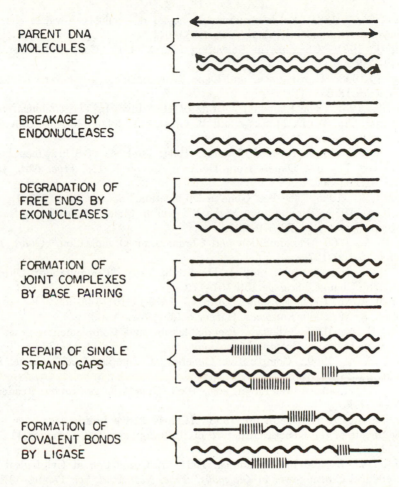

PARENT DNA
MOLECULES

BREAKAGE BY
ENDONUCLEASES

DEGRADATION OF
FREE ENDS BY
EXONUCLEASES

FORMATION OF
JOINT COMPLEXES
BY BASE PAIRING

REPAIR OF SINGLE
STRAND GAPS

FORMATION OF
COVALENT BONDS
BY LIGASE

Fig. 14.20. Diagrammatic representation of the hybrid DNA model of re-combination. (Bodmer, W. F., and A. J. Darlington, 1969. "Linkage and Recombination at the Molecular Level." In *Genetic Organization, Vol. 1,* by Caspari, E. W., and A. W. Ravin, eds., Academic Press, New York, p. 257, Fig. 11.)

Additional Reading

Amati, P., and M. Meselson, 1965. "Localized Negative Interference in Bacteriophage λ," *Genetics* **51**:369–379.

Avery, O. T., C. M. MacLeod, and M. McCarty, 1944. "Studies on the Chemical Nature of the Substance Inducing Transformation of Pneumococcal Types. Induction of Transformation by a Desoxyribonucleic Acid Fraction Isolated from Pneumococcus Type III," *J. Exptl. Med.* **79**:137–158.

Baker, W. K., 1968. "Position-Effect Variegation," *Adv. Genet.* **14**:133–169.

Balbiani, E. G., 1881. "Sul la structure du noyau des cellules salivaires chez les larves de *Chironomus*," *Zool. Anz.* **4**:637–641.

Belling, J., 1931. "Chiasmas in Flowering Plants," *Univ. Calif. Publ. Bot.* **16**: 311–338.

Belling, J., 1933. "Crossing-over and Gene Rearrangement in Flowering Plants," *Genetics* **18**:388–413.

Benzer, S., 1957. "The Elementary Units of Heredity." In *The Chemical Basis of Heredity*, W. D. McElroy and B. Glass, eds. Johns Hopkins University Press, Baltimore.

Boyce, R. P., and P. Howard-Flanders, 1964. "Release of Ultraviolet Light-induced Thymine Dimers from DNA in *E. coli* K 12," *Proc. Nat. Acad. Sci.* **51**:293–300.

Bridges, C. B., 1936. "The Bar 'Gene' a Duplication," *Science* **83**:210–211.

Brink, R. A., 1958. "Paramutation at the R Locus in Maize," *Cold Spring Harbor Symp. Quant. Biol.* **22**:379–391.

Brink, R. A., 1960. "Paramutation and Chromosome Organization," *Quart. Rev. Biol.* **35**:120–137.

Brink, R. A., E. D. Styles, and J. D. Axtell, 1968. "Paramutation: Directed Genetic Change," *Science* **159**:161–170.

Campbell, A. M., 1962. "Episomes," *Adv. Genetics* **11**:101–145.

Campbell, A. M., 1969. *Episomes*. Harper & Row, New York.

Carlson, E. A., 1959. "Allelism, Pseudoallelism, and Complementation at the Dumpy Locus in *Drosophila melanogaster*," *Genetics* **44**:347–373.

Carlson, E. A., 1959. "Comparative Genetics of Complex Loci," *Quart. Rev. Biol.* **34**:33–67.

Caspari, E. W., and A. W. Ravin, eds., 1969. *Genetic Organization*. Academic, New York.

Chase, M., and A. H. Doermann, 1958. "High Negative Interference over Short Segments of the Genetic Structure of Bacteriophage T4," *Genetics* **43**:332–353.

Creighton, H. B., and B. McClintock, 1931. "A Correlation of Cytological and Genetical Crossing-over in *Zea mays*," *Proc. Nat. Acad. Sci.* **17**:492–497.

Darlington, C. D., 1937. *Recent Advances in Cytology*, 2nd ed. Blakiston, Philadelphia.

Davern, C. I., 1971. "Molecular Aspects of Genetic Recombination." In *Progress in Nucleic Acid Research and Molecular Biology*, Vol. 11, J. N. Davidson and W. E. Cohn, eds. Academic, New York.

Delbrück, M., and W. T. Bailey, Jr., 1946. "Induced Mutations in Bacterial Viruses," *Cold Spring Harbor Symp. Quant. Biol.* **11**:33–37.

Fincham, J. R. S., 1966. *Genetic Complementation*. Benjamin, Menlo Park, Calif.

Fincham, J. R. S., and J. A. Pateman, 1957. "Formation of an Enzyme Through Complementary Action of Mutant Alleles in Separate Nuclei of a Heterokaryon," *Nature* **179**:741–742.

Griffith, F., 1928. "The Significance of Pneumococcal Types," *J. Hygiene* **27**: 113–159.

Hayes, W., 1952. "Recombination in *Bact. coli* K 12: Unidirectional Transfer of Genetic Material," *Nature* **169**:118–119.

Hayes, W., 1968. *The Genetics of Bacteria and Their Viruses,* 2nd ed. Wiley, New York.

Heitz, E., and H. Bauer, 1933. "Beweise für die Chromosomennatur der Kernschleifen in Knäuelkernen von *Bibio hortulanus* L.," *Zeits. Zellforsch. u. mikr. Anat.* **17**:67–82.

Hershey, A. D., and M. Chase, 1951. "Genetic Recombination and Heterozygosis in Bacteriophage," *Cold Spring Harbor Symp. Quant. Biol.* **16**:471–479.

Hershey, A. D., and R. Rotman, 1948. "Linkage among Genes Controlling Inhibition of Lysis in a Bacterial Virus," *Proc. Nat. Acad. Sci.* **34**:89–96.

Hershey, A. D., and R. Rotman, 1949. "Genetic Recombination Between Host-Range and Plaque-Type Mutants of Bacteriophage in Single Bacterial Cells," *Genetics* **34**:44–71.

Holliday, R., 1964. "A Mechanism for Gene Conversion in Fungi," *Genet. Res.* **5**:282–304.

Hotchkiss, R. D., 1971. "Toward a General Theory of Genetic Recombination in DNA," *Adv. Genetics* **16**:325–348.

Hotchkiss, R. D., and M. Gabor, 1970. "Bacterial Transformation, with Special Reference to Recombination Process," *Ann. Rev. Genetics* **4**:193–224.

Jacob, F., and S. Brenner, 1963. "Sur la régulation de la synthèse du DNA chez les bactéries: l'hypothese du replicon," *C.R. Acad. Sci.* (Paris) **256**:298–300.

Jacob, F., and E. L. Wollman, 1961. *Sexuality and the Genetics of Bacteria.* Academic, New York.

Janssens, F. A., 1909. "Spermatogénèse dans les Batraciens. V. La Théorie de la chiasmatypie. Nouvelles interprétation des cinèses de maturation," *Cellule* **25**:387–411.

Lederberg, J., and E. M. Lederberg, 1952. "Replica Plating and Indirect Selection of Bacterial Mutants," *J. Bact.* **63**:399–406.

Lederberg, J., and E. L. Tatum, 1946. "Novel Genotypes in Mixed Cultures of Biochemical Mutants of Bacteria," *Cold Spring Harbor Symp. Quant. Biol.* **11**:113–114.

Lewis, E. B., 1941. "Another Case of Unequal Crossing-over in *Drosophila melanogaster,*" *Proc. Nat. Acad. Sci.* **27**:31–34.

Lewis, E. B., 1951. "Pseudoallelism and Gene Evolution," *Cold Spring Harbor Symp. Quant. Biol.* **16**:159–174

Lindegren, C. C., 1953. "Gene Conversion in Saccharomyces," *J. Genet.* **51**:625–637.

Lindegren, C. C., 1955. "Non-Mendelian Segregation in a Single Tetrad of *Saccharomyces* Ascribed to Gene Conversion," *Science* **121**:605–607.

Lissouba, P., J. Mousseau, G. Rizet, and J. L. Rossignol, 1962. "Fine Structure of Genes in the Ascomycete *Ascobolus immersus,*" *Adv. Genet.* **11**:343–380.

Lissouba, P., and G. Rizet, 1960. "Sur l'existence d'une unité génétique polarisée ne subissant que des échanges non réciproques," *Compt. Rend.* **250**:3408–3410.

Luria, S. E., and M. Delbrück, 1943. "Mutations of Bacteria from Virus Sensitivity to Virus Resistance." *Genetics* **28**:491–511.

Meselson, M., 1964. "On the Mechanism of Genetic Recombination," *J. Mol. Biol.* **9**:734–745.

Meselson, M., 1967. "The Molecular Basis of Genetic Recombination." In *Heritage from Mendel*, R. A. Brink, ed. University of Wisconsin Press, Madison, Wisc.

Meselson, M., and J. J. Weigle, 1961. "Chromosome Breakage Accompanying Genetic Recombination in Bacteriophage," *Proc. Nat. Acad. Sci.* **47**:857–868.

Mitchell, M. B., 1955. "Aberrant Recombination of Pyridoxine Mutants of *Neurospora*," *Proc. Nat. Acad. Sci.* **41**:215–220.

Morgan, T. H., 1911. "Random Segregation Versus Coupling in Mendelian Inheritance," *Science* **34**:384.

Muller, H. J., A. A. Prokofyeva-Belgovskaya, and K. U. Kossikov, 1936. "Unequal Crossing-over in the Bar Mutant as a Result of Duplication of a Minute Chromosome Section." *C.R. Acad. Sci.* (USSR) **2**:87–88.

Olive, L. S., 1959. "Aberrant Tetrads in *Sordaria fimicola*," *Proc. Nat. Acad. Sci.* **45**:727–732.

Oliver, C. P., 1940. "A Reversion to Wild-Type Associated with Crossing-over in *Drosophila melanogaster*," *Proc. Nat. Acad. Sci.* **26**:452–454.

Painter, T. S., 1933. "A New Method for the Study of Chromosome Rearrangements and Plotting of Chromosome Maps," *Science* **78**:585–586.

Pontecorvo, G., 1956. "Allelism," *Cold Spring Harbor Symp. Quant. Biol.* **21**:171–174.

Pontecorvo, G., 1958. *Trends in Genetic Analysis.* Columbia University Press, New York.

Pontecorvo, G., and E. Käfer, 1958. "Genetic Analysis Based on Mitotic Recombination." *Adv. Genetics* **9**:71–104.

Pritchard, R. H., 1955. "The Linear Arrangement of a Series of Alleles in *Aspergillus nidulans*," *Heredity* **9**:343–371.

Pritchard, R. H., 1960. "Localized Negative Interference and Its Bearing on Models of Gene Recombination," *Genet. Res.* **1**:1–24.

Setlow, R. B., and W. L. Carrier, 1964. "The Disappearance of Thymine Dimers from DNA: An Error-Correcting Mechanism," *Proc. Nat. Acad. Sci.* **51**:226–231.

Stent, G. S., 1971. *Molecular Genetics.* Freeman, San Francisco.

Stern, C., 1931. "Zytologisch-genetische Untersuchungen als Beweise für die Morgansche Theorie des Faktorenaustausches," *Biol. Zentralbl.* **51**:547–587.

Stern, C., 1936. "Somatic Crossing-over and Segregation in *Drosophila melanogaster*," *Genetics* **21**:625–730.

Sturtevant, A. H., 1925. "The Effects of Unequal Crossing-over at the Bar Locus in *Drosophila*," *Genetics* **10**:117–147.

Sturtevant, A. H., and T. H. Morgan, 1923. "Reverse Mutation of the Bar Gene Correlated with Crossing-over," *Science* **57**:746–747.

Taylor, J. H., 1957. "The Time and Mode of Duplication of Chromosomes," *Amer. Nat.* **91**:209–221.

Taylor, J. H., 1959. "The Organization and Duplication of Genetic Material," *Proc. 10th Inter. Cong. Genetics* **1**:63–78.

Taylor, J. H., 1967. "Patterns and Mechanisms of Genetic Recombination." In *Molecular Genetics*, Part II, J. H. Taylor, ed. Academic, New York.

Whitehouse, H. L. K., 1969. *Towards an Understanding of the Mechanism of Heredity*, 2nd ed. Arnold, London.

Whitehouse, H. L. K., 1970. "The Mechanism of Genetic Recombination," *Biol. Rev.* **45**:265–315.

Whitehouse, H. L. K., and P. J. Hastings, 1965. "The Analysis of Genetic Recombination or the Polaron Hybrid DNA Model," *Genet. Res.* **6**:27–92.

Winkler, H., 1930. *Die Konversion der Gene.* Fischer, Jena.

Zinder, N., and J. Lederberg, 1952. "Genetic Exchange in *Salmonella*," *J. Bact.* **64**:679–699.

Questions

14.1. Distinguish between Janssen's chiasmatype hypothesis and the so-called "classical" theory of crossing over.

14.2. What are heteromorphic homologues? What role did they play in the final proof that the genes are on the chromosomes?

14.3. Diagram the copy-choice and the breakage-reunion theories of genetic recombination.

14.4. How would you distinguish experimentally between alleles, pseudo-alleles, and closely linked but nonallelic genes?

14.5. What is position effect? If a phenotype thought to be due to position effect was found, how could you demonstrate that it was in fact a position effect rather than a point mutation?

14.6. What is a complementation test? How is it used?

14.7. What is a cistron? A muton? A recon?

14.8. How can it be demonstrated most decisively that bacterial resistance to an antibiotic originates by random mutation and not as an adaptive response or a mutation induced by exposure to the antibiotic?

14.9. Transformation, transduction, and conjugation are three types of genetic recombination discovered in bacteria. Compare and contrast these phenomena with one another.

14.10. How does genetic recombination in bacteria differ from genetic recombination in the eukaryotes?

14.11. How does genetic recombination in bacteriophages differ from genetic recombination in bacteria?

14.12. How would you distinguish gene conversion from point mutation?

14.13. What inferences about the nature of synapsis and crossing over have been drawn from the phenomenon of negative interference?

14.14. Suppose that an *r*II mutant in T4 bacteriophages failed to complement any A or B mutant in the trans test. What explanations might emerge from recombination tests of such mutants?

14.15. Back mutations are occasionally observed that are almost, but not quite, like the original wild type. What types of analysis can be used to determine if such a back mutant is due to a suppressor mutation at a different gene locus, at a second site within the same gene, or at the same nucleotide pair as the original mutation?

14.16. What would be the effect of crossing over between sister chromatids?

14.17. Distinguish between transduction and abortive transduction.

14.18. If mitotic crossing over occurred in human cells in tissue culture, what information might be gleaned from this phenomenon?

CHAPTER FIFTEEN

Genetics and Development

The genes have two major functions: (1) the transmission of hereditary information and (2) the expression of the hereditary information. DNA synthesis or replication, an autocatalytic process, makes possible the transmission of genetic information from cell to cell and from generation to generation. RNA synthesis or transcription, a heterocatalytic process, mediates the processes leading to the phenotypic expression of the genes. In this chapter we shall consider further the heterocatalytic function of the genes, particularly the manner in which they are called into action to control development and differentiation. Historically, embryology has been the field of study devoted to unraveling the problems of differentiation and development. In the early days of Mendelism, mutant genes were used to a limited extent as tools in embryological research to get at the relation between gene and phenotype, but otherwise genetics and embryology pursued rather independent paths. More recently, with the advances in understanding of the nature of gene action, genetics and developmental biology have become ever more intimately intertwined. Many of the long-standing problems of developmental biology can now be restated and studied in molecular terms, and it now seems probable that the next major advances in biology will come in developmental genetics.

15.1. The Control of Differentiation

One of the fundamental questions in embryology can be traced back to Aristotle: the question of preformation versus epigenesis. Preformation implies that the fertilized egg contains a fully formed individual and that maturation is simply a matter of growth. Early microscopists claimed to see a complete little man or homunculus in the head of the sperm, but this claim of the "spermists" was disputed by the "ovists," who thought they saw the fully formed individual in the egg. Epigenesis, on the other hand, suggests that the fertilized egg is simpler in structure than the adult organism, and that development involves not just growth but the differentiation of new structures as the result of a complex series of interactions in the developing embryo. Although the debate on this question lasted for centuries, the fact, of course, is that there are elements of truth in both points of view. The genes and, to some extent, the cytoplasm can be thought of as having a preformed program that will determine the course of development of the embryo, controlling, for instance, whether the fertilized egg will develop into a mouse or an elephant. On the other hand, the adult characteristics of the mouse or the elephant are not present as such in the egg but result from a sequence of epigenetic processes that take place in the developing embryo.

Early embryology was of necessity descriptive, but once it was recognized that development was not simply a matter of growth or unfolding, the search for the mechanisms of differentiation began. One concept that played a role in this search was Haeckel's "law" that "ontogeny recapitulates phylogeny." The implication is that the stages in the development of the individual retrace the stages in the evolution of the species. This law was based on such discoveries as the fact that human embryos have transient gill slits even though they never have functional gills. However, these gill slits resemble those seen in fish embryos rather than those in adult fish, and thus do not really support Haeckel's law although they do provide suggestive evidence for the common ancestry of man and fishes. Under this theory, proposed shortly after Darwin's work was published, evolution became a causal explanation of development. Experimental embryology, the study of the immediate causes of differentiation and development, had its beginnings toward the end of the last century, but Haeckel's theory, which provided a ready-made explanation for development, probably inhibited rather than encouraged the causal analysis of development.

Once the details of mitosis and meiosis had been worked out, and the primacy of the nucleus as the carrier of hereditary determinants was established, another question of fundamental significance became obvious. If mitotic cell division gives rise to cells all with identical genomes, and if the genes control the metabolic processes of the cells, how does differentiation of the cells occur, giving rise to cells with a variety of structures and functions ranging from the neurones to the red blood cells. Aside from a few exceptional cases (such as in *Ascaris*), chromosome segregation does not occur in somatic cell division, and hence cannot serve as the basis for differentiation. On the contrary, a variety of evidence indicates that differentiated somatic cells carry a complete diploid complement of chromosomes. Furthermore, in insects with polytene chromosomes, such as *Drosophila*, the same banding patterns have been recognized in diverse cell types. Thus, not only the numbers but the internal structure of the chromosomes are the same in different tissues. When chromosome aberrations occur, they are found in all cell types studied. The Bar duplication in *Drosophila*, for example, affects the morphogenesis of the eye, but the duplication is also present in the salivary gland chromosomes, which are not involved at all in the differentiation of the eye. The nuclei of differentiated cells in a given animal have all been found to contain the same amount of DNA despite their diverse functions, and this amount is equal to twice the amount of DNA in the gametes. Hence, differentiation can hardly be explained as being due to selective loss of unused portions of the DNA.

Therefore, differentiation appears to be the result of the regulation of gene activity so that in each type of differentiated cell a specific set of genes is active in transcription and the other genes are "turned off" or inactive. One of the fundamental propositions of developmental biology today is that the process of differentiation involves a series of interactions between nucleus and cytoplasm that serve to stimulate the activity of some genes and suppress that of others. Not only are different groups of genes active in different tissues, but during the course of development different sets of genes are called into play in sequence in the same tissue. A number of lines of evidence support the variable gene activity theory of differentiation.

15.2. Tests for Totipotent Nuclei

If all differentiated cells carry a complete genome, then theoretically it would be expected that the nucleus from a differentiated cell, in the proper circumstances, should be able to support the differentiation and development of a complete organism. A variety of experiments has been conducted to test this hypothesis. The most convincing demonstration of this point was an experiment by Steward and his associates in which single diploid cells from the secondary phloem of the root of the wild carrot, *Daucus carota*, were isolated and cultured. From these isolated differentiated root cells, complete sexually mature plants were obtained.

In amphibia, techniques have been developed by Briggs and King whereby eggs "activated" to divide can be enucleated and then injected with nuclei from cells at different stages of development. These experiments revealed that blastula nuclei injected into enucleated eggs supported normal differentiation and larval development. Embryos containing late gastrula nuclei ceased to develop at a certain stage and showed characteristic abnormalities. Therefore, the gastrula nuclei were not equivalent to the zygote nucleus or to blastula nuclei in their ability to support development. Some sort of nuclear differentiation had apparently occurred.

The nuclei transplanted from different embryonic regions gave rise to embryos with an excess of cells like those of the donor nucleus and a deficiency of other cell types. For example, a nucleus from the endodermal region in the gastrula gave rise to abnormal embryos with an excess of endodermal cells.

Serial transplantations revealed that the nuclear differentiation was relatively stable. In such experiments a late gastrula endodermal nucleus, for example, is transferred to an enucleated egg and then allowed to develop to the blastula stage, when it is sacrificed to provide a number of nuclei for transplantation into a new group of enucleated eggs. In this way a clone of embryos all derived from the same original endoderm nucleus can be obtained. These embryos can then be permitted to develop except for one that is again sacrificed at the blastula stage to provide nuclei to perpetuate the clone. These serial transplantation experiments revealed that the type of abnormality and the stage at which development ceased were quite uniform and characteristic for each nucleus transplanted, but differed among different nuclei even if they were all derived from the same region, the endoderm (Fig. 15.1).

Therefore, these results, obtained with the genus *Rana*, suggest that in the later stages of development, some form of stable nuclear differentiation has taken place, and stand in contrast to the results with the wild carrot, where a complete, sexually mature plant can be obtained from a single differentiated cell. Statements have been made that in the clawed frog, *Xenopus*, Gurdon has obtained complete, sexually mature frogs by transplantation of nuclei from differentiated cells to enucleated eggs, but this claim is not entirely justified because the nuclei came from endoderm cells in the tadpole rather than from adult frogs, and these cells retain some of

SERIAL TRANSPLANTATION OF ENDODERM NUCLEI

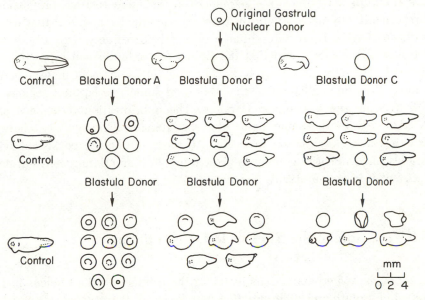

Fig. 15.1. Serial transplantation of *Rana pipiens* endoderm nuclei. (King, T. J., and R. Briggs, 1956. "Serial Transplantation of Embryonic Nuclei," *Cold Spring Harbor Symp. Quant. Biol.* 21:271–290, p. 279, Fig. 3.)

their embryonic characteristics, e.g., the presence of yolk. In theory, there is no reason to suppose that nuclei from differentiated animal as well as plant cells should not be able to support complete normal differentiation and development. It seems merely a matter of time before the secret to reversing nuclear differentiation in animal cells is unlocked. It should be remembered, however, that plant development differs from that of animals. Animals have a well-defined and limited embryonic period, but plants, with a meristematic region at either end, continue growth and differentiation in both root and shoot throughout the adult phase of the plant. Moreover, plant tissues seem more labile than animal tissues, for roots are readily produced by mature stems, buds from leaf margins, and so on. In addition, even though plant cells do undergo differentiation, they do not match the striking cytoplasmic differences to be seen in such differentiated animal cells as muscle, nerve, pancreas, intestinal epithelium, and red blood cells. Thus, plant cells show greater plasticity and less cytoplasmic differentiation than animal cells, and it is not surprising that the problems of development seem somewhat more complex and difficult in animals than in plants.

A more direct test of the genomic equivalence of differentiated cells has been made by DNA-DNA hybridization experiments. In these experiments DNA from mouse brain, kidney, thymus, spleen, and liver were shown to have the same polynucleotide sequences in the same relative proportions. Furthermore, DNA-RNA hybridization experiments have shown that newly synthesized RNA from differentiated cells hybridizes only with 10% or less of the available DNA, and that the RNA from different cell types hybridizes with distinctive DNA fractions. The same DNA segment may be active in more than one cell type so that there are some overlapping regions as well as some that are nonoverlapping, The hybridization tests have also revealed that in the same tissue the spectrum of gene activity changes with time as differentiation proceeds. All these results are consistent with the idea that the entire genome is present in all differentiating cells but only a part of it is active at any one time.

15.3. The Role of the Nucleus in Differentiation

A few of the experiments demonstrating the primacy of the nucleus in the control of growth and differentiation may be cited. If fertilized or activated eggs are enucleated, a certain amount of cleavage may take place, but development soon ceases before gastrulation. Thus, early cleavage is apparently dependent not on the genome of the embryo, but on cytoplasmic factors formed under the control of the maternal genome. However, that phase soon passes, and no further development is possible in the absence of the embryonic genome. Results of this sort have been obtained with sea urchin eggs by Harvey, with salamander eggs by Fankhauser, and with frog eggs by Briggs, Green, and King, among others. The physical methods of enucleation characteristic of these experiments subject the embryo to rather drastic treatment, but recently less severe "chemical enucleation" with actinomycin has produced essentially the same results in sea urchin embryos. The actinomycin binds to the DNA, preventing RNA synthesis, and thus, in effect, blocking gene action, but it does not block DNA synthesis if the dosage is properly adjusted.

Another type of experiment to study the role of the nucleus in development involves hybrids between species whose normal development differs enough from the start so that it can be determined at once whether differentiation is following a maternal, a paternal, or a hybrid pattern. In viable sea urchin hybrids, early morphogenesis follows the maternal pattern, and the expected hybrid influence appears only at or after gastrulation.

Gynogenesis is the development of an egg governed only by the female pronucleus. A sperm may enter the egg, activating its development, but then the sperm degenerates, taking no part in subsequent development. This phenomenon is typical of the nematode, *Rhabdites pellio,* and sometimes occurs in species hybrids in which the paternal chromosomes are eliminated. Irradiation of the sperm in amphibia prior to fertilization produces gynogenesis experimentally, for the female pronucleus controls development. In amphibia, however, a haploid set of chromosomes seems unable to support complete normal development, and the gynogenetic larvae ultimately die. All forms of parthenogenesis, natural and artificial, may be regarded as gynogenesis. In all of these situations in which only a viable maternal nucleus is present, only maternal traits are observed.

Androgenesis is the development of an embryo containing only a paternal nucleus. If an egg is enucleated prior to fertilization, an embryo will develop in which the cytoplasm is maternal and the nucleus is paternal in origin. When sperm and egg differ genetically, it is possible to study which component seems to be in control. In the amphibia, androgenetic haploids fail to complete normal development, but in the wasp, *Habrobracon,* in which the egg pronucleus has been destroyed by irradiation, the male pronucleus does support normal development. In androgenetic hybrids, a typical pattern of development has been observed. Early development appears to be under the control of the cytoplasm. For example, the rate of cleavage in sea urchin hybrids was typically maternal. Later, however, about the time of gastrulation, the sperm nucleus comes into action and takes control. A striking example of this is seen in the Mexican axolotl (*Ambystoma mexicanum*). The pigmentation of the egg and of the early developmental stages is maternal, because the egg pigment is not synthesized after the egg has matured. The larval pigmentation becomes apparent only later, when the melanophores, derived from the neural crest, have differentiated. When an androgenetic hybrid from an enucleated egg of the black strain of the axolotl was fertilized by sperm from the white strain, the larvae that developed were white, showing the predominance of the nucleus over the cytoplasm in this case.

Another experiment of considerable interest was performed with unicellular green algae of the genus *Acetabularia*. Though consisting of only a single cell with a single nucleus, these plants may reach a length of 5 cm. In structure, this differentiated cell somewhat resembles a toadstool, with a rootlike process or rhizoid, a long stalk, and a cap or hat, which is the fruiting body. The rhizoid region contains the nucleus.

If an individual cell is decapitated, a new cap of the same type will be regenerated. This experiment can be repeated indefinitely, and decapitation is always followed by regeneration of a new cap. Thus the nucleated frag-

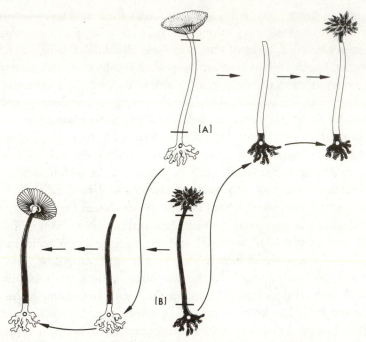

Fig. 15.2. Grafting experiments with *Acetabularia mediterranea* (upper left) and *Acetabularia crenulata* (lower right). The new type of cap formed is characteristic of the species contributing the nucleus. See text. (Saunders, John W., Jr., *Patterns and Principles of Animal Development*, copyright © 1970 by John W. Saunders, Jr. Reprinted with permission of Macmillan Publishing Co., Inc.)

ment has full regenerative power. Removal of the rhizoid region containing the nucleus, however, produces anucleate fragments that are long-lived but have widely varying morphogenetic capacities, apparently depending on the amounts of nucleus-dependent morphogenetic substances present at the time the nucleus is removed. Grafting experiments involving two different species, *Acetabularia mediterranea* and *A. crenulata*, cast further light on the relationship between nucleus and cytoplasm (Fig. 15.2). Each species has a distinctive type of cap. If the cap and the rhizoid with the nucleus are removed from a *mediterranea* cell and the remaining stalk is then grafted to a rhizoid containing the nucleus from *crenulata*, the new cap formed resembles that of *crenulata*. In the reciprocal experiment, the regenerated cap resembles that of *mediterranea*. While the first cap regenerated is sometimes intermediate in character, repeated decapitations lead to the regeneration of caps typical of the nucleus present. Therefore, the type of cap is determined not by the cytoplasm from which it is being generated, but by the nucleus. Presumably, nuclear products are moving up through the stalk to the regenerating cap.

The most plausible explanation of the results with *Acetabularia* is that cap regeneration is dependent on protein synthesis, that this protein synthesis is mediated by RNA present in the cytoplasm, and that this RNA in turn is ultimately dependent for its molecular pattern on the DNA present in the nucleus. The results described above are all consistent with this explanation.

The *Acetabularia* experiments indicate the role of the nucleus in differentiation at the cellular level. We shall now consider other systems at this level before discussing differentiation in multicellular organisms.

15.4. Differentiation in Viruses

One of the simplest systems open to study is that of the bacterium infected by a bacteriophage. Prior to infection the metabolic machinery of the bacterium is tooled up for the maintenance and reproduction of bacteria. After infection by a phage, this machinery is taken over and diverted from its original function to the production of phage particles. This switch can be regarded as differentiation with a vengeance. The process has been studied most thoroughly in the T-even bacteriophages that infect *E. coli*. How far it is safe to extrapolate from this system to differentiation in higher animals remains to be seen, but the insights provided by this system are nonetheless revealing.

Let us consider the events in the life cycle of the T-even phages (Fig. 15.3). The complete, intact virus particle is inert. No metabolism or reproduction occurs until the virus has infected a host cell. Infection involves a chance contact between phage and bacterium, but phage infectivity is itself quite specific and depends on the specific characteristics of the bacterial cell wall and the bacteriophage tail fibers. When the phage is adsorbed to the cell wall, a phage enzyme is released that digests a hole through the cell wall. The tail sheath then contracts, causing the core to penetrate the cell wall like a microsyringe and inject the phage DNA into the bacterium.

After infection there is an *eclipse period* during which no infective phage can be found in the cell. This finding is hardly unexpected, since only the phage DNA was injected. In this form the phage is referred to as *vegetative* phage. Later, when intracellular infective particles start to be

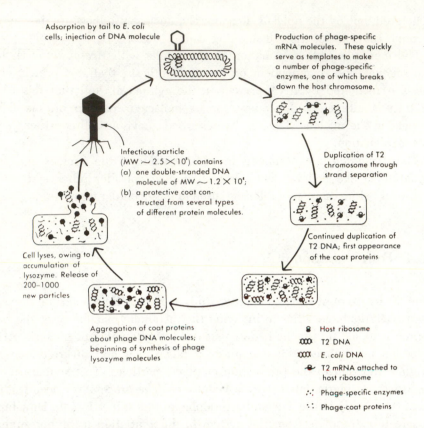

Adsorption by tail to E. coli cells; injection of DNA molecule

Production of phage-specific mRNA molecules. These quickly serve as templates to make a number of phage-specific enzymes, one of which breaks down the host chromosome.

Duplication of T2 chromosome through strand separation

Infectious particle (MW $\sim 2.5 \times 10^8$) contains
(a) one double-stranded DNA molecule of MW $\sim 1.2 \times 10^8$;
(b) a protective coat constructed from several types of different protein molecules.

Continued duplication of T2 DNA; first appearance of the coat proteins

Cell lyses, owing to accumulation of lysozyme. Release of 200–1000 new particles

Aggregation of coat proteins about phage DNA molecules; beginning of synthesis of phage lysozyme molecules

Host ribosome
T2 DNA
E. coli DNA
T2 mRNA attached to host ribosome
Phage-specific enzymes
Phage-coat proteins

Fig. 15.3. The life cycle of the T-even bacteriophages. (From James D. Watson, *Molecular Biology of the Gene*, Second Edition, Copyright © 1970 by J. D. Watson; W. A. Benjamin, Inc., Menlo Park, California.)

formed, their rate of increase is linear rather than exponential. This finding suggests that the new phage particles are not being formed by self-duplication of existing particles, which would be exponential, but by the assembly of independently synthesized component parts.

The discovery that the DNA from the T-even phages contained 5-hydroxymethylcytosine (HMC) rather than the cytosine found in *E. coli* DNA made it possible to distinguish between phage DNA and bacterial DNA. Hence, the kinetics of DNA synthesis within the bacterial cell could be studied, and the amount of phage DNA present could be estimated in terms of phage equivalents. Experiments have shown that synthesis of phage DNA starts about 6 min after infection (Fig. 15.4) and increases rapidly, so that at 12 min, when the first infective particles appear, 50 to 80 phage equivalents of HMC-DNA are present in the infected bacterium.

However, even before phage DNA synthesis begins, other events tran-

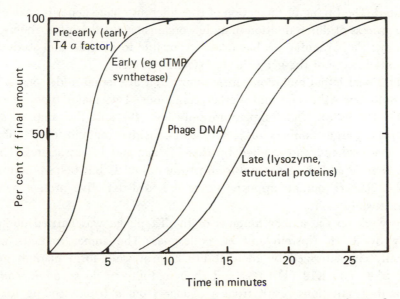

Fig. 15.4. The sequence of synthesis of various components of T4 phage within infected *E. coli* cells at 37°C. (From James D. Watson, *Molecular Biology of the Gene*, Second Edition, Copyright © 1970 by J. D. Watson; W. A. Benjamin, Inc., Menlo Park, California.)

spire. Very shortly after infection with T2 or T4 phage, synthesis of host-specific bacterial DNA, RNA, and proteins ceases. The apparent reason is that within 1 or 2 min after phage penetration, phage-specific messenger RNA has been detected. This mRNA takes over control of the existing protein-synthesizing machinery of the bacterial cell, namely the ribosomes and the transfer RNA. The early proteins synthesized in the first few minutes after infection are different from the proteins in the mature phage particles. These early proteins are enzymes responsible for breaking down the bacterial DNA and setting up the metabolic machinery for the synthesis of viral DNA, which, as noted earlier, starts some 5 or 6 min after infection. The synthesis of the protein coats of the mature infective phage begins only about 9 or 10 min after infection and increases to some 30 or 40 phage equivalents by 12 min, when the first mature phage particles start to appear. The DNA and protein components are synthesized independently within the cell and are then withdrawn irreversibly from the pool for assembly into mature particles. By the time of lysis, about 30 min after infection, some 200 to 300 particles will have been formed. Their release is mediated by a lysozyme, formed under the control of the viral genome, that lyses the bacterial cell wall.

Therefore, the viral life cycle shows a sequence of events that is obviously regulated. The timing of the action of the different phage genes is analogous to the process of differentiation in higher organisms. Furthermore, the use of conditional lethal mutations has made it possible to identify and study the function of many of the genes in the bacteriophages.

Conditional lethal mutations are mutants that are lethal under one set of conditions but wild type under other conditions. Two major types of conditional lethals are the temperature-sensitive (*ts*) and the amber (*am*) mutations. Large numbers of temperature-sensitive mutants of T4 phage have been isolated and studied by R. S. Edgar and his associates. These mutants are unable to grow and form plaques at 42°C but behave normally at 25°C. The *ts* mutants appear to give rise to a defective protein, which is sensitive to heat.

The work on the amber mutants of the T4 phage was carried out primarily by R. H. Epstein and his colleagues. The amber mutants were originally characterized by their ability to grow and form plaques on a strain of *E. coli* K12 (DR 63) and their inability to do so on *E. coli* B. The amber mutations result from a change from a triplet coding for an amino acid to UAG, which is the polypeptide chain-terminating codon. These so-called nonsense mutations therefore give rise to incomplete polypeptide chains of varying length. In the permissive hosts the effects of UAG are suppressed, because "suppressor" mutations produce tRNA molecules that translate the chain-terminating UAG triplet as a compatible amino acid so that the polypeptide is not terminated at that point but a functional polypeptide is formed instead. As a footnote to the history of phage genetics, it may be noted that the "amber" mutations were named in honor of the mother of H. Bernstein (the German equivalent of amber) in fulfilment of a promise made when he agreed to aid in the search for this type of mutant.

Conditional lethal mutants have been of considerable value in the study of gene function in phage. Most of the earlier work had been done with mutants affecting such traits as host range or plaque type, and only a limited portion of the genome was open to study. The conditional lethal mutations, on the other hand, are widely distributed in the genome of the phage, and probably affect all the genes, even those whose functions are indispensable. Hence, these mutants theoretically should permit identification and mapping of all of the genes in the phage. Furthermore, complementation tests can be performed to find the limits of the genes as functional units. Moreover, the conditional lethal nature of the mutants permits mutants with functions essential to life to be isolated and maintained under the permissive conditions and then to be analyzed under the restrictive conditions so that the nature of the block can be studied.

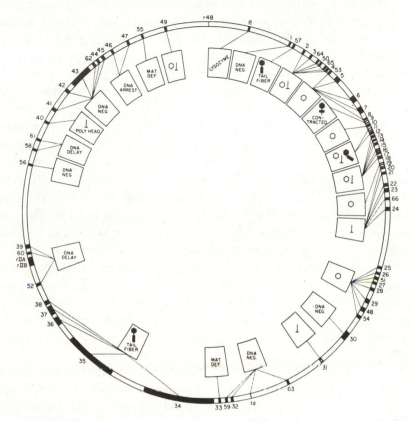

Fig. 15.5. The genetic map of T4 bacteriophage, with cistrons and their phenotypes indicated. The symbols indicate phage components produced by the defective mutants at the sites indicated. ◯ indicates free heads; ⊥, free tails. (Edgar, R. A., and W. B. Wood, 1966. "Morphogenesis of Bacteriophage T4 in Extracts of Mutant-Infected Cells," *Proc. Nat. Acad. Sci. U.S.* 55:498–505, p. 499, Fig. 1.)

Both *ts* and *am* mutations can occur at only a limited number of the total mutable sites within a gene, but they are widely and apparently randomly distributed throughout the T4 phage genome. Both *ts* and *am* mutants have been found in the same gene. As a result of this work, some 70 genes of the T4 phage have now been mapped, which is estimated to be about half of the total possible number of genes on the T4 chromosome (Fig. 15.5). This estimate is based on the observation that the T4 chromosome is about 55 μ or 5.5×10^5 Å long. At 3.4 Å per base pair, there are 1.6×10^5 base pairs. On the assumption that the average gene contains about 1000 base pairs, thus coding for a polypeptide with 300 to 350 amino acids, the phage DNA double helix should contain about 160 genes.

The analysis of conditional lethal mutants has revealed two major groups of genes, one with "early" functions and one with "late" functions. Some genes swing into action immediately after infection, producing early messenger RNA. This early messenger codes for virus-specific proteins, primarily the enzymes responsible for the synthesis of new viral DNA. The late messenger RNA, on the other hand, is concerned with the synthesis of the protein structural components of the phage, their assembly into mature phage particles, and lysis. This sequence of events, (1) synthesis of early enzymes necessary for nucleic acid synthesis, (2) nucleic acid synthesis itself, and (3) synthesis of the other virus-specific proteins including the structural components, appears to be common to all viruses, large and small, DNA and RNA, and single-stranded as well as double-stranded. In the much-studied T4 phage, moreover, the functionally related genes tend to map in clusters. Thus most of the genes concerned with early functions (DNA synthesis) tend to map separately from those with late functions (maturation). This functional segregation is not complete because, for example, gene 31, affecting the head protein, lies between two genes controlling DNA synthesis. However, the maturation genes also tend to occur in functionally related clusters, with some groups of genes involved in formation of the head or capsid, others in the formation of the tail, the contractile sheath, the tail plate, the tail fibers, and so on.

These genes have been delimited by means of complementation tests and mapped. Their functions have been determined by growth of the conditional lethal mutants under restrictive conditions. Studies of the *am* and *ts* mutants have revealed over 40 genes concerned with phage morphogenesis (Fig. 15.6). When these mutants are grown under restrictive conditions, no mature phage particles are formed, but the lysate from such a culture contains recognizable phage components except that one is usually defective or missing. Thus lysates from mutants may have heads but no tails, tails without heads, or heads and tails but no tail fibers. Interestingly enough, mixtures of lysates from mutants with different defects show complementation *in vitro* to give high yields of active phage (Fig. 15.7), but mixtures of lysates with the same defect do not complement. This sort of experiment suggests that to some degree the components of the phage coat are capable of self-assembly.

The tobacco mosaic virus (TMV) is a small RNA virus of very simple structure. Under the proper conditions, in the absence of the RNA, TMV protein spontaneously forms long hollow rods like those of the virus, but they are of indeterminate length. Fraenkel-Conrat and Williams went a step further and isolated the RNA of TMV from the protein as well as the TMV protein free of any RNA. Neither isolate alone was infectious (al-

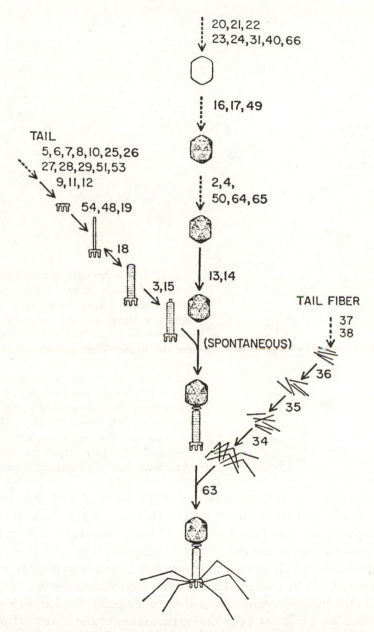

Fig. 15.6. The path of morphogenesis in phage T4. Steps in dashed arrows have not yet been demonstrated *in vitro;* those with solid arrows have. The numbers refer to the genes controlling the steps indicated. (Wood, W. B., R. S. Edgar, J. King, I. Lielausis, and M. Henninger, 1968. "Bacteriophage Assembly," *Fed. Proc.* 27:1160–1166, p. 1163, Fig. 3.)

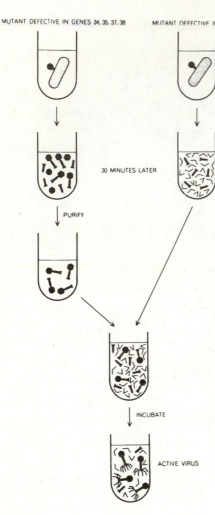

MUTANT DEFECTIVE IN GENES 34, 35, 37, 38   MUTANT DEFECTIVE IN GENE 23

30 MINUTES LATER

PURIFY

INCUBATE

ACTIVE VIRUS

Fig. 15.7. *In vitro* complementation with lysates of phages with different defects. One mutant virus strain (left) carries defective genes for the tail fibers. The other strain (right) has a defective gene for the head. When extracts of the two cultures are mixed and incubated at 30°C, infectious virus particles are produced *in vitro* by the assembly of active virus from the components provided by the two strains. (From Wood, W. B. and R. S. Edgar, "Building a Bacterial Virus." Copyright © July 1967 by *Scientific American*, Inc. All rights reserved.)

though the free RNA was sometimes infectious at a very low level of efficiency compared to the intact virus.) However, when the two isolates were combined *in vitro,* they reconstituted intact, infectious virus particles indistinguishable from natural TMV. In this case the protein cylinders corresponded in length to the RNA molecule. These experiments suggest that the assembly of phage particles occurs spontaneously due to forces related to the molecular structure of their components. With more complex viruses such as T4, it has been shown that some of the structural protein components are capable of self-assembly. However, the entire process of maturation is not simply a matter of spontaneous self-assembly from the component parts, but instead is subject to genetic control, since a number of genes controlling maturation processes have been identified. Gene 20, for example, is responsible for normal head length. Mutants at this locus give rise to "polyheads" of excessive length, while mutants in gene 66 produce abnormally short heads. A mutant in gene 21 results in empty protein heads lacking DNA, and a mutant in gene 31 causes aberrant head protein subunits that aggregate in clumps so that no head is formed at all. Comparable genes have been found that regulate other aspects of morphogenesis in phages. Therefore, the genes in the T4 phage appear to be responsible, not just for the synthesis of the protein subunits of the mature phage particle, but also for the control of various steps in the assembly process. In fact, the life cycle of the phage within the host gives evidence that a number of genes acting in appropriate sequence regulate the entire process, from its earliest to its final stage when lysis occurs. It is for this reason that this seemingly simple system holds so much interest for the student of the genetic control of differentiation. Many details of the process are now known, more than can be profitably considered here, but the references can be consulted for additional information about these fascinating studies.

15.5. Regulation of Gene Activity in Bacteria

Although bacteria are usually considered to be relatively simple organisms, it has long been realized that biochemically they are rather versatile in their ability to synthesize essential metabolites from relatively simple precursors. Moreover, bacterial genes do not all function at full capacity all

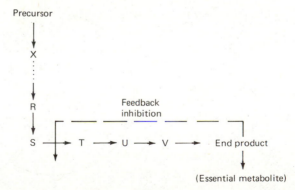

Fig. 15.8. Feedback inhibition. The end product (E) of a synthetic sequence inhibits the activity of an enzyme catalyzing an early step in the sequence (A → B). When the end product is depleted, the inhibition ends, and synthesis of E is resumed.

of the time in mediating the synthesis of proteins. Instead, gene action in bacteria is subject to regulation and control that ensures that the cell functions at a high level of metabolic efficiency. These regulatory mechanisms are built into the bacterial genome, and hence are subject to genetic analysis. The pioneering studies of Jacob and Monod have contributed greatly to our present understanding of the regulatory mechanisms at work in bacteria.

The simplest form of control seems to be *feedback inhibition*. In this situation, the end product of a biosynthetic pathway inhibits the activity of the first enzyme in the synthetic sequence (Fig. 15.8). Hence, if the end product is not being used, the synthetic pathway is shut down, and the

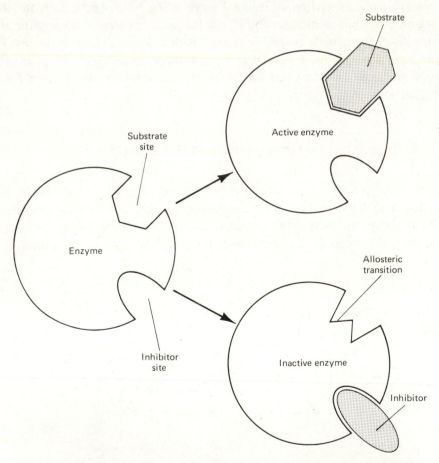

Fig. 15.9. Allosteric inhibition. Enzyme activity is blocked when an inhibitor (or end product) is attached to an inhibitor site on the enzyme molecule, which so alters the structure of the attachment site for the substrate that the substrate can no longer attach to the enzyme and the reaction is blocked.

level of activity is dependent on the rate of utilization of the end product. Feedback inhibition is distinguished by the fact that only the end product or its analogues are effective, that it affects only one early enzyme, and that it acts at once, independently of the presence of existing enzymes. For example, if the end product of a metabolic pathway is tryptophan, addition of tryptophan to the medium leads to an immediate cessation of further synthesis of tryptophan, even though the necessary enzymes are still present. An interesting aspect of end-product inhibition is that the end product does not inhibit by competing with the normal substrate for attachment to the enzyme. Instead, the enzymes involved in feedback inhibition appear to have a protein structure with two distinct specificities, one related to its substrate, the other to the end-product inhibitor. This type of configuration has been called *allosteric*. The small end-product molecule is thought, by its attachment to the larger enzyme molecule, to change the enzyme's configuration so that it no longer can become attached to its normal substrate (Fig. 15.9). The concept of allosteric inhibition can be broadened to include the possibility of activation by allosteric effectors, and evidence for this sort of action has been found. Since the inhibitors and effectors are sterically different from the substrates, this type of system could be used to coordinate the activities of different synthetic pathways, and the model is useful in attempts to explain not only sequential gene action in microorganisms, but also differentiation in higher organisms.

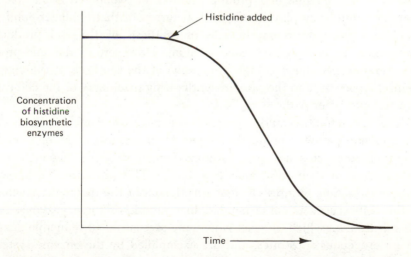

Fig. 15.10. Enzyme repression illustrated with histidine synthesis. When histidine is added to a growing culture of bacteria, the level of the enzymes involved in histidine biosynthesis declines. In essence, the genes responsible for the synthesis of these enzymes are turned off. If this histidine is removed, synthesis of these enzymes is resumed.

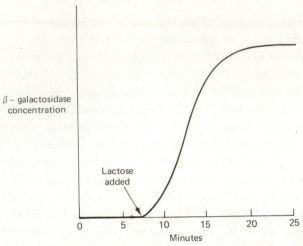

Fig. 15.11. Enzyme induction. When a new substrate, lactose, is added to the culture, synthesis of β-galactosidase, which cleaves lactose into its component sugars, increases to a high level.

Repression is a different type of control mechanism that is also triggered by the end product of a synthetic pathway. Unlike feedback inhibition, however, which has no direct effect on enzyme synthesis, repression involves shutting down the synthesis of all the enzymes involved in the pathway (Fig. 15.10). Since these two mechanisms do the same job, and are triggered by the same end product, it may be wondered what role this seeming redundancy plays in the cell. It turns out that feedback inhibition provides a fine adjustment in rate of synthesis of the end product to fluctuations in end-product concentration. However, if the end-product concentration becomes too high, repression of the synthesis of the enzymes themselves occurs, and the protein-synthesizing machinery of the cell can be diverted into other paths.

Feedback inhibition and repression are negative control systems, with the synthetic pathway being shut down by the excess of its end product. However, some enzymes are synthesized in appreciable amounts only in the presence of their substrate (Fig. 15.11). This phenomenon, known as *enzyme induction*, ensures efficient functioning of the metabolic machinery of the cell—energy is not expended in synthesis of these enzymes unless the substrate on which they act is actually present. Most inducible enzymes act on exogenous substrates, and are exemplified by the enzyme system in *E. coli* that acts on lactose. Induction and repression are similar in several respects. The inducing compound is very specific in its effects although chemical analogues not affected by the enzymes may also act as inducers. Furthermore, if several enzymes are involved in the pathway, all of them will be induced. Finally, *constitutive* mutants have been detected in both inducible and repressible enzyme systems, that is, mutants that cause the continuous synthesis of the enzymes even in the absence of the inducing substrate or in the presence of the repressor.

Our understanding of the mechanisms of induction and repression is due in large measure to work on the control of lactose fermentation in *E. coli* K12 by Jacob and Monod, for which they were awarded Nobel Prizes. As the result of their work on the control of genetic expression in *E. coli,* they recognized two kinds of genes, *structural* genes and *regulatory* genes. A structural gene determines the structure and hence the function of a polypeptide or protein. A regulatory gene, however, produces a cytoplasmic substance that controls the activity of a structural gene or a cluster of structural genes.

15.6. The Operon Concept

We have already noted that in repression the synthesis of a group of enzymes in a pathway is turned off by some sort of repressor. The evidence suggests that the repressor combines with an entity called an operator. The operator is essential for the translation of the genetic information in the structural genes into protein. There is a single operator for the entire group of structural genes controlling the coordinated enzyme system, and the operator plus the structural genes it controls are linked together to form a single integrated unit called an *operon* (Fig. 15.12). The function of the operator locus is to control the initiation of the synthesis of messenger RNA. The repressor, by combining with the operator, prevents the formation of mRNA, and hence synthesis of any proteins specified by the genes in the operon ceases. Apparently, an operon produces a single, large mRNA molecule, which codes for all the proteins formed by all the different structural genes in the operon. If this is the case, the coordinate nature of repression and of induction is understandable, and it also suggests that each structural gene carries the information necessary for polypeptide chain initiation and termination. Moreover, if polycistronic mRNA is formed from a fixed starting point, the operator, then the process of transcription is polarized. This conclusion was borne out by the discovery of polar mutations, in which genes between the operator and the polar mutant have normal function, but those distal to the polar mutant from the operator have impaired activity. Both *amber* and *ochre* chain-terminating mutants and frame-shift mutants are polar mutants.

A. COORDINATED ENZYME INDUCTION

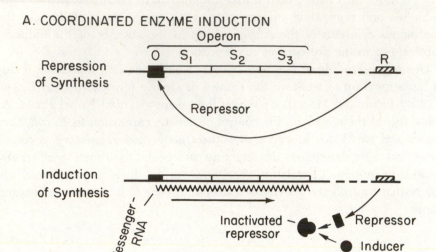

B. COORDINATED ENZYME REPRESSION

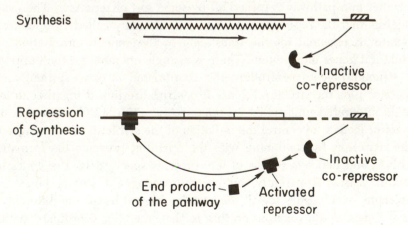

Fig. 15.12. The operon model in relation to enzyme induction and enzyme repression. S_1, S_2, S_3 = three adjacent structural genes; O = the operator gene; R = the regulator gene. A. Enzyme induction: The product of R is a repressor that combines with O to prevent the transcription of the structural genes. The inducer combines with and inactivates the repressor so that the operator gene is no longer repressed, and transcription occurs. B. Enzyme repression: The product of R is an inactive co-repressor that is activated by combination with the end product of the pathway. The activated repressor then combines with the operator gene to repress further synthesis. (Hayes, W. 1968. *The Genetics of Bacteria and Their Viruses*, 2nd ed. John Wiley & Sons, Inc., New York, p. 723, Fig. 137.)

The operon concept has proved very fruitful because it can account for both induction and repression within a single scheme. In induction, the regulator gene, which is not closely linked to the operon it regulates, is postulated to produce a repressor substance (R) that, in the absence of substrate, combines with the operator locus and thus represses the synthesis of mRNA. When a substrate such as lactose is present, the repressor combines with the substrate and is thereby inactivated so that the operator locus now can initiate the transcription of mRNA, and the enzymes for lactose utilization can be formed.

In repression, on the other hand, the R substance produced by the regulator gene is inactive alone, but is activated by union with the end product of the metabolic pathway. This activated repressor then combines with the operator locus to repress further enzyme synthesis.

If the product of the regulator gene is called R and the small molecules or allosteric effectors that combine with R are called F, the interactions between them take the form $R + F \rightleftharpoons RF$. Hence, in induction R represses enzyme synthesis, but in repression RF acts as the repressor. This scheme is now supported by a variety of lines of evidence, among them the fact that mutations have been found not only in structural genes, but in operators and in regulator genes as well. The operator was originally thought to be the site both of repressor action and of the initiation of mRNA synthesis. However, these functions are now known to be separate, because a promoter locus immediately adjacent to the operator locus has been shown to be responsible for the initiation of mRNA synthesis. The operon concept has been further supported by the isolation of the repressor of the *lac* operon and the demonstration of its protein nature.

An important question is the generality of the operon model for bacterial repression-derepression systems of gene action. Is it applicable to the regulation of the differentiation of cells in higher organisms? The answer seems to be that coordinately regulated polycistronic systems of the sort just discussed, that is, operons, are quite rare in eukaryotic organisms and are not especially common even in bacteria. A few clusters of linked genes controlling functionally related steps in intermediary metabolism have been identified in *Neurospora* and yeast (*Saccharomyces*), but most of the functionally related loci are scattered throughout the genome among different chromosomes. Therefore, the applicability of the operon model would seem to have limitations. Perhaps the most important aspect of the work on the operon in relation to differentiation is the demonstration of the existence of regulator genes, for now we know that genes in one part of the genome code for proteins that govern the activity of structural genes in a different, nonadjacent part of the genome. Since more than one regulator

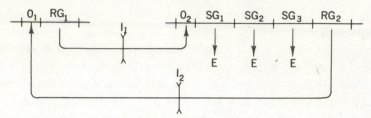

Fig. 15.13. The compound operon model. SG = structural genes; O = operator genes; RG = regulator genes; I = inducers; E = enzymes. RG_1 controls controls the operon containing SG_1, SG_2, and SG_3 plus another regulatory gene, RG_2. RG_1 belongs to an operon sensitive to the repressor synthesized by RG_2. The action of RG_1 is antagonized by I_1, which activates SG_1, SG_2, SG_3, and RG_2. The action of RG_2 is antagonized by I_2. (Davidson, E. H., 1968. *Gene Activity in Early Development*. Academic Press, New York, p. 287, Fig. 95b.)

gene has been found to influence the operation of a single operon, it is becoming clear that regulator genes form a significant part of the genome and that complex systems of control may develop (Fig. 15.13).

15.7. Flagellar and Ciliary Antigens

Another system in bacteria of interest to the student of differentiation is the flagellar antigen system in the genus *Salmonella*. Bacteria in this genus normally have flagella and are motile. The flagella are formed from subunits of a globular protein that appears to be formed as the result of the activity of a single gene. Thus, their antigenic properties or "serotype" can be determined by an appropriate antiserum and reflect the amino acid sequence and arrangement specified by that gene. In some cases, the flagella in a single bacterial strain alternate back and forth between two distinct serotypes. In any one bacterial cell, the flagella are all of one antigenic type, but when multiplication occurs, cells of the other antigenic type arise with a frequency too great to be attributed to mutation. Indeed, the evidence indicates that two separate structural genes are involved, only one of which can be expressed at any one time. Therefore, two phases exist. When the structural gene responsible for one type of flagellar antigen is expressed, the other gene is suppressed. This diphasic system offers a simple model for the study of differentiation, but many of the details remain to be worked out.

A comparable system of ciliary antigens has been studied in the eukaryotic protozoan, *Paramecium*. In this case a number of antigenic types have been identified, each under the control of a single gene. These genes are scattered throughout the genome. Again only one gene at a time is ordinarily expressed, but in cells with identical genotypes grown under different environmental conditions, as many as a dozen different antigens have been identified. Thus, within its genotype the *Paramecium* cell carries the capacity to produce a number of different ciliary antigens, but the activities of these dispersed genes are coordinated in such a way that only one is expressed while all the others are repressed; their expression is mutually exclusive. These systems, in which the activation and repression of a related group of genes depends on the environmental conditions, offer still another model of considerable interest to the student of differentiation and development.

15.8. Differentiation in Multicellular Organisms

The differentiation and development of a multicellular organism start from a single cell, the fertilized egg or zygote. This zygote contains all the information necessary for the development of the mature organism with its highly differentiated cells, tissues, and organs. This information is contained in the DNA of the chromosomes, but not just in the DNA alone—organelles such as mitochondria, chloroplasts, and cell membranes appear to serve as structural templates for the formation of additional structures of the same type. Moreover, certain intact cyclical metabolic sequences must also be present in the fertilized egg. For example, certain amino acids must be present initially to permit the synthesis of the enzymes necessary for the synthesis of more of these same amino acids. As we have already seen, early development is dependent, not on the genome of the zygote, but on a pre-programmed system formed under the control of the maternal genome. However, the embryonic genome assumes control at a fairly early stage, and plays a primary role in the subsequent events during development.

DNA Content of Different Species

Since the role of DNA has been shown to be so important in the governance of metabolic activity, a correlation might be expected between the metabolic complexity of an organism and the total amount of chromosomal DNA. To make such a determination, it is necessary to be able to distinguish between chromosomal and extrachromosomal DNA. When such measurements have been made, it has been shown that the DNA content of various types of differentiated cells from a single individual are equivalent

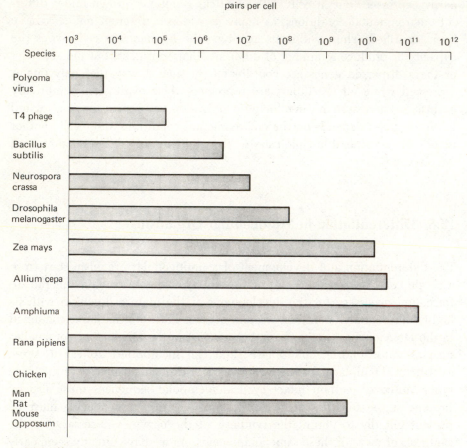

Fig. 15.14. Variations in genome size in different groups of organisms expressed as minimum and maximum numbers of nucleotide pairs per haploid cell in the species examined. All mammals have about the same genome size.

and remain so during the course of development. However, the DNA content of cells from a variety of different species varies greatly, and seems to show little relationship to the apparent relative complexity of these species (Fig. 15.14). However, if the minimum amount of DNA observed for each major group of organisms is regarded as the amount necessary to achieve that level of organization, a more coherent picture emerges, for the groups regarded as showing greater complexity do seem to require a greater minimum amount of DNA than the less complex groups (Fig. 15.15).

The function of this added amount of DNA poses a problem. It has been estimated that the chromosome of *Escherichia coli* contains about 2000

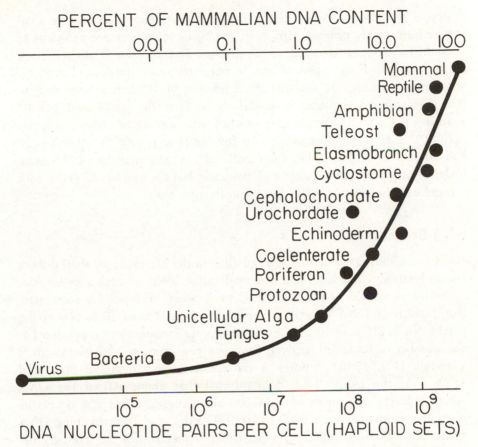

Fig. 15.15. The minimum amount of DNA per haploid genome in different groups of organisms. The values are more reliable for mammals, amphibia, teleosts, bacteria, and viruses than for the other groups. The ordinate has no numerical scale, and the shape of the curve has little meaning. The figure shows that increased complexity of biological organization is accompanied by an increase in DNA content. (Britten, R. J., and E. H. Davidson, 1969. "Gene Regulation for Higher Cells: A Theory," *Science* 165:349–357, p. 352, Fig. 3.)

genes, while the chromosomes of man may contain up to 400,000 genes. Nevertheless, the enzymatic capabilities of *E. coli* are at least equal to those of *Homo sapiens*. Moreover, the amount of DNA needed to code for all the various proteins identified in mammalian cells is less than a thousandth of all the DNA present in those cells. Although additional kinds of protein molecules will undoubtedly be identified, it is unlikely that so far, only one out of every thousand kinds of protein molecules has been recognized. It might be thought that the increased size of the genome is due to redundancy, a repetition of some of the nucleotide sequences, and as we shall see, redundancy is present. However, redundancy alone is not sufficient to account for the increased size of the genome. In the calf and other species, more than half of the DNA occurs in nonrepetitive sequences. Therefore, some other explanation for this great amount of DNA has been sought. The best present explanation seems to be that this additional genetic material has a regulatory function rather than serving as structural genes coding the amino acid sequences in functional enzymes. In other words, the amount of DNA coding for functional enzymes may be of the same order of magnitude in bacteria and mammals, but the number of genes with a regulatory function is vastly greater in the mammals.

DNA Redundancy

Before considering the nature of regulation in the Metazoa, we shall discuss the redundancy that has been observed in the DNA of higher organisms. Although about 45% of the DNA in the calf occurs in repetitive sequences, the quantity of DNA in repeated sequences ranges from·15 to 80% of the total DNA in different species. Furthermore, the frequency of repetition (or the number of copies of a given sequence per genome) also shows great variation (Fig. 15.16). Studies of the reassociation kinetics of denatured DNA from the mouse have demonstrated that about 10% of the DNA forms repeated sequences of 1,000,000 or so copies, about 25% occurs as repetitive sequences with 1,000 to 100,000 copies, and the remaining DNA forms one or a few copies of a particular sequence. On the average, the repeated sequences consist of several hundred nucleotides. Such repetition could conceivably result from polyploidy, polyteny, or linear repetition of the sequence. Polyploidy has been eliminated by direct cytological examination, and polyteny, of the sort seen in *Drosophila* and other Diptera, has not been observed either. Furthermore, these repeated sequences have been measured in gametes, and polytenic gametes would pose difficult genetic problems with respect to the accumulation and transmission of mutations from one generation to the next. Thus, the only possibility remaining is linear repetition of the sequences. If this is the situation, then the organisms with the greatest amount of DNA in their genomes should have the greatest total length of chromosomal material. The evidence, sketchy though it is, suggests this to be the case. The repeated sequences are not always exact copies of one another, but seem to form families of repeated sequences, ranging in similarity from perfect copies to nucleotide sequences matching one another in only about two-thirds of their

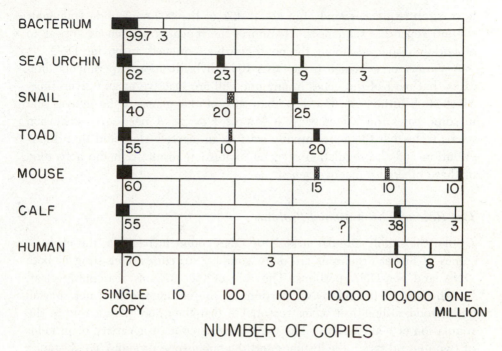

Fig. 15.16. Patterns of repetition of DNA sequences in various organisms. The dark bands indicate repetitive classes of DNA; the width of the band indicates what percentage of the total DNA is present in that class of repetitive DNA. The numbers below the bands give the percentage. The position of the band indicates the number of copies of the repeated DNA sequence. (From Britten, R. J. and D. E. Kohne, "Repeated Segments of DNA." Copyright © April 1970 by *Scientific American,* Inc. All rights reserved.)

nucleotides. In some cases, the repeated sequences are located in the same region of a single chromosome. For example, in the amphibian, *Xenopus,* the 28S and 18S ribosomal RNAs are each encoded by about 450 cistrons arranged in an alternating sequence in a single chromosome region. Another form of multiplicity of the genome has been discovered during oögenesis, for the cistrons governing 28S and 18S rRNA are replicated differentially compared with the rest of the genome. The replicas of these cistrons are freed from the chromosome and then form some 1000 nucleoli lying free in the germinal vesicle, all engaged in the synthesis of rRNA.

In most cases, however, the repeated sequences are scattered throughout the genome. If bovine DNA is broken into fragments, for instance, more than 95% of the fragments about 20,000 nucleotides in length carry repeated sequences, about 75% of the fragments some 5000 nucleotides long have such sequences, and only about 45% of the 500-nucleotide fragments contain repetitive sequences. Therefore, the repeated sequences appear to be widely dispersed among the nonrepetitive sequences in bovine DNA.

A variety of evidence from DNA-RNA hybridization experiments indicates that most of the genes in any one cell are inactive at any given time. Most such estimates show less than 10% of the genes to be active, and in some cases the values were as low as 1 or 2%. A necessary conclusion seems to be that the genes usually exist in an inactive state and that differentiation and the regulation of gene expression result from the activation of these otherwise inactive genes.

The Role of Chromosomal Proteins

Unlike the simple circular strands of DNA found in bacteria, the chromosomes of higher organisms are more complex structures, consisting of both DNA and non-DNA portions. The role of the DNA is reasonably clear, but the role of the non-DNA portions of the chromosomes is not so well understood, although it seems reasonable that they may play a role in the regulation of gene activity. The chromosomes contain a variety of proteins of two general types, the histones and the nonhistone proteins. The histones, which are basic proteins, are associated with the DNA in the chromosomes in approximately equal amounts, the measured ratios of histone to DNA ranging between 0.8 and 1.3. By contrast, the nonhistone proteins vary greatly in amount at different stages of development. In undifferentiated cells the nonhistone proteins may represent only one-tenth the amount of DNA or histone present, but as differentiation progresses the chromosomes may come to contain considerably more nonhistone protein than either histones or DNA. These results have led to the concept that the histones are responsible for the inhibition of DNA-dependent RNA synthesis. In other words, when associated with the DNA, they prevent transcription and hence keep the genes inactive. Several lines of experimental evidence support this hypothesis; in essence they show that removal of the histones leads to derepression and an increase in RNA synthesis while addition of the histones results in resuppression of RNA synthesis. One of the puzzling aspects of the role of the histones is the relatively small number of different types of histone molecules, which is incompatible with the degree of specificity needed if the histones are to regulate the transcription of individual gene loci.

The function of the nonhistone acidic proteins is even less clear, but it seems possible that they have some role to play in regulating the function of the active genes in the otherwise repressed genome.

Regulation of Genes in Multicellular Organisms

Now we can approach the question of how the regulation of gene action occurs in higher organisms. Regulatory processes may occur at various levels in biological systems. Three major levels of control are at the level of transcription by control of the synthesis of mRNA, at the level of translation, through control of the synthesis of polypeptides, or at the enzyme level, by the regulation of enzyme activity. Examples of these types of regulation of gene action by artificial means are the following. The antibiotic, actinomycin D, has been shown to inhibit protein and RNA synthesis but not DNA synthesis. In other words, it blocks the synthesis of mRNA and hence prevents transcription. Another antibiotic, puromycin, intervenes at the translation step, blocking the growth of normal polypeptide chains. Perhaps the most readily understood example of regulation at the enzyme level is that of end-product inhibition or feedback inhibition. A number of cases of feedback inhibition in mammalian systems have been described for purine and pyrimidine metabolism as well as for other synthetic pathways. It may be noted that the three levels of regulation cited above—transcription, translation, and enzyme function—are not necessarily all-inclusive, for other sites or levels of control may also influence gene expression. The complexities of differentiation in multicellular organisms have only recently become the subject of investigation at the molecular level, and much remains to be learned.

The large amounts of DNA in the cells of higher organisms, the dispersal of functionally related structural genes throughout the genome rather than in clusters as in bacterial operons, the redundancy of many of the sequences, and the presence of complete genomes in differentiated cells with only a small fraction of the genes actually expressed, all seem to require an explanation. These facts are not easily accounted for by the operon theory. A recent effort by Britten and Davidson to develop a theory for the regulation of gene action in higher organisms is shown diagrammatically in Fig. 15.17. Britten and Davidson postulate the following: 1. *Producer genes,* the equivalent of structural genes in the operon model, which are transcribed to form all RNA molecules except those involved in genomic regulation. Producer genes would include not only genes generating messenger RNA as a template for polypeptide synthesis, but also such genes as those giving rise to transfer RNA. 2. *Receptor genes,* DNA sequences linked to producer genes that permit transcription of the producer genes when a complex is formed between a receptor gene and an activator RNA molecule. 3. *Activator RNA,* RNA molecules that combine specifically

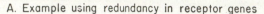

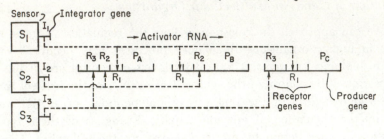

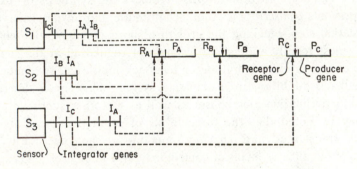

Fig. 15.17. Models for the regulation of gene action in higher organisms. See text. (Britten, R. J., and E. H. Davidson, 1969. "Gene Regulation for Higher Cells: A Theory," *Science* 165:349–357, p. 350, Fig. 1.)

with receptor genes to activate transcription of the linked producer genes. A single-stranded RNA bound to double-stranded DNA is postulated, but it is also suggested that the activators could just as well be protein molecules coded by the activator RNA. 4. *Integrator genes*, which are responsible for the synthesis of activator RNA. Linked integrator genes can cause, through their activator RNAs, the coordinated action of a number of different producer genes in response to a specific initiating event even though the producer genes do not share the same receptor gene sequence. 5. *Sensor genes*, which serve as binding sites for agents such as hormones, resulting in the activation of specific patterns of gene action in the cells. Binding of the inducing agent to the sensor gene is sequence-specific, but may require an intermediary protein molecule, which complexes with the inducing agent and then binds to the sensor gene DNA. 6. *Batteries of genes*, sets of producer genes mobilized by the activation of a sensor gene and its associated integrator genes.

The model postulates that differentiation is due to the activation of various batteries of producer genes. The initiating event may be the presence of a hormone, which is complexed with a sensor gene. The sensor gene then activates transcription of its linked integrator genes, which produce activator RNA. The activator RNA then binds to receptor genes, which in turn promote the transcription of the producer genes. The web of possible interactions with this model is more complex than with the operon model. The observed redundancy of DNA sequences can then be attributed to redundancy either in the receptor genes or in the integrator genes. The histones are postulated to serve in repression but not in activation or regulation. The increased amounts of DNA in higher organisms are thought to be due to the increased numbers of genes involved in the regulation of gene activity rather than to increased numbers of producer genes. Moreover, the model provides for the activation in a tissue of batteries or groups of producer genes that are not physically linked, the situation that prevails in eukaryotes. It also provides for the activation of the same producer genes in a number of different tissues and in different combinations. In addition, it provides an explanation of how a single inducing agent such as a hormone could lead to the activation of a number of different batteries of genes.

While much work undoubtedly remains to be done to establish the validity of the model, evidence already exists compatible with some of its postulates. A number of agents are known, such as various plant and animal hormones and embryonic inducing agents, that cause major shifts in producer-gene activity in specific tissues. RNA molecules confined to the nucleus and binding to chromosomal DNA have been identified and seem to fit the requirements for the postulated activator RNA molecules. Furthermore, some genes with regulatory functions have been identified. In the model, mutations in regulatory genes would be expected to have pleiotropic effects. The Notch series of deficiencies in *Drosophila* fit this requirement in that they cause defects in gut formation, failures in mesodermal differentiation, and abnormalities in the ectoderm, both in the neural and epidermal derivatives. These effects on all three germ layers are consistent with mutation in integrator genes. The Activator-Dissociator system in maize studied by McClintock is another control system compatible with the model. In this case, the expression of the producer genes, for example those governing anthocyanin synthesis, is under the control of the contiguous linked *Ds* element, the equivalent of the receptor genes in the model, and *Ds* in turn is responsive to the *Ac* element, distant from it in the genome and corresponding to the integrator genes in the model.

Two types of evidence have recently come to the fore in the demonstration of differential gene activity in cells in different developmental stages or

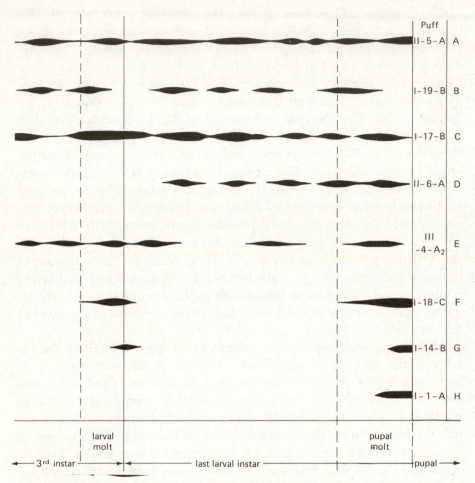

Fig. 15.18. The sequence of puffing at four sites (A, B, C, D) in a salivary gland chromosome of *Chironomus tentans* larvae. Two bands shown do not puff during this period, which, starting from the top, shows the changes observed before and during the molt leading to pupation. (Clever, U., 1966. "Gene Activity Patterns and Cellular Differentiation," *Amer. Zool.* 6:33–41, p. 36, Fig. 2.)

in different tissues. One involves DNA-RNA hybridization techniques to compare RNAs from different sources. These studies have shown that different mRNAs are synthesized in different tissues and in the same tissue at different developmental stages, indicating that different sets of genes are active in different tissues or in the same tissue at different stages of development (Fig. 15.18). Similarly, the puffing patterns in the giant salivary gland chromosomes of *Drosophila* and other Diptera have been found to be tissue- and stage-specific. The puffs are local swellings in the

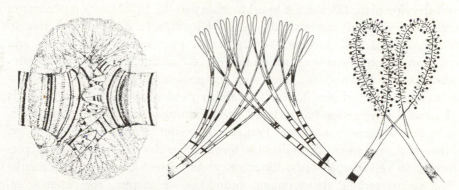

Fig. 15.19. The structure of a large salivary gland chromosome puff. At left, the Balbiani ring as seen in the light microscope. Center, a few of the fibrils at high magnification in the light microscope. Right, with the electron microscope, two puff fibrils can be seen together with granules believed to be mRNA. (From Beermann, W. and U. Clever, "Chromosome Puffs." Copyright © April 1964 by *Scientific American*, Inc. All rights reserved.)

giant chromosomes, usually formed by the uncoiling of a single band from its tightly condensed condition (Fig. 15.19). Along with the uncoiling of the nucleohistone fiber, the puff is marked by an accumulation of acidic, nonhistone proteins and of RNA. Although the exact relation between puffing and gene activity cannot be stated with complete certainty, it is worth noting that the steroid insect hormone ecdysone induces the formation of puffs in certain bands of the salivary gland chromosomes of *Chironomus* larvae (midges). The RNA synthesis that occurs in these puffs can be blocked by actinomycin D, an inhibitor of DNA-dependent RNA synthesis. Thus there seems little reason to doubt that the puffs represent active gene loci. In mammals, the effects of the hormone, estrogen, have been widely studied, and it, too, has been shown to have extensive and rapid effects on patterns of RNA synthesis. These and studies with other hormones, both plant and animal, suggest that hormones produce their effects by activating genes to synthesize RNA, which in turn leads to protein synthesis and the acquisition of new competence and functions by the cells.

This chapter must end, but not because we have exhausted the subject matter in developmental genetics. Rather, there are fascinating topics that we have scarcely touched. Developmental biology is a rapidly changing discipline, influenced, as most other areas of biology have been, by the recent advances in molecular biology. Furthermore, new techniques have made possible new approaches to the old basic problems of differentiation and development. DNA-RNA hybridization and the hybridization of somatic cells from different species have made it feasible to study gene action in ways never before possible. To studies of phenocopies and visible and lethal mutations have been added techniques for the study of temperature-sensitive mutants and the effects of compounds of known effect such as actinomycin D and puromycin. The ability to identify the protein products of normal gene action has reduced the necessity of using mutations in developmental genetic analyses. Perhaps the major change is the ability to formulate and study in molecular terms such classical problems as preformation versus epigenesis, totipotency versus restricted potency, regulative versus mosaic development, induction, organizers, competence, regeneration, immunity, and so on in molecular terms. Striking progress has already been made and is apt to continue at an accelerating rate.

Additional Reading

Beale, G. H., 1957. "The Antigen System of *Paramecium aurelia*," *Int. Rev. Cytol.* 6:1–23.

Beermann, W., 1963. "Cytological Aspects of Information Transfer in Cellular Differentiation," *Amer. Zool.* 3:23–32.

Beerman, W., and U. Clever, 1964. "Chromosome Puffs," *Sci. Amer.* 210(Apr.): 50–58.

Bell, E., ed., 1965. *Molecular and Cellular Aspects of Development.* Harper & Row, New York.

Bonner, J., 1965. *The Molecular Biology of Development.* Oxford University Press, London.

Briggs, R., E. U. Green, and T. J. King, 1951. "An Investigation of the Capacity for Cleavage and Differentiation in *Rana pipiens* Eggs Lacking 'Functional' Chromosomes," *J. Exptl. Zool.* 116:455–499.

Briggs, R., and T. J. King, 1959. "Nucleocytoplasmic Interaction in Eggs and Embryos." In *The Cell*, Vol. I, J. Brachet and A. E. Mirsky, eds. Academic, New York.

Britten, R. J., and E. H. Davidson, 1969. "Gene Regulation for Higher Cells: A Theory," *Science* 165:349–357.

Britten, R. J., and D. E. Kohne, 1968. "Repeated Sequences of DNA," *Science* 161:529–540.

Brown, D. D., and I. B. Dawid, 1969. "Developmental Genetics," *Ann. Rev. Genet.* 3:127–154.

Callan, H. G., 1967. "The Organization of Genetic Units in Chromosomes," *J. Cell Sci.* 2:1–7.

Campbell, A., 1967. "Regulation in Viruses." In *Molecular Genetics*," Part II, J. H. Taylor, ed. Academic, New York.

Changeux, J. P., 1965. "The Control of Biochemical Reactions," *Sci. Amer.* **212** (Apr.) :36–45.

Clever, U., 1963. "Von der Ecdysonkonzentration abhängige Genaktivitätsmuster in den Speicheldrüsenchromosomen von *Chironomus tentans,*" *Devel. Biol.* **6**:73–98.

Clever, U., 1968. "Regulation of Chromosome Function," *Ann. Rev. Genet.* **2**:11–30.

Crick, F., 1971. "General Model for the Chromosomes of Higher Organisms," *Nature* **234**:25–27.

Davidson, E. H., 1965. "Hormones and Genes," *Sci. Amer.* **212** (June):36–45.

Davidson, E. H., 1968. *Gene Activity in Early Development.* Academic, New York.

Epstein, R. H., A. Bolle, C. M. Steinberg, E. Kellenberger, E. Boy de la Tour, R. Chevalley, R. S. Edgar, M. Susman, G. H. Denhardt, and A. Lielausis, 1963. "Physiological Studies of Conditional Lethal Mutants of Bacteriophage T4D," *Cold Spring Harbor Symp. Quant. Biol.* **28**:375–394.

Edgar, R. S., and R. H. Epstein, 1965. "Conditional Lethal Mutations in Bacteriophage T4," *Proc. 11th Int. Cong. Genetics* **2**:1–16.

Fankhauser, G., 1934. "Cytological Studies on Egg Fragments of the Salamander *Triton.* IV. The Cleavage of Egg Fragments Without the Egg Nucleus," *J. Exptl. Zool.* **67**:349–393.

Fraenkel-Conrat, H., ed., 1968. *Molecular Basis of Virology.* Van Nostrand Reinhold, New York.

Fraenkel-Conrat, H., and R. C. Williams, 1955. "Reconstitution of Active Tobacco Mosaic Virus from Its Active Protein and Nucleic Acid Components," *Proc. Nat. Acad. Sci.* **41**:690–698.

Fraser, D., 1967. *Viruses and Molecular Biology.* Macmillan, New York.

Georgiev, G. P., 1969. "Histones and the Control of Gene Action," *Ann. Rev. Genet.* **3**:155–180.

Gottlieb, F. J., 1966. *Developmental Genetics.* Van Nostrand Reinhold, New York.

Gurdon, J. B., 1967. "Control of Gene Activity During the Early Development of *Xenopus laevis.*" In *Heritage from Mendel,* R. A. Brink, University of Wisconsin Press, Madison.

Gurdon, J. B., 1968. "Transplanted Nuclei and Cell Differentiation," *Sci. Amer.* **219** (Dec.):24–35.

Hadorn, E., 1961. *Developmental Genetics and Lethal Factors,* U. Mittwoch, tr. Wiley, New York.

Hadorn, E., 1968. "Transdetermination in Cells," *Sci. Amer.* **219** (Nov.):110–120.

Hämmerling, J., 1953. "Nucleocytoplasmic Relationships in the Development of *Acetabularia,*" *Int. Rev. Cytol.* **2**:475–498.

Harvey, E. B., 1936. "Parthenogenetic Merogony or Cleavage Without Nuclei in *Arbacia punctulata,*" *Biol. Bull.* **71**:101–121.

Hayes, W., 1968. *The Genetics of Bacteria and Their Viruses,* 2nd ed. Wiley, New York.

Hoyer, B. H., and R. B. Roberts, 1967. "Studies of Nucleic Acid Interactions Using DNA-Agar." In *Molecular Genetics,* Part II, J. H. Taylor, ed. Academic, New York.

Jacob, F., and J. Monod, 1961. "Genetic Regulatory Mechanisms in the Synthesis of Proteins," *J. Mol. Biol.* **3**:318–356.

Jacob, F., and J. Monod, 1961. "On the Regulation of Gene Activity," *Cold Spring Harbor Symp. Quant. Biol.* **26**:193–211.

King, T., and R. Briggs, 1956. "Serial Transplantation of Embryonic Nuclei," *Cold Spring Harbor Symp. Quant. Biol.* **21**:271–290.

Lederberg, J., and T. Iino, 1956. "Phase Variation in *Salmonella*," *Genetics* **41**:743–757.

Levine, M., 1969. "Phage Morphogenesis," *Ann. Rev. Genet.* **3**:323–342.

Locke, M., ed., 1966. *Major Problems in Developmental Biology*. Academic, New York.

Loomis, W. F., ed., 1970. *Papers on Regulation of Gene Activity During Development*. Harper & Row, New York.

Luria, S. E., and J. E. Darnell, 1967. *General Virology*, 2nd ed. Wiley, New York.

McCarthy, B. J., and B. H. Hoyer, 1964. "Identity of DNA and Diversity of mRNA Molecules in Normal Mouse Tissues," *Proc. Nat. Acad. Sci.* **52**:915–922.

McClintock, B., 1951. "Chromosome Organization and Genic Expression," *Cold Spring Harbor Symp. Quant. Biol.* **16**:13–47.

McClintock, B., 1961. "Some Parallels Between Gene Control Systems in Maize and in Bacteria," *Amer. Nat.* **95**:265–277.

Markert, C. L., and H. Ursprung, 1971. *Developmental Genetics*. Prentice-Hall, Englewood Cliffs, N.J.

Martin, R. G., 1969. "Control of Gene Expression," *Ann. Rev. Genet.* **3**:181–216.

Means, A. R., and T. H. Hamilton, 1966. "Early Estrogen Action: Concomitant Stimulations Within Two Minutes of Nuclear RNA Synthesis and Uptake of RNA Precursor by the Uterus," *Proc. Nat. Acad. Sci.* **56**:1594–1598.

Miller, O. R., and B. R. Beatty, 1969. "Visualization of Nucleolar Genes," *Science* **164**:955–957.

Monod, J., J. P. Changeux, and F. Jacob, 1963. "Allosteric Proteins and Cellular Control Systems," *J. Mol. Biol.* **6**:306–329.

Nanney, D. L., 1968. "Ciliate Genetics: Patterns and Programs of Gene Action," *Ann. Rev. Genet.* **2**:121–140.

"Nuclear Physiology and Differentiation," *Genetics* **61** (Suppl., 1969):1–469.

Pitot, H. C., 1967. "Metabolic Regulation in Metazoan Systems." In *Molecular Genetics*, Part II, J. H. Taylor, ed. Academic, New York.

Poulson, D. F., 1940. "The Effects of Certain X Chromosome Deficiencies on the Embryonic Development of *Drosophila melanogaster*," *J. Exptl. Zool.* **83**:271–326.

Stern, C., 1954. "Two or Three Bristles," *Amer. Sci.* **42**:213–247.

Steward, F. C., M. Mapes, and K. Mears, 1958. "The Growth and Organized Development of Cultured Cells. II. Organization in Cultures Grown from Freely Suspended Cells," *Amer. J. Botany* **45**:705–708.

Teas, H. J., ed., 1969. *Genetics and Developmental Biology*. University of Kentucky Press, Lexington.

Tomkins, G. M., and D. W. Martin, 1970. "Hormones and Gene Expression," *Ann. Rev. Genet.* **4**:91–106.

Ursprung, H., 1967. "Developmental Genetics," *Ann. Rev. Genet.* 1:139–162.

Wagner, R. P., and H. K. Mitchell, 1964. *Genetics and Metabolism*, 2nd ed. Wiley, New York.

Wood, W. B., and R. S. Edgar, 1967. "Building a Bacterial Virus," *Sci. Amer.* **217** (July):60–70.

Questions

15.1. What is the variable gene activity theory of differentiation?

15.2. How can different types of differentiated cells in the same organism be shown to have equivalent genomes?

15.3. What types of experiments have shown the primacy of the nucleus in the control of differentiation?

15.4. Why is the phage life cycle a useful system to use in the study of genetic control of differentiation?

15.5. What are the advantages of studying conditional lethal mutants as compared with other types of mutations?

15.6. Distinguish between feedback inhibition and repression.

15.7. What is an allosteric molecule? Why has this concept been of particular interest to students of development?

15.8. What data led to the formation of the operon model of gene action?

15.9. What led to the conclusion that not all the DNA in man and other vertebrates is structural DNA?

15.10. Construct a model relating DNA redundancy to quantitative traits.

15.11. What proportion of the genes appear to be active at any given time in a single cell of a higher organism? What is the evidence for this conclusion?

15.12. DNA-containing viruses often inhibit the synthesis of host DNA. Since some cancers are known to be induced by viruses, and cancer cells appear to result from abnormal differentiation of normal cells, how might a virus interfere with the activity of a normal cell at the molecular level to produce a cancer cell?

15.13. In bacteria, genes with related functions have sometimes been found to be adjacent to one another. In higher forms, on the other hand, genes with related functions generally appear to be dispersed throughout the genome. What possible explanations can you suggest for this difference?

15.14. If a cell is enucleated, but protein synthesis continues unabated for several hours, what are the molecular implications?

15.15. How is gene activity regulated?

CHAPTER SIXTEEN

Sex Determination and Sexual Differentiation

In our earlier discussion of sex linkage, two types of sex determination were described, the heterogametic (XY) male type, and the heterogametic (ZW) female type. Prior to the discovery at the turn of the century that sex was an inherited trait, it had long been believed that sex was environmentally determined. The demonstration of the chromosomal basis for sex determination seemed to settle the question; sex was determined at fertilization by the chromosomes present in the zygote, and subsequent development then led to the formation of the specified sex. However, it was soon realized that matters were not quite this simple, that two interrelated phenomena were involved, sex determination and sexual differentiation. Although sex is usually determined at fertilization by the sex chromosome mechanism, fertilization only initiates development. A period of differentiation is necessary before a sexually mature adult is produced. Although a normal fertile male or female is the usual end product of sexual differentiation, the process may be modified by a variety of factors, causing aberrant sexual forms. A review of both normal and abnormal forms of sex determination and sexual differentiation should provide some insight into the processes by which the sexes become differentiated from one another.

16.1. Genetic Sex Determination

In most species with separate sexes, sex determination is under genetic control, with one sex heterogametic and the other homogametic. In many cases, there is a large sex chromosome, called the X chromosome, and its smaller synaptic partner, the Y chromosome. The genetic contents of the X and Y chromosomes are usually different, with the Y being relatively inert genetically. In these cases the sex chromosomes are usually heteromorphic, and can be distinguished from one another cytologically. Among the insect orders Hemiptera, Orthoptera, and Coleoptera, many males are XO rather than XY, the Y chromosome having presumably been lost during the course of evolution. In the lower vertebrates, on the other hand (fishes, amphibia, and reptiles), sex chromosomes seem to exist, but are morphologically indistinguishable in many cases. The sex chromosome mechanism in these groups may be thought of as being in a relatively primitive stage, with the sex chromosomes not yet much differentiated from one another. Where the XO male exists, males have an odd number of chromosomes compared to the even number present in XX females. Among species with heterogametic (ZW) females, the birds, the Lepidoptera (moths and butterflies), the Trichoptera (caddis flies), and some fishes, there are also some ZO female types, especially among the insects.

A third type of genetic sex determination is male haploidy (or haplodiploidy.) In this case the haploid males develop from unfertilized eggs by parthenogenesis, and the fertilized eggs give rise to diploid females. Male haploidy has arisen independently in five different groups of insects as well as once in the arachnids and once in the rotifers. The sex ratio may fluctuate quite widely under male haploidy, depending on the proportion of fertilized eggs laid. It forms a self-correcting system: If the males are rare in the population and some females are not inseminated, their unfertilized eggs will all develop into males, tending to correct the deficiency of males. On the other hand, if males are in excess, the available females will all be inseminated and will produce only females, again tending to balance the sex ratio. Despite this tendency, however, the sex ratio with male haploidy varies more widely than with heterogametic sex determination, which, in theory at least, should produce equal numbers of males and females.

The Hymenoptera (bees, wasps, ants, sawflies, etc.) are the largest and most familiar group showing male haploidy. With few exceptions all species, from the most primitive sawflies to the most advanced social and parasitic species among the wasps and bees, show male haploidy, which must therefore be of ancient origin in this group. Meiosis is normal in the females, but of course synapsis and crossing over cannot occur in the haploid males. There is some disagreement in the literature over the details of spermatogenesis in Hymenoptera, but in general it can be stated that no reduction in chromosome number occurs and all of the sperm produced by any one male are genetically identical. A male hymenopteran is truly an orphan, for he has no father; all of his genes are inherited from his mother.

In the social Hymenoptera such as the honeybee, the males or drones are all fertile, functional males. However, there are two kinds of females, the queens and the workers. The workers are diploid and genetically the same as the queens, but they are ordinarily sterile and do not reproduce. The queen lays the eggs for the entire colony. The differences in fertility and morphology between queens and workers have been traced to the kind of food they receive as larvae: The larvae destined to become queens are fed "royal jelly." We get here a bit of insight into the relationship between sex determination and sexual differentiation. Even though both workers and queens are genetically determined females, only one type differentiates into a functional, reproductive female. The difference in sexual differentiation between workers and queens is not genetic or chromosomal, but environmental, in this case a difference in nutrition.

Among the Bryophyta, the mosses and liverworts, the life cycle shows an alternation of generations between a diploid sporophyte and a haploid gametophyte generation (Fig. 16.1). The gametophyte is larger and more complex structurally than the sporophyte. In other words, the green plant body in mosses and liverworts is the haploid gametophyte. The smaller diploid sporophyte generation remains associated with the gametophyte and is almost completely dependent on the gametophyte for nourishment. Thus, the Bryophytes stand in direct contrast to the higher plants, the Tracheophyta, in which the diploid sporophyte forms the plant body, and the gametophyte is much reduced in size.

In the Bryophytes, the gametophyte generation reproduces by eggs produced in archegonia and sperms produced by antheridia. Some species

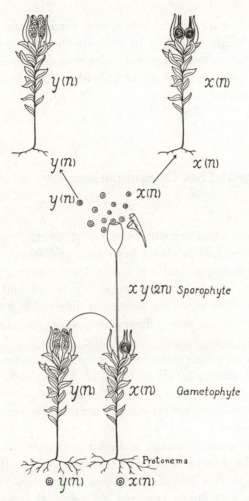

Fig. 16.1. Sex determination in a dioecious moss. (Morgan, T. H., 1928.
The Theory of the Gene, **2nd ed. Yale University Press, New Haven, Conn.,**
p. 125, Fig. 78.)

are monoecious, a single gametophyte bearing both archegonia and
antheridia; other species are dioecious, with archegonia and antheridia
borne on separate male and female gametophytes. In dioecious species, the
$2n$ sporophyte has been shown to be heterozygous for a pair of sex chromosomes and produces two types of spores. Those spores with the large X
chromosome plus a set of autosomes become female gametophytes; those
with the smaller Y chromosome plus a set of autosomes develop into male

gametophytes. Thus, this type of sex determination depends on a *hetero-zygous sporophyte*.

In the monoecious species, both types of gametes are produced by an individual with a single genotype. In this case the differentiation of the sex organs must be due to environmental influences comparable to those in a vertebrate that cause some cells in an individual to differentiate into nerve and others into muscle.

16.2. Environmental Sex Determination

In addition to monoecious bryophytes, many other examples can be cited in which a single individual produces both male and female gametes. Among the higher plants, maize is similar to the mosses in having male and female gametes produced by separate structures, the tassel and the ear. In the earthworm, ovaries and testes are located in different segments of the same individual. Moreover, in many flowering plants, the stamens and pistil are present in the same flower, and in the hermaphroditic snails of the genus *Helix*, a single gonad produces both eggs and sperm from cells some-times in close proximity. In all of these cases, the differentiation of the gonads and the gametes cannot be dependent on any sex chromosome mechanism, but must depend on environmental differences, in some cases very subtle differences.

A somewhat different situation is found in species such as the snail, *Crepidula*, and the marine annelid, *Ophryotrocha*. In these species the individual develops first into a functional male, then passes through a transitional phase, and finally becomes a functional female. Here the en-vironmental factors influencing sexual differentiation appear to be related to size and age.

In the marine echiuroid worm, *Bonellia*, the larvae are free-swimming and sexually undifferentiated. When the larvae settle to the ocean floor to take up a more sedentary existence, their sex is determined by the place where they happen to land. If they land on the proboscis of a female, they take up their abode within the female and differentiate into minute males, parasitic on the female. If they settle on the bottom, they differentiate into females several hundred times larger than the males (Fig. 16.2). It has been shown experimentally that extracts of the female's proboscis are sufficient

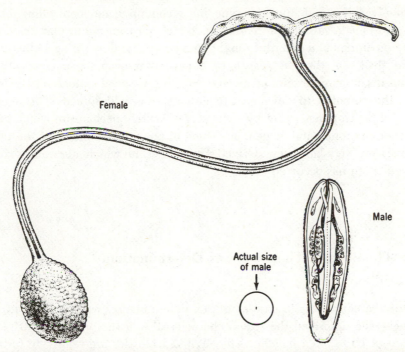

Fig. 16.2. Sex determination in the echiuroid marine worm, *Bonellia viridis*. The female is slightly enlarged and the male magnified because it is barely visible with the naked eye. (Dobzhansky, T., 1955. *Evolution, Genetics, and Man*. John Wiley & Sons, New York, p. 271, Fig. 11.8.)

to induce male differentiation, and that even the addition of hydrochloric acid to the sea water is sufficient to switch the differentiation of the larvae from female to male.

This seemingly haphazard method of sex determination may nevertheless have adaptive value. Any individual settling in an area unoccupied by *Bonellia* differentiates into a large female, and by inhibiting the development of other females in her immediate vicinity, minimizes competition by other females of her own species and ensures herself of a mate if a larva swims within her sphere of influence. Furthermore, there is no wastage of gametes by isolated males, because there are no isolated males. The sequential sexual differentiation in *Crepidula* also seems adaptive, because the transition from male to female is much slower in mated, sedentary males than in wandering, unmated males.

These cases of environmental sex determination are of considerable interest because they indicate the bisexual potentiality of the individual. To speak of environmental sex determination is not to imply that the genes are not involved. Instead, the genotype must contain genes controlling both male and female sexual differentiation. Which set comes into play depends on the environmental stimuli brought to bear on the developing individual. As we shall see, there is reason to believe that even in species with a chromosomal sex-determining mechanism, this bisexual potentiality exists. Thus, the potential to develop into either sex is widespread. The actual course of development is set by some sort of switch mechanism, which may be either environmental or genetic. Much of our understanding of the phenomena involved has been derived from cases in which aberrant sexual differentiation has occurred.

16.3. The Balance Theory of Sex Determination

One of the early, classic studies of sex determination was carried out by Bridges, who described the sexual characteristics of the progeny of triploid females of *Drosophila melanogaster*. Triploid females are relatively fertile; when mated to normal diploid males, they produce offspring with a variety of combinations of sex chromosomes and autosomes. Not all combinations give rise to viable offspring, but the following combinations survived:

Phenotype	No. of X chromosomes	No. of sets of autosomes (A)	Ratio X/A
Superfemale	3	2	1.5
Triploid female	3	3	1.0
Normal diploid female	2	2	1.0
Intersex	2	3	0.67
Normal diploid male	1	2	0.5
Supermale	1	3	0.33

Despite what their names might seem to imply, the supermales and superfemales are sterile and of rather low viability, but they may show somewhat accentuated male or female characteristics (Fig. 16.3). The intersexes are also sterile and exhibit more or less intermediate sexual characteristics or a mixture of male and female traits. An intersex may have sex combs, the genitalia may show a mixture of male and female parts, and the gonads may be rudimentary or resemble ovaries or testes or sometimes one of each. Some of the intersexes tend toward maleness, while others tend more toward femaleness.

The striking thing about these results is that the sexual phenotype is not determined by the sex chromosomes alone. If this were so, the 2X:3A in-

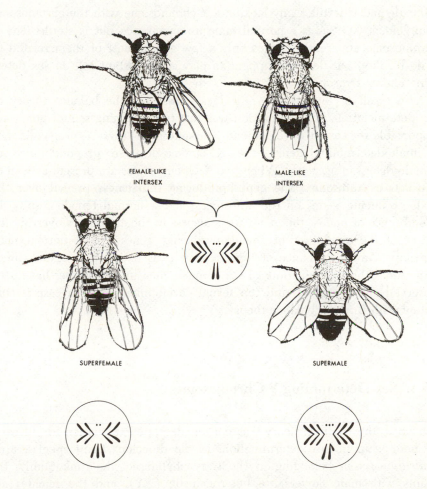

FEMALE-LIKE
INTERSEX

MALE-LIKE
INTERSEX

SUPERFEMALE

SUPERMALE

Fig. 16.3. Supersexes and intersexes in *Drosophila melanogaster* together with their chromosome complements. See text. (King, R. C., 1965. *Genetics*, 2nd ed. Oxford University Press, New York, p. 183, Fig. 12.1.)

dividuals with two X chromosomes should be females. Instead, the sex of the individual seems to depend on the ratio of X chromosomes to autosomes. A ratio of X/A of 1.0 results in the differentiation of females, whether 2X/2A or 3X/3A. A ratio of 0.5X/A causes males to develop. Any ratio between 0.5 and 1.0 leads to the formation of intersexes with intermediate sexual characteristics. Ratios greater than 1.0 or less than 0.5 lead to the development of female or male supersexes with somewhat exaggerated sexual characteristics.

A further question of considerable interest is the role of the Y chromosome. A fly with the chromosome complement XXY:AA is phenotypically a female and is fertile. A fly lacking a Y chromosome with the chromosome complement XO:AA is a normal male in appearance but is sterile. The Y chromosome apparently carries only a few genes, most of them related to male fertility, but does not appear to play an important role in sex determination in *Drosophila*.

As a result of these studies, Bridges formulated the balance theory of sex determination. His assumption was that two opposing sets of genes are responsible for sexual development. The autosomes carry a preponderance of male-determining genes, the X chromosomes a preponderance of female-determining genes. When two X chromosomes are present as well as two sets of autosomes, the genes producing femaleness prevail over the male-producing genes on the autosomes, and a normal female results. In the absence of one X, the genes for maleness in the autosomes override the effect of the single set of female-producing genes and a normal male develops. Any disturbance of this balance leads to aberrant sexual differentiation. Other combinations of sex chromosomes and autosomes have been observed in *Drosophila*, and the sexual development in each case is consistent with Bridges' balance theory.

16.4. Sex-Determining Y Chromosomes

However, the balance theory was not an adequate explanation for all cases of heterogametic sex determination. In the dioecious plant species, *Melandrium album*, belonging to the Caryophyllaceae, the pink family, the plants with male flowers are heterogametic (XY), and the females are homogametic (XX). As in *Drosophila*, a number of combinations of sex chromosomes and autosomes have been recovered. The chromosome constitution and the sex of a number of types of individuals are shown below for *Melandrium* and *Drosophila* for purposes of comparison:

Chromosome complement	Drosophila	Melandrium
2X:2A	♀	♀
3X:3A	♀	♀
4X:4A	♀	♀
2X:3A	☿	♀
2X:4A	♂	♀
3X:4A	☿	♀
XY:2A	♂	♂
XXY:2A	♀	♂
XXY:3A	☿	♂
XXXY:3A	♀	♂

☿ = intersex.

Inspection of the table reveals that individuals with similar chromosome complements in the two species are sometimes of opposite sex. The explanation is that in *Melandrium*, in contrast to *Drosophila*, the Y chromosome is strongly male-determining. Hence, in the combinations shown above for *Melandrium*, individuals with a Y chromosome are male and those lacking a Y are female, no matter what combination of X chromosomes and autosomes are otherwise present. At higher ratios of X to A chromosomes the plants are still essentially male, but they may produce an occasional hermaphroditic flower (e.g., XXXY:4A). However, the XXXXY:4A plants are hermaphroditic. Sex determination in *Melandrium* appears to be dependent on the relation between the X and Y chromosomes, and independent of the ratio of the X chromosomes to the autosomes. The sex-determining mechanism in *Melandrium* was at first regarded as an interesting exception to Bridges' X/A balance theory, but this Y-determining mechanism has recently drawn considerably more interest with the discovery that the Y chromosome is male-determining in man, mice, and other mammals.

The study of mammalian chromosomes lagged for technical reasons compared to cytological studies with *Drosophila*, maize, and *Tradescantia*, and for many years man was thought to have 48 chromosomes. Only in the mid-1950s did improved techniques make human cytological studies worthwhile. One of the first discoveries, by Tjio and Levan, was that the normal diploid somatic number of chromosomes in man is 46. Since the 1950s human cytology has progressed very rapidly, and a number of types of chromosome anomalies have been discovered in man. Among these are XO:2A and XXY:2A individuals with 45 and 47 chromosomes, respectively. In *Drosophila*, you will recall, XO individuals are sterile males. In man, however, XO individuals are phenotypically females, but their ovaries are poorly developed and they are said to have gonadal dysgenesis or Turner's syndrome. In *Drosophila*, XXY flies are fertile females, but in man XXY individuals are sterile males suffering from a condition known as Klinefelter's syndrome. Still other aberrant chromosome combinations are known, and all are consistent with the hypothesis that the Y chromosome is sex-determining in man. Hence, sex determination in man and other mammals is apparently similar to sex determination in *Melandrium* and different from the mechanism in *Drosophila*.

16.5. The Genetics of Sex Determination

In *Drosophila*, sex determination depends on the balance between two systems of polygenes, one on the autosomes favoring maleness, one on the X chromosomes favoring femaleness. The Y chromosome, though not essential for the differentiation of secondary sexual characteristics in the male, carries a number of fertility factors essential for normal spermiogenesis, an exclusively male trait.

In *Melandrium* and man, sex determination also appears to be multigenic, with the genes on the X chromosome favoring femaleness but the genes for maleness residing on the Y chromosome rather than in the autosomes.

The origin and evolution of the sex chromosome mechanism is a problem worth considering. A prevalent theory is that the most primitive sex-determining mechanism was dependent on segregation of alleles at a single gene locus. This idea gains some support from the fact that in the lower vertebrates (fishes, amphibia, and reptiles) heteromorphic sex chromosomes have not been distinguished cytologically. In species with an XY mechanism, the X and Y chromosomes have been postulated as having homologous "pairing" segments, within which synapsis and crossing over occur, and nonhomologous "differential" segments, within which the sex-differentiating genes accumulate and no crossing over can occur without

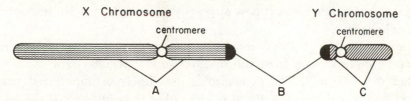

Fig. 16.4. Possible types of sex linkage. A. The differential segment of the X chromosome; X linkage. B. The homologous (pairing) segments of the X and Y chromosomes; "partial" sex linkage. C. Differential segment of the Y chromosome; Y linkage or holandric inheritance. (Moody, P. A., 1967. *Genetics of Man.* W. W. Norton & Co., New York, p. 184, Fig. 12.2.)

disrupting the normal sex-determining mechanism. According to this hypothesis, three types of sex linkage are possible, genes in the homologous pairing segments of the X and Y, genes on the differential segment of the X, and genes on the differential segment of the Y (Fig. 16.4). In species with distinctive X and Y chromosomes, the differential segments appear to make up a major portion of the sex chromosomes. In the lower vertebrates, the differential portions appear to be quite limited in size and may even be restricted to a single gene locus. Furthermore, sex determination in these groups appears to be rather unstable, for shifts from male heterogamety to female heterogamety have been recorded within a single species, and intersexuality and even sex reversal are not uncommon.

At this point, therefore, it seems desirable to cite some cases of the effects of single gene loci on sex. In *Drosophila,* Sturtevant discovered the transformer gene (*tra*) on the third chromosome, an autosome. This recessive gene, when homozygous, causes XX flies to develop into sterile males rather than normal females. These XX *tra/tra* flies are phenotypically normal males, except that the testes are rudimentary and produce no sperm. XY *tra/tra* flies are normal, fully fertile males. Thus, in this case a single gene substitution overrides the effects of the normal sex-determining mechanism, the genes on the X chromosomes producing femaleness. The balance is upset.

In the wasp *Habrobracon,* in which haplodiploidy is the usual mechanism of sex determination, diploid biparental males were sometimes observed. Analysis reveals a number of sex determinants (X^a, X^b, X^c, . . . , X^n) in this species. When heterozygous, these determinants interact in a complementary way to produce diploid females (X^aX^b, X^aX^c, X^bX^c, etc.) Haploid wasps (X^a, X^b, X^c, etc.) and diploid homozygotes (X^aX^a, X^bX^b, X^cX^c, etc.), lacking the complementary gene action, develop into males. The diploid males, however, are semisterile. These sex determinants, which behave like alleles, may nevertheless represent small homologous chromosome segments rather than a single locus. It now appears that a similar mechanism operates in the honeybee.

Support for the idea that complementary gene action causes femaleness in *Habrobracon* came from the discovery of a peculiar type of intersex called a gynandroid. The gynandroids develop from unfertilized partho-genetic eggs with two functional nuclei rather than just one. If the mother is X^aX^b, for example, the egg will develop into a mosaic individual that is mostly male, part of the body being haploid X^a and the other part haploid X^b. However, along the zone of contact between haploid X^a male tissue and haploid X^b male tissue, the tissue of the genitalia becomes feminized, and small femalelike reproductive appendages develop. The most likely explanation is that the diffusion or transport of gene products between X^a and X^b cells produce the complementary gene action necessary for female differentiation in a limited sector of the wasp.

16.6. Sex Determination in Microorganisms

In the mold, *Neurospora*, sexual reproduction occurs only between strains of opposite mating type (heterothallic.) The two mating types, designated as *A* and *a*, are determined by a pair of alleles. Another fungus, *Glomerella*, also a member of the ascomycetes, has a somewhat more complex mechanism for determining mating behavior. *Glomerella* is homothallic, meaning that mating can occur within a single strain, but a number of different strains have been identified, nearly all of which will mate with each other. However, fertility between different strains varies considerably, and can be measured by the numbers of perithecia formed along the boundary between the strains. No single strain shows maximum fertility in all possible crosses; instead, fertility appears to depend on complementary gene action between the strains. Furthermore, at least two gene loci control the mating reaction rather than one as in *Neurospora*.

The yeasts, such as *Saccharomyces cerevisiae*, reproduce asexually by bud-ding and also by the formation of four haploid ascospores by meiosis from a single diploid yeast cell. These spores germinate into small, haploid cells that reproduce mitotically for a time. These haploid cells function as potential gametes, for two haploid cells of opposite mating type may fuse to form a large diploid cell (Fig. 16.5). There are just two mating types, known as plus and minus, which seem to be controlled by alleles at a single gene locus. Hence the diploid is heterozygous for mating type,

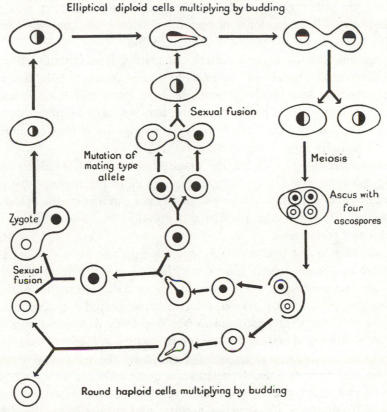

Fig. 16.5. Life cycle of the yeast, *Saccharomyces cerevisiae*. (Catcheside, D. G., 1951. The *Genetics of Micoorganisms*, Sir Isaac Pitman & Sons, London, p. 111, Fig. 24.)

and two of the ascospores will be plus and two will be minus. Although yeast is structurally simple, its life cycle includes a diploid phase, spore formation, a haploid phase, and sexual fusion. These qualities plus its metabolic versatility have made the yeasts useful organisms in genetic research.

The one-celled green alga, *Chlamydomonas*, has two flagella and a single chloroplast. The single cell is haploid and can reproduce asexually by mitotic cell division to form clones. Members of the same clone do not reproduce sexually, but if individuals from different clones are mixed together, they sometimes pair, fuse, and form diploid zygotes. The next two divisions are meiotic, producing four haploid cells. Tests have revealed that there are just two mating types in *Chlamydomonas*, known as mt^+ and mt^-, which are controlled by a pair of alleles at a single nuclear gene locus. Thus, of the four haploid cells resulting from meiosis, two are mt^+

and two are mt^-, giving a 1:1 ratio. Furthermore, mating will occur only between cells from clones of opposite mating type. The mt^+ and mt^- individuals are morphologically indistinguishable.

Among the Ciliates, several varieties of mating type determination have been discovered. Some are genetically quite conventional, resembling *Chlamydomonas* and involving just one gene locus and two alleles. Still others have many alleles at the locus and hence a large number of possible mating types rather than just two. Some species have two or more loci involved in mating-type determination.

In another group of Ciliates, the mating type is not determined directly by the genotype, but rather the genotype specifies the mating-type potential, while the actual mating type produced is dependent on the cytoplasmic environment. These results parallel the situation in ciliary antigen determination in *Paramecium*.

Conjugation in the bacterium, *E. coli*, has already been discussed. The sex factor in *E. coli*, which functions either as an episome or attached to bacterial chromosome, seems to have no exact parallels in higher organisms.

Therefore, a variety of evidence seems to support the concept that the primitive sex-determining mechanism was commonly dependent on the segregation of alleles at a single gene locus. Individuals of opposite mating type tended to be similar morphologically. In the course of evolution, individuals of different sex diverged from one another in morphology and function, and other gene loci became involved in sex determination and sexual differentiation. These genes accumulated primarily on the differential segments of the sex chromosomes, and the sex chromosomes, as the homology between them became reduced, tended to diverge structurally as well as functionally. Crossing over between the differential segments was reduced or eliminated in order to stabilize the sex-determining system. These changes in the genetic system would be favored by natural selection. The result is the evolution of a stable chromosomal mechanism of sex determination giving a 1:1 sex ratio, with the genes responsible for the heterogametic sex accumulated on the Y chromosome and, in some cases, the autosomes, and the genes for the homogametic sex accumulated on the X chromosomes. In some species, more than one pair of chromosomes has become involved in sex determination, and complex sex chromosome systems have evolved. These more complex situations (X_1X_2Y, X_1X_2O, XY_1Y_2, etc.) seem to have been derived from the simpler XY or XO systems either by the inclusion of the autosomes in the sex chromosome mechanism by translocation, or by the "fragmentation" of a single X or Y chromosome into two or more separate chromosomes.

16.7. Abnormal Sexual Differentiation

Some of the most revealing insights into sex determination and differentiation have been gained from sexually aberrant forms. We have already mentioned the supersexes and intersexes in *Drosophila,* the gynandroid in *Habrobracon,* and some examples of hermaphroditism. Still other aberrant sexual forms of considerable interest have been discovered.

The *gynandromorph* is a sexual mosaic, made up of male and female tissues. The most striking gynandromorphs are insects, for instance, the solitary wasp, in which the individuals are bilaterally half male and half female (Fig. 16.6). The most probable cause is the loss of one X chromosome from a daughter nucleus at the first mitotic division after fertilization. The resulting individual is thus an XX/XO mosaic. One conclusion from the clear-cut separation of male and female tissues is that the insects do not have circulating sex hormones comparable to those in mammals. The sexual differentiation of the cells is autonomous and dependent on the chromosome complement in each cell. Gynandromorphs of this type would not be found in mammals.

Sex reversal seems to be relatively easy to achieve in the anuran amphibia. When reared at elevated temperatures, genetically female frogs differentiate into males. This result raises the interesting possibility of a fertile mating between two genetic females, a normal female (XX) and a sex-reversed male (also XX). One can predict that the progeny of such a cross, reared under normal conditions, will all be XX and hence all females and this, in fact, is the case.

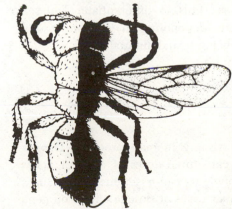

Fig. 16.6. Gynandromorph of the solitary wasp, *Pseudomethora canadensis.* Right: male, black coloration, 13 antennal segments, 2 ocelli, wings, male-type legs. Left: female, bright red coloration, 12 antennal segments, no ocelli, no wings, female-type legs. (Stern, C., 1968. *Genetic Mosaics and Other Essays.* Harvard University Press, Cambridge, Mass., p. 32, Fig. 8.)

A famous case of sex reversal is that of a normal hen, reported to have laid fertile eggs, that later assumed the appearance and behavior of a rooster and actually fathered chicks. In the hen, only one gonad develops into a functional ovary; the other remains in an undifferentiated state. The early gonad consists of two regions, an outer cortex from which the ovary forms, and an inner medulla from which the testis differentiates. In this remarkable chicken the functional ovary was apparently destroyed by disease, and the undifferentiated gonad, released from the inhibiting effect of the female sex hormones from the ovary, differentiated a testis from the medulla. The male hormones produced by the testis then caused the bird to develop male secondary sexual characteristics.

When twins of opposite sex are born in cattle, one is a normal male and the other is usually a female intersex referred to as a "freemartin." In these cases it has been shown that the twins, due to a fusion of blood vessels, share a common circulation. Since the male hormones come into play earlier than the female hormones, the sexual differentiation of the female is modified in the male direction before her own hormone system can come into play. She develops into a sterile intersex, but sex reversal is not complete.

Complete sex reversal has not been reported in man. The occasional bearded lady in the circus, who, if she has the right genotype, may also start to become bald, may suffer from a masculinizing tumor of the adrenal cortex. The steroid hormones, produced in excess, cause the masculine traits to appear. If the condition is corrected, the affected individual reverts to a more feminine appearance.

These cases of sex reversal again indicate the bisexual potentiality of the developing individual. The usual sex-determining mechanism is modified by temperature or hormones in such a way that the alternative set of genes is called into play, and the expected sex fails to differentiate. The sex reversal may be complete or incomplete, depending on the timing and strength of the action of the factors influencing sexual differentiation.

16.8. Dosage Compensation

We have already noted that a simple trisomy in man produces Down's syndrome or Mongolian idiocy. In this case three doses of a set of genes rather than the customary two are enough to cause serious disturbances in normal development. Many other similar cases of dosage effects on gene expression could be cited. These results suggest that a certain dosage of all genes is required for normal differentiation and development. For autosomal genes, two alleles are ordinarily required. If, because of a deficiency, only one normal allele is present, or if three or more doses of the normal allele are present on an otherwise diploid background, development is apt to be aberrant, seemingly because of the deficiency or excess of the normal gene product in relation to the proteins synthesized under the control of genes present in double dose.

For the genes on the sex chromosomes, however, the situation is somewhat different. For species with the XY mechanism, the males carry one dose of the genes in the differential segment of the X, while the females have two doses. Despite the difference in dosage, most of these genes in *Drosophila,* for example, have the same expression in males and females. At first this occasioned no surprise, but as dosage effects came under study, it was realized that the similarity in expression of sex-linked genes in males and females implied some sort of dosage compensation to equalize the single dose-double dose difference between the sexes.

In *Drosophila* carrying the X-linked apricot eye-color gene (w^a) in the hemizygous or homozygous condition, XO and XY males and XX, XXY, and XXYY females all have the same eye color. Hence the Y chromosome seems to play no part in dosage compensation. Furthermore, an XX *tra/tra* individual homozygous for apricot (a homozygous apricot female transformed into a male) has the same eye color as an XY male. Hence the male or female phenotype as such seems to be unimportant in dosage compensation. If it were, these phenotypic males with two doses of apricot should have a darker eye color than the XY males with one dose. The conclusion is that a complex system of modifying genes on the X chromosome itself is responsible for dosage compensation in *Drosophila,* the net effect of which is to enhance the expression of the single X in the male to equal the effect of two X's in the female.

This conclusion is supported by the fact that in the salivary glands of males and females, the DNA of the X chromosomes appears in the expected 1:2 ratio, but the males' single X appears to have as much associated RNA and protein as the two synapsed X's in the females.

In man and other mammals, however, dosage compensation is effected in quite a different way. In 1949, Barr and Bertram reported a consistent difference between the interphase nuclei of male and female cats. The female nuclei contained a small, stainable body, the "sex chromatin" or "Barr body," which was absent from male nuclei. It was soon discovered that the Barr body was actually a condensed and heavily stained or heteropycnotic X chromosome.

A number of karyotypes have been examined for Barr bodies. In

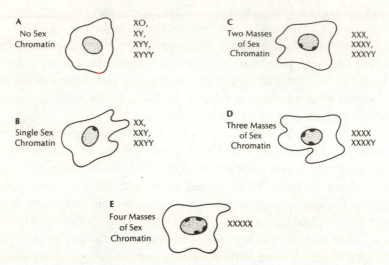

Fig. 16.7. The relationship between the sex chromatin (Barr bodies) in interphase nuclei and the sex chromosome pattern in the karyotype. (Barr, M. L., 1966. "The Sex Chromosome in Evolution and in Medicine," *Canad. Med.* Assn. J. 95:1137–1148, p. 1144, Fig. 4.)

general, the number of Barr bodies equals one less than the number of X chromosomes present, although they may sometimes appear fused. Thus, the XY male lacks a Barr body. The XO individual with Turner's syndrome is phenotypically female, but also lacks a Barr body. The XX female has one Barr body, as does the phenotypically male XXY person with Klinefelter's syndrome. The Barr body is not a female phenotypic trait, but rather is related to X chromosome number (Fig. 16.7). The XXX and XXXY conditions show two Barr bodies, while XXXX individuals have three. Therefore, only one X chromosome resembles the autosomes in staining properties and in replication time.

A decade later, an unusual mosaicism was described in the mouse. Females carrying a translocated autosomal dominant wild-type allele on one of their X chromosomes and the recessive allele for brown coat color on a normal autosome showed patches of brown and wild-type (agouti) fur. Males with the same translocation and genotype were not mosaic, but wild type. Furthermore, XO individuals, which in the mouse are fertile females, were also not mosaics, but wild type like the males. These results suggest that this position-effect mosaicism is in some way related to whether one or two X chromosomes are present. However, not only translocated autosomal genes but many ordinary X-linked mutants were observed to cause mosaicism in heterozygous mammalian females.

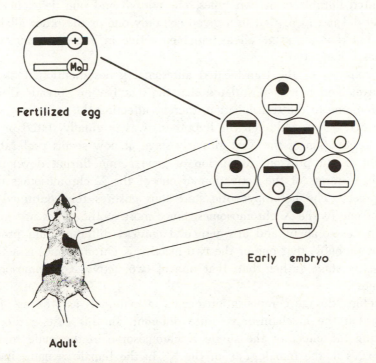

Fertilized egg

Early embryo

Adult

Fig. 16.8. Diagrammatic representation of the Lyon hypothesis. The inactive X chromosome is shown as circular, the active one as rod-shaped. (Lyon, Mary F., 1963. "Attempts to Test the Inactive-X Theory of Dosage Compensation in Mammals," *Genetical Research* 4:93–103, p. 93, Fig. 1.)

Mary Lyon proposed the most widely accepted explanation of these cytological and genetic observations, which has come to be known as the Lyon hypothesis. In essence, she postulated that only one X chromosome is needed for normal development in mammals and that the other X becomes genetically inactive (the Barr body). This inactivation is a normal process, randomly affecting one X in some cells and the other X in other cells (Fig. 16.8). The time of determination comes early in development and is irreversible in a given cell lineage. Therefore, the two cell types will occur in groups or patches. The theory accounts for the observations on the expression of a number of X-linked traits in heterozygous mammalian females, ranging from tortoise-shell females in cats to women heterozygous for anhidrotic ectodermal dysplasia, who are mosaic for patches of normal skin and skin with virtually no sweat glands. One of the first studies to support Lyon's concept showed that females heterozygous for glucose-6-phosphate dehydrogenase deficiency (G-6-PD) produced single-cell clones of cultured fibroblasts of two types, one normal and one deficient. Thus, not only did it appear that in a given cell only one or the other allele was active, but that the same allele remained active in a given clone, as the hypothesis suggested.

The behavior of the translocated autosomal genes mentioned above is easily explained if it is postulated that the inactivation spreads from the X chromosome to the translocation, thereby affecting the expression of the translocated genes. Although the hypothesis was originally stated in terms of the inactivation of the X at an early stage, it now seems preferable, in the light of our previous discussion of gene action during development, to speak in terms of the activation of one of the X chromosomes during development. The heteropycnotic state thus ensures the continued inactivity of one of the X chromosomes. Since most of the genes are inactive early in development and are activated only as differentiation proceeds, it seems probable that one of the two inactive X chromosomes is activated at an early stage rather than that one of two active X chromosomes is inactivated.

Therefore, dosage compensation occurs in mammals as well as in *Drosophila*, but the mechanism is quite different. In this case, rather than enhancing the effect of the single X chromosome in the male to match the two X's in the female, one of the X's in the female remains inactive, and the gene activity in the females is brought to the male level. Since the mechanisms of dosage compensation are so different, it should be clear that *Drosophila* females heterozygous for sex-linked genes will not show mosaicism. The fact that X chromosome mosaicism in heterozygous mammalian females appears to be the normal condition provides considerable food for thought, especially as to its possible adaptive significance.

16.9. Reproduction in Higher Plants and Animals

If one compares the genetic systems of higher plants and higher animals, one is struck by the high incidence of polyploidy among plants and its rarity in animals. Still another difference lies in the much greater frequency of various types of asexual reproduction in plants than in animals. Parthenogenesis is practically the only form of such reproduction in animals above the Coelenterata, but this and various other forms of apomixis are much more common in plants than in animals. Still another generalization is that hermaphroditism is much more frequent in plants than animals. It seems probable that some of these phenomena are related.

In hermaphroditic species, the possibility of self-fertilization exists, and hence the possibility of harmful consequences arising from inbreeding. Various devices have evolved in many hermaphroditic species to minimize the chances of selfing, ranging from different maturation times for male and female gametes to structural or behavioral modifications that favor reciprocal cross-fertilization. Self-sterility is one simple genetic mechanism that assures outcrossing. In essence, systems of self-sterility alleles exist that prevent the growth of any pollen carrying an allele similar to either of the alleles present in the female parent. In tobacco, for example, neither s_1 nor s_2 pollen will grow down the style of an s_1/s_2 plant, but pollen with s_3 or other s alleles will grow. Thus, all the plants will be heterozygous at this locus and at others as well, and the outcrossing would seem to ensure some degree of heterosis.

It has been estimated that about one-third of the angiosperms, the flowering plants, are polyploid, with most of them being allopolyploids derived from hybridization between different species followed by a doubling of the chromosome number. Thus, polyploidy appears to have played a significant role in the evolution of higher plants. However, among the animals polyploidy is quite rare and thus could only have played at best a minor role in the evolution of higher animals. Muller first proposed the classical explanation for the rarity of polyploidy in animal species. He pointed out that the sex-determining mechanism would be thrown out of kilter in polyploid dioecious animal species. In *Drosophila,* for example, which is dependent on the balance between X chromosomes and autosomes for sex determination, an XXY + 3A triploid fly is a sterile intersex. If formed, an XXYY + 4A tetraploid fly should be a normal fertile male and a 4X + 4A fly a female, but the progeny from a cross of two such flies would be XXXY + 4A intersexes and sterile. Hence, it seems unlikely that fertile, polyploid strains of *Drosophila* could be established.

However, in a number of plant genera such as *Melandrium*, stable, autotetraploid, dioecious strains have been developed with no difficulty. In *Melandrium* the XXXY + 4A are fertile males, the 4X + 4A plants are females, and their progeny are males and females in equal numbers. Therefore, contrary to Muller's hypothesis, these species do form normal dioecious tetraploid lines. The difference seemingly lies in the presence of the strongly male-determining Y chromosome in *Melandrium*. Hence, Muller's theory may account for the absence of polyploidy in some animal groups, but probably is inadequate to account for its rarity throughout the animal kingdom.

Muller also predicted that polyploidy would be most common in hermaphroditic or parthenogenetic animal species. This prediction seems to be borne out among parthenogenetic species, but relatively few polyploid hermaphroditic animals have been found. Therefore, Muller's theory again seems inadequate. These results suggest that sexual reproduction is itself in some way responsible for the low frequency of polyploidy in animals.

Still another suggestion, by Stebbins, is that differentiation and development in animals are so much more complex that polyploidy disturbs normal development far more in animals than it does in plants, and hence is strongly selected against. Some evidence for this comes from triploid salamanders, which have fewer but larger cells than their diploid relatives, and are less vigorous and less well adapted to survive than the diploids. Therefore, in addition to the anomalies in the sex-determining mechanism created by polyploidy, developmental anomalies may constitute a second block to its establishment in animals.

In any event, it should be added that both polyploidy and asexual reproduction do not seem to have led to major new evolutionary breakthroughs in the groups in which they occur. Parthenogenetic species seem able to adapt well to a particular ecological niche, but their genetic system limits their ability to adapt to new conditions and may lead to their extinction. This limitation is surmounted in some species, for example, the aphids, by the presence of parthenogenetic summer generations followed by a sexual generation in the fall. Many plant species that reproduce primarily asexually retain the ability to reproduce sexually and occasionally revert to sexuality. Apomictic reproduction in plants is genetically controlled and generally behaves as a recessive to sexual reproduction although it is often controlled by more than one locus. Since most species are not completely obligatory apomicts, crosses between sexual and apomictic forms or between two apomicts do occur. Apomixis appears to result from a rather delicate balance involving both genetic and environmental factors. Crosses of this sort tend to break up the gene combinations favoring apomixis, and cause a reversion to sexual reproduction or else result in sterility or some other abnormal genetic outcome.

Even in plants, polyploidy may lead to disruption in sexual reproduction, and hence apomixis tends to be favored among polyploids. However, the polyploids, like the apomicts, seem not to have produced any major new trends in plant evolution. However, both polyploidy and apomixis represent mechanisms by which heterosis may be perpetuated from generation to generation.

16.10. The Conversion of Maize from a Monoecious to a Dioecious Condition

As we have seen, maize is normally monoecious, but many genes have been identified that change the plants into staminate or pistillate types. One such mutant is tassel seed (*ts*), in which silks (pistils) rather than stamens are produced in the tassel. Thus a *ts/ts* plant is pistillate or female. The silkless (*sk*) mutant prevents the formation of silks on the ears, but the tassels are normal and the *sk/sk* plants are effectively staminate or male. The dioecious stock that was synthesized was the following:

$$ts/ts; \; sk/sk \times +/ts; \; sk/sk$$

Since only one locus is segregating, male and female plants are produced in equal numbers from this cross. The other factor making the cross work is that the *ts* gene in the *ts/ts*; *sk/sk* individuals promotes the formation of silks on the ears as well as the tassels, overriding the effects of the male-promoting *sk* gene.

The ease with which maize is converted from monoecious to dioecious, and the ease with which apomictic species may revert to sexuality, give some indication of the flexibility in modes of reproduction that exist in plants. The genetic system in plants seems much more easily modifiable than the mode of reproduction in animals.

The study of sex determination and sexual differentiation has not only considerable inherent interest, but is also of significance for the insight that it provides into the basic phenomena of development.

Additional Reading

Allen, C. E., 1935. "The Genetics of Bryophytes," *Botan. Rev.* 1:269–291.

Bacci, G., 1965. *Sex Determination.* Pergamon, Oxford.

Baltzer, F., 1925. "Untersuchungen über die Entwicklung und Geschlectsbestimmung der *Bonellia*," *Pubbl. Staz. Zool. Napoli* 6:223–286.

Baltzer, F., 1935. "Experiments on Sex-Development in *Bonellia*," *The Collecting Net* **10**:3–8.

Barr, M. L., 1959. "Sex Chromatin and Phenotype in Man," *Science* **130**:679–685.

Barr, M. L., 1966. "The Significance of the Sex Chromatin," *Int. Rev. Cytol.* **19**:35–95.

Barr, M. L., and E. G. Bertram, 1949. "A Morphological Distinction Between Neurones of the Male and Female, and the Behavior of the Nucleolar Satellite During Accelerated Nucleoprotein Synthesis," *Nature* **163**:676–677.

Beadle, G. W., 1945. "Genetics and Metabolism in *Neurospora*," *Physiol. Rev.* **25**:643–663.

Beale, G. H., 1954. *The Genetics of* Paramecium aurelia. Cambridge University Press, London.

Bridges, C. B., 1925. "Sex in Relation to Chromosomes and Genes," *Amer. Nat.* **59**:127–137.

Coe, W. R., 1943. "Sexual Differentiation in Molluscs. I. Pelecypods," *Quart. Rev. Biol.* **18**:154–164.

Coe, W. R., 1944. "Sexual Differentiation in Molluscs. II. Gastropods, Amphineurans, Scaphopods, and Cephalopods," *Quart. Rev. Biol.* **19**:85–97.

Crew, F. A. E., 1921. "Sex Reversal in Frogs and Toads: A Review of the Recorded Cases of Abnormality with an Account of a Breeding Experiment," *J. Genet.* **11**:141–181.

Crew, F. A. E., 1926. "Abnormal Sexuality in Animals. I. Genotypical," *Quart. Rev. Biol.* **1**:315–359.

Davidson, R. G., H. M. Nitowsky, and B. Childs, 1963. "Demonstration of Two Populations of Cells in the Human Female Heterozygous for Glucose-6-Phosphate Dehydrogenase Variants," *Proc. Nat. Acad. Sci.* **50**:481–485.

Dronamraju, K. R., 1965. "The Function of the Y-Chromosome in Man, Animals, and Plants," *Adv. Genetics* **13**:227–310.

Ephrussi, B., 1953. *Nucleo-cytoplasmic Relations in Micro-organisms.* Oxford University Press, London.

Falkow, S., E. M. Johnson, and L. S. Baron, 1967. "Bacterial Conjugation and Extrachromosomal Elements," *Ann. Rev. Genet.* **1**:87–116.

Fincham, J. R. S., and P. R. Day, 1963. *Fungal Genetics.* Davis, Philadelphia.

Ford, C. E., K. W. Jones, P. E. Polani, J. C. de Almeida, and J. H. Briggs, 1959. "A Sex-Chromosome Anomaly in a Case of Gonadal Dysgenesis (Turner's Syndrome)," *Lancet* **1**:711–713.

Goldschmidt, R. B., 1934. "Lymantria," *Biblio. Genet.* **11**:1–186.

Hartmann, M., 1956. *Die Sexualität*, 2nd ed. Fischer, Stuttgart.

Hess, O., and G. F. Meyer, 1968. "Genetic Activities of the Y Chromosome in *Drosophila* During Spermatogenesis," *Adv. Genet.* **14**:171–223.

Jacob, F., and E. L. Wollman, 1961. *Sexuality and the Genetics of Bacteria.* Academic, New York.

Jacobs, P. A., and J. A. Strong, 1959. "A Case of Human Intersexuality Having a Possible XXY Sex-Determining Mechanism," *Nature* **183**:302–303.

Jones, D. F., 1934. "Unisexual Maize Plants and Their Bearing on Sex Differentiation in Other Plants and in Animals," *Genetics* **19**:552–567.

Lillie, F. R., 1917. "The Freemartin: A Study of the Action of Sex Hormones in the Foetal Life of Cattle," *J. Exptl. Zool.* **23**:371–452.

Lyon, M. F., 1961. "Gene Action in the X-Chromosome of the Mouse (*Mus musculus* L.)," *Nature* **190**:372–373.

Lyon, M. F., 1962. "Sex Chromatin and Gene Action in the Mammalian X-Chromosome," *Amer. J. Human Genet.* **14**:135–148.

Lyon, M. F., 1968. "Chromosomal and Subchromosomal Inactivation," *Ann. Rev. Genet.* **2**:31–52.

Moore, K. L., ed., 1966. *The Sex Chromatin*. Saunders, Philadelphia.

Mittwoch, U., 1967. *Sex Chromosomes*. Academic, New York.

Morgan, T. H., and C. B. Bridges, 1919. "The Origin of Gynandromorphs." Carnegie Inst. Wash. Publ No. 278, pp. 1–122.

Muller, H. J., 1925. "Why Polyploidy is Rarer in Animals Than in Plants," *Amer. Nat.* **59**:346–353.

Muller, H. J., 1932. "Some Genetic Aspects of Sex," *Amer. Nat.* **66**:118–138.

Muller, H. J., 1950. "Evidence for the Precision of Genetic Adaptation," *Harvey Lectures* **43**:165–229.

Muller, H. J., and W. D. Kaplan, 1966. "Dosage Compensation of *Drosophila* and Mammals as Showing the Accuracy of the Normal Types," *Genet. Res.* **8**:41–59.

Nesbitt, M. N., and S. M. Gartler, 1971. "The Application of Genetic Mosaicism to Developmental Problems," *Ann. Rev. Genet.* **5**:143–162.

Overzier, C., ed., 1963. *Intersexuality*. Academic, New York.

Russell, L. B., 1961. "Genetics of Mammalian Sex Chromosomes," *Science* **133**:1795–1803.

Sager, R., 1955. "Inheritance in the Green Alga *Chlamydomonas reinhardi*," *Genetics* **40**:476–489.

Sager, R., and F. J. Ryan, 1961. *Cell Heredity*. Wiley, New York.

Seiler, J., 1965. "Sexuality as Developmental Process," *Proc. 11th Intern. Cong. Genet.* **2**:199–207.

Sonneborn, T. M., 1957. "Breeding Systems, Reproductive Methods and Species Problems in Protozoa." In *The Species Problem*, E. Mayr, ed. American Association for the Advancement of Science, Washington, D.C.

Stebbins, G. L., 1950. *Variation and Evolution in Plants*. Columbia University Press, New York.

Stent, G. S., 1963. *Molecular Biology of Bacterial Viruses*. Freeman, San Francisco.

Stern, C., 1960. "Dosage Compensation-Development of a Concept and New Facts," *Can. J. Genet. Cytol.* **2**:105–118.

Sturtevant, A. H., 1945. "A Gene in *Drosophila melanogaster* That Transforms Females into Males," *Genetics* **30**:297–299.

Tjio, J. H., and A. Levan, 1956. "The Chromosome Number of Man," *Hereditas* **42**:1–6.

Warmke, H. E., 1946. "Sex Determination and Sex Balance in Melandrium," *Amer. J. Botany* **33**:648–660.

Westergaard, M., 1958. "The Mechanism of Sex Determination in Dioecious Flowering Plants," *Adv. Genetics* **9**:217–281.

White, M. J. D., 1973. *Animal Cytology and Evolution*, 3rd ed. Cambridge University Press, London.

Whiting, P. W., 1943. "Multiple Alleles in Complementary Sex Determination of *Habrobracon*," *Genetics* **28**:365–382.

Whiting, P. W., 1945. "The Evolution of Male Haploidy," *Quart. Rev. Biol.* **20**:231–260.

Winge, Ø., 1934. "The Experimental Alteration of Sex Chromosomes into Autosomes and Vice Versa as Illustrated by *Lebistes*," *Compt. Rend. Trav. Lab. Carlsberg, Ser. Physiol.* **21**:1–49.

Witschi, E., 1929. "Studies on Sex Differentiation and Sex Determination in Amphibians. II. Sex Reversal in Female Tadpoles of *Rana sylvatica* Following Application of High Temperature," *J. Exp. Zool.* **52**:267–291; "III. Rudimentary Hermaphroditism and Y Chromosome in *Rana temporaria*," *J. Exp. Zool.* **54**:157–223.

Questions

16.1. In species with unequal numbers of chromosomes in males and females, which sex will ordinarily be the heterogametic sex?

16.2. How is the sex ratio regulated in species with haploid males and diploid females?

16.3. How do you explain the differentiation of both ovaries and testes in a single hermaphroditic individual?

16.4. a. In *Drosophila melanogaster,* what will be the sexual characteristics of individuals of the following karyotype:

   1. XXY:2A
   2. XXY:3A
   3. XO:2A

   b. What would be the sexual characteristics of the same three karyotypes in *Melandrium album?*

   c. What would be the sexual characteristics of individuals with the same three karyotypes in *Homo sapiens?*

16.5. How many types of sex-linkage have been postulated?

16.6. How is it possible to have diploid males in species with haplodiploidy as the sex-determining mechanism?

16.7. In species with sex determined by alleles at a single locus, what advantage might there be to having multiple alleles at this locus rather than just two? If this is so, why are there only two sexes in man and other higher plant and animal species?

16.8. Distinguish among hermaphroditism, gynandromorphism, sex reversal, and intersexuality.

16.9. Why are gynandromorphs observed in insects but not in mammals? How might you attempt to find sex chromosome mosaicism in mammals?

16.10. Distinguish among sex-linked, sex-limited, and sex-influenced traits.

16.11. Distinguish between sex determination and sexual differentiation. What are the implications of the "bisexual potentiality" of individuals?

16.12. In *Drosophila melanogaster*, what sex ratio would be expected from a cross between males homozygous and females heterozygous for the transformer (*tra*) gene?

16.13. What situations appear to be most favorable for the occurrence of sex reversal?

16.14. How is dosage compensation attained in man and other mammals? How is dosage compensation achieved in *Drosophila?*

16.15. Why are mosaics for sex-linked traits, such as the tortoise-shell cat, seen in female mammals but not in female *Drosophila?*

16.16. At the molecular level, what do you think the Barr body represents?

16.17. What evidence supports the Lyon hypothesis?

16.18. Why is polyploidy more common among higher plants than among higher animals?

16.19. Diagram a cross between a tetraploid female and a diploid male in *Drosophila*. What would be the chromosomal makeup and phenotype of the progeny?

16.20. If you were to screen species of animals in a search for polyploidy, where would you be most apt to find it?

16.21. Tortoise-shell cats can result only when the animal is heterozygous for the sex-linked black gene (*B*) and its yellow allele (*b*). These alleles are found only on the X chromosomes and not on the Y. Rarely, a sterile, tortoise-shell male cat has been found. What must be his chromosome complement?

CHAPTER SEVENTEEN

Extrachromosomal Inheritance

As Mendelism and the chromosomal mechanism of heredity were becoming established, there remained a lingering doubt that this could be the only mechanism of heredity. The Galton-Pearson school, for example, which studied the inheritance of quantitative traits, felt that Mendelian segregation and independent assortment could not account for the inheritance of continuously varying traits. Among the embryologists of the day were other skeptics, who believed that Mendelism could account for the inheritance of trivial differences in coat color or eye color, for example, but that the fundamental traits that distinguish, say, a man from a mouse were controlled by some other mechanism. Moreover, the possibility that acquired characters might be inherited has always had considerable emotional appeal, but the chromosome theory seems incompatible with Lamarckianism. Hence, those seeking a mechanism for the inheritance of acquired traits have also sought a nonchromosomal mechanism of heredity. Many of these questions have been resolved, but the question of whether there are any extrachromosomal mechanisms of inheritance has persisted. Our purpose in this chapter is to examine the evidence for patterns of heredity that seem inexplicable under the chromosome theory of heredity.

17.1. Exceptions to Mendelian Inheritance

The search has been directed toward the cytoplasm for separate systems of heredity comparable to the chromosome system in the nucleus. Such a system should be self-duplicating, capable of mutation, and should have a detectable phenotypic effect. In essence, hereditary phenomena are sought that seem to be exceptions to the usual rules of Mendelian inheritance. Then, breeding tests are run to eliminate all possible variations of chromosomal behavior that might account for the results. Since additional unusual chromosomal phenomena are constantly being discovered, the problem of establishing clear-cut cases of extrachromosomal inheritance becomes increasingly difficult. A number of criteria for the detection of cytoplasmic or nonchromosomal inheritance have proven useful, most of which are suggestive but not necessarily conclusive proof of non-Mendelian behavior. The following are some of the tests most commonly used:

1. The results of reciprocal crosses differ. For autosomal genes, the outcome from reciprocal crosses is ordinarily the same. However, the cytoplasmic contribution of male and female gametes is usually quite different, leading to the expectation that reciprocal crosses would differ.

2. The trait in question cannot be mapped in one of the linkage groups— If the genes controlling the trait are on the chromosomes, it should be possible to map them.

3. If segregation fails to occur or if aberrant segregation ratios are obtained at the time of meiosis, nonchromosomal heredity may be involved.

4. On the other hand, if segregation is obtained subsequent to mitotic cell division, when chromosome segregation is not occurring, then the segregation of cytoplasmic factors may be responsible.

5. Although this test is not always possible, nuclear substitution may be informative. If a trait persists despite the substitution of a nucleus known to govern an alternative state for that trait, it suggests nonchromosomal control.

6. The presence of nonchromosomal DNA may indicate the presence of a cytoplasmic hereditary system.

Now we shall consider some of the hereditary phenomena that seem to violate Mendelian principles and discuss the explanations for these phenomena.

Sex Linkage

One obvious exception to the rule that reciprocal crosses should not differ under Mendelian heredity is sex linkage. Sex-linked traits, of course, do not always behave in the same way in reciprocal crosses, but it is generally possible, by appropriate crosses, to establish the sex-linked nature of the trait.

Preferential Segregation

Another cause of differences between the progeny from reciprocal crosses is preferential chromosome segregation during meiosis. Nonrandom segregation affects the constitution of female gametes because in oögenesis or megasporogenesis only one of the four haploid nuclei formed becomes a functional egg or megaspore nucleus. Hence, genes on the preferentially segregated chromosome have a higher frequency of transmission in females. In males, however, all four haploid products of meiosis form gametes, and no such effect is observed. Because the frequency of transmission of genes on a preferentially segregating chromosome differs depending on the sex of the parent from which it comes, the progeny from reciprocal crosses will differ.

Maternal Effects

In some apparent cases of cytoplasmic inheritance, the phenotype of the progeny is determined by the genotype of the mother rather than by their own genotype. A classical example of this sort is found in the inheritance of the direction of coiling in the shells of fresh-water snails of the genus *Limnaea.*

Coiling is usually dextral (i.e., the opening of the shell is to the right), but occasionally it is sinistral. In reciprocal crosses between true-breeding dextral and sinistral strains, the direction of coiling in the F_1 is always like that of the mother. At first glance this appears to be cytoplasmic inheritance, but all the F_2 progeny from both types of cross are dextral, and in the F_3 a ratio of three dextral to one sinistral is obtained, no matter what type of coiling was seen in the F_1 (Fig. 17.1).

With one exception, the Mendelian explanation is quite simple and straightforward. A single locus with two alleles is involved, with the + or dextral allele acting as a dominant to the *s* or sinistral allele. The cross can be outlined as follows, where the phenotypes D for dextral and S for sinistral are shown above the line, and the genotypes below the line.

P_1 $\qquad \dfrac{D}{++} \, ♀ \times \dfrac{S}{ss} \, ♂$ $\qquad\qquad\qquad\qquad \dfrac{S}{ss} \, ♀ \times \dfrac{D}{++} \, ♂$

$\qquad\qquad\qquad\qquad\qquad\quad \downarrow \qquad\qquad\qquad\qquad\qquad\qquad\qquad\qquad \downarrow$

F_1 $\qquad\qquad\qquad\qquad \dfrac{D}{+s} \qquad\qquad\qquad\qquad\qquad\qquad\qquad\qquad \dfrac{S}{+s}$

$\qquad\qquad\qquad\qquad\qquad\quad \downarrow \qquad\qquad\qquad\qquad\qquad\qquad\qquad\qquad \downarrow$

F_2 $\quad \dfrac{D}{++} \quad \dfrac{D}{+s} \quad \dfrac{D}{+s} \quad \dfrac{D}{ss} \qquad\qquad \dfrac{D}{++} \quad \dfrac{D}{+s} \quad \dfrac{D}{+s} \quad \dfrac{D}{ss}$
$\qquad\quad \downarrow \qquad \downarrow \qquad \downarrow \qquad \downarrow \qquad\qquad\quad \downarrow \qquad \downarrow \qquad \downarrow \qquad \downarrow$

F_3 $\qquad \underline{D} \qquad \underline{D} \qquad \underline{D} \qquad \underline{S} \qquad\qquad \underline{D} \qquad \underline{D} \qquad \underline{D} \qquad \underline{S}$

Hence, the direction of coiling is not determined by the genotype of the individual in the shell, but by the genotype of its mother. It is apparent

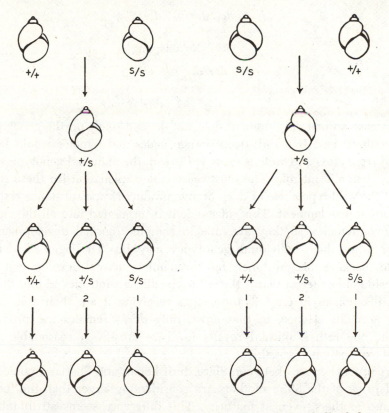

Fig. 17.1. Inheritance of the direction of coiling of the shell of the snail, *Limnaea peregra.* $+$ = dextral; s = sinistral. See text. (Sturtevant, A. H. and G. W. Beadle, 1939. *An Introduction to Genetics.* W. B. Saunders Co., Philadelphia, p. 330, Fig. 113.)

that the egg develops under the control of the maternal genotype, and the cleavage pattern, which is ultimately responsible for the direction of coiling, is already established in the egg prior to fertilization. It should be noted that the shells of the young are not necessarily like the shell of the mother, but are like her genotype, for an $S/+s$ mother has only dextral progeny and a D/ss mother has only sinistral progeny. Therefore, the Mendelian segregation is completely normal; the unusual aspect is the lag of a generation in the expression of the genes.

Another similar but somewhat more complex case of this sort is seen with the *deep orange* (*dor*) sex-linked eye-color mutant in *Drosophila*. The following crosses are relevant:

1. $dor/+$ ♀ $\times dor/Y$ ♂
                 ↓
1dor/dor ♀ : 1$dor/$ + ♀ : 1dor/Y ♂ : 1+/Y ♂

2. dor/dor ♀ $\times dor/Y$ ♂
                      ↓
No offspring

3. dor/dor ♀ $\times +/Y$ ♂
                      ↓
$dor/+$ ♀ progeny only

In this case normal development does not depend solely on the genotype of the fertilized egg. If it did, deep orange males and females would be expected from cross 2 as well as cross 1. Instead, the maternal genotype seems to play a significant role. The most reasonable explanation for these results is that the wild-type allele of deep orange produces some substance essential for normal development. This substance is incorporated into all the eggs of the *dor*/+ females, so they are able to complete normal development no matter what their individual genotypes may be. The eggs of *dor/dor* females, lacking this substance, fail to complete normal development. It is of considerable interest that if the wild-type allele is introduced into the egg at fertilization as in cross 3, those eggs receiving it are then able to develop normally. Hence, in that cross, only *dor*/+ females are produced. Despite the rather unusual results in these crosses, a reasonable Mendelian explanation is possible.

In reciprocal crosses between Shire draft horses and Shetland ponies, the F_1 offspring of the Shire mothers are considerably larger than the foals of Shetland mothers, even at maturity. This difference seems attributable to the persistence of the effects of the more favorable uterine environment provided by the Shire mothers. This sort of nutritional effect will disappear within a generation or at most within a few generations and hence cannot be regarded as a clear-cut case of extrachromosomal heredity—if it were, the differences would be expected to persist undiminished through successive generations.

Phenocopies are phenotypic changes induced by environmental stimuli that simulate the effects of a mutant gene. For example, wild-type *Drosophila* larvae exposed to suitable concentrations of silver nitrate ($AgNO_3$) develop into adults with a light body color similar to that of the yellow or straw mutants. However, progeny of such flies, reared under normal conditions, have the normal wild-type body color. Thus, no genetic effect is apparent; the agent must have had an extrachromosomal influence during development, but this effect does not persist into the next generation.

In *Phaseolus*, however, exposure of the plants to chloral hydrate produces abnormal leaf development, and the effect persists during several generations of sexual reproduction. However, the percentage of affected plants gradually declines, dropping from about 75% affected in the first generation after treatment to 50% by the fourth generation and to 0% in the seventh generation even though only affected plants are used as parents. This phenomenon has been called *Dauermodification*. Unlike the phenocopies, this phenomenon does not seem due to a temporary modification of gene action. Rather, some sort of extrachromosomal inherited change must be involved that is, however, unable to persist indefinitely. It has been suggested that the treatment induces abnormal extrachromosomal determinants, which gradually lose out in time in competition with their normal homologues. In other words, selection operating on a heterogeneous group of cytoplasmic determinants is the suggested explanation.

17.2. Particulate Cytoplasmic Inheritance

The maternal effects that we have considered thus far have been traced to the effect of the mother's genotype or to environmental influences that are more or less transient. True cytoplasmic inheritance would be expected to cause persistent rather than transient non-Mendelian hereditary changes. A number of such persistent effects have been discovered, many of them associated with cytoplasmic particles. Complexity emerges here because while some of the cytoplasmic particles clearly seem to be normal cytoplasmic constituents and others quite clearly seem to be infective agents not essential to the life of the cell, in many cases the distinction between normal cytoplasmic particle and infective particle is very difficult to make, strange though this may seem. However, examples of some of these types of extrachromosomal phenomena will give some insight into what has been learned about cytoplasmic inheritance and also into the nature of the interactions between nucleus and cytoplasm.

In man, syphilis, caused by a spirochaete, may be transmitted by an infected mother to her infant; it thus *appears* to be a congenital disease. Even though this may seem to be a case of maternal inheritance, however, its infectious nature is quite obvious.

Mammary Tumors in Mice

Mammary cancer in the mouse was at first thought to be a case of cytoplasmic inheritance because it was transmitted by a mother to her offspring. However, a series of experiments, primarily the work of John Bittner, revealed that the development of mammary tumors in the mouse depends on a combination of factors. First is an inherited susceptibility to mammary cancer; some strains are quite susceptible, while others have a low susceptibility. In some susceptible strains the susceptibility seems to be due to a single dominant gene, but in other susceptible strains a more complex system of genes seems to be involved. A second factor is the hormonal influence on the mammary glands. Males of susceptible strains seldom if ever develop mammary tumors, but if they are treated with female hormones, their incidence of mammary tumors rises markedly. Studies have revealed hereditary patterns of hormonal activity, which are inherited independently of the genes for tumor susceptibility. The third and most surprising discovery was that an agent transmitted through the milk from mother to offspring was causing the tumors. The milk factor was discovered when baby mice were taken from their mothers at birth and reared by foster mothers. When mice from high- and low-incidence lines were cross-fostered, the results showed that mice from low-incidence lines nursed by mothers from high-incidence strains had an increased tumor incidence compared with their controls, but the mice from high lines had a reduced incidence of tumors when nursed by low-line mothers. Further study has revealed that the agent is a nucleoprotein particle with viral-like properties. It has also been transmitted by injection with blood from animals with mammary tumors and is sometimes transmitted by males from cancerous strains to females from noncancerous strains by coitus.

Carbon Dioxide Sensitivity in Drosophila

Drosophila melanogaster can ordinarily withstand prolonged anesthesia with CO_2 without harm. However, strains have been discovered that are paralyzed and killed by brief exposure to CO_2. This sensitivity is transmitted in a non-Mendelian fashion, primarily through the egg but sometimes through the sperm. Sensitive flies were found to harbor a viruslike agent called sigma (σ). Sigma is not ordinarily contagious, but nevertheless can be transmitted by injection. In the injected flies, sigma seems to be present in an unstable state: It is present at high concentrations in the tissues of the new host but is seldom transmitted, never in the sperm. However, sometimes it becomes stabilized in the oöcytes of injected females and from then on behaves like the naturally occurring forms of the virus. A number of mutant types of sigma have been discovered, and it has also been shown that certain genes in the host inhibit viral growth.

Sex Ratio in Drosophila

The sex-ratio trait in *Drosophila* leads to deviations from the one-to-one ratio of males to females. Some cases are known to be controlled by nuclear genes, but others are maternally inherited. The latter have been found in a number of species of *Drosophila: bifasciata, prosaltans, willistoni, paulistorum, equinoxialis, borealis,* and *nebulosa.* In these species the female produces progeny that are predominantly or exclusively females. Oöplasm from dying eggs from SR females produces the sex-ratio condition when injected into normal females. The SR trait can be "cured" if affected flies are subjected to high temperatures. Strain differences in susceptibility to the SR agent have been demonstrated, so the genotype of the host must play a role in the expression of the trait. The agent, surprisingly enough, has been identified in *D. willistoni* and *D. nebulosa* as a spirochaete. This spirochaete is thus a parasite that causes the death of males but not females.

Kappa in Paramecium

Sonneborn's work on the killer character in *Paramecium* is one of the notable studies of extrachromosomal inheritance. He found that some strains of *P. aurelia* cause the death of members of other strains when they are grown together in mixed culture. The killer strains contained cytoplasmic particles, which were called kappa particles. Kappa replicates within the cytoplasm of *Paramecium;* the number of particles per *Paramecium* may range from 100 to a few thousand. Kappa particles have been shown to contain both DNA and protein, and have been compared in structure and complexity to *Rickettsia.*

Originally it was thought that killer strains release a toxin called paramecin into the medium, which kills sensitive individuals but has no effect on killer individuals. Now the toxic effect is believed to be due to the release of mature particles into the medium, where they are transformed into P particles that act as toxic agents. It has been found that in some cases contact with a single particle suffices to kill a sensitive cell.

Several mutant forms of kappa have been described, one of which protects the organism against P particles but does not form P particles itself, while another, pi, neither protects against nor forms P particles. In addition, two other closely related classes of particles have now been discovered, called lambda and mu. Lambda strains, like those with kappa, kill sensitive cells when cultured with them. Mu, however, is called mate-killer, because the killing takes place only after conjugation between a killer and a sensitive cell.

Since the sensitive strains live and reproduce normally in the absence of kappa, lambda, and mu, it is clear that these particles are not normal cell constituents but must be a form of intracellular parasite. However, an intimate relationship must develop between the particles and their hosts, because the reproduction and maintenance of these particles is dependent on the presence of certain nuclear genes in the *Paramecium*. The gene K is required for kappa maintenance (and also s_1 and s_2), L for lambda, and M_1 and M_2 for mu.

Unlike the mate-killers, killers with kappa can conjugate with sensitive cells without killing them (Fig. 17.2). The killer and sensitive exconjugants have the same genotype, but their cytoplasms differ. Ordinarily, the exconjugants with killer cytoplasm produce killer clones, and the exconjugants with cytoplasm from the sensitive strain produce sensitive clones. However, autogamy in the Kk killer exconjugant will produce kk cells with kappa particles in the cytoplasm. When this happens, the kappa particles are irreversibly lost within about five generations. In some cases the sensitive exconjugants were found to give rise to killer clones. Further analysis revealed that the cytoplasmic bridge formed during conjugation persisted longer in these cases, permitting the passage of kappa into the cytoplasm of the sensitive strain. An added fact of interest is that, while kk will not support the maintenance of kappa at all, KK individuals have, on the average, twice as many kappa particles as the heterozygous Kk type.

Reproduction of kappa normally occurs in synchrony with the rate of fission of the paramecia so that the number of particles per cell remains about the same. However, the rate of multiplication of the ciliate can be raised to outstrip that of the kappa particles, and sensitive KK paramecia free of kappa can be obtained. If such sensitive KK cells (also carrying s_1 and s_2) are placed in suspensions containing free kappa particles, some of them acquire kappa and are converted into killers. Thus kappa, under

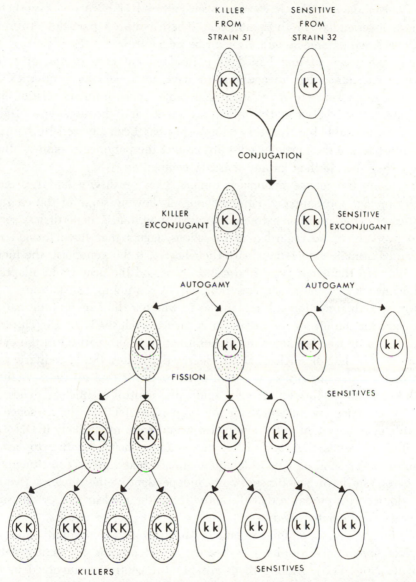

Fig. 17.2. The transmission of kappa and the killer trait in *Paramecium aurelia*. The dots represent the kappa particles. See text. (King, R. C., 1965. *Genetics*, 2nd ed. Oxford University Press, New York, p. 299, Fig. 16.2.)

these conditions, is infectious. It is worth noting that a killer can be "cured" of its kappa particles by treatment with aureomycin, chloromycetin, nitrogen mustard, X-rays, or high temperature. Another interesting point is that a single *Paramecium* can carry two different mutant forms of kappa.

Finally, lambda particles have now been cultured *in vitro* on a synthetic medium independent of the paramecia. These cultured particles are able to kill sensitive paramecia and also can be infective.

Thus, what was at first hailed as a fundamental new system of cytoplasmic inheritance has turned out to have many of the properties of infectious organisms. However, it is remarkable for the intimate relationship that has evolved between the infectious agents and the genotype of the host. It seems probable that this relationship has been favored by natural selection because the particles not only protect the paramecia against other killers, they also rid them of any sensitive competitors.

Therefore, the mouse mammary tumors, CO_2 sensitivity and sex ratio in *Drosophila*, and kappa in *Paramecium*, all showing some of the characteristics of cytoplasmic or extrachromosomal inheritance, nevertheless seem more properly to be regarded as infectious agents that have formed remarkably intimate associations with their hosts. To the geneticist, the interplay between the genotypes of host and agent and the origin and adaptive significance of this relationship are problems of continuing fascination.

However, the question still remains as to whether the genetic information necessary for the life of the organism is carried exclusively by the chromosomal DNA or whether some of this information is also carried in the cytoplasm. Even though nuclear DNA clearly dominates the hereditary pattern of cells and organisms, it is still open to doubt that all the structures in the cell can be formed *de novo* under the control of nuclear genes. In view of what is now known about the properties of DNA, the presence of a stable genetic system in the cytoplasm would seem more likely if DNA or, possibly, self-replicating stable RNA were found associated with cytoplasmic structures. We shall now consider the evidence that various constituents of the cytoplasm have genetic continuity independent of the nucleus. Even if they do, it seems probable that the activities of the nuclear and cytoplasmic systems will be integrated.

A number of cytoplasmic organelles have physical continuity and are therefore potential candidates as bearers of cytoplasmic hereditary determinants (Fig. 17.3). Among them are (1) the centrioles, involved in the formation of the aster and possibly the spindle during cell division, (2) the basal granules or blepharoplasts, involved in the formation of cilia, flagella, and sperm tails, (3) the mitochondria, which play a fundamental role in respiratory metabolism, (4) the chloroplasts, associated with photosynthesis, (5) the kinetoplasts of trypanosomes and some other Protozoa, and (6) the endoplasmic reticulum and the cell membranes. For some of these organelles, some knowledge of their structure and function is accumulating, but such information about others is still rather limited.

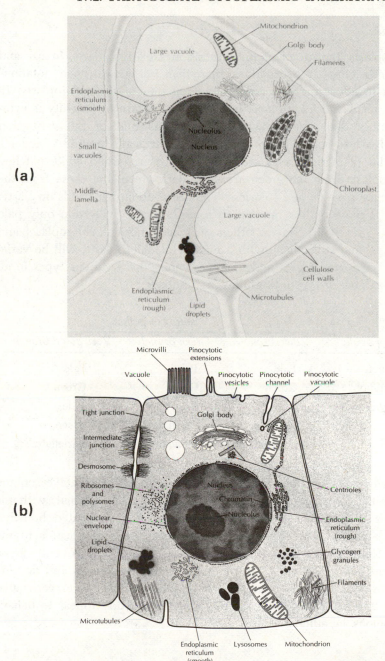

Fig. 17.3. (a) Diagram of a generalized plant cell. (b) Diagram of a generalized animal cell. (From *Cell Structure and Function*, Second Edition, by Ariel G. Loewy and Philip Siekevitz. Copyright © 1963, 1960 by Holt, Rinehart and Winston, Inc. Reprinted by permission of Holt, Rinehart and Winston, Inc.)

Chloroplasts

The chloroplasts, the site of photosynthesis, are relatively large in size and complex in structure. In maize alone, more than 100 nuclear gene mutations are known to affect the appearance and function of the chloroplasts. In some cases of plastid inheritance, however, the pattern of heredity is cytoplasmic rather than nuclear. One such case is that of the variegated four-o'clock, whose scientific name, *Mirabilis jalapa,* calls to mind a Model-T Ford. In this plant the leaves are mottled, with green and white patches. The chloroplasts from the green areas are dark green, but those from the white patches are pale and colorless. The distribution of plastids or plastid precursors is more or less random so some branches will have only pale chloroplasts and white leaves, others will have normal chloroplasts and green leaves, and still others, with a mixture of chloroplasts, will be variegated. It then is possible to make crosses among these various types in all possible combinations, with the following results:

| | Source of pollen | | |
Source of ovule	Pale branch	Green branch	Variegated branch
Pale branch	Pale	Pale	Pale
Green branch	Green	Green	Green
Variegated branch	Pale, green, or variegated	Pale, green, or variegated	Pale, green, or variegated

It is obvious that the source of the pollen has no effect whatever on the phenotype of the progeny. The inheritance is strictly maternal, with the three types of seedlings grown from seeds from the variegated branches being explained by the distribution of the two types of plastids to the embryos. Embryos on the variegated branches may receive only green plastids and hence become green plants, or only pale plastids and develop into albinos. If they receive both types of plastids, they form variegated plants like the maternal parent. In this case the plastids appear to behave as autonomous hereditary units.

Iojap in Maize

An unusual relationship between a nuclear gene and the chloroplasts has been discovered in maize. Iojap is a recessive mutation on chromosome 7 that when homozygous causes plants with contrasting green and white stripes to develop. The following series of crosses will illustrate the nature of the effect of iojap (*ij*):

1.　　　　P₁　　　　green ♀ (+/+) × striped ♂ (*ij*/*ij*)
　　　　　　　　　　　　　　　　　↓
　　　　　F₁　　　　　　　　　green (+/*ij*)
　　　　　　　　　　　　　　　　　↓
　　　　　F₂　　　1 green (+/+):2 green (+/*ij*):1 striped (*ij*/*ij*)

Thus in this case iojap appears to behave as an ordinary Mendelian recessive. However, in the reciprocal cross, the results are quite different:

2.　　　　P₁　　　　striped ♀ (*ij*/*ij*) × green ♂ (+/+)
　　　　　　　　　　　　　　　　　↓
　　　　　F₁　　　green (+/*ij*); striped (+/*ij*); white (+/*ij*)
　　　　　　　　　　　　　　　　　　　　　　　　　(lethal)

These three kinds of F₁ progeny may be obtained in widely varying ratios, with the lethal white type relatively rare. Further crosses were made as follows:

3.　　　　green F₁ ♀ (+/*ij*) × green ♂ (+/+)
　　　　　　　　　　　　　　↓
　　　　　all green (+/+:+/*ij*)

4.　　　　striped F₁ ♀ (+/*ij*) × green ♂ (+/+)
　　　　　　　　　　　　　　↓
　　　　　three possible types of progeny:
　　　　　(1) all green (+/+ and +/*ij*)
　　　　　(2) all white (+/+ and +/*ij*)
　　　　　(3) a mixture of green, white, and some striped
　　　　　　　 individuals (+/+ and +/*ij*)

The following are the most significant points to note. In the F₂ of cross 1, striped individuals appear, apparently as a result of the influence of the homozygous *ij* allele on the chloroplasts. However, once the change is induced by the nuclear genes, it is irreversible, since plants with the normal +/+ genotype, entirely lacking the iojap allele, may have only abnormal plastids, as shown by the white (+/+) progeny in cross 4.

The variety of progeny with the same genotypes in cross 4 can best be explained by the segregation of normal and abnormal plastids during embryo sac formation—some plants receiving only normal plastids, others only the abnormal plastids, and others a mixture. Thus the change induced by a nuclear gene then behaves as an autonomous extrachromosomally inherited entity. The phenotype is now independent of the nuclear genotype.

Since striped *ij*/*ij* plants contain both normal and abnormal plastids, the change was obviously induced in only some of the plastids by the action of the iojap allele. For this reason, it has been suggested that the effects of the gene are indirect rather than the result of direct induction of plastid mutations. The exact nature and site of action of the iojap gene have not yet been identified. However, it is clear that the chromosomal genes have induced an autonomous heritable change in an extrachromosomal system.

Despite the fact that nuclear genes are known to affect the chloroplasts, there is also good evidence that they are autonomous, self-duplicating cytoplasmic organelles. It is possible to "cure" a plant cell of its chloroplasts by treatment with the antibiotic streptomycin; once the cell has completely lost its plastids, it can no longer form them *de novo*. A variety of chemical agents, heat treatment, and growth in the dark have all been found to cause the disappearance of chlorophyll from cells. If the chloroplasts and their proplastid precursors are both eliminated, regeneration does not occur. On the positive side is the fact that chloroplasts have been found to have their own DNA distinctive in base composition from that in the nucleus. Moreover, they have their own separate protein-synthesizing system, including distinctive ribosomes, tRNAs, and activating enzymes. Since DNA is present, it is not surprising that a number of plastid mutations have been discovered, both spontaneous and induced. Much of this work has been done with lower forms such as *Euglena, Spirogyra,* and *Chlamydomonas.* Enough work has been done with higher plants, however, so that it now seems safe to assume that DNA is generally present in chloroplasts.

Mitochondria

The mitochondria are structures found in the cytoplasm of all eukaryotic species, in other words, in just about everything but viruses, bacteria, and the blue-green algae. Their primary role in oxidative metabolism and in the formation of energy-rich phosphate bonds in ATP (adenosine triphosphate) has been gradually worked out over the past few decades. More recently it has been discovered that the mitochondria, like the chloroplasts, contain their own distinctive DNA and apparently constitute another genetic system in eukaryotic cells separate from the chromosomal system. Mitochondrial DNA has sometimes been found in circular strands of naked DNA double helices (Fig. 17.4) and thus seems to resemble bacterial or viral DNA rather than the chromosomal DNA of eukaryotes, which is normally associated with large amounts of protein. Furthermore, isolated mitochondria can carry on protein synthesis independently, and their protein-syn-

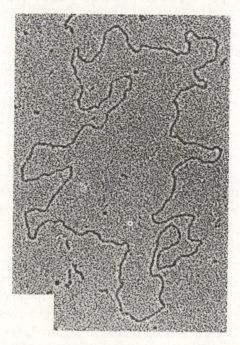

Fig. 17.4. Circular mitochondrial DNA in the dimer form. (Courtesy J. Vinograd, California Institute of Technology.)

thesizing system is distinct from that found elsewhere in the cytoplasm. New mitochrondria are formed from previously existing mitochondria rather than *de novo*, and thus have genetic continuity. Even in anaerobic yeast, which appear to lack mitochondria but form them when switched to aerobic respiration, it is now thought that promitochondria comparable to the proplastids are actually present and responsible for the genesis of these mitochondria. Self-replicating mutant mitochondria have been identified, and it is now clear that the mitochondria represent another genetic system in the cell. However, the amount of DNA in the mitochondria is so small, representing less than 1% of the nuclear DNA in mammalian cells, that it can only code for a part of the protein in the mitochondria. The synthesis of the cytochrome *c* found in mitochondria, for example, is known to occur in the microsomal fraction of the extramitochondrial cytoplasm under the control of nuclear genes. Therefore, while both mitochondria and chloroplasts seem to have a semiautonomous existence, and their DNA forms the basis for genetic systems separate from that in the nucleus, it is nevertheless clear that there is interaction and collaboration between these systems and the nuclear genes. They are not completely independent of one another. The discovery of DNA in plastids and mitochondria added a new dimension to the study of genetics: It established the existence of coordinate hereditary systems in the cytoplasm in addition to the much better understood chromosome mechanism in the nucleus. Furthermore, it now seems possible that still other cytoplasmic organelles with physical continuity may also turn out to have a measure of genetic autonomy, and the study of cytoplasmic inheritance seems likely to become an increasingly significant aspect of genetics.

Petite in Yeast

While working with baker's yeast (*Saccharomyces cerevisiae*), Ephrussi and his collaborators found a type of mutant that, when grown on a solid medium containing a fermentable sugar, produced small colonies in comparison with normal strains. These "petite" colonies are small because they lack a number of respiratory enzymes, including cytochromes *a* and *b* and some dehydrogenases. Under anaerobic conditions both normal and petite strains grow at the same rate because then their metabolism is based on fermentation rather than respiration. Under aerobic conditions, however, the normal strains switch to respiration and grow some five times faster than the petites, which continue to rely on their anaerobic metabolic system. Mitochondria are present in the petite strains, but they are aberrant in appearance and lack an inner membrane. While some of the enzymes are absent entirely, others are present at very low concentrations.

The petite mutants were recovered at the surprisingly high frequency of 10^{-2} to 10^{-3}. Furthermore, if normal yeast cells were grown in the presence of the dye, acriflavine, petite mutations were induced in essentially 100% of the cells. These induced changes were shown to be due to mutation rather than to some selective process, and proved remarkably stable; in fact, they seemed to be irreversible. Results such as these have not ordinarily been obtained with nuclear genes. The stability and irreversibility of the mutants and the fact that several enzymes were affected suggested that the change was due to a loss rather than a structural alteration in the genetic material. However, as noted above, the mitochondria were aberrant rather than absent.

When normal yeast is crossed with a petite strain, only normal progeny are obtained. Normal diploid colonies are formed by vegetative growth; following meiosis, only normal haploid colonies are derived from the ascospores. Thus the petite trait does not appear in the expected Mendelian ratios, but disappears even though chromosomal marker genes in these same crosses segregate in the expected 1:1 ratio. Even if the progeny are repeatedly backcrossed to the petite strain, the petite trait fails to reappear. The petite trait is stable and heritable with vegetative reproduction, but shows non-Mendelian behavior in crosses even though nuclear genes in these same crosses segregate normally. The usual interpretation of these results is that a cytoplasmic factor, usually referred to as the rho (ρ) particle, is responsible. When fusion between petite and normal cells occurs, the normal rho particle is contributed by the normal strain to the diploid zygote. Following replication, it is transmitted to all diploid daughter cells during vegetative growth and to all the ascospores produced by meiosis. Cells lacking the particle are rho minus (ρ^-) and form petite colonies. Petite strains behaving in this way are called "neutral petites."

When two (ρ^-) strains of independent origin are crossed, the progeny are always (ρ^-). In other words, no complementation occurs to produce functional mitochondria. The same defect must thus be present in all the (ρ^-) strains. Unfortunately, further genetic analysis of these diploid (ρ^-) cells is not possible, because diploid petite cells are unable to sporulate. Apparently, sporulation is an aerobic process, requiring a respiratory metabolic system that is lacking in the petite cells.

A different type of petite, called the "suppressive petite," has also been discovered. The suppressive petites differ from the neutral petites in that some of the zygotes formed by a cross of petite and normal fail to produce normal diploid daughter cells during vegetative multiplication. In other words, the production of rho seems to be suppressed in some of the progeny, or some of the zygotes fail to transmit rho to their diploid daughter cells, so that they are petite rather than normal. Different strains show different proportions of suppressed cells, ranging from nearly zero to as high as 99%. Moreover, if the suppressed zygotes are placed on a sporulating medium immediately after fusion but before vegetative growth begins, sporulation can be induced and ascospores then give rise to haploid suppressive petite colonies. The unsuppressed normal zygotes can also sporulate, of course, and produce only normal ascospores. Thus, this group of cells acts like neutral petites in the mode of inheritance of the respiratory deficiency. Therefore, in the suppressive petites, segregation of a sort does occur, but it is irregular and non-Mendelian.

Further analysis of the suppressive petite strains led to the suggestion that a suppressive factor (SF) acts to inhibit the replication of normal rho particles. It is apparently passively transmitted, and those cells failing to receive it are normal while those that do, give rise to suppressive petite colonies. The incidence of suppression would then depend on the rate of formation of SF.

In addition to the neutral and suppressive non-Mendelian petites, segregational petite mutants have been discovered, which segregate in the expected Mendelian 2:2 ratio in ascospores and are linked to known chromosomal gene markers. Thus, respiratory mutants may be nuclear as well as cytoplasmic. Unlike the mitochondria of the cytoplasmic petites, the mitochondria of the segregational petites appear normal.

If a segregational petite is crossed with a vegetative petite, the diploid cells derived from the zygote show normal growth. The diploid cells receive from the vegetative petite parent the wild-type allele of the mutant gene in the segregational petite, and the previously inactive but normal-appearing mitochondria from the segregational petite parent. These nuclear and cytoplasmic components interact to produce full normal activity.

Studies of mitochondrial DNA in cytoplasmic (ρ^-) petite strains have revealed that there is effective loss of the mitochondrial DNA. It is assumed that the DNA is itself the rho factor, whose normal function is to control the synthesis of the inner membrane system of the mitochondrion. However, the amount of DNA in the mitochondrion is quite small and thus has a limited coding capacity, estimated to be the equivalent of about 30 proteins. Since the mitochondria carry on replication, transcription, and translation, this amount of DNA seems insufficient to code for the total synthesis of a mitochondrion. The evidence indicates that nuclear genes as well as the mitochondrial DNA collaborate in mitochondrial formation, and that the mitochondria have a limited self-reproductive capacity. Quite clearly, the mitochondria are not completely autonomous cytoplasmic bodies.

Poky *in* Neurospora

In *Neurospora* as in yeast, there is evidence that new mitochondria are formed by the division of existing mitochondria during vegetative growth. The biogenesis of mitochondria appears to depend on genetic information carried by both nuclear and mitochondrial DNA. In support of this statement is the fact that some respiratory mutants in *Neurospora* show typical Mendelian behavior, while others are non-Mendelian or cytoplasmic in inheritance. Moreover, the evidence indicates that the soluble mitochondrial proteins and specifically cytochrome *c* are not synthesized by the mitochondrial protein-synthesizing system but elsewhere in the cell. In many respects the findings concerning respiratory mutants in *Neurospora* are similar to the results with yeast.

The aptly named *poky* mutant in *Neurospora* is one of a group of maternally inherited (*mi*) mutants causing a slow rate of mycelial growth. *Poky* (also known as *mi-1*) shows a typical pattern of maternal inheritance in crosses with the normal type. If the protoperithecial or maternal parent is *poky,* all of the progeny will be *poky;* if the maternal parent is normal, all of the progeny will be normal, and *poky* cannot be recovered in subsequent generations. In these same crosses, biochemical chromosomal marker genes show normal Mendelian segregation without regard to the direction of the cross. *Poky* is deficient in cytochromes *a* and *b,* but has much more than the normal amount of cytochrome *c.*

Poky (*mi-1*), *mi-3,* and *mi-4* can be distinguished phenotypically, and therefore must represent different cytoplasmic mutations. In reciprocal crosses between *poky* and *mi-3* or *mi-4,* the mitochondria in the progeny resemble those of the maternal parent, and there is no evidence for genetic recombination of any kind. Heterokaryon formation in *Neurospora* provides another test of nonallelic genes. In the *poky* + *mi-4,* heterokaryon growth was normal. However, the growth was due to some form of physiological complementation between two types of defective mitochondria, because only *poky* or *mi-4* progeny could be recovered from the heterokaryon. Apparently no normal mitochondria had been formed.

An interesting note is that mitochondria from a cytoplasmic mutant, *abn-1* (similar to *mi-1*), when injected into the hyphae of a normal strain, gave rise to maternally inherited abnormalities, similar to but not identical with *abn-1,* and hence suggesting interaction between the normal and abnormal mitochondria.

Biochemical analysis of the defects in mitochondrial structural protein in *mi-1* and *mi-3* indicate that they differ from wild type by a single amino acid substitution. Thus, the change in mitochondrial DNA that causes a single amino acid replacement in one protein apparently causes multiple enzyme deficiencies. Such a result suggests interdependence of the different proteins in the membrance structure of the mitochondria.

17.3. The Origin of Chloroplasts and Mitochondria

In the last century there was speculation that the mitochondria originated from a primitive microorganism such as a bacterium that invaded a nucleated cell and developed into an endosymbiont. Similarly, it has been suggested that the chloroplasts originated from algae. The information recently acquired about these organelles has conferred new respectability on these speculations.

For example, mitochondria and bacteria are similar in several respects. They are about the same size and the mitochondria resemble the rod-shaped bacteria in shape. The mitochondrial outer membrane differs from the bacterial cell wall, but the inner membrane is analogous to the bacterial cell membrane. Furthermore, the respiratory assemblies are similar in the bacterial membrane and the inner mitochondrial membrane, as are the lipids. Even more interesting is the circular nature of mitochondrial DNA. Until this was discovered, circular DNA had been observed only in microorganisms and not in higher groups. Moreover, nearest-neighbor analysis indicated that mitochondrial DNA was more like bacterial DNA than the chromosomal DNA of the cells in which the mitochondria occurred. Add to this the presence of a reasonably complete protein-synthesizing system, and the mitochondrion takes on many of the aspects of an autonomous, self-replicating system closely allied to a free-living bacterium.

As intensive work on the nature of chloroplasts and mitochondria has proceeded, it has also become clear that these two organelles resemble one another at many levels of organization, from the biochemical to the structural. Both contain their own DNA and RNA and have the capacity for independent growth and division, both are enclosed by a double semipermeable membrane and are highly membranous structures, and the metabolic systems in both show many similarities. Although there is evidence that the chloroplasts may differentiate from proplastids, as yet there is no evidence for a relationship between proplastids and mitochondria.

Nevertheless, the observations lend credence to the idea that a phylogenetic relationship may exist. The present form of the theory is that one organelle may have evolved from the other, or that both may have evolved from a common ancestor. The most intriguing hypothesis is that the ancestor was a free-living microorganism that invaded the cell, or was ingested, and became a symbiont. Under this theory the host cell is envisioned as an anaerobe getting energy by fermentation (glycolysis), which is an extra-mitochondrial process, and the symbiont as an autotrophic microorganism. The adaptive advantage of the combination lies in its ability to obtain energy either from the substrate or from light. Subsequent modification of the symbionts then could have led to the evolution of the more efficient oxidative respiration system now associated with the mitochondria.

While the theory that mitochondria and chloroplasts represent symbiotic intracellular microorganisms is certainly intriguing, and we have already discussed several cases of this sort, it can hardly be regarded as being established. In opposition to the theory is the fact that the mitochondria and chloroplasts have limitations on their autonomy. Their limited amount of DNA is not sufficient for the complete biosynthesis of the organelle, and nuclear genes and protein-synthesizing systems outside the organelle are responsible for the synthesis of essential components in the mitochondria and chloroplasts. Thus, the theory must be viewed with reservations. The important advances lie in the new insights into the relationship between nucleus and cytoplasm and the proof of extranuclear genetic systems.

17.4. Episomes

Some hereditary particles have been found to exist in two alternative states: either in an *autonomous* state in the cytoplasm, where they replicate independently of the chromosomes, or in an *integrated* state incorporated into the chromosome and replicating in synchrony with it (see Fig. 14.14). Particles with these properties are known as **episomes**, and include such things as the sex factor, the colicinogenic factors and the factors for antibiotic resistance in bacteria, and the lysogenic bacteriophages. The episomes are apparently not essential to the life of the bacteria, because they may or may not be present. If they are absent, they can be acquired only from an external source. The F⁻ *E. coli* cells, for example, can only receive the sex factor (F) during conjugation or by transduction. A lysogenic phage may be acquired either by infection, or by conjugation, or by transduction.

The episome concept grew out of the studies of lysogenic phages. If a temperate phage infects a sensitive bacterium, it usually goes through the typical cycle of a vegetative phage, replicating more rapidly than the bacterial genome and forming new phage particles that are released upon lysis. However, more rarely the temperate phage becomes an integral part of the bacterial chromosome in a form known as a "prophage," a process known as lysogeny. Interestingly enough, each type of temperate virus has its own specific site of attachment in the bacterial chromosome. The evidence is that the prophage is not substituted for some portion of the bacterial genome, but rather represents an addition to the genetic material of the bacterium. The prophage replicates in synchrony with the bacterial chromosome, but does not mediate the synthesis of viral protein. It also suppresses such synthesis by homologous viruses that invade the cell, thus conferring immunity on the host cell.

The prophage may be induced to change into the vegetative phage by exposure to X-rays, ultraviolet light, or certain chemicals; or, at a much lower rate, it may change spontaneously. When such a change is induced, it has been found that a bacterial gene may now be incorporated in the genome of the phage. Thus, not only may a phage act as a genetic element in the bacterium, but a bacterial gene may act as a part of the phage genome (e.g., the *gal*⁺ gene with phage λ).

Conjugation in bacteria occurs between "male" F^+ and "female" F^- cells, with the male cells donating genetic material to the females (see Chapter 14). Males occasionally revert to females, apparently because of the loss of the sex factor F, but females never become males spontaneously —they do so only by receiving F from the males during conjugation. In F^+ males, F exists in an autonomous state with a number of replicas, for it can spread through an F^- population very rapidly. When F is in the autonomous state in F^+ cells, it is transmitted at a very high frequency but independently of the bacterial chromosome. The bacterial genes may be transmitted by F^+ cells, but at a considerably lower frequency than F itself. The sex factor can be eliminated from F^+ cells by treatment with acridine dyes, converting them to F^- recipient female cells.

The F^+ strains sometimes mutate to Hfr, in which F is incorporated into the bacterial chromosome. As a result, the Hfr strains act as donors of bacterial genetic material to the F^- females at a high frequency. However, the sex factor, at the distal end of the bacterial chromosome, is no longer transmitted freely during conjugation, and F^- cells are rarely converted into males.

The Hfr males with F in the integrated state do not carry the sex factor in the autonomous extrachromosomal state, suggesting that, like the two phases of the temperate phage, the autonomous and integrated phases of F are mutually exclusive. Furthermore, although acridine dyes can eliminate the F factor from F^+ males, making them F^-, the F factor is not affected by acridine treatment when it is integrated into the bacterial chromosome. Thus, in some way the integrated F factor is protected against the action of the dye.

When F is in the integrated state it can acquire, apparently by genetic recombination, small segments of the bacterial chromosome. If the F then changes from the integrated to the autonomous state, free F factors carrying bacterial genes may exist in the cytoplasm. In this manner the lac^+ bacterial gene has been introduced from F^+ lac^+ cells into F^- lac^- cells. The process has been called sex-duction or F-duction and is an instance of cytoplasmic inheritance of the gal^+ gene. The process shows many parallels to transduction by bacteriophages.

The episomes may be either present or absent from the cell, they may be chromosomal or cytoplasmic, they may be endogenous or exogenous in origin, and they may be pathogenic or harmless. Thus, they seem to span a range from normal cell constituent to intracellular parasite. While it is customary to think in terms of discrete categories, nuclear genes, self-replicating cytoplasmic organelles, or intracellular parasites, the episomes show that in reality these three types of entity may be intimately related. Not only can the episomes exist in the integrated or in the autonomous state, but in the process of changing from one state to the other they form a link between these different categories of hereditary material.

For example, the temperate bacteriophages can exist in all three states. However, relatively few genes control the ability of the phage to become integrated on the bacterial chromosome and the expression of strictly viral characteristics. A single mutation can cause the virus to lose its ability to become attached to the bacterial chromosome, and with its power of lysogeny lost, the virus no longer acts as an episome but only as an obligatory infectious virus particle. On the other hand, mutation in one of the genes controlling vegetative reproduction of the phage may cause it to lose its ability to exist as a vegetative phage; henceforth it can only persist in the integrated prophage state. In this condition it is virtually indistinguishable from the bacterial genome.

Just as viral genes can be incorporated into the bacterial genome, bacterial genes may be incorporated into the viral genome by recombination during the prophage state and subsequent reversion to the vegetative state. The virus (the episome) can then serve as the vector for the infective transfer of bacterial genes from the former host to a new host cell (transduction). The distinctions obviously begin to break down when nuclear genes become infective and the genome of an infectious particle becomes inseparable from the genome of its host.

A similar spectrum exists with respect to the pathogenicity of episomes. The sex factors are not pathogenic. The colicinogenic factors, responsible for the production of colicins, proteins that act as antibiotics, are potentially lethal but are not infectious. The temperate phages are not only potentially lethal but also potentially infectious. The phage genome, by mutation or recombination, may become defective and lose its infectious properties, but retain its ability to synthesize viral protein and hence remain potentially pathogenic. The virus then would resemble the colicinogenic factors, and it is possible to speculate that by mutation they too could be rendered similar to the sex factors and nonpathogenic.

These considerations, therefore, should make us wary in dealing with the various types of hereditary determinants. While the various types of entities will ordinarily retain their identity, the work on episomes opens up new and intriguing possibilities for the behavior of the genetic material. In the course of evolution, extrachromosomal genetic material could have originated by the detachment of episomic elements from the chromosomes. Conversely, the chromosomes could have evolved to their present form by the acquisition of episomic material from the cytoplasm. Still other possibilities exist, because the chromosomes are highly organized structures and must have evolved from simpler, more primitive structures. It is not inconceivable that the episomes resemble the primitive form of the genetic material from which the more highly organized genetic systems have evolved.

17.5. Controlling Elements in Maize

The episome concept grew out of studies involving bacteria, and most of our knowledge about episomes has been gained from microorganisms. However, some work with higher forms has revealed systems showing resemblance to the episomes. One such system is the Activator-Dissociator or *Ac-Ds* system in maize studied by Barbara McClintock. *Ac* and *Ds* are chromosomal elements with unusual properties. *Ds* can induce chromosome breaks and mutations at whatever position it occupies in a chromosome, and may be transposed from one locus to another. This transposition can be detected by the changed linkage relations of its effects. The other element, *Ac*, is required for the expression of *Ds* activity, and the timing of *Ds*-induced mutations or chromosome breaks is modified by the number of *Ac* elements present. The greater the dosage of *Ac*, the later in development the *Ds*-induced events occur (Fig. 17.5). Like *Ds*, *Ac* can also change its position from locus to locus and can induce chromosome breakage or mutation. Unlike *Ds*, *Ac* can act autonomously. McClintock called *Ac* and *Ds* and similar entities *controlling elements* because of their influence on the activity and behavior of the chromosomal genetic material.

Ds was originally recognized because it caused chromosome breakage, and a breakage-fusion-bridge cycle of chromosome behavior resulted (Fig. 17.6). This cycle can be detected cytologically or, with the use of suitable marker genes, by its effects on the phenotype when recessive genes are expressed because of the loss of an acentric fragment. It was later found, however, that chromosome breakage is not an essential part of the phenomenon: Mutation and transposition may occur even in the absence of detectable chromosome breakage. *Ds* was found to undergo "changes in state" such that it no longer induced breakage but only mutation. It could affect not only its own locus but also nearby closely linked genes. In some cases, *Ds*-induced events were so frequent that the loci being affected appeared mutable. For this reason, reservations were expressed as to whether *Ds* was in fact inducing mutations or was perhaps in some way repressing gene action. On the other hand, germinal mutations induced by *Ds* were sometimes stable.

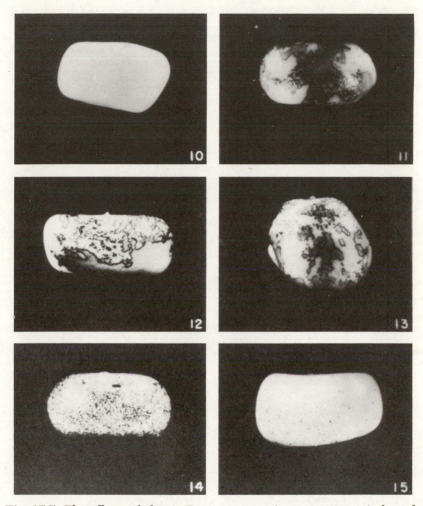

Fig. 17.5. The effects of the *Ac-Ds* system on pigment patterns in kernels of maize. The variegation results from chromosome breaks at *Ds* and the subsequent loss of dominant genes that have masked the expression of their recessive alleles. 10. No *Ac* is present, no breaks occur at *Ds*, and the kernel is colorless because a dominant color inhibitor *I* is present. 11–13. One *Ac* is present, the breaks at *Ds* occur early in development, and the colored sectors are large. 14. Two *Ac* factors are present, the *Ds* breaks occur somewhat later in development, and the colored sectors are correspondingly smaller. 15. Three *Ac* factors are present, and the *Ds* breaks occur so late that only a few colored specks appear. Thus, the dosage of *Ac* affects the timing of the breaks at *Ds*–the greater the number of *Ac* factors, the later in development the breakage occurs. (McClintock, B., 1951. "Chromosome Organization and Genic Expression," *Cold Spring Harbor Symp. Quant. Biol.* 16:13–47, pp. 22–23, Fig. 10.15.)

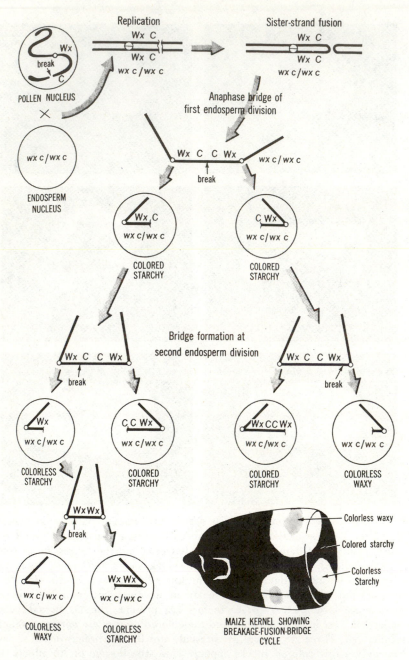

Fig. 17.6. The chromatid type of breakage-fusion-bridge cycle in maize. The endosperm markers on chromosome 9 are *Wx* = starchy, *wx* = waxy; *C* = colored, *c* = colorless. (Baker, W., 1965. *Genetic Analysis.* Houghton Mifflin Co., p. 115, Fig. 8.9.)

The exact nature of the controlling elements remains to be determined. Since other systems have been identified (e.g., *Dt-a*, and *Spm-a*$_1$$^{m \cdot 1}$), the *Ac-Ds* system is not unique. Their significance appears to lie in their possible role in the control of gene action during differentiation and development. They resemble episomes in that they may be present or absent and may occupy different positions on the chromosomes. They presumably exist in an extrachromosomal state during transposition, but this state has not been identified. At present the evidence is not sufficient to regard the controlling elements as being homologous with the episomes. Because of their properties, the controlling elements have sometimes been likened to blocks of heterochromatin. More recently, the controlling elements have been compared to the operon model of bacterial gene action, with *Ac* as the regulator and *Ds* as the operator. This approach suggests that the controlling elements may play a significant role in regulating normal developmental processes. However, the analogy runs into difficulty because some of the mutationlike events are stable rather than reversible like the regulator-repressor systems in bacteria.

17.6. Paramutation

Paramutation is a phenomenon that does not fit comfortably into any category, but since it shows some resemblance to controlling elements and has been studied primarily in maize, it can best be treated at this point. Normally, when two different alleles coexist in a heterozygote and then segregate, they can be recovered in their original intact form. However, the R^r gene in maize, which controls anthocyanin production in the plant and in the seed, invariably has a reduced ability to form anthocyanin after passing through a heterozygote with its stippled R^{st} allele. The changed form of the allele, known as $R^{r'}$, is transmitted through the gametes. When made homozygous, it reverts toward the standard R^r allele, but not completely. The R^r allele is said to be paramutable; the R^{st} allele is paramutagenic. The change $R^r \rightarrow R^{r'}$ is not a typical mutational event. It occurs only with particular heterozygous combinations of alleles, but then, instead of being a rare event, it occurs invariably. Furthermore, even though the change persists over a number of generations, there is a tendency toward reversion to the standard R^r allele. Although paramutation has been compared to gene conversion, which in *Neurospora* involves intragenic recombination, and to interallelic interactions resembling the effects of episomes or controlling elements, the discovery that the functional state of R may change even in plants hemizygous for the R locus would seem to rule out these possibilities. At present, the significance of paramutation seems to lie in the insight it may give into the mechanism of control of gene action.

Still other phenomena of cytoplasmic inheritance could be cited, such as cytoplasmic male sterility in maize, streptomycin resistance in the alga *Chlamydomonas*, the "minutes" in *Aspergillus*, the many studies with the plant genus *Epilobium*, and the barrage and senescence phenomena in the ascomycete *Podospora*. However, the mechanisms involved in these and other phenomena are even less well understood than those already discussed. By this time it should be clear that one of the exciting new frontiers in genetics is the study of extrachromosomal heredity. It is important because the existence of self-replicating genetic systems in the cytoplasm has now been established, the genetics of which are only now coming under study. Moreover, the nature of nuclear-cytoplasmic interactions can now be studied more profitably than ever before. Finally, the control of gene action during differentiation and development, which involves subtle interactions between nucleus and cytoplasm and the environment, is now open to attack in ways never before possible.

Additional Reading

Beale, G. H., 1954. *The Genetics of* Paramecium aurelia. Cambridge University Press, London.

Bittner, J. A., 1958. "Genetic Concepts in Mammary Cancer in Mice," *Ann. N.Y. Acad. Sci.* **71**:943–975.

Borst, P., and A. M. Kroon, 1969. "Mitochondrial DNA: Physico-Chemical Properties, Replication, and Genetic Function," *Int. Rev. Cytol.* **26**:107–190.

Brink, R. A., 1960. "Paramutation and Chromosome Organization," *Quart. Rev. Biol.* **35**:120–137.

Campbell, A. M., 1962. "Episomes," *Adv. Genet.* **11**:101–145.

Campbell, A. M., 1969. *Episomes*. Harper & Row, New York.

Caspari, E., 1948. "Cytoplasmic Inheritance," *Adv. Genet.* **2**:1–68.

Chevremont, M., 1962. "Localization and Synthesis of Deoxyribonucleic Acids in the Cytoplasm of Somatic Cells of Vertebrates: The Role of Mitochondria," *Biochem. J.* **85**:25–26.

Correns, C., 1909. "Vererbungsversuche mit blass (gelb) grünen und buntblättrigen Sippen bei *Mirabilis jalapa, Urtica pilulifera* und *Lunaria annua*," *Z. Ind. Abst. Verebungslehre* **1**:291–329.

Duvick, D. N., 1965. "Cytoplasmic Pollen Sterility in Corn," *Adv. Genet.* **13**:1–56.

Ephrussi, B., 1953. *Nucleo-Cytoplasmic Relations in Microorganisms*. Oxford University Press, London.

Falkow, S., E. M. Johnson, and L. S. Baron, 1967. "Bacterial Conjugation and Extrachromosomal Elements," *Ann. Rev. Genet.* **1**:87–116.

Gahan, P. B., and J. Chayen, 1965. "Cytoplasmic Deoxyribonucleic Acid," *Int. Rev. Cytol.* **18**:223–247.

Gibor, A., and S. Granick, 1964. "Plastids and Mitochondria: Inheritable Systems," *Science* **145**:890–897.

Goldschmidt, R., 1955. *Theoretical Genetics,* University of California Press, Berkeley.

Goodenough, U. W., and R. P. Levine, 1970. "The Genetic Activity of Mitochrondria and Chloroplasts," *Sci. Amer.* **223** (Nov.):22–29.

Hayes, W., 1968. *The Genetics of Bacteria and Their Viruses,* 2nd ed. Wiley, New York.

Jacob, F., and E. L. Wollman, 1961. *Sexuality and the Genetics of Bacteria.* Academic, New York.

Jinks, J. L., 1964. *Extrachromosomal Inheritance.* Prentice-Hall, Englewood Cliffs, N.J.

Jollos, V., 1939. *Dauermodifications. Handb. Vererbw.* **ID**:1–106.

Kirk, J. T. O., and R. A. E. Tilney-Bassett, 1967. *The Plastids.* Freeman, San Francisco.

Levine, R. P., and U. W. Goodenough, 1970. "The Genetics of Photosynthesis and of the Chloroplast in *Chlamydomonas reinhardi,*" *Ann. Rev. Genet.* **4**:397–408.

L'Héritier, P., 1958. "The Hereditary Virus of *Drosophila,*" *Adv. Virus Res.* **5**:195–245.

L'Héritier, P., 1970. "*Drosophila* Viruses and Their Role as Evolutionary Factors," *Evol. Biol.* **4**:185–209.

McClintock, B., 1956. "Controlling Elements and the Gene," *Cold Spring Harbor Symp. Quant. Biol.* **21**:197–216.

Margulis, L., 1970. *Origin of Eukaryotic Cells.* Yale University Press, New Haven, Conn.

Margulis, L., 1971. "Symbiosis and Evolution," *Sci. Amer.* **225** (Aug.):48–57.

Merrell, D. J., 1947. "A Mutant in *Drosophila melanogaster* Affecting Fertility and Eye Color," *Amer. Nat.* **81**:399–400.

Michaelis, P., 1954. "Cytoplasmic Inheritance in *Epilobium* and Its Theoretical Significance," *Adv. Genet.* **6**:287–401.

Mitchell, M. B., and H. K. Mitchell, 1952. "A Case of "Maternal" Inheritance in *Neurospora crassa,*" *Proc. Nat. Acad. Sci.* **38**:442–449.

Nanney, D. L., 1957. "The Role of the Cytoplasm in Heredity." In *The Chemical Basis of Heredity,* W. D. McElroy and B. Glass, eds. John Hopkins, Baltimore.

Nanney, D. L., 1963. "Cytoplasmic Inheritance in Protozoa." In *Methodology in Basic Genetics,* W. J. Burdette, ed. Holden-Day, San Francisco.

Oehlkers, F., 1964. "Cytoplasmic Inheritance in the Genus *Streptocarpus,*" *Adv. Genet.* **12**:329–370.

Poulson, D. F., 1963. "Cytoplasmic Inheritance and Hereditary Infections in *Drosophila.*" In *Methodology in Basic Genetics,* W. J. Burdette, ed. Holden-Day, San Francisco.

Poulson, D. F., and B. Sakaguchi, 1961. "Nature of 'Sex-Ratio' Agent in *Drosophila,*" *Science* **133**:1489–1490.

Rhoades, M. M., 1946. "Plastid Mutations," *Cold Spring Harbor Symp. Quant. Biol.* **11**:202–207.

Rhoades, M. M., 1955. "Interaction of Genic and Non-genic Hereditary Units and the Physiology of Non-genic Inheritance." In *Handbuch der Pflanzenphysiologie,* Vol. I, pp. 19–57, W. Ruhland, ed. Springer Verlag, Berlin.

Ris, H., and W. Plaut, 1962. "Ultrastructure of DNA-Containing Areas in the Chloroplast of *Chlamydomonas,*" *J. Cell Biol.* **13**:383–391.

Roodyn, D. B., and D. Wilkie, 1968. *The Biogenesis of Mitochondria.* Methuen, London.

Sager, R., 1965. "Non-chromosomal Genes in *Chlamydomonas,*" *Proc. 11th Inter. Cong. Genet.* **3**:579–589.

Sager, R, 1965. "Genes Outside the Chromosomes," *Sci. Amer.* **212** (Jan.):70–79.

Sager, R., 1972. *Cytoplasmic Genes and Organelles.* Academic, New York.

Sager, R., and F. J. Ryan, 1961. *Cell Heredity.* Wiley, New York.

Seecof, R. L., 1962. "CO_2 Sensitivity in *Drosophila* as a Latent Virus Infection," *Cold Spring Harbor Symp. Quant. Biol.* **27**:501.

Sonneborn, T. M., 1959. "Kappa and Related Particles in *Paramecium,*" *Adv. Virus Res.* **6**:229–356.

Sturtevant, A. H., 1923. "Inheritance of the Direction of Coiling in *Limnaea,*" *Science* **58**:261–270.

Swift, H., 1965. "Nucleic Acids of Mitochondria and Chloroplasts," *Amer. Nat.* **99**:201–228.

Waddington, C. H., 1960. "Evolutionary Adaptation." In *Evolution after Darwin. I. The Evolution of Life,* S. Tax, ed., University of Chicago Press, Chicago.

Waddington, C. H., 1961. "Genetic Assimilation," *Adv. Genet.* **10**:257–293.

Wilkie, D., 1964. *The Cytoplasm in Heredity.* Methuen, London.

Questions

17.1.   What criteria are useful in trying to distinguish between Mendelian and non-Mendelian heredity?

17.2.   What types of Mendelian inheritance may lead to differences between the progeny from reciprocal crosses?

17.3.   Why is it sometimes difficult to make a distinction between a normal cytoplasmic particle and an infective particle?

17.4.   Is it correct to say that segregation does not occur in extrachromosomal inheritance? If it occurs, what differences would you expect between Mendelian and non-Mendelian segregation?

17.5.   What methods are available to introduce a genome into a foreign plasmon?

17.6.   How would screening for mutations in cytoplasmic genes differ from screening for nuclear gene mutations?

17.7.   In *Limnaea,* what will be the direction of coiling of the shells among progeny from the cross $+s$ ♀ with sinistral shell × $+s$ ♂ with dextral shell?

17.8. In *Limnaea*, what will be the direction of coiling of the shells among progeny from the cross ss ♀ with dextral shell × $++$ ♂ with dextral shell?

17.9. How could you produce a true-breeding stock to carry the deep orange gene?

17.10. What factors influence susceptibility to mammary tumors in mice?

17.11. Distinguish between a phenocopy and a Dauermodification.

17.12. a. What will be the phenotype of the progeny from a cross in *Mirabilis jalapa* involving pollen from a pale branch and ovules in flowers on a variegated branch?

b. What will be the phenotypes of the progeny from the reciprocal cross?

17.13. Genic male sterility due to recessive genes has been found at a number of loci in maize (*Zea mays*), but cytoplasmic male sterility in maize has also been discovered. How would you distinguish genic from cytoplasmic male sterility in maize?

17.14. In maize, if a striped F_1 ♀ from the P_1 cross, striped ♀ (ij/ij) × green ♂ ($+/+$), is crossed with a green ♂ ($+/+$), three outcomes are possible:

1. All green progeny
2. All white progeny
3. A mixture of green, white, and striped individuals

In each case the genotypes of the progeny with respect to the iojap locus are 50% $+/ij$ and 50% $+/+$. How do you account for these results?

17.15. What do you consider to be the most convincing evidence for the existence of extranuclear genetic systems?

17.16. What is the evidence to support the idea that chloroplasts and mitochondria are symbionts?

17.17. What is an episome?

17.18. How do episomes bridge the gaps between chromosomal genetic systems, extrachromosomal genetic systems, and pathogens?

17.19. Distinguish between paramutation, gene conversion, and mutation.

17.20. Segregational (Mendelian) petites as well as neutral and suppressive (non-Mendelian) petites have been discovered in yeast. How can you explain this variety of types of petite mutants?

17.21. Is it possible to have a *Paramecium* that is not a killer but is nonetheless not sensitive to kappa?

17.22. Compare controlling elements in maize with episomes in bacteria.

PART THREE

GENETICS OF POPULATIONS

CHAPTER EIGHTEEN

Population Genetics

The concept of evolution has been traced back to the time of the Greeks. However, modern evolutionary theory stems from the work of Lamarck (Fig. 18.1) and Darwin in the nineteenth century. Lamarck's concepts of evolution and the inheritance of acquired characteristics, set forth in his *Philosophie Zoologique* in 1809, were not generally accepted, but Darwin's *The Origin of Species* (1859), in which his evidence for evolution and natural selection was presented, won general acclaim and relatively quick acceptance. Thus the publication of *The Origin of Species* marks the true beginning of modern scientific interest in the problem of evolution.

And what is the problem? All mammals, for instance, are warm-blooded, air-breathing vertebrates with hair, mammary glands, and four limbs. Nevertheless, the mammals occur in a variety of forms, ranging from mice and moles to horses, cattle, elephants, whales, giraffes, and men. Despite the diversity of mammalian types, all mammals conform to the same basic body plan. The question, then, is why so many obviously different species of mammals show so many similarities. The same sort of situation exists in many other groups of animals and plants. The Linnaean system of classification, with its hierarchy of genera, families, orders, and so on, is based on a hierarchy of resemblance.

Darwin (Fig. 18.2) proposed that the similarities observed among different species were due to their descent, with modification, from a common

Fig. 18.1. Jean Baptiste Pierre Antoine de Monet, Chevalier de Lamarck (1744–1829). (From *Portraits of Geneticists*, 1967. *Genetics.*)

Fig. 18.2. Charles Robert Darwin (1809–1882). (From a water color by George Richmond.)

ancestry. In other words, the resemblances reflected a genetic relationship between the species: the more alike they were, the closer the relationship, and the more recently they had diverged from one another. In support of his theory of evolution, Darwin presented evidence from such fields as paleontology, comparative embryology and anatomy, taxonomy, and biogeography.

Lamarck proposed the inheritance of acquired characters as the mechanism of evolution; Darwin proposed another mechanism—natural selection. The theory of natural selection was based on a few readily made observations and the conclusions that followed from these observations. In the first place, the number of individuals in a species, although subject to fluctuations, must be relatively constant. Nevertheless, the reproductive capacity of any species, including those with the lowest reproductive rates, is far greater than is necessary to maintain the population at its existing level. These two observations lead to the conclusion that many of the zygotes formed never complete development to the adult stage.

The next observation is the uniqueness of the individual members of a species. Variation in natural populations is universal; no two individuals are exactly alike. Given the elimination of excess individuals from the population and the fact of variation, it follows that some individuals will be better adapted than others to survive and reproduce under the existing conditions. To the extent that the variations between individuals are genetic, the genes conferring greater fitness will be transmitted to the next generation at a higher frequency than the other genes, and the adaptedness of the population to the existing conditions will thereby improve.

Therefore, as Darwin saw, a keystone of the theory of natural selection was knowledge of the nature and transmission of biological variation. Knowledge about heredity, however, was scant although Mendel was Darwin's contemporary. Mendel was familiar with Darwin's work, but there is no evidence that Darwin ever became aware of Mendel's experiments. It is interesting to speculate about the possible course of events if Mendel had ever written directly to Darwin or sent him copies of his papers. Darwin was intensely interested in the mechanism of heredity, because he realized its importance to the theory of natural selection. As it was, he developed a theory of heredity called pangenesis, now discredited, which had Lamarckian implications, and in fact he more or less retreated into a Lamarckian position in the later editions of *The Origin of Species*.

The development of modern genetics has supplied the missing information about the nature and transmission of variation and has thus laid a secure foundation under the theory of natural selection, which is now more firmly established than ever before. However, in the early days of this century many geneticists regarded the theory of natural selection as outmoded and passé. This attitude stemmed, at least in part, from the realization that natural selection was not the only means by which evolutionary change could occur. The modern theory of the mechanism of evolution is often referred to as the modern synthesis or neo-Darwinism, and is primarily the result of the work of R. A. Fisher, J. B. S. Haldane, and Sewall Wright. They developed a mathematical theory of the genetics of populations that has implications not only for evolution and speciation, but also for anthropology, human genetics, medicine, animal and plant breeding, and any other areas involved with genetic changes in populations.

The problem attacked by Fisher, Haldane, and Wright was how, within the framework of Mendelian genetics, evolution takes place. How does a population with one set of characteristics arise from an ancestral population that had a somewhat different set of characteristics? They thought of evolution as nothing more than the changes occurring in the kinds or frequencies of genes in populations, and tried to explain how new hereditary variations originate and increase in frequency in populations. Evolution thus was a population phenomenon, and an understanding of population genetics was essential to an understanding of evolution.

Fig. 18.3. William Ernest Castle (1868–1962). (From *Portraits of Geneticists*, 1967. *Genetics.*)

18.1. Castle-Hardy-Weinberg Equilibrium

The first step toward an understanding of the genetics of populations was made by Castle (Fig. 18.3) in 1903 and later, in 1908, by Hardy and by Weinberg, who independently formulated what is now known as the Castle-Hardy-Weinberg law. This law dealt with populations in which the numbers were large and there was no mutation, migration, or natural selection. Under these hypothetical conditions it could be shown that gene frequencies remained constant from one generation to the next. The concept of gene frequency needs to be made more explicit. Let us suppose we have an isolated oceanic island in the Pacific on which there are no rats. Let us further suppose that a ship strikes a reef off the island, losing all hands, except for a number of rats that make their way safely to shore. Most of the rats are agouti or gray in color, the typical wild rat, but several are black. In the rat, gray coat color is due to an autosomal dominant gene A, and black to the homozygous condition of the recessive a. Since each rat is diploid, it carries ashore with it two A genes, one received from its mother, the other from its father. Let us further suppose that 50 rats reach the island. Thus, a given population with N individuals contains $2N$ A genes.

The possible genotypes for the rats are AA, Aa, or aa. If 20 rats are AA, 20 are Aa, and 10 are aa, the frequency of the A and the a genes among the survivors is readily obtained. Let p equal the frequency of A

and q the frequency of a, and let D equal the frequency of AA, H the frequency of Aa, and R the frequency of aa genotypes. Then

$$f(A) = p = \frac{2D+H}{2N} = \frac{40+20}{100} = 0.6 = \frac{D+\frac{1}{2}H}{N}$$

$$f(a) = q = \frac{2R+H}{2N} = \frac{20+20}{100} = 0.4 = \frac{R+\frac{1}{2}H}{N}$$

Therefore, of the A genes introduced into the island, 60% are dominant A and 40% are recessive a. It is thus possible to think not only in terms of the frequency of the different genotypes but also in terms of the frequency of the different alleles at the A locus introduced onto the island.

18.2. Random Mating

Having reached the island, the rats start to interbreed at random. Random mating implies that any individual of one sex is equally likely to mate with any individual of the opposite sex in the population. If, among the stranded rats, each genotype consists of equal numbers of males and females, the possible matings are as follows:

$$\delta\ \delta \qquad 10\,AA \qquad 10\,Aa \qquad 5\,aa$$
$$\female\ \female \qquad 10\,AA \qquad 10\,Aa \qquad 5\,aa$$

From this array it is possible to determine the expected frequency of each of the nine possible matings and the genotype ratios resulting from each kind of mating, but such a procedure is obviously rather cumbersome. Instead, since mating and fertilization are random processes, it follows that the genotype frequencies in the next generation can be determined directly from the gene frequencies in the parents. The frequency or probability of a given zygote is the product of the frequencies of the two genes combining to form it. Therefore, the expected genotype frequencies can be calculated as follows:

♀ \ ♂	$f(A)$ p	$f(a)$ q
$f(A)$ p	$p^2_{f(AA)}$	$pq_{f(Aa)}$
$f(a)$ q	$pq_{f(Aa)}$	$q^2_{f(aa)}$

or

$$p^2\ +\ 2pq\ +\ q^2\ = 1$$
$$f(AA) + f(Aa) + f(aa) = 1$$

In this case,

$$p^2 + 2pq + q^2 = (0.6)^2 + 2(0.6)(0.4) + (0.4)^2 = 1$$
$$0.36(AA) + 0.48(Aa) + 0.16(aa) = 1$$

Let us suppose that conditions on the island are favorable, and the population doubles in size in this generation. The island population can then be expected to consist of 36 homozygous gray (AA), 48 heterozygous gray (Aa), and 16 black (aa) rats. Therefore, the relative proportions of the three genotypes will be expected to change somewhat from the original 20 AA:20 Aa:10 aa proportions. The gene frequencies, however, can be shown to have remained constant:

$$f(A) = p = \frac{D + \frac{1}{2}H}{N} = \frac{36 + 24}{100} = 0.6$$

$$f(a) = q = \frac{R + \frac{1}{2}H}{N} = \frac{16 + 24}{100} = 0.4$$

With continued random mating and no mutation, selection, or further migration, not only the gene frequencies but the genotype frequencies can be expected to remain constant on the island if the population remains large. Thus, the population is in equilibrium. This C-H-W equilibrium reveals a most significant fact: The available variability in a population is not lost as a consequence of crossing, as Darwin believed, but tends to persist at the existing frequencies. Therefore, the C-H-W equilibrium is a conservative factor in evolution, and evolution may even be defined as a shift in the C-H-W equilibrium. The four factors that may bring about such changes in gene frequency are mutation, natural selection, migration or gene flow, and random genetic drift.

18.3. Mutation

Let us suppose that on another island there is a population of rats all of which are homozygous for the dominant A gene ($p = 1$). Spontaneous mutations from dominant A to recessive a occasionally occur, but they are rare events. Let us suppose that the mutation rate from A to a is 2/100,000 gametes per generation. Since the effect of this mutation rate is to increase the frequency of the a gene in the population, mutation can be thought of as a pressure operating on the population to raise the gene frequency of a. Over a very long period of time, if no other evolutionary factors such as selection were operative, all of the A genes in the population could be transformed by mutation to a. If evolution can be defined as a change in the kinds or frequencies of genes in populations, it should be clear that, in theory, mutation alone can bring about evolutionary change independent of any other evolutionary mechanism.

However, back mutations are also known to occur. As soon as the a genes start to appear in the population, they will have a certain probability of mutating back to A. Since the rate of reversion is usually lower than the rate from "wild type" to mutant, let us suppose that the mutation rate from a to A is 1/100,000 gametes per generation. Therefore, the mutation rates are opposed to one another, and the question becomes one of assessing the nature of the effect of these opposing mutation rates.

Let us suppose that the

$$\text{rate of } A \rightarrow a = u = 2 \times 10^{-5}$$

$$\text{rate of } a \rightarrow A = v = 1 \times 10^{-5}$$

The effect of u is to cause a decrease in the frequency of A or p, but v will cause an increase in p. The extent of the effect is dependent on the frequencies of the A and a genes in the population capable of undergoing mutation. Hence the increase in $A = vq$ and the decrease in $A = up$, while the net change in the frequency of A, or Δp, is

$$\Delta p = vq - up$$

On the island where $p = 1$ and $q = 0$,

$$\Delta p = -(2 \times 10^{-5}) \quad \text{and} \quad p_1 = 0.99998$$

On the island where $p_0 = 0.6$ and $q_0 = 0.4$,

$$\Delta p = (1 \times 10^{-5})(0.4) - (2 \times 10^{-5})(0.6)$$
$$= (4 \times 10^{-6}) - (12 \times 10^{-6}) = -8 \times 10^{-6}$$

Therefore, since $p_1 = p_0 + \Delta p$,

$$p_1 = 0.6 - (8 \times 10^{-6}) = 0.599992$$
$$q_1 = 0.400008$$

Obviously, mutation pressure in this case has a very slight effect on gene frequencies, causing an increase in a of only 8 per million genes.

Where two opposing forces are at work, it is appropriate to ask whether an equilibrium is possible. In this case, an equilibrium exists when

$$\Delta p = vq - up = 0$$

or when

$$vq = up$$

At this point the actual number of mutations in each direction is equal, so that the net result is no change in gene frequency; however, this is a dynamic rather than a static equilibrium.

When the above equation is solved for the equilibrium value of p (or $\hat{p}$), the following relationship emerges:

$$up = v(1-p)$$
$$up = v - vp$$
$$p(u+v) = v$$
$$\hat{p} = \frac{v}{u+v}$$

Similarly,

$$\hat{q} = \frac{u}{u+v}$$

The equilibrium value of p is, therefore, completely independent of the initial value of p and depends only on the mutation rates. The same equilibrium point will be reached from any initial values of p and q, including 0 and 1. Although the equilibrium is dynamic, it is also stable. For the mutation rates given above, the equilibrium point is

$$\hat{p} = \frac{1 \times 10^{-5}}{(2 \times 10^{-5}) + (1 \times 10^{-5})} = 0.333$$

$$\hat{q} = 0.667$$

Thus, at equilibrium the number of a genes is twice as great as the number of dominant A genes, but since they mutate only half as often, the total number of mutations in each direction is the same. Furthermore, it should be noted that in this case, at any value of p less than 0.333, A will increase in frequency as a result of mutation despite the fact that u, the rate from A to a, is twice as great as v. The reason, of course, is that Δp is dependent on gene frequencies as well as on the mutation rates.

Therefore, mutation plays an independent role in its effect on gene frequency change, and must be studied as a separate force interacting with other evolutionary forces. The equilibrium due to opposing mutation rates is dependent on these rates, and is independent of initial gene frequencies and different from the Castle-Hardy-Weinberg equilibrium.

18.4. Natural Selection

The expression "natural selection" has always had bloodthirsty connotations, for it has been related to "survival of the fittest," "competition," and similar terms. In fact, however, the modern interpretation of natural selection equates it with differential reproduction by different genotypes. Many factors may lead to differential reproduction, and while survival and competition may be important in many cases, in others cooperative behavior, even altruism, may lead to differential reproduction in favor of the genotypes showing this behavior. Thus, any trait, of whatever nature, that has a genetic component and permits the genotypes manifesting the trait to leave proportionately more progeny than other genotypes will tend to increase in frequency in the population as the relevant genes increase in frequency. The problem is to measure the effects of natural selection on gene frequencies.

18.5. Fitness and the Selection Coefficient

Let us suppose that we have equal numbers of the two homozygous types of rats, AA and aa, but that for every 100 gray AA rats that survive and reproduce, only 90 black aa rats survive and reproduce. We may then say that the fitness, W, of the black rats is only 90% as great as that of the gray rats. If the fitness of gray is set equal to one, the fitness of the black rats is

$$W = 1 - s$$

where s is known as the **selection coefficient**. In this case

$$0.9 = 1 - s$$

or

$$s = 0.1$$

In other words, s is a measure of the disadvantage of the less fit type. Now let us examine how selection operates in a population of diploid individuals.

18.6. Zygotic Selection

Let us suppose that the following fitness relations exist:

Genotype	AA	=	Aa	>	aa
W	1.0		1.0		0.90
s	0		0		0.1

Genotype	AA	Aa	aa	Total
Frequency before selection	p_0^2	$2p_0q_0$	q_0^2	1
Frequency after selection	p_0^2	$2p_0q_0$	$q_0^2(1-s)$	$1 - sq_0^2$

The change in frequency, then, is

$$\Delta p = p_1 - p_0$$

$$p_1 = \frac{D + \frac{1}{2}H}{D + H + R} = \frac{p_0^2 + p_0 q_0}{1 - s q_0^2}$$

Therefore

$$\Delta p = \frac{p_0^2 + p_0 q_0}{1 - s q_0^2} - p_0$$

$$= \frac{p_0^2 + p_0 q_0 - p_0 + s p_0 q_0^2}{1 - s q_0^2}$$

$$= \frac{s p_0 q_0^2}{1 - s q_0^2}$$

or, if sq^2 is small,

$$\Delta p \cong s p q^2$$

If $s = 0.1$, $p_0 = 0.6$, and $q_0 = 0.4$, then

$$\Delta p \cong (0.1)(0.6)(0.4)^2 \cong 0.0096 \cong 0.01$$

Therefore,

$$p_1 = 0.61$$

The form of the equation, $\Delta p = s p q^2$, reveals certain relevant facts about selection. It is, for example, not effective if p or q is zero. In other words, selection cannot act in a homozygous population; there must be alternatives present on which selection can operate. Of course, Δp is also zero if s is zero. Furthermore, selection is most effective at intermediate gene frequencies. For the same value of s, Δp will vary as the values of p and q vary, with Δp very small when either p or q is small. For example, if p is 1×10^{-6} and s is also small, say 1×10^{-3}, it has been estimated that it would take 11,738 generations to increase the frequency of p from 1×10^{-6} to 2×10^{-6}. A mutation rate of 1×10^{-6} could produce the change in a single generation.

The relationship between mutation and selection should be noted. New genes are introduced into a population by mutation. At low frequencies selection is relatively ineffective, so recurrent mutation may lead to an increase in the frequency of the gene to a level at which selection can operate effectively. The ultimate fate of the gene in the population will be determined primarily by selection rather than mutation. Since most mutations are deleterious, mutation pressure and selection pressure generally tend to oppose or balance one another. The deleterious genes originate and increase in frequency in the population due to recurrent mutation, but natural selection tends to eliminate these genes. Thus, an equilibrium can be achieved between mutation and selection. For example, let us consider the equilibrium frequency of a recessive lethal gene. In this case the fitness relations are

	AA	=	Aa	>	aa
W	1.0		1.0		0
s	0		0		1.0

At equilibrium,

$$\Delta p = spq^2 + vq - up = 0$$

The vq term is ordinarily negligible in this case because of the low gene frequency of the lethal and the small value of v. Therefore, at equilibrium,

$$spq^2 = up$$
$$sq^2 = u$$

and in this case, since $s = 1$,

$$\hat{q}^2 = u$$

In other words, the frequency of the homozygous recessive lethal individuals born into the population gives a direct estimate of the mutation rate. Thus, for example, if the aa type has a frequency of 1×10^{-5}, the mutation rate is 1×10^{-5}.

If the recessive trait is deleterious but not lethal, the relationship is

$$\hat{q}^2 = \frac{u}{s}$$

In the case of albinism in man, the incidence is estimated to be about 1 in 20,000, and s has been estimated to be 0.5. The mutation rate from the gene for normal pigment to the gene for albinism can be estimated to be

$$u = (0.5)(5 \times 10^{-5}) = 2.5 \times 10^{-5}$$

It must be stressed that the validity of these estimates is very much dependent on the validity of the assumptions made, for example, about relative fitness and about equilibrium.

A balance between mutation and selection is not the only way that a deleterious gene might be maintained in equilibrium in a population. If the heterozygote has an adaptive advantage over both homozygotes, a stable equilibrium will also be established. Let us suppose that the following fitness relations exist:

Genotype	AA	<	Aa	>	aa
W	$1 - s_1$		1		$1 - s_2$
Frequency before selection	p^2		$2pq$		q^2
Frequency after selection	$p^2(1 - s_1)$		$2pq$		$q^2(1 - s_2)$

In this case

$$\Delta p = \frac{pq(s_2q - s_1p)}{1 - s_1p^2 - s_2q^2}$$

At equilibrium Δp will equal zero when $s_2q - s_1p = 0$. Therefore, at equilibrium,

$$s_2q = s_1p$$
$$s_2(1-p) = s_1p$$

and the equilibrium value of p can be shown to be

$$\hat{p} = \frac{s_2}{s_1 + s_2}$$

while

$$\hat{q} = \frac{s_1}{s_1 + s_2}$$

In other words, an equilibrium will be established in which the gene frequencies depend on the magnitude of the selection coefficients. The gene for sickle cell anemia, a fatal recessive disease in man, is maintained at a high frequency in some populations in the malarial regions of Africa, because the heterozygotes for this gene seem less seriously affected by malaria than the homozygotes for the normal allele.

18.7. Migration or Gene Flow

Rather than forming one large, randomly mating population, a species is typically subdivided into a number of breeding populations, each more or less isolated from the others. If the isolation is complete, each population will pursue its own independent path. However, within a species isolation is seldom complete, and a certain amount of movement of individuals from one breeding group to another occurs. In this way genes flow back and forth between populations, and the tendency toward divergence among isolated breeding populations is counteracted by migration. Migration may be regarded as a pressure similar to selection or mutation. **Migration pressure** results from the invasion of a population by individuals that interbreed with members of the population and thus add their genes to the gene pool going to form subsequent generations. In this sense all hybridization may be treated as migration.

Visualize a small rocky offshore island inhabited by a population of rats.

Let us suppose that in each generation some rats make their way to the island from the mainland. If the gene frequencies differ on the island and the mainland, this influx of genes will have an effect on the gene frequencies on the island. The magnitude of this effect is obtained from the equation

$$\Delta p = -m(p - p_m)$$

where p is the frequency of, for example, the gray gene (A) on the island, p_m is the frequency of the gray gene among the immigrants, and m is the migration coefficient or coefficient of replacement. In reality, m is a measure of the proportion of gametes going to make up the next generation that is derived from the immigrants.

If, for example, the island is of black lava and the black (a) gene has a frequency of 0.9 $(q = 0.9; p = 0.1)$ while the black gene on the mainland has a frequency of 0.1 $(q_m = 0.1; p_m = 0.9)$ and m is 10%, the change due to migration will be

$$\Delta p = -0.1(0.1 - 0.9)$$
$$= +0.08$$
$$p_1 = p + \Delta p = 0.18$$

and there is a marked increase in the frequency of the gray gene on the island.

An equilibrium will exist when $p = p_m$. Furthermore, it is worth noting that if m is rather large and p greatly different from p_m, very rapid changes in gene frequency are possible. It is for this reason, of course, that breeders so often introduce a new sire or some other source of new genetic material into the population with which they are working. In natural populations, migration permits the spread of favorable genes or gene combinations throughout the species.

18.8. Genetic Drift

Mutation, selection, and migration have been referred to as pressures because it is possible to predict the direction and magnitude of the changes in gene frequency due to their action. Evolution, in turn, shows evidence of directive forces at work, leading, for instance, to better adaptation. At the same time, evolution sometimes seems to show a random aspect. Mutation is a random process in the sense that mutations do not occur as an adaptive response by the organism to changes in its environment. Thus, insects exposed to DDT, for example, do not show an increased number of mutations conferring DDT resistance. The "spontaneous" mutations are random in this sense, and also in the sense that it is not possible to predict which particular gene is about to mutate.

The other source of the random aspect of evolution is **genetic drift**. The significance of migration to evolution is dependent on the population structure of the species. In the case of genetic drift, the important factor is population size: the smaller the population, the greater the potential for genetic drift.

To illustrate genetic drift, let us consider a very simple example. Suppose we have a single, self-fertilized, heterozygous (Aa) plant that each generation produces a single surviving offspring. N, the population size, is constant and equals one. The possible kinds of progeny are AA, Aa, aA, and aa. There is thus a 50% chance that the sole surviving offspring will be an Aa heterozygote like the parent, but also a 50% chance that it will be homozygous, either for AA or aa. In the latter case, the gene frequency will have changed from $p = 0.5$ to either $p = 1.0$ or $p = 0$. The A gene thus may be either fixed or lost from the population (of one) due solely to chance, which alone determines which of the four possible kinds of progeny happens to survive. This case is extreme; most populations consist of more than one individual. In populations with equal numbers of males and females, the rate of decrease in heterozygosity due to random genetic drift (k) has been shown to be

$$k = \frac{1}{2N}$$

where N is the effective size of the breeding population. N is a measure of the individuals actually contributing their genes to the next generation and thus may be somewhat different from a simple census figure for adults. If N, for example, were 50 in a population with 100 heterozygous loci, the expected rate of decrease in heterozygosity would be 1/100. The rate of decrease in heterozygosity may also be thought of as the rate of fixation or the rate at which homozygosity is being achieved. Hence, one of the 100 loci would be expected to become homozygous for one allele or the other in the next generation. Obviously, as N becomes large, genetic drift becomes a very minor factor in the dynamics of the population. It should be added that when the population is fluctuating in size, the effective size is considerably closer to the minimum size than to the maximum. Similarly, if the sexes are unequal in number with, say, many females, but only a few males serving as sires, the effective size is less than the total for both sexes, and is most closely related to the number of the rarer sex. The effective size of the breeding population is thus a statistical concept that permits estimates of the amount of genetic drift to be made.

This brief introduction to population genetics should convey some conception of the way that mutation, selection, migration, and drift produce changes in gene frequencies in populations.

Additional Reading

Castle, W. E., 1903. "The Laws of Heredity of Galton and Mendel, and Some Laws Governing Race Improvement by Selection," *Proc. Amer. Acad. Arts Sci.* **39**:223–242.

Crow, J. F., and M. Kimura, 1970. *An Introduction to Population Genetics Theory.* Harper & Row, New York.

Darwin, C., 1859. *The Origin of Species.* (Many editions and printings.)

Demerec, M., ed., 1955. "Population Genetics," *Cold Spring Harbor Symp. Quant. Biol.* **20**:1–346.

Dobzhansky, T., 1962. *Mankind Evolving.* Yale University Press, New Haven, Conn.

Dobzhansky, T., 1970. *Genetics of the Evolutionary Process.* Columbia University Press, New York.

Ewens, W. J., 1969. *Population Genetics.* Methuen, London.

Falconer, D. S., 1960. *Introduction to Quantitative Genetics.* Ronald, New York.

Fisher, R. A., 1930. *The Genetical Theory of Natural Selection.* Clarendon, Oxford. (Reprinted 1958, Dover, New York.)

Grant, V., 1963. *The Origin of Adaptations.* Columbia University Press, New York.

Haldane, J. B. S., 1932. *The Causes of Evolution.* Longmans, Green, London. (Reprinted 1966, Cornell University Press, Ithaca, New York.)

Hardy, G. H., 1908. "Mendelian Proportions in a Mixed Population," *Science* **28**:49–50.

Hogben, L., 1946. *An Introduction to Mathematical Genetics.* Norton, New York.

Huxley, J., ed., 1940. *The New Systematics.* Oxford University Press, London.

Huxley, J., 1943. *Evolution. The Modern Synthesis.* Harper & Row, New York.

Ingram, V. M., 1963. *The Hemoglobins in Genetics and Evolution.* Columbia University Press, New York.

Jukes, T. H., 1966. *Molecules and Evolution.* Columbia University Press, New York.

Kempthorne, O., 1957. *An Introduction to Genetic Statistics.* Wiley, New York.

Lamarck, J. B. P. de., 1809. *Philosophie Zoologique.* Dentu, Paris.

Lerner, I. M., 1954. *Genetic Homeostasis.* Oliver and Boyd, Edinburgh.

Lerner, I. M., 1958. *The Genetic Basis of Selection.* Wiley, New York.

Lewontin, R. C., 1967. "Population Genetics," *Ann. Rev. Genet.* **1**:37–70.

Lewontin, R. C., ed., 1968. *Population Biology and Evolution.* Syracuse University Press, Syracuse, N.Y.

Li, C. C., 1955. *Population Genetics.* University of Chicago Press, Chicago.

Malécot, G., 1966. *Probabilités et hérédité.* Presses Université de France, Paris.

Malécot, G., 1969. *The Mathematics of Heredity,* D. M. Yermanos, tr. Freeman, San Francisco.

Mather, W. B., 1964. *Principles of Quantitative Genetics.* Burgess, Minneapolis.

Mayr, E., 1942. *Systematics and the Origin of Species.* Columbia University Press, New York.

Mayr, E., 1963. *Animal Species and Evolution.* Harvard University Press, Cambridge, Mass.

Mayr, E., 1970. *Populations, Species, and Evolution.* Harvard University Press, Cambridge, Mass.

Merrell, D. J., 1962. *Evolution and Genetics.* Holt, Rinehart and Winston, New York.

Mettler, L. E., and T. G. Gregg, 1969. *Population Genetics and Evolution.* Prentice-Hall, Englewood Cliffs, N.J.

Provine, W. B., 1971. *The Origins of Theoretical Population Genetics.* University of Chicago Press, Chicago.

Rasmuson, M., 1961. *Genetics on the Population Level.* Heinemann, London.

Rensch, B., 1960. *Evolution above the Species Level.* Columbia University Press, New York.

Robinson, R., 1971. *Lepidoptera Genetics.* Pergamon, Elmsford, N.Y.

Schmalhausen, I. I., 1949. *Factors of Evolution,* I. Dordick, tr. Blakiston, Philadelphia.

Simpson, G. G., 1944. *Tempo and Mode in Evolution.* Columbia University Press, New York.

Simpson, G. G., 1953. *The Major Features of Evolution.* Columbia University Press, New York.

Spiess, E. B., 1968. "Experimental Population Genetics," *Ann. Rev. Genet.* **2**: 165–208.

Stebbins, G. L., 1950. *Variation and Evolution in Plants.* Columbia University Press, New York.

Tax, S., ed., 1960. *Evolution after Darwin.* Vol. I. *The Evolution of Life.* Vol. II. *The Evolution of Man.* Vol. III. *Issues in Evolution.* University of Chicago Press, Chicago.

Wallace, B., 1968. *Topics in Population Genetics.* Norton, New York.

Weinberg, W., 1908. "Über den Nachweis der Vererbung beim Menschen." *Jahreshefte Verein f. vaterl. Naturkunde in Württemberg* **64**:368–382. (Partial translation in C. Stern, 1943, "The Hardy-Weinberg Law," *Science* **97**:137–138.)

Wright, S., 1931. "Evolution in Mendelian Populations," *Genetics* **16**:97–159.

Wright, S., 1967. "The Foundations of Population Genetics." In *Heritage from Mendel*, R. A. Brink, ed. University of Wisconsin Press, Madison.

Wright, S. *Evolution and the Genetics of Populations.* Vol. I, 1968; Vol. II, 1969. University of Chicago Press, Chicago.

Questions

18.1. A wild population of a butterfly species consists of 99% orange butterflies and 1% yellow. Genetic tests proved the yellow form to be due to an autosomal recessive gene. Estimate the percentage of heterozygotes in this population. What assumptions have you made in order to reach this estimate?

18.2. A wild population of a moth species was found to have 96% black or melanic moths and 4% mottled white moths. Crosses proved the melanic form to be due to an autosomal dominant.

a. Estimate the frequency of the dominant gene.

b. Estimate the frequency of the heterozygotes in this population.

c. How could you check on the accuracy of your estimate of the frequency of heterozygotes?

18.3. If 6% of the males in a population have the same type of sex-linked recessive red-green color blindness, estimate the frequency of women in the population with this same type of color blindness. What proportion of the women will have normal color vision but nevertheless have half of their sons color blind because of this gene?

18.4. If 10 white mice homozygous for the autosomal recessive gene for albinism were introduced into a pure agouti population of 90 individuals, estimate the equilibrium frequency of albino mice in the population.

18.5. The frequency of a dominant gene is 0.2. What proportion of the population would you expect to show the dominant trait?

18.6. The so-called "wild-type" phenotype is the result of the interaction of a number of "wild-type" genes, each of which is ordinarily dominant to other alleles at its locus.

a. What is a "wild-type" gene?

b. Why are "wild-type" genes usually dominant?

18.7. Would you expect a sex-linked dominant trait to be more frequent in men or in women? If a dominant sex-linked trait affects 5% of the men in a population, what proportion of the women do you estimate would be affected?

18.8. In a population of mice, an autosomal recessive trait has a frequency of 1 in 100. In a nearby population this same trait has a frequency of 1 in 144. Assume a C-H-W equilibrium in both populations. Could this difference in gene frequency have arisen due to a single generation of selection against only the homozygous recessive individuals?

18.9. Suppose that an Indian tribe had 90.25% of blood group O and 9.75% of blood group A. Intermarriage with members of a neighboring tribe with 100% O resulted in 20% of the genes in the next generation coming from the neighboring tribe. What are the expected frequencies of the blood groups in the next generation?

18.10. Suppose that you saw an albino squirrel in a neighbor's yard. What information would you have to obtain and what assumptions would you have to make in order to make an estimate of the frequency of the gene of albinism?

18.11. Under what conditions will the following relationships hold?

a. $\dfrac{p}{q} = \dfrac{v}{u}$

b. $s_1 p = s_2 q$

c. $\dfrac{sq^2}{u} = 1$

18.12. From a population 80% normal and 20% heterozygous for an autosomal recessive lethal, what percentage of the next generation will be expected, with random mating, to be homozygous for the recessive lethal?

18.13. Breeding populations of mice are usually quite small. In 100 populations of five pairs each, all heterozygous for a given locus, in how many of these populations would that locus be expected to become homozygous in the next generation due to chance?

18.14. Thalassemia major is a severe anemia, usually fatal in childhood and rather frequent in populations in the Mediterranean region. It results from homozygosity for a particular gene. Thalassemia minor (a very mild anemia) results from heterozygosity for this gene. If, among people of southern Italian or Sicilian ancestry now living in Rochester, New York, thalassemia major occurs in about 4 in 10,000 births, what is the expected incidence of thalassemia minor?

18.15. If the situation in Problem 18.14 were an equilibrium being maintained by mutation, what must be the mutation rate to the thalassemia gene?

18.16. If the situation in Problem 18.14 were a balanced polymorphism maintained by superiority of the heterozygotes over the homozygotes, what must be the fitness of the homozygous normals as compared to the heterozygotes?

18.17. If 1 in 50,000 males is born with the sex-linked trait, hemophilia, and all die before reproducing, what is the expected frequency of hemophilic females?

18.18. About 1 in 20,000 children born is homozygous for the autosomal recessive gene for albinism. (Let the frequency of albinism equal 0.000049 rather than 0.000050.) If albinos on the average have only 60% as many children as normally pigmented individuals, what is the expected frequency of albinos in the next generation?

18.19. If the mutation rate $A \rightarrow a$ is 1.7×10^{-6} and the rate $a \rightarrow A$ is 0.8×10^{-6}, what will be the equilibrium frequencies of these alleles?

18.20. On an offshore island, a light-colored deermouse population had a low frequency (4%) of the recessive dark-colored type found exclusively on the mainland. On the island the dark type was only 75% as fit as the light type. What amount of gene flow is necessary to maintain an equilibrium at these frequencies on the island?

18.21. In a population of wild mice consisting of five breeding pairs with 60 loci heterozygous, how many of these loci would be expected to become homozygous in the next generation due to chance?

CHAPTER NINETEEN

Inheritance of Quantitative Traits

Thus far, the variation discussed has been dependent on one or a few gene pairs, and individuals with different traits have been readily separable into different groups. The variation, in other words, is discontinuous. In such traits as comb shape in fowl, eye color in *Drosophila,* and blood groups in man, the expression of the trait is clear-cut. Each individual, for example, can unequivocally be classified as having type A, B, AB, or O blood.

Many traits of interest, however, are continuous in nature rather than discontinuous. Evolutionary change, for instance, generally seems to involve changes in quantitative traits, and most economically important characters in domesticated species of plants and animals, such as yield in corn, milk production, efficiency of gain in pigs, and so on, vary quantitatively. Moreover, it is clear in many cases that heredity plays a role in the expression of the trait even though this role may be poorly defined. There is, for example, an obvious relationship between the heights of father and son, but the correlation is by no means perfect. Unlike flower color, where a plant will have, for instance, red or white flowers depending on the segregation of alleles at a single locus, the determination of a quantitative trait is far more complex. If we wished to study yield in corn, we would find that it de-

Fig. 19.1. Continuous variation. Soldiers arranged according to height form a frequency distribution known as the normal curve. See Appendix. (Blakeslee, A. F., "Corn and Men." Reprinted with permission from *The Journal of Heredity* 5:511–518, 1914.)

pended on the number of ears per plant, the length of the ears, the number of rows per ear, the number of kernels per row, the size of the kernels, and so on. Each of these traits may well be affected by genes at many different loci, and furthermore, the environmental conditions of temperature, moisture, and soil type will also play an important role in the yield ultimately produced.

Such traits as height (Fig. 19.1), weight, or growth rate are continuously variable in the true sense (see Appendix). Such characters as egg production in birds, litter size in mice, scales in the lateral line in fish, and bristle number in *Drosophila* are also universally regarded as quantitative traits, and are treated statistically and genetically as such. They are more properly regarded as examples of meristic variation, or variable traits in which the differences can be counted. Hence, they are discontinuous in every sense of the word. The rationale for grouping them with the continuously varying traits is that they tend to have a normal frequency distribution, and hence can be treated in a similar fashion statistically. Furthermore, the assumption is made that the complex of processes necessary for scale formation in fish, for instance, varies continuously, but that before an additional scale can be formed, some threshold must be exceeded. Hence, the discontinuity in meristic traits is thought to reflect an underlying physiological continuity. In a statistical sense, there seems to be no particular problem in treating continuous variation and meristic variation in a similar manner, for the measurement process itself makes the continuous variable discontinuous. From a genetic standpoint, one may wonder whether comparable genetic systems control such traits as egg production in birds and scale number in fish in contrast to length or weight. If yield in corn were measured by the number of kernels produced per plant and also by the weight of corn per plant, the conclusions reached would probably be quite similar. However, if yield in tomatoes were measured as number of tomatoes per plant instead of by weight, the conclusions could be very different—a few large tomatoes from one plant may considerably outweigh 10 times as many small tomatoes from another. In both cases, however, it seems probable that the genetic mechanism underlying the determination of numbers of kernels or tomatoes may be somewhat simpler than that for the complex of variables that enter into yield as measured by total weight. One reason for discussing this point is that much of the recent experimental work on quantitative inheritance has been done on bristle number in *Drosophila*. If, in fact, the genetics of meristic traits differ significantly from the genetics of continuously variable traits, then extrapolations from this work to work on continuous variation may be of dubious value.

The origins of the study of quantitative variation are several. A century before Mendel, Kölreuter described the results of a study of size inheritance in tobacco. He observed that the F_1 was intermediate in size between the parents, but the highly variable F_2 posed problems for which he had no answers. Mendel himself deliberately avoided quantitative variation. Although he worked with size inheritance in his peas, the dwarf plants ranged in height from 9 to 18 in., and the tall parents from 6 to 7 ft. There was never any doubt as to whether they should be classed as tall or dwarf, and he did not concern himself with the continuous variation in size within each class. Mendel's success was in large measure due to his avoidance of quantitative traits, but he laid the groundwork for the subsequent development of the theory of quantitative inheritance.

Animal and plant breeders long have been concerned with the inheritance of quantitative characters. Notable among them was Robert Bakewell (1726–1795), who played a major role in establishing the British breeds of livestock. Right up to the present day, much of the interest in and research on the genetics of quantitative traits has come from animal and plant breeders.

Darwin himself drew a great deal upon the knowledge accumulated by breeders, and his formulation of the theory of evolution based on the gradual accumulation of numerous hereditary changes by natural selection made the study of the inheritance of this type of change of primary interest. The study of evolution is a second area where interest in quantitative inheritance has remained high.

In the latter part of the last century, Darwin's cousin, Sir Francis Galton, became interested in the study of heredity. Like Mendel, he recognized the importance of a quantitative approach to heredity, but where Mendel counted, he chose to measure. And where Mendel broke through to the laws of segregation and independent assortment, Galton, from his study of continuous variation, came up with a "law of ancestral inheritance," which failed to provide the insight he sought (although it later led to a bitter battle between the early Mendelian geneticists and the followers of Galton). However, Galton's study of the relation between height of parents and children led him to develop the concepts of correlation and regression (see Appendix), and marked the beginning of the development of modern statistics. Thus, his efforts were not entirely wasted.

The primary adversaries in the controversy in Britain between the Mendelians and the biometricians were William Bateson, on the one side, and Karl Pearson on the other. The essence of the dispute was whether a Mendelian explanation was possible for the inheritance of quantitative characters. Pearson argued that Mendelian principles did not apply to continuous variation while Bateson held that a number of Mendelian factors might be involved in traits showing continuous variation. The question was ultimately resolved, but the bitterness and animosity that it generated between the Mendelians and the biometricians in England persisted, and undoubtedly delayed the application of statistics in genetics.

In addition to the bitterness, the dispute was notable for the confusion that prevailed. When Bateson argued for a Mendelian explanation for continuous variation, Pearson responded by showing that the parent-offspring correlations he had observed were too high to be consistent with the Mendelian interpretation. It was then pointed out that Pearson had assumed complete dominance and that if dominance were incomplete, correlations of the order observed would be obtained. de Vries solved the problem in quite a different way by arguing that Mendelian principles were universal because continuous variation was never inherited. And so it went. More fundamental, perhaps, was the failure to distinguish between genotype and phenotype, to make a distinction between environmentally induced and inherited variation. Associated with this was the erroneous concept of the unit character, the idea that a given trait is associated with a single factor or gene. Before we consider how the conflicts were finally resolved, it may be helpful to study the results of Johannsen's experiments with the garden bean, and then the results of a cross involving a quantitative trait.

19.1. Johannsen's Experiments

As his experimental material Johannsen chose the garden bean (*Phaseolus vulgaris*), which, like Mendel's garden pea, is normally self-pollinating.

Table 19.1 Differences Between Pure Lines of the Garden Bean (*Phaseolus vulgaris*) and the Ineffectiveness of Selection Within Pure Lines

Pure line	Avg. wgt. of lines (centigrams)	No. beans weighed	Weight of mother beans					
			20	30	40	50	60	70
			Average weight of offspring beans (number of beans weighed shown in parentheses)					
I	64.2	(145)					63.1 (54)	64.9 (91)
II	55.8	(475)			57.2 (86)	54.9 (195)	56.5 (120)	55.5 (74)
III	55.4	(282)				56.4 (144)	56.6 (40)	54.4 (98)
IV	54.8	(307)				54.2 (32)	53.6 (163)	56.6 (112)
V	51.2	(255)			52.8 (107)	49.2 (29)		50.2 (119)
VI	50.6	(141)		53.5 (20)	50.8 (111)		42.5 (10)	
VII	49.2	(305)	45.9 (16)		49.5 (262)		48.2 (27)	
VIII	48.9	(159)		49.0 (20)	49.1 (119)	47.5 (20)		
IX	48.2	(241)		48.5 (117)		47.9 (124)		
X	46.5	(533)		42.1 (28)	46.7 (412)	46.9 (93)		
XI	45.5	(418)		45.2 (114)	45.4 (217)	46.2 (87)		
XII	45.5	(83)	49.6 (14)			45.1 (42)		
XIII	45.4	(712)		47.5 (93)	45.0 (219)	45.1 (205)	45.8 (95)	
XIV	45.3	(106)		45.4 (21)	46.9 (51)		42.8 (34)	
XV	45.0	(188)	46.9 (18)			44.6 (131)	45.0 (39)	
XVI	44.6	(273)		45.9 (147)	44.1 (90)	41.0 (36)		
XVII	42.8	(295)	44.0 (78)		42.4 (217)			
XVIII	40.8	(357)	41.0 (54)	40.7 (203)	40.8 (100)			
XIX	35.1	(219)		35.8 (72)	34.8 (147)			

From *Genetics in Relation to Agriculture*, by Babcock, E. B., and R. E. Clausen. Copyright 1918. Used with permission of McGraw-Hill Book Company. (Data from Johannsen.)

Table 19.2 The Lack of Effect During Six Generations of Selection in Line I of the Garden Bean (*Phaseolus vulgaris*)

Harvest years	Total number of beans	Mean weight of mother beans of the select strains		Differ- ence b − a	Mean weight of progeny seeds of select strains		Difference B − A
		a-*minus*	b-*plus*		A-*minus*	B-*plus*	
1902	145	60	70	10	63.15 ± 1.02	64.85 ± 0.76	+1.70 ± 1.27
1903	252	55	80	25	75.19 ± 1.01	70.88 ± 0.89	−4.31 ± 1.35
1904	711	50	87	37	54.59 ± 0.44	56.68 ± 0.36	+2.09 ± 0.57
1905	654	43	73	40	63.55 ± 0.56	63.64 ± 0.41	+0.09 ± 0.69
1906	384	46	84	38	74.38 ± 0.81	73.00 ± 0.72	−1.38 ± 1.08
1907	379	56	81	25	69.07 ± 0.79	67.66 ± 0.75	−1.41 ± 1.09

From *Genetics in Relation to Agriculture*, by Babcock, E. B., and R. E. Clausen. Copyright 1918. Used with permission of McGraw-Hill Book Company. (Data from Johannsen.)

Any bean resulting from continuous self-pollination over many generations will be homozygous at practically every locus, its progeny will all have genotypes similar to one another and to the parent, and the F_2 progeny in turn will be similar to the preceding generations. Such a line of descent is called a pure line.

Johannsen started with 19 different pure lines of beans. When the individual beans were weighed, there was a wide range of variation in weight. However, when the average weights for the different lines were compared, they were found to be significantly different (Table 19.1). Therefore it was possible to select among the different lines. The progeny from a line with large beans weighed more on the average than the progeny from a line with small beans.

On a given plant, the size range of the seeds is great due to number and position in the pod, time of seed set, and so on. Johannsen also selected for several generations the largest and smallest seeds within a single line, and reared their progeny. Even though in some instances the beans differed by as much as 40 centigrams in weight (line VII), the average weight of their progeny was the same. In other words, selection within a homozygous line was futile (Tables 19.1 and 19.2). He was dealing only with environmental variation; there was no genetic variability, and hence the selection was ineffective.

The conclusions from these experiments are so widely understood and accepted today that they no longer seem revolutionary. Nevertheless, they led Johannsen to make the distinction between variation due to heredity and that due to environmental factors, and to coin the words genotype for the hereditary constitution of the individual and phenotype for the sum total of the traits expressed as a result of the interaction between heredity and environment. He showed that a given trait exhibits both hereditary and environmental variation, and that the only way to determine the magnitude and direction of their effects is by experimental breeding tests. Furthermore, the environmental effects are so great that they tend to mask any possible genetic discontinuity, leading to continuous variation in weight, and suggesting that the genes responsible must individually be of small effect. Previously, selection was thought to be effective on all variation, but Johannsen showed clearly for the first time that selection could be effective only if genetic variability was present, and had no effect whatever on environmental variation. Another misconception laid to rest was the idea that selection was effective indefinitely. Johannsen's work showed that selection was effective only so long as hereditary variation remained. Once this was exhausted, the effectiveness of selection ceased, for it could not create new variation.

19.2. The Inheritance of Corolla Length in *Nicotiana longiflora*

East (1916) published the results of a study of corolla length in the flowers of tobacco (*Nicotiana longiflora*). As parents he used two inbred varieties with average corolla lengths of 40.5 mm and 93.3 mm. When the varieties were crossed, the mean length of the flowers of the F_1 was intermediate between the means of the parents. There was some variation in size in all three groups, which roughly fitted a normal distribution in each case, but there was no overlap in size between groups (Fig. 19.2). The parents and the F_1 all had about the same amount of variability as measured by the coefficient of variation (see Appendix). The F_2 had about the same mean as the F_1, but it had a wider range of variability, and hence a more platy-kurtic normal curve and a larger coefficient of variation than the previous generations. Variation in the F_3 was less than in the F_2 on the average, but was greater than in the parents or in the F_1. Moreover, the mean length of the different F_3 populations was correlated with the corolla length of the F_2 plant from which they were descended. Although, as noted above, a Mendelian explanation for results such as these had been sought earlier, it was not until about 1910 that Nilsson-Ehle and East independently proposed the multiple-factor hypothesis to explain the evidence they presented on the inheritance of quantitative characters.

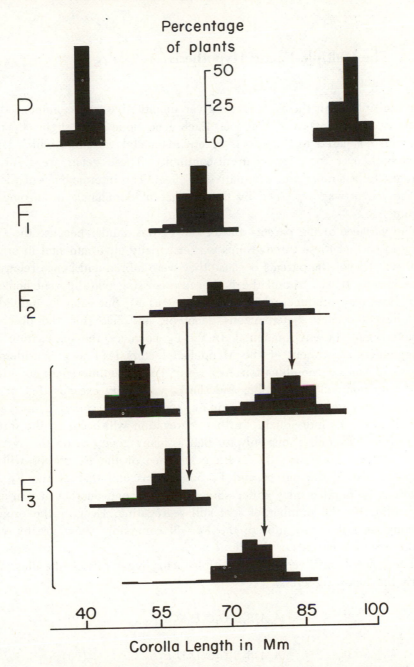

Fig. 19.2. The inheritance of corolla length in *Nicotiana longiflora*. The data are grouped into 3-mm classes, giving a spurious appearance of discontinuity. See text. (Reprinted from Kenneth Mather and John L. Jinks: *Biometrical Genetics.* © 1971 K. Mather and J. L. Jinks. Used by permission of Cornell University Press.)

19.3. The Multiple-Factor Hypothesis

The multiple-factor theory is that a given quantitative trait is influenced by genes at a number of different loci. Each gene alone has a slight effect on the trait compared to the effects of environmental variation, which tends to smooth over the Mendelian discontinuity. These genes were further thought to be similar and cumulative in effect. The intermediacy of the F_1 to the parent was explained by the absence of dominance or incomplete dominance.

The variance of the parents and of the F_1 was similar because the variation in each of these three groups was essentially environmental in origin. This was true of the parents because they were inbred and hence relatively homozygous. It was true of the F_1 because crossing two different homozygous lines may produce a highly heterozygous F_1, but each F_1 individual will have the same heterozygous genotype, and here too the observed variation must be environmental. In the F_2, however, the segregation and independent assortment of the Mendelian factors lead to a considerable degree of genetic recombination. As a result, genotype differences are added to the environmental variation, and the F_2 variability exceeds that of the two previous generations. Furthermore, each locus becomes homozygous in half of the F_2 individuals. Further segregation will occur in the formation of the F_3 families, but only for half as many gene loci on the average as in the F_2. Therefore, the average variance of the F_3 groups will lie between that of the parent and F_1 populations and that of the F_2, but different F_3 families may differ somewhat from one another in variance depending on the number of loci still segregating. Even if the original parents are not highly inbred, the F_2 will commonly show greater variability than the parents or the F_1.

The concepts involved in the multiple-factor hypothesis can be illustrated with the following cross:

$$P_A \qquad A_1A_1B_1B_1C_1C_1D_1D_1E_1E_1\ldots \qquad (10)$$
$$\times$$
$$P_B \qquad A_2A_2B_2B_2C_2C_2D_2D_2E_2E_2\ldots \qquad (20)$$
$$\downarrow$$
$$F_1 \qquad A_1A_2B_1B_2C_1C_2D_1D_2E_1E_2\ldots \qquad (15)$$
$$\downarrow$$

recombinant genotypes

$$F_2 \qquad A_1A_1B_1B_1C_1C_1D_1D_1E_1E_1\ldots \qquad (10)$$
$$A_1A_2B_1B_1C_1C_1D_1D_1E_1E_1\ldots \qquad (11)$$
$$A_2A_2B_1B_1C_1C_1D_1D_1E_1E_1\ldots \qquad (12)$$
$$A_2A_2B_1B_2C_1C_1D_1D_1E_1E_1\ldots \qquad (13)$$

$$\vdots$$

$$A_2A_2B_2B_2C_2C_2D_2D_2E_1E_2\ldots \qquad (19)$$
$$A_2A_2B_2B_2C_2C_2D_2D_2E_2E_2\ldots \qquad (20)$$

The subscripts to the gene symbols can be used to indicate the phenotypic effect of each allele. If the effects are additive, P_A can then be thought of as having a phenotypic value of 10, and P_B a phenotypic value of 20. The F_1 is intermediate at 15, and a wide range of genotypes and phenotypes will be obtained in the F_2, ranging over all values from 10 to 20. The frequencies of the various F_2 phenotypes will, however, be quite different. On the assumption of equal, additive effects and no linkage, these frequencies can be obtained from the binomial $(p+q)^{10}$, where $p = q = \frac{1}{2}$ and p is the chance of having an allele with subscript 1 and q the chance of an allele with subscript 2.

Phenotypic value	Frequency
10	1
11	10
12	45
13	120
14	210
15	252
16	210
17	120
18	45
19	10
20	1

As can readily be seen, the frequency distribution is symmetrical and approximates the normal curve. With phenotypic values so close together, one can also see that environmental variations could easily blur the distinctions between adjacent classes, making the frequency distribution even more like the normal curve.

One of the first questions related to the theory was whether genes of similar effect actually existed. In Chapter 4 we discussed the duplicate factors in the shepherd's purse, *Capsella bursa-pastoris*, where genes at two different loci have similar effects. The case of seed color in wheat studied by Nilsson-Ehle is even more instructive. In crosses of red-grained and white-grained wheat varieties, he found that the F_2 ratios in some instances were 3 red:1 white, in others 15 red:1 white, and in still others 63 red:1 white. Red thus appeared to be dominant to white, but the red seeds were found to vary in color from pale to dark red. The results suggested that the different ratios were due to segregation for color taking place at one, two, or three loci. In the 3:1 cross two shades of red were found, giving a 1:2:1 ratio in reality. In the 15:1 cross the F_2 showed the following frequencies:

Phenotype	Frequency	Number of R genes
white	1	0
pale red	4	1
red	6	2
dark red	4	3
very dark red	1	4

Thus a dihybrid cross of the type, $R_1R_1R_2R_2 \times r_1r_1r_2r_2$, where both loci have similar, additive effects, gave a symmetrical frequency distribution but the classes could still be distinguished reasonably well.

$$P_1 \qquad R_1R_1R_2R_2 \times r_1r_1r_2r_2$$

$$F_1 \qquad R_1r_1R_2r_2$$

F_2	R_1R_2	R_1r_2	r_1R_2	r_1r_2
R_1R_2	$R_1R_1R_2R_2$ [4]	$R_1R_1R_2r_2$ [3]	$R_1r_1R_2R_2$ [3]	$R_1r_1R_2r_2$ [2]
R_1r_2	$R_1R_1R_2r_2$ [3]	$R_1R_1r_2r_2$ [2]	$R_1r_1R_2r_2$ [2]	$R_1r_1r_2r_2$ [1]
r_1R_2	$R_1r_1R_2R_2$ [3]	$R_1r_1R_2r_2$ [2]	$r_1r_1R_2R_2$ [2]	$r_1r_1R_2r_2$ [1]
r_1r_2	$R_1r_1R_2r_2$ [2]	$R_1r_1r_2r_2$ [1]	$r_1r_1R_2r_2$ [1]	$r_1r_1r_2r_2$ [0]

In the trihybrid cross the expected F_2 frequency distribution is (1:6:15:20:15:6:1), and the six classes of red begin to grade into one another. Thus, this case was similar in many respects to the results of other crosses involving quantitative traits. The F_1 was uniform and intermediate between the parents, while the F_2 showed a normal frequency distribution of a continuously varying trait. Since seed color was shown to be Mendelian, a Mendelian explanation for quantitative traits now seemed quite reasonable.

The later discovery that wheat is an allohexaploid suggested, however, that this case may be oversimplified. It seems plausible, at least, that each of the three sets of chromosomes may carry a homologous locus for red.

19.4. Estimates of the Number of Gene Pairs Controlling a Quantitative Trait

Nilsson-Ehle's work suggests a method for determining the number of loci segregating in relation to a particular quantitative trait. When only one locus was involved, $\frac{1}{4}$ of the F_2 was like one or the other parent; with two loci, $\frac{1}{16}$ was like one of the parents; for three loci, $\frac{1}{64}$ was like one of the parents. The relationship between the frequency of recovery of the parental types in the F_2 and the number of segregating loci obviously is as follows:

Number of gene loci	Proportion of F_2 as extreme as one of the parents
1	1 in 4
2	1 in 16
3	1 in 64
4	1 in 256
.	.
.	.
.	.
n	1 in 4^n

In a cross by East between Black Mexican sweet corn and Tom Thumb pop corn, a few individuals in the F_2 had ear lengths in the range of the parents (Fig. 19.3). Since the F_2 numbered 221, it can be estimated that about four gene pairs were all that were segregating for ear length in this cross.

A somewhat different approach makes use of a comparison of the variance (see Appendix) in the F_1 and F_2. The F_1 variance, as noted above, is primarily environmental. The F_2 variance is both environmental and genetic, and it seems reasonable to assume that the environmental variance in the

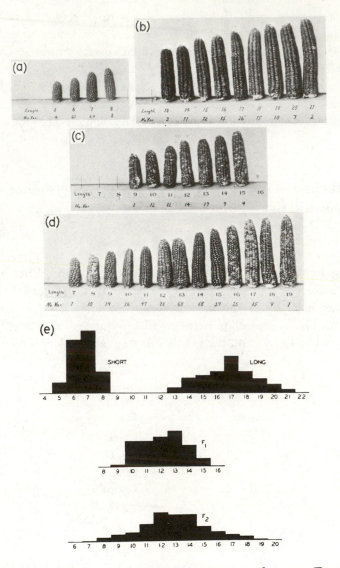

Fig. 19.3. The inheritance of ear length in a cross between Tom Thumb pop corn and Black Mexican sweet corn. The length is in centimeters. The numbers below the lengths are the numbers of ears of that length. (a) Tom Thumb parent. (b) Black Mexican parent. (c) F_1. (d) F_2 (e) Histograms of the data. The horizontal axes are in centimeters; the vertical axes show the frequencies of each size class. Again, the trait varies continuously, but the data have been grouped for ease of handling. (East, E. M., and H. K. Hayes, 1911. "Inheritance in Maize," *Conn. Ag. Expt. Stat. Bull.* 167:1–142. Plates XVIII, XIX, XX opposite p. 120; and Sturtevant, A. H., and G. W. Beadle, 1939. *An Introduction to Genetics.* W. B. Saunders Co., Philadelphia, p. 265, Fig. 101.)

F_1 and F_2 will be about the same. Therefore, subtraction of the F_1 variance from the total variance in the F_2 should give an estimate of the F_2 genetic variance:

$$\text{genetic } s^2(F_2) = \text{total } s^2(F_2) - s^2(F_1)$$

The F_2 genetic variance can be estimated as

$$s^2(F_2) = \frac{Ne^2}{2}$$

where N is the number of loci involved and e is the contribution of each effective allele. The value of e can be estimated from

$$e = \frac{D}{2N}$$

where D is the difference between means in the parental lines. Substitution in the equation for the F_2 genetic variance above then gives

$$\text{genetic } s^2(F_2) = \frac{D^2}{8N}$$

or

$$N = \frac{D^2}{8\left[s^2(F_2) - s^2(F_1)\right]}$$

This estimate assumes equal and additive effects for the genes, no dominance, no epistasis, no linkage, and also that all the plus alleles are in one parent and the minus alleles in the other. Since these conditions are rarely met, the estimate must be regarded as only a very crude approximation. Nevertheless, the conclusion that the greater the number of loci involved, the smaller the variability likely to be observed in the F_2 is of importance and is perhaps contrary to what one might first expect.

19.5. Transgressive Variation

Sometimes, when different strains are crossed, new true-breeding lines can be selected from among their descendents that are either larger or smaller than the original strains. Such lines may even be derived when two strains of about the same size serve as parents. This type of result is called *transgressive variation*, because the new extremes in size "transgress" beyond the size limits observed in the parental strains. The explanation with the multiple-factor hypothesis is very straightforward. If both parents carry some factors for both increasing and decreasing size, then recombination may produce a greater concentration of plus or minus genes in the F_2 than in the parents. For example,

$$P_1 \qquad A_1A_1B_1B_1C_2C_2D_2D_2\ldots \qquad (12)$$
$$\times$$
$$A_2A_2B_2B_2C_1C_1D_1D_1\ldots \qquad (12)$$
$$\downarrow$$
$$F_1 \qquad A_1A_2B_1B_2C_1C_2D_1D_2\ldots \qquad (12)$$
$$\downarrow$$
$$F_2 \qquad A_1A_1B_1B_1C_1C_1D_1D_1\ldots \qquad (8)$$
$$\cdot$$
$$\cdot$$
$$\cdot$$
$$A_2A_2B_2B_2C_2C_2D_2D_2\ldots \qquad (16)$$

In this case, parents and F_1 all have the same average phenotype, but in the F_2 true-breeding lines both smaller and larger than the parents are obtained. Here, as elsewhere, similar phenotypes result from different genotypes. From a practical standpoint, it demonstrates that progress by selection may still be possible even though the variability may appear to be limited. Such cases also show that the assumption that all plus alleles are in one parent is often false.

19.6. The Nature of Multiple Factors

Since it has generally been impossible to identify and study individually by Mendelian methods these postulated multiple factors, in the past there have been doubts as to their existence or else suggestions that they are in some way different from genes of large effect. Nevertheless, the evidence now seems reasonably clear that these two kinds of genes, sometimes termed "polygenes" and "oligogenes," are in reality essentially alike in all respects.

The results from reciprocal crosses are ordinarily essentially the same, indicating that quantitative inheritance is nuclear and not cytoplasmic. The slight differences that may be observed in reciprocal crosses are no greater than is the case for qualitative traits. We have already seen that the pattern of variability in the F_1, F_2, and F_3 generations conforms to expectations based on segregation at a number of independent loci. The inference thus seems reasonable that the mechanism of quantitative inheritance is chromosomal. If so, linkage should be demonstrable. One way to show that Mendelizing multiple factors exist is to demonstrate linkage between a "major" gene for a qualitative trait and the polygenes. The first such experiment was reported by Sax in 1923, who worked, like Johannsen, with the garden bean (*Phaseolus vulgaris*). The qualitative trait was seed coat color; the quantitative trait was seed weight. From a cross between a line with large, colored seeds and one with small, white seeds, the F_2 was 3:1 colored to white, but seed weight was a continuous variable. The F_3 progeny were reared to distinguish F_2 colored homozygotes from F_2 colored heterozygotes. The weights of the beans from the three F_2 classes could then be compared, with the following result:

Genotype	N	Mean seed weight (centigrams)
PP	45	30.7 ± 0.6
Pp	80	28.3 ± 0.3
pp	41	26.4 ± 0.5

From the standard errors it is clear that the differences in weight for the three color genotypes are significant. One possible explanation for these results is that the genes responsible for the size differences, not detectable alone, were detected in this case by their linkage to color genes. However, an alternative possibility is that the color genes are directly responsible for the size differences as a secondary pleiotropic effect. Comparable results were obtained by Castle in studies of coat color genes and body weight in mice and by Lindstrom with size and skin color in tomatoes.

A further refinement of this type of experiment was that of Rasmusson (1935), who worked with flower color in the garden pea and with variation in flowering time as the continuous variable. In this case he was able to show that the linkage between the color genes and the genes affecting flowering time must have been broken by crossing over. Thus, the existence of independent loci affecting a quantitative trait was demonstrated.

19.7. Modifying Factors

The so-called modifying factors may be regarded as multiple factors that are detectable only through their quantitative effects on the expression of a major gene. A most interesting example of modifiers is to be found in the work of Castle with the hooded trait in rats, which is due to a homozygous recessive. Castle successfully selected over a number of generations in plus and minus directions to increase or decrease the amount of pigmentation in

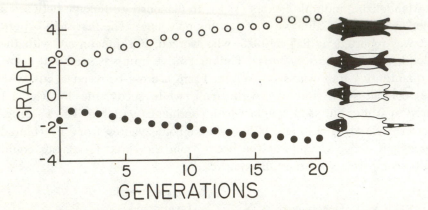

Fig. 19.4. The effect of 20 generations of selection on the hooded trait in inbred lines of rats. Shown above are four of the arbitrary grades used to measure pigmentation. Below are the results of selection: open circles show the average score for the line selected for increased pigmentation; filled circles show the line selected for decreased pigmentation. (Data from Castle.)

the hooded rats (Fig. 19.4). At first he believed that he was, by his selection, actually changing the hooded allele itself. However, the changes in expression of hooded when the selected plus and minus strains were outcrossed to wild rats soon convinced him that selection of plus or minus modifiers at loci other than the hooded locus was responsible for his results. Another case of the effects of modifiers is that of brachyury in the house mouse, mentioned in Chapter 4 on gene interaction. In that case different sets of modifiers caused brachyury to be expressed as a dominant in the European species and as a recessive in the Asiatic species when in the heterozygous condition.

In this connection it may be well to mention the category of genes variously known as suppressors, inhibitors, or repressors. These are genes that suppress the expression of a gene at another locus (often of the mutant allele to produce the wild type), but generally have no other detectable phenotypic effect. The suppressors may be regarded as a type of modifying factor, and in many cases it is possible, by suitable linkage tests, to map a suppressor to a particular locus.

The experimental evidence thus suggests that the inheritance of quantitative traits is governed by a number of chromosomal genes at different loci. When Mather reintroduced the term polygene in place of the earlier expression multiple factor, he suggested that the polygenes might in some way differ from the oligogenes. In particular, he felt that they might be localized primarily in the heterochromatic regions of the chromosomes. The evidence for such a theory has not been forthcoming, but the term polygene has continued to be used as a synonym for multiple factor.

It now seems clear that many different genes are involved in the determination of quantitative characters, but their effects are seldom equal or additive as originally postulated in the multiple-factor hypothesis. Linkage, dominance, epistasis, pleiotropy, developmental restrictions, and genotype-environment interactions may all modify the outcome of crosses involving quantitative traits. In quantitative inheritance, dominance is used to refer to all degrees of allelic interaction leading to a heterozygous phenotype that is other than halfway between parents. Epistasis is used in a very broad sense to refer to all types of nonallelic genetic interactions. In view of the one-gene, one-polypeptide concept that has emerged in relation to gene action, it seems reasonable to suppose that many of the genes seemingly responsible for a quantitative trait may be so because of a pleiotropic effect related to their primary function. The developmental restrictions that may prevent the recovery of certain types of progeny in a quantitative cross are those that would produce individuals so grotesque as to be unable to survive. Alternatively, the pattern of differentiation may be such that not all types theoretically recoverable will develop. To give a hypothetical and perhaps far-fetched example of the latter case, it is unlikely that we will ever observe a man 6'6" with legs only 2' long, not necessarily because he would die but because if he had the growth hormones needed to support growth to this size, his legs would grow too. Under the circumstances, perhaps the most surprising aspect of the studies in quantitative inheritance is that so large a portion of the variance in so many cases turns out to be additive genetic variance.

It is interesting that there has been relatively little effort to incorporate the recent discoveries about DNA into the theory of quantitative inheritance. One of the finest recent books on the subject fails to mention DNA at all. Nevertheless, it seems likely that further progress in the study of quantitative inheritance awaits the development of more realistic models based on our rapidly growing knowledge of DNA.

19.8. Heritability

Since individual genes cannot ordinarily be identified and studied in quantitative inheritance, genetic studies of quantitative traits focus on the study of the variance. The main effort is to partition the variance into the various components responsible for the observed phenotypic variation. The total phenotypic variance can be thought of as

$$V_P = V_G + V_E$$

where V_P = phenotypic variance, V_G = genetic variance, and V_E = environmental variance. The genetic variance can be further subdivided:

$$V_G = V_A + V_D + V_I$$

where V_A = additive variance, V_D = dominance variance, and V_I = variance due to epistatic interactions.

In addition to these effects, interactions may also be observed between the genotype and the environment since a given environmental change may not produce a comparable effect in all genotypes. Thus, the phenotypic variance may include a V_{GE} term. If we were to write a fairly complete equation for the components of the phenotypic variance, therefore, it would be

$$V_P = V_A + V_D + V_I + V_E + V_{GE}$$

Of these various components, the additive genetic variance is of particular interest because it is the chief cause of the resemblances between relatives and hence affords a measure of the breeding value of individuals and of the response of populations to selection. Furthermore, it is much easier to estimate than the other variance components. For this reason the ratio of the additive genetic variance to the total phenotypic variance is of special importance; it has been called the **heritability**, symbolized by H. (Originally Wright used h^2.) Hence

$$H = \frac{V_A}{V_P}$$

Heritability estimates have been made for a number of traits in many different species. Despite the fact that different estimates of heritability for the same trait may vary considerably, there is an inclination to seek some common figure as an average estimate of the heritability of a particular trait. This is unfortunate because, as can be seen from the equations above, heritability is not just the property of a given trait but also depends on the population and the environment in which the trait is observed. All the components of variance will affect the heritability, and a change in any component will influence the estimate. For example, a relatively homozygous population will show lower heritabilities than a relatively heterozygous population. Also, variable environmental conditions will reduce heritabilities while they will increase under more uniform conditions.

Therefore, generalizations about heritability must be made with caution, because each heritability estimate is unique for a particular population in a particular environment.

In quite a few instances, the F_1 in a quantitative cross, rather than being intermediate between the parents, has a mean somewhat lower than the arithmetic mean expected with additive gene effects, and has a positively skewed F_2 frequency distribution. Such results have been thought to indicate nonadditive gene action. For example, the following model shows the results if the effects of the genes are squared rather than added:

Genes for increased size	Additive effect	$(Effect)^2$	
0	50	50	
1	56	51	
2	62	54	
3	68	59	F_1 mean
4	74	66	
5	80	75	
6	86	86	

This and similar models for nonadditive polygene action are probably improvements over the additive model, because they suggest that the contribution of a particular gene in absolute terms will vary according to the rest of the genotype, and thus the model attempts to take gene interactions into account. The additive model, of course, implies that a given gene makes a fixed contribution to the phenotype in a variety of genotypes. However, it is undoubtedly still a considerable oversimplification to regard the complex gene interactions leading to the development of any quantitative trait in terms of some simple multiplicative model.

Although the interpretation of such data in terms of gene interaction seems reasonable, other possibilities should perhaps be pointed out. Similar results would be obtained, for instance, if the gene effects were additive but the genes for small size were partially dominant. They might also arise as a consequence of a metrical bias, a skewness due to an inherent relation between the scale of measurement and the phenotype, with no particular genetic significance. Were the scale of measurement to be changed, the metrical bias might be eliminated. Still another possibility is that there may be a physiological lower limit to the expression of the trait, causing a clustering of individuals toward the lower end of the frequency distribution.

Therefore, at the present time, the models are useful in the description of the data and even in making predictions in relation to the inheritance of quantitative traits. However, they fall considerably short of giving a realistic picture of the genetic framework underlying the development of a quantitative character. It is possible that the models work as well as they do because many of the effects related to the various types of interaction in essence cancel one another out.

In studies of the effects of individual chromosomes on quantitative traits, as for example on egg size, bristle number, and DDT resistance in *Drosophila,* the results indicate that all chromosomes contribute to the effect, with some chromosomes or chromosome segments having greater impact than others. These results imply that while many loci may be involved, some are more important than others, and that it may be possible to identify and analyze the effects of some of these more significant loci. Such an approach has been possible in some of the recent work on quantitative inheritance by Wehrhahn and Allard with wheat and by Thoday et al. and by Robertson with *Drosophila.* In the converse type of study on the effect of known major genes on quantitative traits, Castle, working many years ago with the effect on body weight of coat color genes in mice, gained insight into the variety of effects and interactions possible. As just one example of his results, he found that the brown gene in most genotypes increased the average size of the mice, but in the presence of the leaden mutant, it decreased the average size of the mice. In addition to the trend toward the identification and study of individual genes affecting quantitative inheritance, there is also presently considerable attention being directed toward studying the effects of linkage in quantitative traits. When it is possible to couple studies such as these with advances in developmental genetics tying the genotype to the phenotype, it seems inevitable that our understanding of the genetic basis for quantitative traits will be considerably advanced.

Additional Reading

Bateson, W., 1902. *Mendel's Principles of Heredity, A Defence.* Cambridge University Press, London.

Castle, W. E., 1919. "Piebald Rats and the Theory of Genes," *Proc. Nat. Acad. Sci.* **5**:126–130.

Castle, W. E., 1921. "An Improved Method of Estimating the Number of Genetic Factors Concerned in Cases of Blending Inheritance," *Science* **54**:223.

Castle, W. E., 1941. "Influence of Certain Color Mutations on Body Size in Mice, Rats, and Rabbits," *Genetics* **26**:177–191.

Charles, D. R., and Goodwin, R. H., 1943. "An Estimate of the Minimum Number of Genes Differentiating Two Species of Goldenrod with Respect to Their Morphological Characters." *Amer. Nat.* **77**:53–69.

East, E. M., 1910. "A Mendelian Interpretation of Variation That Is Apparently Continuous," *Amer. Nat.* **44**:65–82.

East, E. M., 1916. "Studies on Size Inheritance in *Nicotiana*," *Genetics* **1**:164–176.

East, E. M., and H. K. Hayes, 1911. "Inheritance in Maize," *Conn. Agr. Exp. Stat. Bull.* **167**:1–142.

Emerson, R A., and E. M. East, 1913. "The Inheritance of Quantitative Characters in Maize," *Nebraska Agr. Exp. Stat. Res. Bull.* **2**.

Falconer, D. S., 1960. *Introduction to Quantitative Genetics*. Ronald, New York.

Fisher, R. A., 1918. "On the Correlation Between Relatives on the Supposition of Mendelian Inheritance," *Trans. Roy. Soc. Edinburgh* **52**:399–433.

Galton, F., 1889. *Natural Inheritance*. Macmillan, New York.

Johannsen, W., 1903. *Über Erblichkeit in Populationen und in reinen Linien.* Fischer, Jena.

Johannsen, W., 1909. *Elemente der exakten Erblichkeitslehre,* Fischer, Jena.

Kempthorne, O., 1957. *An Introduction to Genetic Statistics*. Wiley, New York.

Lindstrom, E. W., 1924. "A Genetic Linkage Between Size and Color Factors in the Tomato," *Science* **60**:182–183.

Lindstrom, E. W., 1928. "Linkage of Size, Shape, and Color Genes in *Lycopersicon*," *Zeit. ind. Abst. Vererb. Suppl.* **2**:1031–1057.

Mather, K., 1943. "Polygenic Inheritance and Natural Selection," *Biol. Rev.* **18**:32–64.

Mather, K., 1944. "The Genetical Activity of Heterochromatin," *Proc. Roy. Soc. B* **132**:308–332.

Mather, K., and J. Jinks, 1971. *Biometrical Genetics,* 2nd ed. Cornell University Press, Ithaca, N.Y.

Mather, W. B., 1964. *Principles of Quantitative Genetics*. Burgess, Minneapolis.

Nilsson-Ehle, H., 1909. "Kreuzungsuntersuchungen an Hafer und Weizen." *Lunds. Univ. Aarskr. N.F.* **5**(2):1–122.

Rasmusson, J. M., 1935. "Studies on the Inheritance of Quantitative Characters in *Pisum*. I. Preliminary Note on the Genetics of Flowering," *Hereditas* **20**:161–180.

Robertson, A., 1967. "The Nature of Quantitative Genetic Variation." In *Heritage from Mendel,* R. A. Brink, ed. University of Wisconsin Press, Madison.

Sax, K., 1923. "The Association of Size Differences with Seed-Coat Pattern and Pigmentation in *Phaseolus vulgaris*," *Genetics* **8**:552–560.

Serebrovsky, A. S., 1928. "An Analysis of the Inheritance of Quantitative Transgressive Characters," *A.I.A.V.* **48**:229–243.

Thoday, J. M., J. B. Gibson, and S. G. Spickett, 1964. "Regular Responses to Selection. 2. Recombination and Accelerated Response," *Genet. Res.* **5**:1–19.

Wehrhahn, C., and R. W. Allard, 1965. "The Detection and Measurement of the Effects of Individual Genes Involved in the Inheritance of a Quantitative Character in Wheat," *Genetics* **51**:109–119.

Wright, S., 1968. *Evolution and the Genetics of Populations.* Vol. I. *Genetic and Biometric Foundations*. University of Chicago Press, Chicago.

Questions

19.1. Distinguish among continuous variation, discontinuous variation, and meristic variation.

19.2. If you wished to carry out a successful program of selection for increased size of beans in *Phaseolus vulgaris*, how would you proceed?

19.3. Why, in East's study of corolla length inheritance in tobacco, did the F_2 have a greater coefficient of variation for length than the parents and the F_1? Why was the F_3 coefficient of variation less than that of the F_2 but greater than those of the F_1 and the parents?

19.4. Why does continuous variation in biological material tend to fit a normal distribution curve rather than a bimodal curve or some other type of frequency distribution?

19.5. When the effects of individual genes cannot be identified, how is it possible to estimate the number of gene loci influencing a particular quantitative trait? What assumptions are necessary to make these estimates?

19.6. Because crosses involving quantitative traits usually resulted in an F_1 intermediate to the parents, Darwin believed in blending inheritance rather than particulate inheritance. Why is a theory of blending inheritance inadequate to explain the inheritance of quantitative traits?

19.7. A cross between a pure line of tall plants and a pure line of short plants produced an F_1 that was intermediate in size. Of 512 F_2 offspring, only two were as short as the P_1 parent. Estimate the number of gene loci affecting the difference in height between the parents.

19.8. For F_2 populations of equal finite size, would you expect the range of variation for a quantitative trait to be greater or less as the number of gene pairs influencing that trait increases?

19.9. Suppose that the parental means for a quantitative trait were $P_1 = 50$ and $P_2 = 40$ and the F_1 and F_2 means both equal 45. However, the F_1 variance is 2.5 and the F_2 variance is 5.0. Estimate the number of gene pairs influencing the difference between the means of the parents.

19.10. If you wished to develop a variety of pigeons larger than any of the numerous known varieties raised by pigeon fanciers, how would you proceed?

19.11. In spotted varieties of animals, it has been possible to select lines with either increased or decreased amounts of spotting. Spotting is ordinarily controlled by an autosomal recessive gene. How would you determine whether the changed amount of spotting was due to selection for different recessive spotting alleles at the spotting locus or to selection for modifying factors at other loci?

19.12. If environmental conditions are very carefully controlled so that they are quite uniform, what will be the effect on heritability estimates for quantitative traits? What are the implications of your conclusion for the animal and plant breeder; for the student of human genetics?

19.13. The mean and standard deviation for egg production per year for a flock of hens was 150 ± 20, while the mean and standard deviation for their weight was 4.5 ± 0.4 lb. Which trait is more variable in this flock of hens, egg production or body weight?

19.14. Do you think efforts to identify and study individual gene loci influencing quantitative traits are worthwhile? Why?

19.15. In what ways might the recent discoveries about DNA as the genetic material be useful to the student of quantitative inheritance?

19.16. The following measurements were made on a group of male college students in the United States:

Student No.	Height (in.)	Length of forearm (in.)	Weight (lb)
1	67	18	135
2	68	17.5	165
3	68	18.5	140
4	69	19	145
5	69	18.5	155
6	70	19.5	150
7	71	19	180
8	71	20	165
9	72	19	160
10	73	19.5	170

Calculate the means, the variances, and the standard deviations for each of these traits.

19.17. Calculate the coefficients of variability for the three traits in Question 19.16. Which trait is least variable? Which trait is most variable?

19.18. What is the standard error of the mean for each of the traits in Question 19.16? What inferences can be drawn from a knowledge of the mean and its standard error? What inferences can be drawn from a knowledge of the mean and its standard deviation?

19.19.   Using the data in Question 19.16, plot forearm length against height and weight against height. Calculate the correlation coefficients for each of these combinations. Calculate the regression of weight on height and of forearm length on height. Why is the regression coefficient generally thought to be more useful than the correlation coefficient?

19.20.   Referring to Question 19.19, which did you expect to be more strongly correlated, height and length of forearm or height and weight? Why?

19.21.   A group of male college students in Great Britain was measured for height with the following results:

Student No.	Height (in.)
1	65
2	66
3	66
4	67
5	67
6	67
7	68
8	68
9	69
10	69
11	70
12	71

Does this group of students differ significantly in height from the group of American college students in Question 19.16? What is the appropriate test of significance in this case? What test could be used if the samples were large?

19.22.   From the following data, determine whether inoculation provides any protection against typhoid fever.

	Typhoid fever	
	Attacked	Not attacked
Inoculated	56	6,759
Not inoculated	272	11,396

CHAPTER TWENTY

Inbreeding, Outbreeding, Heterosis, and Selection

The topics in this chapter are of particular importance in relation to the use of genetic principles in animal and plant breeding. The economically important traits are usually quantitative traits governed by genes at many loci. The breeder begins with a population that has a certain array of characteristics and a corresponding array of genes in its gene pool. His problem is to shift the average characteristics of the population in a desired direction. Ordinarily, the solution will require shifts in the gene frequencies in the population. Thus, the breeder is dealing with quantitative inheritance, with population genetics, and, in a very real sense, with evolution, because by his manipulations of the gene pool he brings about descent with modification. Among the tools at his disposal are inbreeding, outbreeding, heterosis, and selection.

20.1. Inbreeding

Simply defined, inbreeding is the mating of related individuals. Since there are many different degrees of relationship, there are also different degrees of inbreeding. The closest possible system of inbreeding is self-fertilization, which occurs in many plants and in a few animal species. In species with compulsory cross-fertilization, the closest possible inbreeding would be brother-sister, mother-son, or father-daughter, with the other degrees (uncle-niece, cousins, etc.) more remote.

You may be surprised to learn that you yourself are the product of inbreeding. The proof is relatively simple. If each of your ancestors has entered your pedigree just once, you will have 2 parents, 2^2 grandparents, 2^3 great grandparents, and 2^n ancestors in each generation farther back. Some 20 generations back, you would have had 1,048,576 ancestors. If there were four generations per century, this would carry us back 500 years, to about 1450 A.D. Ten generations more, back to about 1200 A.D., and in theory you would have had 2^{30} or 1,073,741,824 ancestors. Since the human population on earth at that time was not even close to a billion people, obviously some of your ancestors appear more than once in your pedigree and you are to some extent inbred. However, you are not nearly as inbred as most farm animals, and it has been shown that unless inbreeding is as close as second cousins, it does not cause an appreciable increase in homozygosity.

Inbreeding is commonly thought to be harmful. The so-called Knight-Darwin law states that "Nature abhors perpetual self-fertilization." If a normally cross-fertilized plant such as corn is self-pollinated, a rapid decline in quality, vigor, and yield ensues in the next few generations. The inbred lines are so poor that no self-respecting farmer would plant them. Furthermore, many species of plants are self-sterile, so that outbreeding is compulsory for them. One interpretation of the prevalence of separate sexes is that it ensures cross-fertilization and averts to some extent the deleterious effects of inbreeding. In man, you are probably aware of the taboos and the legal and religious bans against incest. In the United States all states have laws against marriage of relatives closer than first cousins, and some states—Minnesota, for example—forbid marriage if the individuals are closer than second cousins.

In contrast to the supposed harmful effects of inbreeding stands Cleopatra, long regarded as a benchmark of feminine beauty and intelligence. She was the product of very close inbreeding (Fig. 20.1) since the Ptol-

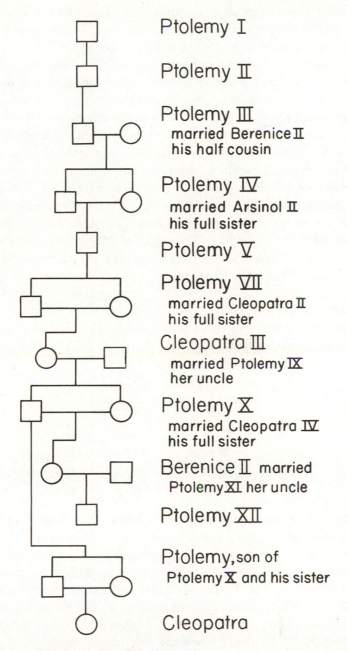

Ptolemy I

Ptolemy II

Ptolemy III
married Berenice II
his half cousin

Ptolemy IV
married Arsinol II
his full sister

Ptolemy V

Ptolemy VII
married Cleopatra II
his full sister

Cleopatra III
married Ptolemy IX
her uncle

Ptolemy X
married Cleopatra IV
his full sister

Berenice II married
Ptolemy XI her uncle

Ptolemy XII

Ptolemy, son of
Ptolemy X and his sister

Cleopatra

Fig. 20.1. A classical example of consanguineous marriage. A fragment of the genealogy of Cleopatra VII, mistress of Julius Caesar and Marc Antony, legendary for her beauty and intelligence. (Reprinted with permission from *The Journal of Heredity* 8:354, 1917.)

emies, the royal family of Egypt, regarded themselves so highly that marriage to mere commoners was unthinkable. Along with Cleopatra, we must place wheat, oats, tobacco, rice, peas, beans, and tomatoes, all naturally self-pollinated and all of normal vigor and productivity.

We are then faced with the question of whether inbreeding is harmful or not. On the one hand we have Cleopatra, tomatoes, wheat, oats, rice, and so on. On the other hand, we have clear evidence of the harmful effects of inbreeding when normally cross-fertilizing species such as corn are inbred. The answer, of course, is that inbreeding, in itself, is neither harmful nor beneficial. Its effect is to increase the level of homozygosity. The harmful effects sometimes observed following inbreeding are the result of bringing deleterious recessive genes to expression in the homozygous condition, genes that were formerly concealed in the heterozygous state. If favorable genes are present, they too may be fixed in the homozygous condition by inbreeding. The effect of inbreeding, then, is to increase the level of homozygosity for all loci and for all genes, dominant and recessive, favorable and deleterious.

The effect of inbreeding can be studied most readily if we consider the effect of self-fertilization, the most rapid inbreeding system, on an initially 100% heterozygous population. Selfing will reduce the amount of heterozygosity by one-half each generation. Thus, at a given locus the following will transpire:

Generation	AA (%)	Aa (%)	aa (%)
0	—	100	—
1	25	50	25
2	37.5	25	37.5
3	43.75	12.5	43.75
4	46.875	6.25	46.875

Theoretically, after just four generations of selfing, only 1 in 16 individuals will be heterozygous. After seven generations, more than 99% of the population will be homozygous, and the approach to complete homozygosity will become asymptotic. The proportion of heterozygotes in the population relative to the initial population will be $(\frac{1}{2})^n$, where n is the number of generations of selfing. The rapidity of this transition is due to the fact that, with selfing, any homozygote formed from then on gives rise only to homozygotes like itself while the remaining heterozygotes produce 50% homozygous types each generation. For example, if the population size is held at 1600, with each individual leaving one offspring, the expected numbers will be distributed as follows:

Generation	Genotype		
	AA	Aa	aa
0	—	1600	—
1	400	800	400
2	400 + 200	400	200 + 400
3	600 + 100	200	100 + 600
4	700 + 50	100	50 + 700

With other systems of mating, the increase in proportion of homozygotes occurs less rapidly, as shown in Fig. 20.2. Even brother × sister mating requires more than twice as many generations to reach an equivalent level of homozygosity as selfing, while continued second-cousin matings do not raise the level of homozygosity above 51%.

The effect of inbreeding, then, is to increase homozygosity at all loci. The most efficient system of inbreeding is self-fertilization. When selfing is

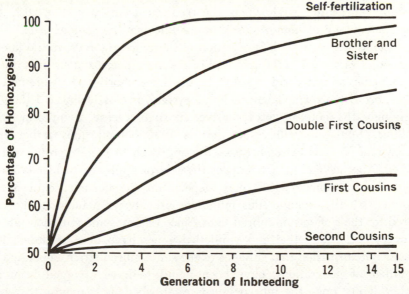

Fig. 20.2. Changes in the percentage of homozygosity in successive generations with various systems of inbreeding. (Sinnott, E. W., L. C. Dunn, and T. Dobzhansky, 1958. *Principles of Genetics*, 5th ed. McGraw-Hill Book Co., New York, p. 258, Fig. 19.3.)

possible, "pure lines" can be developed quite rapidly. The greater the number of heterozygous loci in the initial population, the greater the possible number of homozygous lines. For instance, for one heterozygous locus, *Aa*, two pure lines are possible: *AA* and *aa*. For two loci, *AaBb*, four lines (*AABB*, *AAbb*, *aaBB*, and *aabb*) are possible. For *n* heterozygous loci, 2^n pure lines are possible. In species that are not self-fertilizing, brother × sister matings are commonly used to develop inbred lines. The approach to homozygosis is slower, as shown above, but the theoretical number of possible inbred lines is the same as with selfing. If 100 loci are heterozygous in the starting population, 2^{100} pure lines are possible, an exceedingly large number. Therefore, it is not surprising that inbred lines developed from the same heterozygous parent population differ from one another although the individuals within a line are quite uniform. Because deleterious recessive genes are brought to expression, many of the lines will lack vigor or may even die out. Therefore, inbreeding is customarily combined with selection, the less vigorous and fertile lines being culled. In this way fully vigorous and fertile lines have been developed that have a high degree of homozygosity and are very uniform in their characteristics. The numerous inbred lines of mice, widely used in research, are perhaps the best example of the diversity of pure lines that can emerge following inbreeding.

A word of caution is desirable about the level of homozygosity reached following inbreeding as shown in the theoretical curves in Fig. 20.2. The actual percentage of homozygosity in the line may be somewhat less than that theoretically expected. If selection of the more vigorous, fertile lines is practiced, it may lead to the selection of the more heterozygous lines due to the phenomenon of hybrid vigor, to be discussed shortly. The extreme case of this sort is that of a balanced lethal system of closely linked lethals, in which only the heterozygotes survive. In such a case, a uniform, true-breeding population will result that is 100% heterozygous rather than homozygous. In less extreme cases, the approach to homozygosity will be delayed. Another factor that reduces the rate of approach to homozygosity is linkage, for homozygosity may depend on the occurrence of a rare crossover. Finally, spontaneous mutations are always a factor to be considered for their effects on inbred lines. Since chromosome mutation rates of the order of 1% for lethal and semilethal genes have been recorded, it is perhaps not surprising that sublines of the same inbred strain, separated for a number of generations, have been found to have diverged from one another. Thus, once the inbred line is established, it is unwise to regard it as fixed in the homozygous condition. Instead, the inbreeding should be continued, and the line checked regularly for evidence of mutations.

Earlier, it was pointed out that all of us are inbred to some extent.

However, rather than consider everyone inbred, we are confronted with the question of when the mating of related individuals should be regarded as inbreeding. The customary answer is when the individuals are more closely related than the members of the population from which they are derived. In each case, therefore, a baseline population is used as a reference point from which to measure the amount of inbreeding. The amount of inbreeding is measured by the coefficient of inbreeding, F. It should be kept in mind that the value of F depends on the system of mating and also on the reference population. Therefore, if the reference populations are quite different, two similar values of F might have rather different implications in terms of the actual amount of homozygosity achieved in the two inbred populations.

We have already stated that the effect of inbreeding is to increase homozygosity. This increased homozygosity results from the likelihood that an individual's parents carry an allele identical by descent from a common ancestor. Two types of identity of alleles at a locus may be recognized. If the alleles are functionally identical, indistinguishable by any functional test, they may be termed alike in state. If the alleles are identical due to the replication and transmission of one ancestral gene to subsequent generations, they are identical by descent. Thus two kinds of homozygotes can be recognized, those carrying two alleles alike in state, and those carrying two alleles identical by descent. The coefficient of inbreeding, F, may be defined as the probability that two alleles at any locus in an individual are identical by descent. Therefore, F refers to an individual and is a means of expressing the relationship between his parents.

The inbreeding coefficient was originally defined by Sewall Wright as the correlation coefficient between uniting gametes. This definition, which seems somewhat different from the one given above, is nevertheless equivalent to it. The difference between the definitions stems from the different approaches used to study the effects of inbreeding. Wright's approach was to think in terms of the effects on gene and genotype frequencies of matings between related individuals. This approach requires knowledge about the pedigrees of the individuals in the population. The other approach, stemming from Malécot, treats the problem as a sampling problem related to population size, and does not require such knowledge of the individual pedigrees. Each approach has certain advantages, depending on the circumstances. Our primary concern is to show the consequences of the random sampling of gametes in small populations or in matings between related individuals. Under these conditions random changes in gene frequency will occur, leading to a dispersive process. As a result, subpopulations will become differentiated from one another, but individuals within a given population will tend to become more alike genetically. Furthermore, the dispersive process will tend to increase the frequency of homozygotes above what it would be if all of the subpopulations were to form one large panmictic or random mating population.

The following derivation of F, following Malécot and Falconer, is given to afford some insight into the meaning of F. Let us assume a diploid population of hermaphrodites capable of both self- and cross-fertilization in which there are N individuals. In the initial baseline population none of the alleles at a given locus is regarded as identical by descent, so there are $2N$ different kinds of gametes possible. In the initial population, therefore, the inbreeding coefficient, is zero, or

$$F_0 = 0$$

In the next generation the probability that two random gametes giving rise to an individual carry genes identical by descent is $1/2N$. In other words, this is the probability that two uniting gametes carry alleles derived by replication of a single parental gene. Therefore

$$F_1 = \frac{1}{2N}$$

In the next generation the probability again is $1/2N$ that uniting gametes will carry newly replicated alleles identical by descent. In addition, the rest of the zygotes $(1 - 1/2N)$ will carry alleles that are independent in origin in generation 1, but may be identical in origin in generation 0. This probability is given by the inbreeding coefficient of generation 1. Therefore the inbreeding coefficient of the second generation consists of two components, one due to the new replication and the other to earlier replications:

$$F_2 = \frac{1}{2N} + \left(1 - \frac{1}{2N}\right)F_1$$

This relation holds for subsequent generations and can be generalized as

$$F_t = \frac{1}{2N} + \left(1 - \frac{1}{2N}\right)F_{t-1} \tag{1}$$

Thus, F_t consists of two terms, one measuring the "new" inbreeding, the other the inbreeding in previous generations. It is worth noting that if, for some reason, there is no new inbreeding and that term $(1/2N)$ equals zero, the effects of the earlier inbreeding will remain intact, as expressed by the term $(1 - 1/2N)F_{t-1}$.

If the new inbreeding is symbolized as ΔF, then

$$\Delta F = \frac{1}{2N}$$

and equation (1) can be written as

$$F_t = \Delta F + (1 - \Delta F)F_{t-1}$$

which in turn gives rise to

$$\Delta F = \frac{F_t - F_{t-1}}{1 - F_{t-1}} \qquad (2)$$

Hence, ΔF is seen to be a proportion, with the increase of the inbreeding coefficient in one generation in the numerator and the remaining amount of inbreeding needed to reach complete inbreeding in the denominator. Therefore ΔF is a measure of the rate of inbreeding.

To extend the inbreeding coefficient over more than one generation, the panmictic index (P) is useful. $P = 1 - F$, and thus measures the amount of random mating still prevailing in the population. Obviously, if $F = 0$, $P = 1$ and complete panmixia exists. In a base population, for example, by definition $F = 0$ and $P = 1$. Substitution of P in equation (2) gives

$$\Delta F = \frac{P_{t-1} - P_t}{P_{t-1}}$$

which in turn simplifies to

$$\frac{P_t}{P_{t-1}} = 1 - \Delta F$$

For the next generation

$$\frac{P_t}{P_{t-2}} = (1 - \Delta F)^2$$

and extension back to P_0, the original base population, gives

$$P_t = (1 - \Delta F)^t P_0$$

When $P_0 = 1$, conversion back to the inbreeding coefficient gives

$$1 - F_t = (1 - \Delta F)^t$$

or

$$F_t = 1 - (1 - \Delta F)^t$$

The relationship, $\Delta F = 1/2N$, holds strictly only in the situation where random mating includes the possibility of self-fertilization. If self-fertilization is not possible, the gene replications responsible for the increment of new inbreeding must have occurred in the grandparents rather than the parents. In this case,

$$\Delta F = \frac{1}{2N + 1}$$

but if the population size is reasonably large,

$$\Delta F = \frac{1}{2N+1} \cong \frac{1}{2N}$$

We have already stated that the effect of inbreeding is to increase homozygosity. If we think in terms of two alleles at a single locus in a base population from which a number of inbred lines or subpopulations are derived without selection, the overall gene frequency for all lines should average out to equal that of the base population. The variance about this mean depends on the gene frequency and on the population size of the subpopulations. This variance (see Appendix) among lines at generation t is:

$$\sigma_p^2 = \sigma_q^2 = p_0 q_0 \left[1 - (1 - \frac{1}{2N})^t \right]$$

The genotype frequencies for the entire population are obtained from their mean value for all the subpopulations. This mean value for homozygous recessives, for example, can be written as $\overline{q^2}$, and can be obtained from the variance in gene frequencies among the subpopulations. This variance equals the mean of the squared values of q minus the square of the mean, or

$$\sigma_q^2 = \overline{q^2} - \overline{q}^2$$

Hence $\overline{q^2} = \overline{q}^2 + \sigma_q^2$

Since there is no change in gene frequency without selection, q equals q_0. Therefore, in the entire population the increase in the frequency of homozygotes is as follows:

Genotype	Frequency in total population
AA	$p_0^2 + \sigma_q^2$
Aa	$2pq - 2\sigma_q^2$
aa	$q^2 + \sigma_q^2$

Now, as shown above,

$$\sigma_q^2 = p_0 q_0 \left[1 - (1 - \frac{1}{2N})^t \right]$$

$$= p_0 q_0 [1 - (1 - \Delta F)^t]$$

$$= p_0 q_0 F$$

Hence, the genotype frequencies can also be given in terms of the inbreeding coefficient.

	Frequency			
Genotype	C-H-W		Inbreeding effect	
AA	p_0^2	+	$p_0 q_0 F$	
Aa	$2p_0 q_0$	−	$2p_0 q_0 F$	$= 2p_0 q_0 (1 - F)$
aa	q_0^2	+	$p_0 q_0 F$	

If F equals zero, the usual Castle-Hardy-Weinberg frequencies exist. Otherwise the heterozygotes decrease by an amount equal to $2p_0 q_0 F$, which is distributed equally to the two homozygous classes, and the second term in each frequency denotes the change due to inbreeding.

Earlier we pointed out that two classes of homozygotes could exist, those alike in state and those identical by descent, and that the inbreeding coefficient is defined as the probability that the two genes at a given locus in an individual are identical by descent. The frequency of these two classes of homozygotes can be derived algebraically from the expression above to produce the following relation:

	Origin			
Genotype	Independent		Identical by descent	
AA	$p_0^2 (1 - F)$	+	$p_0 F$	
Aa	$2p_0 q_0 (1 - F)$			
aa	$q_0^2 (1 - F)$	+	$q_0 F$	

Thus it can be seen that the frequency of both types of homozygotes identical by descent is dependent on the initial gene frequencies and that together they equal the inbreeding coefficient, because $p_0 F + q_0 F = F(p_0 + q_0) = F$. The rest of the genotypes, with genes of independent origin, are present in the Castle-Hardy-Weinberg proportions expected with random mating.

Another way to look at the above relations is in terms of the frequency of heterozygotes, which are reduced in frequency from $2p_0 q_0$ to $2p_0 q_0 (1 - F)$. $(1 - F)$ is the panmictic index, P, which therefore can be thought of as a measure of the amount of heterozygosity in the inbred population relative to that in a random mating or panmictic population. In other words, if $H_0 = 2p_0 q_0$ and $H_t = 2p_0 q_0 (1 - F)$ at generation t, then

$$H_t = H_0 (1 - F)$$

and

$$P = (1 - F) = \frac{H_t}{H_0}$$

In addition to its effect in increasing the proportion of homozygotes in a population in relation to expectations based on the C-H-W equilibrium, inbreeding may also lead to fixation of a gene in a population (and hence to the loss of its alleles). The rate of fixation—the proportion of unfixed or heterozygous loci that become fixed or homozygous in each generation— equals ΔF. Hence, when the dispersal of gene frequencies from the initial frequency reaches a steady state, fixation then occurs at the rate of $1/2N$ per generation. In the steady state the frequency distribution of gene frequencies is flat, ranging between zero and one. The process of fixation has been likened to an open-ended trough filled with sand. The grains in the trough represent the various gene frequencies and will become uniformly dispersed if the trough is shaken. The grains falling off at one end or the other (0 or 1) represent the loss or fixation of an allele. The analogy also makes clear the irreversible nature of fixation.

The inbreeding coefficient also makes it possible to quantify the effect of inbreeding in bringing deleterious recessive genes to expression in the homozygous condition. The frequency of recessives with inbreeding was given above as $q^2 + pqF$. If $q = 0.01$ and $F = 0.5$, the frequency of recessive homozygotes will be about 0.0050, or some 50 times more frequent than the number observed with random mating (0.0001). Even though the individual deleterious recessives may have low frequencies, such genes may occur at many different loci; thus it should hardly be surprising that, in normally outbred populations, striking reductions in viability and fitness result from inbreeding. In normally crossbred populations, then, the effects of inbreeding are ordinarily an increased uniformity of genotype and phenotype within the inbred line and a decrease in viability, both effects being due to the increase in homozygosity. Once brought to expression by inbreeding, the deleterious recessives may be eliminated by either natural or artificial selection. When inbred lines are derived from crossbreeding parental populations, the combination of inbreeding plus selection may in time result in inbred lines with reasonably good viability and fertility.

20.2. Assortative Mating

Inbreeding has been defined as the mating of related individuals and is thus determined by the degree of genetic relationship between the genotypes of individuals. If not accompanied by selection, it does not lead to changes in gene frequency.

Assortative mating is similar to inbreeding in that no gene frequency change results from assortative mating. Assortative mating, however, is ordinarily used to refer to matings based on phenotypic similarities or differences. Positive assortative mating is the tendency for individuals similar in some trait to mate. Positive assortative mating will ordinarily lead to an increase in the proportion of homozygotes in the population, while negative assortative mating will tend to produce a greater proportion of heterozygotes as compared to a population mating at random. In neither case will gene frequencies change, unless some form of selection also occurs. Therefore, inbreeding and assortative mating are similar in that they do not lead to gene frequency change, but they are dissimilar in that inbreeding is genotypic and assortative mating is phenotypic. Inbreeding and assortative mating should be distinguished from selective mating, in which certain parental types tend to leave greater numbers of progeny than others.

20.3. Outbreeding

Earlier, we defined inbreeding as the mating of individuals more closely related than the average for the population from which they are derived. In a similar way, outbreeding may be defined as the mating of individuals less closely related than the average for the population. A crossbreeding population tends to retain its heterozygosity and hence its variability, the deleterious recessives are protected in the heterozygous condition from the full force of selection, and viability and fertility tend to remain high.

We noted that in some species self-fertilization is the customary method of reproduction. In other species, however, crossbreeding is mandatory; for example, in all species with separate sexes. Even in hermaphroditic species, various devices are known to reduce or prevent selfing. They range from systems of self-sterility alleles to differences in time of maturation of the male and female gametes or physical separation of the male and female reproductive tracts. Systems of balanced lethals and balanced chromosome polymorphisms for translocations or inversions also ensure more or less permanent heterozygosity, in many instances associated with hybrid vigor.

Crossbreeding is useful to the animal or plant breeder as a means of introducing new genetic material into the population. If the cross is too wide, for example, between two different species, the breeder may be plagued by sterility, low viability, or both. The mule is the classic example of the production of a superior economic animal by crossing the horse and the donkey, but the mule is sterile. On the other hand, if a sheep is crossed with a goat, the zygote starts to develop but soon dies. Thus, the limits to successful outcrossing are set by sterility and viability barriers. Within these limits, crossbreeding, followed by inbreeding and selection, has been used to develop new and more productive or otherwise useful kinds of animals or plants. Outcrossing is also widely used to achieve the hybrid vigor or heterosis that so often results, giving rise to maximum vigor and yield.

20.4. Heterosis

Heterosis may be defined as the increased vigor, size, yield, developmental rate, longevity, or disease resistance in crossbred individuals as compared to their more inbred parents. The word, coined by Shull and intended to carry no theoretical implications, is thus a term describing a phenomenon and is usually considered to be synonymous with hybrid vigor. Heterosis has become a part of modern life: It is now widely used in the commercial production of corn, pigs, chickens, and many other species of plants and animals. In the corn fields scattered across the United States it is common to see one or another variety of hybrid corn advertised, for virtually all of the field corn and much of the sweet corn raised today is hybrid corn.

The production of hybrid corn involves the development of inbred lines with typically reduced size and vigor, which are crossed to produce a heterotic F_1. The traits of interest are quantitative traits, and heterosis is usually defined as being present when the mean of the F_1 individuals exceeds the mean of the larger of the parental lines. It is worth noting that the presence or absence of heterosis thus may depend on the trait selected for study. Furthermore, heterosis may be present in one trait but absent in another in the same F_1 population, because the mean for one trait exceeds the means of the parents but the mean for another trait is intermediate to the parental means. Therefore, heterosis is not necessarily characteristic of the organism as a whole, but may be trait-specific.

Maximum heterosis will be observed in the F_1 generation, with the vigor of the F_2, F_3, and successive generations gradually diminishing when the members of these generations are inbred. By the F_8, the vigor observed will again be reduced to about that of the original inbred lines (Fig. 20.3 and Table 20.1). Therefore, maximum heterosis can ordinarily be achieved only by repeating the crosses between the inbred lines. However, heterosis will be maintained if the F_1 can be propagated asexually.

Table 20.1 The Production of Grain (in bushels per acre) of Two Inbred Strains of Maize and Their Hybrid, from the F_1 to the F_8 Generation, Successively Self-fertilized

Year grown	P_1'	P_1''	F_1	F_2	F_3	F_4	F_5	F_6	F_7	F_8
1917	5.5	21.5	64.5	56.0						
1918	23.5	26.9	120.6	128.4	15.1					
1920	27.8	16.2	127.6	47.6	34.8	29.0	9.9			
1921	13.1	19.6	72.8	54.5	49.2	32.7	15.3	23.1		
1922	26.1	20.0	160.4	83.2	73.7	67.5	48.8	35.7	22.7	
1923	21.0	13.2	61.0	45.0	40.5	47.0	15.8	23.0	26.2	27.2
Average	19.5	19.6	101.2	69.1	42.7	44.1	22.5	27.3	24.5	27.2

From Jones, D. F., 1924. "The Attainment of Homozygosity in Inbred Strains of Maize," *Genetics* 9:405–418, p. 415, Table 4.

Fig. 20.3. The effect of crossing two inbred lines of maize followed by self-fertilization through the F_8 generation. Representative plants from the two inbred lines at left, the F_1 hybrid next right, followed by successive self-fertilized generations from this hybrid through the F_8. All plants grown in adjoining rows in the same year. (Jones, D. F., 1924. "The Attainment of Homozygosity in Inbred Strains of Maize," *Genetics* 9:405–418, p. 414, Fig. 2.)

It should be noted that heterosis has been redefined by Dobzhansky in terms of fitness, heterosis being present when the fitness of the more heterozygous types exceeds that of their more inbred parents. Under such a definition, the mule cannot be regarded as an example of heterosis because of its sterility, but instead is said to manifest luxuriance. This definition leads to still another difficulty. Optimum fitness in birds, for example, may be associated, not with maximum clutch size, but with an intermediate clutch size, which permits the parents to rear more young than if they are overburdened with too many mouths to feed. Thus in this case also, the original definition of heterosis and this newer definition lead to diametrically opposite conclusions. A further problem with this definition is the difficulty of measuring fitness. What is usually measured is viability, fecundity, fertility, and so on, or some combination of these traits, and fitness is inferred from these measurements. Hence fitness as a trait is not easy to measure directly, and the definition of heterosis is made to depend on a rather poorly defined trait. Since the original definition of heterosis is well established and not subject to some of these ambiguities, it seems advisable to retain it, and to seek other terms and methods for dealing with the relationship between genotype and fitness.

20.5. Theories of the Mechanism of Heterosis

The first major theory to explain the origin of heterosis was Jones' 1917 theory of linked favorable dominant genes. Jones' theory was a further development of the multiple-factor hypothesis proposed earlier by Nilsson-Ehle and East. Jones assumed that there are numerous gene loci on all the chromosomes governing such quantitative traits as size, vigor, yield, and so on. The genes responsible for favorable or "normal" growth and development are generally dominant or at least partially dominant, while the more deleterious genes are recessive. The distribution of these dominants and recessives on the chromosomes is such that no one chromosome can be expected to carry favorable dominants exclusively. Therefore, when inbred lines are formed in normally outcrossing species, they will be homozygous for both dominant and recessive genes. The homozygous recessives will reduce the vigor of the inbred lines below that of the original parental population. However, each different inbred line will tend to be homozygous for a different group of recessives so that, when two inbred lines are crossed, the dominants in one strain will mask the deleterious recessives in the other, and vice versa. Therefore, there will be a favorable dominant gene at virtually all loci, and the F_1 hybrids will show maximum vigor. The decline in vigor in the F_2 and subsequent generations with inbreeding is due to segregation, which leads again to an increasing proportion of loci homozygous for deleterious recessives.

If favorable dominants are responsible for maximum vigor, then in theory it might seem possible to develop an inbred line homozygous for a favorable dominant at every locus. In this way a true-breeding line with maximum vigor could be obtained, and the need to repeat the crosses between inbred lines each generation would be ended. Unfortunately, no such inbred line has ever been developed. The reason seems to be that many gene loci are involved in heterosis, and hence each chromosome will carry a number of such loci, which fall in a single linkage group. Moreover, each chromosome will contain both favorable and unfavorable genes scattered along its length. Only if, by opportune crossovers at just the right places in the F_1 hybrids, it were possible to obtain gametes carrying favorable dominants exclusively, and these then combined to form a zygote, could a true-breeding line with maximum vigor be established. The probabilities of such an event are apparently so vanishingly small that it has never been observed. Moreover, Singleton has shown that even without linkage the chances of recovering such a line are extremely slight. For example, if only 30 effectively unlinked loci were responsible for heterosis in corn, 4^{30} plants would be required to give a 50:50 chance of recovering one that carried all 30 favorable dominants. It has been estimated that the experimental corn field necessary would be 2000 times as large as the total land area of the earth. In summary, the essence of Jones' theory is the covering up or masking of the effects of the deleterious recessives in the F_1 by their more favorable dominant or partially dominant alleles.

The second major genetic theory to explain heterosis was first presented in detail by East in 1936 and was based on a concept of allelic interaction. It had earlier been termed superdominance by Fisher and is now best known as the overdominance theory of heterosis, a term coined somewhat later by Hull. As presented by East, the theory postulated that a number of genes in different linkage groups were involved, as in Jones' theory. However, because he felt that there was no theoretical reason for genes for large size to be dominant, and also that heterosis was due to the action of "normal" genes rather than to the covering up of "defective" genes, he postulated the absence of dominance. Instead he suggested that hybrid vigor was due to heterozygosity itself, with the degree of stimulus proportional to the number of heterozygous factor pairs. Rather than the heterozygote equalling the dominant homozygote as in Jones' theory ($AA = Aa > aa$), the heterozygote was thought to be more vigorous than either homozygote ($a_1a_1 < a_1a_2 > a_2a_2$) due to an allelic interaction between a_1 and a_2.

The dominance and the overdominance theories of heterosis are not mutually exclusive. There is some experimental evidence in support of both theories. Although easy to visualize, the theories are not easy to subject to a critical test. The reason for this difficulty is the practical problem of holding the remainder of the genotype constant so that only the differences in vigor at a single locus are being compared. If a few closely linked genes or small chromosome segments are heterozygous, the heterozygote may be superior to both homozygotes, but the superiority may not be due to overdominance but rather to dominance at these loci. In other words, the test should involve

I
$$\begin{matrix} B & | & | & B \\ A & | & | & a \\ C & | & | & C \end{matrix} \quad \text{vs.} \quad \begin{matrix} B & | & | & B \\ A & | & | & A \\ C & | & | & C \end{matrix} \quad \text{vs.} \quad \begin{matrix} B & | & | & B \\ a & | & | & a \\ C & | & | & C \end{matrix}$$

or

II
$$\begin{matrix} B & | & | & b \\ A & | & | & a \\ C & | & | & c \end{matrix} \quad \text{vs.} \quad \begin{matrix} B & | & | & b \\ A & | & | & A \\ C & | & | & c \end{matrix} \quad \text{vs.} \quad \begin{matrix} B & | & | & b \\ a & | & | & a \\ C & | & | & c \end{matrix}$$

and not

III
$$\begin{matrix} B & | & | & b \\ A & | & | & a \\ c & | & | & C \end{matrix} \quad \text{vs.} \quad \begin{matrix} B & | & | & B \\ A & | & | & A \\ c & | & | & c \end{matrix} \quad \text{vs.} \quad \begin{matrix} b & | & | & b \\ a & | & | & a \\ C & | & | & C \end{matrix}$$

In the latter case, the dominance theory is a sufficient explanation for the superiority of the Aa heterozygote without resorting to overdominance. Technically, however, it is difficult to be certain that the test for an overdominant locus is indeed of type I or II rather than III. Unfortunately, it has become commonplace to describe almost any case of heterosis with the heterozygotes superior to both homozygotes as a case of overdominance. Such usage makes the word virtually synonymous with the phenomenon of heterosis and diminishes its usefulness in describing a theoretical mechanism for heterosis.

In view of the difficulties in distinguishing between the dominance and overdominance theories experimentally, the attempt to make the distinction may not seem worth the effort. However, the implications of the two theories for breeding practices are quite different. If the dominance theory is correct, efforts should be made to accumulate the maximum number of favorable genes for size, yield, growth rate, and so on in the inbreds. The ultimate hope would be to develop a pure line carrying only favorable genes. On the other hand, if the overdominance theory is correct, the

Table 20.2   Data from a Cross Between Two Lines of Tomatoes

Line	n (ripe fruit)	Wt./tomato (g)	yield (g)
4102	4.4 ± 0.69	138 ± 12.81	607 ± 86
4110	109.1 ± 11.34	17 ± 0.58	1868 ± 149
F_1 hybrid	44.5 ± 2.52	55 ± 2.93	2428 ± 150

From Powers, L., 1944. "An Expansion of Jones's Theory for the Explanation of Heterosis," *Amer. Nat.* 78:275–280, p. 277, Table 1.

primary concern in the development of inbred lines should be their combining ability with other inbred lines rather than the innate qualities of the inbred lines themselves. Breeding efforts would then be directed toward the improvement of the combining ability of the inbreds.

One final mechanism for heterosis will be described that shows how dependent the phenomenon is on the trait selected for study and how heterosis may be due to the complementary interaction of nonallelic genes at different loci. The data given in Table 20.2 are from Powers' (1944) work with tomatoes. He crossed one line (4102) that had a few large fruits with another (4110) that had numerous small tomatoes. The F_1 hybrid between the two lines was intermediate to the parental lines both in the average size and number of the fruits it bore. In fact, the F_1 means for these two traits were both somewhat lower than the arithmetic mean of the parental values, approximating the geometric means instead. One possible interpretation of this result is that the genes for fewer fruit and for smaller fruit were partially dominant. Nevertheless, despite the lack of heterosis for either fruit size or number, when total yield was the trait studied, a marked heterotic effect on yield was observed in this cross. The heterosis in this case seems most readily attributable to complementary interactions between nonalleles, between the genes affecting size on the one hand and number on the other. Thus, the heterosis may be said to be due to epistasis in the broad sense of the word.

Therefore, it seems reasonable to say that no one theory of heterosis is likely to account adequately for all cases of heterosis. The most probable explanations include (1) dominance, the masking of deleterious recessives (or conversely, the combined action of favorable dominants), (2) overdominance or allelic interaction, and (3) nonallelic or epistatic interactions, complementary gene action in the usual sense.

Although heterosis has been exploited in pig breeding and in hybrid chickens and in other species, the most successful commercial application of heterosis has been in the production of hybrid corn. The methods used

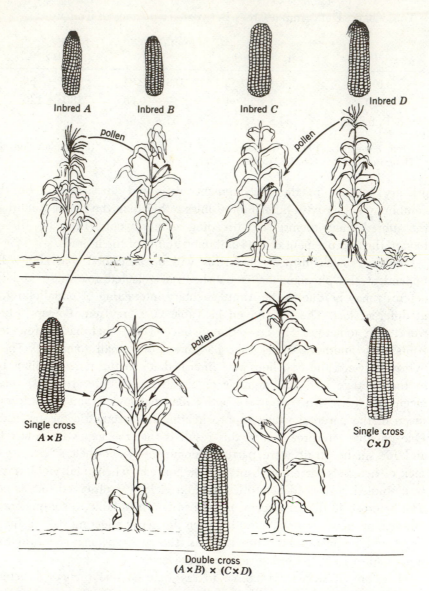

Fig. 20.4. The double-cross method for the commercial production of hybrid corn. See text. (Dobzhansky, T., 1955. *Evolution, Genetics, and Man.* John Wiley & Sons, New York, p. 216, Fig. 9.7.)

are outlined in Fig. 20.4. Rather than a single cross between two inbred lines, a double cross is made in order to obtain adequate seed production. Four different inbred lines (*A, B, C,* and *D*) are initially crossed (*A × B* and *C × D*). The vigorous single-cross hybrids, *AB* and *CD*, are then crossed with one another to give the vigorous double-cross hybrid, *ABCD,* which contains genes from all four inbred lines. In this way the seed used for planting is produced on a vigorous hybrid F_1 plant rather than on one of the inbred parents. Originally, in order to ensure cross-pollination, extensive detasseling operations were necessary to produce hybrid corn. More recently, however, this costly and time-consuming step has been eliminated in some cases by the use of cytoplasmic male sterility factors together with genes that restore fertility.

20.6. Selection

A key tool in animal and plant breeding is selection. By choosing some individuals to perpetuate the population and excluding others, the breeder hopes to shift the characteristics of the population in a desired direction. Because he imposes his will on the mating system, this selection is usually referred to as artificial selection. Artificial selection should be thought of as being superimposed on natural selection rather than as supplanting it completely because individuals homozygous for a recessive lethal, for example, will continue to die in a population under artificial selection. Thus, both natural and artificial selection may be at work in a population. Ideally, the breeder wishes that natural selection would reinforce the effects of his artificial selection. In practice, however, the goals sought by artificial selection may be counter to the direction of natural selection, and a balance between the two may lead to a cessation of progress. For example, if increased size is the goal of artificial selection, and sterility increases as size increases, then at some point progress will cease, or else the line will die out. A further consideration for the breeder using almost any form of artificial selection is to avoid, so far as possible, the deleterious effects of inbreeding by keeping the rate of inbreeding at a low level.

When a breeder assumes control of a population in order to practice artificial selection, he has to make a series of judgments as to which individuals should be retained for breeding purposes and which should be culled. The available relevant information may be of three different types. The first is the individual's phenotype, which includes not just external appearance but such traits as rate of gain, milk production, disease resistance, and so on. The second type is information from the pedigree of the individual in question. The closer the relationship, the more useful the information to be obtained from the relatives. For example, information about the parents is of greater significance than information about the great-grandparents. A Kentucky Derby winner in the remote blood lines of a horse will have little bearing on how fast he can run. In some cases such data are indispensable. The potential of a Holstein bull, for instance, for increasing the milk production of his daughters cannot be determined from his own phenotype. Instead, his worth can be estimated from the production records of his dam or his sisters or half-sisters. This type of information is obtained from the ancestors or collateral relatives of the individual. The third type of information is obtained from the descendants of the individual. The data from actual progeny tests will furnish the most reliable measure of an individual's breeding value. However, the advantages of progeny testing may be more than outweighed by the increased length of generation, so progress may not be as rapid as with other methods of selection.

Phenotypic Selection

Within this framework, a number of breeding practices can be identified. Undoubtedly, the selection of individuals by their phenotypes has been widely practiced and has often led to rapid progress. Although individual selection is the easiest to use, on occasion it may not be reliable. More than one prize-winning bull has consistently lowered the milk production of his daughters below that of their mothers. *Individual selection* is the expression used for phenotypic selection where records on individual matings are kept. When a number of individuals are selected and allowed to mate at random, as for instance when certain plants are saved for seed, the phenotypic selection is called *mass selection*.

Pedigree Selection

Selection based on the characteristics of the relatives is often called pedigree selection, of which there are several types. When the trait under selection has a low heritability, individual phenotypic selection tends to be relatively unreliable because environmental effects obscure the genotypic value of the individual. Under these circumstances family selection is used, in which whole families are accepted or rejected on the basis of their mean phenotypic value. The families ordinarily will consist of either full sibs or half-sibs. For family selection to be effective, family size should be large enough that individual environmental variations tend to cancel out, and the average phenotypic value for the family gives a good indication of the mean genotypic value. At times, all members of a family may resemble one another, not because of the similarity in their genotypes but because they have shared a common environment different from that of other families. If this is the case, selection is operating on environmental variation rather than genetic differences between families, and progress will be far less than expected. When, as noted above, selection can be based only on relatives because the trait cannot be measured in the individual (e.g., milk production in bulls), it is referred to, somewhat ambiguously, as sib selection. Hence, sib selection is a form of family selection in which measurements on the selected individuals do not enter the family mean. It should be understood that if heritability of the trait is reasonably high, individual phenotypic selection is preferable to family selection. Family selection is useful for traits of low heritability in large families that show relatively little effect of a common environment during development.

Family selection in animals is similar to line selection in plants. In self-pollinated plants a single plant and its progeny can be used as the starting point for development of a pure line, and testing and selection can then be carried on among the various pure lines. In cross-pollinating species, artificial self-pollination may be used to develop pure lines. In some cases uncontrolled cross-pollination is permitted, and selection is practiced only on the maternal line without regard to the male parent, a method of selection that is less efficient but is also less laborious than hand-pollination.

Pedigree selection based on a detailed knowledge of the characteristics of the individuals in the pedigrees is, of course, common in animal breeding, especially for large animals such as race horses. Its success can be judged from the fact that the Kentucky Derby has been won by two trios of grandfather, father, and son: Reigh Count in 1928, his son, Count Fleet, in 1943, and his grandson, Count Turf, in 1951; and by Pensive in 1944, his son, Ponder, in 1949, and his grandson, Needles, in 1956.

Progeny Testing

Despite the appeal of being able to base decisions on which individuals produce the best offspring rather than on which ones have the best phenotype or the most distinguished ancestors or relatives, progeny testing is, in general, so cumbersome that its inherent advantages are not used as often as might be expected. The long generation length and the need to maintain large numbers of individuals while their breeding value is being determined from their progeny make progeny testing difficult. It may, however, be useful when other methods of selection have failed to produce progress.

20.7. The Response to Selection

The response to selection will depend on the nature of the trait under selection and on its mode of inheritance. Selection for a recessive trait requires simply the identification and mating together of a homozygous recessive male and female in order to establish a pure line for the trait. Selection for a dominant trait is a bit more involved, but still relatively simple. Probably the easiest method under most circumstances is to test-cross individuals with the desired dominant trait to homozygous recessives in order to distinguish between homozygous and heterozygous dominants. Once homozygous dominant males and females have been identified, a pure line for the dominant trait can be established.

Unfortunately, most of the traits of interest to breeders are not this simply inherited, but are instead quantitative traits controlled by a number of genes at different loci. Even so, if the gene effects were additive, and the heritability equalled 1.0, the breeders' problems would not be difficult. However, the effects of the genes are not ordinarily simply additive, but may be complicated by dominance, epistatic interactions, linkage, and environmental influences. As a consequence, the breeders' attempts to bring about gene frequency changes to shift the average characteristics of a population in a desired direction often result in rather slow progress instead of rapid gains toward the goal.

The response to selection for a quantitative trait is dependent on the heritability of the trait under selection and on the intensity of selection. The intensity of selection is measured by the selection differential, which is the difference between the mean of the parents selected to propagate the next generation and the mean for the entire population from which these parents were drawn. The intensity of selection must always be balanced against the need to maintain an adequate population size and to minimize the deleterious effects of inbreeding.

The gain or response to selection for a given selection differential will then be dependent on the heritability. If the heritability is 1.0, the observed variation among phenotypes being due solely to additive genetic differences among genotypes, the mean of the progeny will equal the mean of the selected parents. If $H = 0$, as in a homozygous population in which the observed phenotypic variation is environmental, no progress will be made. For intermediate values of H, intermediate gains will be made. In fact, the ratio of the actual response to selection to the selection differential may be used to estimate the heritability, that is, $H = R/S$, where R is the response to selection and S is the selection differential. Conversely, if a heritability estimate is available and the selection differential is known, it is possible to predict the expected gain or response to selection, that is, $R = HS$.

Since heritabilities seldom equal one, the typical response to artificial selection is a group of progeny whose mean falls somewhat below that of their parents but above that of the entire population in the previous generation. Galton called this type of response, regression toward the mean. The regression coefficient, so widely used today, was developed as the result of Galton's search for a means to describe the regression of offspring on parents in his studies of quantitative inheritance.

Thus far, we have dealt with selection as though it were carried on only for one trait at a time. Often, however, the breeder must be concerned with several traits at once, which greatly complicates his problem. Nevertheless, as a minimum, most animal breeders are to some extent concerned with their animals' outward appearance and their fertility as well as their productivity in economic terms although of course these traits may be interrelated. Under these circumstances, a selection index is frequently used, because it has been found that simultaneous selection is the most efficient way to bring about improvement in the desired directions. The index is constructed by weighting each trait according to its relative importance and its heritability. Correlations between the traits introduce further complexity. However, the outcome is a single value, the selection index, on which the selection is based. Simultaneous selection for a number of different traits undoubtedly slows the rate of progress toward any single goal, but it is seldom possible for a breeder to concentrate his efforts on a single trait.

Additional Reading

Allard, R. W., 1960. *Principles of Plant Breeding.* Wiley, New York.

Allard, R. W., S. K. Jain, and P. L. Workman, 1968. "The Genetics of Inbreeding Populations." *Adv. Genet.* 14:55–131.

Carson, H. L., 1967. "Inbreeding and Gene Fixation in Natural Populations." In *Heritage from Mendel,* R. A. Brink, ed., University of Wisconsin Press, Madison.

Dobzhansky, T., 1952. "Nature and Origin of Heterosis." In *Heterosis,* J. W. Gowen, ed. Iowa State University Press, Ames.

East, E. M., 1910. "A Mendelian Interpretation of Variation That Is Apparently Continuous," *Amer. Nat.* 44:65–82.

East, E. M., 1936. "Heterosis," *Genetics* 21:375–397.

East, E. M., and D. F. Jones, 1919. *Inbreeding and Outbreeding*. Lippincott, Philadelphia.

Falconer, D. S., 1960. *Introduction to Quantitative Genetics*. Ronald, New York.

Fisher, R. A., 1922. "On the Dominance Ratio," *Proc. Roy. Soc. Edinburgh* **42**:321–341.

Fisher, R. A., 1965. *The Theory of Inbreeding*, 2nd ed. Oliver and Boyd, Edinburgh.

Gowen, J. W., ed., 1952. *Heterosis*. Iown State University Press, Ames.

Harpstead, D. D., 1971. "High-Lysine corn," *Sci. Amer.* **225** (Aug.):34–42.

Hayes, H. K., F. R. Immer, and D. C. Smith, 1955. *Methods of Plant Breeding*, 2nd ed. McGraw-Hill, New York.

Hull, F. H., 1945. "Recurrent Selection for Specific Combining Ability in Corn," *J. Amer. Soc. Agron.* **37**:134–145.

Jones, D. F., 1917. "Dominance of Linked Factors as a Means of Accounting for Heterosis," *Genetics* **2**:466–479.

Lerner, I. M., 1954. *Genetic Homeostasis*. Oliver and Boyd, Edinburgh.

Lerner, I. M., 1958. *The Genetic Basis of Selection*. Wiley, New York.

Lerner, I. M., and H. P. Donald, 1966. *Modern Developments in Animal Breeding*. Academic, New York.

Libby, W. J., R. F. Stettler, and F. W. Leitz, 1969. "Forest Genetics and Forest Tree Breeding," *Ann. Rev. Genet.* **3**:469–494.

Lush, J. L., 1945. *Animal Breeding Plans*, 3rd ed. Iowa State University Press, Ames.

Malécot, G., 1969. *The Mathematics of Heredity*, D. M. Yermanos, tr. Freeman, San Francisco.

Mangelsdorf, P. C., 1951. "Hybrid Corn: Its Genetic Basis and Its Significance in Human Affairs." In *Genetics in the 20th Century*, L. C. Dunn, ed. Macmillan, New York.

Mangelsdorf, P. C., 1951. "Hybrid Corn," *Sci. Amer.* **185** (Feb.):39–47.

Nilsson-Ehle, H., 1909. "Kreuzungsuntersuchungen an Hafer und Weizen." *Lunds. Univ. Aarskr. N.F.* **5** (2):1–122.

Powers, L., 1944. "An Expansion of Jones's Theory for the Explanation of Heterosis," *Amer. Nat.* **78**:275–280.

Richey, F. D., 1946. "Hybrid Vigor and Corn Breeding," *J. Amer. Soc. Agron.* **38**:833–841.

Richey, F. D., 1950. "Corn Breeding," *Adv. Genet.* **3**:159–192.

Robertson, A., 1967. "Animal Breeding," *Ann. Rev. Genet.* **1**:295–312.

Sheppard, P. M., 1958. *Natural Selection and Heredity*. Hutchinson, London.

Shull, G. H., 1908. "The Composition of a Field of Maize," *Rept. Amer. Breeders' Assoc.* **4**:296–301.

Shull, G. H., 1948. "What is 'heterosis'?" *Genetics* **33**:439–446.

Singleton, W. R., 1941. "Hybrid Vigor and Its Utilization in Sweet Corn Breeding," *Amer. Nat.* **75**:48–60.

Sprague, G. F., 1967. "Plant Breeding," *Ann. Rev. Genet.* **1**:269–294.

Winters, L. M., 1948. *Animal Breeding*, 4th ed. Wiley, New York.

Wright, S., 1921. "Systems of Mating," *Genetics* **6**:111–178.

Wright, S., 1922. "The Effects of Inbreeding and Crossbreeding on Guinea Pigs," U.S. Dept. Agr. Bull. 1090.

Wright, S., 1931. "Evolution in Mendelian Populations," *Genetics* **16**:97–159.

Wright, S., 1969. *Evolution and the Genetics of Populations*. Vol. 2. *The Theory of Gene Frequencies*. University of Chicago Press, Chicago.

Questions

20.1. Under what circumstances can inbreeding occur without having detrimental effects?

20.2. Under what system of inbreeding is homozygosity approached most rapidly?

20.3. In a population heterozygous for two alleles at each of five loci, how many different pure lines are possible? At 10 loci? At 15 loci? At 20 loci?

20.4. Define inbreeding. What implications does this definition have if you attempt to compare inbreeding coefficients for different populations?

20.5. What is the effect of inbreeding on the proportion of heterozygotes in a population? How can this effect be expressed quantitatively?

20.6. Discuss the similarities and differences between inbreeding and positive assortative mating.

20.7. Discuss the usefulness of inbreeding and outbreeding as tools for the animal and plant breeder.

20.8. Heterosis has sometimes been defined as being present when the F_1 mean exceeds the average for both parental lines rather than the average of the larger parental line. What is the effect of using this definition?

20.9. What is the relationship between heterosis and fitness?

20.10. Why is it so difficult to test the various theories of the mechanism of heterosis experimentally?

20.11. Would it make any difference from a practical standpoint if over-dominance rather than the dominance theory of heterosis were shown to be primarily responsible for heterosis in a cultivated plant species?

20.12. Is hybrid vigor or heterosis a characteristic of organisms or of particular traits of organisms? Explain.

20.13. In the commercial production of hybrid corn, what is the advantage of the double-cross method involving four inbred lines and two generations as compared to a single cross involving just two inbred lines and one generation?

20.14. If progeny testing provides the most reliable information on which to base a selection program for the improvement of domesticated species, why is it not more widely used?

20.15. How would you attempt to study the biochemical basis of heterosis?

20.16. If a parental population averages 84 cm in height but only the largest 10% of the plants are used for breeding and these average 96 cm, what will be the expected mean height of the next generation if the heritability estimate for height in this case is 0.25?

20.17. Suppose that a population of mice averages 30 g in weight. The largest 20% of the mice are used as parents in an attempt to select for increased size; these mice average 35 g. What is the selection differential? The progeny of these larger mice also averaged 30 g. What is the heritability of weight in this population? How do you account for these results?

20.18. In another mouse population, the average weight was 29 g and the largest 20% of the mice averaged 33 g. The progeny of these larger mice averaged 31 g. What is the heritability estimate for weight in this population?

20.19. A mouse breeder selected for handedness in an inbred line of mice and failed to make any progress toward developing lines with a preferred paw for bar-pressing. He thereupon concluded that this behavioral trait was not under genetic control. Comment.

20.20. Many effective mutagenic agents are available to plant and animal breeders, but induced mutations have not played a very significant role in plant or animal breeding programs. Why has this been the case? What circumstances would be most favorable for the use of induced mutations in breeding programs?

20.21. Disease resistance in plants and animals is frequently due to genes at one or a few loci, and these genes are often dominant. What effect do these facts have on the ability of breeders to develop resistant strains of plants and animals?

CHAPTER TWENTY-ONE

Domesticated Animals and Cultivated Plants

The most appropriate opening statement about the origin of agriculture is that surprisingly little is known about it. Man is now dependent on domesticated animals and plants for his existence, yet in most cases their time and place of origin are obscure. So uncertain is the information that controversies have repeatedly arisen among the experts over some of the more fundamental aspects of domestication. For example, some authorities have believed that agriculture originated in fertile river valleys, but differ as to whether it was the Nile, the Tigris and Euphrates, or the valley of the Indus. Others have argued for an origin in dry upland areas, and still others for a forest or jungle origin, or for mountain valleys. Some believe that agriculture was invented only once and then diffused throughout the world. The contrasting opinion is that there were several independent centers of origin. Some feel that the sequence was from hunter to

herdsman to farmer; others feel that this is improbable because the herdsman is constantly on the move, and hence could not guard his crops. An
alternative is that a sedentary fishing population was better able to make
the transition from seed-gathering to agriculture. And not everyone feels
that agriculture started with seed crops, but instead that tubers and
melons may have been the first cultivated crops. It seems generally agreed
that the dog was probably the first animal to be domesticated, but no
one seems very certain as to why. And so it goes.

The taxonomy of domesticated species is equally confused, with a variety
of scientific names often used by different authorities for the same domesticated species. The taxonomic muddle reflects the ignorance and confusion
about the origins of domesticated species. However, recent advances in
genetic and cytogenetic studies of domestic animals and plants have begun
to cast some light on the relationships of domesticated species to one
another and to their wild ancestors. In some cases, the wild ancestors are
unknown and probably extinct. The genetic information is probably the
most reliable type of information now available. Combined with archeo-
logical studies, it offers hope that further insight will be gained into the
origin of agriculture and of the domesticated species now so widely
disseminated to all parts of the world.

21.1. Domestication

Although there seems to be general agreement as to which species are
domesticated, it is rather difficult to define domestication itself by specifying those characteristics by which domesticated species can be distinguished from wild species. Several factors are involved in domestication,
none of which alone is sufficient to define it. Usually we tend to think of
a domesticated species as one whose reproduction is controlled by man.
Such control is essential if man is to carry on artificial selection. However,
the cat is certainly a domesticated species, but for the most part cats
continue to breed pretty much as they please. Domesticated species are
often thought of as living in close association with man, but the English
sparrow, the house mouse, and the Norway rat all live close to and seem
dependent on man yet are generally regarded as wild species. Sheep, on
the other hand, may live far from human habitation, but are domesticated.
Some species, such as maize, have become dependent on man for
existence and would die without his intervention. The ear of corn is no

longer adequate for seed dispersal, so unless corn is planted by man, its reproductive ability is very limited. Other domesticated species readily become feral, given the opportunity, and may even become weeds or pests. In animals, tameness often characterizes domesticated species. Yet stories abound of tame wolves and seals and bears—all wild species; on the other hand, a feral house cat is about as ferocious a small package as one would care to encounter. We tend to think of the domesticated species as those converted to man's use, but, of course, man also puts many wild species to use as well. In addition, many species have been domesticated for esthetic or even religious or ceremonial purposes. Many species and varieties of cultivated plants have been developed for the beauty of their flowers. Other plants have been cultivated as oddities or curios. The same reasons seem to apply to animals. A canary is valued for esthetic reasons; only its owner is apt to regard a Pekingese dog as anything but a freak.

Nevertheless, domestication generally does involve man's control of the reproduction of his crops or animals to some extent. Even if deliberate crosses or matings are not made, he may eliminate the types he considers less desirable and thus, through artificial selection, influence the gene pool of his domesticated populations. Selection for tameness in animals, for example, seems almost inevitable, for man is not apt to tolerate very long any individual that is likely to do him harm, and the wilder, less tractable animals are also more likely to escape.

The environment of domesticated species is man-made. Man tends to protect his animals from predators and parasites, and to weed and dust his crops. As a result, interspecific competition is minimized, and the force of natural selection is moderated. Furthermore, competition becomes primarily intraspecific, with man intensifying the intraspecific competition through artificial selection. In the sheltered environment provided by man, unusual variants may survive and be preserved that would otherwise be quickly eliminated in nature. Such variants are often saved as curiosities, but they may also have utility in the new environment. The ear in corn is a case in point. Although poorly adapted for seed dispersal, it is easy to pick and to store.

Another result of domestication is an increase in the amount of inbreeding. Migration and hence hybridization of domesticated plants and animals are reduced when man settles down to an agricultural existence. Artificial selection, by reducing the numbers of individuals used as parents, also enhances the amount of inbreeding. In domesticated animals often only a few males are retained as sires, which of course leads to an increase in the inbreeding coefficient.

Paradoxically, perhaps, domestication may also lead to outbreeding involving far wider crosses than would ever be observed in nature, for man through his travels and migrations may bring together and cross individuals that otherwise would never interbreed.

The most effective breeding system for rapid evolutionary change is one that combines moderately close inbreeding and fairly rigorous selection with occasional wide outcrosses. In this way many distinct but uniform varieties can be produced. Obviously, domestication has taken place under circumstances well suited to enable man to mold the domesticated species to his needs rather quickly.

21.2. The Origin of Agriculture

The earliest available evidence bearing on the origin of agriculture is less than 10,000 years old. The transition from a hunting and gathering mode of life to a planting and harvesting agricultural life was undoubtedly gradual rather than abrupt, and thus no exact date for the origin of agriculture seems possible. One widely discussed theory of the geographer Sauer postulates that during the retreat of the ice from the last glacial period, good fishing along the numerous lakes and streams made it possible for the formerly nomadic hunting tribes to settle in one favorable site for long periods. Agriculture requires such permanence, for a farmer must plant his crops, watch over them as they mature, and then harvest them. It was further suggested that the wastes accumulated from such settlements in dump heaps or piles of refuse formed a very fertile, open environment for plants to grow in. The fishermen observed the luxuriant plant growth in this rich soil and learned to plant the seeds of desirable species there to have a nearby supplement to their diet rather than having to go on foraging expeditions to gather the plants as food.

Another suggestion by Sauer is that agriculture based on roots and tubers may have originated even earlier than that based on the seed grains. The root or tuber can be dug up, part of it eaten, and the rest replanted to grow until the next visit to the site. Such primitive agriculture does not even require a permanent settlement. However, cereals seem to have been cultivated very early, and all of the important civilizations have been based on cereals rather than roots or tubers.

An interesting point is that if agriculture is, in fact, less than 10,000 years old, then, since most domesticated species have, at best, one generation per year, all the great diversity of varieties and breeds in domesticated plants and animals has evolved in less than 10,000 generations. Hence, evolutionary change in domesticated species has evidently been extremely rapid. An alternative possibility is that perhaps the history of domesticated species extends somewhat further back into the past than is presently thought to be the case. If so, then more time has been available for these evolutionary changes.

The first comprehensive studies of the origin of cultivated plants were made by de Candolle around 1880. While de Candolle tended to stress the similarities between domesticated species and their wild relatives and presumed ancestors, Darwin had been more interested in the differences that had arisen in domesticated species as further evidence for evolution. de Candolle made the rather straightforward assumption that cultivated species had originated in the area where their wild relatives are found. Darwin, on the other hand, was struck by the fact that no obvious wild prototypes could be found for many cultivated plant species. He pointed out that this fact could be most readily explained by the assumption that many cultivated species had undergone such profound changes during domestication that they no longer closely resembled their wild progenitors. In his book on domestication Darwin set forth his theory of pangenesis, which was his attempt to develop an adequate theory of heredity and variation. However, only with the rediscovery of Mendelism and the growth of genetics did it become possible to study the genetic basis for the variation in domesticated species and for their differences from their wild relatives. Mendel himself apparently realized that his work with the garden pea, a cultivated plant, had implications with respect to domestication, but he did not delve further in this direction.

The modern era in the study of the origin of cultivated plants stems from the work of the Russian geneticist Vavilov in his *Studies on the Origin of Cultivated Plants* (1926). He made collecting expeditions to many parts of the world, seeking out plants to be brought back for study and also for use in developing improved varieties of crops. In addition to the more traditional methods of study, he relied on genetic and cytogenetic analysis of his material. Vavilov found that centers of diversity could be identified for cultivated plant species where many different varieties of the same species existed. He argued that the center of diversity was also the center of origin for the cultivated species. Initially, Vavilov recognized five major centers of diversity (actually six), with the proviso that still others probably would be recognized. The centers, shown on the map in Fig. 21.1, were (1) southwestern Asia, (2) southeastern Asia, (3) the Mediter-

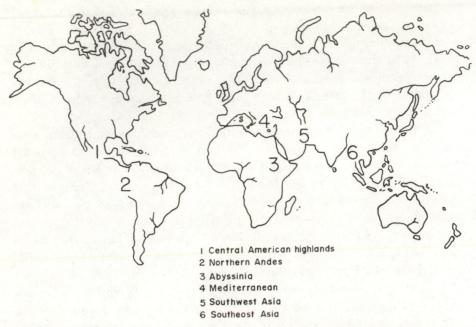

1 Central American highlands
2 Northern Andes
3 Abyssinia
4 Mediterranean
5 Southwest Asia
6 Southeast Asia

Fig. 21.1. Vavilov's centers of diversity for cultivated plants. (From *Plants and Civilization*, 2nd edition, by Herbert G. Baker. © 1965, 1970 by Wadsworth Publishing Company, Inc., Belmont, California 94002. Reprinted by permission of the publisher.)

ranean region, (4) Ethiopia, and (5) the New World, where he recognized Mexico and Peru as separate centers.

Although in at least one instance it has been suggested that agriculture was invented just once, probably in southeastern Asia, and has since spread throughout the world, the general tendency since Vavilov's time has been to increase the number of centers of origin rather than to decrease them. The origin of agriculture, of course, may be a separate question from the origin of domesticated species, for the idea of agriculture may have arisen once and then spread, with different species being domesticated in different regions as the idea itself spread. However, the present consensus seems to be that agriculture arose independently in several parts of the world, and that the centers of origin of domesticated species are in fact also centers of origin for agriculture. Following is a representative list of the areas thought to have served as centers of origin for cultivated plants (Fig. 21.2):

1. India-Burma
2. Southeastern Asia
3. Ethiopia

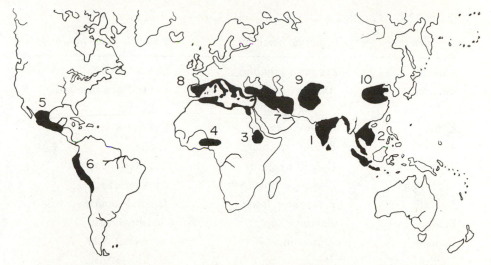

Fig. 21.2. Major centers of origin for cultivated plants: 1. Indian; 2. Indonesian; 3. Ethiopian; 4. Nigerian; 5. Mexican; 6. Peruvian; 7. Near Asiatic; 8. Mediterranean; 9. Central Asiatic; 10. Chinese. (Darlington, C. D., 1963. *Chromosome Botany and the Origins of Cultivated Plants*, 2nd ed. George Allen & Unwin Ltd.)

4. Central Africa
5. Central America
6. South America
7. Southwestern Asia
8. Mediterranean
9. Central Asia
10. China

A list of some of the more important and familiar economic plants is give in Table 21.1. The table contains the more commonly accepted taxonomic names and centers of origin for these crops. They are then regrouped in Table 21.2 by centers of origin. The present trend is to acknowledge our debt to the pioneering efforts of de Candolle and Vavilov, but to stress that, except for a few crops such as wheat, cotton, and corn, for which some progress is now being made, our knowledge of the origin of most cultivated plant species is superficial and sketchy at best. The situation in the better-studied cases has turned out to be far more complex than was originally suspected. In view of this dearth of knowledge, it must be stressed that the information in Tables 21.1 and 21.2 must be regarded as quite tentative. The usefulness of the tables lies more in the general impressions they convey than in any specific details.

Table 21.1   The More Common Useful Cultivated Plants

Common name	Scientific name	Probable place of origin
Almond	*Prunus amygdalus*	Southwest Asia
Amaranth	*Amaranthus* spp.	Central America, India-Burma
Apple	*Pyrus malus*	Central Asia
Apricot	*Prunus armeniaca*	China
Artichoke	*Cynara scolymus*	Mediterranean
Asparagus	*Asparagus officinalis*	Mediterranean
Avocado	*Persea americana*	Central America
Bamboo	*Bambusa* spp.	Southeast Asia
Banana	*Musa* spp.	Southeast Asia
Barley	*Hordeum* spp.	Southwest Asia
Bean		
Broad	*Vicia faba*	Mediterranean
Castor	*Ricinus communis*	Ethiopia
Jack (horse)	*Canavalia* spp.	South America
Kidney	*Phaseolus vulgaris*	Mexico
Lima	*Phaseolus lunatus*	Peru
Mung	*Phaseolus aureus*	India
Scarlet runner	*Phaseolus coccineus*	Central America
Soy	*Glycine max*	China
Beet	*Beta vulgaris*	Mediterranean
Blackberry	*Rubus* spp.	Europe
Blueberry	*Vaccinium* spp.	North America
Breadfruit	*Artocarpus communis*	Southeast Asia
Buckwheat	*Fagopyrum esculentum*	Central Asia
Cabbage	*Brassica oleracea capitata*	Mediterranean
Broccoli	*Brassica oleracea italica*	Mediterranean
Brussels sprouts	*Brassica oleracea gemmifera*	Mediterranean
Cauliflower	*Brassica oleracea botrytis*	Mediterranean
Kohlrabi	*Brassica oleracea gongylodes*	Mediterranean
Carrot	*Daucus carota*	Southwest Asia
Cashew nut	*Anacardium occidentale*	South America
Cassava (manihot, manioc, tapioca, yuca)	*Manihot esculenta*	South America

Table 21.1   The More Common Useful Cultivated Plants (*Cont.*)

Common name	Scientific name	Probable place of origin
Celery	*Apium graveolens*	Mediterranean
Cherry	*Prunus avium*	Southwest Asia
Chestnut	*Castanea sativa*	Mediterranean
Chickpea	*Cicer arietinum*	Southwest Asia
Cinnamon	*Cinnamonum zeylanicum*	Ceylon
	Cinnamonum cassia	Southeast Asia
Clove	*Eugenia aromatica*	Molucca Islands
Cocoa	*Theobroma cacao*	South America
Coconut	*Cocos nucifera*	Southeast Asia
Coffee	*Coffee arabica*	Ethiopia
Cotton		
	Gossypium herbaceum (2×)	Africa
Tree	*Gossypium arboreum* (2×)	Asia
Upland	*Gossypium hirsutum* (4×)	Central America
Sea Island (Egyptian)	*Gossypium barbadense* (4×)	South America
Cowpea (Blackeyed pea)	*Vigna sinensis*	India-Burma
Cranberry	*Vaccinium macrocarpum*	North America
Cucumber	*Cucumis sativus*	India-Burma
Currant	*Ribes sativum*	Europe
Date palm	*Phoenix dactylifera*	Southwest Asia
Eggplant	*Solanum melongena*	India-Burma
Endive	*Cichorium endivia*	Mediterranean
English walnut	*Juglans regia*	Central Asia
Fig	*Ficus carica*	Southwest Asia
Flax	*Linum usitatissimum*	Southwest Asia
Garlic	*Allium sativum*	Mediterranean
Ginger	*Zingiber officinale*	Southeast Asia
Gooseberry	*Ribes grossularia*	Europe
Gourd	*Cucurbita* spp.	Central America
Calabash	*Lagenaria siceraria*	Africa
Grape	*Vitis vinifera*	Southwest Asia
Grapefruit	*Citrus paradisi*	West Indies
Guava	*Psidium guajava*	South America

Table 21.1   The More Common Useful Cultivated Plants (*Cont.*)

Common name	Scientific name	Probable place of origin
Hemp	*Cannabis sativa*	Central Asia
Hops	*Humulus lupulus*	Mediterranean
Horseradish	*Rorippa armoracia*	Europe
Indigo	*Indigofera tinctoria*	India-Burma
Jerusalem artichoke	*Helianthus tuberosus*	North America
Jujube	*Zizyphus jujuba*	China
Jute	*Corchorus olitorius*	India-Burma
Kapok	*Ceiba pentandra*	Central America
Kola nut	*Cola acuminata*	Central Africa
Lemon	*Citrus limon*	Southeast Asia
Lentil	*Lens esculenta*	Southwest Asia
Lettuce	*Lactuca sativa*	Mediterranean
Licorice	*Glycyrrhiza glabra*	Southwest Asia
Lime	*Citrus aurantifolia*	Southeast Asia
Loganberry	*Rubus* spp. (4×)	California
Maize	*Zea mays*	Mexico
Mango grain	*Bromus mango*	South America
Mango	*Mangifera indica*	Southeast Asia
Maté	*Ilex paraguariensis*	South America
Millet	*Panicum miliaceum*	Central Asia
Italian (foxtail)	*Setaria italica*	Central Asia
Raggee (finger)	*Eleusine coracana*	Ethiopia
Pearl	*Pennisetum glaucum*	Africa
Japanese	*Echinochloa frumentacea*	China
Mulberry		
White	*Morus alba*	China
Paper	*Broussonetia* spp.	China
Muskmelon	*Cucumis melo*	Southwest Asia
Mustard	*Brassica* spp.	Asia
Nutmeg	*Myristica fragrans*	Southeast Asia
Oats	*Avena sativa*	Europe
Hulled	*Avena strigosa*	Mediterranean
Naked	*Avena nuda*	China

Table 21.1   The More Common Useful Cultivated Plants (*Cont.*)

Common name	Scientific name	Probable place of origin
Okra	*Hibiscus esculentus*	Ethiopia
Olive	*Olea europaea*	Mediterranean
Onion	*Allium cepa*	Mediterranean
Opium poppy	*Papaver somniferum*	Southwest Asia
Orange	*Citrus sinensis*	China
Oyster plant	*Tragopogon porrifolius*	Mediterranean
Papaya	*Carica papaya*	South America
Parsley	*Petroselinum sativum*	Mediterranean
Parsnip	*Pastinaca sativa*	Europe
Pea	*Pisum sativum*	Southwest Asia
Peach	*Prunus persica*	China
Peanut (ground nut)	*Arachis hypogaea*	South America
Pear	*Pyrus communis*	Central Asia
Pecan	*Carya pecan*	North America
Pepper (bell, chili, red)	*Capsicum* spp.	Central America
Pepper (spice)	*Piper nigrum*	India-Burma
Persimmon	*Diospyros kaki*	China
Pigeon pea	*Cajanus cajan*	India (Africa)
Pineapple	*Ananas comosus*	South America
Plantain	*Musa paradisica*	Southeast Asia
Pomegranate	*Punica granatum*	Southwest Asia
Plum	*Prunus domestica*	Southwest Asia
Potato	*Solanum tuberosum*	South America
Pumpkin	*Cucurbita* spp.	Central America
Quince	*Cydonia oblonga*	Southwest Asia
Quinine	*Cinchona calisaya*	South America
Quinoa	*Chenopodium quinoa*	South America
Radish	*Raphanus sativus*	China
Raspberry	*Rubus* spp.	North America
Rhubarb	*Rheum rhaponticum* (food)	Central Asia
	Rheum officinale (drug)	China

Table 21.1 The More Common Useful Cultivated Plants (*Cont.*)

Common name	Scientific name	Probable place of origin
Rice	*Oryza sativa*	India
Rubber	*Hevea brasiliensis*	South America
Rye	*Secale cereale*	Europe
Sesame	*Sesamum indicum*	Central Africa
Sisal hemp	*Agave sisalana*	Central America
Sorghum	*Sorghum* spp.	Africa
Spinach	*Spinacia oleracea*	Southwest Asia
Squash	*Cucurbita* spp.	Central America
Strawberry	*Fragaria grandiflora* (4×)	South America × North America
Sugar beet	*Beta vulgaris esculenta*	Europe
Sugar cane	*Saccharum officinarum*	Southeast Asia
Sunflower	*Helianthus annuus*	North America
Sweet potato	*Ipomoea batatas*	Central America
Swiss chard	*Beta vulgaris cicla*	Mediterranean
Tangerine	*Citrus reticulata*	Southeast Asia
Taro	*Colocasia esculenta*	India-Burma
Tea	*Thea sinensis*	China
Tobacco	*Nicotiana tabaccum*	South America
Tomato	*Lycopersicum esculentum*	South America
Tumeric	*Curcuma longa*	Southeast Asia
Turnip	*Brassica rapa*	China
Vanilla	*Vanilla planifolia*	Central America
Watermelon	*Citrullus vulgaris*	Central Africa
Wheat		
Einkorn	*Triticum monococcum* (AA)	Southwest Asia
Emmer (macaroni, durum)	*Triticum dicoccum* (AABB)	Southwest Asia
Bread	*Triticum aestivum* (AABBDD)	Southwest Asia
Yam	*Dioscorea* spp. (food)	Southeast Asia
	Dioscorea spp. (drug)	Central America

Table 21.2   Places of Origin of the More Common Useful Cultivated Plants

Old World

Central Asia
1. Apple
2. English walnut
3. Hemp
4. Millet
5. Millet, Italian
6. Mustard
7. Pear
8. Rhubarb (food)

China
1. Apricot
2. Bean, soy
3. Buckwheat
4. Jujube
5. Millet, Japanese
6. Mulberry
7. Oats, naked
8. Orange
9. Peach
10. Persimmon
11. Radish
12. Rhubarb (drug)
13. Tea
14. Turnip

India-Burma
1. Amaranth
2. Bean, mung
3. Chickpea
4. Cinnamon
5. Cowpea
6. Cucumber
7. Eggplant
8. Indigo
9. Jute
10. Mango
11. Pepper
12. Pigeon pea
13. Rice
14. Taro

Southeast Asia
1. Bamboo
2. Banana
3. Breadfruit
4. Cinnamon
5. Clove
6. Coconut
7. Cotton
8. Ginger
9. Lemon
10. Lime
11. Nutmeg
12. Plantain
13. Sugar cane
14. Tangerine
15. Tumeric
16. Yam (food)

Southwest Asia
1. Almond
2. Barley
3. Carrot
4. Cherry
5. Date palm
6. Fig
7. Flax
8. Grape
9. Lentil
10. Licorice
11. Muskmelon
12. Opium poppy
13. Pea
14. Pomegranate
15. Plum
16. Quince
17. Spinach
18. Wheat

Ethiopia
1. Bean, castor
2. Coffee
3. Cotton
4. Kola nut
5. Millet, raggee
6. Okra

Central Africa
1. Calabash
2. Millet, pearl
3. Sesame
4. Sorghum
5. Watermelon

Europe
1. Blackberry
2. Currant
3. Gooseberry

Table 21.2 Places of Origin of the More Common Useful Cultivated Plants (*Cont.*)

4. Horseradish	2. Asparagus	12. Garlic
5. Oats	3. Bean, broad	13. Hops
6. Parsnip	4. Beet	14. Kohlrabi
7. Raspberry	5. Broccoli	15. Lettuce
8. Rye	6. Brussels sprouts	16. Oats, hulled
9. Sugar beet	7. Cabbage	17. Olive
	8. Cauliflower	18. Onion
	9. Celery	19. Oyster plant
Mediterranean	10. Chestnut	20. Parsley
1. Artichoke	11. Endive	21. Swiss chard

New World

Central America	15. Vanilla	4. Cassava
1. Amaranth	16. Yam (drug)	5. Cocoa
2. Avocado		6. Cotton
3. Bean, kidney	*North America*	7. Guava
4. Bean, scarlet	1. Blueberry	8. Mango grain
runner	2. Cranberry	9. Maté
5. Cotton	3. Jerusalem artichoke	10. Papaya
6. Gourd	4. Loganberry	11. Peanut
7. Grapefruit	5. Pecan	12. Pineapple
8. Kapok	6. Raspberry	13. Potato
9. Maize	7. Strawberry	14. Quinine
10. Pepper, capsicum	8. Sunflower	15. Quinoa
11. Pumpkin		16. Rubber
12. Sisal hemp	*South America*	17. Strawberry
13. Squash	1. Bean, jack	18. Tobacco
14. Sweet potato	2. Bean, lima	19. Tomato
	3. Cashew nut	

Compared to the land surface of the earth, the centers of origin for cultivated plants are quite restricted in size. They are located in regions of tropical, semitropical, or at worst, mild climatic conditions. Most cultivated plants appear to have originated in the Old World in the region between 20° and 40° North latitude. Three centers of primary importance appear to lie in southwestern Asia, southeastern Asia, and in the New World. In southwestern Asia, wheat was the most important grain, in southeastern Asia it was rice, and in the New World, maize. Some of the centers of origin are clearly of secondary importance in terms of the numbers of species that originated there. Tropical Africa and tropical America, for example, have contributed rather few species to modern agriculture compared with Asia. Furthermore, Europe (apart from the Mediterranean region) and North America, now highly productive agricultural regions, can claim to be the center of origin for only a bare handful of cultivated plants.

21.3. The Process of Domestication

The initial domestication of a wild plant species would almost have to occur within the area of its natural distribution. Furthermore, the evidence suggests that the evolutionary changes associated with domestication are facilitated in plants if gene flow from wild relatives continues to feed new genetic variation into the domesticated species. Thus, the argument for centers of diversity being equivalent to centers of origin would seem to be well founded. On the other hand, it is also true that cultivated plants often thrive better in regions remote from their site of origin because then they can grow in the absence of their naturally occurring pests and parasites. Therefore, the present-day distribution and frequency of a particular domesticated species may be quite misleading with respect to the actual origin of that species. Man's migrations have tended to disperse many cultivated plants (and "weeds," too, for that matter) to the far corners of the earth with amazing rapidity. As they have spread, selection, both natural and artificial, has continued to adapt them to the new environmental conditions to which they are exposed. In this way the numerous varieties of a single domesticated species have evolved.

In view of the tremendous recent scientific and technological progress in so many fields, it is rather surprising to discover that most cultivated plant species were domesticated in prehistoric times. The increased knowledge of genetics has been applied to the improvement of existing cultivated species. There has been little effort to search out and convert to man's uses additional wild species. Furthermore, it is interesting to note that the important agricultural plant species are so few in number compared with the total number of all plant species, and that nearly all of them are either monocotyledonous or dicotyledonous angiosperms. The species base for modern agriculture therefore is rather narrow, and domestication occurred so long ago that the origin of most cultivated plants is obscure and can be studied only by inference. In a few cases the origin is known. Although berries of different kinds were gathered for a long time, they were not cultivated until rather recently. Raspberries and blackberries were apparently first cultivated during the Middle Ages; gooseberries and currants first began to be mentioned as cultivated plants in the sixteenth century. The present cultivated strawberry (*Fragaria grandiflora*) arose as an allopolyploid following hybridization in Europe between the North American species, *F. virginiana,* and the South American species, *F. chiloensis,* in the mid-eighteenth century. The hexaploid loganberry apparently originated in California in the 1880s following hybridization between a diploid raspberry and an octoploid California blackberry.

The history of the tomato is rather curious. A New World species, the wild plants in South America bear small fruit the size of currants. However, cultivated tomatoes became an important part of the diet in Mexico in pre-Columbian times. Introduced into Europe and the United States, it was grown primarily as an ornamental plant, with the fruits thought to be poisonous and sometimes referred to as "love apples." Only in Italy was it quickly adopted as an important vegetable, becoming an essential component in many Italian dishes. Its adoption by the rest of Europe and in the United States has taken place only within the last century. If you had offered a guest tomato juice for breakfast a century ago, he might have suspected you of trying to poison him.

The history of domestication of the beet, *Beta maritima,* illustrates a situation that is paralleled in a number of other cases. *Beta maritima* was first domesticated for its foliage, which was used as greens and known as chard. Later selection gave rise to the mangelwurzel, whose coarse yellowish roots were used as cattle fodder. The garden beet, with its deep red, rather sweet root used as a table vegetable, is first mentioned in the sixteenth century. The discovery in the mid-eighteenth century that beets contained sugar led to deliberate efforts to select for increased sugar content. These efforts were so successful that the percentage of sugar was ultimately increased from about 5% to over 20% (Fig. 21.3), and thus the sugar shortage was eased in France during the Napoleonic wars. Sugar beets have been grown commercially in the United States only since about 1875, but they are now the most important crop derived from *Beta maritima.* The cabbage, *Brassica oleracea,* is another example of the diversity of uses made of a single original domestication.

The Para rubber tree (*Hevea brasiliensis*) is another recently domesticated species. The tree is native to the Amazonian tropical rain forest, but more than 97% of the rubber produced today comes from Southeast

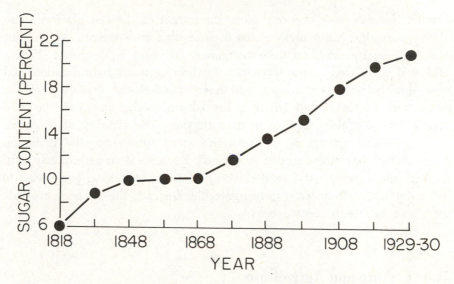

Fig. 21.3. The increase in the sugar content of sugar beets due to selective breeding. (Schwanitz, F., 1966. *The Origin of Cultivated Plants.* Harvard University Press, Cambridge, Mass., p. 159, Fig. 57.)

Asia and to a lesser degree from Africa. It was not introduced into Southeast Asia until about 1875, after apparently being smuggled out of Brazil. Some of these original trees are thought to be still living. Here is a case in which a New World species is now being grown primarily in the Old World, but it is counterbalanced by coffee, an Old World species from Africa now grown primarily in the New World. In both cases, the species probably do better because they are removed from their pests and diseases.

Thus, in a few cases it is possible to learn something directly about the origin of domesticated species. In other cases, information can be gained from a genetic and cytological comparison of the cultivated species with its supposed wild ancestors. More difficult to study are the species whose wild ancestors are unknown and probably extinct. Among them are such species as maize (*Zea mays*), the onion (*Allium cepa*), the broad bean (*Vicia faba*), the lentil (*Lens esculenta*), ginger (*Zingiver officinale*), and the date palm (*Phoenix dactylifera*). The reasons for extinction may differ. Drastic environmental changes may have eliminated the ecological conditions favorable to the wild species. Another possibility is that essentially the entire species population has been domesticated. A third possibility is that the species has been so greatly modified under domestication that its affinities to its wild ancestors are no longer obvious. Probably all three factors are involved to varying degrees for most such cases. Rather dramatic climatic changes have occurred since the retreat of the last glaciers some 10,000 years ago. Noteworthy is the expansion of arid regions. Agriculture itself has greatly modified the environment for most wild species. Where wild and cultivated forms have remained in contact, hybridization and gene flow between them may well have speeded the evolution of the domesticated species, but led to a breakdown in the fitness of the wild population, ultimately leading to its extinction. The rapidity of evolution in domesticated species in the relatively short time since the invention of agriculture has already been mentioned. Even so, it seems unlikely that, if their wild ancestors still existed, this evolution could have progressed to the point where they were unrecognizable. Instead, the chances are that they have, in fact, become extinct.

21.4. Culture and Agriculture

One of the more interesting aspects of the study of cultivated plants is the light shed on cultural contacts between different regions. For instance, the evidence is reasonably clear for an early and rather direct contact between India and Central Africa that did not involve the Fertile Crescent region of southwestern Asia. The question of contact between the Old World and the New prior to Columbus has stimulated much discussion. The evidence from cultivated plants suggests that it was minimal. New World agriculture was based on maize, squashes (*Cucurbita*), and beans (*Phaseolus*), quite different from the species complexes used in the Old World. Furthermore, domesticated animals, closely associated with Old World agriculture from prehistoric times and used for work, food, and fiber, were not thus used in the New World, nor was the wheel in use in agriculture here. The best hypothesis seems to be that the American Indians reached the New World via the Bering Straits region, probably more than 10,000 years ago and hence prior to the origin of agriculture in the Old World. Agriculture developed independently in the two regions. The tendency for some to argue in favor of pre-Columbian cultural contacts between Old and New Worlds turns out to be based on relatively few plant species that were apparently common to both the Old and New Worlds prior to Columbus. Of these the coconut palm (*Cocos nucifera*) and the calabash gourd (*Lagenaria siceraria*) probably were distributed naturally by ocean currents from the Old World to the New, the coconut across the Pacific from Southeast Asia, the calabash across the Atlantic from Africa. The peanut (*Arachis hypogaea*) and maize (*Zea mays*)

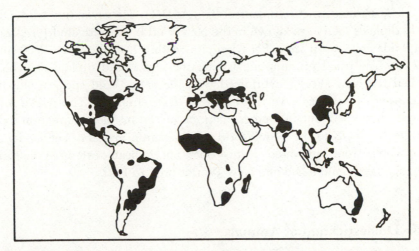

Fig. 21.4. Present-day areas of cultivation of maize. Although indigenous only to the Americas, maize is now cultivated on all continents. (Edlin, H. L., *Man and Plants*, © 1967, Aldus Books Limited, London; artist, Edward Poulton.)

probably were restricted to the New World before 1492, but were then widely and rapidly disseminated over the Old World via the Portuguese trade routes (see Fig. 21.4).

The sweet potato (*Ipomoea batatas*) was present in both the Old and New Worlds before 1492, for it is found in both America and Polynesia. It was domesticated in tropical America, and was evidently carried to Polynesia by man. There are two reasons for making this statement. Unlike the coconut and the gourd, the sweet potato is not adapted to survive long immersion in sea water. Furthermore, it bears variations of the same name in South America and throughout the Pacific, suggesting that its name went with it on its travels, which is possible only if it traveled with man. It is more important in western and southern Polynesia than in central Polynesia, where Southeast Asian plants predominate. Whether the Polynesians reached South America and brought back the sweet potato or the South American Indians carried it to Polynesia remains a moot point. In either case, it does not appear that the cultural contact was important or extensive. Rather, it seems similar in nature to the brief contact made with the New World by the Vikings about 1000 A.D.

Cotton constitutes more of a puzzle. There are distinctive Old World diploid species and the genus *Gossypium* with 13 pairs of large chromosomes (AA), and New World diploid species with 13 pairs of small chromosomes (DD). The New World cultivated species are tetraploid, with 26 pairs of chromosomes, 13 large pairs (AA) and 13 small pairs (DD), so that their chromosome complement is (*AADD*). However, no wild Old World diploids (*AA*) are known in the New World, yet cultivated (*AADD*) tetraploids are known from Peruvian sites dating from between 2000 and 3000 B.C. The usual presumption is that the Old World diploid somehow reached the New World, hybridized with the native wild species and, by chromosome doubling, gave rise to the *AADD* tetraploid. The wild (*AA*) diploid then died out. Whether man or an errant floating cotton boll brought the seeds to the New World may remain a mystery. The evidence as a whole from cultivated plants does not indicate extensive cultural exchange between the Old and New Worlds prior to 1492.

21.5. Domestication of Animals

Our ignorance of the time and place and often the original reasons for the domestication of animals is equally great. The number of domesticated animal species is even smaller than the number of cultivated plant species. The great majority are mammals, mostly Artiodactyla, the cloven-hoofed mammals, with birds the next largest group. Hence, nearly all domesticated animal species are warm-blooded vertebrates. About the only invertebrates of any significance are the honeybee and the silkworm, and recently, perhaps, the oyster. Most domesticated animals originated in the Old World, with only the llama and the alpaca, the Muscovy duck, the guinea pig, and the turkey domesticated in the New World. None of these except the turkey plays a very significant role in modern agriculture. Again it is worth noting that most of these species were domesticated in prehistoric times. The efforts of modern animal breeders have been devoted to the improvement of existing domesticated species rather than to the domestication and improvement of additional wild species.

Some of the more important domesticated animal species are shown in Table 21.3. For some of the more familiar species—dog, cattle, pig, and goose—the place of origin is so uncertain that "Old World" is about the only satisfactory designation. This difficulty may stem, in part, from the possibility that these species were domesticated more than once in different parts of the Old World. This is thought to be true for the goose and the pig, which are both thought to have been domesticated in China as well as in Europe, and probably at points between as well. Doubts about the ancestry of the dog and of cattle complicate the picture for these species. At various times, the wolf (*Canis lupus*), the golden jackal (*Canis*

Table 21.3   The More Common Useful Domesticated Animals

Common name	Scientific name	Probable place of origin
Alpaca	*Lama pacos*	South America
Camel		
Dromedary (one-hump)	*Camelus dromedarius*	Southwest Asia
Bactrian (two-humps)	*Camelus bactrianus*	Central Asia
Cat	*Felis catus*	North Africa
Cattle	*Bos taurus*	Old World
Dog	*Canis familiaris*	Old World
Donkey	*Equus asinus*	Africa
Duck	*Anas platyrhyncha*	Asia
Muscovy duck	*Cairina moschata*	Central America
Elephant		
Africa	*Loxodonta africana*	Africa
Indian	*Elephas maximus*	India
Fowl	*Gallus gallus*	Southeast Asia
Goat	*Capra hircus*	Southwest Asia
Goose	*Anser anser*	Old World
Guinea pig	*Cavis porcellus*	South America
Honeybee	*Apis mellifera*	Old World
Horse	*Equus caballus*	Asia
Llama	*Lama glama*	South America
Mink	*Mustela vison*	North America
Pig	*Sus scrofa*	Old World
Pigeon	*Columba livia*	Eurasia
Rabbit	*Oryctolagus cuniculus*	Spain
Reindeer	*Rangifer tarandus*	Northern Eurasia
Sheep	*Ovis aries*	Southwest Asia
Silkworm	*Bombyx mori*	China
Turkey	*Meleagris gallopavo*	Mexico
Water buffalo	*Bubalus bubalis*	India
Yak	*Bos grunniens*	Tibet
Zebu cattle	*Bos indicus*	India

aureus, or a hypothetical wild dog has been suggested as the ancestor of the domesticated dog, *Canis familiaris.* Lack of evidence for the hypothetical wild dog species has led to the abandonment of this theory. The differences between dogs and the golden jackal and the similarities in behavior and anatomy between dogs and wolves have led to the present tendency to believe that domesticated dogs were derived from one of the smaller subspecies of wolves in the Near East or southern Asia at least 10,000 or 12,000 years ago.

Domesticated cattle are thought to be derived from the large wild aurochs, *Bos primigenius.* However, another smaller short-horned species (*B. longifrons*) has also been implicated, but knowledge is so uncertain that it has also been suggested that *longifrons* is merely the female of *primigenius.*

The reasons for domestication of animals must have varied, for they are used for food, milk, blood, fuel, as a draft animal or a mount, or in a combination of such uses. The dog, which seems to have been domesticated early, may originally have been a sort of camp follower, gradually developing a symbiotic relationship with man. His original role was probably that of a scavenger, from which he gradually evolved into his varied roles of watch dog, aid in the chase, companion, and emergency source of food. Some species today have considerable ceremonial or religious significance in certain cultures, and this element may well have played a role in the domestication of these species. The taboos that exist in certain present-day societies against eating beef or pork are matched in our own by our aversion toward eating, wittingly at any rate, horse meat or dog meat.

The circumstances of domestication also must have varied. Some have argued that the domestication of animals became possible only after man had learned to cultivate plants and was able to settle in one place with a more or less stable food supply for himself and his animals, but this can hardly have been universally true. It is not true, for example, of the reindeer even today, and probably was not an essential feature in the domestication of other large herd animals. For such species man may have changed from a hunter to a protector of the herd, taking up a nomadic existence in order to stay near and watch over his major food source, and gradually assuming ever greater control over the herd. There can be little doubt that the cultivation of plants led to changes in the management practices for some herd animals but not for all. However, the ecology and behavior of other species is such that their domestication seems much more probable after man became a farmer.

Pigs and chickens, for example, do not seem well adapted to a nomadic existence.

Present evidence suggests that sheep and goats were domesticated very early, preceded only by the dog, and that cattle, pigs, chickens, geese, horses, and so on were domesticated at various times and places thereafter. It seems probable that one useful method of domestication was to rear immature animals, especially females, as pets. Young animals are more tractable and more easily tamed than adults, and having lost their fear of man, they may continue to live in association with him, even while rearing their young.

21.6. Animal Breeding

Although animals were domesticated long ago, systematic efforts at animal improvement are relatively recent. The Arabs were interested in the pedigrees of their horses more than a thousand years ago, but little is known about whether this information was used in any significant way in breeding. Some of the first known efforts toward animal improvement were those of English royalty in developing improved war horses and later, race horses. To this end they imported Arabian horses and others from the continent to be crossed with their own horses.

The Englishman Robert Bakewell (1725–1795) is generally regarded as the founder of modern animal breeding. He was a practical man rather than a theoretician, and his principles must be inferred from his practice since he never published them. His influence seems to have been responsible for the development of pedigree breeding in the latter part of the eighteenth century. He operated on the principles, "like begets like" and "breed the best to the best." Thus, unlike most of his contemporaries, he did not avoid inbreeding, but instead used it, together with selection, for fixing the type of his animals. Since he apparently also used occasional wide outcrosses to introduce new genes into his flocks or herds, he also took advantage of the benefits from crossbreeding as well as inbreeding in developing superior groups of animals. On the other hand, he did not allow unlimited crossbreeding to dissipate the gains he had made. The development of these obviously improved and distinctive groups of animals led to the next trend in animal breeding, the development of distinctive breeds and the formation of breed registry societies that kept records on all the members of a given breed.

Breeds of animals as we know them today are a relatively recent phenomenon. They developed after a local breed became so numerous and widespread that it became impossible to keep track of the pedigrees and verify the ancestry of the animals without some formal central system of keeping records. The first herdbook, for the Thoroughbred horse, appeared in 1791. The Shorthorn herdbook was started in 1822, and was followed by a herdbook for Herefords in 1846, and for other breeds still later. The first swine herdbook, for American Berkshires, did not appear until 1876. Initially, any superior individual that approximated the type of the breed was accepted for registration. However, the herdbooks were later closed, and no further animals were registered except "purebreds," both of whose parents were already registered.

Therefore, a breed developed along the following lines. A breeder first developed a type of animal superior in some way to the usual type, but not very distinct in pedigree from it. The new type of animal was kept together in closed herds and inbred so that members of the herd became distinct from other herds in ancestry as well as in phenotype and could be identified as a separate breed. If the new breed was popular, additional herds would be established to the point where a central herdbook was needed to keep track of all of the pedigrees. Finally, a breeders' society was formed to supervise the herdbook, maintain the purity of the breed, and thus protect the economic interests of the breeders.

The development of the different breeds, while leading to improvement in various domesticated animals over the earlier stock, also led to the belief that "purebred" animals were superior to animals bred in any other manner. This emphasis on purity of breed has, however, tended to inhibit further progress in animal breeding because, in some cases at least, the benefits to be gained from heterosis following crossbreeding have been ruled out by the purebred philosophy.

Another outgrowth of the development of distinct breeds was the livestock show. Such shows were useful in bringing breeders together so that they could compare the merits of their respective animals and could better define their goals or ideals in breeding. They were not an unmixed blessing, however, for breeders tended to emphasize, in their breeding, traits that would produce prizewinners in the show ring. Such animals were not necessarily the best on performance in economic traits as distinct from show-ring traits. However, performance tests had long been a consideration in animal breeding. The early breeders of the Thoroughbred horse were interested primarily in how fast their horses could run, and Bakewell is known to have conducted tests for efficiency of weight gain by his pigs. Nevertheless, emphasis on performance did not become widespread until the end of the nineteenth century, when tests of milk and butterfat production in dairy cattle and laying trials in poultry began to come into use.

As means became available for measuring performance in economic traits, emphasis in the breeding associations began to shift from protection and maintenance of the breed toward breed improvement, and it came to be recognized that the basic criterion of success should be efficiency of production. The primary method used was individual selection within the pure breeds. The pedigrees were available, but the proper interpretation and use of the pedigrees in making breeding decisions remained obscure.

The development of the science of genetics and subsequently of population genetics supplied a theoretical background that ultimately gave animal and plant breeders insight into the nature of the problems confronting them and their possible solutions. Mendelian segregation and independent assortment immediately accounted for the observations that sibs differed, and that "reversions" to an ancestral type sometimes occurred. Furthermore, the fundamental distinction between genotype and phenotype, so essential to success in breeding, was clarified by the work of Johannsen. It then was realized that the phenotype of an individual reflected environmental as well as genetic effects, but that only the genetic component was important in breeding. This, in turn, led to a better understanding of the need for and use of pedigrees, performance records, progeny tests, and sib tests.

The Castle-Hardy-Weinberg law, the first step in demonstrating the consequences of Mendelian inheritance in populations, showed that, contrary to prior belief, variation remains constant in a breeding population rather than being continually eroded by blending inheritance. It then became possible for the breeders to concentrate more on the effects of selection without so much concern about the effects of their breeding practices in exhausting the available variability.

Another advance resulting from the new knowledge of genetics was a better understanding of the consequences of different mating systems. The systematic use of inbreeding to produce inbred lines that were crossed to form commercial hybrids with heterotic vigor resulted in the spectacular success of hybrid corn and the extension of the technique to poultry and other species. At last, theories became available to account for heterosis and for inbreeding depression, and a more rational use of the various mating systems was possible.

The use of the selection index, incorporating and weighting performance in several traits of interest to the breeder, marked another step in improved

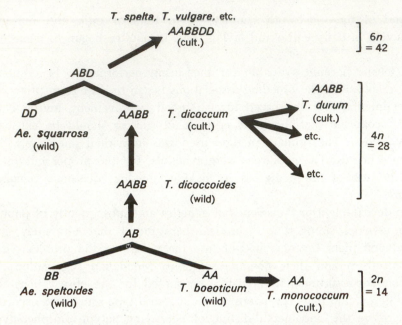

Fig. 21.5. Diagram of the history of the origin of diploid, tetraploid, and hexaploid cultivated wheat. *Ae.* = *Aegilops. T.* = *Triticum.* (From *Plants and Civilization,* 2nd edition, by Herbert G. Baker. © 1965, 1970 by Wadsworth Publishing Company, Inc., Belmont, California 94002. Reprinted by permission of the publisher.)

breeding practice. This index plus some knowledge of heritabilities made it possible to predict expected progress. When progress did not agree with expectations based on additive effects, the effects of dominance (including overdominance), epistasis, and linkage were investigated. Thus, the development of genetic theory provided a rationale for breeding practice that was previously lacking.

The selection index is useful because the breeder, although concerned with yield, often must deal with such diverse traits as disease resistance, palatability, winter hardiness, and so on. With such diverse goals, it is obviously necessary that some way be found to assist the breeder's decision-making process.

Along with the use of selection, inbreeding, crossbreeding, and heterosis, certain special techniques in breeding have been developed. Artificial insemination has been widely used in cattle breeding as a means of getting maximum dissemination of the genes of superior sires, and has been used to varying degrees in other species—including the honeybee, whose nuptial flight had always posed a considerable practical problem for the bee breeder. Artificial insemination is also being used in some cases in man.

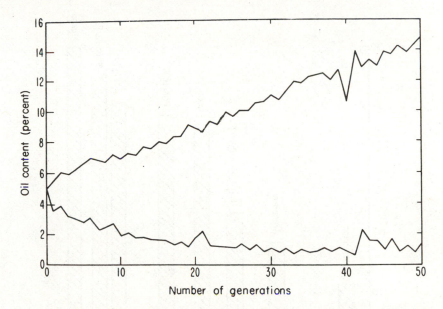

Fig. 21.6. The effect of 50 generations of selection for high and low oil content in maize. (Schwanitz, F., 1966. The *Origin of Cultivated Plants*. Harvard University Press, Cambridge, Mass., p. 118, Fig. 50.)

Some experimental work with ova transplantation has been done, but it is not yet being widely used. It offers the prospect, at least, that scrub cattle, for example, could produce pedigreed progeny. In theory, a movie actress could have her children in this way without interruption of her career or jeopardy to her figure, but ova transplants in man appear to be lagging behind heart and kidney transplants.

Artificially induced mutations have not yet been found particularly useful in animal breeding, but selection of mutations for increased production of antibiotics and other organic compounds by bacteria and fungi has been quite successful, and induced mutations have been utilized in some crop plants such as barley and oats. The usefulness of this new tool cannot yet be properly assessed, but it seems likely to be most useful where large numbers of progeny can be screened at little expense in search of potentially favorable mutations.

A large proportion of cultivated plants are polyploid, either autopolyploid or allopolyploid (e.g. wheat, Fig. 21.5). Spontaneous polyploids have evidently been preserved in the past because of the increased size or gigantism that often accompanies polyploidy. Autopolyploids are particularly common in ornamental plants such as narcissus, petunia, and cyclamen, where fertility is not a major concern, but larger, more intensely colored

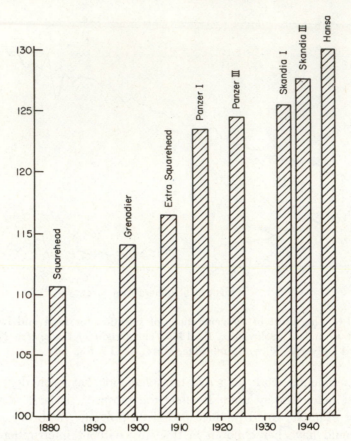

Fig. 21.7. The effect of selection on yield of winter wheat in Sweden. (Schwanitz, F., 1966. *The Origin of Cultivated Plants.* Harvard University Press, Cambridge, Mass., p. 163, Fig. 58.)

flowers are desired. The fertility of allopolyploids is usually high, and because they are in essence permanent hybrids, heterosis contributes to their vigor and yield. The discovery that colchicine induces polyploidy in plants provided a means for producing polyploids at will. The impact of this discovery has not been as great as was at first anticipated, but improved varieties of red clover and alfalfa have been produced, and work with ornamental plants has also met with success. Parallel efforts to induce polyploidy in animals have not been conducted to any great extent because of the problems associated with sex determination. Even though triploid rabbits have been reported, it is unlikely that polyploidy will play a significant role in animal breeding.

In the matter of sex determination, crosses in the silkworm and in chickens have been arranged in such a way that the segregation of sex-

Fig. 21.8. The Santa Gertrudis breed of cattle and its ancestors, the Brahma and Shorthorn breeds. (Courtesy *The Cattleman Magazine*.)

linked genes gives rise to males and females of different colors. In this way the sexes can be identified and sorted very efficiently at an early age. This method of autosexing in chicks is no longer common since the discovery that young chicks can be sexed by examination of the cloaca. However, because in many domesticated species one sex is of economically greater value than the other, efforts continue to influence the sex ratio in such a way as to produce a greater proportion of the more valuable sex. As yet, these efforts have not been notably successful. It is at least conceivable that induced parthenogenesis could be one solution to this problem.

The very success of animal and plant breeders has led to certain hazards. The creation of a few highly successful breeds or varieties of a domesticated species often has led to their widespread adoption commercially. Since these populations have only a limited representation of the total gene pool of the species, there is a danger that much of the available genetic variation so essential to further breeding progress will be lost simply because no one bothers to grow the older, less desirable varieties. For this reason, deliberate attempts are now being made in many domesticated species to preserve and maintain as much of this variation as possible.

It is perhaps worth citing a few of the notable successes of animal and plant breeders so that you will have some appreciation of the progress

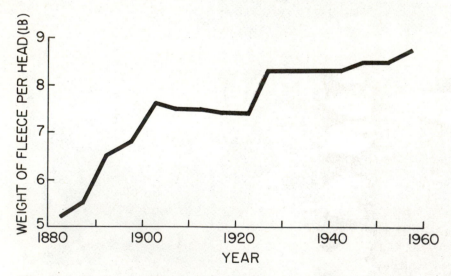

Fig. 21.9. Increased yield of wool per sheep in Australia. (Frederick B. Hutt–*Animal Genetics*. Copyright © 1964 The Ronald Press Company, New York.)

that has been made. The development of hybrid corn has already been discussed, but another long-term selection experiment for oil content is worth citing. In the high-selection line oil content has been increased from 5% to about 15%, while in the low-selection line the oil content has dropped below 2% (Fig. 21.6). Wheat breeders have successfully increased the yield of winter wheat some 25% by crossing winter-hardy varieties with high-yielding types such as Squarehead and selecting for winter-hardy, high-yielding recombinants (Fig. 21.7). Selection for early ripening has extended the range over which spring wheat can be grown considerably farther north. In addition, resistance to pests, to drought, and to wheat rust have been incorporated into varieties with good milling qualities.

Plant breeders are continually introducing new winter-hardy varieties of ornamental as well as crop plants, extending the range of many cultivated species far north of the limits tolerated by their wild ancestors or relatives.

The King Ranch in Texas faced the opposite problem. The existing

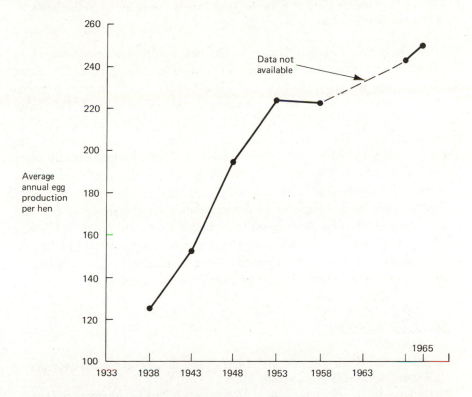

Fig. 21.10. The effects of selection for egg production in poultry. Note that egg production has doubled, from 125.6 in 1933 to 249.6 in 1965. (Data of Lerner and Lowry.)

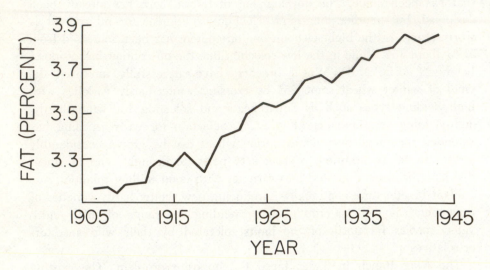

Fig. 21.11. Average yield of butterfat from Friesian cows in the Netherlands. (Data from Roelofs.)

beef cattle breeds failed to thrive in the hot, subtropical environment of southern Texas. Consequently, they developed a new breed, the Santa Gertrudis, by selection among the descendants from crosses between English Shorthorn and Indian Brahma cattle (Fig. 21.8). This new breed of beef cattle thrives both in hot, dry areas and in humid, subtropical regions. They are resistant to ticks, insects, and parasites, are not adversely affected by high temperatures, and gain weight rapidly feeding on grass alone. The Santa Gertrudis is one of the few new animal breeds to be developed in this century. The figures show a sampling of progress made by breeders to give some notion of the progress that has been achieved in such diverse traits as fleece in sheep (Fig. 21.9), egg production by hens, (Fig. 21.10), and butterfat in dairy cattle (Fig. 21.11). Improved management and nutrition have undoubtedly contributed to these gains to some extent, but the primary gains are genetic.

Additional Reading

Ames, O., 1939. *Economic Annuals and Human Cultures*. Botanical Museum of Harvard University, Cambridge, Mass.

Anderson, E., 1960. "The Evolution of Domestication." In *Evolution after Darwin*. Vol. 2. *The Evolution of Man*, S. Tax, ed, University of Chicago Press, Chicago.

Anderson, E., 1967. *Plants, Man and Life*. University of California Press, Berkeley.

Bailey, L. H., 1949. *Manual of Cultivated Plants*, rev. ed. Macmillan, New York.

Baker, H. G. 1970. *Plants and Civilization*, 2nd ed. Wadsworth, Belmont, Calif.

de Candolle, A., 1886. *Origin of Cultivated Plants*. Hafner, New York (Reprint 1959).

Darlington, C. D., 1963. *Chromosome Botany and the Origins of Cultivated Plants*, 2nd ed. Hafner, New York.

Darwin, C., 1868. *The Variation of Animals and Plants under Domestication*. Murray, London.

Edlin, H. L., 1967. *Man and Plants*. Aldus Books, London.

Epstein, H., 1972. *The Origin of the Domestic Animals of Africa*. 2 vols. Africana, New York.

Frankel, O. H., E. Bennett, R. D. Brock, A. H. Bunting, J. R. Harlan, and E. Scheiner, eds., 1970. *Genetic Resources in Plants*. I.B.P. Handbook No. 11. Davis, Philadelphia.

Galinat, W. C., 1971. "The Origin of Maize," *Ann. Rev. Genet.* **5**:447–478.

Hafez, E. S. E., ed., 1969. *The Behavior of Domestic Animals*, 2nd ed. Ballière, Tindall and Cox, London.

Hamilton, W. D., 1964. "Genetical Evolution of Social Behavior," *J. Theor. Biol.* **7**:1–52.

Harlan, J. R., 1961. "Geographic Origin of Plants Useful to Agriculture." In *Germ Plasm Resources*, AAAS Publ. **66**:3–19.

Harlan, J. R., 1971. "Agricultural Origins: Centers and Non-centers," *Science* **174**:468–474.

Harris, D. R., 1967. "New Light on Plant Domestication and the Origins of Agriculture: A Review." *Geogr. Rev.* **57**:90–107.

Harris, D. R., 1972. "The Origins of Agriculture in the Tropics," *Amer. Sci.* **60**:180–193.

Hanks, J. G., 1970. "The Origins of Agriculture." *Economic Botany* **24**:131–133.

Hedrick, U. P., ed., 1919. *Sturtevant's Notes on Edible Plants*. N.Y. Dept. Ag. 27th Ann. Rept. Vol. 2. Pt. II.

Hill, A. F., 1937. *Economic Botany*. McGraw-Hill, New York.

Hutchinson, J., ed., 1965. *Essays on Crop Plant Evolution*. Cambridge University Press, London.

Indian Society of Genetics and Plant Breeding, 1951. "Symposium of the Origin and Distribution of Cultivated Plants in South Asia," *Indian J. Genet. and Plant Breeding* **11** (1).

Janick, J., R. W. Schery, F. W. Woods, and V. W. Ruttan, 1969. *Plant Science, an Introduction to World Crops*. Freeman, San Francisco.

MacNeish, R. S., 1965. "The Origins of American Agriculture," *Antiquity* **39**:87–94.

Mangelsdorf, P. C., 1971. *Corn, Its Origin, Evolution and Improvement*. Harvard University Press, Cambridge, Mass.

Mangelsdorf, P. C., R. S. MacNeish, and W. C. Galinat, 1964. "Domestication of Corn," *Science* **143**:538–545.

Purseglove, J. W., 1968. *Tropical Crops: Dicotyledons*. Wiley, New York.

Reed, C. A., 1959. "Animal Domestication in the Prehistoric Near East," *Science* **130**:1629–1639.

Robinson, R., 1971. *Genetics for Cat Breeders*. Pergamon, Elmsford, N.Y.

Sauer, C. O., 1952. *Agricultural Origins and Dispersals*. American Geographical Society, New York.

Schwanitz, F., 1966. *The Origin of Cultivated Plants*. Harvard University Press, Cambridge, Mass.

Scott, J. P., 1968. "Evolution and Domestication of the Dog," *Evol. Biol.* 2:243–275.

Scott, J. P., and J. L. Fuller, 1965. *Genetics and the Social Behavior of the Dog*. University of Chicago Press, Chicago.

Stakman, E. C., E. R. Bradfield, and P. C. Mangelsdorf, 1967. *Campaigns Against Hunger*. Harvard University Press, Cambridge, Mass.

Ucko, P. J., and G. W. Dimbleby, eds., 1969. *The Domestication and Exploitation of Plants and Animals*. Aldine, Chicago.

Vavilov, N. I., 1926. *Studies on the Origin of Cultivated Plants*. Institute de Botanique Appliquée et D'Amelioration des Plantes, Leningrad.

Vavilov, N. I., 1949–1950. "The Origin, Variation, Immunity and Breeding of Cultivated Plants," K. S. Chester, tr, *Chronica Botanica* 13:1–366.

Zeuner, F. E., 1963. *A History of Domesticated Animals*. Harper & Row, New York.

Questions

21.1. How would you define domestication?

21.2. How have selection pressures changed in domesticated species as compared with their wild relatives and ancestors?

21.3. Do you think that selection pressures on human populations have changed as the result of the invention of agriculture? In what ways? In the 10,000 years since the advent of agriculture, do you think that the domesticated species or man himself have undergone more genetic changes? Do you think man has evolved at all in the last 10,000 years?

21.4 What are the putative centers of origin for agriculture? What condiditions seem to favor the development of an agricultural society?

21.5. What are the implications of the rates of genetic change observed in domesticated species for possible rates of evolutionary change in other species?

21.6. Modern civilizations evolved only after the origin of agriculture. Do you believe that agriculture was an essential prerequisite to the development of these societies, or that they could also have arisen in groups whose subsistence depended on hunting and gathering? Why do you think grains such as rice, wheat, and maize form the basis for the most successful agricultural systems rather than other types of domesticated species?

21.7. Many cultivated species are now grown throughout the world wherever conditions are favorable. In some cases these species thrive better in other parts of the world than in their original native habitat. Why might this be so?

21.8. If you wished to investigate the origin of New World cultivated tetraploid cotton, how would you proceed? What techniques and what disciplines might aid you in your study? What practical benefits might accrue as the result of your study?

21.9. Suppose that you were a psychologist interested in aggression. The literary image of such species as the rabbit is that of a timid, fearful creature, while the weasel is portrayed as fearless, aggressive, and bloodthirsty. Do you believe it is possible, by selective breeding, to develop an aggressive rabbit and a docile timid weasel? How would you proceed to attempt to do so? In planning your project, how many years would you estimate it might require?

21.10. As an agricultural scientist you are interested in developing new domesticated species that will more or less fill in gaps among the existing array of domesticated species. What new types of domesticated species do you think are most needed at the present time?

21.11 From your answer to question 21.10, choose one species and develop a breeding program that could lead to the creation of a new species useful to man.

21.12. What hazards are inherent in agriculture based on one or a few crops?

21.13. Distinguish between artificial selection and natural selection. Does natural selection cease to function in domesticated species of plants and animals?

21.14. In addition to artificial selection, what techniques are available to the breeder? Which of all the methods available have been most widely used and most successful?

21.15. How would you select for egg production in roosters? In other words, how would you determine which rooster to use as a sire who would most consistently raise the egg production of his daughters?

CHAPTER TWENTY-TWO

Genetics and the Origin of Species

22.1. The Evolutionary Process

In the process of evolution, an ancestral population with one set of hereditary characteristics (or gene frequencies) gives rise to a descendent population with a different set of hereditary characteristics (or gene frequencies). This form of evolution (species A → species B → species C → species D) is sometimes referred to as *phyletic evolution*. However, it is a gradual, continuous process rather than discrete as implied above, because it is impossible to determine a point in time that separates, say, species A from species B. A phylogenetic sequence of distinct species will be observed only if there are gaps in the fossil record.

A second, more complex, mode of evolution is known as *speciation*, in the limited sense. In this case a single ancestral species A gives rise to two contemporary species, B and C, that differ from the ancestral species A and also differ from, and are reproductively isolated from, each other. Any theory of the mechanism of evolution must account for both phyletic evolution and speciation.

Analysis of wild populations of plants and animals has revealed that they normally contain a rather surprising amount of genetic variability. All such variation is ultimately traceable to an origin by mutation. These mutations persist in populations for various reasons. If recessive, a mutation may be present in a population in the heterozygous state even though deleterious. It will be eliminated by natural selection only when it is expressed in the homozygous condition. Whole-chromosome mutation rates for new deleterious mutations are of the order of 0.010 to 0.015 per chromosome per generation. Thus, mutation constantly feeds new deleterious mutations into the population, and eventually an equilibrium will be established between mutation and selection. Because natural selection has constantly winnowed and sifted out and preserved the more favorable alleles in the population, most new mutations will be deleterious compared with the existing alleles. Therefore, some of the variation observed in wild populations is due simply to the effect of mutation pressure in generating deleterious mutants that have not yet been eliminated by natural selection, in other words, to a balance between mutation and selection.

An occasional rare mutant will confer greater fitness than the existing alleles and will be favored by natural selection. During the time the new allele is replacing the old, both types will be fairly frequent in the population, and *transient polymorphism* is said to exist.

In some cases, more than one allele will be maintained in the population by natural selection, a situation referred to as *balanced polymorphism*. Various types of selection pressure may maintain balanced polymorphism. Commonly cited is overdominance, where the heterozygote is fitter than either of the homozygotes $(A_1A_1 < A_1A_2 > A_2A_2)$. Frequency-dependent selection, in which the less frequent allele is always favored by selection, will also result in a balanced equilibrium of gene frequencies. Seasonal selection, with selection favoring one allele at one season and another in a different season, will also tend to maintain more than one allele in a population. If one allele is favored in males and another in females, the opposing selection pressures will tend to counterbalance one another. Self-sterility alleles and other self-incompatibility systems in hermaphroditic species also lead to balanced polymorphism. Negative assortative mating and density-dependent selection are other means by which, in theory at least, balanced polymorphism can be maintained.

Some idea of the amount of concealed genetic variation may be obtained from Table 22.1. There it can be seen that the frequency of chromosomes with lethal or semilethal effects in wild *Drosophila* populations ranges from 10 to 33%. This means that up to one-third of the chromosomes in

Table 22.1   Chromosomes with Lethal or Semilethal Effects in Three *Drosophila* Species

Species	Percent L and SL		
	II	III	IV
D. pseudoobscura	33	25	26
D. persimilis	26	23	28
D. prosaltans	33	10	—

From Dobzhansky, T., and B. Spassky, 1953. "Genetics of Natural Populations. XXI. Concealed Variability in Two Sympatric Species of *Drosophila*," *Genetics* 38:471–484, p. 476, Table 4; and Dobzhansky, T., and B. Spassky, 1954. "Genetics of Natural Populations. XXII. A Comparison of the Concealed Variability in *Drosophila prosaltans* with that of Other Species," *Genetics* 39:472–487, p. 478, Tables 5 and 6.

these natural populations, when made homozygous, reduced the viability of the homozygote to less than half the expected value. In addition, among the more nearly normal chromosomes, many more were subvital and others caused sterility or reduced the rate of development. Extrapolation from such findings as these leads to the realization that no diploid individual, whether fly or human, is apt to be entirely free of harmful mutations. The demonstration of wild-type iso-alleles, genes with similar but not identical effects, indicates the existence of an additional pool of variability consisting of genes that are not markedly detrimental in contrast to those cited above.

The presence of this array of genetic variability in natural populations means that the old concept of "the wild type," individuals homozygous for the normal or wild-type allele at every gene locus, is misleading and erroneous. Although individuals from natural populations may appear similar in phenotype, each possesses a unique and distinctive genotype. Furthermore, Mendelian segregation and recombination are constantly shuffling and reshuffling the genes in the population and dealing them out as new genotypes. In sexually reproducing diploid species, the possible gene combinations are so far in excess of those actually produced that each new individual is in essence a unique adaptive experiment being exposed to natural selection. The genotypes better adapted to the existing environmental and biotic conditions will survive and reproduce and leave a proportionately greater sample of their genes in the gene pool of the next generation than the less fit genotypes.

In a broad sense, the species may be considered as the fundamental unit of evolution. Species have been defined as "groups of actually or potentially interbreeding natural populations that are reproductively isolated from other such groups." The implication is that significant gene flow may occur within a species but not between species. In this sense, all members of a given species form one inclusive breeding population. In reality, the species population is subdivided into a number of breeding populations or *demes*, each isolated to some degree from the others and subjected to a unique combination of environmental conditions. Therefore, in a very real sense, the actual unit of evolution is the breeding population. Although the breeding population is composed of individuals, each of whom is a unique adaptive experiment, the individuals are not the units of evolution because they lack continuity due to death. The ongoing entity is the deme's pool of genes, which individuals hold only in temporary custody. Such a view of the biological world may seem alien at first and may require some re-orientation in thinking, but it is necessary to appreciate the implications of modern evolutionary theory.

Viewed in this light, each deme becomes a unique evolutionary experiment. Since no two sets of environmental and biotic conditions will ever be exactly alike, the selection pressures in different demes will differ, and hence the demes will tend to diverge from one another in adaptive response to these selection pressures. Furthermore, the mutations occurring in different populations are unlikely to be the same. Since mutation at a particular locus is a rare and random process, a somewhat different spectrum of mutations will accumulate in different finite populations. Added to this is the fact that any new mutation has a high probability of being lost by chance. Thus, the mutations persisting in different populations are unlikely to be the same. Therefore mutation, like natural selection, can be expected to cause genetic divergence between demes. Finally, if the breeding populations are small, random genetic drift will lead toward the random fixation or loss of genes in the populations. The net result of the effects of selection, mutation, and drift in different demes is to cause them to diverge from one another genetically.

This divergence may lead in time to the patterns of variation within a species that are recognized as *ecotypes*, or as *clines*, or as *geographical races* or *subspecies*. Ecotypes are especially well adapted to some particular segment of the total environment occupied by a species. In widely distributed plant species, for instance, desert or alpine ecotypes may evolve in those portions of the species' range (Fig. 22.1). Where the environment changes gradually, clines or continuous character gradients may be observed that reflect the adaptation of the species' populations to these gradual changes. If some isolation between populations exists, different geographic races or subspecies may differentiate, and the species is then said to be

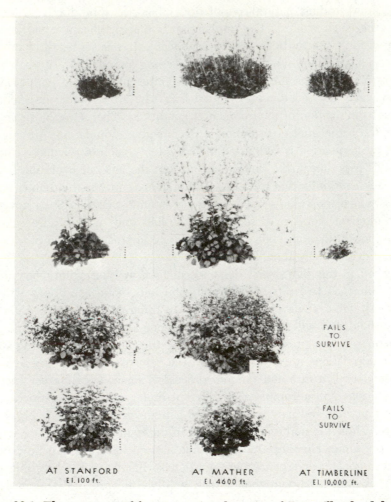

Fig. 22.1. The responses of four ecotypic subspecies of *Potentilla glandulosa* to three contrasting environments at Stanford (left), Mather (center), and Timberline (right). The four subspecies are *nevadensis* from 10,000 ft (top row), *Hanseni* from 4600 ft (second row from top), *reflexa* from 2500 ft (third row from top), and *typica* from 900 ft (bottom row). (Clausen, J., 1951. *Stages in the Evolution of Plant Species.* Cornell University Press, Ithaca, N.Y., p. 46, Fig. 17.)

polytypic. The taxonomic value of the subspecies concept may be debatable, but the existence of species composed of geographically distinctive populations can be observed in many species of plants and animals, including man.

While mutation, selection, and genetic drift will, in general, tend to cause different populations of the same species to diverge from one another, the fourth evolutionary factor, migration or gene flow, tends to prevent such divergence. The gene flow resulting from the migration of individuals between demes leads to increased similarity in gene frequencies among different populations of the same species. Gene flow is thus a factor in maintaining the coherence or identity of the entire species population as a single entity.

All forms of hybridization may be treated under the rubric of migration or gene flow. The relatively wide crosses that sometimes occur, especially in plants, may play a significant evolutionary role. Occasional hybridization between related species may permit the influx or introgression of genes from the gene pool of one species into that of another. The resulting burst of variability may give rise to new adaptive types better able to exploit new ecological niches than either of the parental species. This phenomenon is known as *introgressive hybridization*. Even wider crosses may succeed if followed by chromosome doubling to form an allopolyploid. The evolutionary importance of allopolyploidy in plants is evidenced by the fact that at least one-third of the existing species of angiosperms appear to be allopolyploid in origin.

If continued over a long enough period of time, the evolutionary processes outlined above will lead to phyletic evolution. As a population continues to adapt to changing conditions or to improve its adaptation to relatively stable conditions, in time the population will have changed sufficiently to be considered a new species, taxonomically distinct from the original species from which it descended. It should be reemphasized that no discontinuity is implied. The same mechanism that leads to the evolution of ecotypes or geographic races may result, given longer periods of time, in the formation of groups recognized as species, or genera, of families, or orders, or even higher taxonomic categories. The only known discontinuous evolutionary process, by which a distinctly different species originates at a single step, is allopolyploidy. Aside from this, the various theories of evolution involving "saltation," "macroevolution," or "systemic mutations" have little evidence in their support.

The process of speciation is more complex because we must account not only for the divergence between populations of the same species but also for the origin of the reproductive isolation between them. We have already considered the factors that lead to differentiation among populations of the same species. However, new species are recognized only when populations that formerly belonged to the same species have diverged to the point where they are reproductively isolated from one another. At that point, significant gene flow between them ceases, and they henceforth follow independent evolutionary paths. Thus, migration is the force that tends to counteract the forces leading to speciation, and reproductive isolation is the criterion by which species status is determined. Therefore the origin of reproductive isolation becomes a question of prime importance.

The evidence indicates that for genetic divergence between populations of the same species to occur, there must be a minimum of gene flow in the early stages and hence a maximum amount of isolation between them. Since at this time the populations are parts of the same species and not reproductively isolated, the isolation must be geographical, topographical, or physical in the sense that members of the different populations do not meet one another as potential mates. Spatial isolation differs in kind from reproductive isolation, because the latter is due to genotypic differences.

Reproductive isolating mechanisms may be classified according to the timing and nature of their effects. The *prezygotic* isolating mechanisms take effect prior to fertilization and thus prevent the formation of hybrid zygotes. For example, *ecological* isolation results from differences in habitat preferences or requirements. Thus, populations living in the same general area (sympatric populations) and otherwise likely to interbreed are kept apart because they occupy different types of habitat. Sympatric populations will remain distinct due to *seasonal* isolation if their sexual cycles do not coincide. If the reproductive season in one group does not overlap that of the other, no interbreeding will occur even though otherwise such matings might readily occur and result in fully viable, fertile progeny. *Sexual* or *ethological* isolation is due to differences in behavior that inhibit the completion of courtship and mating. Morphological differences in the reproductive organs of plants or animals may lead to *mechanical* isolation between different related groups. *Physiological* isolation may prevent fertilization if, for example, sperm are adversely affected in the reproductive tract of an alien female, or if pollen fail to grow properly on a foreign style. By preventing fertilization, the prezygotic isolating mechanisms tend to prevent the wastage of gametes on the less well adapted hybrids that otherwise would be produced.

Even if fertilization occurs between members of different groups, *postzygotic* isolating mechanisms may intervene. Often the zygotes fail to complete normal development so that *hybrid inviability* is a barrier to gene flow. When vigorous but sterile hybrids develop, *hybrid sterility*, due to abnormal gonads or aberrant meiosis, is an effective isolating mechanism. In some cases, vigorous, fertile F_1 hybrids are produced, but the F_2 or backcross generations have reduced vigor or fertility or both, a phenomenon known as *hybrid breakdown* and apparently due to disharmonious gene combinations. Thus, even if hybrid zygotes are formed, a barrier to gene flow will exist if they are not normal, fertile individuals.

Two theories of the origin of reproductive isolation have been proposed, which are complementary rather than alternative. Muller suggested that reproductive isolation arises as an incidental by-product of the genetic divergence that occurs in spatially isolated populations. Since, when the populations first become isolated, their gene pools are similar and they are not reproductively isolated, physical isolation of the populations in the early stages is essential if the evolving divergence is not to be swamped by hybridization and gene flow between them. However, with a long enough period of physical isolation and sufficient genetic divergence, the chances are that when the populations again come in contact, the reproductive patterns will have diverged to the point where some degree of incompatibility will have crept in.

Dobzhansky has postulated that reproductive isolation may actually be reinforced by natural selection. If the hybrids resulting from mating between members of the diverging populations are not well adapted to existing conditions, the gametes invested in them are wasted, and any genes tending to inhibit or prevent hybridization will be favored by natural selection. In this way sexual isolation between the populations will be strengthened. Once reproductive isolation becomes established, two species will exist where before there was only one.

The evidence seems clear that, apart from polyploidy, evolution is a gradual process. Nevertheless, many gaps or discontinuities appear in the fossil record. The reasons for the gaps are related to the nature of the fossil record rather than to the evolutionary process itself. The fossil record, by its very nature, is incomplete. Ordinarily only the hard parts of organisms form fossils, and then only if they are buried quickly so that rapid decay does not occur. Thus, whole groups may be missing from the record because they lack hard parts or else live in a habitat where quick burial is very unlikely. The fossil record is only a small sample of all of the organisms that have ever lived. The gaps are most frustrating because they seem to exist so often at the crucial points in a phylogeny where a new higher taxonomic group has emerged. However, the conditions favoring rapid evolution occur in relatively small, relatively isolated breeding populations subjected to strong new selection pressures. Therefore, only one population from a number of existing populations may pursue the evolutionary path that leads to a new group, and the transition may require, geologically speaking, a relatively short period of time. The numbers of individuals in this transitional group compared to the numbers in closely related groups before, during, and after the change may be relatively quite small, so that their chances of being fossilized and, if fossilized, of being discovered are rather low. The gaps in the fossil record, therefore, do not necessarily require that a different mechanism of evolution be postulated to account for the origin of higher categories. Furthermore, even though the origin of a new order or family may represent a major adaptive shift (e.g., browsing to grazing, or plantigrade to digitigrade locomotion), there is no reason to believe that during the transition the population is poorly adapted. They did not starve during the switch from browsing to grazing, nor hobble around until digitigrade locomotion had been perfected. The evolving unit is the breeding population, and it must remain reasonably well adapted to its environment no matter what its evolutionary course, or it will become extinct.

22.2. Rates of Evolution

Evolutionary change is often thought to require millions of years, and rates of evolution are calculated in terms of geological time. However, we have already noted the dramatic changes that have occurred in domesticated species in less than 10,000 years. These changes have been mediated by artificial selection, and may be regarded as artificial evolution, but they demonstrate the great potential for change that exists in wild species. It is entirely reasonable to suppose that, under appropriate conditions giving rise to strong new selection pressures, comparable changes could occur without the intervention of man. Indeed, the evolution of insects resistant to insecticides occurred in less than a decade, and the evolution of industrial melanism in nearly 100 different species of moths required less than a century. These examples require relatively simple genetic changes, but a landlocked population of the marine seal, *Phoca vitulina*, cut off in a freshwater lake in northern Canada about 5000 years ago, is now recognized as a distinct subspecies. Hence, the evolution of this degree of difference has required only about 1000 generations.

The evolutionary sequence leading to modern man has also been rather rapid. The numerous changes involved in the evolution from the genus *Australopithecus*, creatures with cranial capacities ranging from 300 to 600 cc, through *Homo erectus* to *Homo sapiens*, with a cranial capacity averaging 1350 cc, have occurred in about 2 million years. The assumption of an average generation length of 20 years leads to a crude estimate of 100,000 generations for the evolution of *Homo sapiens* from *Australopithecus*. The fossil record of modern man, *Homo sapiens*, extends back only some

50,000 years or about 2500 generations. Certainly there seems little reason to believe that human evolution has miraculously ceased, or that man has necessarily reached some evolutionary pinnacle.

The fossil record of the horse family, the Equidae, is one of the most complete sequences available. In this record eight successive genera have been recognized as forming a phylogenetic sequence lasting some 60

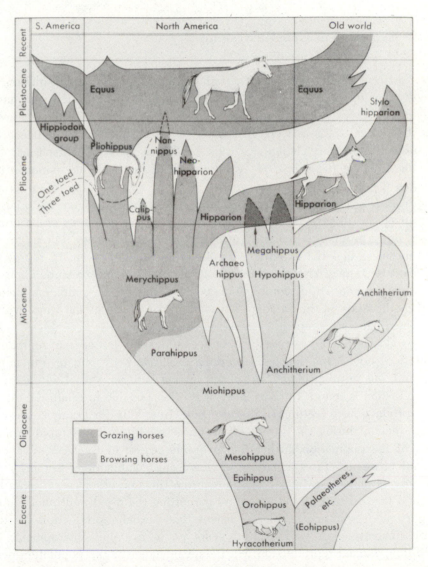

Fig. 22.2. A simplified phylogeny of the horse family. (Simpson, G. G., 1951. *Horses.* Oxford University Press, New York, p. 148, Fig. 10.)

million years. Simple division indicates that evolution in the horses has proceeded at the rate of 7.5 million years per genus in the line leading from the tiny *Hyracotherium* to *Equus* (Fig. 22.2). Evolutionary rates in some of the invertebrate groups appear to be somewhat slower, but in either case the available time spans and the observed evolutionary changes are compatible with the concept of gradual rather than cataclysmic evolution.

22.3. Evolution of Genetic Systems

The fossil record is a record of certain anatomical traits, and evolution is usually thought of primarily in terms of the morphological changes that have occurred. However, the phenotype consists of physiological, biochemical, and behavioral traits as well, all subject to natural selection and hence to adaptive evolutionary change. Even the hereditary mechanism itself may be modified during the course of evolution. All of the available evidence indicates the universality of the genetic code in a wide range of different organisms, but the genetic systems into which the hereditary material is organized are diverse. The DNA may have a simple structure or be organized into a group of structurally more complex chromosomes. The origin of chromosomes and the evolution of mitosis and meiosis seem to represent a means for regularizing the duplication, distribution, and recombination of the genetic material. Recombination occurs even in the simpler systems, and modification in the amount of recombination is one of the ways evolution has influenced genetic systems.

Sexual reproduction permits genetic recombination, but asexual reproduction does not. Some species reproduce exclusively by one or the other method, but in others some sort of alternation between sexual and asexual reproduction is found. In these cases it is often possible to relate the mode of reproduction to features of the life cycle. In aphids, for example, asexual reproduction during the spring and summer months permits rapid expansion of the population during the favorable season of the year. The fall sexual phase generates a variety of new gene combinations just prior to the advent of harsh winter conditions. From this and other cases it can be argued that asexual reproduction, by permitting the rapid multiplication of a particular genotype, is more efficient than sexual reproduction for the rapid increase of a well-adapted genotype in a stable environment. In addition, asexual reproduction will be advantageous if the population density is so low that chances of meeting a suitable mate are very slight. Thus, forms with asexual reproduction are well adapted to invade and

colonize previously unoccupied areas. An asexually reproducing population is dependent on the chance occurrence of rare favorable mutations to
adapt to changing environmental conditions. Microorganisms, with large
numbers and high reproductive rates, seem able to adapt to changing
conditions in this way without great difficulty. Hence, asexual reproduction
is not necessarily to be regarded as simply the mode of reproduction among
primitive forms that have not succeeded in evolving a sexual form of reproduction. It undoubtedly has adaptive significance. In many species of
higher plants and animals, various forms of apomixis and parthenogenesis
have clearly evolved from sexual forms, and it has been argued that even
in the lower forms asexual reproduction may also have been secondarily
derived from some form of sexual reproduction involving genetic recombination. If this is the case, asexual reproduction must clearly have adaptive advantages over sexual reproduction in some cases.

The prevalence of sexual reproduction among the more complex or
"higher" species, however, suggests that it has played an important role in
the evolution of the higher groups of plants and animals. The essence of
ordinary sexual reproduction is genetic recombination, which generates an
array of unique genotypes each generation. These genotypes are tested by
natural selection, and some are undoubtedly found wanting, but the
chances that others are adaptively superior to previously existing genotypes
are greater than would be true for asexual reproduction. Thus, it is perhaps
not surprising that evolutionary changes have been greatest in sexually reproducing lines.

In sexually reproducing species, the sexes may be separate, or both kinds
of gametes may be produced by a single individual known as an hermaphrodite. Hermaphroditism is common in plants and in the lower animals and
hence appears to be the ancestral condition in multicellular forms from
which the separation of the sexes has been derived.

Another fundamental difference between genetic systems is that between
haploidy and diploidy. In the simplest eukaryotes, the flagellates and green
algae, the haploid phase is dominant, for the zygote formed at fertilization immediately undergoes meiosis, which restores the haploid condition.
Haploidy thus appears to be the ancestral state, but the evolution of both
higher plants and animals is marked by a trend toward the predominance of
the diploid phase in the life cycle. In other words, after fertilization a number of mitotic divisions of the diploid zygote nucleus occur before meiosis
takes place. In this way the diploid phase of the life cycle is extended
until in some groups meiosis immediately precedes fertilization rather than
immediately following it, and the organisms are diploid through almost all
of the life cycle.

The usual explanation for this trend is that diploidy permits the retention of more genetic variability in the population than haploidy. In haploids the whole genotype is expressed and therefore exposed to selection. Any genes unfavorable at the moment will be eliminated, and the variability in the population will be minimal. In diploids, recessive genes may be carried along unexpressed in the heterozygous condition. Each generation a portion of this variability will be released by genetic recombination, and the population retains some flexibility to respond to changing conditions.

The amount of genetic recombination itself may be under the control of natural selection. High chromosome numbers and high chiasma frequencies increase recombination. In hermaphrodites, self-fertilization may be inhibited by systems of self-sterility alleles or by other means so that recombination comparable to that achieved by the separation of the sexes is attained. Permanent hybridity, due to balanced lethals, inversion or translocation heterozygosity, or allopolyploidy, stabilizes gene combinations that often confer heterosis.

Conversely, linkage of genes on the same chromosome and a decrease in chromosome number both tend to restrict recombination, as does a reduction in chiasma frequency. The integrity of gene complexes or blocks of genes may be preserved by such means as positive interference, inversion or translocation heterozygosity, or by a complete absence of crossing over as in *Drosophila* males. Self-fertilization in hermaphrodites is a different avenue for limiting genetic recombination, and asexual reproduction is the most effective method of all. It is worth noting that if one of the mechanisms limiting recombination (self-fertilization, asexual reproduction, or limitations on crossing over) becomes established in a group, the other two will probably be absent. The implication is that selection acts on the genetic system to reduce recombination, and when one mechanism to this end is established, the other options are no longer necessary.

22.4. The Evolution of Genes

The universality of the genetic code, the fact that the nucleotide triplets code for the same amino acids in all organisms, confers a certain stability on the evolutionary process. The meaning of the codons does not change from species to species. Instead, at the molecular level, evolution occurs when the replacement of one codon by another leads to the substitution of a different amino acid in the peptide being synthesized, and this substitution, in turn, has an effect on fitness.

That such a change may affect fitness is illustrated by sickle cell anemia in man. The human hemoglobin molecule is composed of four heme groups, each attached to a separate polypeptide chain. There are two types of chains, alpha (α) with 141 amino acids of known sequence, and beta (β) with 146. Normal adult hemoglobin A is symbolized as $\alpha_2^A\beta_2^A$. Persons with the severe and often fatal disease, sickle cell anemia, were found to have hemoglobin S, in which both beta chains were abnormal ($\alpha_2^A\beta_2^S$). Hemoglobin S differs from hemoglobin A only in the substitution of valine for glutamic acid at position 6 in the beta chains. The other 145 amino acids are the same. The messenger RNA coding for glutamic acid is GAA or GAG, that for valine is GUA or GUG. Thus, a single base substitution of U for A at the second position in the mRNA codon (preceded, of course, by a corresponding change in the DNA), causes a drastic reduction in fitness, and severe disease results from a seemingly trivial change affecting just 2 of the 574 amino acids in the hemoglobin molecule. Moreover, this mutation is not merely a detrimental change, but has evolutionary significance because of the selective advantage of the heterozygotes over both homozygotes in malarial regions.

Therefore, changes in the DNA code that lead to changes in the polypeptides being synthesized are the basis for evolutionary change. The changes in the code may involve simple base substitutions, but they may also involve changes in the number and sequence of the bases and thus may include duplications and deletions of DNA segments. These changes or mutations are essentially random, at least in relation to fitness, but only those mutations that permit the production of an array of proteins that form a viable, well-adapted organism will pass through the screen of natural selection and be perpetuated. A haploid mammalian cell has been estimated to contain about 3 billion bases. The theoretically possible different sequences in which the four bases (A, C, G, and T) could be arranged is therefore equal to 4 raised to about the three billionth power, a number much greater than the estimated number of elementary particles in the universe. Obviously, in a given species, most of these sequences will never be formed. Among those that are formed, most will undoubtedly be deleterious compared to the prevalent types, and only rarely will a favorable new DNA sequence occur. Moreover, minimal perturbations in an existing code such as a single base substitution would seem generally better able to support the development of a viable organism than more drastic changes in the number or sequence of the bases, except perhaps for duplications. If true, this conclusion helps to explain why evolution is generally gradual, and also suggests the possibility that phylogeny can be studied at the biochemical level by determining the degree of homology between the DNA or protein molecules of different species.

The spectacular progress in molecular biology has made it possible to study evolution at the molecular level. Some of the earliest attempts were directed toward comparisons of the base composition of the DNA in a number of diverse species. Measured in terms of guanine plus cytosine (G + C), the range observed lay between about 20 and 75%, with less variation in higher plants and animals than in lower forms. More recently, DNA hybridization experiments have given a more direct measure of the homologies involved. The essence of the technique is the separation of the two complementary strands of a DNA double helix and the mixing of this single-stranded DNA with either single-stranded DNA or with RNA from a different species. The extent of the reaction leading to the formation of double-stranded hybrid molecules indicates the proportion of sequences that are held in common by the two species. Furthermore, the exactness of the fit or degree of similarity in the sequences can be estimated from melting-point determinations. With this technique, those species judged most similar on other grounds tend to show higher levels of renaturation than less similar species, and the DNA hybridization technique is now used to draw phylogenetic inferences about the relationship between species and the rates of divergence in the genes during evolution.

Ultimately it may be possible to learn the entire DNA sequence for a given cell. This sequence could then be compared with that in another individual or another species to determine the fraction of the DNA complement common to both, and the nature of the differences between them. The DNA sequence could also be compared with the RNA sequences in the cell to establish the source of the various RNA molecules transcribed during the life of the cell. It would also be possible to predict the polypeptide sequences that could be produced by this cell. As yet determinations of this sort have not been possible, but the base sequences have been determined in simple nucleotides such as transfer RNA.

Another biochemical approach to evolution has been through comparative studies of proteins such as hemoglobin and cytochrome *c*. These studies became possible after the techniques were developed by which the amino acid sequences in these proteins could be determined. In this way the effects of mutations on human hemoglobin as well as the differences between human hemoglobin and that of other mammals could be pinpointed. In most cases the abnormal human hemoglobins are attributable to the substitution of a single amino acid in the chain, resulting from a single base change in the codon. The hemoglobins in different species, however, differed by several amino acid substitutions, ranging from one between human and orangutan hemoglobin to 17 between human and horse hemoglobin.

Interestingly enough, myoglobin, found in muscle, and the α, β, γ, δ and ϵ chains of human hemoglobin appear to be related molecules, which originated following duplication of the gene controlling synthesis of this type

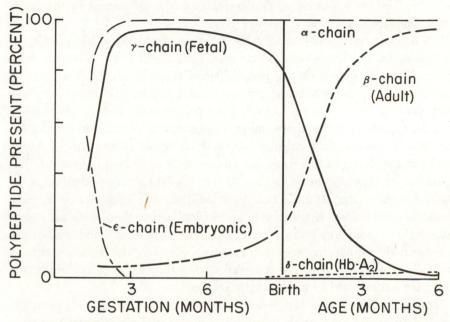

Fig. 22.3. Changes in human hemoglobins during development. (Huehns, E. R., N. Dance, G. H. Beaven, F. Hecht, and A. G. Motulsky, 1964. "Human Embryonic Hemoglobins," *Cold Spring Harbor Symp. Quant. Biol.* 29:327–331, p. 329, Fig. 3.)

of polypeptide and divergence in the function of the duplicated genes due to subsequent mutations. Myoglobin has a molecular weight of about 17,000 and is a monomer of the globin-heme configuration, having just one heme group associated with one polypeptide chain. The hemoglobin of the lamprey, a primitive vertebrate, is also a monomer with a molecular weight of 17,000. However, as noted above, normal adult hemoglobin A is a tetramer ($\alpha_2^A\beta_2^A$). Fetal hemoglobin or hemoglobin F, with a higher oxygen affinity than hemoglobin A, is produced in the fetus, but its production drops after birth to almost nil by 6 months (Fig. 22.3). At the same time, A, which is low in the fetus, increases to adult levels. Hemoglobin F is also a tetramer of formula $\alpha_2^A\gamma_2^F$. Both A and F have a dimer of the α polypeptide, but it appears that the gene for the γ (gamma) chain is shut down at the time the gene for the β chain is being turned on. The δ (delta) chain is a part of hemoglobin A_2 ($\alpha_2^A\delta_2^A$), which is a normal component of adult human blood, but the ratio of A to A_2 is about 25 to 1. The ϵ (epsilon) chain has been found only in the hemoglobin of the early human embryo, probably as $\alpha_2\epsilon_2$. The picture that is beginning to emerge, therefore, is one that suggests that evolution of the hemoglobin molecule has proceeded by means of gene duplication followed by gene divergence due to base substitution. The net result of this differentiation is a refinement in the functional adaptive roles these molecules play.

While some comparative studies of different species have been made for the hemoglobins, much more information of this sort is available for cytochrome c, used in respiratory metabolism. Cytochrome c from species as diverse as man, baker's yeast, chickens, rattlesnakes, and tuna fish have been studied. Horse cytochrome c has 104 amino acids, and the cytochrome c of many other species shows many similarities. A significant observation was that at some sites a considerable flexibility exists in the type of amino acid present, but at other sites the amino acids are always identical. The sequence of 11 amino acids at sites 70 through 80 has been invariant in all the species thus far studied, and there are probably about 20 or so additional invariant sites. A reasonable conclusion is that these sites are essential to the normal function of the cytochrome c molecule, and amino acid substitutions cannot be tolerated at these sites. At the variable sites, a number of substitutions have occurred during the course of evolution, and these have been used in phylogenetic studies.

The field known as biochemical systematics has centered primarily around phenolic compounds, alkaloids, and other similar "secondary compounds," secondary in the sense that they seem not to be involved in the basic metabolism of the cells. For this reason, they appear to be better able to tolerate variations in structure than compounds more essential to the life of the cell. A great deal of this work has been done with plants, and it marks, in a sense, a continuation and extension of some of the earliest work done by Scott-Moncrief, Lawrence, and others on flower pigments as a means of studying genetics on a biochemical level.

One of the hazards of biochemical studies, as of morphological studies, is the problem of homology versus analogy. A similarity in DNA or protein structure may be indicative of homology and thus have phylogenetic significance, but since the genetic code is universal, such similarities may also be related to the functional requirements of the organism and convey no useful phylogenetic information whatever. For example, the human hemoglobin β chain is more like that of the pig than that of the lemur, another primate, and human cytochrome c is more like that of the kangaroo, a marsupial, than that of another placental mammal, the horse. Taken at face value, this evidence would suggest that man is closer to the pigs than to the other primates and to the pouched mammals than to the placentals. While, on other grounds, the first conclusion might not be so difficult to accept, the second does seem pretty far-fetched. Thus, unfortunately, even at the molecular level phylogenetic studies may be complicated by the possibility of parallel or convergent evolution or by chemical analogy rather than homology.

It seems reasonably safe to say that the DNA code in its present form has existed at least since the time of origin of the metazoa, and is thus more than half a billion years old. The origin of the code is at present a matter of speculation. The two major kinds of macromolecules in cells are proteins and nucleic acids. The nucleic acids store genetic information in their nucleotide sequences and transmit it to daughter cells by replication. Moreover, after a mutation the nucleic acids catalyze their self-duplication in their new mutated form, a fact crucial to biological evolution. The coded information is translated into proteins, which carry on the functions of the cells. A mutation in the nucleic acid code leads to the synthesis of a modified protein, whose effects are subjected to natural selection. The indefinitely cumulative nature of this process accounts for the seemingly almost limitless potential for evolution by natural selection. The transfer of information now is from DNA to RNA to protein. The fact that reverse translation does not occur from protein to nucleic acid suggests that even the most primitive code must have been some form of nucleic acid and that earlier speculations about the origin of life in terms of protein alone were in error. In other words, it seems unlikely that the first form of life was some sort of metabolic system capable of carrying on assimilation, growth, and division and that only later did the genetic material evolve within the protoplasm or become associated with it. The three essential properties for life and evolution are (1) self-duplication, (2) mutation and the ability of the mutant form to duplicate itself, and (3) the ability to organize and utilize nonliving materials into structures and functions that support the continued existence of life. The nucleic acids alone seem to have this property; the proteins form an accessory system that supports and maintains life.

The nature of the primitive genetic material is conjectural. In some viruses, RNA rather than DNA serves as the hereditary material. One might therefore postulate that RNA, which can be self-replicating and also mediates protein synthesis, is a logical candidate for the primitive genetic material. Alternatively, one could suggest that the RNA viruses represent a degeneration or simplification of the DNA viruses. One speculation is that the complex triplet coding system used today was preceded by a simplified triplet code in which only the first two bases per triplet specified the amino acids and the third base could be, interchangeably, any one of the four bases used in the code. Such a code, essentially a doublet code, would thus have only 16 meaningful permutations and at best could code for only 16 amino acids. The thread of doublets in the existing code supports this possibility, for the code is degenerate, more than one triplet coding for a single amino acid. In such triplets the first two bases are the same, but the third position may be occupied by two or more alternatives. According to this scheme, primitive proteins were composed of fewer different amino acids than existing proteins. Mutations in the code could be tolerated because the functions of the proteins were not greatly modified by changes in their amino acid sequences. The synonyms among the codons minimized the possibility of unreadable triplets. Modification of the meaning of the code permitted coding for additional new amino acids, which would be incorporated into the system if favored by natural selection. This change would be simplest if the code for each new amino acid involved triplets used for a functionally similar existing primitive amino acid. This hypothesis would then help to account for the fact that similar codons now code for similar amino acids. As proteins and cells became more complex, changes in the meaning of the code could no longer be tolerated, and the code achieved its present fixed, universal meaning. It has also been argued that the present form of the code minimizes the adverse effects of single base substitutions. The code has evolved toward greater stability so that there are no unreadable triplets, and similar codons, differing by a single base change, code for similar amino acids. In this way, the effects of mutation are minimized, but it also appears that selection has favored a code that gives maximum accuracy of translation.

One final issue of considerable interest at present is the question of whether there is any inherent, inevitable relationship between a given amino acid and its codon. In other words, do an amino acid and its codon "recognize" one another or interact by pairing with one another? A priori, one might expect this to be the case, but the case is yet to be made. The evolution of the existing complex translation mechanism, involving DNA, messenger RNA, activating enzymes, transfer RNA, and the ribosomes, is only now becoming a subject of speculation. It should be obvious from our discussion that even though the broad principles governing biological evolution are beginning to be understood, we still have much to learn.

Additional Reading

See also references at the end of Chapter 18.

Alston, R. E., 1967. "Biochemical Systematics," *Evol. Biol.* 1:197–305.
Anderson, E., 1949. *Introgressive Hybridization.* Wiley, New York.
Anfinsen, C. B., 1959. *The Molecular Basis of Evolution.* Wiley, New York.
Blair, W. F., ed., 1961. *Vertebrate Speciation.* University of Texas Press, Austin.

Bleibtreu, H. K., ed., 1969. *Evolutionary Anthropology*. Allyn and Bacon, Boston.

Blum, H. F., 1969. *Time's Arrow and Evolution*, 3rd ed. Princeton University Press, Princeton, N.J.

Bryson, V., and M. J. Vogel, eds., 1965. *Evolving Genes and Proteins*. Academic, New York.

Calvin, M., 1969. *Chemical Evolution*. Oxford University Press, London.

Clark, W. E. L., 1957. *History of the Primates*. Phoenix, Chicago.

Clark, W. E. L., 1967. *Man-Apes or Ape-Men?* Holt, Rinehart and Winston, New York.

Clausen, J., 1951. *Stages in the Evolution of Plant Species*. Cornell University Press, Ithaca, N.Y.

Clausen, J., and W. M. Hiesey, 1958. "Experimental Studies on the Nature of Species. IV. Genetic Structure of Ecological Races." Carnegie Inst. Wash. Publ. No. 615, pp. 1–312.

Darlington, C. D., 1958. *The Evolution of Genetic Systems*, 2nd ed. Cambridge University Press, London.

De Ley, J., 1968. "Molecular Biology and Bacterial Phylogeny," *Evol. Biol.* 2:103–156.

Dobzhansky, T., 1970. *Genetics of the Evolutionary Process*. Columbia University Press, New York.

Fitch, W. M., and E. Margoliash, 1970. "The Usefulness of Amino Acid and Nucleotide Sequences in Evolutionary Studies," *Evol. Biol.* 4:67–109.

Florkin, M., 1949. *Biochemical Evolution*. Academic, New York.

Fox, S. W., ed., 1965. *The Origins of Prebiological Systems*. Academic, New York.

Howells, W., 1967. *Mankind in the Making*, rev. ed. Doubleday, Garden City, N.Y.

Ingram, V. W., 1963. *The Hemoglobins in Genetics and Evolution*. Columbia University Press, New York.

Jepsen, G. L., E. Mayr, and G. G. Simpson, eds., 1949. *Genetics, Paleontology, and Evolution*. Princeton University Press, Princeton, N.J.

Jukes, T. H., 1966. *Molecules and Evolution*. Columbia University Press, New York.

Korn, N., and F. Thompson, eds., 1967. *Human Evolution. Readings in Physical Anthropology*. Holt, Rinehart and Winston, New York.

Kraus, B. S., 1964. *The Basis of Human Evolution*. Harper & Row, New York.

Lasker, G. W., 1961. *The Evolution of Man*. Holt, Rinehart and Winston, New York.

Leone, C. D., ed., 1964. *Taxonomic Biochemistry and Serology*. Ronald, New York.

McCarthy, B. J., 1966. "The Evolution of Base Sequences in Polynucleotides," *Prog. Nucl. Acid Res.* 4:129–160.

McCarthy, B. J., and E. T. Bolton, 1963. "An Approach to the Measurement of Genetic Relatedness among Organisms," *Proc. Nat. Acad. Sci.* 54:1636–1641.

Mayr, E., ed., 1957. *The Species Problem*. American Association for the Advancement of Science, Washington, D.C.

Muller, H. J., 1967. "The Gene Material as the Initiator and the Organizing Basis of Life." In *Heritage from Mendel*, R. A. Brink, ed., University of Wisconsin Press, Madison.

Simpson, G. G., 1951. *Horses.* Oxford University Press, New York.
Simpson, G. G., 1964. "Organisms and Molecules in Evolution," *Science* **146**: 1535–1538.
Simpson, G. G., 1967. *The Meaning of Evolution*, rev. ed. Yale University Press, New Haven, Conn.
Stebbins, G. L., 1959. "The Role of Hybridization in Evolution," *Proc. Amer. Phil. Soc.* **103**:231–251.
Stebbins, G. L., 1960. "The Comparative Evolution of Genetic Systems." In *Evolution after Darwin*. Vol. I. *The Evolution of Life*, S. Tax, ed. University of Chicago Press, Chicago.
Washburn, S. L., and P. C. Jay, eds., 1968. *Perspectives on Human Evolution.* Holt, Rinehart & Winston, New York.
White, M. J. D., 1973. *Animal Cytology and Evolution*, 3rd ed. Cambridge University Press, London.
White, M. J. D., 1969. "Chromosomal Rearrangements and Speciation in Animals," *Ann. Rev. Genet.* **3**:75–98.
Young, L. B., ed., 1970. *Evolution of Man.* Oxford University Press, New York.

Questions

22.1. Which type of genetic polymorphism would you expect to be more common in natural populations, balanced polymorphism or transient polymorphism? Under what circumstances would you expect the frequencies to approach closest to one another?

22.2. What mechanisms may be responsible for balanced polymorphism?

22.3. What is the basic unit within which evolutionary changes occur? What is the basic evolutionary unit that gives rise to higher taxonomic groups such as families and orders?

22.4. What evolutionary factors give rise to ecotypes, clines, and geographic races?

22.5. Distinguish between polymorphic and polytypic. Are these conditions mutually exclusive within a species?

22.6. How may new genes be added to the gene pool of a deme?

22.7. What is the only known mechanism by which new species can originate at a single step? Why are they considered new species?

22.8. Which type of isolating mechanism is less costly to the species involved, prezygotic or postzygotic?

22.9. What is the statistical argument for the gaps in the fossil record just where it would be most desirable to find crucial "missing links"?

22.10. Evolutionary rates are sometimes estimated from a comparison of the number of amino acid substitutions in proteins such as hemoglobin or cytochrome c in different animal taxa. What flaws can you see in this procedure that may lead to erroneous conclusions?

22.11. What is the genetic system?

22.12. Compare the adaptive advantages of asexual versus sexual reproduction.

22.13. How may the amount of genetic recombination be influenced by natural selection?

22.14. Does the apparent universality of the genetic code necessarily have any evolutionary implications? Explain.

22.15. What advantages are these to studying evolution at the molecular level?

22.16. The analysis of chromosomes from natural populations of *Drosophila* has revealed that from 10% to more than 50% of the chromosomes carry recessive lethal and semilethal genes. If natural selection is constantly eliminating these deleterious genes whenever they are expressed, what reasonable explanations can you suggest for such a high frequency for these chromosomes?

22.17. In England and on the continent of Europe, dark populations of moths belonging to a number of different species and genera have appeared in industrial areas within the last 100 years or so. Genetic analyses have shown that virtually all of these cases of industrial melanism are controlled by single dominant genes. Since recessives and multiple factors giving dark coloration are also known in these species, how do you account for the rapid spread of the dominants rather than the recessives or the multiple-factor systems for melanism?

22.18. If introgression is occurring between two species, how will the phenotypes of populations appear in areas where the species are sympatric as compared to phenotypes of allopatric populations?

22.19. If there is no introgression but the species are in competition where they are sympatric, what will be the appearance of sympatric as compared with allopatric populations?

22.20. If Dobzhansky's theory of the origin of sexual isolation is operative, would you expect to find a greater degree of sexual isolation between allopatric populations or sympatric populations of two different species?

22.21. Suppose that you have just purchased an abandoned farm and find that the farm is occupied by some dusky rodents that you are unable to identify with the aid of available taxonomic keys or even the aid of a competent taxonomist. How would you proceed to determine whether you are dealing with a new species, a distinct subspecies, or just a color variant?

22.22. What is a "species"?

PART FOUR

GENETICS AND MAN

CHAPTER TWENTY-THREE

Populations, Pedigrees, and Twins: Methodology in Human Genetics

The pioneer in the study of human genetics was Sir Francis Galton (Fig. 23.1), who was Charles Darwin's first cousin. In the course of studying the inheritance of quantitative traits in man, Galton developed biometrical methods that laid the foundations of modern statistics. In addition he initiated twin studies, recognizing their importance in relation to the "nature versus nurture" question. In England William Bateson and A. E. Garrod, a physician, led in developing the Mendelian approach to human genetics at the beginning of this century, and R. A. Fisher and J. B. S. Haldane were dominant figures in the development of human genetics during the first half of the century. Consequently, England has been a major center of activity in human genetics, but the Scandanavian countries, Germany, and the United States have also been centers of research. In the

Fig. 23.1. Sir Francis Galton (1822–1911).

United States, C. B. Davenport was probably the most influential of the early workers in human genetics. It is of interest that both Galton and Davenport were interested in the genetic improvement of man, for which Galton coined the word, eugenics. Much of the early research in human genetics was justified by the hope that human qualities could thereby be bettered. The early human geneticists were overly enthusiastic and overly optimistic about the benefits to be derived from their work. The perversion of eugenic concepts by Hitler and others, and the early misplaced confidence in the ease with which man could be "improved," have made eugenics a source of embarrassment to many human geneticists. After World War II, for example, the British journal, *The Annals of Eugenics,* was renamed *The Annals of Human Genetics.* Furthermore, the development of genetic counseling has probably been inhibited to some extent by the fate of the eugenics movement. Human genetics has progressed a long way from the early days of the century, much of the progress having come in the last 25 years, but advances in the field have emotional, moral, and legal aspects that are unique compared to the more or less scientific approach that can be taken with respect to the application of genetic knowledge to other species.

The study of the genetics of man is particularly important, not only because the benefits derived from new discoveries may be so great, but also because the insights gained into the human condition may be so revealing. Research in human genetics, compared with genetic research with *Drosophila,* or corn, or mice, or microorganisms, is faced with certain difficulties, but it also has certain definite advantages. The disadvantages are perhaps more obvious. In man, controlled matings between individuals of known genotype are not condoned. Furthermore, in corn or *Drosophila,* a large number of progeny can be obtained from a single mating, but human families tend to be small. Even for simple Mendelian traits, simple Mendelian ratios will not be observed in human families because an excess of affected children above expectation will appear in families with any affected children at all. Such families are counterbalanced by families in which the parents are of the appropriate genotype, but no affected children have been produced, and the family thus escapes detection. Therefore, only by appropriate corrections for biases due to small family size is it possible to study even simple Mendelian inheritance in man.

Because human families are small, it is customary to group families showing similar traits in order to get enough data for statistical analysis. Since similar traits have turned out to be dominant in one pedigree and recessive in another, or sex-linked in one and autosomal in another, such lumping together of families carries with it the hazard that similar traits with distinctly different modes of inheritance may be grouped. If that happens, the analysis of the data may become hopelessly confused.

A further problem is a lack of environmental control comparable to that possible with experimental organisms. For some traits, such as albinism, this constitutes no real problem, because the trait is expressed under all conditions. However, for many traits, expressivity may vary considerably depending on environmental conditions, or there may even be a complete lack of penetrance.

The generation length in man is also a great handicap. Observations that can be made in several hours in bacteria, or weeks in *Drosophila,* or months in corn, may require decades in man. This time lapse is also responsible, in large measure, for the difficulty in getting accurate, complete data of the desired type, because many subjects of crucial importance may have died long ago. Furthermore, the human geneticist is quite dependent on the cooperation, interest, and goodwill of his subjects, and this too may complicate the matter of obtaining reliable information. Even with the best of cooperation, good data may be hard to get since so few records are ordinarily kept. It is always interesting to discover how few college students know with any certainty the eye color of all four of their grandparents when their grandparents were of college age.

The advantages of man as an object of genetic study are perhaps less obvious. Man is undoubtedly the most closely studied species of all. All sorts of physical, physiological, biochemical, and other traits have been scrutinized. Despite their sometimes frustrating inadequacies, many more records on individuals are kept for man than for any other species. Birth, death, school, medical, and many other types of records are maintained. Even though controlled matings are unavailable in man, it may be possible to seek out and find matings of the desired type. With the development of population genetics and of statistical methods appropriate for the analysis of data in human genetics, considerable progress has been made. One result is that an ever-increasing array of hereditary traits in man has been identified, especially in the area of medical genetics. In terms of the number of recognized hereditary traits, man is one of the best-known organisms.

Because controlled matings comparable to those carried out with other species are not available in man, other, more indirect approaches to human heredity are necessary. Several broad types of approach are possible: the study of populations, the study of families, the study of twins, and more recently, the study of cells in culture.

23.1. The Study of Populations

The epidemiological approach entails the study of all the factors related to the development of specific diseases in populations. Three factors must be considered in the etiology of any disease: a possible causative agent or pathogen, the environmental conditions, and the characteristics of the host. In infectious diseases, all three kind of factors are significant. In hereditary disease, only the environment and the condition of the host are important. In the past, epidemiologists have tended to concentrate on the identification of infectious agents in disease and their control through manipulation of the environment. The use of antibiotics and chemotherapeutic agents has dramatically reduced the death rate from infectious diseases so that greater emphasis is now being concentrated in medicine on those diseases in which the host's genotype plays an important role in the etiology of the disease.

The epidemiologist's interests are very broad, for he explores all paths in his efforts to identify the circumstances responsible for a disease. For example, for a disease of unknown etiology, he may wish first to gain some idea of its prevalence, its frequency in the general population. He may also determine if there are regional variations in frequency. Such variations could be due to environmental differences—in climate, for example—or to genetic differences among the human populations, or to differences in rate of exposure to a pathogen in different regions. He may also seek seasonal variations in incidence, or differences related to social class, race, parental age, birth order or parity, epidemics, or other environmental variables. Depending on his findings, he will further narrow his search in order to isolate and, if possible, control the conditions leading to the disease. Because of its broad, more or less shotgun approach, epidemiology may merely tend to incriminate heredity in the etiology of a disease rather than lead to a well-defined statement of its etiological role.

If the genetics of the disease have been worked out, the techniques of population genetics can be applied to the study of different human populations. The study of gene frequency differences in different populations has opened new vistas not only in medicine, but in anthropology as well. Some genes were found to be more or less universal in distribution, but to differ in frequency; others were more restricted in their distribution. The historical or evolutionary significance and the adaptive significance (if any) of these genes have become a matter of interest and speculation.

23.2. The Study of Families

In order to determine the mode of inheritance of a trait, it is ordinarily necessary to analyze data from family histories. In this way, it can be established if dominant or recessive, autosomal or sex-linked inheritance is involved. Possible biases stemming from the manner of selection of families or from pooling data from families that are actually dissimilar have already been mentioned. However, the analysis of family pedigrees may aid in disentangling these clinically similar conditions, and furthermore may aid in the recognition of subclinical conditions and in the detection of heterozygotes.

Retrospective family studies usually start with an affected individual (the proband, or propositus) and work outward from him through the existing family. One problem with this approach may be the inability to obtain relevant data on deceased relatives or others who are not readily available. The problems inherent in retrospective studies are minimized to some extent by *prospective* or *longitudinal* studies, which work forward in time rather than backward. The data are apt to be more reliable in such studies, but considerable time may be required to complete the study unless the family histories of affected individuals identified a generation or more ago can be brought up to date.

Another way to use family data is the **contingency** method. The essence of this method is the determination of the frequency of affected individuals among the different groups of relatives of a proband. If the frequency of the trait increases the closer the relationship to the proband, a hereditary effect is indicated. Such data may not necessarily lead to a conclusion as to mode of inheritance, but they may provide empirical risk figures useful in counseling on the probability that the trait will appear in sibs, offspring, or other relatives of an affected person. Furthermore, it may indicate an etiological relationship not otherwise suspected between seemingly distinct traits. Such was the case for spina bifida and anencephaly: An increased frequency of spina bifida has been found among sibs of propositi with anencephaly.

If the progeny from marriages between relatives (or consanguineous matings) show an increased frequency of a trait, rare recessive genes may be implicated. Therefore, the study of the effects of consanguinity on incidence offers still another approach to the analysis of family data.

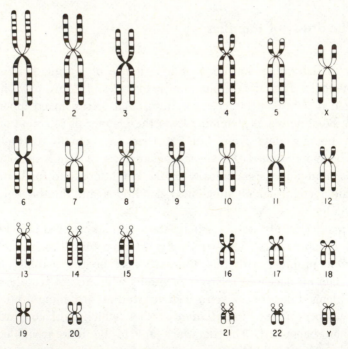

Fig. 23.2. Diagram of the banding structure of the human chromosomes revealed by Giemsa stain. (Crossen, P. E., 1972. "Giemsa Banding Patterns of Human Chromosomes," from *Clinical Genetics* 3:176, Fig. 8. By courtesy of Munksgaard Publishers, Copenhagen.)

One point that probably cannot be overemphasized is the necessity for adequate, carefully chosen, and thoroughly studied controls. If the control groups are inadequate, the conclusions may be in error, no matter how carefully the rest of the work is done.

23.3. Studies of Cells and Chromosomes

Recent developments in human cytogenetics have made a new technique available to the human geneticist by which the human karyotype can be studied in detail and chromosome anomalies recognized. Progress in these studies has been rapid and dramatic in its impact on human genetics (Fig. 23.2).

The culturing of human cells by techniques comparable to those used with microorganisms offers the possibility of studying human genetics separate from man. These studies have not yet progressed as rapidly as human cytogenetics, but it seems probable that subjects such as linkage, mutation, and human biochemical genetics can be investigated intensively *in vitro* in the near future. Recently, hybridization between cells from different species in tissure culture—for example, human cells and mouse cells—has added a new technique to the armamentarium of the human geneticist, which is useful in linkage and other types of studies.

23.4. The Study of Twins

Another technique useful in human genetics is the twin method. The study of twins does not ordinarily reveal a great deal about the mode of inheritance of a particular trait. However, it is singularly useful in the evaluation of the relative contributions of hereditary and environmental factors in the determination of a trait. In other words, human twins permit us to explore the nature-nurture question in man.

Quite apart from their scientific value, twins have long caused curiosity in others and consternation to their parents. The initial shock when confronted by two identical bundles of joy often leads parents of twins to give them similar names, which further compounds the confusion. Twins are the object of superstition among primitive peoples. In some tribes in Africa they were regarded as an omen of fertility and plenty, and both mother and twins were given preferential treatment. In others, however, twins were regarded as visitations from the devil, and mother and twins were sacrificed to eliminate the evil influence. In the Philippines, the bestial situation of a woman bearing a litter was corrected by killing one of the twins. Another superstition, a potential source of trouble for the mother, was that one man can father only one child at a time. Lest we smile too benignly at these superstitions, it may be worth pointing out that our society seems to have a few of its own. One such belief is that twins are mentally and physically inferior to singly born children. Another is that one twin may often be sterile, apparently from an analogy with the sterile freemartin in twin cattle. There seems to be little evidence to support either of these beliefs.

There are two different types of twins. The fraternal, or dizygotic or two-egg, twins are genetically no more alike than ordinary single-born sibs. They develop from two different eggs fertilized by two different sperm cells. The unusual feature is that the mother ovulates two eggs in the same menstrual cycle, both of which are fertilized and undergo development. On the average, dizygotic twins, like ordinary siblings, have 50% of their genes in common. They may be of the same sex or of opposite sexes.

Identical, or monozygotic or one-egg, twins, on the other hand, develop from a single fertilized egg. They are therefore identical in genotype, having 100% of their genes in common, and they are always of the same sex. The unique feature is that at an early stage of development, the single embryo separates into two halves, each of which differentiates into a separate individual.

The causes of identical twinning in man remain obscure. In experimental animals, however, it has been possible to induce twinning at early stages by techniques that lead to the subdivision of the developing embryo. This has been done by separating the cells at the two-cell stage either mechanically or chemically, after which each forms a complete embryo rather than half an embryo. At later stages constriction of the embryo with a hair loop will also result in the formation of two normal embryos. In mammals that normally produce litters, the young are usually fraternal, the result of multiple ovulations and fertilizations. The nine-banded armadillo, however, normally produces monozygotic quadruplets, all of the same sex. Development of the armadillo zygote to the blastula stage appears normal, but then four primitive streaks form rather than one and around these the four embryos differentiate.

The time of separation in monozygotic twinning is apparently variable. The formation of an XY male and an XO individual with Turner's syndrome, who by the usual criteria appeared to be monozygotic twins,

Fig. 23.3. Conjoined twins, Simplicio and Lucio Godino, on the tennis court; Lucio about to serve. (Walker, N, F., 1952. "A Discussion of the Zygosity and Asymmetries of Two Pairs of Conjoined Twins." Courtesy of Instituto Mendel (Rome): *Acta Geneticae Medicae et Gemellologiae.*)

suggests that the separation in this case was at the two-cell stage, and that the Y chromosome had been lost from one of the two cells. However, in most instances the separation probably occurs somewhat later. In fact, sometimes the separation starts but is not completed, and the two differentiating embryos remain attached to one another, forming what are known as Siamese twins (Fig. 23.3). (The original "Siamese" twins, Chang and Eng, were actually Chinese.) The point of attachment and the degree of connection are variable, and hence the survival of these rare anomalies depends on the nature of the attachment between them and on the skill of the surgeons, who today usually try to separate them so that they can lead normal lives. In many cases, however, they do not long survive.

Frequency of Twinning

The frequency of multiple births varies in different human populations, being highest among Negroes, about 1 in 70, intermediate in Caucasians, about 1 in 90, and lowest in Orientals, the Japanese rate, for example, being about 1 in 145. The figures are usually given only for those births in which at least one twin is born alive.

An empirical rule of thumb, known as Hellin's law, can be used to estimate the expected frequency of the higher multiple births. The relationship observed was that if the frequency of twin births was $1/b$, the frequency of triplets would be $1/b^2$, of quadruplets $1/b^3$, and so on. Actual data for 21 countries for 10 years on 120,061,398 pregnancies showed 1,408,-912 twins or 1/85.2, 15,738 triplets or $1/(87.3)^2$, and 179 quadruplets or $1/(87.5)^3$. For the Japanese population, $1/(145)^2$ triplets would be expected and $1/(141)^2$ were observed. By this rule, quintuplets would be very rare, expected in less than 1 in 50,000,000 pregnancies. However, it may be noted that the widespread use of birth-control pills and fertility drugs may well have significant effects on these frequencies. The surprising thing, perhaps, is that Hellin's rule works as well as it does, since it deals with two apparently biologically unrelated phenomena, monozygotic and dizygotic twinning.

The relative frequency of monozygotic and dizygotic twins has also been found to vary in different populations. These frequencies can be estimated because of the fact that all unlike-sexed twins must be dizygotic. If we assume the probability of a boy to be a, and of a girl b, and that the formation and development of the two zygotes are independent events, then the proportions of boy-boy, boy-girl, and girl-girl pairs can be estimated from the binomial expansion of $(a+b)^2$, or a^2 (boy-boy), $2ab$ (boy-girl), and b^2 (girl-girl). For the simple assumption that the sex ratio at birth is 1:1, then $a = \frac{1}{2}$ and $b = \frac{1}{2}$, and $a^2 = \frac{1}{4}$ (boy-boy), $2ab = \frac{1}{2}$ (boy-girl), and $b^2 = \frac{1}{4}$ (girl-girl). Thus, the total for all dizygotic twins should be twice the observed number of boy-girl pairs, and the number of monozygotic pairs is then determined by subtracting the estimate for the number of dizygotic twin pairs from the total number of all twin pairs.

The sex ratio at birth in most human populations shows a slight excess of males. Therefore a more precise estimate for the proportion of monozygotic twins can be obtained by using the known values of a and b for the probabilities of boys and girls at birth. In this case the proportion of boy-girl pairs is

$$\frac{\text{boy-girl pairs}}{\text{all DZ pairs}} = \frac{2ab}{a^2 + 2ab + b^2} = \frac{2ab}{1}$$

Hence

$$\text{all DZ pairs} = \frac{\text{boy-girl pairs}}{2ab}$$

Finally,

$$\text{all MZ pairs} = \text{all twin pairs} - \frac{\text{boy-girl pairs}}{2ab}$$

This method of estimating the proportions of monozygotic and dizygotic twins is known as Weinberg's differential method, after the pioneer human geneticist who also played a role in developing the Castle-Hardy-Weinberg law. When the method has been applied, it has revealed that in the United States about one-third of the sets of twins born are monozygotic. In Japan, with its lower overall rate of twinning, nearly two-thirds of the twin pairs are monozygotic. Consequently, it appears that Japanese women ovulate more than one egg at a time less often than women in the United States, and that the difference in rates of twinning is due primarily to the production of fewer dizygotic twins in Japan.

Inheritance of Twinning

The role of heredity in twinning is not clear-cut, but there is evidence that twin births seem to cluster in certain families. The incidence of twins also shows a relationship to the age of the mother and to birth order or parity. The tendency to have dizygotic twins increases to about age 39 and then drops sharply; the tendency to have monozygotic twins increases only very slightly with age (Fig. 23.4). Furthermore, when the effect of age is controlled, the tendency to have twins can be shown to increase with parity. Thus, the effect of age and of birth order on twinning are similar but independent, the twinning rate increasing with age and with parity, but the effect is more pronounced with dizygotic than with monozygotic twins.

Empirical risk figures give the chances that a mother, having had one set of twins, will have another. Such data show that a mother of dizygotic twins has about three times the chance of having another pair of twins

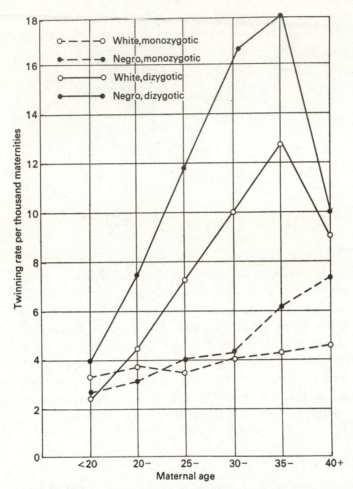

Fig. 23.4. The relationship between maternal age and twinning in American whites and Negroes. (Bulmer, M. G., 1970. *The Biology of Twinning in Man.* Clarendon Press, Oxford, p. 88, Fig 4.8.)

as a mother in the general population, but that the incidence of twinning is not significantly increased in subsequent pregnancies of mothers of monozygotic twins.

Family pedigree studies in Norway revealed wide variation in the incidence of twins, ranging from no twins at all in 800 births in one family group to 107 twin pairs in 2,840 births (3.91%) in another. Application of the contingency method to such families reveals an increased incidence of twins among the relatives of mothers of dizygotic twins but not of monozygotic twins. Therefore, the available evidence suggests a tendency toward the inheritance of fraternal twinning but not identical twinning.

An oddity of the data, for which there is no adequate explanation, is that a tendency for dizygotic twinning is found not only among the relatives of the mother of dizygotic twins, but among the relatives of the father as well. In passing, we might note that Alexander Graham Bell, better known in another connection, was successful in selecting for an increased incidence of twins in sheep, still another bit of evidence that twinning is influenced by heredity.

Diagnosis of Zygosity

In twin studies it is essential to determine whether each set of twins is monozygotic or dizygotic. If the twins are of opposite sex, they are obviously dizygotic. If like-sexed twins differ in a trait known to be genetically determined and fully penetrant, they cannot be identical. However, if the twins are alike in sex and in other hereditary traits, they are not necessarily identical. Hence, some method is needed to determine zygosity in these more difficult cases.

In the past, the fetal membranes were regarded as useful for this purpose. The embryo is enclosed within two membranes, an inner sac or amnion and an outer chorion, and is attached to the uterine wall by the placenta. About the only safe generalization in this respect is that if both twins are enclosed within a single chorion, they are monozygotic. However, twins with separate chorions may be either monozygotic or dizygotic, so the chorion cannot be used to determine zygosity in all cases. The possibilities are shown below:

	Twins	
Membranes	MZ	DZ
1 chorion, 1 amnion	+	−
1 chorion, 2 amnions	+	−
2 chorions, 2 amnions	+	+

Diagnosis based on the placentas is even less reliable, as can be seen from the relationships below, since both types of twins may have either type of placenta.

	Twins	
Membranes	MZ	DZ
1 placenta, 1 chorion	+	−
1 placenta, 2 chorions	+	+
2 placentas, 2 chorions	+	+

A more reliable method of twin diagnosis is the similarity method. Its essence is the use of a number of simply inherited, polymorphic traits with complete penetrance and uniform expression. The blood groups are particularly useful for this purpose, because they are simply inherited, fully and uniformly expressed, and the different alleles supply the variability needed for segregation. Since fraternal twins, on the average, hold half of their genes in common, it is not surprising if they are the same for a given trait. There is a fairly high probability that they will be alike. However, for each additional trait studied, the probability that fraternal twins will be alike diminishes until a point is reached at which it is highly improbable that these twins would be similar in so many traits if they were actually dizygotic. Thus, the similarity method provides a probability statement on the zygosity of twins. However, if the twins differ in even one of the traits, they are considered dizygotic. Hence, it is possible to prove dizygosity but not monozygosity by the similarity method, although monozygosity becomes highly probable as the number of relevant traits increases.

Skin grafting is probably the most reliable test of zygosity, but it is used only in exceptional cases. Even though fraternal twins may have some antigens in common, they will differ with respect to others. Therefore, when a skin graft is made, antibodies will form and the graft will eventually be rejected. Identical twins, with identical genotypes, will have identical antigens, and since they do not form antibodies against their own antigens, a tissue transplant between identical twins will not be rejected. Thus, the best guarantee of success at present in tissue and organ transplants is to have identical twins as the donor and recipient. Because surgical procedures are required, this test is not used in routine diagnoses of zygosity.

The Use of the Twin Method

The usefulness of twins in the study of the relative importance of heredity and environment in the formation of the phenotype is ordinarily based on a comparison of monozygotic twin pairs reared together with like-sexed dizygotic twin pairs reared together. Monozygotic twins have identical genotypes with 100% of their genes in common, but dizygotic twins on the average have only 50% of their genes in common. The environment for both types of twins is much the same. The environment, as used in this discussion, encompasses all nongenetic influences, both prenatal and postnatal, including the internal environment of the body and any cytoplasmic differences. Intrauterine environment, maternal age, birth order, nutrition, training, culture, and so on, are all rather similar for monozygotic and dizygotic twins reared together. Therefore, the extent to which mono-

Fig. 23.5. Discordance for harelip in monozygotic twin brothers. (Grebe, H., 1952. "Diskordanzursachen bei erbgleichen Zwillingen." Courtesy of Instituto Mendel (Rome): *Acta Geneticae Medicae et Gemellologiae.*)

zygotic twins are more alike than dizygotic twins must be due primarily to heredity.

If both of a pair of twins are alike for a given trait, they are said to be *concordant*. If the twins differ for the trait, they are *discordant*. If concordance in identical twins is complete, the trait in question must be completely determined by heredity. For simply inherited traits such as the blood groups, all monozygotic twin pairs will be concordant, since environmental variations do not seem to influence the blood type. On the other hand, if monozygotic and dizygotic twin pairs show similar degrees of concordance, the trait must be strongly influenced by environmental conditions and the genetic component must be small or nonexistent. Comparisons of identical with fraternal twin pairs for many different traits have revealed that many traits are influenced to varying degrees by both heredity and environment, but relatively few are determined completely by the environment.

Discordance in even one pair of monozygotic twins is evidence for an environmental influence on the trait in question (Fig. 23.5). Furthermore, discordance in a single pair of monozygotic twins is critical evidence against any theory of incomplete penetrance being due solely to different modifiers of a primary gene, since identical twins will have identical modifiers as well as the same major gene.

Another method used in twin studies is the comparison of monozygotic twins reared together with monozygotic twins reared separately from one another. Monozygotic twin pairs reared apart are relatively rare, so such data are hard to gather, but enough cases have now been studied to provide some useful information. To the extent that it is possible to measure, quantify, and evaluate environmental conditions, it becomes possible to study the effect of environmental variables on the development of individuals with identical genotypes. Of course, it must be remembered that the prenatal environment of monozygotic twins is the same in either case, and the postnatal environment as well until the twins are separated. Hence, if early environmental conditions are important in the differentiation of the trait, these separated monozygotic twins may be alike due to their early environments rather than their identical genotypes.

Still another technique possible with twins is to compare, for example, like-sexed dizygotic twins with like-sexed sibs in the same family. To the extent that the dizygotic twins are more alike than ordinary brothers or sisters, the greater similarity must be environmental, since genetically they are comparable.

A different method of study of the relative effects of nature and nurture, not involving twins, is to compare natural children with their parents, and foster children with their foster parents. If natural children are no more like their parents than foster children, then the trait in question is presumably environmentally determined. The higher the correlation between natural children and their parents as compared to foster children and their foster parents, the more important the role of heredity.

Identical twins offer a unique opportunity in man to carry on experiments with adequate controls. If it were desired, for instance, to evaluate a new method of teaching reading or roller skating, a group of identical twins could be tested, with one of each pair learning the new way, and his co-twin serving as a control. In this way, the genotypes in the experimental and the control groups are known to be comparable, and the usual sampling problems involved in getting adequate controls for such studies are eliminated. Studies of this sort, with a co-twin control, useful as they would be, are relatively uncommon.

Twin studies have been subjected to criticism, and reservations about their value have often been expressed because of the biases possible in twin research. While many of these criticisms are valid, they do not necessarily invalidate the use of the twin method. As long as the goals of twin research remain modest—namely, to determine, for a given trait or series of traits, the relative roles of hereditary and environmental influences —the conclusions reached can be accepted. However, twin studies, as such, reveal very little about the mode of inheritance or, for that matter, about the nature of the environmental influences at work. Moreover, efforts to make quantitative estimates of heritability, or penetrance, or variance ratios from twin data are not particularly useful because they are not generally applicable to the population at large. In fact, they may not even be valid for twins. The practice, for instance, of estimating penetrance from monozygotic twin data is of dubious value. The concordance rate in monozygotic twins gives an upper limit to the penetrance, but no more, because different sets of identical twins have different genotypes and there is really no way of estimating the number of monozygotic twin pairs who carry a susceptible genotype but none of whom develops the trait. Therefore, it is possible to estimate the number of monozygotic twin pairs concordant for the trait, the number of monozygotic twin pairs discordant for the trait, but not the number of monozygotic twin pairs concordant in that both fail to show the trait even though they may have the appropriate genotype. The picture is complicated even more by the possibility that phenocopies may be responsible for some of the discordant pairs.

The basic data from twin studies are the concordance rates in monozygotic and dizygotic twins. Twin data have been used to calculate heritabilities, correlations, variance ratios, and the like, but such estimates, as in the case of penetrance, are not particularly reliable. The primary usefulness of twin studies is to demonstrate environmental influences on traits commonly thought to be hereditary or, conversely, to show a genetic component in a trait usually thought to be environmental. Catching the measles, for instance, may be thought of as an environmental trait dependent only on exposure to the virus. However, the fact that the concordance rate is slightly higher in monozygotic than in dizygotic twins reared together

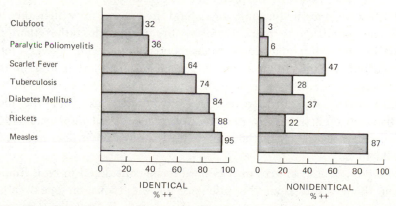

Fig. 23.6. Concordance and discordance rates in twins affected by various pathological conditions. (Data from Herndon and Jennings, poliomyelitis; Idelberger, clubfoot; and von Verschuer.)

suggests that a genetic difference in susceptibility to measles may also be involved (Fig. 23.6). The twin data are particularly useful for research on complex quantitative traits showing continuous variation, such as size or intelligence. The basic question is usually whether monozygotic twins are more alike than dizygotic twins. If they are, a genetic component to the trait is indicated.

One possible source of bias in twin studies lies in the method of ascertainment of the twins. For example, the problem may arise because of the greater likelihood of detecting twin pairs with both twins affected or, if case reports are used, because of a greater likelihood that discordant identical twin pairs will get into the literature than the more routine concordant identical twins. In any event, care must be exercised in this regard or the concordance rates may be badly biased.

Other criticisms are that prenatal and natal influences (position in the uterus, implantation, crowding, order or manner of delivery, and the like) may cause identical twins to differ for nongenetic reasons, or else that these effects may be different in monozygotic and dizygotic twins so that even at this stage, the assumption of equivalent environmental conditions is erroneous. This point is argued even more strongly with regard to postnatal environment, since identical twins, because they are so much alike, are often treated more similarly than are fraternal twins. While this observation seems to argue against the assumption that monozygotic and dizygotic twins reared together have comparable environments, to some extent the differences in treatment and environmental conditions are actually a reflection of the greater similarity in the genotypes of monozygotic twins as compared to dizygotic twins. Thus even these environmental differences arise as an expression of the genotype.

The effect of mirror imaging, or lateral inversion, in which the left side of one monozygotic twin resembles the right side rather than the left side of the other, may also cause monozygotic twins to vary more than expected. On the other hand, it has been suggested that if mutual circulation were established in fraternal twins *in utero* comparable to that sometimes observed in twin cattle, the dizygotic twins would be more alike than expected. Errors of diagnosis of zygosity will also tend to bias the results, but improved methods of twin diagnosis have now minimized this problem.

All of these possible sources of bias need to be considered in twin studies. Some of them result from the influence of the twins on one another. Nevertheless, twin studies need not be abandoned. If their limitations are recognized, and the method is not pushed beyond these limits, useful information can be obtained. The co-twin control approach offers a unique and relatively unexploited opportunity to study environmental effects in man.

Additional Reading

Allen, G., 1952. "The Meaning of Concordance and Discordance in Estimation of Penetrance and Gene Frequency," *Amer. J. Human Genet.* 4:155–172.

Boyd, W. C., 1950. *Genetics and the Races of Man.* Little, Brown, Boston.

Boyer, S. H., ed., 1963. *Papers on Human Genetics.* Prentice-Hall, Englewood Cliffs, N.J.

Bulmer, M. G., 1970. *The Biology of Twinning in Man.* Clarendon, New York.

Burdette, W. J., ed., 1962. *Methodology in Human Genetics.* Holden-Day, San Francisco.

Cavalli-Sforza, L. L., and W. F. Bodmer, 1971. *The Genetics of Human Populations.* Freeman, San Francisco.

Clarke, C. A., 1969. *Selected Topics in Medical Genetics.* Oxford University Press, London.

Dahlberg, G., 1926. *Twin Births and Twinning from a Hereditary Point of View.* Bekförlages A.B., Stockholm.

Dahlberg, G., 1948. *Mathematical Methods for Population Genetics.* Wiley-Interscience, New York.

Dahlberg, G., 1948. "Genetics of Human Populations," *Adv. Genet.* 2:67–98.

Gedda, L., 1951. *Studio dei Gemelli.* Edizioni Orizzonte Medico, Rome.

Gedda, L., 1961. *Twins in History and Science,* M. Milani-Comparetti, tr. Charles C. Thomas, Springfield, Ill.

Harrison, G. A., J. S. Weiner, J. M. Tanner, and N. A. Barnicot, 1964. *Human Biology.* Oxford University Press, London.

Hsia, D. Y., ed., 1966. *Lectures in Medical Genetics.* Year Book Publishers, Chicago.

Kallmann, F. J., 1953. *Heredity in Health and Mental Disorder.* Norton, New York.

Koch, H. L., 1966. *Twins and Twin Relations.* University of Chicago Press, Chicago.

Krooth, R. S., G. A. Darlington, and A. A. Velazquez, 1968. "The Genetics of Cultured Mammalian Cells," *Ann. Rev. Genet.* 2:141–164.

Li, C. C., 1961. *Human Genetics.* McGraw-Hill, New York.

McKusick, V. A., 1970. "Human Genetics," *Ann. Rev. Genet.* 4:1–46.

Moody, P. A., 1967. *Genetics of Man.* Norton, New York.

Nance, W. E., 1969. "Twins: An Introduction to Gemellology," *Medicine* 38: 403–414.

Neel, J. V., and W. J. Schull, 1954. *Human Heredity.* University of Chicago Press, Chicago.

Newman, H. H., 1940. *Multiple Human Births.* Doubleday, Garden City, N.Y.

Newman, H. H., F. N. Freeman, and K. J. Holzinger, 1937. *Twins, a Study of Heredity and Environment.* University of Chicago Press, Chicago.

Penrose, L. S., 1959. *Outline of Human Genetics.* Wiley, New York.

Penrose, L. S., 1959. "Genetical Analysis in Man." In *Biochemistry of Human Genetics,* G. E. W. Wolstenholme and C. M. O'Connor, eds. Little, Brown, Boston.

Price, B., 1950. "Primary Biases in Twin Studies," *Amer. J. Human Genet.* 2: 293–352.

Roberts, J. A. F., 1970. *An Introduction to Medical Genetics*, 5th ed. Oxford University Press, London.

Scheinfeld, A., 1967. *Twins and Supertwins*. Lippincott, Philadelphia.

Shields, J., 1962. *Monozygotic Twins*. Oxford University Press, London.

Smith, S. M., and L. S. Penrose, 1955. "Monozygotic and Dizygotic Twin Diagnosis," *Amer. Human Genet.* **19**:273–289.

Steinberg, A. G., 1959. "Methodology in Human Genetics," *J. Med. Educ.* **34**: 315–334.

Stern, C., 1973. *Human Genetics*, 3rd ed. Freeman, San Francisco.

Sutton, H. E., 1965. *An Introduction to Human Genetics*. Holt, Rinehart and Winston, New York.

Sutton, H. E., 1967. "Human Genetics," *Ann. Rev. Genet.* **1**:1–36.

Whittinghill, M., 1965. *Human Genetics*. Van Nostrand Reinhold, New York.

Questions

23.1.   What are the advantages in using man as a subject of genetic study? What handicaps confront the student of human genetics that geneticists working with other species do not have to face?

23.2.   If the incidence of a particular disorder is found to be significantly different in different human populations, suggest possible causes for these differences in frequency.

23.3.   What type of study is most useful in indicating whether or not heredity plays a significant role in the etiology of a particular human trait?

23.4.   What type of study provides the most reliable information as to the mode of inheritance of a particular human trait?

23.5.   If you were responsible for setting up a program of research into the relationship between heredity and diabetes, what types of studies do you believe would be most informative?

23.6.   Suggest a problem that would be especially suitable for study in human cell culture.

23.7.   What is the basis for Weinberg's differential method for estimating the proportion of monozygotic twins among all twins born?

23.8.   What types of information can be gained from comparisons of
   a. Monozygotic twins reared together and dizygotic twins reared together?
   b. Monozygotic twins reared together and monozygotic twins reared apart?
   c. Dizygotic twins reared together and dizygotic twins reared apart?

d. Dizygotic twins reared together and ordinary sibs reared to-
    gether?
e. Natural children with their natural parents and foster children
    with their foster parents?
f. Foster children with their foster parents and with their natural
    parents?

23.9. In which of the comparisons in Question 23.8 would you not wish
    to use opposite-sex dizygotic twins?

23.10. If 51% of all children born are boys, and 32% of all twins born are
    of opposite sex, what proportion of all twin pairs born will be
    identical twins?

23.11. What is the evidence that heredity influences the tendency for
    twinning?

23.12. If you wished to screen the school population of a large city for
    identical twins, how would you proceed?

CHAPTER TWENTY-FOUR

Heredity and Disease

The idea that diseases are inherited can be traced back to the Greeks, who believed that some hereditary constitutions are favorable to the development of a disease like tuberculosis but others are not. The concept of contagion also has a long history; for centuries it was commonly thought that malaria was due to exposure to a miasma emanating from swamps and marshes. However, until a century ago heredity was generally considered to be the primary cause of disease. The idea that a particular infectious agent could cause a specific disease did not become well established until the late 1800s, when Pasteur and Koch and their students demonstrated the importance of microorganisms as pathogenic agents. During the past century, bacteria, protozoa, viruses, yeasts, and rickettsia have all been identified as pathogens in specific diseases. These exciting new discoveries reversed the previous trend of thought favoring heredity, and by the early days of this century it was felt that the primary requirements for the control of disease were identification of the pathogen and of the vehicles or routes of infection. Then methods of treatment could be devised to eliminate the pathogen from the host or to prevent it from infecting the host. The success of vaccination and other methods of immunization seemed to confirm this view of disease and to increase the optimism of the medical workers.

The birth of genetics came at a time when the old concept of the inheritance of disease seemed outmoded: New pathogens were regularly being identified, and immunologists were making steady progress in understanding the acquisition of both active and passive immunity. Nevertheless, the rediscovery of Mendelism eventually led to a far better understanding of the nature of disease, and to a reconciliation of the hereditary and contagious theories of disease, for as is often the case in such situations, there were elements of truth on both sides.

The concept of disease that has emerged may be likened to a three-legged stool, with one leg representing the gene complex of the host, the second the gene complex of the pathogen, and the third the environmental complex.

24.1. Hereditary Diseases

In hereditary diseases, a single gene substitution of the sort to be discussed in the next chapter may produce the disease. There is no pathogen, but the defective gene itself may be thought of as the pathogen. Although hereditary diseases have often been regarded as incurable, such fatalism is no longer always necessary, for treatments are being found for hereditary defects. The use of insulin in diabetes mellitus, of low-phenylalanine diets for phenylketonuric infants, of pituitary transplants or extracts in dwarf mice whose pituitary glands lack eosinophil cells are all examples of effective treatments for hereditary diseases. As the nature of the genetic defect becomes better understood, the possibility of successful treatment increases. The essence of the treatment is to manipulate the environment so as to minimize the effect of the defective gene. Such treatments, since they tend to preserve deleterious genes in the population, must be regarded as dysgenic.

In the long run, it is possible that efforts to repair genetic damage will be made at the DNA level by the substitution of normal DNA segments for the segments of defective DNA. Some success in this direction has already been achieved with cultured human fibroblasts derived from a patient with galactosemia. Galactosemia is an inborn error of metabolism characterized by the absence of the enzyme galactose-1-phosphate uridyl transferase (GPU transferase) and the inability to take up and metabolize galactose. Merril, Geier, and Petricciani have reported that bacteriophage λ particles have infected these human fibroblasts, and have been transcribed and translated within these eukaryotic cells. Moreover, they are stably transmitted through many generations of fibroblast multiplication. Even more exciting is the fact that transducing phages derived from λ and carrying the *Escherichia coli* galactose operon have infected the galactosemic fibroblasts, in which the GPU transferase activity has then increased from 10 to 75 times that of the controls. In other words, through transduction with a phage, a bacterial gene has been transferred to a human cell, where the bacterial gene remedies the human cell's biochemical deficiency. Although many questions remain (for example, it is not known whether the phage genome resides in the nucleus, the cytoplasm, or the mitochondria of the cell, nonetheless, if the work is confirmed, it must be regarded as a striking breakthrough of far-reaching implications.

A rather different approach has been used in Fabry's disease, a sex-linked enzyme deficiency in the breakdown of the glycosphingolipids. In this case, the missing enzyme, ceramide trihexosidase, has been either administered by intravenous injection, or provided by the transplant of a third kidney from an unaffected person, which results in a more permanent alleviation of the condition. Thus, the treatment of human genetic diseases is becoming increasingly sophisticated.

24.2. Environmental Diseases

Some diseases are environmental in origin. Rickets, for example, is due to a deficiency of vitamin D, and kwashiorkor, found mostly in the tropics, is due to a diet deficient in protein. Some goiters develop because of an iodine deficiency. In each case, the disease can be alleviated by supplying the missing dietary component. In such diseases there is no pathogen and the genotype of the affected person is normal; the environment is deficient and will not support normal development.

24.3. Infectious Diseases

With infectious diseases, all three legs of the stool come into play and the interrelationships become very complex: The course of the disease will depend on the genotype of the host, the genotype of the pathogen, and on the environmental conditions under which pathogen and host come into contact. Some of the early work with plant diseases led to an appreciation of these factors. For example, plant breeders sought to develop varieties resistant to plant pathogens such as wheat rust rather than to develop treatments for the diseased plants. Their efforts were successful, and led to the development of varieties of wheat resistant to wheat rust. Thus was demonstrated the fact that some genotypes are more resistant to a particular disease organism than others. However, in time the resistant varieties of wheat began to show signs of infection by wheat rust. Analysis of the situation revealed that these infections were due to new and different strains of rust that had originated by mutation. Therefore, whether or not the disease appeared depended not only on the genotype of the host but on the genotype of the pathogen. The wheat breeder had to return to his experimental plots to breed new varieties resistant to the new strains of rust. This cycle of stepwise, parallel evolution of wheat and wheat rust seems destined to continue unless some means of breaking the chain can be found. In the meantime, the wheat breeders seem assured of a job for some time to come.

Although the idea that the hereditary constitution of the host plays a role in disease has had a long history, the realization that the pathogen is also subject to variation is much more recent. Although antibiotics have produced many medical triumphs, the microorganisms, when exposed to antibiotics, have shown a disturbing tendency to become resistant. Analysis has shown such cases to be due to an evolutionary change in the microbial populations. The susceptible organisms die, but the resistant mutant types survive to become progenitors of the subsequent generations. Thus, natural selection is operating to produce resistant populations of microorganisms. In some cases, mutants not only resistant to but dependent on the antibiotic have been discovered. If such a mutant is responsible for an infection, treatment with the antibiotic may actually cause the patient's condition to worsen rather than improve.

In the development of epidemics and pandemics, it is becoming increasingly clear that not only the genotype of the host but the genotype of the pathogen is of considerable importance. The most serious outbreaks occur when a large proportion of the hosts have genotypes susceptible to a particular pathogen. Often the pathogen is a new mutant strain to which the population has not been previously exposed. The periodic outbreaks of influenza in man are traceable to new virulent strains of influenza virus that tend to spread through the human populations all over the earth.

A host's resistance to disease is often relative rather than absolute. For instance, degrees of susceptibility to mouse typhoid (*Salmonella typhimurium*) vary widely in different inbred strains of mice, and may range from as high as 92% survival in a resistant strain to 2% survival in a susceptible strain for a given inoculum. Nevertheless, a large enough inoculation will be 100% lethal even to animals from the resistant strain.

The environmental conditions also play a role in the etiology of disease. Respiratory infections, for example, are much more common in the colder months than in the summer. Rheumatic fever, which is apparently triggered by a group A beta hemolytic streptococcal infection in susceptible persons, is more frequent in temperate than in tropical regions. It is, for example, more common among Puerto Rican immigrants to northern cities in the United States than it is among Puerto Ricans in Puerto Rico. Studies of rheumatic fever or of heart damage resulting from rheumatic fever are complicated by the fact that the etiology involves exposure to infection of a person with a susceptible genotype under appropriate environmental conditions. Small wonder that questions remain about the etiology of the disease.

With three variables, the host genotype, the pathogen genotype, and the environment, the etiology and course of infectious diseases become more difficult to study than the hereditary diseases to be described in the next chapter. The clinical symptoms may show wide variations in expression, and genetic analysis may be difficult even though heredity clearly is involved in the etiology of the disease.

24.4. The Major Causes of Death

Tables 24.1 and 24.2 show the 10 leading causes of death in the United States in 1900 and in 1970. It should be noted that the various categories are usually a group of similar disease entities under a single heading rather than separate, distinct diseases. We shall consider briefly the role heredity plays in the various causes of death, because these are the diseases that kill a sizable proportion of all who die and are the ones the practicing physician encounters most frequently.

A noteworthy change between 1900 and 1970 is the reordering of the major causes of death due in large part to control of infectious diseases. In 1900 the three major killers were infectious diseases, accounting for nearly a third of all deaths. In 1970, the fifth major cause of death is the first to be clearly due to infection and accounts for just 3% of all deaths. In 1900 heart disease, stroke, and cancer combined caused 17.9% of all deaths; now they cause 66.1% of all deaths. The death rate per 100,000 has been cut almost in half during this period. In part, this change reflects different age structures in the populations then and now, but the

Table 24.1   Ten Leading Causes of Death in the United States—1900

Rank	Cause of death	Rate per 100,000 population	Percent of deaths
1.	Pneumonia and influenza	202.2	11.8
2.	Tuberculosis	194.4	11.3
3.	Diarrhea, enteritis, and ulceration of the intestines	142.7	8.3
4.	Diseases of the heart	137.4	8.0
5.	Intracranial lesions of vascular origin	106.9	6.2
6.	Chronic and other forms of nephritis	88.6	5.2
7.	Congenital malformations and other deaths in early infancy	74.6	4.3
8.	Accidents	72.3	4.2
9.	Cancer and other malignant tumors	64.0	3.7
10.	Senility	50.2	2.9
	All other causes	585.8	34.1
	All causes	1,719.1	100.0

From *Vital Statistics in the U.S. 1900–1940.* U.S. Government Printing Office, Washington, D.C.

Table 24.2   Ten Leading Causes of Death in the United States—1970

Rank	Cause of death	Rate per 100,000 population	Percent of deaths
1.	Diseases of the heart	359.0	38.2
2.	Malignant neoplasms	161.8	17.2
3.	Cerebrovascular diseases	100.9	10.7
4.	Accidents	54.1	5.8
5.	Pneumonia and influenza	30.8	3.3
6.	Certain diseases of early infancy	21.6	2.3
7.	Diabetes mellitus	18.6	2.0
8.	Arteriosclerosis	15.8	1.7
9.	Cirrhosis of liver	15.6	1.7
10.	Bronchitis, emphysema, and asthma	14.8	1.6
	All other causes	145.6	15.5
	All causes	938.6	100.0

Source: National Center for Health Statistics, Public Health Service, U.S. Dept. of Health, Education and Welfare.

average life expectancy in the United States has increased some 20 years since 1900, approximately from 50 to 70.

This change in the major causes of death has far-reaching implications for the practice of medicine and for the training of physicians. As the control of infectious diseases has improved, the common causes of death have become diseases with an important hereditary component. Rather than combatting the effects of infectious organisms, physicians are devoting more and more of their efforts to combatting the effects of inadequate or defective genotypes. Since this is true, it seems essential that they have insight and understanding of genetic principles, but medical education in the United States for the most part is seriously deficient in this respect.

In 1970 three of the ten leading causes of death (1, 3, 8) involved the circulatory system and accounted for half of all deaths. Major factors in the development of circulatory diseases are hypertension (high blood pressure), arteriosclerosis (hardening of the arteries), and heart damage resulting from rheumatic fever.

Both twin data and family history studies indicate that a hereditary predisposition toward hypertension may exist, but it is also true that environmental stress may aggravate the situation. Hypertension may also develop as a secondary consequence of other diseases. If excessive, hypertension may cause an enlarged heart and heart failure, coronary disease, brain hemorrhage (stroke), or kidney disease.

Arteriosclerosis, which causes a thickening and hardening of the arterial walls, is responsible for a significant proportion of the deaths due to heart disease. A common cause of this condition is the formation of fatty deposits in the walls of the arteries (atherosclerosis). The danger from arteriosclerosis arises from the reduced flow of blood and from the formation of blood clots at such points. These, in turn, may produce a heart attack or coronary thrombosis if formed in the coronary arteries of the heart, or a stroke if the clot or thrombus is formed in the arteries of the brain. At least two hereditary metabolic errors of fat metabolism may result in atherosclerosis, *hypercholesterolemia* and essential *hyperlipemia*. In the first, cholesterol is deposited in the walls of the blood vessels; in the second, dietary lipids are not removed from the blood and stored but instead form deposits in the arteries. In persons so affected, fatty deposits may be seen under the skin or on the eyelids. In both cases dominant or partially dominant genes are involved, and many cases of heart attacks at an early age (before 40) are attributable to the predisposing effect of the gene for hyperlipemia.

Congenital heart defects may be environmental in origin, due, for instance, to German measles (rubella) in the mother during pregnancy. They may also be genetic, as in the case of the trisomy for the 21st chromosome causing Mongolian idiocy (Down's syndrome). A heart defect is a typical part of the syndrome.

Another major factor in heart disease is rheumatic fever, the second most important cause of death among children 5 to 19 years old. As mentioned above, environmental conditions and an infection seem to be involved in the etiology, but there is evidence that hereditary differences in susceptibility are also a factor.

The second major cause of death today is cancer. A number of different types of neoplasms are included in this category. Some of the rare types of cancer mentioned in the next chapter are clearly hereditary. For such common types as breast cancer, the role of heredity is more difficult to demonstrate. Nevertheless, by analogy with the work done on cancer in mice and other animals, it seems safe to conclude that in many cases the genotype will influence the tendency to develop a particular type of tumor. Even if viruses are implicated in human cancer, as seems probable by analogy with tumors in other vertebrates, individuals with different genotypes may well show different susceptibilities to the virus. Environmental agents may be carcinogenic, as for example the coal-tar derivatives in cigarette smoke. However, not all heavy smokers develop lung cancer, so that genotypic differences in reaction to carcinogens may well exist.

When we reach accidents, the fourth major cause of death, you may feel that here, surely, almost by definition, the cause of death is purely environmental. Nonetheless, to the extent that the accidents result from faulty judgment, a defective sensory apparatus, or poor reflexes, and to the extent that these inadequacies are due to heredity, even accidents may result from an unfavorable genotype.

Pneumonia and influenza are the first major category involving infectious disease, but the mouse data cited above and twin data on infectious diseases in man both suggest hereditary differences in susceptibility. For example, the concordance rates are considerably higher in identical than in fraternal twins when one of the pair is infected by tuberculosis or paralytic poliomyelitis. The tenth category is a catchall, which has apparently moved into tenth place because of the increase in deaths due to emphysema.

The diseases of early infancy again include a variety of different causes of death. Birth injuries are environmental, but many genetic traits are also grouped in this category, many of them deleterious recessives brought to expression in the homozygous condition.

The last category not yet considered is diabetes mellitus. The role of heredity in this trait is discussed in Chapter 25. However, the wide range of variation in age of onset and in severity of the disease indicate the role of the environment in the development of clinical symptoms. The fact that

2.0% of all deaths in the United States are due to diabetes mellitus again raises the question as to why the gene or genes responsible have such a high frequency and suggests that they do not markedly influence reproductive fitness. Moreover, it shows that despite the use of insulin and special diets, diabetes mellitus is being treated rather than cured. Although the diabetic's life may be considerably prolonged, he still may die of the effects of the disease.

This brief treatment of the role of heredity in disease should serve to bring out the growing importance of hereditary factors in the etiology of the more common medical problems confronting the medical profession (and the people suffering from these diseases) today. Moreover, the complexities of many of these conditions stand in marked contrast to the genetically simple and clear-cut hereditary diseases to be discussed in the next chapter.

24.5. Cancer

Cancer represents one of the more perplexing problems in medicine today. It is not a single disease, but an array of different diseases, all characterized by uncontrolled cell division. In normal differentiated cells, cell division is under some form of control. The transition from controlled to uncontrolled cell division in a differentiated cell is heritable and can occur in nearly all cell types. Thus the malignant or cancer cells may be skin cells, kidney cells, bone cells, nerve cells, liver cells, blood cells, and so on, and continue to show many of the morphological and functional traits of their normal precursors. Aggregates of cancer cells are called tumors. Some tumors grow rapidly and others more slowly. Some cancer cells tend to remain at their site of origin, but others spread rapidly through the organism and invade all sorts of normal tissues. Unless the tumor can be removed by surgery or by radiation *in situ,* or unless the spread of the invasive cancers can be arrested, death is the almost inevitable result.

The question of particular interest to the study of genetics is what is the nature of the heritable change that causes a normal differentiated cell to become malignant. In addition to more rapid cell division, cancer cells are characterized by a loss of contact inhibition (of movement and division) and of their specific affinity for cells of their own type. Both changes suggest that the cell surface of a cancer cell has in some way been modified. In addition, cancer cells show increased glucose consumption and enhanced lactic acid production compared with normal cells, but the significance of this finding is still problematical.

The recent discovery that the use of immunosuppressive drugs leads to a significant increase in the incidence of spontaneous tumors has led to the belief that most tumor cells that arise are attacked and destroyed by the host's immune system and that only the exceptional cancer cell escapes from the host's immune response. It has also, of course, led to the hope that cancer can be brought under immunological control.

It has long been known that both radiation and various chemicals are carcinogenic, but the effects of the treatments are so drastic that the exact nature of their carcinogenic effect is obscured by their numerous other effects. The problem is further compounded by the fact that these carcinogens are not 100% effective; rather, only a fraction of the cells become cancerous. It is clear that the transition from a normal cell to a cancerous cell is a permanent hereditary change, because *in vitro* a single isolated cancer cell transmits its cancerous properties to all of its descendants, and these, on transplantation to an appropriate host, will cause the formation of tumors.

Various theories have been proposed to explain the nature of the hereditary change from normal to cancer cell. One theory is that cancer is due to an irreversible differentiation similar to that seen in the differentiation of nerve cells or muscle cells, which henceforth multiply only as nerve cells or muscle cells. While descriptive, this theory is not informative because we do not yet understand irreversible differentiation in any type of cell.

A second hypothesis is that of somatic mutation. Since the incidence of cancer increases with age, it is usually postulated that several somatic mutations must accumulate before the cell becomes malignant. The assumption here is that the cancer results from the loss of some essential gene function.

The third theory, that viruses cause cancer, is currently being very actively pursued. Many cases of viruses as carcinogenic agents have been investigated in experimental animals, so there can be little doubt that viruses can transform normal cells into cancer cells. In this case, added genetic material brings about the transition to the cancerous state. Since the proteins and nucleic acids of viruses are open to manipulation and study at the molecular level, it is becoming possible to study cancer in molecular terms. Where this research will lead remains to be seen, but it seems to hold great promise.

24.6. Immunology

The field of immunology is advancing rapidly at the present time. The increased knowledge of the nature of the immune response will not only have a significant impact in medicine on the understanding of immunization, transplanation, allergy, autoimmune disease, degenerative disorders, cancer, and so on, but it will also contribute to our understanding of the nature of gene action, molecular recognition, cellular differentiation, and other fundamental questions in the biology of higher organisms.

Antigens are macromolecules or else are composed of macromolecules (e.g., viruses and bacteria), and are usually proteins although some polysaccharides and nucleic acids are also antigenic. Antigenicity is determined by the ability of a substance to stimulate specific antibody synthesis in vertebrates. Antibodies are proteins called immunoglobulins found in the blood plasma. The immune response involves not only antibody synthesis but cellular immunity as well (Fig. 24.1). Many questions remain about the role of the cells; only recently, the thymus, whose function was previously unknown, has been found to play a key role in the development of active immunity. It now appears that a class of lymphocytes in the bone marrow migrates to the thymus, where they develop new surface antigens and immunocompetence. They then migrate to peripheral lymphoid tissues and "search" for antigen, with which they are now capable of reacting specifically. These "thymus-derived" cells are responsible for the various phenomena of cell-mediated immunity such as the graft-versus-host reaction. A second class of lymphocytes from the bone marrow serves as precursors of

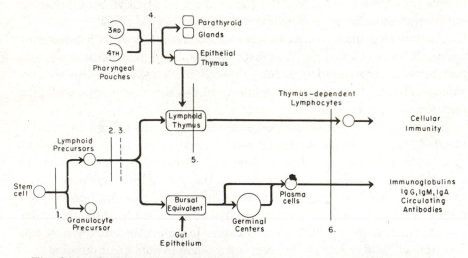

Fig. 24.1. The basis for the immune response and possible blocks leading to immunological disorders. (Hoyer, J. R., M. D. Cooper, A. E. Gabrielsen, and R. A. Good, 1968. "Lymphopenic Forms of Congenital Immunologic Deficiency: Clinical and Pathological Patterns." In R. A. Good and D. Bergsma, eds. *Immunologic Deficiency Diseases in Man.* Birth Defects Original Article Series Vol. 4, National Foundation Press, New York, p. 101, Fig. 8.)

the cells that secrete antibody. These "bone marrow-derived" cells migrate directly to distinct anatomical sites in the lymphoid tissues. Thymus-derived and bone marrow-derived cells cooperate in the antibody response. An antigen or hapten is bound to the thymus-derived cell, and this complex then serves as an efficient stimulus to antibody production by the bone marrow-derived plasma cells.

The immune response is a protective reaction in vertebrates that counteracts the harmful effects of pathogenic organisms and their toxins. This is accomplished by a union between the microorganism or foreign macromolecule and its corresponding specific antibody. These complexes are then engulfed and destroyed by scavenger white blood cells such as the macrophages. The study of antigens and their interaction with the immune system is known as immunology.

An essential part of the immune defense system is the ability of an organism to distinguish between its own macromolecules and foreign macromolecules. The number of distinct antigens that can induce specific antibody formation in unknown, but it must be very large—well over 10,000, possibly in the millions. It is not even known whether a given species can synthesize a large but limited number of antibodies or has the capacity to synthesize virtually an infinite number. At present an experimental test of these alternatives is not practical. The antibodies are very specific in their reaction with the type of antigen that induced their formation; cross-reactions with unrelated protein molecules practically never occur. Moreover, a single antigenic determinant can induce the formation of a heterogeneous array of antibody molecules. Thus, it appears that the immune system responds to particular molecular configurations at a number of different sites on the surface of a molecule or microoraganism by producing a variety of antibodies against that particular antigenic determinant.

All antibodies belong to the group of related serum proteins called immunoglobulins. Three major and several minor classes of antibodies have been identified. The major classes differ from one another primarily in size. Of the major types, the immunoglobulin known as gamma globulin (γG) is predominant, accounting for 80% of all of the plasma immunoglobulin. Far more is known about gamma globulin than the others, so the rest of our discussion will be limited to this immunoglobulin.

The steps leading to our present knowledge of gamma globulin go back to 1890, when von Behring and Kitasato at Robert Koch's Institute in Berlin demonstrated the existence of antibodies by making an animal immune to tetanus with an injection of blood serum from an animal that had survived the disease and become immune to it. Then, in 1917, Landsteiner showed that animals could make antibodies against small organic molecules of known structure, which he called "haptens," and that antibodies recognize different antigens by differences in their shape. Moreover, since the antibodies could specifically bind synthetic haptens or molecules that the species had never previously encountered, it seemed probable that an animal could form antibodies against just about any kind of foreign antigen. More recently, starting with the work of Edelman and Porter in 1959, the structure of the immunoglobulin molecules has been subjected to intensive study. The tremendous heterogeneity among antibody molecules, however, posed an almost insurmountable obstacle to any sort of detailed chemical analysis of antibody structure such as had been possible with the hemoglobin and insulin molecules.

This heterogeneity problem was by-passed by the discovery that individuals with multiple myeloma had large quantities of a single variety of immunoglobulin in their blood plasma. Multiple myeloma or myelomatosis is characterized by multiple tumors in the bone marrow, by excessive production of serum gamma globulins known as myeloma proteins, and by the excretion in the urine, in many cases, of characteristic proteins known as Bence Jones proteins after the physician Henry Bence Jones, who first reported them in 1847. Multiple myeloma is a form of cancer in mice as well as men involving the plasma cells, the type of cell responsible for antibody production. Apparently, the malignant proliferation of a single plasma cell results in the formation of large numbers of identical plasma cells, which produce large quantities of immunoglobulin molecules with identical amino acid sequences. Because they are present in the blood in such high concentration, they are rather easy to isolate. Moreover, each myeloma patient has a different type of myeloma protein or immunoglobulin, with a distinctive amino acid sequence, so comparative studies of the different sequences are possible. The Bence Jones protein in the urine turned out to be excess homogeneous light chains of the myeloma protein made by the tumor, but not incorporated into whole immunoglobulin molecules. Because the Bence Jones light chains were all alike, were readily available in large quantities, and were smaller than the complete immunoglobulin molecule, attention was first directed toward an analysis of their structure and later to an analysis of the complete immunoglobulin molecule.

These detailed analyses have revealed the following facts. Each gamma globulin is formed from two identical light (short) polypeptide chains (MW $\cong$ 23,000) and two identical heavy (long) chains (MW $\cong$ 55,000) and hence has a molecular weight of about 150,000. Each light chain is linked to a heavy chain by a covalent disulfide bridge ($-$S$-$S$-$), and the two heavy chains are connected by two disulfide bridges as shown in Fig.

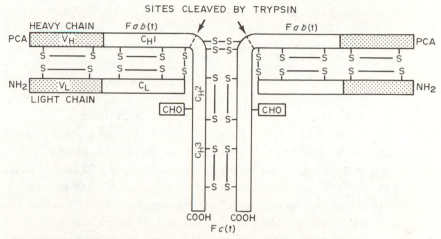

Fig. 24.2. Structure of a gamma globulin molecule. Sulfur to sulfur bonds are shown as —S—S—. Light chains have variable and constant regions (V_L and C_L). Heavy chains have a variable region (V_H) and a constant region composed of three homology regions (C_H1, C_H2, and C_H3). (CHO) indicates carbohydrate. The chains have amino (NH_2), carboxyl (COOH), and pyrollidonecarboxylic acid (PCA) ends. V_H and V_L are homologous, and C_L, C_H1, C_H2, and C_H3 are homologous to one another. Fab(t) and the "crystallizable" fragment Fc(t) are the antibody-binding fragments produced by cleavage of the protein with trypsin. (After Edelman.)

24.2. Furthermore, it is now known that each light chain and each heavy chain has both a constant and a variable segment, with the specificity dependent on the amino acid sequences in the variable segments of both the light and heavy chains.

The light chains contain some 214 amino acid residues in a polypeptide chain. The amino acids are numbered from the amino-terminal or N-terminal end, the end of the molecule made first, to the carboxyl terminal or C-terminal end. The constant region, common to all the light chains, includes residues 108 to 214 in the C-terminal half, while the variable region, specific for each different type of antibody, includes residues 1–107 in the N-terminal half of the molecule. In the heavy chains the variable region includes about 110 amino acids, also at the N-terminal end, and the constant region is about three times as long, some 330 amino acid residues at the C-terminal end. The variable regions in both light and heavy chains are of approximately the same length.

The specificity of the antibody resides in the variable segment and is apparently determined by the specific sequence of the amino acids. Thus far, no two Bence Jones proteins have been found to be identical in the variable segment. Of the 107 residues in this segment, at least 40 have been found to vary. If only two different amino acids could occupy each of these 40 sites (and this is an underestimate), the possible number of different sequences is 2^{40} or more than 10 billion. A similar sort of variation is present in the variable segments of the heavy chains. Clearly, the system has the potential to generate a wide variety of specific antibodies.

The first injection of an antigen causes the production of a small amount of antibody, which gradually disappears from the blood plasma, but a second injection a month or so later produces a response 50 to 100 times greater than the first. Apparently many more plasma cells are engaged in antibody production the second time, and the organism is resistant to the disease or to the adverse effects of the antigen. Although the amount of circulating antibody may again fall to a low level, henceforth subsequent challenges with the antigen elicit a strong, rapid immune response. One of the more intriguing questions about the immune system is the nature of the molecular memory that enables an animal, once exposed to a specific antigen, to recognize it on subsequent exposure, and to respond so differently the second time from the way it did the first. At present, the explanations for immunological memory are mostly conjectural.

Of equal or even greater interest to the geneticist is the question of how a vast array of antigens induces the synthesis of a corresponding array of specific antibodies. Antibodies are proteins, and protein synthesis is controlled by DNA, so the question then becomes how the DNA codes for the synthesis of such a variety of different antibodies more or less on demand, depending on the types of antigens encountered. Since antibody specificity is known to be due to the sequence of amino acids in the variable portion of the polypeptide chains but DNA constancy in the nuclei of the cells of higher organisms is a postulate of molecular genetics, a paradox seems to exist.

Initially, the most plausible explanation for antibody formation seemed to be the "instructive" theory, that the antigen acts as a template in pluripotent cells, causing them to develop the capacity to synthesize antibody specific to that particular antigen. Despite its plausibility, the "instructive" theory turned out to be incorrect. It is now generally agreed that the "selective" theory or the "clonal selection" theory of Burnet and Jerne is more in accord with the facts (Fig. 24.3). The essence of the selective theory is that, prior to the appearance of any antigen, a large number of plasma cells or plasma cell precursors already exist in the organism, each able to make just one specific type of antibody. Contact with an antigen

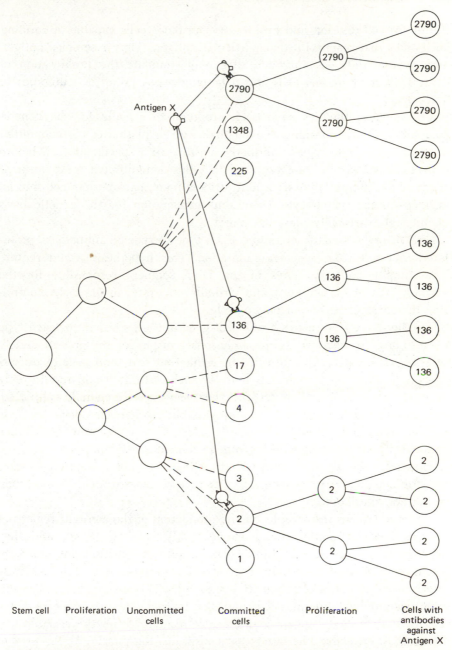

| Stem cell | Proliferation | Uncommitted cells | Committed cells | Proliferation | Cells with antibodies against Antigen X |

Fig. 24.3. The clonal selection theory postulates that the stem cell precursors of the antibody-producing cells become committed during development to the production of a unique immunoglobulin, indicated by the numbers. These cells then can interact with various antigens, and a single antigen may be recognized by more than one antibody-producing cell. Interaction between the cell and the antigen stimulates proliferation of that cell type and synthesis of antibody.

stimulates cell division and proliferation in those cells capable of forming antibodies that react with that particular antigen. Thus a complex antigen with several different antigenic sites may stimulate the proliferation of several different plasma cells and a heterogeneous group of antibodies to that antigen.

A number of theories have been proposed to account for the genetic basis of antibody variability. Given that a given plasma cell is committed to produce a single type of antibody sensitive to a specific antigen before it ever encounters that antigen, and that this committment must occur at the time of differentiation of a nonrestricted bone marrow stem cell into an antigen-specific lymphocyte, then some explanation for the genetic basis of antibody variability must be sought.

One theory is somatic mutation. Each type of specific antibody is postulated to result from a somatic mutation. This hypothesis would require high somatic mutation rates in the DNA segment responsible for the variable part of the antibody and virtually zero rates in the DNA controlling the constant portion of polypeptide.

Another theory postulates a different gene in the germ line for each type of antibody; in other words, an evolutionary origin for the present variety of antibodies, with each controlled by a different structural gene or cistron. Although it has been estimated that 10^5 such cistrons would require only about 1% of the mammalian genome, this theory is not generally supported.

It has also been suggested that the variability originates not at the level of the DNA but through a variable translation mechanism, which can produce a variety of polypeptides from a common genetic message.

However, the theory presently attracting the most attention invokes somatic crossing over between a few or several genes to explain the origin of antibody variability.

Structural genes responsible for the production of the constant regions of various immunoglobulin chains have already been identified, and they segregate as expected in diploid organisms. An immunoglobulin light chain, consisting of a constant region and a variable region, is evidently synthesized as a single polypeptide molecule. This means, then, that this light-chain polypeptide must be controlled by two genes, the gene responsible for the constant segment and a a gene responsible for the variable segment. It is, of course, the latter that constitutes the puzzle. At the present time intensive work is in progress to test the concept of somatic recombination involving two genes or two sets of genes as the mechanism that produces the great variety of genes responsible for the variable regions of the immunoglobulin molecules and hence the specificity of antibodies. In light of the rate of progress in immunology in the past decade, it seems probable that evidence bearing on this theory of somatic recombination will soon be forthcoming.

Additional Reading

Allison, A. C., 1968. "Genetics and Infectious Disease." In *Haldane and Modern Biology*, K. R. Dronamraju, ed. Johns Hopkins University Press, Baltimore.

Bartalos, M., ed., 1968. *Genetics in Medical Practice*. Lippincott, Philadelphia.

Benacerraf, B., and H. O. McDevitt, 1972. "Histocompatibility-Linked Immune Response Genes," *Science* 175:273–279.

Blumberg, B. S., ed., 1961. *Genetic Polymorphisms and Geographic Variations in Disease*. Grune and Stratton, New York.

Carter, C. O., 1969. *An ABC of Medical Genetics*. Little, Brown, Boston.

Cold Spring Harbor Symp. Quant. Biol., 1967. Vol. 32, pp. 1–619. "Antibodies."

Edelman, G. M., 1970. "The Structure and Function of Antibodies," *Sci. Amer.* 223 (Aug.):34–42.

Edelman, G. M., and W. E. Gall, 1969. "The Antibody Problem," *Ann. Rev. Biochem.* 38:415–466.

Goldschmidt, E., ed., 1963. *The Genetics of Migrant and Isolate Populations*. Williams and Wilkins, Baltimore.

Good, R. A., and D. W. Fisher, eds., 1971. *Immunobiology*. Sinauer Assoc., Stamford, Conn.

Haldane, J. B. S., 1949. "Disease and Evolution," *La Ricerca Sci.* 19 (Suppl.):68.

Herzenberg, L. A., H. O. McDevitt, and L. A. Herzenberg, 1968. "Genetics of Antibodies," *Ann. Rev. Genet.* 2:209–244.

Jones, F. A., ed., 1961. *Clinical Aspects of Genetics*. Pitman, London.

Knudson, A. G., 1965. *Genetics and Disease*. McGraw-Hill, New York.

McKusick, V. A., 1969. *Human Genetics*, 2nd ed. Prentice-Hall, Englewood Cliffs, N.J.

Merril, C. R., M. R. Geier, and J. C. Petricciani, 1971. "Bacterial Virus Gene Expression in Human Cells," *Nature* 233:398–400.

Neel, J. V., M. W. Shaw, and W. J. Schull, eds., 1965. "Genetics and the Epidemiology of Chronic Diseases," *Public Health Service Publ.* 1163:1–395.

Osborne, R. H., ed., 1961. "Genetic Perspectives in Disease Resistance and Susceptibility," *Ann. N.Y. Acad. Sci.* 91:595–818.

Platt, R., and A. S. Parkes, eds., 1967. *Social and Genetic Influences on Life and Death*. Plenum, New York.

Porter, I. H., 1968. *Heredity and Disease*. McGraw-Hill, New York.

Porter, R. R., 1967. "The Structure of Antibodies," *Sci. Amer.* 217 (Oct.):81–90.

Roberts, J. A. F., 1970. *An Introduction to Medical Genetics*, 5th ed. Oxford University Press, London.

Temin, H. M., 1971. "The. Protovirus Hypothesis: Speculations on the Significance of RNA-Directed DNA Synthesis for Normal Development and for Carcinogenesis," *J. Nat. Cancer Inst.* **46**:3–7.

Temin, H. M., 1971. "Mechanism of Cell Transformation by RNA Tumor Viruses," *Ann. Rev. Microbiol.* **25**:609–648.

Questions

24.1. In an infectious disease, what factors may influence the course of the infection?

24.2. What methods are available to alleviate or correct the harmful consequences of genetic defects?

24.3. Why is the development of resistant strains of crop plants unlikely to provide a permanent solution to the problem of diseases in crop plants?

24.4. Do you believe there is any parallel between the sporadic outbreaks of wheat rust in cultivated wheat and the periodic influenza epidemics in human populations? Explain.

24.5. Which of the major causes of death showed the least change in frequency between 1900 and 1970? Which showed the greatest change?

24.6. Not only was there a change in the pattern of leading causes of death between 1900 and 1970, but the death rate itself dropped How great was the decline in the death rate? What was the effect of this decline on average life span? Is it possible to calculate the average life span from these data?

24.7. Since the current major causes of death represent constellations of disease entities with complex etiologies rather than simply inherited genetic defects, what possible practical benefits will accrue from making genetics an integral part of medical education?

24.8. Why has the control of cancer seemed such an intractable problem?

24.9. What agents are known to cause a normal cell to become malignant?

24.10. What theories have been proposed to explain the origin of cancer?

24.11. How may genetics contribute to the solution of the cancer problem?

24.12. DNA constancy is one of the tenets of molecular genetics. Another is that DNA, through mRNA, controls protein synthesis. Antibodies are protein molecules and therefore, presumably, their synthesis is controlled by DNA. The specificity and variety of antibody molecules poses a problem for the developmental geneticist. What theories have been proposed to explain the origin of antibody specificity?

24.13. Given the variety and heterogeneity of immunoglobulins, determination of the molecular structure of any one type of immunoglobulin was virtually impossible. What discovery made it possible to determine the amino acid sequences for single immunoglobulins?

24.14. What are the genetic implications of the "instructive" theory of antibody formation? What are the genetic implications of the "clonal selection" theory of antibody formation?

CHAPTER TWENTY-FIVE

Human Traits and Heredity

In this chapter we shall examine some of the variety of simply inherited structural and functional traits in man. The classification of traits in this manner is rather arbitrary and somewhat artificial, since structure and function are intimately related. Nevertheless, for didactic purposes, some organizational scheme is needed that gives an indication of the wide spectrum of inherited human characteristics. For the most part, only simply inherited traits are discussed here, those inherited as Mendelian autosomal dominants (D) or recessives (R) or as sex-linked dominants or recessives (XD or XR). Not too long ago, the list of known simply inherited human traits was relatively short, but it has now grown to the point where McKusick in 1966 published a sizable volume that was just a catalog, with references, of the simply inherited traits in man. Progress has been so rapid that it is already in its third edition, with each edition double the previous one in size.

The purpose of this chapter is to heighten your awareness of how pervasive the effects of heredity are in human development. Perhaps by seeing a few of the many ways that defective genes can cause defective development, you will begin to appreciate how intricate and delicately interwoven are the processes that govern the differentiation of a "normal" human being. Although some of these traits are quite rare, others are fairly common, and the sum total of the frequencies of known simply inherited genetic traits in

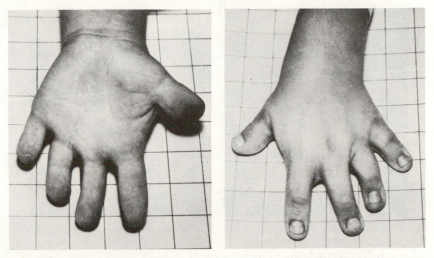

Fig. 25.1. Brachydactyly. (Hoefnagel, D., and P. S. Gerald, 1966. "Hereditary Brachydactyly," *Annals of Human Genetics* 29:377–382, Plate 1, facing p. 380.)

man is of considerable magnitude: It is estimated that presently 2% of all infants born are or will be affected by detrimental traits of simple genetic origin. As the control of infectious and environmental diseases has improved, hereditary diseases have become increasingly important medical and human problems.

25.1. Structural Traits

We shall consider here the effects of heredity on the skeletal system, on the epidermal system, including skin, hair, and teeth, and on the muscular system.

Skeletal System

The first trait in man shown to be inherited as a Mendelian dominant was *brachydactyly* (D), in which all the digits are short due to the apparent absence of the middle phalanges or their fusion with the terminal phalanges (Fig. 25.1). Several different forms of brachydactyly, all D, have now been identified.

The presence of extra fingers and toes is called *polydactyly,* which may be preaxial or postaxial. Subtypes of both preaxial and postaxial polydactyly

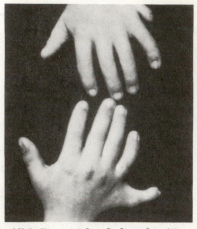
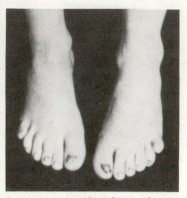

Fig. 25.2. Postaxial polydactyly. (From *Genetics*, Fourth Edition, by A. M. Winchester, p. 15. Copyright © 1972, Houghton Mifflin Company. Reprinted by permission. Photo by C. Nash Herndon.)

are known. In one form of postaxial polydactyly (Fig. 25.2), the extra digit is well formed, and the trait is D with good penetrance; in the other types of polydactyly, however, the genetics are less clear-cut, though probably D with reduced penetrance and variable expressivity. More severe R traits lead to severely deformed or absent hands and feet. For example, the hands and feet are completely absent in individuals with *acheiropody* (R) (Fig. 25.3).

Fig. 25.3. Acheiropody. (Koehler, O., 1936. "Die hand-und fusslosen brasilianischen Geschwister," *Z.f. menschl. Vererbungs-und Konstitutionslehre* 19:670–690, p. 670, Fig. 1.)

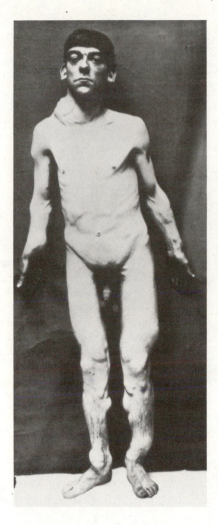

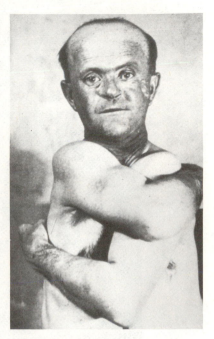

Fig. 25.4. Cleidocranial dysostosis. The "folding man" syndrome. (Kelley, A., "Hereditary Absence of Clavicles." Reprinted with permission from *The Journal of Heredity* 20:352–355, 1929.)
Fig. 25.5. Diaphyseal aclasis (multiple exostoses). (Stocks, P., 1925. "Hereditary disorders of bone development," *Treasury of Human Inheritance* 22–23:1–182, Plate A.)

The traits just cited affecting the limbs in man illustrate a situation often observed in human genetics. The less severe traits are transmitted as Mendelian dominants, while similar but more deleterious conditions are recessive.

Cleidocranial dysostosis, the "folding man" syndrome, is also a complete D (Fig. 25.4). Lacking clavicles or collar bones, affected individuals can almost make their shoulders touch; they also have persistent open skull sutures and other bone defects.

In *diaphyseal aclasis* or multiple exostoses (Fig. 25.5), bony outgrowths or exostoses are formed in the distal region of the diaphysis or main shaft of the long bones. The bony or cartilaginous growths are usually benign,

Fig. 25.6. Henri de Toulouse-Lau-trec (1864–1901). Once thought to have osteogenesis imperfecta, Toulouse-Lautrec is now thought to have suffered from a clinically similar but different disease. (From McKusick, Victor A.: *Heritable Disorders of Connective Tissue*, ed. 2, St. Louis, The C. V. Mosby Co.)

and heredity is D, but more males than females appear to be affected due to reduced expression and penetrance in females.

Osteogenesis imperfecta is a generalized connective tissue disorder that is sometimes known as "brittle bones" because of the ease and frequency with which bone fractures occur, even in the fetus in some cases (Fig. 25.6). The most significant functional defects are brittle bones and deafness (otosclerosis), but the blue sclerae ("whites") of the eyes are a striking feature. Thin skin, loose-jointedness, and hernia also occur as manifestations of the syndrome. Because of the wide range of expression of the trait, several clinical types have been described. However, they may simply reflect the variable expression of the same D gene, although the very severe form, *osteogenesis imperfecta congenita*, lethal at a very early age, seems to be R.

A variety of stories is told about the fragility of the bones of persons afflicted with the trait. Whittling may fracture the forearm, or writing may fracture the phalanges. A patient may have both femora broken when his girl sits in his lap, or simply by his own weight when standing. Ivar the Boneless, who masterminded the Scandinavian invasion of England in the late ninth century A.D., is reputed to have had *osteogenesis imperfecta*. He was unable to walk and had to be carried into battle on a shield. This

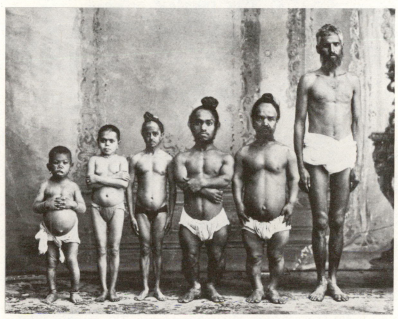

Fig. 25.7. Dwarfs in India. Left to right: cretinous dwarf, two ateliotic dwarfs, two achondroplastic dwarfs, and a man of normal height. Also see Figs. 25.17 and 25.18. (Rischbieth, H. and A. Barrington. 1912. "Dwarfism." *Treasury of Human Inheritance.* Vol. I, parts 7 & 8. Eugenics Laboratory Memoirs, No. 15. Published by the Cambridge University Press, London, for the Francis Galton Laboratory of National Eugenics, University of London. Plate LL (77).)

story, like some of the others, may be apocryphal, but the fragility of the bones is not.

Classical *achondroplasia* (D) is a type of dwarfism characterized by short limbs and a large head (Fig. 25.7). Seven out of eight cases are estimated to result from new mutations, because obstetric problems related to the small pelvis in the females and a reduced rate of reproduction in both sexes lower the reproductive fitness.

The *Marfan syndrome* or *arachnodactyly* affects primarily the skeleton, the eyes, and the cardiovascular system (Fig. 25.8). The name arachnodactyly is derived from the characteristically long, thin extremities. Ectopia lentis or displaced lens is the customary ocular defect, and an aneurysm or dilation of the ascending aorta is the typical cardiovascular defect. This syndrome is D with fairly high penetrance and variable expressivity. The trait is of special interest because of the recent suggestion that Abraham

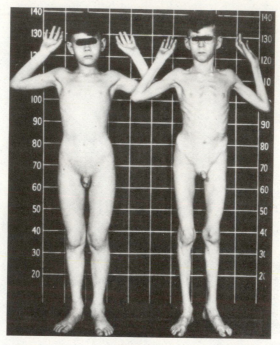

Fig. 25.8. Arachnodactyly or Marfan's syndrome. The normal brother on the left is 10 years old; the brother on the right with Marfan's syndrome is only 8 years old. (From McKusick, Victor A.: *Heritable Disorders of Connective Tissue*, ed. 4, St. Louis, The C. V. Mosby Co.)

Lincoln had Marfan's syndrome. Since Marfan first described the trait in 1896, 31 years after Lincoln was assassinated, the diagnosis is obviously open to question. Lincoln was tall, with unusually large hands and feet, but he was renowned for his athletic prowess. If he was affected, his was a mild case. It should be added that his intelligence, his sense of humor, and his warmth of personality are not incompatible with the syndrome.

Integumentary System

Skin. In "complete" *albinism* (R), the melanocytes or pigment cells are present but fail to produce normal amounts of melanin (Fig. 25.9). Melanin is not entirely absent in human albinos, in contrast to albino mice, rats, and rabbits. However, human albinos are characterized by extremely light skin and hair and by reduced pigmentation of the eyes. Other ocular defects are usually present, and photophobia and squinting are characteristic.

Partial albinism, piebaldism, or white forelock are three names given to a D trait in which individuals have unpigmented patches on various parts

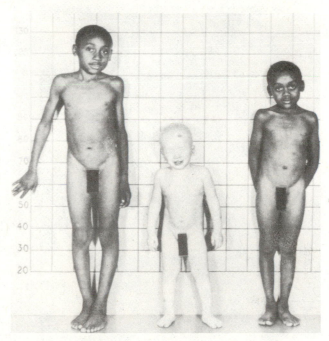

Fig. 25.9. Young Negro albino boy with his normally pigmented brothers. (Goodman, R. M., and R. J. Gorlin, 1970. *The Face in Genetic Disorders.* C. V. Mosby Co., St. Louis, Mo., p. 123, Fig. 52B.)

of the body, usually including the forehead, hence the white forelock (Fig. 25.10).

Xeroderma pigmentosum (R) is characterized by extreme sensitivity of the skin to sunlight. Onset is marked by the development of frecklelike lesions on the skin, which generally become malignant carcinomata causing death at an early age (Fig. 25.11).

Neurofibromatosis or von Recklinghausen's disease (D) is marked by cafe-au-lait or coffee-colored spots on the skin and by the formation of tumors at the nerve endings (neurofibromas) in the skin and also along the nerve trunks. These tumors, which may become quite large, may have secondary effects due to pressure, and so on, and occasionally they may become malignant (Fig. 25.12).

Ichthyosis is a disorder marked by a dry, scaly skin, somewhat resembling fish skin (Fig. 25.13). The mild, common form, *ichthyosis vulgaris*, is D. Individuals with the rarer R form, *ichthyosis congenita*, may die of complications very early in life due to infection or loss of fluids from the deep fissures lying between the horny plates that cover the entire surface of the body. Less severe cases are, however, compatible with life.

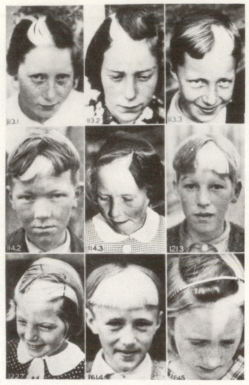

Fig. 25.10. Partial albinism, a dominant trait with the white forelock its most obvious effect. These children are members of a Norwegian family with 42 individuals known to have the trait, which is traceable through four generations from an affected great grandfather. Except for the lack of pigment, these people are normal. (Sundför, H. "A Pedigree of Skin-Spotting in Man." Reprinted with permission from *The Journal of Heredity* 30:67–77, 1939.)

A rather unusual trait is ***anhidrotic ectodermal dysplasia*** (XR), in which sweat glands are absent, the teeth are absent or deficient, and the hair is fine and sparse. Being unable to sweat, affected males suffer in warm weather. Heterozygous females may show minor manifestations, but there seems to be some question about the expression in carrier females. This trait would seem useful in studies of the Lyon hypothesis of the inactivated X chromosome. In some pedigrees, clinically rather similar conditions are R.

The "India rubber men" of circus sideshows, who can stretch their skin for half a foot or more, have the ***Ehlers-Danlos*** syndrome (D) (Fig. 25.14). In addition to hyperelastic skin, they have hyperextensible joints and fragile skin and blood vessels. The basic defect seems to lie in the connective tissue.

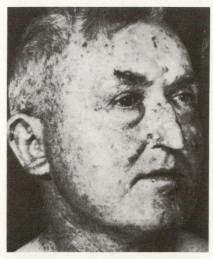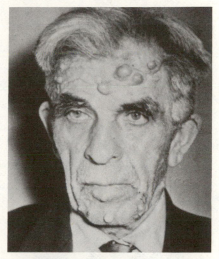

Fig. 25.11. Xeroderma pigmentosum. This patient is less severely affected than some, but shows progressive atrophic scarring and malignant degeneration. (Goodman, R. M., and R. J. Gorlin, 1970. *The Face in Genetic Disorders.* C. V. Mosby Co., St. Louis, Mo., p. 167, Fig. 74.)

Fig. 25.12. Neurofibromatosis or von Recklinghausen's disease. (Goodman, R. M., and R. J. Gorlin, 1970. *The Face in Genetic Disorders.* C. V. Mosby Co., St. Louis, Mo., p. 119, Fig. 50B.)

*Hair. **Premature graying*** of the hair, with the hair starting to turn gray in the late teens and being white by the mid-twenties, seems to be D. Aside from their striking appearance, individuals seem otherwise normal.

Hair color in man, like coat color in mammals, is undoubtedly governed by genes at a number of different loci. For this reason, the genetics of hair color are not simple. However, ***red hair*** seems to result from homozygosity for a recessive gene at a single locus, because two red-headed parents almost invariably have all red-haired children, and in families where neither parent has red hair but red-headed children are born, the corrected ratio of nonred to red is approximately 3:1. The various shades of red from auburn to reddish blonde can be explained by the brunet to blonde range of residual genotypes on which the red genes are being expressed.

In some cases ***baldness*** results from disease, but most cases of baldness are hereditary. However, the expression of the trait is sex-influenced, with many more men than women being bald. This difference is not due to sex-linkage, however, but to the differences in the sex hormones of men and women. The autosomal gene responsible for baldness acts as a dominant in males, with both heterozygous and homozygous men becoming bald. In women, however, heterozygotes retain their hair while homozygotes may have thin hair or, occasionally, partial baldness. These conclusions are supported by pedigree analysis and also by studies of unusual hormonal

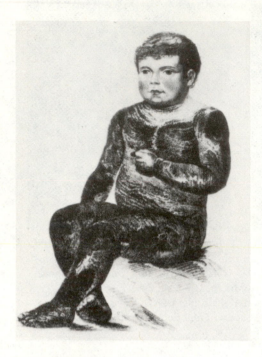

Fig. 25.13. Martin's 1818 case of hereditary ichthyosis. Jane Holden, aged 3 years. (Penrose, L. S., and C. Stern, 1958. "Reconsideration of the Lambert Pedigree (*Ichthyosis hystrix gravior*)," *Ann. Human Genetics* 22:258–283, Plate 5 facing p. 278.)

situations in men and women. For example, Hippocrates long ago observed that eunuchs never become bald. More recently, the administration of male hormones to a group of males who had failed to mature sexually and none of whom were bald caused some of them to start losing their hair. Furthermore, the development of masculinizing tumors in the adrenal cortex, often responsible for the beard of the bearded ladies in the circus, has also been found to induce baldness in such women if they have the appropriate genotype. Various types of patterns of baldness have been identified, and heredity also plays a role in the pattern of baldness, but this role is not yet well worked out. However, the relatively late age of onset, the variability in expression, and the sex-influenced nature of baldness all contribute to making its study more complex than that of a simple autosomal dominant expressed at birth. Nevertheless, the hypothesis that baldness results from an autosomal gene, dominant in the male and recessive or subrecessive in the female, is in accordance with the facts.

Although *hairy ears* were formerly thought to be an example of Y-linked inheritance, the trait passing from a father to all of his sons, this hypothesis has recently been under attack. As with baldness, there is an age effect; the trait may not appear in some men until late middle age. Moreover, the trait may be sex-limited, with expression restricted to the males. If so, the

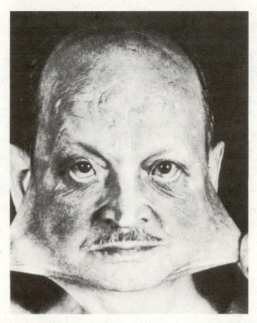

Fig. 25.14. Ehlers-Danlos syndrome. (Goodman, R. M., and R. J. Gorlin, 1970. *The Face in Genetic Disorders.* C. V. Mosby Co., St. Louis, Mo., p. 61, Fig. 21A.)

pedigrees can be interpreted in more than one way. At present, no unequivocal example of Y-linkage is known in man.

Teeth. Absence or reduction of the **upper lateral incisors** is due to a D. An XD is responsible for **enamel hypoplasia,** which causes the teeth to wear down rather quickly to mere stumps because the enamel is so thin. The differences in expression between affected males and females have been cited in support of the Lyon hypothesis. The males have uniformly thin enamel, but in females the teeth appear vertically grooved, because the enamel is much thicker in some places than in others.

Muscular System

The ability to "curl" the tongue by rolling up the lateral edges was originally reported to be inherited as a D, but this was later cast into doubt by a report of discordance in monozygotic twins. Even if the trait were strongly affected by nongenetic influences, a "high frequency of discordance" in identical twins seems rather surprising. For this reason, additional twin studies and pedigree analyses seem desirable. About 70% of the people of European ancestry can curl their tongues, but the ability

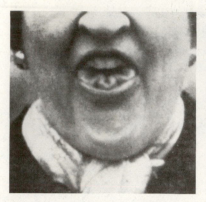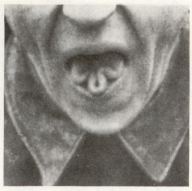

Fig. 25.15. Clover-leaf tongue in a mother and her son. (Whitney, D. D., "Cloverleaf Tongues." Reprinted with permission from *The Journal of Heredity* 41:176, 1950. Copyright 1950, by the American Genetic Association.)

to form the tongue into a cloverleaf configuration (Fig. 25.15) is much rarer. It, too, seems to be D.

Peroneal muscular atrophy (Charcot-Marie-Tooth syndrome) occurs in three forms, D, R, and XR, with the R type more severe than the D type and the XR type intermediate in effect. The onset is marked by weakness and atrophy of the peroneal muscles in the calf, while later other distal muscles of the legs and arms are involved. The basic defect appears to be degeneration in the spinal cord of the roots of the motor nerves to the distal parts of the extremities.

Myotonias are characterized by tonic spasms of the muscles following voluntary contraction so that they are slow to relax. In *paramyotonia congenita* (D) the paralysis is expressed mainly when the muscles are exposed to cold. The trait is not progressive so that affected persons can live reasonably normal lives if they avoid such "hazards" as swimming in cold water or eating ice cream.

A number of types of hereditary muscular dystrophy have been identified. Of these, progressive *pseudohypertrophic muscular dystrophy* (Duchenne type) has its onset before age 6; the muscles waste away so much that the patient is chairbound by 12, and dead before 20. This type is XR, and only affected males are observed since homozygous recessive females would have to receive a recessive gene from each parent, but boys with the gene do not survive long enough to reproduce. An R type of muscular dystrophy (type II) quite similar to the Duchenne XR type has been identified (Fig. 25.16). In this form also, pseudohypertrophy of the muscles occurs before they start to waste away. *Facio-scapulo-humeral*

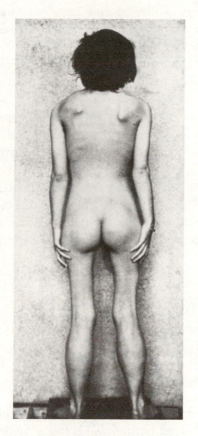

Fig. 25.16. Autosomal limb-girdle muscular dystrophy. A girl of 15 with prominent calves but wasting of the thighs and gluteal muscles and of the upper arms and shoulder muscles. (Stevenson, A. C., 1953. "Muscular Dystrophy in Northern Ireland," *Annals of Eugenics* 18:50–93, Plate 3.)

progressive muscular dystrophy (D) is marked by the wasting of the muscles of the face, shoulders, and arms. The age of onset, between 7 and 20, is later than in the recessive types above, and the course of the D disease is not so severe, as is often the case with clinically similar D and R traits. It has been estimated that more than 200,000 people in the United States are currently affected by one or another of the various types of hereditary muscular dystrophy, half of them children between the ages of 3 and 13.

25.2. Functional Traits

Sense Organs

Eyes. In persons affected with *ocular albinism*, pigmentation is normal except in the eyes. The trait is X-linked; in affected males the iris is only slightly pigmented, while the fundus (the internal surface at the back of the eye) lacks pigment. Furthermore, nystagmus (a constant, involuntary movement of the eyeball), head nodding, and impaired vision are characteristic. In heterozygous females, the fundus shows a mosaic of pigmentation, as would be expected under the Lyon hypothesis.

Congenital night blindness causes people with normal vision in daylight to be blind or nearly so at night. Night blindness may also be due to a deficiency of vitamin A, and this form responds to supplementary vitamin A in the diet, but the congenital type is refractory to such treatment. One hereditary type of night blindness is D; another is X-linked and distinguishable from the D type by being associated with myopia or nearsightedness.

Color blindness is one of the most familiar and common X-linked traits in man: About 8% of Western European males are color blind. However, the genetics are not as simple as was formerly thought, because several types of color blindness have now been described. These can be identified as follows:

1. Total color blindness
2. Deutanomaly   2a. Deutanopia
    (green weakness)   (green blindness)
3. Protanomaly   3a. Protanopia
    (red weakness)   (red blindness)
4. Tritanomaly   4a. Tritanopia
    (blue weakness)   (blue blindness)

Total color blindness is also known as "day-blindness," an apt expression because the cones are absent and affected persons see better at night. This trait is a rare autosomal recessive.

The protan and deutan defects are the most common types of color blindness. Of the 8% of males affected, some 6% have a deutan defect and 2% a protan defect. The relative frequency of affected males and females suggests that two loci on the X chromosome are involved rather than just one. If one locus were involved, the frequency of homozygous color-blind females expected would be $q^2 = (0.08)^2 = 0.0064 = 0.64\%$. With two loci this frequency would be $q_d^2 + q_p^2 = (0.06)^2 + (0.02)^2 = 0.0036 + 0.0004 = 0.40\%$. Actual data on these relative frequencies agree more closely with expectations based on the two-locus theory. Furthermore, women known to carry both protan and deutan types of color blindness because their sons have one or the other type do not have defective color vision. Therefore, by the complementation test for allelism, these defects are not allelic. The simplest hypothesis is that at each locus there is a series of three alleles with dominance in the order shown:

normal > deutanomaly > deutanopia
normal > protanomaly > protanopia

Less is known about the tritan defects, both because they are rare and because deficiencies in blue vision are of less practical importance than those involving red or green. However, D and X-linked forms of tritanopia have been reported, while tritanomaly is thought to be X-linked.

Retinitis pigmentosa, a hereditary cause of blindness, produces a gradual progressive degeneration of the retina as well as deposits of pigment in the retina. This trait is notable for its genetic heterogeneity, for a number of different forms of the disease have been recognized. In some, only the retinal degeneration occurs, and D, R, and XR pedigrees of this sort have been reported. In addition, retinitis pigmentosa may occur as one component in a syndrome. Nearly 10 such syndromes have already been reported, and retinitis pigmentosa has been found to be associated with deafness, idiocy, and other defects, again with various D, R, and X-linked modes of inheritance. Thus, the phenotype results from a number of different genetic defects.

Retinoblastoma is a D that causes tumors to develop in the retina in early childhood. The condition is fatal unless the tumorous eyes are removed by surgery.

Ears. **Darwin's points** are not inherited in a clear-cut fashion, but the pedigrees suggest a D with incomplete penetrance and variable expressivity. They are of interest because they are thought to be a vestigial hereditary trait, suggesting the longer, pointed ears of our more primitive mammalian ancestors.

A number of types of hereditary deafness have been described. In some, such as **otosclerosis**, the middle ear is affected, in this case by deposition and hardening of bony tissue. In others, the inner ear is affected, and the auditory nerve itself may be involved. Deafness is also often found as part of a syndrome. As might be expected, various modes of inheritance are found in different pedigrees, and all degrees of severity have been observed. The age of onset also varies in different pedigrees, and sometimes within the same pedigree, but the deafness usually becomes progressively worse. Those forms of deafness that are congenital or present at birth prevent the development of speech unless special training is given. Hence, they are sometimes called deaf-mutism even though the basic defect is deafness.

Taste and Smell. Studies of these senses are undoubtedly hampered by the difficulties in devising adequate objective tests. However, hereditary *anosmia* (inability to smell) due to incomplete development of the olfactory lobes has been reported, although the mode of inheritance remains in doubt.

Taste sensitivity to phenylthiocarbamide (PTC) is the most thoroughly studied trait involving taste. The ability to taste PTC is inherited as a D, with PTC tasting bitter to about 7 out of 10 Europeans. The difference between tasters and nontasters is not absolute, but rather is a difference in the taste threshold between the two groups. Taste sensitivity decreases with age and is undoubtedly influenced by other conditions, but the bimodal distribution of taste thresholds is quite clear-cut, and the number of anomalous intermediate cases is relatively few. Additional special tests for this group have been devised.

The response to PTC points up some interesting problems. First, heredity influences our perception of the world, and not all of us perceive the world in the same way: To many people, PTC is a distasteful, noxious substance, but to others it causes little or no reaction. Second, the taster and nontaster alleles seem to be in polymorphic equilibrium ($T = 0.45$; $t = 0.55$), and the polymorphism is evidently of some duration since it also exists among chimpanzees. However, the adaptive significance of the PTC polymorphism remains in doubt. The only relevant finding seems to be that nontasters are more frequent than expected among persons with a hypothyroid condition called nodular goiter, but the exact significance of this fact is not known.

Alimentary System

Cystic fibrosis of the pancreas (R) is the most common lethal inherited disease of childhood, with an incidence estimated as 1 in 3700 or higher in Caucasians, with a gene frequency of 0.016 and a heterozygote frequency of about 3%. The effects of the gene are noted primarily in the pancreas, the small intestine, and the bronchioles of the lungs. The exocrine function of the pancreas, the secretion of pancreatic juice into the digestive tract, is affected, as are the intestinal and bronchial glands. As a result, affected children fail to thrive, have fatty, foul-smelling stools, and suffer from recurrent respiratory infections. No adequate explanation for the high incidence of such a deleterious recessive is yet available.

Urogenital System

Two major types of *polycystic disease* of the kidneys have been identified. The age of onset in one is at or near the time of birth; for the other the age of onset ranges between 30 and 60. The congenital form, usually causing death shortly after birth, is R, while the adult form is D. The resultant renal insufficiency may lead to associated traits such as hypertension and uremia. In the recessive form, cystic changes may also occur in the liver and pancreas. In still other families, polycystic kidneys may appear as one feature of a complex syndrome.

Adrenal virilism is due to a R type of *adrenal hyperplasia* that causes excessive secretion of adrenal androgens (hormones similar to male sex hormones). The condition leads to precocious growth and sexual maturity in males and to the masculinization of females. Counterbalancing hormone treatments will help affected individuals to keep their sexual and physical development within normal limits.

Endocrine System

The failure of the pancreas to secrete sufficient insulin, a hormone needed for the proper utilization of sugar in the body, leads to the development of *diabetes mellitus*, a disease with an incidence of about 1% in the population of the United States. It is marked in untreated patients by increased glucose levels in the blood, the presence of glucose in the urine, excessive thirst and urine production, and eventually by dehydration. Because it is so common and because the age of onset and the severity of the disease may be quite variable, the mode of inheritance of diabetes mellitus remains obscure. D, R, and multiple-factor hypotheses have been advanced. While heredity plays an undoubted role in many cases, in others the symptoms may arise in association with other conditions.

Pituitary diabetes insipidus, which resembles diabetes mellitus primarily in the abnormal thirst and frequent urination, responds to treatment with the antidiuretic hormone, vasopressin, from the posterior pituitary. This rare trait is inherited as an autosomal D, although X-linked recessive families have also been reported.

Nephrogenic diabetes insipidus is marked by the failure of the renal tubules to reabsorb water, a condition that does not respond to antidiuretic hormone. The best present information suggests autosomal D inheritance, with penetrance and expressivity reduced more in females than in males. It too is marked by excessive thirst and the excretion of large volumes of dilute urine. Unlike pituitary diabetes insipidus and diabetes mellitus, both of which result from a hormone deficiency and can be controlled by administration of the proper hormone, the primary defect in nephrogenic diabetes insipidus seems to lie in the renal tubules themselves, and no specific treatment has been developed.

Renal glycosuria is a benign trait inherited as an autosomal D and is marked by excretion of glucose in the urine even though the blood sugar

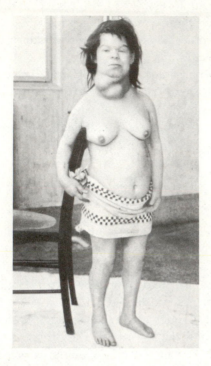

Fig. 25.17. Cretinism. Note the goitrous enlargement of the thyroid gland. (Rischbieth, H. and A. Barrington. 1912. "Dwarfism." *Treasury of Human Inheritance.* Vol. I, parts 7 & 8. Eugenics Laboratory Memoirs, No. 15. Published by the Cambridge University Press, London, for the Francis Galton Laboratory of National Eugenics, University of London. Plate GG (63).)

is normal. Apparently, the glucose in the glomerular filtrate, which is normally almost completely reabsorbed in the proximal convoluted tubules of the kidney, is not efficiently reabsorbed in persons with this trait. However, unlike the various types of diabetes cited above, which may lead to diseased states of varying severity, renal glycosuria, though it simulates diabetes mellitus in one respect, does not require treatment.

Five forms of *familial cretinism with goiter* have been identified. Cretinism is a characteristic form of dwarfism associated with mental deficiency (Fig. 25.17). Both physical and mental development are retarded because of a thyroid deficiency, and the goiter results from compensatory growth of the thyroid gland. The synthesis of the thyroid hormone requires a series of steps, and four types of cretinism result from blocks at different points in the synthesis of the hormone. In the fifth type an abnormal thyroglobulin is synthesized and can be detected in the blood serum instead of the normal thyroid hormone. The most probable genetic explanation is that these various forms of cretinism are due to different autosomal recessive genes, but many cases of cretinism are of environmental origin. Administration of the thyroid hormone will generally help to prevent the development of cretinism and mental retardation.

Fig. 25.18. Three brothers. The two ateliotic dwarfs are 24 and 22 years old and about 40 in. tall. (Rischbieth, H. and A. Barrington. 1912. "Dwarfism." *Treasury of Human Inheritance.* Vol. I, parts 7 & 8. Eugenics Laboratory Memoirs, No. 15. Published by the Cambridge University Press, London, for the Francis Galton Laboratory of National Eugenics, University of London. Plate BB (45).)

Ateliotic dwarfs are normal in proportions and in intelligence but are midgets in size (Fig. 25.18). The simplest form of ateliotic dwarfism is due to an R that causes a deficiency of only the growth hormone among the anterior pituitary hormones, the other hormones being present. Another type of ateliotic dwarf suffers from a deficiency of all of the anterior pituitary hormones and thus is affected by hypothyroidism, hypoadrenalism, and hypogonadism as well. The two types are sometimes distinguished as the sexual and asexual forms of ateliotic dwarfism.

Hemopoietic System

A considerable number of hereditary diseases of the hemopoietic system are known in man, far too many to be considered in any detail. We have already discussed the blood groups and the structure of the hemoglobins in earlier chapters. Hence, only a few examples will be cited to show that the hemopoietic system, too, is subject to hereditary diseases. For convenience, we shall consider in turn abnormalities of the erythrocytes, leucocytes, and platelets.

Sickle cell anemia, discussed previously in Chapter 22, is representative of a group of diseases resulting from mutations in structural genes that lead to the synthesis of abnormal hemoglobins. In the case of sickle cell anemia, valine is substituted for glutamic acid at position 6 in the β chains. Rather than normal adult hemoglobin (hemoglobin A, $a_2{}^A\beta_2{}^A$), patients with sickle cell anemia have only hemoglobin S ($a_2{}^A\beta_2{}^{6 \text{ Val}}$ or $a_2{}^A\beta_2{}^S$) and are homozygous for the gene producing hemoglobin S (Hb S). The heterozygotes, with one normal and one sickling gene, show the sickling trait of the erythrocytes under reduced oxygen tension, but are not otherwise seriously affected ordinarily. They have about 40% Hb S and 60% Hb A, so the genes apparently act independently of one another to produce nearly equal amounts of $a_2{}^A\beta_2{}^S$ and $a_2{}^A\beta_2{}^A$. The sickling trait occurs in up to 45% of some African tribes, and is associated with reduced mortality and increased fertility of the heterozygotes in areas of endemic falciparum malaria.

Many additional types of abnormal hemoglobins have been discovered, and for many the substituted amino acid has been identified.

Thalassemia is a chronic hemolytic anemia in which the red cells are deficient in hemoglobin, abnormal in shape, and destroyed at an unusually rapid rate. It is also known as Cooley's anemia, after its discoverer, and as Mediterranean anemia because it was once thought to be confined to persons whose origins were in the Mediterranean region. Again, homozygotes for the gene for thalassemia are severely affected with *thalassemia major*, while the heterozygotes with a single dose of the gene show only mild symptoms referred to as *thalassemia minor*. The homozygotes usually die at an early age.

In thalassemia, the defect appears to be the low rate of production of normal adult hemoglobin rather than the formation of an abnormal hemoglobin comparable to Hb S. Fetal hemoglobin (Hb F $- a_2{}^A\delta_2{}^F$) and Hb A$_2$ ($a_2{}^A\delta_2{}^{A2}$) are produced in normal amounts. Therefore, it has been suggested that classical Mediterranean anemia is due to the failure of an operator gene to initiate the formation of the β chains necessary for Hb A ($a_2{}^A\beta_2{}^A$). Thus β-chain thalassemia would be accompanied by elevated amounts of Hb F and Hb A$_2$. Absence of the a chain would have even earlier and more severe consequences, because it is utilized in Hb F and thus may endanger the fetus. Nevertheless, a-chain thalassemia may occur because hemoglobin H is known to be a tetramer, β_4, as is hemoglobin Bart's, γ_4.

Therefore, it appears that sickle cell anemia and other diseases due to abnormal hemoglobins result from mutations in structural genes, but that thalassemia and related conditions result from mutations in operator genes. These two types of mutations are codominant and are independently inherited and expressed. When two such mutants coexist in a single individual, an anemia often develops.

Hereditary spherocytosis, a disease marked by the presence in the blood of red cells more nearly spherical than normal, has three major clinical effects: anemia, jaundice, and an enlarged spleen. These abnormal red cells are osmotically and mechanically fragile, and are destroyed at an accelerated rate. The enhanced rate of destruction, primarily in the spleen, is compensated to some extent by an increased rate of red cell formation in the bone marrow. The clinical course of the disease depends on how well this balance is maintained. The basic defect has not yet been identified, but is thought to be in the red cell membrane, which is abnormally permeable to sodium. Under favorable metabolic conditions, as in the plasma, the defect in the spherocytic red cells is evidently compensated for by an increased rate of metabolism. However, conditions are less favorable during circulation of these red cells through the spleen, and degenerative changes occur. The damaged red cells are then slowed in their passage through the spleen, and this in turn leads to further damage. Repeated passage through the spleen eventually produces irreversible damage in the spherocytic red cells, which are then trapped in the spleen where they undergo hemolysis.

Normal red cells remain normal when transfused to patients with hereditary spherocytosis; hereditary spherocytes have a reduced life span even when transfused to a normal subject. Thus, the red cell defect seems to be intrinsic to the red cell itself. Hereditary spherocytosis appears to be inherited as a D with an incidence of about 2 per 10,000, so affected persons are nearly always heterozygous.

Perhaps the most unusual aspect of this hereditary disease is that the clinical symptoms disappear when the spleen is surgically removed. However, the inherited defect in the red cells remains, because if they are transfused from a splenectomized patient to a normal subject, they again undergo rapid hemolysis in the spleen. Thus, splenectomy cures the symptoms without curing the defect. Nevertheless, splenectomy is by far the most effective treatment for any patient with hereditary spherocytosis.

Glucose-6-phosphate dehydrogenase deficiency (G6PD) is a hereditary biochemical defect in the red blood cells with several interesting aspects. The trait was first discovered when previously innocuous drugs caused acute hemolytic anemia in certain susceptible persons. Antimalarial drugs, such as primaquine, some sulfonamides, and a variety of others, have been found to act in this way. Moreover, in some persons, eating the *Vicia fava* bean precipitated hemolytic anemia. These reactions were traced to a deficiency of the enzyme glucose-6-phosphate dehydrogenase in the erythrocytes, but the exact sequence of events remains in some doubt. Further analysis revealed more than 20 variants of G6PD, some of them quite common. For instance, the gene that causes sensitivity to *Vicia fava* is quite frequent among Caucasians of Mediterranean origin. A high incidence of G6PD deficiency is also found among Negroes, but they have a different variant and manifest primaquine sensitivity but not favism. All of the variants are apparently due to different alleles at a single X-linked locus. Many cause G6PD deficiency, but there are several variants with normal activity and even some with enhanced G6PD activity. The G6PD locus has been mapped at about 5 map units from the locus for deutan color blindness and also seems to be quite close to the locus for hemophilia A on the X chromosome. This trait provides further support for Lyon's hypothesis of the inactivated X chromosome, because heterozygous females have two types of red cells, one with normal enzyme activity and the other deficient in G6PD activity. Lastly, a comparison of the distribution of the more frequent G6PD alleles with the incidence of falciparum malaria has led to the conclusion that these alleles in some way confer protection against malaria.

In the **Pelger-Huët nuclear anomaly,** the polymorphonuclear leucocytes, which are usually segmented into two to five distinct chromatin masses connected by thin threads of chromatin, are either unsegmented or show at most two or rarely three subdivisions. All three types of polymorphonuclear leucocytes, the neutrophils, eosinophils, and basophils, are affected, but the anomaly is not seriously debilitating. However, some reduction in fitness has been postulated on the assumption that affected persons are less able to combat bacterial infections because their leucocytpes are less active in phagocytosis. The anomaly behaves as a clear-cut D.

"Essential" or "familial" *thrombocytopenia* is an X-linked condition characterized by a reduced number of platelets in the blood. The bleeding symptoms in this trait are relatively mild, but **Aldrich's syndrome,** also X-linked and also marked by thrombocytopenia, produces eczema, proneness to infection, and bloody diarrhea, and is nearly always fatal before the age of 10.

Thus red cells, white cells, and the blood platelets each may be affected by different mutant genes. Still other genes are known to cause abnormalities in components of the blood plasma. In the past half-century much has been learned about blood clotting, and the picture of events that has emerged is complex, since a number of different factors must be present for normal clotting to occur. Hemophilia is an example of a disease due to a hereditary deficiency of a clotting factor in the plasma.

Hemophilia has probably been known as a hereditary disease longer than any other; both the Jews and the Arabs recognized its hereditary nature many centuries ago. Furthermore, in the second century A.D. the Jews permitted a special dispensation with respect to circumcision of a male child if two of his brothers had previously died of uncontrollable bleeding following circumcision or if two of his male cousins on his mother's side had died as the result of circumcision. Thus, even then, it was realized that the disease is hereditary, that it affects primarily the males, but is transmitted by females. However, hemophilia as a disease entity was not described in the medical literature until 1803, and more than a century passed before its X-linked recessive mode of inheritance was understood.

Hemophilia is usually recognized early in childhood because the tendency to hemorrhage appears following circumcision or minor injury. Uncontrollable bleeding may follow extraction of teeth or the removal of tonsils. Affected boys bruise easily and may bleed internally, either in the internal organs or intracranially, which is often fatal. Bleeding into the joints will in time lead to crippling deformities. Thus the lot of a hemophiliac is not an easy one, and in the past it is estimated that more than half of them died before the age of 6 and only a quarter survived beyond the age of 20. The survival rate has improved somewhat in recent years as the result of better understanding of the nature of the disease.

One of the more important advances is the recognition that a clotting defect may be due to any one of a number of different causes. *Classic hemophilia* (now also called *hemophilia A*) is now known to be due to a deficiency of a serum protein known as antihemophilic globulin (AHG) or factor VIII. A second important type of hemophilia is *hemophilia B* or *Christmas disease*. The latter name originated because the disease was first reported in an English family named Christmas during Christmas week, and this combination was apparently irresistible. Christmas disease is due to a deficiency of plasma thromboplastin component (PTC) or factor IX. Clinically, classic hemophilia and Christmas disease are indistinguishable although some cases of the latter are relatively mild. However, the two types were first distinguished when it was found that the blood from hemophilic A individuals would correct the clotting defect of hemophilic B individuals and vice versa. Interestingly enough, hemophilia B is also X-linked, but is not allelic with hemophilia A, and the two loci are apparently rather far apart on the X chromosome. Classic hemophilia (A) is five or more times more frequent than Christmas disease (B). In both types the heterozygous female carriers can often be shown to have reduced amounts of AHG or PTC, but the overlap between carriers and normal women is sufficiently great to make proper classification very difficult. A number of additional inherited defects in the clotting mechanism have now been identified.

Classic hemophilia occupies a special niche among hereditary human diseases because of its far-reaching political ramifications. Queen Victoria of England was a carrier and, because the royal families of Europe intermarry, the gene spread to other royal families, with especially dire consequences to the royalty in Russia and Spain. The power and influence of the notorious monk, Rasputin, over the Czar and Czarina of Russia was gained in large measure because of his supposed ability to control the bleeding in the Czarevitch or Crown Prince. The combination of the oppressive rule of the mad monk and the lack of a normal heir to the throne were major contributing factors in precipitating the Russian revolution. The last Queen of Spain, like the last Czarina of Russia a granddaughter of Victoria of England, bore two hemophilic heirs to the Spanish throne, both of whom, oddly enough, died of hemorrhaging following relatively minor

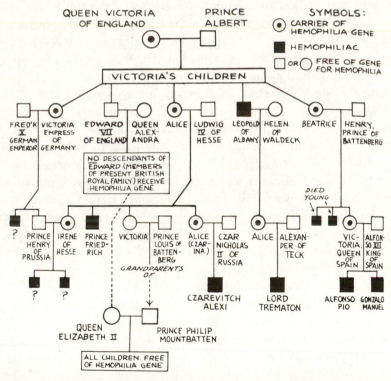

Fig. 25.19. Partial pedigree of Queen Victoria showing the transmission of hemophilia among the royal families of Europe. Some individuals free of the hemophilia gene, such as Princess Margaret, have been omitted. (From *Your Heredity and Environment* by Amram Scheinfeld. Copyright © 1965 by Amram Scheinfeld. Copyright 1939, 1950 by Amram Scheinfeld. Reprinted by permission of J. B. Lippincott Company.)

injuries in separate automobile accidents. Ironically, the British royal family is free of the gene, because Prince Philip is not hemophilic and Queen Elizabeth is descended from Queen Victoria through a line of three nonhemophilic males. The pedigree of Queen Victoria is shown in Fig. 25.19. Since there is no evidence for hemophilia among her ancestors, this gene seems to have originated as a mutation, most likely in an X chromosome she received from one of her parents. J. B. S. Haldane, one of the first to estimate human mutation rates, calculated the rate to hemophilia to be 1 in 50,000 gametes. He also once commented that Queen Victoria's greatest contribution to the world was her gene for hemophilia.

One other type of serum protein deficiency warrants mention at this point, the disease known as *agammaglobulinemia*. Patients with this trait lack plasma cells and suffer from repeated severe bacterial infections, such as pneumonia or septicemia. They seem to respond normally to virus infections, but not to bacterial infections, for they have a marked deficiency of gamma globulins in their blood plasma and are unable to form antibodies following antigenic stimulation. This form of agammaglobulinemia, the most common type, behaves as an X-linked recessive, and is thus primarily an affliction of males. Death used to occur within the first decade, but therapy with antibiotics plus regular administration of gamma globulin has been successful in drastically reducing the number of infections.

Inborn Errors of Metabolism

To Sir Alexander Garrod goes the distinction of being the first to suggest that hereditary diseases are due to metabolic blocks resulting from the deficiency or absence of an enzyme controlling a single metabolic step. This suggestion was made in 1902 in relation to *alkaptonuria*. Garrod was studying patients with this condition and found that they excreted large amounts of homogentisic acid, but that none was excreted by normal persons. He suggested that in these patients the homogentisic acid produced during the normal metabolism of phenylalanine and tyrosine could not be oxidized further due to a metabolic block, and therefore it accumulated in the blood and was excreted in the urine. The observation that the trait tended to appear in sibs from consanguineous unions of normal parents led him, after conferring with Bateson, to postulate recessive inheritance, and alkaptonuria was, therefore, the first recessive trait to be recognized in man. A half-century later his hypothesis was confirmed when homogentisic acid oxidase activity was shown to be absent from the liver of an alkaptonuric patient. Alkaptonuria is usually first detected because the urine turns dark on standing due to the presence of the homogentisic acid. Later, dark pigmentation of the cartilages appears, and by middle age a deforming arthritis sets in, particularly affecting the spine.

By 1908, when Garrod presented his famous Croonian lectures, he had added *albinism*, *pentosuria*, and *cystinuria* to his list of "inborn errors of metabolism." Despite his insight, biochemical genetics did not develop for another 30 years, when Beadle and Tatum introduced their one-gene, one-enzyme concept, which has subsequently been modified to a one-gene, one-polypeptide concept of gene action.

Albinism, discussed earlier, is now regarded as due to a deficiency of the enzyme tyrosinase in the melanocytes.

Like albinism and alkaptonuria, essential pentosuria (L-xylosuria) is R, but the exact location of the block remains in doubt. Although pentose appears in the urine, the condition is benign and no treatment is required.

Cystinuria, the last of Garrod's quartet of recessive traits, is now known to be due, not a metabolic block as such, but to a hereditary defect in an amino acid transport mechanism. In cystinuria there is impaired intestinal absorption and renal tubule reabsorption of not only cystine but three other dibasic amino acids, lysine, arginine, and ornithine. The only clinical effect appears to be the accumulation of urinary stones of nearly pure cystine, the least soluble of these amino acids. Thus, cystinuria is not due to a block in a metabolic pathway as Garrod first postulated. Nevertheless, the defect in the transport mechanism may still represent some sort of defect in an enzyme system.

Since Garrod's original work, a large number of hereditary biochemical disorders in man have been identified, many more than we can possibly consider. However, it may be worthwhile to describe some of the variety of types of biochemical processes that are affected.

Protein Metabolism. Albinism and alkaptonuria are traits resulting from blocks at different points in the metabolism of phenylalanine and tryosine.

Another block in this metabolic pathway causes **phenylketonuria**, which is characterized by phenylpyruvic acid in the urine and causes severe mental retardation. Phenylketonuria (PKU) has a frequency of about 1 in 10,000 in the United States and accounts for about 1% of all mental defectives in institutions. In addition to the mental defect, the patients are deficient in pigment and are often blond and blue-eyed; they suffer from dry skin and eczema. They tend to be below average height and weight and are often microcephalic. Various aspects of behavior are markedly affected—victims show abnormal stance, gait, and sitting posture. Abnormal EEG (electroencephalogram) patterns and epilepsy are common, and they typi-

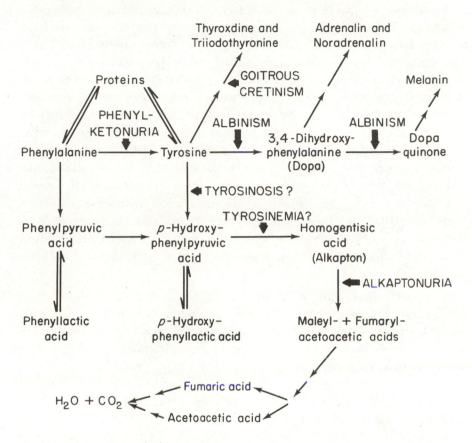

Fig. 25.20. Diagram of phenylalanine and tyrosine metabolism in man. The heavy arrows indicate probable blocks, because of a deficiency of the indicated enzyme, in several of the synthetic steps. In each case the metabolic defect or inborn error of metabolism is due to the presence of a recessive allele in the homozygous condition; hence the normal dominant allele is responsible for the production of the enzyme.

cally show stereotyped purposeless movements of the hand, fingers, or the entire body. They are difficult patients to care for, since they are hyperactive, irritable, and subject to uncontrollable temper tantrums. Thus, in contrast to such a relatively benign condition as cystinuria, for which two glasses of water at bedtime is the recommended treatment to help prevent cystine crystal formation, phenylketonuria has drastic effects, both physically and mentally.

Phenylketonuria is R, and is due to the hereditary deficiency of the liver enzyme phenylalanine hydroxylase, which mediates the conversion of the essential amino acid phenylalanine to tyrosine (Fig. 25.20). As a result, phenylalanine accumulates in the blood and the symptoms, including the irreversible brain damage, are ultimately attributable to this excess phenylalanine. Minor metabolic pathways permit a detour around the block. For example, phenylalanine is converted by its transaminase to phenylpyruvic acid, which is excreted. However, the excess phenylalanine seems to have far-reaching metabolic repercussions, even in pathways seemingly unrelated to phenylalanine. With the discovery of the metabolic basis for the disease, treatment has become possible through rigid dietary control of phenylalanine intake. If started soon enough after birth, the low phenylalanine diet not only alleviates many of the symptoms of phenylketonuria, but also results in an improvement in intelligence. Unless dietary treatment is begun in the first few weeks of life, irreversible neurological damage and intellectual impairment occur. Therefore, routine diagnosis of newborn infants for PKU with the simple $FeCl_3$ test for phenylpyruvic acid in their urine would seem almost mandatory and is now required in some states.

Carbohydrate Metabolism. Diabetes mellitus and pentosuria are two hereditary diseases affecting carbohydrate metabolism that have already been discussed. To these may be added *galactosemia* (R), which is characterized by an inability to carry out the normal conversion of galactose to glucose and is due to the deficiency of an enzyme catalyzing one of the steps in this conversion. Affected infants have excessively high levels of galactose in the blood, and galactose appears in the urine. They soon show nutritional failure and enlargement of the liver, and cataracts of the eyes and mental retardation also develop. The disease may be fatal unless milk and milk products, which contain galactose, are withheld from the diet. When this is done within the first few weeks of life, the symptoms disappear and no long-term damage results.

Lipid Metabolism. **Amaurotic family idiocy** is a term applied to a group of hereditary diseases that have rather similar clinical and histological features, but differ in age of onset. Hence, reference may be found to the infantile, late infantile, juvenile, late juvenile, adult, or congenital forms of the disease. However, despite the similarities, the basic metabolic defect in these different forms of amaurotic idiocy seems not to be the same. The best studied type of amaurotic family idiocy is the infantile type, which is also known as **Tay-Sachs disease** or **ganglioside lipidosis**. The first symptoms appear between 4 and 6 months when growth retardation and an abnormal sensitivity to sound may be observed. Loss of motor control, dementia, and blindness develop progressively, with death usually occurring between the ages of 2 and 4 years. A central, cherry-red spot in the retina is characteristic of the disease. The clinical symptoms result from an accumulation of gangliosides, a type of lipid, in abnormal amounts in the nervous system, leading to degeneration and death of the cells. The exact nature of the biochemical defect remains unknown, but presumably it results from a block in the normal metabolism of the gangliosides.

Tay-Sachs disease is R with complete penetrance. Although the disease has been reported in most ethnic groups, its incidence is far higher among Jews than among non-Jewish peoples. In the United States, for example, the frequency is about 1 in 5,000 births in the Jewish population but only 1 in 500,000 births in the rest of the population. Furthermore, among Jews a high frequency of Tay-Sachs disease occurs in the Ashkenazi Jews, whose ancestors came from northeastern Europe, but a much lower frequency is found among the Sephardic Jews, whose ancestors came from Spain and the Mediterranean area.

Among the Jews in New York City, the gene frequency of this recessive is estimated to be approximately 0.016, which leads to the estimate that 1 in 30 is a heterozygous carrier as compared to 1 in 300 heterozygotes among non-Jewish Americans. A most puzzling question is why the gene has achieved such a high frequency among Ashkenazi Jews but not in other populations. Heterozygous advantage has been suggested as one possible explanation, but then the question becomes why the gene fails to show heterozygous advantage in the other populations. At present no satisfactory explanation is available.

Purine Metabolism. Gout, a disease with a long and honorable (or dishonorable) history traceable back to Hippocrates, results from a disorder in purine metabolism that causes increased amounts of uric acid in the blood and recurrent attacks of kidney stones or of painful arthritis. The usual picture called to mind is that of a stout old English gentleman sitting in his club in obvious pain, with his foot elevated on a cushion (Fig. 25.21).

The basic defect is **hyperuricemia**, but this will be detected only if kidney stones or arthritis or both develop. Thus asymptomatic individuals may complicate genetic studies, as will individuals with uric acid levels elevated due to environmental causes. Moreover, the disease is far more common in men than women, who normally have lower levels of uric acid in their blood than men. All of these factors complicate the genetic picture, but at present gout is tentatively thought to be an autosomal D condition (although genetic heterogeneity and even a polygenic explanation have been invoked). It is well to end our discussion with this sort of trait, for it serves as a reminder that even for a well-known disease with a long history of being considered hereditary, it is not always possible yet to provide a simple genetic explanation.

This recital of human hereditary ailments and ills may seem most dis-

Fig. 25.21. "Marriage à la mode. The contract." William Hogarth (1697–1764). A noble earl, whose family traces back to William the Conqueror, Duke of Normandy, is proudly displaying his genealogical tree to a wealthy citizen, who is giving up a large dowry in order to gain entry into a noble family through marriage of his daughter to the earl's son. Hogarth uses the earl's gout as the mark of a profligate nobleman.

heartening. And it must be added that only a sampling of a much larger array of hereditary diseases has been presented. Nevertheless, it should not be forgotten that human genes also made possible Jack Benny's beautiful blue eyes, Einstein's soaring genius, and the consummate artistry of Michelangelo, Shakespeare, and Beethoven.

Additional Reading

Bartalos, M., and T. A. Bararuki, 1967. *Medical Cytogenetics.* Williams and Wilkins, Baltimore.

Gates, R. R., 1946. *Human Genetics,* 2 vol. Macmillan, New York.

Harris, H., 1963. *Garrod's "Inborn Errors of Metabolism"* (reprinted with a supplement by H. Harris). Oxford University Press, London.

Hoefnagel, D., and P. S. Gerald, 1966. "Hereditary Brachydactyly," *Ann. Human Genet.* 29:377–382.

Hsia, D. Y., 1966. *Inborn Errors of Metabolism,* 2nd ed. Year Book Publishers, Chicago.

Kalmus, H., 1965. *Diagnosis and Genetics of Defective Color Vision.* Pergamon, Elmsford, N.Y.

Knudson, A. G., 1969. "Inborn Errors of Metabolism," *Ann. Rev. Genet.* 3:1–24.

McKusick, V. A., 1964. *On the X Chromosome of Man. AIBS,* Washington, D.C.

McKusick, V. A., 1966. *Heritable Disorders of Connective Tissue,* 3rd ed. Mosby, St. Louis.

McKusick, V. A., 1969. *Human Genetics,* 2nd ed. Prentice-Hall, Englewood Cliffs, N.J.

McKusick, V. A., 1971. *Mendelian Inheritance in Man,* 3rd ed. Johns Hopkins University Press, Baltimore.

Moore, K. L., ed., 1966. *The Sex Chromatin.* Saunders, Philadelphia.

Ratnoff, O. D., 1972. "Hereditary Disorders of Hemostasis." In *"The Metabolic Basis of Inherited Disease,* J. S. Stanbury, J. B. Wyngaarden, and D. S. Frederickson, eds. McGraw-Hill, New York.

Rimoin, D. L., and R. N. Schimke, 1971. *Genetics Disorders of the Endocrine Glands.* Mosby, St. Louis.

Scheinfeld, A., 1965. *Your Heredity and Environment.* Lippincott, Philadelphia.

Sorsby, A., 1951. *Genetics in Ophthalmology.* Mosby, St. Louis.

Sorsby, A., ed., 1953. *Clinical Genetics.* Mosby, St. Louis.

Stanbury, J. B., J. B. Wyngaarden, and D. S. Frederickson, eds., 1972. *The Metabolic Basis of Inherited Disease,* 3rd ed. McGraw-Hill, New York.

Thompson, J. S., and M. W. Thompson, 1966. *Genetics in Medicine.* Saunders, Philadelphia.

Waardenburg, P. J., A. Franceschetti, and D. Klein, 1961. *Genetics and Ophthalmology,* 2 vol. Van Gorcum, Assen., Netherlands. (Charles C Thomas, Springfield, Ill.)

Questions

25.1. What explanation can you suggest for the observation that in clinically similar hereditary traits in man, the less severe conditions tend to be inherited as Mendelian dominants while the more severe conditions are inherited as recessives?

25.2. Many traits, originally described as a single entity, have been found on further study to be an array of clinically similar but genetically distinct conditions. In the light of this fact, what precautions seem necessary if large studies involving a number of different family groups are to be meaningful?

25.3. How could you distinguish between Y-linked traits and sex-limited (but not Y-linked) traits in man?

25.4. What type of hereditary deleterious trait in man is apt to cause the greatest emotional, social, and financial burden?

25.5. In a human population, 4% of the males have a deutan defect in color vision and 1% have a protan defect.

a. If only one locus on the X is responsible for color blindness, what is the expected frequency of color-blind females?

b. If two loci on the X are responsible, what is the expected frequency of color-blind females?

c. Suppose, in a sample of 10,000 women, that 16 were color blind. Would this deviation from the expected numbers under the one locus hypothesis be statistically significant so that the hypothesis could be rejected?

d. What other type of test might you use to determine whether the protan and deutan genes are allelic?

25.6.   Cystic fibrosis of the pancreas is an autosomal recessive lethal with an incidence estimated to be of the order of 1 in 3700. If this high frequency were being maintained by a balance between selection and mutation, what mutation rate would be needed to maintain the gene at this frequency? What other mechanisms can you suggest that might be responsible? How would you test these possibilities?

25.7.   In a disease such as diabetes, with a late and variable age of onset, variations in expression and prenetrance, and a complex etiology, what types of studies do you think will be most fruitful in determining the role of heredity?

25.8.   Occasionally a deleterious recessive trait that is rare in most human populations will have a relatively high frequency in one or a few human populations. What possible explanations can you propose for this situation? Criticize each of your proposals. How would you test your theories?

25.9.   McKusick's roster of known human hereditary traits has grown about fourfold in three editions. How long would you guess this logarithmic rate of growth will continue? What limits are there to the number of new hereditary human traits that can be identified?

25.10.   Do you believe the frequency of hereditary deleterious conditions in man is the same, greater, or less today than it was in 1900? In 1500? In 10,000 B.C.? Explain.

CHAPTER TWENTY-SIX

Genetics and Behavior

26.1. The Origins of Behavior Genetics

In the space of a brief chapter only a few of the highlights of the relationship between genetics and behavior can be mentioned. As in so many other areas, Darwin was a pioneer in the study of behavior and its relation to heredity and evolution. The lineal descendants of Darwin's approach to the study of behavior work in the discipline known as *ethology*. Modern ethology stems primarily from the work of Lorenz and Tinbergen in Europe during the 1930s and 1940s. For the most part ethologists have been trained as zoologists, and have been interested in comparative studies of animal behavior in such groups as fish, birds, and insects. Their data are derived from field observation and experimentation, and much of their attention has been directed toward "instincts" or "innate behavior patterns" and the evolution of behavior.

In the United States the main thrust for research in animal behavior has come from people trained in psychology, who have experimented under carefully controlled laboratory conditions with mammals, especially the laboratory rat. The problem of greatest interest to them has been learning, and their orientation has been more toward human behavior than in the case of the ethologists. Out of their experiments with mammals, they have

hoped, by analogy, to develop theories of behavior, especially human behavior. Because of their interest in learning, these psychologists have been
far more interested in envionmental influences on behavior than in the role
of heredity. The most extreme environmentalist stand was taken by the
American psychologist J. B. Watson, who attributed individual differences
in behavior solely to environmental factors, a school of thought known as
behaviorism.

The studies of human behavior initiated by Freud, which led to the
development of psychoanalysis and to a considerable emphasis on the
importance of early experience as a determinant of later behavior, also led
to a preoccupation with the environmental aspects of behavior, even
though Freud himself recognized the importance that hereditary constitutional factors might have in human personality and behavior. The work
of Freud and his disciples, revolutionary though it has been, has been
based more on insight than on experiment, and the major defect in much
of the research in psychoanalysis has been the lack of adequate controls.

The great pioneer in physiological studies of behavior was Pavlov, whose
study of conditioned reflexes opened still another approach to behavior.
Since attention was directed more toward the conditioning than toward
the reflex, here too environmental influences on behavior were of primary
concern.

Therefore, aside from ethology, these other lines of research—psychology,
psychoanalysis, and physiology—tended to stress the environment as the
most important determinant of behavior almost to the exclusion of heredity.
Most workers in the fields of psychology and psychiatry, especially in the
United States, did not consider the relationship between genetics and
behavior to be particularly close or important. In addition, the concepts
developed rested on a very narrow experimental base. Pavlov worked almost exclusively with dogs and cats; the experimental psychologists concentrated to a striking degree on the domesticated Norway rat. Since mammals constitute less than 1% of all known animal species, most behavioral
research has been concentrated on just a few species of one of the less
numerous groups of animals. Furthermore, studies of learning and conditioning have far outnumbered studies of such topics as sensory capacities or
reflexes and simple reaction patterns. Under the circumstances, it is perhaps
not surprising that the work of the physiologists and psychologists has resulted in some rather distorted views of the nature of animal behavior.

Although behavior genetics has come to be recognized as a separate
discipline only in the last two decades, its origins can be traced back to the
early days of this century in the work of such pioneers as Yerkes, who
studied the effects of the waltzing gene on mouse behavior, and Sturtevant,

who observed the effects of *Drosophila* mutants on sexual behavior. Other landmarks were the selection experiments of Tryon and of Heron during the 1920s and 1930s for maze-learning ability in rats, and the work of Scott and Fuller, begun in the 1940s, on genetics and behavior in dogs. Much of the more recent work in behavior genetics has been conducted with two species well known genetically, the mouse, *Mus musculus,* and the fruit fly, *Drosophila melanogaster.*

In the study of behavior, the nature-nurture controversy flourished, just as it did in many other areas of biology. Here it took the form of instinctive versus learned behavior. Because of the predominance of the behaviorists, with their interest in conditioning and learning, and of the Freudians, who stressed the effects of early experience, the environmentalist or "nurture" viewpoint tended to prevail. This attitude was further reinforced by the perversion of genetic concepts under Hitler and by the ideas of the "white supremacists," for such abuses made more rational men reluctant even to consider the possible role of heredity in the ontogeny of human behavior. Here too, however, the dichotomy between nature and nurture is a false one. No trait develops in an environmental void, and no animal develops without a genotype. It is possible to study the relative importance of heredity and environment in the differentiation of a trait for a given set of genetic and environmental conditions. In some cases, one or the other may loom large in its influence on the course of events, but it is never possible to exclude one or the other from the scene entirely. However, because so much attention has been devoted in the past to environmental influences on behavior, we shall be more concerned with the role of heredity in behavior.

In order to indicate the influence of heredity on behavior, let us consider a case in which it is rather difficult to escape the conclusion that the genotype has something to do with behavior, namely, nest building in birds. It is true that young birds grow up in the nest and could, therefore, learn something about its location and its final structure. However, they have no opportunity to observe its actual construction since it is completed before the eggs from which they hatch are laid. Yet, in most cases, the following year the young birds build nests, without ever having seen one built and without any instruction from older birds. In the process they must first pick a suitable site and then accumulate the necessary nest-building materials and work them into an appropriate structure. The sites and the materials chosen are quite characteristic for each species, and differ widely among species. The delicately woven basket of the Baltimore oriole hanging at the tip of an elm branch is very different from the barn swallow's nest, fashioned of mud and located in a barn or under a bridge.

While it can be argued that learning could be involved in some aspects of nest building behavior such as site selection, for many aspects of the process the conclusion seems inescapable that the complex behavior patterns seen in nest building must in some way be built right into the genotype. If true, this conclusion raises the interesting question of the chain of events that connects the DNA controlling polypeptide synthesis on the one hand with the complex nest-building behavior on the other. Not only is the nest built, but it is usually concealed and often camouflaged. If this complicated sequence of events is largely governed by heredity, it must have evolved to its present state. Thus the next question could well be how such behavior patterns could have evolved.

The birds evolved from the reptiles, which lay shelled eggs on the ground or in the sand. The eggs of such a modern-day reptile as the turtle are laid in the sand, where they are accessible to such predators as raccoons and skunks. It is not uncommon for an entire clutch of turtle eggs to be uncovered and consumed. Hence, any variation in behavior that causes eggs to be laid in less accessible places will tend to enhance the probability of survival of the offspring. Therefore, any tendency by ancestral birds to lay their eggs in shrubs or trees where they would be less vulnerable to terrestrial predators would be favored by natural selection. Moreover, any behavior tending to keep the eggs in the trees, such as placing a few sticks in a crotch, would also increase the chances of survival of the offspring and hence be favored by natural selection. The crude, rather rickety nest of twigs of the mourning dove seems hardly to have progressed beyond this point. If this explanation seems oversimplified or far-fetched, it should be remembered that vast expanses of time, millions of years, were available during which evolutionary changes could occur. It is also worth pointing out that birds living on oceanic islands without terrestrial predators frequently nest on the ground even though their close relatives elsewhere nest in trees. This reversion to ground nesting may represent a relaxation of natural selection, or there may even be positive adaptive advantages to ground nesting on oceanic islands. However, when man has introduced such species as the rat on these islands, they have wreaked havoc on the ground-nesting birds, who have not been able to adapt to this sudden new selection pressure. Viewed in these terms and over a long time span, the evolution of the diversity and complexity to be seen in nesting behavior in birds may no longer be quite so difficult to comprehend.

That nest-building behavior may actually be strongly influenced by the genotype is evidenced by the results of studies by Dilger with two species of lovebirds and the F_1 hybrids between them. One species (*Agapornis personata fischeri*) carries strips of nesting material to the nest in its bill; the other species (*Agapornis roseicollis*) tucks the nesting material into its rump feathers to carry it to the nest. The F_1 hybrids at first almost invariably tried to tuck the nesting material into their feathers, but were never successful. In some cases the tucking movements were incomplete, but often the movements were normal and the bird either failed to let go of the strip after tucking it in, or seized the strip in the center rather than at the end, or else tucked it into an unsuitable place. Thus, the F_1 hybrids seemed to act in a thoroughly confused manner, unable to decide whether to carry the strips they cut in their feathers or in their beaks. In their initial efforts toward nest building they spent much of their time trying to tuck material into their feathers, but invariably it fell out before they reached the nesting site. Six percent of the time they succeeded in carrying the nesting strips to the nesting site in their bills. With the passage of time, the birds gradually increased the proportion of strips carried in their bills, but even after three years they continued to carry out some of the activities associated with tucking even though it was never successful.

These experiments showed that these behavior patterns related to carrying nesting materials were to a large degree innate. The hybrids, carrying genes producing conflicting behavior patterns, had great difficulty in learning to suppress one totally unsuccessful pattern in order to use the other. Nevertheless, over a three-year period, they did gradually learn to carry more of the strips in their bills even though they never learned to suppress completely their tucking activity. In fact, they did gradually learn to become somewhat more adept at tucking material into their feathers, even though they never successfully carried it to the nest. Since the hybrids were sterile, it was impossible to study F_2 or backcross birds, or to learn what effect this behavioral conflict might have on breeding success. Nevertheless, this fascinating study should give some inkling of the role that heredity can play in the ontogeny of behavior. It is also noteworthy that even in this case some modification of behavior with experience, that is, learning, did occur.

Many behavior patterns in animals are even more stereotyped than this example in the lovebirds, while others are much more easily modified by experience. It would be well to consider all forms of behavior as falling on a continuum, with the variables being the relative influence of the genotype and the environment, rather than trying to separate types of behavior into "instinctive" or "learned" categories. However, some species, with man a prime example, show far more ability than others to modify their activities as the result of experience. Moreover, man is able not only to learn but to reason; he often arrives at a rational solution to a problem and acts upon the basis of his reasoning without the benefit of prior experience. Lest we think man unique in this respect, it is well to recall that a monkey, confronted by a banana and a stick and with no prior experience, will usually in short order use the stick to bring the banana within reach. There is also the story, probably apocryphal, of the experimenter who wished to test the intelligence of a chimpanzee and therefore set before him a problem for which there were three possible solutions. The chimp somewhat confounded the experiment and the experimenter by using a fourth solution that had not occurred to his human counterpart.

It is worth noting that the ability to learn and the ability to reason must also have evolved as the result of natural selection. If we assume that the more stereotyped patterns of behavior are more primitive, as seems quite reasonable, and if learning contributes to fitness as compared to blind instinct, then the ability to learn and to reason will have adaptive value and will be favored by natural selection. The instincts will then tend to be suppressed or modified by experience. Moreover, the limits on the ability to learn are influenced by the genotype. Chimpanzees reared by humans with all the care, attention, and affection given to human infants never learn to talk or to read or to behave entirely like humans, although they do "ape" human behavior and may become caricatures of man. Human behavior is beyond the capacity of the chimpanzee's genotype; he cannot learn to be human.

26.2. Genetics and Human Behavior

Now we shall consider more specifically the role of heredity in relation to human behavior. The frequency with which birth injury, a poor home environment, an unhappy or broken marriage, and a wide variety of other environmental factors are cited as causes of abnormal or subnormal human behavior is a reflection of the preoccupation of students of human behavior with environmental influences. In part, this tendency seems to stem from a feeling that environmental factors can be modified and the aberrant behavior thereby corrected, but that hereditary disorders present a less hopeful picture. As we have already seen, this is not necessarily true. Whatever the reason for this attitude, it does seem desirable next to consider the major forms of abnormal human behavior, then to cite a few examples of human behavioral aberrations with a simple genetic basis.

Human mental aberrations fall into two major categories, mental disease and mental retardation. Although usually less striking in their outward manifestations than the physical disabilities already discussed, mental disorders pose far more serious medical problems than physical disabilities. Medical research in the area of mental health has lagged behind research in other areas, in terms of both support and progress, yet this is undoubtedly the number-one medical problem in the United States today. About half of all the hospital beds in the United States are occupied by mental patients, with more than half a million people in institutions for long-term care. Not only are there large numbers of institutionalized mental patients, but an even larger number of people with mental problems are not in institutions. The care of all these individuals is often difficult and costly, and they represent an emotional and financial drain both on their families and on society. One hopeful sign in recent years is the decline in the patient population in psychiatric hospitals, due in large measure to the discovery of drugs that make it possible for the patient to shorten his stay in the hospital and resume life on the outside sooner than was possible in the past. However, such treatment controls rather than cures the basic problem.

Mental disease has also been called insanity, lunacy, madness, dementia, and psychosis. The psychotic person has a profound disorganization of his mind, personality, or behavior. He tends to be irrational and to misinterpret reality even though his intellectual capacity may fall within the normal range. The neurotic person, on the other hand, suffers, sometimes acutely, from anxieties and internal conflicts, but remains rational and perceives reality fairly accurately. The difference between the two is reflected in the saying that the psychotic thinks 2 and 2 make 5, while the neurotic knows that 2 and 2 are 4, but it makes him nervous. Although psychosis is generally regarded as more severe than neurosis, the two conditions do not form a continuum, but seem to be independent forms of abnormality. A neurotic does not often become psychotic, nor does a psychotic often change into a neurotic. Thus psychosis is not simply a more severe form of neurosis, and neurotic individuals are not ordinarily regarded as being mentally ill or suffering from mental disease.

The other major type of mental disorder is mental retardation, which has also been called mental deficiency, feeblemindedness, amentia, and oligophrenia. While psychotics may be considered as having a diseased or pathological mentality, the mentally retarded have an underdeveloped mentality. They are not mentally ill but have a subnormal intellectual capacity. A useful operational definition is the following: "Mental retardation refers to subaverage general intellectual functioning which originates in the developmental period and is associated with impairment in adaptive behavior." By the latter is meant, for example, that a mentally retarded person may fail to learn that fire burns and hence must be protected against the consequences of his own actions.

Mental Aberrations of Simple Genetic Origin

We have already discussed phenylketonuria, which is of interest in the present context because it is a simple autosomal recessive that not only causes mental retardation but also has distinctive effects on the personality and behavior of untreated affected individuals.

Huntington's chorea is an autosomal dominant with complete penetrance. The age of onset is variable, but the first symptoms usually do not appear until the 30s or 40s, when the reproductive period is nearing completion. A progressive atrophy of the nervous system leads to the development of physical and behavioral symptoms. Facial grimacing, weaving movements of the body, and choreiform twitching of the arms and legs become progressively more pronounced. In the past these movements were considered evidence that the sufferers were possessed by the devil, and occasionally they were burned as witches. Speech and intellect are gradually impaired, psychotic symptoms develop into dementia, and eventually, after a period of years, death ensues. This neurological disorder has drawn attention for several reasons: its association with witchcraft, the number of excellent studies of the trait, made possible in part by the simple Mendelian dominant mode of inheritance, and the fact that the late age of onset ordinarily permits a new generation to be born, half of whom are destined to face the same inevitable, long-drawn-out deterioration and death. No cure is yet known, nor is it yet possible to distinguish between carriers of the gene and their normal sibs at an early age. Even if it does become possible to identify carriers of the dominant gene for Huntington's chorea before they start reproducing, it will present a most difficult problem in genetic counseling. Even though it may be highly desirable to break this sad chain by informing carriers of the gene of this fact in the hope that they will refrain from reproduction, the psychological impact of this information on a youth some 20 years or more before the age of onset may be devastating.

Down's syndrome, also known as mongolism· or mongolian idiocy, is a syndrome that long constituted an etiological puzzle. It is characterized by mental deficiency and a typical appearance with broad head, snub nose, small ears, and slanted, widely spaced eyes, whence the term mongolism (Fig. 26.1). The bones are shortened, and other congenital anomalies, espe-

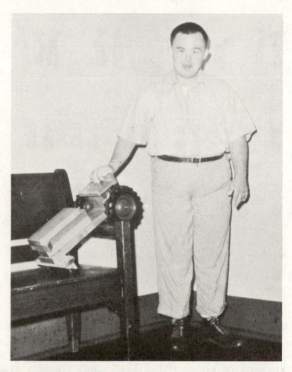

Fig. 26.1. Down's syndrome, mongolism, or trisomy 21. Age 25 years, Stanford-Binet IQ 26. The facial appearance and expression and the short height and stubby fingers are characteristic of the condition. (Courtesy of Raymond M. Mulcahy, Superintendent, Brandon Training School, Brandon, Vermont.)

cially of the heart, are common. Death often results from heart disease or infection.

The pieces of the puzzle began to fall into place with the discovery in 1959 that Down's syndrome resulted from trisomy for the small 21st human chromosome. Instead of the 46 chromosomes usually seen, children with mongolism had 22 pairs of chromosomes plus three of chromosome 21 for a total of 47 (Fig. 26.2). The condition appears far more often among children of older mothers, apparently because nondisjunction occurs more frequently in women approaching the end of their reproductive period.

Subsequently, mongolism was discovered associated with a translocation rather than trisomy. In this situation the extra 21st chromosome had become attached to another chromosome in the complement. This form of the syndrome poses a more difficult problem. Where the mother of a child with trisomy 21 runs a risk, even if past age 45, of only about 1 chance in 50

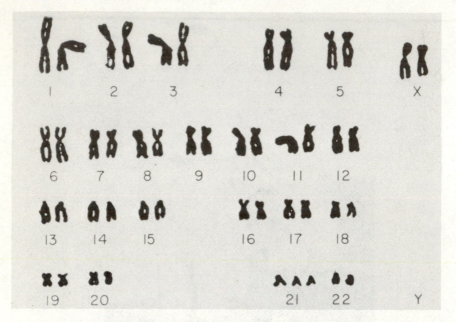

Fig. 26.2. Karyotype of a female with Down's syndrome or trisomy 21. The presence of two X chromosomes results in a female; trisomy for chromosome 21 produces Down's syndrome. (Courtesy of Dr. Victor A. McKusick, Division of Medical Genetics, Johns Hopkins Hospital, Baltimore, Maryland 21205.)

of having a second affected child, the chances for a mother who carries the translocation range between 25 and 33%. The carriers of the translocation can be normal, but then they have only 45 chromosomes, one of which, of course, carries the translocated 21st chromosome. When a gamete bearing the translocation is fertilized by a normal gamete, three sets of genes from the 21st chromosome are present, leading to the developmental imbalance that is manifested as Down's syndrome.

A further diagnostic problem arose when individuals were discovered some of whose body cells had an extra 21st chromosome while others did not. These chromosomal mosaics have intermediate symptomatology and further complicate the diagnosis.

Down's syndrome, trisomy 21, is estimated to have a frequency of about 1 in 500 births. Still other trisomies with equally dire effects have been identified, and there seems no reason to suppose that trisomy for any one of the 23 types of human chromosomes could not occur. In the Jimson weed (*Datura stramonium*), with $n = 12$, twelve different simple trisomics have been identified, one for each of the chromosome types. The type of trisomy can be identified from the shape of the seed capsule, because each type of trisomy has a distinct and constant effect on the phenotype (see Fig. 9.13). This situation seems in all ways comparable to the effects of trisomy in man, because the distinctive phenotype of individuals with trisomy 21 is quite striking. It seems probable that trisomy for the larger human chromosomes is lethal at early stages of development and thus difficult to detect. The frequency of chromosome anomalies in the human population seems surprisingly high, and there is no simple explanation of why this should be so. The drastic consequences of most of these anomalies hardly support any contention that in some way they confer added fitness to the population. Therefore, one might expect natural selection to stabilize the human genetic system in such a way as to reduce the frequency of these anomalies. Although adequate data for comparison probably do not exist, in no other species do deleterious chromosome anomalies seem to arise with comparable frequency.

Mental Retardation

The development of intelligence tests by Binet in France early in this century led to the use of these tests to define different degrees of mental deficiency in terms of the intelligence quotient (IQ). Very briefly, perform-ance on an IQ test is standardized by assuming that, for a given age group, intelligence is normally distributed, and the mean for the group is then arbitrarily set at 100 and the other scores fitted to a normal distribu-tion curve (see Appendix). The intelligence quotient is then obtained by dividing the mental age of the individual as measured by the test by his chronological age (and multiplying by 100). For example, if a 4-year-old scores as well on the test as the average 5-year-old, his IQ is readily obtained:

$$IQ = \frac{\text{mental age}}{\text{chronological age}} \times 100 = \frac{5}{4} \times 100 = 125$$

Although various schemes have been used to categorize the different degrees of mental defect, we shall use the following:

IQ	Designation	Mental competence (in years)	Percent of total population	Percent in institutions
0–19	Idiot	0–2	0.1	25.0
20–49	Imbecile	3–7	0.3	15.0
50–69	Moron	8–12	3.0	1.0

Table 26.1   Incidence of Mental Retardation Among the Co-twins of Institutionalized Mentally Retarded Twins

Study	Monozygotic		Dizygotic		Heritability $H = \dfrac{C_{MZ} - C_{DZ}}{100 - C_{DZ}}$
	No. of pairs	Percent concordance	No. of pairs	Percent concordance	
Rosanoff et al.	126	91	240	53	0.81
Juda	71	97	149	56	0.93

From Rosen, E., R. E. Fox, and I. Gregory, *Abnormal Psychology*, 2nd ed. Published by W. B. Saunders Company, Philadelphia, 1972.

These figures indicate that there are more than 6 million mentally retarded persons in the United States, of whom some 200,000 are in institutions. The more severely retarded individuals are much more apt to be institutionalized. Since persons with IQ scores ranging from 70 to 79 are often considered of borderline intelligence, the dimensions of the problem are rather awesome.

We have already considered such traits as phenylketonuria, amaurotic idiocy, cretinism, and mongolism, in which the etiology for the mental retardation is primarily genetic. Many other individual causes of mental retardation, both genetic and environmental, have been identified, but in addition to such cases, there is also a large undifferentiated group in which differential diagnosis is not possible. In these cases the etiology may be primarily genetic, primarily environmental, or some combination of unfavorable environmental and genetic conditions. It is of some interest to try to determine the role of heredity in these cases.

The first step in such an analysis is a look at twin data. The data of Rosanoff, Handy, and Plesset (1937) and of Juda (1939), reproduced in Table 26.1, are representative of studies of concordance rates for mental deficiency in twins.

The lack of complete concordance in monozygotic twins and the slight increase in concordance above the expected 50% in dizygotic twins both indicate that environmental factors may be important in the etiology of mental retardation. If environmental conditions are, in fact, comparable for both types of twins, the significantly higher concordance rate in monozygotic as compared to dizygotic pairs suggests an important role for heredity. If not, the greater similarity of the monozygotic twins may reflect environmental as well as genetic effects.

The last column in Table 26.1 gives the results of calculating heritabilities from concordance data in twins using a simple but rough method to estimate the contribution of heredity to the variance. The formula is

$$H = \frac{C_{\mathrm{MZ}} - C_{\mathrm{DZ}}}{100 - C_{\mathrm{DZ}}}$$

where C_{MZ} = percent concordant monozygotic twins and C_{DZ} = percent concordant like-sexed dizygotic twins. It can be seen that H varies from a value of zero when $C_{\mathrm{MZ}} = C_{\mathrm{DZ}}$, to 1.0 when $C_{\mathrm{MZ}} = 100$ and C_{DZ} is less than 100.

For traits showing continuous variation such as height or intelligence, heritabilities may be estimated in a comparable manner from the intraclass correlation coefficients (see Appendix) for monozygotic and for like-sexed dizygotic twins. The equation takes the form

$$H = \frac{r_{\mathrm{MZ}} - r_{\mathrm{DZ}}}{1 - r_{\mathrm{DZ}}}$$

Here also, H can be expected to vary from zero to 1.0. If the twins were always concordant, the correlation coefficient would equal 1.0. If there is no relationship between members of a pair, the correlation coefficient would be zero.

It is important to note that if the concordance rates or the correlation coefficients for both monozygotic and dizygotic twins are quite high, H is not a reliable estimate of the contribution of heredity. For instance, if r_{MZ} were 1.0 and r_{DZ} were 0.8, then $H = 1.0$. This value is suspect because it implies that the environment plays no role in the ontogeny of the trait, yet the expected r_{DZ} due to heredity is only 0.5, and the increase to $r_{\mathrm{DZ}} = 0.8$ must be due to the effect of environmental factors.

The heritability estimate is sometimes based on the variances (V) (see Appendix), as follows:

$$H = \frac{V_{\mathrm{DZ}} - V_{\mathrm{MZ}}}{V_{\mathrm{DZ}}}$$

Such comparisons of monozygotic and dizygotic twins should minimize the variation due to birth order, age of mother, and sex. The estimates of heritability obtained in this way should not be accepted without reservation, however. For one thing, it is assumed that the environments of monozygotic and dizygotic twins are equivalent, which may not be true. For another, no account is taken of possible interactions between heredity and environment.

A further consideration is that heritability estimates are distorted by the effects of positive assortative mating, which is frequent in man. Assortative mating will tend to raise dizygotic twin, parent-child, and sib-sib correlations. Therefore, heritability estimates based on comparisons of monozygotic and dizygotic twins have certain limitations. Nevertheless, they are relatively simple to calculate, and in the absence of any better statistical tool, have often been used to obtain a rough estimate of the proportion of the variance due to heredity.

As can be seen from the two heritability estimates in Table 26.1, the twin data indicate quite a high heritability for mental retardation. This conclusion is in agreement with findings based on other types of data.

The Reeds, in their massive study of mental retardation which included over 80,000 individuals, found 54 fertile unions of two mentally retarded persons. 215 children were produced, of whom 85 (39.5%) were retarded. Terman and his co-workers, on the other hand, made an extended longitudinal study of a group of gifted children whose IQ's ranged from 135 to 200. The gifted children, grown up, married spouses who also, on testing, were found to be highly intelligent. This finding is not unexpected because of the known tendency for positive assortative mating in man. The ability of the "gifted children" and of their spouses was evidenced not only by their test scores, but also by their educational achievements and other measures of success. A study of the IQ scores of 1525 children of the "gifted children" and their spouses revealed that the average IQ of the offspring was 132.7, with one-third scoring in the "genius" class (140 or above), while only 2% had IQs below 100. The contrast between 39.5% retarded children from retarded parents and the high ability of the offspring from the gifted parents is striking, but reveals little about the relative contributions of heredity and environment. The mentally retarded parents were likely to give their children not only an unfavorable genotype but an unfavorable environment as well. On the other hand, the children of the gifted were not only likely to receive a favorable genotype but were provided by their highly successful parents with a very favorable environment.

A most significant point in these studies is that 60.5% of the children of two retarded parents had IQs in the normal range. The most reasonable interpretation of this finding is that genetic recombination gave rise to favorable genotypes in these children that enabled them to surmount to some degree the unfavorable environmental conditions into which they were born. On the other hand, 0.5% of the offspring of the gifted were mentally retarded despite the propitious circumstances of their birth. In these cases, apart from those due to irreparable accidental damage, the

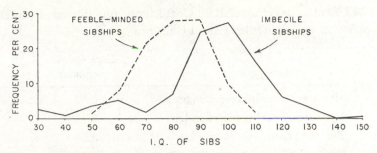

Fig. 26.3. The frequency distributions of the IQs of 562 sibs of the feeble-minded and imbeciles in the IQ range from 30 to 68. (Roberts, J. A. F., 1952. "The Genetics of Mental Deficiency," *Eugenics Review* 44:71–83, p. 78, Fig. 4. Reprinted by courtesy of The Eugenics Society.)

most plausible explanation is genetic segregation giving rise to detrimental genotypes that cannot respond even to the most favorable of environments.

When the mentally retarded population is studied further, it can be separated into two rather distinct groups. The severely retarded group, the idiots and imbeciles, are nearly all physically as well as mentally defective, and the mental retardation is associated with a chromosomal, endocrine, metabolic, neurological, or other abnormality. The more numerous mildly retarded group, the morons, on the other hand, do not ordinarily show such physical symptoms in conjunction with their mental defect. The mildly retarded individuals tend to come from families in which low intelligence in the parents, sibs, and other relatives is quite common. On the other hand, the families of the idiots and imbeciles show a more or less normal distribution of intelligence compared with the general population, except for a small peak in the defective range due to individuals like the propositi (Fig. 26.3).

The most reasonable conclusion seems to be that the idiots and imbeciles are defective due to a rather specific cause, which may range from a homozygous recessive gene to a birth injury causing brain damage. The morons, however, may be thought of as individuals whose intelligence falls at the low end of the normal curve of distribution for intelligence. Both environmental and genetic influences may be at work here, also, but the environmental effects are more subtle, and the genetic influences are multigenic. As the technques of differential diagnosis improve, the etiology for a greater proportion of the cases of mental retardation will become clear. However, it seems probable that for some time to come, differential diagnosis will not be possible for a fairly sizable group of cases, which represent the low end of the normal distribution of intelligence and manifest the effects of unfavorable combinations of genes.

Table 26.2 Intraclass Correlation Coefficients for IQs of Different Kinds of Twins and for Other Paired Children

Comparison	Correlation coefficient
Monozygotic twins reared together	0.87
Monozygotic twins reared apart	0.75
Fraternal twins reared together	0.53
Ordinary sibs reared together	0.49
Unrelated children reared together	0.23

From Erlenmeyer-Kimling, L., and L. F. Jarvik, 1963. "Genetics and Intelligence: A Review," *Science* 142:1477–1479.

Genetics and Intelligence

Heredity can quite definitely be implicated in the etiology of many types of mental deficiency. There is also evidence that it plays an important role in the development of normal intelligence. A number of useful comparisons have been made, involving identical twins reared together and separately, fraternal twins, ordinary sibs, parents and their children, and parents and foster children.

Table 26.2 was compiled from 52 studies by Erlenmeyer-Kimling and Jarvik in 1963 and shows the correlation between IQs of individuals under various genetic and environmental conditions.

The expected correlations based solely on heredity would be 1.0 for identical twins, 0.5 for fraternal twins and ordinary sibs, and 0.0 for unrelated children (Table 26.3).

The difference between monozygotic twins reared together and monozygotic twins reared separately seems to reflect the effect of greater environmental diversity on those reared separately. However, measures of the degree of environmental difference involved are very crude. Even though separated, for example, twins are ordinarily reared within the same culture. Greater diversity of cultural and environmental conditions might give rise to greater disparity in the correlation coefficients.

Fraternal twins reared together are only very slightly more alike than ordinary singleton sibs reared together, and both are close to the expected genetic correlation of 0.5.

The unrelated children reared together show a positive correlation in IQ of 0.23. This may reflect the effect of their common environment, but it may also be due to the efforts of adoption agencies to match the IQs of

Table 26.3 Correlations for Intellectual Ability Among Relatives Compared with Expected Values Based on Mendelian Inheritance

Relationship	Number of studies	Median correlation Observed	Expected
Unrelated—reared apart	4	−0.01	0
Unrelated—reared together	5	+0.23	0
Foster parent-child	3	+0.20	0
Parent-child	12	+0.50	+0.50
Sibs—reared apart	2	+0.42	+0.50
Sibs—reared together	35	+0.49	+0.50
Dizygotic twins—opposite sex	9	+0.53	+0.50
Dizygotic twins—same sex	11	+0.53	+0.50
Monozygotic twins—reared apart	4	+0.75	1.00
Monozygotic twins—reared together	14	+0.87	1.00

From Erlenmeyer-Kimling, L., and L. F. Jarvik, 1963. "Genetics and Intelligence: A Review," *Science* 142:1477–1479.

the children they place with those of the adoptive parents so far as they are able to do so.

Most twin studies involve the comparison of monozygotic twins reared together with dizygotic twins reared together. The other major type of twin study, the comparison of monozygotic twins reared together with monozygotic twins reared apart, is more difficult to carry out because cases of separated identical twins are so rare that it is hard to find enough cases for meaningful statistical treatment. Nevertheless, such comparisons are very useful because they permit a rough estimate of the proportion of the variance attributable to the environment. Limitations in the amount of available data dictate that the comparisons are usually based on quantitative traits and correlation coefficients rather than on qualitative traits and concordance rates. The equation used is

$$E = \frac{r_{\text{MZT}} - r_{\text{MZA}}}{1 - r_{\text{MZA}}}$$

where E = estimate of the proportion of phenotypic variation due to the environment, r_{MZT} = intraclass correlation coefficient for monozygotic twins reared together, and r_{MZA} = intraclass correlation coefficient for monozygotic twins reared apart. As with H, the values of E will ordinarily range between zero and 1.0, and E is meaningful only if r_{MZT} and r_{MZA} are not both large. Two very useful studies by Newman, Freeman, and Holzinger (1937) and

Table 26.4 Resemblance in IQ Scores of Monozygotic Twins Reared Together (MZT) and Apart (MZA) and Dizygotic Twins Reared Together (DZT)

Study		Correlation coefficients				
		MZT	MZA	DZT	H	E
Newman et al. (1937)	n	50	19	51		
	Corrected mean difference in Binet IQ	3.1	6.0	8.5		
	Corrected correlation coefficient	0.88	0.77	0.63	0.68	0.48
Shields (1962)	n	34	37	7		
	Mean difference in combined IQ	7.4	9.5	13.4		
	Correlation coefficient	0.76	0.77	0.51	0.51	−0.04
Wingfield (1928)	Correlation coefficient	0.90		0.57	0.77	
Herrman and Hogben (1933)	Correlation coefficient	0.84		0.48	0.69	

by Shields (1962) have compared the intelligence of monozygotic twins reared together and apart, and have also included dizygotic twins reared together in their data. The studies of Wingfield and of Herrman and Hogben of monozygotic and dizygotic twins reared together have been included to give further data on monozygotic and dizygotic twins reared together (Table 26.4).

In both the Shields and the Newman et al. studies, the mean IQ difference is greater in the MZA than the MZT twins, but both are lower than the differences between DZT. Therefore, some environmental effect is present, but the important influence of heredity on intelligence is clearly demonstrated. The negative value of E in Shields' work stems from the fact that the correlation coefficient for MZT is slightly less than for MZA.

However, his value for MZT (0.76) is low compared with the others given in Tables 26.2 and 26.3. It does suggest some of the problems that may arise from estimating H and E from twin data. The heritabilities reported for intelligence seem rather high, but the reservations expressed earlier about the validity of heritabilities based on twin data are also applicable here. Nevertheless, the conclusion seems inescapable that the twin data taken as a whole support very clearly the idea that heredity is a very important factor in intelligence.

The comparison of MZT and MZA twins, in essence, permits the study of the effect of environmental variation while the genotype is held constant. In theory, the comparison of monozygotic and dizygotic twins makes it possible to study the effect of different genotypes in a common environment. The comparison of fraternal twins with pairs of single-born sibs is comparable to the MZT-MZA studies since both groups have comparable genotypes, but the dizygotic twins are assumed to develop under more similar environmental conditions. These comparisons seldom reveal any very striking differences, probably because in both groups the pairs grow up in a common home environment. The correlations usually turn out to be rather close to the expected theoretical correlation of 0.5.

Further interesting and useful comparisons can be made between parents and their offspring and between parents and their foster children. Two such studies are shown in Table 26.5. Both studies show a slight positive correlation between the intelligence of children and that of their foster parents, and suggest that the influence of the mother may be somewhat greater than that of the father. The correlation between children and their own parents is considerably higher and close to the theoretical value of 0.5. These studies, in which bias due to selective placement or other factors has been minimized, again suggest the important relation between heredity and intelligence.

Mental Disease

The relationship between heredity and psychosis is often complex. Huntington's chorea has already been cited, however, as a simple dominant condition leading to dementia and death, and the effects of phenylketonuria on behavior as well as intelligence have been pointed out. *Wilson's disease* or *hepatolenticular degeneration* is another rare recessive trait with effects on behavior, some patients showing symptoms similar to schizophrenia while others show less pronounced changes in personality or behavior. Pathologically it is marked by degenerative changes in the brain, especially in the basal ganglia, by cirrhosis of the liver, by greenish brown rings in

Table 26.5 Correlation of Intelligence Between Parent and Own Children and Parent and Foster Children

Study	Children and foster parents		Children and true parents	
	N	r	N	r
Burks (1928)				
Father's MA	178	0.07	100	0.45
Mother's MA	204	0.19	105	0.46
Leahy (1935)				
Father's MA	178	0.15	175	0.51
Mother's MA	186	0.20	191	0.51

From Neel, J. V., and W. J. Schull, 1954. *Human Heredity*. University of Chicago Press, Chicago, p. 113, Table 9.4. Copyright 1954 by the University of Chicago.

the cornea of the eyes, and by an increased deposition of copper in the tissues. The basic defect seems to be a disturbance in copper metabolism, and the level of the copper-bearing serum protein, ceruloplasmin, is low. Thus, Wilson's disease seems to represent another simply inherited inborn error of metabolism that influences behavior.

Schizophrenia is the most frequent of all the psychoses; it is estimated that 1% of all persons will become schizophrenic at some point in their lives. The age of onset ranges from about 15 to 45, and the illness tends to be chronic and long-lasting so that roughly half of all resident patients in public mental hospitals are schizophrenic. The condition was formerly called dementia praecox, but since early onset and irreversible mental deterioration did not invariably occur, both words were inappropriate, and the term schizophrenia was adopted. Four subtypes of schizophrenia are generally recognized: simple, paranoid, hebephrenic, and catatonic. The major symptoms of schizophrenia are bizarre and unreal thought patterns, autistic withdrawal from the real world into a self-created world, ambivalent behavior making strong emotional responses difficult, and inappropriate affect or response to a given emotional situation (e.g., tragic news may be greeted with laughter).

The diagnostic problem poses a serious difficulty for genetic research on schizophrenia. In addition to the four subtypes listed above, other types of schizophrenia are sometimes recognized as well. However, entirely different systems of classification are sometimes used that cut across these categories. Not only is there disagreement as to the number and kind of types and subtypes of schizophrenia, but also as to whether there are any separate categories at all—some argue that the symptoms range along a

Table 26.6   Contingency Study of Schizophrenia

Relationship	Incidence of schizophrenia (%)
Unrelated	
General population	0.85
Step sibs	1.8
Marriage partners	2.1
Related	
First cousins	2.6
Nephews and nieces	3.9
Grandchildren	4.3
Half-sibs	7.0
Parents	9.2
Sibs	14.3
Fraternal twins	14.7
Children	16.4
Identical twins	85.8

From F. J. Kallmann, *The Genetics of Schizophrenia,* J. J. Augustin Publisher, Locust Valley, N.Y. 11560. Pp. 144–145, Table 54.

continuum rather than falling into separate groups. The diagnoses are based on the observed behavioral aberrations; even though more reliable physical or biochemical criteria have been repeatedly sought as an aid in diagnosis, none has yet proved useful. Therefore, it is not even clear yet whether schizophrenia is a single disease entity with a wide range of expression or a number of distinct but similar conditions. Past genetic studies of schizophrenia have generally treated all forms of schizophrenia as various manifestations of a single trait. Under the circumstances, it is perhaps not surprising that even though there is evidence implicating heredity in the etiology of schizophrenia, there is little agreement as to the mode of inheritance for a genetic predisposition to schizophrenia, or even whether a genetic predisposition is necessary.

One type of evidence stems from contingency studies of the incidence of schizophrenia among individuals with different degrees of relationship to a schizophrenic individual, as shown in Table 26.6. These data show a rather striking increase in the frequency of schizophrenia as the degree of relationship becomes closer. Among the identical twins, concordance was only 77.6% among those separated for 5 years or more, while it was 91.5% for those who remained together. Although suggestive, the contingency data could

Table 26.7   Estimated Concordance Rates in Monozygotic (MZ) and Dizygotic (DZ) Co-twins of Schizophrenics

Investigator	Zygosity	Number of pairs	Percent concordance	H
Luxenburger	MZ	21	67	
(1930)	DZ	60	3	0.66
Rosanoff et al.	MZ	41	67	
(1934)	DZ	101	10	0.63
Essen-Möller	MZ	7	71	
(1941)	DZ	24	17	0.65
Slater	MZ	41	76 [a]	
(1953)	DZ	115	14 [a]	0.72
Kallmann	MZ	268	86 [a]	
(1953)	DZ	685	15 [a]	0.84
Inouye	MZ	55	76 [a]	
(1961)	DZ	17	22 [a]	0.69

[a] Age corrected.

From Rosen, E., R. E. Fox, and I. Gregory, *Abnormal Psychology,* 2nd ed. Published by W. B. Saunders Company, Philadelphia, 1972.

also be explained as the result of direct, cultural, nongenetic transmission from one family member to another or due to a common unfavorable environment.

However, a number of additional studies of concordance rates in monozygotic and dizygotic twins also support the genetic hypothesis, and, to some extent, counter the argument for direct nongenetic transmission. The data from six such studies are given in Table 26.7. Again, the lack of complete concordance in identical twins demonstrates that environmental factors must play a significant role in the etiology of schizophrenia. However, the difference between the concordance rates in monozygotic and dizygotic twins and the high heritabilities support an important role for heredity. An added fact of some interest is that while the estimates of the incidence of schizophrenia among the children of a schizophrenic range between 7 and 16%, the estimates of the frequency of schizophrenia among the children of two schizophrenic parents range between 40 and 68%.

Various hypotheses have been proposed for the mode of inheritance involved in schizophrenia, some recessive, some dominant, and some multigenic. The theories have been further refined to fit the data by the postulation of variations in penetrance or expressivity or differences in expression between heterozygotes and homozygotes. To date, no single theory seems to have won complete acceptance although there is an inclination toward polygenic or multiple-factor theories. This appears to be an unfortunate choice of words, for "polygenic" and "multiple-factor" are both associated with theories of quantitative inheritance that seem to have been outmoded by the discovery of the DNA code. The specificity of the code would seem to assure that even if many gene loci are involved, these genes will be specific and different in their effects rather than equal and additive. Thus, it seems preferable to call the theory a multigenic theory to avoid the implications of the older terms. The implications of such a theory are that various combinations of genes and environmental circumstances may push the individual past a threshold into an area of behavior that is recognized as overt schizophrenia. Such a theory is, in a sense, a recognition of the complexity of the condition we call schizophrenia. However, it should hardly be regarded as the final word on the subject, but rather as a working hypothesis. With greater refinement in diagnosis and treatment, it may be possible eventually to tease apart the various genetic and environmental factors involved in the etiology of schizophrenia.

Another common mental disease is manic-depressive psychosis, marked by mood changes, either toward depression or euphoria, that are sometimes so intense as to represent a break with reality. The manic is hyperactive, elated, and greatly overestimates his abilities, tending to make impossible plans, or promises he cannot keep. The depressed patient is not only depressed in mood but is slow in movement and thought and is generally overcome by a feeling of worthlessness and hopelessness. Any combination of mania and depression may occur; the mood swings may veer from episodes of mania through remissions or normal periods to episodes of depression, or one type of reaction may predominate, or the sequence may be irregular and unpredictable. In contrast to the schizophrenic, who seems withdrawn from his environment, the manic-depressive seems oversensitive to his environment. Furthermore, the outcome of manic-depressive psychosis is different from schizophrenia, which tends to lead toward mental deterioration; manic-depressive psychosis is cyclical and recurrent, episodes of mania or depression being followed by complete remission, and the disease does not ordinarily lead to progressive deterioration of the mind and personality.

The incidence of manic-depressive psychosis is generally cited as being about 3 or 4 per 1000 or less than 0.5%. An excess of women over men is found among institutionalized patients, the ratio being about three women to two men. However, these figures reflect only part of the picture because many patients with manic-depressive psychosis are never hospitalized or

Table 26.8 Estimated Concordance Rates in Monozygotic (MZ) and Dizygotic (DZ) Co-twins of Manic-Depressives

Investigator	Zygosity	Number of pairs	Percent concordance	H
Rosanoff et al.	MZ	23	70	
(1935)	DZ	67	16	0.64
Luxenburger	MZ	56	84	
(1942)	DZ	83	15	0.81
Kallmann	MZ	27	100 [a]	
(1953)	DZ	58	26 [a]	1.00
Slater	MZ	8	57 [a]	
(1953)	DZ	30	29 [a]	0.39
Da Fonseca	MZ	21	75	
(1959)	DZ	39	38	0.60

[a] Age corrected.

Reprinted by permission from Rosen, E., R. E. Fox, and I. Gregory, 1972. *Abnormal Psychology*, 2nd ed. Philadelphia, W. B. Saunders Company, p. 201, Table 11–1.

even treated. One study in Sweden indicated that the diagnosed patients represented only one-seventh of the total number of manic-depressives in the general population. Moreover, suicidal tendencies are common among depressed persons, and suicide rather than institutionalization may be the outcome of the psychosis. Since suicide rates are roughly four times higher among men than women, these facts may help to account for the excess of female manic-depressives in institutions.

The incidence figures in different countries, different states, and even in different mental hospitals may vary quite widely. To some extent, these figures probably represent real differences in the incidence of manic-depressive psychosis in different populations, either in their predisposition to the psychosis or in the precipitating factors leading to its onset. However, the evidence is clear that these differences also reflect, to some degree, fashions in diagnosis. For example, the diagnosis of manic-depressive psychosis is far less popular today in the United States than it was in the early part of the century, and many patients who earlier would have been called manic-depressive are now being classed as schizophrenics. At present, the criteria for manic-depressive psychosis are more restrictive than in the past.

Again heredity has been implicated in this psychosis by twin data (Table 26.8) and by the high frequency of manic-depressive psychosis among the parents, sibs, and children of manic-depressives (Table 26.9).

The concordance rates are high not just for monozygotic twins but also

Table 26.9 Estimates of Percent Frequency of Manic-Depressive Psychosis Among Relatives of Manic-Depressive Probands

Investigator	Banse (1929)	Röll and Entres (1936)	Slater (1938)	Strömgren (1938)	Sjögren (1948)	Kallmann (1950)	Stenstedt (1952)
Parents	10.8	13.0	15.5	7.5	7.0	23.4	7.4
Sibs	18.1			10.7	3.6	23.0	12.3
Fraternal twins						23.6	
Children		10.7	15.2				9.4

Reprinted by permission from Rosen, E., R. E. Fox, and I. Gregory, 1972. *Abnormal Psychology*, 2nd ed. Philadelphia, W. B. Saunders Company, p. 202, Table 11–2.

relatively high for dizygotic twins, although there is considerable variation in the data. The incidence among relatives also varies rather widely, but it should be noted that, for a given study, the figures tend to be similar in the different groups being compared. Hence, these variations probably represent differences in the diagnostic criteria being used in the different studies. The similar incidence among parents, sibs, and offspring of manic-depressives has led to the suggestion that an autosomal dominant with incomplete penetrance is involved, but multigenic theories have also been proposed. Not only because of the high incidence among relatives but because of the frequent inability to identify any precipitating event in the onset of the manic or depressive episodes, there has long been a willingness to accept hereditary or constitutional factors as being important in the etiology of manic-depressive psychosis. Recently, a useful diagnostic and genetic distinction has been drawn between a unipolar psychosis, marked by recurrent episodes of depression, and a bipolar psychosis, characterized by both manic and depressive phases.

Behavior and Physique

Human behavior and body build have long been thought to be intimately related. Modern studies have concentrated on the relationship between physique, personality, and psychiatric disorders. Various systems and terms have been used to classify human body types, but all show striking similarities. The one proposed by William Sheldon will serve to illustrate some of the findings. Sheldon rated individuals on three dimensions of physique based on the three germ layers of the developing embryo, which he termed endomorphy (visceral development), mesomorphy (muscular development), and ectomorphy (development of the ectodermal derivatives, the skin and nervous system.) Each individual was rated on a scale from 1 to 7 for each of these three components, and his body type or somatotype could then be described with a three-digit code. A 4-6-1, for example, would be medium in endomorphy, high in mesomorphy, and very low in ectomorphy. Sheldon also devised a scale for temperament related to the three components of physique. The endomorph has a viscerotonic temperament marked by sociability or extraversion, love of physical comfort and food, and in general a relaxed approach to life. The mesomorph has a somatotonic temperament and is energetic, highly competitive, self-assertive, and action-oriented. The ectomorph has a cerebrotonic temperament and is introverted, self-conscious, and sensitive, loves privacy, and is restrained in posture and movement.

Attempts to relate somatotype with particular mental disorders have revealed that heavy-set persons with a strong endomorph component seem prone to develop manic-depressive psychosis, the athletic mesomorphs are apt to show paranoid reactions, and the ectomorphs seem likely to become schizophrenic.

These findings pose some very interesting questions about the relationship between physique and behavior. Correlations of this sort prove very little about causation, so interpretation of such results is not simple. Does a person behave as he does because he has a certain physique, or does some common underlying factor influence both behavior and physique? Moreover, biases in the methodology itself are difficult to eliminate since the research worker may be influenced by the appearance of the mesomorphic subject to rate him higher on aggressiveness and energy-level than he otherwise might. Nevertheless, the findings seem of sufficient interest to warrant further exploration of the relation between behavior and physique.

Thus far, we have discussed abnormal human behavior, but there is evidence that the personality of normal individuals is also influenced by heredity. Evidence from the twin studies of Newman, Freeman, and Holzinger (1937) and those of Shields (1962) is of considerable interest in this regard (Table 26.10).

The comparison of identical twins reared together and separately and fraternal twins reared together is also most revealing (Table 26.11). The comparison of the mean differences in scores for monozygotic twins reared together and separately with those of like-sexed dizygotic twins reared together shows that the differences are less for the MZA group than for the DZT, and they are often very close to the MZT mean difference,

Table 26.10 Heritabilities Estimated from Monozygotic and Like-Sexed Dizygotic Twins Reared Together

Trait	Newman et al.	Shields
Height	0.808	0.893 [a]
Weight	0.775	0.563 [a]
Intelligence	0.678	0.510
Personality		
Neuroticism	0.304	0.303
Extraversion	–	0.504

[a] Females only.

From Newman, H. H., F. N. Freeman, and K. J. Holzinger, 1937. *Twins: A Study of Heredity and Environment.* University of Chicago Press, Chicago, p. 117, Table 37; and Shields, J., 1962. *Monozygotic Twins.* Oxford University Press, London, p. 55, Table 4.

except for the Newman et al. study on weight. Clearly, heredity is implicated not only in such traits as height and weight but in IQ and personality as well. The heritabilities in Table 26.10 suggest, however, that height and weight in these studies were less influenced by environmental factors

Table 26.11 Mean Differences Between Monozygotic Twins Reared Together (MZT) and Separately (MZA) and Like-Sexed Dizygotic Twins Reared Together (DZT)

Trait		MZT	MZA	DZT
Height	Newman et al.	1.7	1.8	4.4
(cm)	Shields	1.3	2.1	4.5
Weight	Newman et al.	4.1	9.9	10.0
(lb)	Shields	10.4	10.5	17.3
IQ	Newman et al.	3.1 [a]	6.0 [a]	8.5 [a]
	Shields	7.4	9.5	13.4
Neuroticism	Newman et al.	5.3	5.0	6.7
	Shields	3.0	3.1	4.0
Extraversion	Shields	2.7	2.5	4.7

[a] Corrected.

Data from Shields, J., 1962. *Monozygotic Twins.* Oxford University Press, London, p. 139, Table 31; and Newman, H. H., F. N. Freeman, and K. J. Holzinger, 1937. *Twins: A Study of Heredity and Environment.* University of Chicago Press, Chicago. Copyright 1937 by the University of Chicago.

Table 26.12 Concordance Rates for Criminal Behavior in Monozygotic and Dizygotic Twins Reared Together. Summary of Seven Studies

	Number of pairs	Percent concordance
Monozygotic	143	68
Dyzgotic	142	28

From Stern, 1960.

than intelligence or personality, with personality appearing to be the most susceptible to environmental influence.

At a time when crime in the streets is a matter of increasing public concern, and the usual explanation for criminal behavior is an unfavorable environmental background, it is of some interest to examine data on criminality in twins (Table 26.12). As always, such data are suggestive but by no means conclusive: The greater concordance in identical twins may be due to a possible greater similarity in their environmental experiences as well as or even rather than their greater similarity in genotype. Studies on criminal twins reared separately are needed to help resolve this question. However, one of the interesting facets of these studies was that identical twins were not only more often concordant in criminality but also tended to be more alike in the type of crime committed than fraternal twins. Personality is influenced by heredity, and it in turn seems to influence the type of crime committed, since personalities of different types seem to get involved in forgery, burglary, and murder. Thus, the data are, to some extent, misleading; in some cases the dizygotic twins were concordant in criminality but discordant in the type of crime, while the monozygotic twins were generally very similar in personality but only one had run afoul of the law. Obviously, the subject requires considerably more study, but at least the possibility must remain open that the genotype may play a role in the development of criminal behavior.

A new aspect of this problem has been added because of the recent suggestion that men with an extra Y chromosome are particularly prone to criminal behavior. This suggestion adds a new dimension to legal proceedings because it now becomes conceivable that an XYY male will plead innocent to a charge because of his extra Y chromosome.

26.3. Nest Cleaning in Bees

In concluding this chapter, let us turn away from human behavior and consider the results of a genetic analysis of nest-cleaning behavior in honey-

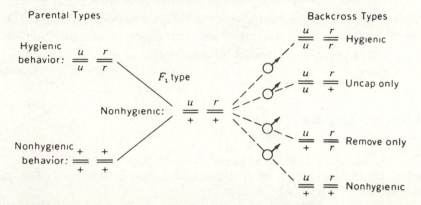

Fig. 26.4. A genetic hypothesis to account for the different responses to broods killed by American foulbrood disease in 63 colonies of honeybees. (Rothenbuhler, W. C., 1964. "Behavior Genetics of Nest Cleaning in Honey Bees. IV. Response of F_1 and Backcross Generations to Disease-Killed Broods," *Amer. Zool.* 4:111–123, p. 122, Fig. 14.)

bees. Two inbred lines were selected, one for resistance, the other for susceptibility to the disease, American foulbrood. These lines differed strikingly in behavior: The resistant line removed foulbrood-killed larvae from the nest, but the susceptible line failed to do so. When crosses were made, the F_1 hybrids behaved like the susceptible parents and did not remove the foulbrood-killed larvae. Therefore the factors involved in nest cleaning appeared to be recessive. Backcrosses of haploid drones from the F_1 queens to hygienic queens led to the conclusion that two loci were involved in the difference in behavior between the two inbred lines. The recessive gene at one locus caused the bees to uncap the cells containing dead larvae, but they did not remove them. The recessive gene at the other locus caused the bees to remove the dead brood, but they would not uncap the cells. Thus, homozygosity for recessives at both loci was necessary for the complete hygienic behavior pattern to be carried out (Fig. 26.4).

We have introduced this elegant experiment at this time to point out that these experiments do not demonstrate that hygienic behavior in the honeybee is governed by genes at just two loci. What it does demonstrate is that the difference in hygienic behavior between these two inbred lines is determined by differences at just two loci. There can be little doubt that other crosses involving other strains would reveal additional loci that influence the nest-cleaning behavior of the honeybee. Where complex behavior patterns patterns exist, they may be influenced or disrupted by genes at many different loci. By the same token, it seems probable that complex human behavior is subject to the influence of genes at many loci.

This fact means that the study of the genetics of human behavior will be complex, but as the function of each additional segment of human DNA becomes understood, we shall move one step closer to an understanding of the relation between genetics and human behavior.

Additional Reading

Bliss, E. L., ed., 1962. *Roots of Behavior*. Harper & Row, New York.

Darwin, C., 1871. *The Descent of Man and Selection in Relation to Sex*. Murray, London.

Darwin, C., 1872. *The Expression of the Emotions in Man and Animals*. Murray, London.

Dilger, W. C., 1962. "The Behavior of Lovebirds," *Sci. Amer.* **206** (Jan.):88–98.

Erlenmeyer-Kimling, L., and L. F. Jarvik, 1963. "Genetics and Intelligence: A Review," *Science* **142**:1477–1479.

Freud, S., 1938. *The Basic Writings of Sigmund Freud*. Random House, New York.

Fuller, J. L., 1960. "Behavior Genetics," *Ann. Rev. Psychol.* **11**:41–63.

Fuller, J. L., and W. R. Thompson, 1960. *Behavior Genetics*. Wiley, New York.

Gottesman, I. I., 1968. "A Sampler of Human Behavioral Genetics," *Evol. Biol.* **2**:276–320.

Heinroth, O., 1911. "Beiträge zur Biologie, namentlich Ethologie und Psychologie der Anatiden." In *Verh. 5th Int. Ornithol.-Kongr.*, pp. 589–702. Deutsche Ornithologische Gesellschaft, Berlin.

Heinroth, O., 1938. *Aus dem Leben der Vögel*. Springer, Berlin.

Heron, W. T., 1935. "The Inheritance of Maze Learning Ability in Rats," *J. Comp. Psych.* **19**:77–89.

Herrman, L., and L. Hogben, 1933. "The Intellectual Resemblance of Twins," *Proc. Roy. Soc. Edinb.* **53**: 105–129.

Hinde, R. A., 1966. *Animal Behaviour. A Synthesis of Ethology and Comparative Psychology*, McGraw-Hill, New York.

Hirsch, J., ed., 1967. *Behavior-Genetic Analysis*. McGraw-Hill, New York.

Howard, H. E., 1920. *Territory in Bird Life*. Murray, London.

Juda, A., 1939. "Zur Ätiologie des Schwachsinns. Neue Untersuchungen an Hilfs-schule-Zwillingen," *Z. ges. Neurol. Psychiat.* **165**:90–97.

Juda, A., 1939. "Neue psychiatrisch-genealogische Untersuchungen an Hilfsschul-zwillingen und ihren Familien. I. Die Zwillingsprobanden und ihre Partner," *Z. ges. Neurol. Psychiat.* **166**:365–452.

Kretschmer, E., 1948. *Körperbau und Charakter*, 19th ed. Springer, Heidelberg.

Lejeune, J., R. Turpin, and M. Gautier, 1959. "Le mongolisme, premier example d'aberration autosomique humaine," *Ann. Genet. Hum.* **1**:41–49.

Lorenz, K., 1935. "Der Kumpan in der Umvelt des Vogels," *J. f. Ornithol.* **83**: 137–213, 289–413.

McGaugh, J. L., N. M. Weinberger, and R. E. Whalen, eds., 1967. *Psychobiology. The Biological Bases of Behavior*. Freeman, San Francisco.

McGill, T. E., ed., 1965. *Readings in Animal Behavior*. Holt, Rinehart and Winston, New York.

Manning, A., 1967. *An Introduction to Animal Behavior*. Addison-Wesley, Reading, Mass.

Manosevitz, M., G. Lindzey, and D. D. Thiessen, 1969. *Behavioral Genetics. Method and Research*. Appleton-Century-Crofts, New York.

Merrell, D. J., 1965. "Methodology in Behavior Genetics," *J. Hered.* **56**:263–266.

Newman, H. H., F. N. Freeman, and K. J. Holzinger, 1937. *Twins, a Study of Heredity and Environment*. University of Chicago Press, Chicago.

Parker, C. E., J. Melnyk, and C. H. Fish, 1969. "The XYY Syndrome," *Amer. J. Med.* **47**:801–808.

Parnell, R. W., 1958. *Behaviour and Physique*. Arnold, London.

Pavlov, I. P., 1927. *Conditioned Reflexes*, G. V. Anrep, tr. Oxford University Press, London.

Penrose, L. S., 1963. *The Biology of Mental Defect*, 2nd ed. Grune and Stratton, New York.

Pratt, R. T. C., 1967. *The Genetics of Neurological Disorders*. Oxford University Press, London.

Reed, S. C., and E. W. Reed, 1965. *Mental Retardation: A Family Study*. Saunders, Philadelphia.

Roe, A., and G. G. Simpson, eds., 1958. *Behavior and Evolution*. Yale University Press, New Haven, Conn.

Rosanoff, A. J., L. M. Handy, and I. R. Plesset, 1937. "The Etiology of Mental Deficiency," *Psychol. Monogr.* **48**:1–137.

Rosen, E., R. E. Fox, and I. Gregory, 1973. *Abnormal Psychology*, 2nd ed. Saunders, Philadelphia.

Rosenthal, D., 1970. *Genetic Theory and Abnormal Behavior*. McGraw-Hill, New York.

Rosenthal, D., and S. S. Kety, eds., 1968. *The Transmission of Schizophrenia*. Pergamon, Oxford.

Rothenbuhler, W. C., 1967. "Genetic and Evolutionary Considerations of Social Behavior of Honeybees and Some Related Insects." In *Behavior-Genetic Analysis*, J. Hirsch, ed. McGraw-Hill, New York.

Scott, J. P., and J. L. Fuller, 1965. *Genetics and Social Behavior of the Dog*. University of Chicago Press, Chicago.

Sheldon, W. H., C. W. Dupertuis, and E. McDermott, 1954. *Atlas of Men*. Harper & Row, New York.

Sheldon, W. H., S. S. Stevens, and W. B. Tucker, 1940. *The Varieties of Human Physique*. Harper & Row, New York.

Shields, J., 1962. *Monozygotic Twins*. Oxford University Press, London.

Slater, E., and V. Cowie, 1971. *The Genetics of Mental Disorders*. Oxford University Press, New York.

Slater, E., and M. Roth, 1969. *Clinical Psychiatry*. Williams and Wilkins, Baltimore.

Stern, C., 1973. *Human Genetics*, 3rd ed. Freeman, San Francisco.

Sturtevant, A. H., 1915. "Experiments in Sex Recognition and the Problem of Sexual Selection in *Drosophila*," *J. Animal Behav.* **5**:351–366.

Terman, L. M., and M. H. Oden, 1959. *The Gifted Group at Mid-Life. Genetic Studies of Genius*, Vol. V. Stanford University Press, Stanford, Calif.

Thoday, J. M., and A. S. Parkes, eds., 1968. *Genetic and Environmental Influences on Behaviour*. Plenum, New York.

Tinbergen, N., 1951. *The Study of Instinct*. Oxford University Press, London.

Tryon, R. C., 1940. "Genetic Differences in Maze-Learning Ability in Rats." In *39th Yearbook Nat. Soc. Stud. Educ.*, Part I, pp. 111–119. Public School Publ. Co., Bloomington, Ill.

Vandenberg, S. G., ed., 1965. *Methods and Goals in Human Behavior Genetics*. Academic, New York.

Vandenberg, S. G., ed., 1968. *Progress in Human Behavior Genetics*. Johns Hopkins University Press, Baltimore.

Watson, J. B., 1924. *Behaviorism*. University of Chicago Press, Chicago.

Whitman, C. O., 1899. "Animal Behavior." In *Biol. Lect. Marine Biol. Lab. Wood's Hole, Mass. 1898*, pp. 285–338.

Whitman, C. O., 1919. "The Behaviour of Pigeons." Carnegie Inst. Wash. Publ. No. 257, pp. 1–161.

Wingfield, A. H., 1928. *Twins and Orphans: The Inheritance of Intelligence*. J. B. Dent & Sons, Ltd., London.

Yerkes, R. M., 1907. *The Dancing Mouse*. Macmillan, New York.

Questions

26.1. Why has the study of environmental influences on behavior received relatively more attention in the past than the study of genetic influences on behavior?

26.2. Cite three types of behavior in invertebrates that are not learned.

27.3. What is the adaptive advantage of stereotyped unlearned behavior patterns? What is the adaptive advantage of learning? Which is more common in the animal kingdom?

27.4. What are the major types of mental health problems?

27.5. Generally speaking, it is more often possible to demonstrate a simple genetic origin for cases of mental deficiency (e.g. phenylketonuria, Down's syndrome) than it is for cases of psychosis. What explanation can you suggest for this difference?

26.6. What is the evidence that IQ scores are influenced by heredity?

26.7. Do you think heritability estimates based on comparisons of concordance rates, correlation coefficients, or variances for various traits in monozygotic and dizygotic twins are as reliable as heritability estimates obtained by the methods given in Chapter 20? Explain. Are the heritability estimates from twins applicable only to the twin population itself, or to the population from which the twins were drawn as well?

26.8. What is the evidence that schizophrenia is influenced by heredity?

26.9. There is evidence that reproductive rates in schizophrenics are lower than among nonschizophrenics and also that heredity is involved in the etiology of schizophrenia. Nevertheless, schizophrenia has a remarkably high frequency of 1% in the general population despite the selection pressure against it. What explanations can you suggest for this apparent paradox? How would you test your hypothesis?

26.10. Does the success of drug therapy in controlling the symptoms of mental illness and shortening the stay of psychotic patients in mental hospitals argue for a greater or a lesser role for heredity in the etiology of mental illness?

26.11. Do you believe that schizophrenia is a single entity or a group of related syndromes? What evidence supports your answer?

26.12. What type of study holds the greatest promise for working out the genetic factor or factors involved in the etiology of schizophrenia?

26.13. What type of research do you believe holds the greatest promise for providing insight into the relationship between genetics and human behavior?

CHAPTER TWENTY-SEVEN

Genetics and Race

Thirty-four separate subspecies or races have been described in the North American song sparrow, *Passarella melodia* (Fig. 27.1). If our discussion were to be confined to a consideration of the nature, origin, and significance of the different subspecies of song sparrows, it could be handled in a relatively objective, dispassionate way. However, the discussion of races in man is apt to arouse such emotional overtones that rational consideration of the subject becomes rather difficult. Nevertheless, it seems desirable to try to deal with the topic simply because it is a matter of such significance and because it is surrounded by so many misconceptions, fears, and erroneous beliefs.

Some people try to make the subject go away by denying that there are human races. One eminent anthropologist, for example, titled a paper "On the Nonexistence of Human Races." Another distinguished anthropologist, on the other hand, went so far as to suggest that the different human races existed as separate entities even prior to the evolutionary emergence of *Homo sapiens,* and that they crossed the imaginary evolutionary line from subhuman to human hominids separately and at different times. Aside from giving taxonomists fits because this idea implied the origin and persistence of subspecific traits prior to the origin of specific traits, this suggestion was seized upon and used by racists in support of their concepts of racial supremacy. It can hardly be gainsaid that supposedly nonexistent entities are the source of a lot of controversy in the world.

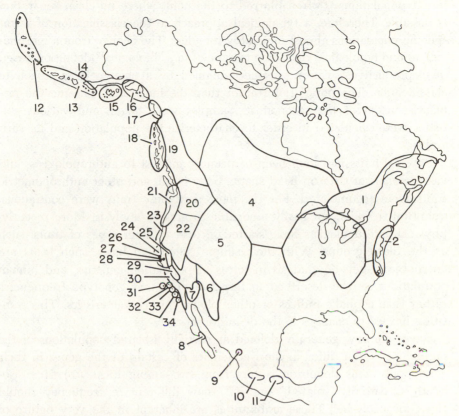

Fig. 27.1. The approximate breeding ranges of the 34 races or subspecies of the polytypic species, the song sparrow, *Passarella melodia*, formerly known as *Melospiza melodia*. (Miller, A. H., 1956. "Ecologic Factors that Accelerate Formation of Races and Species of Terrestrial Vertebrates," *Evolution* 10:262–277, p. 264, Fig. 1.)

In order to discuss race (or subspecies), it is well to define the term, but the variety and vagueness of the many definitions proposed merely help to confuse the issues. Quite apart from its human context, the sub-species concept has been attacked as being useless in the classification of organisms. Taxonomy was long dominated by the type concept, whereby species were described in relation to a single representative specimen, the type specimen, primarily on the basis of its morphological features. Because reproductive isolation bars gene flow between species, discontinuities exist between different species and make it reasonably easy in most cases to distinguish between members of separate species on morphological grounds. In the case of subspecies or races, no such reproductive barriers exist, gene flow can and often does occur, and the distinctiveness between different populations is often blurred to the point where no clean separation is possible. Therefore, a typological approach to the classification of infraspecific populations almost inevitably founders. The change from a morphological and typological definition of species to a "biological" definition based on reproductive isolation was accompanied by a change in approach to classification. The population, rather than the individual, became the primary concern of the taxonomist. Samples were taken, and various statistics were calculated in order to characterize the population and its variability.

Physical traits had long been of primary interest to anthropologists, and vast amounts of data on head shape, body build, and other anthropometric traits were accumulated. For the most part these traits were continuous, quantitative traits, not easily amenable to genetic analysis. More recently, physical anthropologists have started to turn to other types of traits, such as the blood groups. With their simple Mendelian basis, such traits are amenable to genetic analysis in terms of population genetics, and human populations can be described in terms of gene and genotype frequencies rather than cephalic indices or other phenotypic characteristics. This transition has led to changes in the definition of race.

Races are now generally defined as partially isolated populations of the same species that differ in the frequencies of certain of the genes in their gene pools. This definition suffers from two ambiguities: (1) How isolated is "partially isolated," and (2) how different in frequency should the gene pools be? These ambiguities are inherent in the very nature of races, and help to explain why some authorities identify as few as five different human races while others have recognized 34 or even more distinct human races. It also helps to explain why some argue that races as such do not exist, while others recognize three different kinds of races—geographical races, local races, and microgeographical races or micro races.

In our earlier discussion of genetics and the origin of species, the deme or local breeding population was described as the smallest identifiable population unit, and the species itself as the largest and most inclusive breeding population. Each deme is unique and its gene pool differs to some extent from that of every other deme. However, when a number of demes have similar gene pools but differ from other groups of demes, subspecific entities may be recognized as, for example, ecotypes or geographical races. These units are often named primarily as matters of convenience, depending on the type of problem under investigation. Whatever the system used, there is little that is sacred about it, since sharp boundaries are usually lacking between the different units due to the ebb and flow of genes throughout the species. When it is operationally useful to identify different races or subspecies, the taxonomist is apt to do so. If he is dealing with clines, attempts to name subspecies are almost meaningless. Therefore, the system used to categorize subspecific variation depends on the patterns of variation to be observed in the breeding populations of the species and also on the interests of the investigator.

By the usual criteria applied by systematists today, all living men belong to a single species, *Homo sapiens,* since just about every conceivable type of human cross can and does take place. In other words, the gene flow among different human populations is sufficiently great that they form a single species. If anything, reproductive isolation is less today than in the past due to modern man's mobility.

An objective look at man leads to the inescapable conclusion that he is a vertebrate, a placental mammal, and a primate. All men manifest a number of traits that separate them from the other primates. For example, the large toe is in line with the others rather than opposable, yet the foot is arched. The spinal column has a unique lumbar curve, not seen in other primates. The body is nearly hairless, and tactile hairs are absent. Man's chin is distinctive, his canine teeth do not project noticeably beyond his other teeth, and his lips are outrolled with a median furrow. His nose has a bridge and an elongated tip. The human brain, some 2½ to 3 times larger than that of a gorilla, is one of his more unusual features. These traits, plus those that mark man as a primate, a placental mammal, and a vertebrate, are shared by all men. The traits that distinguish different human populations involve such things as hair, eye, and skin color, body size and shape, hair shape and head shape, blood group frequencies, and so on. It seems safe to say that these differences are minor compared to the more fundamental differences cited above. A major question is the origin of the differences observed between different human populations. Some have argued that they are nonadaptive while others have argued that they do have adaptive significance. Although it must be admitted that evidence bearing on the adaptive significance of racial differences in man is surprisingly scant, it seems, by analogy with other species, that adaptive significance will be detected for many such traits.

In polytypic species made up of a number of geographical races, the intraspecific patterns of variation have emerged as the result of the combined effects of mutation, natural selection, random genetic drift, and gene flow. While gene flow among different populations tends to prevent their divergence, mutation and genetic drift are apt to produce random differences among populations. Because the environmental conditions, both biotic and physical, are never exactly alike, natural selection causes gene frequency changes leading to improved adaptation to local conditions. Thus, the members of a geographic race, whether of song sparrows or leopard frogs, turn out to be better adapted to their immediate local environment than members of a different geographic race from another locality would be. Since mutation, selection, drift, and gene flow also govern the course of human evolution, the pattern of human intraspecific variation would seem likely to have a similar nature and origin.

Human populations in different parts of the world fall rather easily into a pattern of geographical races. The so-called primary races can each be identified essentially with a continent:

1. Negroid—Africa
2. Caucasoid—Europe
3. Mongoloid—Asia
4. Australoid—Australia
5. American Indian—The Americas

A more refined classification of geographical races of man and their pre-Columbian distribution is the following:

1. African—Africa south of the Sahara
2. European—Europe, Africa north of the Sahara, the Middle East
3. Indian—The Indian subcontinent

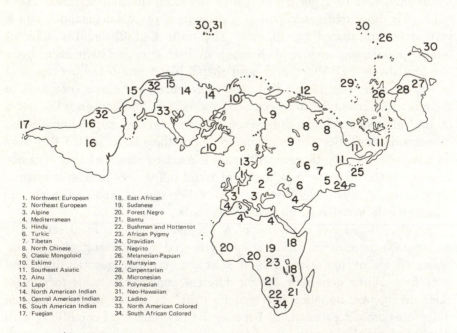

1. Northwest European
2. Northeast European
3. Alpine
4. Mediterranean
5. Hindu
6. Turkic
7. Tibetan
8. North Chinese
9. Classic Mongoloid
10. Eskimo
11. Southeast Asiatic
12. Ainu
13. Lapp
14. North American Indian
15. Central American Indian
16. South American Indian
17. Fuegian
18. East African
19. Sudanese
20. Forest Negro
21. Bantu
22. Bushman and Hottentot
23. African Pygmy
24. Dravidian
25. Negrito
26. Melanesian-Papuan
27. Murrayian
28. Carpentarian
29. Micronesian
30. Polynesian
31. Neo-Hawaiian
32. Ladino
33. North American Colored
34. South African Colored

Fig. 27.2. The geographic distribution of human races in which 34 groups are recognized. (Dobzhansky, T., 1962. *Mankind Evolving.* Yale University Press, New Haven, Conn., p. 263, Fig. 10.)

4. Asiatic—Indonesia, Southeast Asia, Tibet, China, Japan, Mongolia, Siberia
5. Australian—Australia
6. Melanesian-Papuan—Western Pacific from New Guinea to New Caledonia and Fiji
7. Micronesian—Western Pacific from Guam to Marshall and Gilbert Islands
8. Polynesian—Eastern Pacific from New Zealand to Hawaii and Easter Island
9. American Indian—The Americas

Further subdivision gives a classification in which 34 different races are recognized, as shown in Fig. 27.2 and Table 27.1.

Table 27.1   A Classification of *Homo sapiens* into 34 Races

1. *Northwest European*—Scandinavia, northern Germany, northern France, the Low Countries, United Kingdom, and Ireland
2. *Northeast European*—Poland, Russia, most of the present population of Siberia
3. *Alpine*—from central France, south Germany, Switzerland, northern Italy, eastward to the shores of the Black Sea
4. *Mediterranean*—peoples on both sides of the Mediterranean, from Tangier to the Dardanelles, Arabia, Turkey, Iran, and Turkomania
5. *Hindu*—India, Pakistan
6. *Turkic*—Turkestan, western China
7. *Tibetan*—Tibet
8. *North Chinese*—northern and central China and Manchuria
9. *Classic Mongoloid*—Siberia, Mongolia, Korea, Japan
10. *Eskimo*—arctic America
11. *Southeast Asiatic*—South China to Thailand, Burma, Malaya, and Indonesia
12. *Ainu*—aboriginal population of northern Japan
13. *Lapp*—arctic Scandinavia and Finland
14. *North American Indian*—indigenous populations of Canada and the United States
15. *Central American Indian*—from southwestern United States, through Central America, to Bolivia
16. *South American Indian*—primarily the agricultural peoples of Peru, Bolivia, and Chile
17. *Fuegian*—nonagriculture inhabitants of southern South America

Table 27.1 A Classification of *Homo sapiens* into 34 Races (*Cont.*)

18. *East African*—East Africa, Ethiopia, a part of Sudan

19. *Sudanese*—most of the Sudan

20. *Forest Negro*—West Africa and much of the Congo

21. *Bantu*—South Africa and part of East Africa

22. *Bushman and Hottentot*—the aboriginal inhabitants of South Africa

23. *African Pygmy*—a small-statured population living in the rain forests of equatorial Africa

24. *Dravidian*—aboriginal populations of southern India and Ceylon

25. *Negrito*—small-statured and frizzly-haired populations scattered from the Philippines to the Andamans, Malaya, and New Guinea

26. *Melanesian–Papuan*—New Guinea to Fiji

27. *Murrayian*—aboriginal population of Southeastern Australia

28. *Carpentarian*—aboriginal population of northern and central Australia

29. *Micronesian*—islands of the western Pacific

30. *Polynesian*—islands of the central and eastern Pacific

31. *Neo-Hawaiian*—an emerging population of Hawaii

32. *Ladino*—an emerging population of Central and South America

33. *North American Colored*—the so-called Negro population of North America

34. *South African Colored*—the analogous population of South Africa

From Dobzhansky, T., 1962. *Mankind Evolving.* Yale University Press, New Haven, Conn., pp. 363, 365.

Each of these systems has validity; each has some merit; each is also inadequate to describe the full range of intraspecific variation between different human populations because, no matter which system is used, some human populations will not fit comfortably into one or another of the groups. The very nature of human intraspecific variation precludes the separation of human beings into a neat system of racial categories. Furthermore, our knowledge of the genetic structure of human populations makes the old shibboleth of the "pure race" an obvious myth. The notion of a pure Aryan race or a pure Nordic type is simply inconsistent with all that is known about the genetics of human populations. A "pure" race implies genetic homozygosity or isogeneity for all members of the population. Although it seems probable that the isolation between different human populations was somewhat greater in the past than at present, it is extremely doubtful that any human population ever even approached the level of homozygosity to be found in an inbred strain of mice. Hence, the earlier tendency to attempt to identify a few ancestral races of man, and to regard many living individuals as the result of hybridization, seems an oversimplification. While migration and gene flow are increasing today, migrations and gene flow occurred in the past as well. The most reasonable conclusion is that, in the past as at present, *Homo sapiens* was a polytypic species and that the population structure was in a constant state of flux. As a result, the populations identifiable as races were also subject to change. In the list of 34 races given above, the last four are recognizably hybrid in origin and less than 400 years old. The Neo-Hawaiian is of Polynesian, Mongolian, and Caucasian origin; the Ladino or mestizo is of American Indian, Caucasian, and probably some Negroid ancestry; and the North American Colored and South African Colored are of Negroid and Caucasian ancestry. The present inhabitants of Pitcairn Island are descended from Polynesians and the mutineers from the British ship, H.M.S. Bounty. Though few in number and thus not apt to be included in any list of races, they now form a rather distinctive group, of particular interest because so much, relatively, is known about their origin.

Since adaptation to local environmental conditions has been shown to be characteristic of subspecies of other species of plants and animals, it seems reasonable to inquire whether there are adaptively significant differences between different human races. The answer is that there undoubtedly are such differences, but that the evidence bearing on the subject is surprisingly sketchy. There are several reasons for this situation. In the first place, because of the nature of the evolutionary process, with its combination of random and directed events, the mere existence of differences is not sufficient evidence of their adaptive significance. Second, the demonstration of a selective advantage leading to improved fitness is a difficult task in any species, but is particularly difficult in man. In addition, these adaptations may have arisen in the past, when the selection pressures operating on human populations differed from those now at work. Furthermore, in man genetic selection may not only operate in relation to the physical environment and his diseases and parasites, but sexual selection and social and cultural factors may also be involved, which may be much more difficult to identify. Finally, research in this area has probably been inhibited by the realization that the results of such research might be seized upon by racists in support of their views. While such fears are not groundless, they reflect a misunderstanding of the meaning of biological differences.

To say that one person differs from another or that one race differs from another does not mean that one is superior to the other. The mere existence of differences is not in itself evidence of superiority or inferiority, even though many have slipped into this logical fallacy. If such words are to be used, a frame of reference or system of values is required by which such judgments can be made. The question then becomes what set of values, or rather, whose set of values shall prevail.

In the past, many have made such judgments. In the United States Presidents Jefferson and Lincoln, both usually regarded as enlightened humanitarians, believed Negroes to be inferior; in England Sir Francis Galton was convinced of Nordic superiority. However, their beliefs did not lead them to behave like the Nazis. Adolph Hitler made racism an integral part of Nazism, glorifying the blond Nordic race and justifying the extermination of 6 million Jews on the grounds that they were an inferior race. Oddly enough, both the Germans and the Jews have remarkably diverse origins, and their identity as groups is more cultural than biological.

If, in fact, differences among races are adaptive responses to local conditions, then one group will be better adapted to one set of environmental conditions and the other to another. Under these circumstances, to deal in absolute terms such as superior and inferior becomes meaningless.

Although skin pigmentation is only one among many traits in which human populations differ, it looms large in discussions of racial differences, probably because it is so visible, and races have been described in terms of their skin color as yellow, brown, black, red, and white. One might expect that if the adaptive significance of any human trait were known, it would be skin color, but the fact is that even though theories abound, there is no general agreement about the adaptive significance of differences in human skin color.

The most widely accepted theory is that the melanin pigment provides a protective barrier against the harmful effects of sunlight. Accordingly, Negroes are considered better adapted to life under the hot tropical sun than light-skinned people, and their dark skin has been attributed to the action of natural selection. In lighter-skinned individuals the ability to tan is regarded as an individual adaptive response to exposure to sunlight. The geographical distribution of human races is generally in accord with the idea that skins should be darkest in areas with clear skies and bright sunshine and lightest in cool and cloudy regions. Dark peoples live south of the Sahara in the savannas of Africa, in southern India, and in Melanesia, New Guinea, and Australia. The natives in the tropical forests of Africa are lighter than those in the savannas, and the South African Bushmen are yellowish-brown. Cool, cloudy northern Europe is the center for peoples with the lightest skin. The American Indians do not conform to the hypothesis, however, since the amount of pigmentation is essentially the same in populations ranging from Tierra del Fuego through the tropics into temperate North America. The relatively recent arrival of the Indians in the American hemisphere via the Bering Straits some 25,000 years ago or less, and their subsequent spread southward, may account for this discrepancy. The Mongoloids of Siberia and the Eskimos of the far North in America are more pigmented than might be expected under the theory, but this has been explained as a response to the bright glare from Arctic snow. Since these peoples are normally heavily clothed, this explanation seems a bit forced, and fairly recent migration into their present harsh environment may be a factor here as well.

When the physiological effects of sunlight are studied, two general categories can be recognized: One is the effect of adding the energy of sunlight to the heat load of the body, the other is the effects of certain wavelengths (the ultraviolet) in causing sunburn; these same wavelengths are also carcinogenic and antirachitic. Since dark skin absorbs about 30% more energy from sunlight than light skin, dark-skinned individuals would seem to be less well adapted to the tropics than light-skinned persons because of their increased heat load.

The skin consists of an outer horny layer of dead epidermal cells, a layer of living epidermal cells, and an underlying dermal layer. Sunburn results from damage to the living cells, primarily in the epidermis. A suntan results in not only an increased deposition of melanin by the epidermal cells but also in a thickening of the outer horny layer. The protection against the action of ultraviolet seems due as much to one as the other. Furthermore, it has been observed that Negro skin, which is refractory to sunburn, and has more melanin than white skin, also has a thicker horny layer than white skin. Therefore, the greater resistance to sunburn may not be due solely to the greater melanin content. In any case, however, dark skin does seem to confer more protection against sunburn than light skin.

Repeated exposure to ultraviolet light over long periods of time tends to induce skin cancers in the exposed areas in whites, but Negroes are much less susceptible to such cancers. Negro immunity to skin cancer and to sunburn is probably due to the same cause—the greater opacity of the outer layer of epidermis. However, since these cancers occur late in life in whites, their effect on fitness may be small.

A third effect of ultraviolet in man is to mediate the synthesis of vitamin D in the skin from a precursor substance. Vitamin D is essential for normal bone development in man, and a deficiency leads to rickets. It has been suggested that the adaptive advantage of light skin is related to its greater ability to carry on vitamin D synthesis in the northern regions. However, vitamin D can also be obtained in the diet so that the connection with fitness is somewhat tenuous. Furthermore, it has not even been clearly established that greater synthesis does occur in light skin than in dark skin. Moreover, this antirachitic theory fails to account for the prevalence of fair hair color and light eye colors in light-skinned populations since hair color and eye color are controlled by different sets of genes than skin color.

Therefore, the adaptive significance of skin color does not seem to be clearly established. Dark skin seems to protect against sunburn and skin cancer, but it increases the heat load in warm climates. The relation between vitamin D synthesis and fitness can be regarded as no more than an interesting speculation at present. Still another, quite different hypothesis is that dark skin provides protective coloration to hunters in such areas as tropical rain forests. This theory, too, is primarily speculative, with no real evidence to back it up. If nothing else, this discussion of the possible adaptive significance of skin pigmentation in man should point up some of the difficulties in studying adaptation and perhaps prevent some hasty jumps to unwarranted conclusions about the adaptive value of human traits. This discussion should not be interpreted, however, as suggesting that skin color has no selective value. Dark skin has evidently evolved independently in several otherwise rather different human populations—for example, in Africa, southern India, and Melanesia. Just because we do not presently have very much insight into the role of skin pigments in man does not necessarily mean that this situation will always prevail or that they lack adaptive significance.

A number of other traits in man have been studied for their adaptive significance. In other warm-blooded species, a relation between climate and the size and proportions of the body has been discovered. These relations have been stated as ecological "rules" or generalizations. Bergmann's rule is that "within a polytypic warm-blooded species, the cooler the climate where a subspecies lives, the greater its body size." Allen's rule states that "within a warm-blooded species, the warmer the climate, the larger the size of protruding body parts, such as ears, legs, and tail." These rules appear to be based on adaptations for either the conservation or dissipation of body heat. As the body increases in size, its surface increases as the square, and its volume as the cube of its linear measurements. In a cold climate, to minimize heat loss, which occurs at the body surface, a small surface-to-volume ratio (or a roly-poly physique) is most effective. To maximize heat loss in a warm climate, a large surface-to-volume ratio (or a long, lean physique) is most efficient. Measurements of the body surface in man are quite difficult. Therefore, the comparisons involve such traits as weight, stature, or sitting height rather than body surface. For example, Fig. 27.3 shows the regression of body weight on mean annual temperature, and it is clear that in various parts of the world the mean body

Table 27.2 Stature (in cm), Weight (in kg), and Stature:Weight Ratio Among Inhabitants of Different Parts of the World

Population	Stature	Weight	Ratio
Caucasian			
Finland	171.0	70.0	2.44
United States (Army)	173.9	70.2	2.48
Iceland	173.6	68.1	2.55
France	172.5	67.0	2.57
England	166.3	64.5	2.58
Sicily	169.1	65.0	2.60
Morocco	168.9	63.8	2.65
Scotland	170.4	61.8	2.76
Tunisia	173.4	62.3	2.78
Berbers	169.8	59.5	2.85
Mahratta (India)	163.8	55.7	2.94
Bengal (India)	165.8	52.7	3.15
Negroid			
Yambasa	169.0	62.0	2.78
Kirdi	166.5	57.3	2.90
Baya	163.0	53.9	3.02
Batutsi	176.0	57.0	3.09
Kikuyu	164.5	51.9	3.17
Pygmies	142.2	39.9	3.56
Efe	143.8	39.8	3.61
Bushmen	155.8	40.4	3.86
Mongoloid			
Kazakh (Turkestan)	163.1	69.7	2.34
Eskimo	161.2	62.9	2.56
North China	168.0	61.0	2.75
Korea	161.1	55.5	2.90
Central China	163.0	54.7	2.98
Japan	160.9	53.0	3.04
Sundanese	159.8	51.9	3.08
Annamites	158.7	51.3	3.09
Hong Kong	166.2	52.2	3.18

From Baker, P. T., 1958. "The Biological Adaptation of Man to Hot Deserts," *Amer. Nat.* 92:337–357, pp. 355–357. Appendix A.

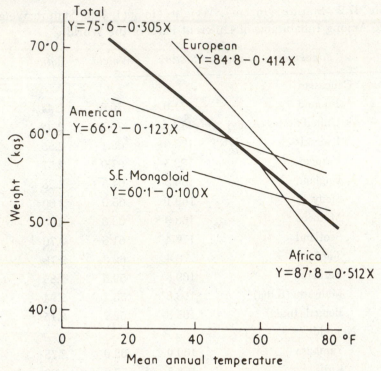

Fig. 27.3. The regression of body weight on mean annual temperature. In all cases, the higher the mean annual temperature, the lower the body weight. (Roberts, D. F. B., 1953. "Body Weight, Race and Climate," *Amer. J. Phys. Anthr.* 11:533–558, p. 548, Fig. 8.)

weight is lower in hot regions than in the cooler areas. Moreover, studies of the ratio of sitting height to total height and of arm span to total height show that man's limbs tend to be longer relative to his torso in the warmer climates. The ratio of height to weight again shows an increasing ratio the warmer the climate (Table 27.2). Thus, in a general way, these human populations seem to conform to these ecogeographical rules in the same way as other warm-blooded species. There is no good evidence as yet as to what extent these differences are environmental and to what extent genetic. It would be of interest to study the physiques of different races reared in climates other than that of their ancestors. No matter what the relative contributions of heredity and environment are, it does seem clear that inhabitants of hot regions tend to be lean and slender, those of cool regions heavier and more thickset.

Other factors are undoubtedly involved in height as well. Both the Pygmies, who are very small, and the Batutsi, Nilotic Negroes who are very tall, live in the hot tropics, the Pygmies as hunters in the rain forest,

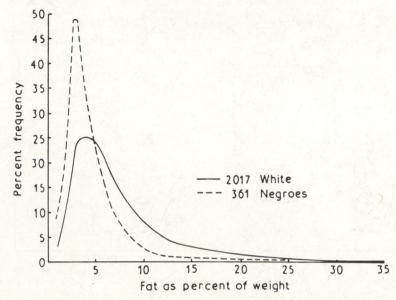

Fig. 27.4. Body fat in American Caucasians and American Negroes as measured by skin-fold thickness. (Reprinted from "Skinfold Measurements in young American males," *Human Biology* 28.2 (May 1956): 154–64, by Russell W. Newman, by permission of the Wayne State University Press.)

the Batutsi as herdsmen in the open grasslands. Both are slender in build, as expected, but they differ by more than a foot in height, on the average. It is tempting to speculate about the adaptive significance of this striking difference in height, but no real evidence is available as yet.

Another aspect of this problem is the measurement of subcutaneous body fat. This body fat not only contributes to the rounded physique of the more northerly peoples, but is also an effective insulator against heat loss. All these considerations led to the hypothesis that the flattened Mongoloid face, well padded with subcutaneous fat, is particularly well adapted to severely cold and windy climates. This theory remains just that, but some data exist suggesting differences in body fat between different populations in the expected directions (Fig. 27.4). However, nutrition is an important factor here, and carefully controlled studies are needed.

Other studies have been made of human response to heat stress and to cold stress. In such studies, of course, acclimatization is an important factor. Even after allowing for acclimatization, however, there appear to be some differences in physiological tolerance in different races.

Another area in which differences among human races have been sought is resistance to disease, not from acquired immunity due to recent ex-

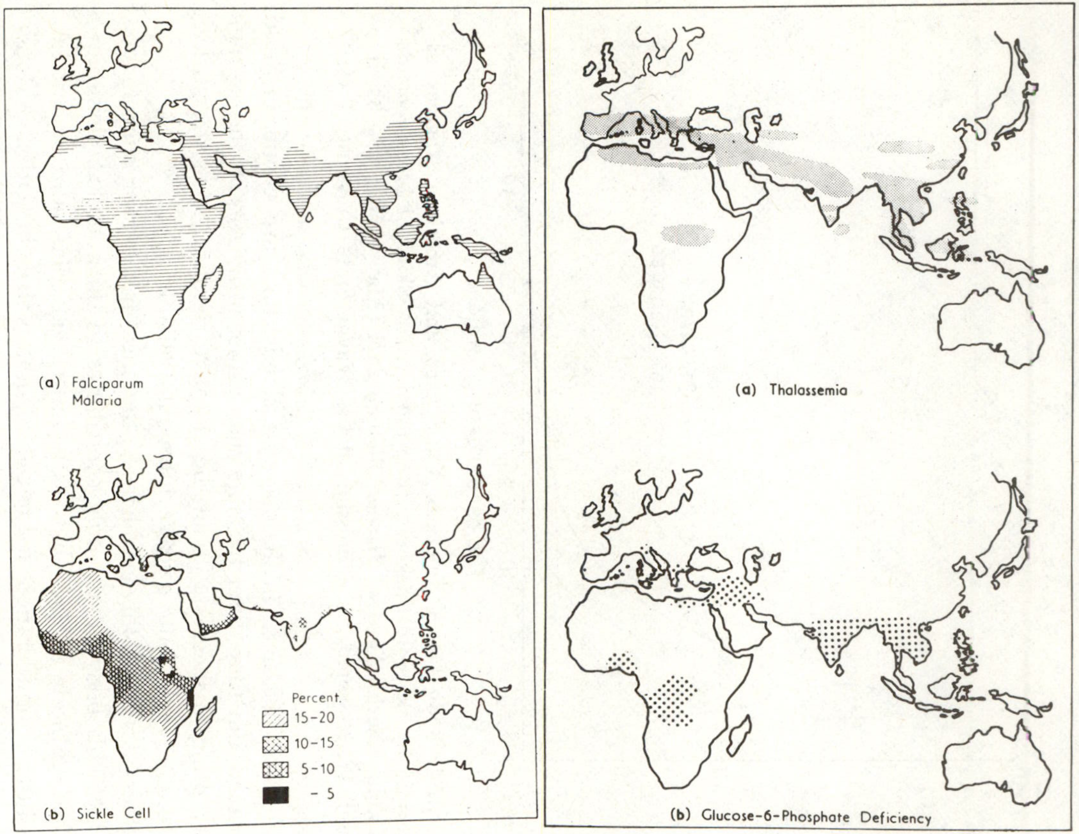

(a) Thalassemia

(b) Glucose-6-Phosphate Deficiency

(a) Falciparum Malaria

(b) Sickle Cell

Percent.
15–20
10–15
5–10
– 5

Fig. 27.5. Left: distribution of (a) falciparum malaria and (b) sickle cell gene in the Old World. Right: distribution of (a) thalassemia and (b) glucose-6-phosphate dehydrogenase deficiency in the Old World. (Left, a and b: Reprinted with permission from A. C. Allison, "Abnormal Hemoglobin and Erythrocyte Enzyme-Deficiency," in G. A. Harrison, ed., *Genetical Variation in Human Populations*, 1961, Pergamon Press, Inc. [out of print]. Right, a: Chernoff, A. I., 1959. "The Distribution of the Thalassemia Gene: A Historical Review," *Blood* 14:899–912, p. 903, Fig. 1. By permission of *Blood*. Right, b: Reprinted from "Metabolic Polymorphisms and the Role of Infectious Diseases in Human Evolution," *Human Biology* 32:28–62, p. 50, Fig. 5 (1960), by A. G. Motulsky, by permission of the Wayne State University Press.)

posure, but resistance based on genetic differences. The three red blood cell polymorphisms in man, for the sickle cell trait, thalassemia, and glucose-6-phosphate dehydrogenase deficiency (G6PD), seem to be cases in point, for their present high frequency seems to coincide with areas where exposure to falciparum malaria has been high, and selection for these genes, which confer some degree of immunity, has been correspondingly high (Fig. 27.5). In other diseases, such as measles, tuberculosis, and poliomyelitis, there is suggestive evidence that genetic factors for resistance exist even though they have not been individually identified. The evidence comes from several sources: (1) from twins, where concordance rates for an infectious disease are usually somewhat higher in monozygotic than dizygotic twins despite similar rates of exposure, (2) from previously unexposed populations in which there is decreased mortality following several generations of exposure despite the virtual absence of any health measures, and (3) from differences in incidence among different families or different populations despite similar rates of exposure. It is not always easy in these situations to distinguish acquired immunity from genetically determined natural immunity. However, experiments in animals have demonstrated that resistant or susceptible strains can be readily established by selection for such diseases as tuberculosis in rabbits, plague in rats, and poliomyelitis in mice, all of which closely parallel similar human diseases.

The high mortality rate when a previously unexposed population is first exposed to a disease also suggests genetic differences in susceptibility. Measles, for example, a serious but seldom fatal childhood disease in Europeans, caused high mortality among Polynesians when it was first introduced among them. Similarly, when populations of Polynesians, Eskimos, American Indians, and South African Bushmen were first exposed to tuberculosis, the mortality rate was high, compared for instance with the urban populations of Europe. The death rate from tuberculosis among urban Europeans fell long before any effective treatment was discovered, and in these more recently exposed populations a parallel drop in the death rate has occurred in a few generations. Such data are at least suggestive if not conclusive evidence that selective factors have operated on the genotypes of these populations. In summary, it seems safe to say that different human populations differ in their patterns of resistance and susceptibility to disease. The differences appear to be adaptive and dependent on the diseases to which the populations have been exposed. To some extent these differences are environmental, dependent on differences in such factors as nutrition, activity, and so on, but it seems almost inevitable that some genetic differences are involved as well.

Thus far, even though it was pointed out earlier that random as well as adaptive differences might occur between different populations, we have discussed the differences between different human races in terms of their possible adaptive significance. This approach seems reasonable by analogy with other species despite the inadequacy of the data in man. For a long time the differences in frequency of the various blood group genes in different human races were cited as adaptively neutral traits in man, but more recently the discovery of a higher incidence of type O blood among persons with duodenal ulcers, and of type A blood in persons with stomach cancer, has made this a rather dubious assumption. The exact reason for this association remains unclear, but it is no longer safe to assume that these genes are adaptively neutral. Different races have rather wide variations, not only in the frequency of blood group genes (Fig. 27.6), but of genes for PTC tasting (Table 27.3), color blindness (Table 27.4), and for other simply inherited human traits. The conclusion to be drawn from these traits as well as those discussed above is that different human populations are, indeed, genetically different in some respects, primarily in the frequencies of genes at certain loci rather than in absolute terms. These differences may have adaptive significance although in most cases we are not at all sure what it might be.

One subject worth mentioning at this time is the biological effect of hybridization or miscegenation (L. *miscere* to mix + *genus* race). Two

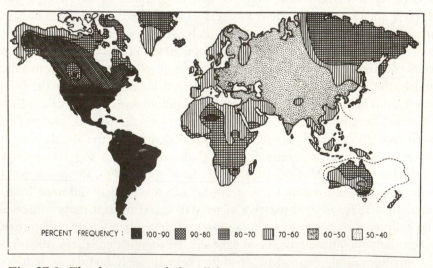

PERCENT FREQUENCY : ■ 100-90 ▨ 90-80 ▥ 80-70 ▥ 70-60 ▨ 60-50 ▨ 50-40

Fig. 27.6. The frequency of the allele responsible for blood group O in human populations. (Mourant, A. A., A. C. Kopec, and K. D. Sobczak. 1958. *The ABO Blood Groups.* Blackwell Scientific Publications, Oxford, p. 270, Map 6.)

Table 27.3 Distribution of High and Low Gene Frequencies for PTC Nontasters Within and Between Garn's (1961) major races

Race and population	Author	Gene frequency
African, Batutsi and Bahuta	Hiernaux (1954)	0.114
African, Bantu, Kenya	Lee (1934)	0.285
Amerindian, Caraja, Brazil	Junqueira et al. (1957)	0.000
Amerindian, Eskimo, Labrador	Sewall (1939)	0.639
Asian, Chinese, Singapore	Lugg and White (1955)	0.142
Asian, Malayan	Thambipillai (1955)	0.400
Australian, natives, southern	Simmons et al. (1954)	0.702
Australian, natives, central	Simmons et al. (1957)	0.707
European, Lapps, Finland	Allison and Nevanlinna (1952)	0.253
European, Danish	Mohr (1951)	0.584
Indian, Hindu, Riang	Kumar and Sastry (1961)	0.403
Indian, Mala-Vedan	Büchi (1958)	0.732
Melanesian, New Hebrides	Simmons et al. (1956)	0.303
Melanesian, Pygmies, New Guinea	Simmons et al (1956)	0.713
Micronesian, Turkese	Simmons et al. (1953)	0.428
Micronesian, Kapinga	Simmons et al (1953)	0.529
Polynesian, Easter Islanders	Simmons et al. (1957)	0.281
Polynesian, Cook Islanders	Simmons et al. (1955)	0.404

From Spuhler, J. N., and G. Lindzey, 1967. "Racial Differences in Behavior." In *Behavior-Genetic Analysis*, J. Hirsch, ed. McGraw-Hill Book Co., New York, p. 384, Table 19.2.

theoretical possibilities would seem to exist based on studies in other species The first is that human racial crosses will produce heterosis in the hybrid progeny, who will be more vigorous and fertile than their parents. The other is that if each race represents a population well adapted to a particular local environment, then their hybrid progeny, not being well adapted to any particular environment, will be less fit than their parents. Although a considerable amount of crossing between human races has occurred, very little reliable evidence bearing on this subject is available. It does seem safe to say that there is no evidence for "hybrid breakdown" in F_2 and subsequent generations such as is sometimes observed in other species, and there is a bit of evidence suggesting heterosis among the Pitcairn Islanders. However, until systematic studies of the effects of human hybridization are made, any conclusions about its possible beneficial or adverse effects are premature.

Table 27.4 Frequency of Color Blindness (Protan and Deutan Groups) in Males

Population	Race (Garn, 1961)	Author	No.	Frequency, %
Fiji Islanders	Melanesian	Geddes (1946)	200	0.0
Brazilian Indians	Amerindian	Mattos (1958)	230	0.0
Bagandas	African	Simon (1951)	537	1.9
Navaho Indians	Amerindian	Spuhler (1951)	163	2.4
Australian natives	Australian	Mann and Turner (1956)	378	3.3
Marshall Islanders	Micronesian	Mann and Turner (1956)	268	4.1
Turks, Istanbul	European	Garth (1936)	473	5.3
Chinese, Peking	Asian	Chang (1932)	1164	6.9
Tonga Islanders	Polynesian	Beaglehole (1939)	67	7.5
Belgians	European	Francois (1956)	1243	8.6
Russians	European	Flekkel (1955)	1343	9.3
V.N.B. Brahmins, Bombay	Indian	Sanghvi (1949)	100	10.0
Americans	European	Shoemaker (1943)	803	11.4
Todas, India	Indian	Rivers (1905)	320	12.8
Dutch, Brazil	European	Saldanha (1960)	97	15.5
Kotas, India	Indian	Sarkar (1958)	28	61.0

From Spuhler, J. N., and G. Lindzey, 1967. "Racial Differences in Behavior." In *Behavior-Genetic Analysis*, J. Hirsch, ed. McGraw-Hill Book Co., New York, p. 388, Table 19.4.

The most sensitive area in discussions of racial differences has to do with intelligence. For this reason, it seems worthwhile to consider the nature of the evidence and of the controversy. The controversy stems not so much from the evidence, on which there seems to be fairly good agreement, but on the interpretation to be placed upon it. Most of the data deal with differences between Negroes and whites in the United States. A number of studies agree in showing that the average IQ of American Negroes is about 85, or 15 points below the average of 100 for American whites (Fig. 27.7). There can be no doubt that the difference is statistically significant. The controversy arises over the question of whether any of this difference is innate or whether all of it is due to the undoubted educational, economic, social, and cultural disadvantages of the Negroes. In other words, we are again confronted by the old nature-nurture controversy, in still another and more confusing guise.

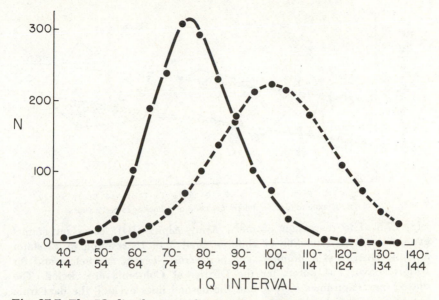

Fig. 27.7. The IQ distributions of normative white and southeastern Negro school children. One of the more extreme differences reported is shown here. The differences usually range between 10 and 20 IQ points. (From "Biogenetics of Race and Class" by I. I. Gottesman, in *Social Class, Race, and Psychological Development* edited by Martin Deutsch, Irwin Katz and Arthur R. Jensen. Copyright © 1968 by Holt, Rinehart and Winston, Inc. Reprinted by permission of Holt, Rinehart and Winston, Inc.)

The problem here is more subtle and complex than in the other cases of hereditary-environmental interaction that we have considered. In the first place, although the IQ score is a single numerical value, there is no conclusive evidence that intelligence itself is unitary. What we call intelligence, as measured by IQ tests, may be made up of a number of more or less independent components. Furthermore, even though the ultimate goal in the construction of an IQ test may be to measure innate ability and to rid the test as much as possible of biases due to environmental factors, this goal can never be entirely realized. For one thing, the ability a person exhibits, as manifested by his responses to the test items, is the product of his genotype interacting with his environment and all that this encompasses. Hence, the genotype does not determine ability as such but may be thought of as setting limits to the potential development of the individual. The particular level of ability reached by that genotype within these limits will then depend on the environmental conditions within which it develops. To minimize environmental effects, attempts have been made to devise culture-free or culture-fair tests or to control the groups being compared for educational or socioeconomic status. However, an entirely

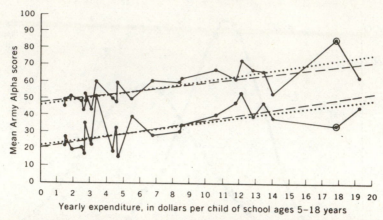

Fig. 27.8. The regression of mean Army Alpha (IQ) scores on annual expenditures for education of children aged 5 to 18 residing in 23 states and the District of Columbia. The points for Negroes are below and for whites above. The points for the District of Columbia are circled. The dotted regression lines include and the dashed lines exclude the data from the District of Columbia. (From "Racial Differences in Behavior" by J. N. Spuhler and G. Lindzey, in *Behavior-Genetic Analysis*, J. Hirsch, ed. Used with permission of McGraw-Hill Book Company. Copyright 1967 by McGraw-Hill, Inc.)

culture-free test or even a culture-fair test hardly seems possible, nor can one be certain that persons of different races assigned to the same socio-economic group have actually had entirely comparable environments. Therefore, the IQ test can hardly be regarded in any sense as a measure of the genetic endowment for intelligence. This is not to say that heredity plays no role in intelligence or even in IQ scores; we have already considered the evidence that it does. Rather, what is implied is that unless comparable environmental conditions prevail, such tests can tell us very little about possible genetic differences between individuals or races in intelligence.

When "culture-free" tests have been used, Negroes have gotten slightly lower scores than they got on the more conventional IQ tests such as the Stanford-Binet. Attempts to control for educational attainment or socio-economic status (SES) reduce but do not eliminate the difference in IQ between whites and Negroes. Other minority groups in the United States—American Indians, Mexican-Americans, and Puerto Ricans—also score below the norm of 100, although not as low as the Negroes, but Chinese- and Japanese-Americans average out at about 100. A comparison of IQ scores of whites and Negroes from different areas or by states always shows the Negroes to be lower than the whites in a given region. However, these IQ scores can be shown to be correlated with such factors as annual expenditure per child for education (Fig. 27.8). In the South, where the amounts

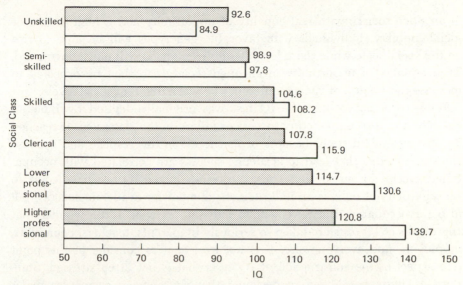

Fig. 27.9. The relationship between social class and intelligence. The white bars show the mean IQ of fathers in the various occupational groups. The gray bars show the mean IQ of their children. The IQs of the wives are not shown, but correlate well with those of the husbands. In each case the children show some regression toward the overall mean IQ of 100. (Data from Burt.)

are very low, both whites and Negroes have very low average IQ scores. In some instances, northern Negroes had higher averages than southern whites. These findings have been subjected to various interpretations, but in view of the reservations about IQ tests set forth above, they raise more questions than they answer. What they do seem to show, however, is that IQ scores may be correlated with an environmental factor such as dollars per student, and furthermore, that despite the obvious disadvantages of the minority groups, the distributions of IQ scores for different racial groups show broad areas of overlap. Removal of the environmental disadvantages of minority groups would seem inevitably destined to move these distributions closer together.

Another type of comparison often made is the distribution of IQs of children of different social class or by occupation of the father or similar criteria of socioeconomic status (Fig. 27.9). The results generally show an association between average IQ and social class: the higher the SES, the higher the mean IQ. These differences must reflect a considerable environmental component, but there is a willingness by investigators in this case to entertain the possibility that genetic differences may be involved as well. If the barriers between classes are high or there is a strong caste system, the observed differences are apt to be largely environmental. However, in an open society with real equality of opportunity and a high degree of social mobility, individuals with favorable genotypes will be able to rise on the social scale while those less favorably endowed will move downward. True equality of opportunity in an open society can then be expected to increase the amount of genetic difference between different social levels.

It is worth noting that the IQ test was originally devised by Binet in France as an aid in predicting success or failure in the classroom. It reached its present exalted status because of its success in this respect, and because the same factors that lead to academic success are correlated with occupational success in our society. Thus, the results of the IQ test have come to be regarded not only as an indicator of potential for academic success, but also for occupational achievement in later life as well. Little wonder that the results of IQ tests are taken so seriously by parents, teachers, counselors, employers, and everyone else. It is not irrelevant to consider at this point one aspect of the construction of IQ tests that is not often stressed. Standard IQ tests have been so designed that the average scores for males and females are equal. When the tests were being devised, girls were found to do consistently better than boys on such things as rote memory, language usage, and esthetic and social questions, while boys scored higher in mathematics, abstract reasoning, mechanical ability, and spatial visualization. Rather than have a test in which one of the sexes consistently outscored the other, the test items were selected and balanced in such a way that neither sex was favored. These differences in the mental processes of boys and girls have been found to persist into later life. Under these conditions, it is obvious that IQ tests cannot be used to prove man's intellectual superiority over woman or vice versa. What it does suggest is that men and women differ significantly in the ways they think. What it also suggests is the care with which the test items were selected and the tests constructed to give a certain type of result, in this case, equality to the sexes. Since men and women in the same population draw their genes from the same gene pool, have 22 pairs of autosomes in common, and differ only in that the female has two X chromosomes while the male is XY, the differences between them in mental processes must be developmental and environmental rather than genetic.

This digression on the differences between the sexes has some bearing on the problem of racial differences in intelligence. It shows, for instance, that individuals sharing the same sorts of genes for intelligence and similar environmental conditions, aside from that fundamental difference, their sex, can nevertheless differ markedly in their thought processes. Little wonder then if individuals who differ genetically, environmentally, and culturally also differ in their test scores. The selection of the test items to achieve equality of the sexes raises the question of whether a test could not be so constructed that the results showed equality of the races or even reversed the present results.

At this point, however, it seems worthwhile to pause over the meaning of an IQ score. Earlier we spoke of the fact that words like superior and inferior can be used only with reference to individual or racial differences, if some system of values is superimposed. For no human trait is this more apt to be done than it is for IQ. It is almost automatically assumed that a high IQ is good, a low IQ bad; that a person with a high IQ is superior, not just in having a high test score, but superior, period. It is for this reason that the difference in average IQ scores between races has become such a sensitive issue. Nor is it only the racists who think this way; people with a wide spectrum of beliefs seem to accept this standard.

In a biological sense, the only criterion of success is reproductive fitness. If this criterion is applied, there turns out to be no significant correlation between IQ and reproductive success within the normal IQ range (Fig.

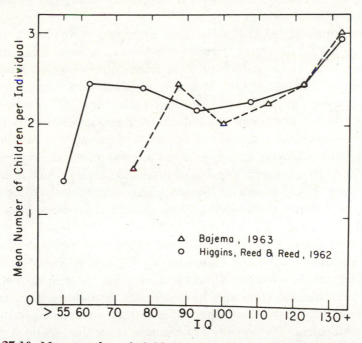

Fig. 27.10. Mean number of children per adult (including childless adults) at various IQ levels in two samples of American whites. (Arthur R. Jensen, "How much can we boost IQ and scholastic achievement?," *Harvard Educational Review* 39, Winter 1969, p. 94. Copyright © 1969 by President and Fellows of Harvard College.)

27.10) Among the retarded, however, reproductive success is diminished. This reduction is due to the failure of a sizable proportion of individuals with low IQs to find marriage partners. Thus, selection would seem to be at least maintaining or perhaps increasing average IQ levels. This conclusion is contrary to the formerly held belief that persons with low IQs were out-reproducing those with higher IQs, and that the level of human intelligence must therefore be declining.

However, IQ tests are imperfect instruments for measuring the full range of human abilities. Such factors as creativity, motivation, and personality, all difficult to define and equally difficult to measure, are important elements in human achievement, and are not highly correlated with IQ, yet the tests do not measure them. Such special abilities as musical talent, artistic talent, mathematical aptitude, and athletic ability are still other facets of the picture that are, with the possible exception of mathematics, inadequately represented in the IQ test. Babe Ruth and Jascha Heifitz were both great artists, each in his own way, yet who is to say which was greater (though many undoubtedly would be willing to try), or who would be willing to say that someone with a higher IQ score than either was superior to both.

It is sometimes stated that some race or ethnic group has a special aptitude for music, or art, or science, or athletics, or some other field of human endeavor. These statements stem from cases where an unusually high number of individuals with some special talent appear in a particular group. It is unsafe to jump to the conclusion that these differences in frequency of talented individuals are indicative of genetic differences among the populations because a special concatenation of genetic and environmental circumstances is necessary to produce a great musician, or artist, or scientist. Since all races seem capable of producing highly talented individuals, the observed differences in frequency may just as well reflect social, economic, or cultural differences as any genetic differences among the populations.

The most reasonable attitude would seem to involve recognition of the range of human diversity, and efforts to enable each individual to realize his full potential, in whatever direction it may lie. All men do not have the same genotype; therefore men have different capabilities. They are not created equal but different, and they differ in their capabilities and potential for development. The equality to be sought is not the genetic similarity of identical twins, but an equality of opportunity whereby each individual has the chance to realize his potential to the fullest extent possible. If this were done, many of the problems now associated with race, social class, and creed would be greatly diminished. If our attitude toward human differences could change to parallel that of the Frenchman with respect to the difference between the sexes, we could all join in saying "Vive la différence!"

Additional Reading

Barnicot, N. A., 1959. "Climatic Factors in the Evolution of Human Populations," *Cold Spring Harbor Symp. Quant. Biol.* **24**:115–129.

Bleibtreu, H. K., ed., 1969. *Evolutionary Anthropology.* Allyn and Bacon, Boston.

Blum, H. F., 1961. "Does the Melanin Pigment of Human Skin Have Adaptive Value?" *Quart. Rev. Biol.* **36**:50–63.

Boyd, W. C., 1950. *Genetics and the Races of Man.* Little, Brown, Boston.

Coon, C. S., 1962. *The Origin of Races.* Knopf, New York.

Coon, C. S., S. M. Garn, and J. B. Birdsell, 1950. *Races, a Study of the Problems of Race Formation in Man.* Charles C Thomas, Springfield, Ill.

Dobzhansky, T. 1962. *Mankind Evolving.* Yale University Press, New Haven, Conn.

Dunn, L. C., and T. Dobzhansky, 1946. *Heredity, Race, and Society.* Penguin, New York.

Garn, S. M., 1961. *Human Races.* Charles C Thomas, Springfield, Ill.

Garn, S. M., ed., 1968. *Readings on Race,* 2nd ed. Charles C Thomas, Springfield, Ill.

Garn, S. M., and C. S. Coon, 1955. "On the Number of Races of Mankind," *Amer. Anthr.* **57**:996–1001.

Gottesman, I. I., 1968. "Biogenetics of Race and Class." In *Social Class, Race, and Psychological Development,* M. Deutsch, I. Katz, and A. R. Jensen, eds. Holt, Rinehart and Winston, New York.

Grant, V., 1963. *The Origin of Adaptations.* Columbia University Press, New York.

Harrison, G. A., J. S. Weiner, J. M. Tanner, and N. A. Barnicot, 1964. *Human Biology.* Oxford University Press, London.

Harvard Educational Review. Reprint Series No. 2. *Environment, Heredity, and Intelligence.*

Higgins, J. V., E. W. Reed, and S. C. Reed, 1962. "Intelligence and Family Size: A Paradox Resolved," *Eug. Quart.* **9**:84–90.

Korn, N., and F. Thompson, eds., 1967. *Human Evolution.* Holt, Rinehart and Winston, New York.

Lasker, G. W., 1961. *The Evolution of Man.* Holt, Rinehart and Winston, New York.

Livingstone, F. B., 1964. "On the Nonexistence of Human Races." In *The Concept of Race,* A. Montagu, ed. Macmillan, New York.

Mayr, E., 1942. *Systematics and the Origin of Species.* Columbia University Press, New York.

Mayr, E., 1963. *Animal Species and Evolution.* Harvard University Press, Cambridge, Mass.

Miller, A. H., 1956. "Ecologic Factors That Accelerate Formation of Races and Species of Terrestrial Vertebrates," *Evol.* **10**:262–277.

Shapiro, H. L., 1962. *The Heritage of the Bounty,* rev. ed. Doubleday, Garden City, N.Y.

Spuhler, J. N., and G. Lindzey, 1967. "Racial Differences in Behavior." In *Behavior-Genetic Analysis,* J. Hirsch, ed. McGraw-Hill, New York.

UNESCO, 1952. *The Race Concept.* Columbia University Press, New York.

Questions

27.1. How would you distinguish between a geographic race or sub-species and a species?

27.2. Compare the evolutionary processes leading to the origin of races with those leading to the origin of species.

27.3. What is the evidence that some of the differences between different geographic races in man are adaptive?

27.4. In discussions of IQ and race, heritability estimates are sometimes used. Is it appropriate to use a heritability estimate obtained from twin data or from one race for other populations? Explain.

27.5. Do you believe that the present trend in human evolution is toward further divergence among different human populations or toward a reduction in the genetic differences among different human populations? Explain.

27.6. Discuss the effect of each of the following evolutionary factors in present-day human populations as compared to human populations more than 10,000 years ago: mutation, natural selection, gene flow, random genetic drift.

CHAPTER TWENTY-EIGHT

Genetics and Ecology

28.1. The Nature of Ecological Genetics

Population genetics stems primarily from the work of three men: Fisher, Haldane, and Wright. Their approach was mathematical, an effort to study the statistical consequences of Mendelian heredity in populations and to determine how evolutionary factors produced changes in gene frequency. This approach established a theoretical model with which actual populations, both natural and experimental, could be compared.

Population ecology is a different aspect of the study of populations, in which the theories were developed quite independently of the theoretical developments in population genetics. Population ecology deals with the distribution and abundance of individuals and species found in nature. In the past, relatively little was done to relate the theories of population genetics and population ecology to one another or to actual populations. Although population geneticists talked of natural selection and adaptation, it was usually in the abstract rather than in concrete terms related to existing environmental conditions. Although they tended to think more realistically about the environment, population ecologists, on the other hand, generally ignored the genotypes of the organisms with which they dealt.

Recently, these two disciplines have begun to merge in the burgeoning field now known as ecological genetics. The primary concern of this emerging discipline is to study the ways in which natural populations are adapted to their environments, and the mechanisms by which they respond to environmental change. This approach holds promise for advances of considerable theoretical and practical significance.

Research in ecological genetics may involve studies of natural populations in the field, studies of such populations under common environmental conditions, and crosses to determine the nature of the genetic differences between populations. In addition to estimates of gene frequencies, mutation and migration rates, and selection coefficients, more estimates of population size and density, reproductive rates, death rates, and other relevant parameters are needed. In this way, the statics and dynamics of populations can be studied. Some observations can be made only in natural populations; others are best made under experimentally controlled conditions. Thus a combination of field and laboratory research is the most fruitful approach to research in ecological genetics.

The types of problems open to investigation in ecological genetics are quite varied. Some of the earliest work by Turesson, and by Clausen, Keck, and Hiesey, involved studies of the genetic and environmental differences among different ecological races or ecotypes of plants. Some species have distributions that cut across a variety of habitats. Members of the same species living in different habitats usually look rather different. By growing such plants together in a common environment, these early workers showed that the differences persisted to a considerable extent and hence were not solely environmental. Moreover, crosses between different ecotypes revealed that a number of gene loci were involved.

Another problem of interest is the importance of genetic drift in the origin and maintenance of differences among populations, a subject of considerable controversy. In order to resolve the controversy, information on effective population size and on mutation, selection, and migration coefficients in wild populations is essential. Such estimates are not always easy to make, but they are necessary if this question is to be answered.

A number of polymorphisms are known in natural populations. A few examples of such polymorphisms are melanism in the coat color of the hamster, color and pattern polymorphisms in grouse locusts, dominant pattern mutants in the leopard frog and in the wood frog, inversion polymorphism in *Drosophila* species, the ABO and other blood groups in man, shell color and banding pattern in snails, and Batesian mimicry in insects. In contrast to studies of ecotypes, studies of polymorphism deal with problems of genetic adaptation within populations. In most cases the origin and adaptive significance (if any) of these polymorphisms remain obscure. Analysis has not progressed far beyond the descriptive level. Nevertheless, the prevalence of such polymorphisms, many of them controlled by allelic differences at a single locus, suggest a biological significance to genic polymorphism that is, as yet, seldom understood.

The response of populations to changes in their environment is a subject of considerable interest at present. The development of resistance to insecticides by mosquitoes, flies, and other insects, and of resistance to antibiotics by bacteria have both been shown to be due to strong selection pressure favoring genes for resistance in these populations. Industrial melanism in moths resulted from parallel responses by a number of different species to the dirt and soot of industrial pollution. Dominant genes for melanism were favored by natural selection in the polluted regions. Research on the ability of both animal and plant species to withstand polluted conditions is of great importance at a time when pollution of the air, the water, and the soil is increasing rapidly. For many species the alternatives seem to be either to adapt quickly to the rapidly changing conditions or to perish. Basic studies on the genetics of such populations are urgently needed. Equally important are studies of populations living in undisturbed habitats, for only thus will baseline data become available against which to measure the effects of pollution or other environmental disturbances.

A related area of concern in ecological genetics is the ability of populations to withstand the damaging effects to the gene pool of ionizing radiation. A subject that has aroused considerable controversy, it can best be settled through actual studies of exposed populations in the field and in the laboratory.

At the present time, species are being introduced into new areas and new habitats at an ever-increasing rate. Some of the introductions are unintentional, the result of the increased ease of transport from one part of the world to another. The gypsy moth, Japanese beetle, chestnut blight, and Dutch elm disease are only a few instances of the dire consequences of such introductions in the United States.

Other introductions are intentional, either aimed at biological control of introduced pests, or else with the hope of adding useful species to the local scene. The introduction of game birds and mammals and of many species of fishes into new areas requires that thought be given to finding those genotypes within these species that will be best adapted to the new habitat, and at the same time providing sufficient variability so that selection can operate effectively in the new environment to produce optimally adapted populations. Similar precautions are needed in the reforestation projects that are now covering vast regions. On the other hand, some thought must be given to the possible disruptive effects of the introduced species on the existing ecological balance.

These types of problems require an orientation of thought peculiar to ecological genetics—the realization that populations are dynamic units, very precisely adapted genetically to their environments and sensitive to, and able to respond, within limits, to any change in their environmental conditions. No longer can a population be regarded as a set of identical black boxes, all responding identically to their impinging environment. Conversely, no longer can the environment be regarded as a fixed set of parameters, acting uniformly on a variable population. Rather the interplay between the genetically variable population and its ever-changing environment must be the central focus of attention for biologists dealing with natural populations.

28.2. Pollution and Evolution

Because pollution presents a problem of increasing concern, it seems worthwhile to consider some of its effects. In most of the literature on pollution, for example Rachel Carson's *Silent Spring*, attention has centered on the extinction of entire populations of birds or other species. However, the cases in which species have adapted to changed environmental conditions are of considerable biological interest and also pose some new and different problems. They also serve to illustrate the nature and significance of research in ecological genetics, and to show that, rather than leading to extinction, in some cases pollution leads to evolution.

Industrial Melanism in Moths

The Industrial Revolution began in England about 1800. Coal was the primary source of energy in the factories, and consequently the countryside in the industrial areas came to be blanketed under a coat of soot and dirt from the numerous smokestacks. In England, a number of species of moths normally rested in the open on tree trunks or rocks during the day and were inconspicuous because their pigment patterns blended in so well with the bark of the trees or with the lichens growing on the trees and rocks. The soot killed the lichens and darkened the trees so that these light-colored moths were quite conspicuous against their new background, and no longer had protective, cryptic coloration (Fig. 28.1).

Fig. 28.1. Typical and melanic forms of the moth, *Biston betularia*, resting on a lichen-covered tree in unpolluted country (left) and a blackened, lichen-free tree in an industrial area (right). (From the experiments of Dr. H. B. D. Kettlewell, University of Oxford.)

In the course of the last century and a half, the light-colored typical form in more than 80 species of English moths has been replaced in industrial areas by a darker, melanistic form that blends in very well with the dark background. Similar changes have been observed in additional species in the industrial areas of continental Europe and the United States. In the early 1800s, the dark form was a rare, aberrant type of moth, found only occasionally among populations of light moths. Today some populations consist of very close to 100% of the dark type, and the light form is the rare, aberrant type.

Evolution has been defined as "descent with modification"; in other words, an existing population has descended with some hereditary changes from an ancestral population. In this case, the existing populations of dark moths are known to have descended from the ancestral populations of light moths so that, by definition, this transition represents an example of evolution.

In nearly all cases, the melanism has turned out to be due to a single simple dominant gene. Thus, the change is hereditary, not due to a direct environmental effect. The dominant gene has increased in frequency from being rare to being very common. It is the new "wild-type" gene in these populations.

The mechanism by which this change was produced is natural selection. Birds have been found to prey more heavily on dark moths resting on a light background than on light moths, but they capture proportionately more of the light moths resting on a dark background. Thus, in the new environment the dark moths survived better because they were less conspicuous, and they left more progeny than the light moths. The frequency of the gene that causes melanism increased very rapidly as a result of this differential predation and differential reproduction. The evolutionary change was relatively rapid, and it was adaptive in that the population was better able to survive and reproduce in its new polluted environment.

Insecticide Resistance

Organic insecticides, of which DDT is the most familiar, started to come into use in the early 1940s. The first resistant house flies were reported in 1946. Thereafter, the number of species in which insecticide resistance was reported increased rapidly (Table 28.1). Not only house flies, but mosquitoes, lice, fruit flies, and many other insect species were found to have become resistant. Probably many additional species have become resistant but have never been studied and reported. Nearly every species exposed to an insecticide has been able to develop some degree of resistance, and there seems to be no insecticide among the great variety tested to which some species has not become resistant. Furthermore, the resistance is due to an evolutionary change, to a change in the genetic composition of the exposed populations as a result of natural selection. The resistance is inherited, and is not some sort of physiological tolerance acquired by an exposed individual. This can be shown by taking parents from a resistant population and rearing their progeny without any contact with DDT. When first exposed to DDT, they are resistant, unlike the progeny of susceptible individuals.

The mechanism responsible for the difference in resistance is different in different populations, in some cases in different populations of the same

Table 28.1 Number of Insect Species Reported to Be Resistant to DDT

Year	Reported number of resistant species
1948	12
1951	16
1954	25
1957	76
1960	137
1963	157
1966	165+

Data from Crow, 1966.

species. Differences in DDT resistance have been found to be due to differences in (1) absorption of DDT through the insect cuticle; (2) lipid content (DDT is fat soluble, and is thus stored in the fat bodies where it does little harm); (3) rate of detoxification by enzymes; (4) sensitivity of the nervous system; and (5) behavior, the insects having behavior patterns that tend to minimize their exposure to DDT. All of these differences are hereditary and, as might be expected, different genetic mechanisms are responsible for the different modes of resistance; dominant and recessive, autosomal and sex-linked, single factors and multiple factors have all been reported. The diversity of paths taken reveals another aspect of evolution—its opportunism. Natural populations of insects contain a great deal of genetic variability in their gene pools, but the variability in the gene pool of one population is different from the variability in the gene pool of another. When the populations are confronted by the lethal stress of DDT, any kind of gene that in some way permits an individual to survive in the presence of DDT will be passed on to subsequent generations and will increase in frequency because individuals lacking such genes will die, and their genes will be eliminated from the population. When all susceptible individuals die, it is not surprising that resistance has evolved so rapidly in insects, in as little as a year or two. Furthermore, the opportunism of evolution under these conditions should not be surprising since the populations can only work within the framework of their existing variability. The insecticides are not mutagenic and do not cause mutations to occur. Thus, there is no Lamarckian type of response to this new environmental stress.

Although evolution is not usually thought to be a subject adapted to study

Table 28.2 DDT Resistance in Control (C) and Resistant (R) Populations of *Drosophila melanogaster*

Population	$ED_{50}(mg)$	Relative resistance
731 C	0.04	1
91 C	0.16	4
731 R	4.2	105
91 R	11.7	293

ED_{50} = effective dose in milligrams that kills 50% of the females in 24 hr.
Data from Merrell, 1960.

in the laboratory, nevertheless it is possible to study microevolutionary changes such as the development of insecticide resistance in experimental populations. Exposure of genetically variable populations of fruit flies, *Drosophila melanogaster*, to DDT in the lab has resulted in a 100-fold increase in DDT resistance compared to the unexposed controls (Table 28.2). The resistance rose rapidly at first and has now been maintained in these experimental populations for more than two decades. The original dosage in the populations was 0.2 or 0.5 mg, with 75% to 90% mortality. Now dosage in the populations is 150 mg, with minimal mortality. The flies have become so resistant that they now walk about on visible deposits of DDT crystals during testing, and it has become increasingly difficult to get good estimates of the effective lethal dose. In these populations, it is possible to study a number of aspects of the way in which evolutionary changes actually occur rather than to speculate by inference from the fossil record.

In contrast to the genetically variable populations, highly inbred populations did not develop increased resistance when exposed to DDT. This result has certain implications. First, favorable spontaneous mutations did not accumulate fast enough to permit increased resistance to develop. Moreover, there was no indication that DDT was able to induce adaptive mutations. Finally, there was no sign that physiological adaptation of some sort was involved.

In addition, genetically variable exposed and unexposed populations from the same original population diverged from one another not only in DDT resistance, but in other respects as well. In general, the unexposed controls showed a higher fecundity and fertility, but the resistant flies had a longer life span. The environmental conditions for control and exposed populations differed only in the presence of DDT-impregnated strips in the exposed populations, but this difference put a selective premium on fecundity and fertility in the first flies to reach fresh food in the unexposed group. On the other hand, the selective pressure in the other populations favored not only DDT resistance but also an extended life span and the ability to produce eggs over a long period of time. Thus, at least some of the eggs would be laid under conditions favorable to their development, after the DDT strip had become covered by food and debris. Hence, natural selection seems to operate in subtle as well as in more direct and obvious ways.

The Rabbit in Australia

Before the Europeans reached Australia, the only placental mammals there were bats, several species of rats, and the Australian aborigines and their dogs. The mammalian fauna consisted almost exclusively of marsupials, the kangaroos, the wallabies, and a number of other unique and unusual species of pouched mammals. When Europeans colonized the world during the sixteenth, seventeenth, and eighteenth centuries, they brought with them, intentionally or unintentionally, many species of plants and animals, both domesticated and wild. One species introduced into Australia with dire consequences was the European rabbit, *Oryctolagus cuniculus*. In 1859 about two dozen wild English rabbits were deliberately released on a sheep station in Victoria. By 1865, 20,000 rabbits had been killed on that one ranch alone, and they were spreading rapidly, eventually covering all of southern Australia despite desperate efforts to halt their spread by such devices as barrier fences that covered thousands of miles. This represents a new and different sort of pollution. In the course of a century the rabbit population increased from 24 to several hundred million individuals, and their effect on agriculture and ranching was devastating (Fig. 28.2). Efforts at control were futile, and there were no natural enemies or predators.

The myxoma virus, a member of the poxvirus group, was originally found only in the Americas. It caused a relatively benign disease (lesions called fibromas) in South American rabbits (*Sylvilagus brasiliensis*) that was transferred by mosquitoes. In European rabbits, however, the virus caused a very severe, generalized infection that was nearly always fatal. The suggestion was made as early as 1918 by the Brazilian Dr. Aragão that the myxoma virues be used to control the rabbit pests in Australia, but the idea was not tested until 1950.

At that time virulent Brazilian-type myxoma virus was inoculated into rabbits at trial sites in the Murray Valley. The disease escaped from the test sites, and by 1954 had spread throughout the rabbit population of Australia. The initial death rate was estimated to be 99.8% of the rabbits infected, but the death rate after a year or so was found to have fallen to about 90%. Thus, the virus was losing its effectiveness as a means of controlling the rabbit population.

As you should now expect, the rabbits, upon test, were found to have

Fig. 28.2. The population explosion of rabbits in Australia. Swarming rabbits drinking at a waterhole. (Fenner, F., 1954. "The Rabbit Plague," *Sci. Amer.*, Feb. 1954, p. 31. By permission of Black Star Publishing Company.)

evolved an increased level of resistance to the virus through natural selection of the genetically more resistant individuals as the progenitors of the next generation.

Perhaps more surprising is the fact that the virus itself was found to have evolved from the original highly virulent strain to a less virulent type of virus. Why should this be? A parasite that is too virulent rapidly kills its host, and this in turn leads to its own demise. Thus, it is to the advantage of the parasite to reach some sort of accommodation with its host that permits it to survive and spread to other hosts. Therefore, mutant viral strains of reduced virulence permitted infected rabbits to survive longer and to serve as reservoirs of virus that could be picked up by mosquitoes and transmitted to other hosts. Thus natural selection acted to increase the frequency of the less virulent viruses, and both the rabbit and the myxoma virus have evolved in relation to one another. The consequence of this evolution, hand in hand, by rabbit and virus is that, given the reproductive potential of the rabbit (a 5- to 12-fold annual increase), the virus no longer is completely effective in keeping the pest under control.

Because pollution may have such far-reaching and unforseen consequences, the indiscriminate pollution of the environment has been widely condemned. However, most attention has been directed toward the deleterious or lethal effects of the pollutant on existing species. The lethal effects of pesticides on the target species are welcome, but their harmful effects on other species are cause for concern. Because death results when environmental stresses are beyond the ability of organisms to respond adaptively, mortality is usually the most conspicuous effect of pollution. Hence, the death of individuals and the possible extinction of species have been the major source of concern about pollution.

That species might evolve so rapidly in response to changing environmental conditions was at first quite unexpected. The evidence now clearly indicates that whenever a new environmental stress on a genetically variable population permits differential survival and reproduction, the population will evolve adaptively to the stress. Thus, there are now reports not only of insects resistant to DDT, but of rats resistant to rat poison. Consequently, new and even more difficult problems of control may arise because of the ability of pest species to evolve. While there may be some comfort in the realization that other species may also have the ability to adapt in the face of new and rigorous selection pressures, unabated pollution can hardly be justified on this basis.

Biological control is usually regarded as more desirable than control by pesticides, because its effects are more specific and restricted. However, the evolutionary changes in the rabbit and the myxoma virus suggest that even biological control may not provide a permanent panacea. Furthermore, just as the introduction of a new pesticide may have unforeseen consequences, the introduction of a new predator, or parasite, or pathogen may also have unexpected biological ramifications.

In our present state of knowledge, there are no experts on pollution. However, it is becoming clear that, in finding new and unusual ways to foul or disrupt his environment, man runs the risk of affecting the evolution not only of other species but of his own as well. The great present need is for additional reliable information through studies in ecological genetics. Until such data are available, past experience suggests that we should proceed with caution and circumspection in dealing with problems of pollution.

Additional Reading

Anderson, E., 1949. *Introgressive Hybridization*. Wiley, New York.

Baker, H. G., and G. L. Stebbins, eds., 1965. *The Genetics of Colonizing Species*. Academic, New York.

Birch, L. C., 1960. "The Genetic Factor in Population Ecology," *Amer. Nat.* **94**: 5–24.

Clausen, J., 1951. *Stages in the Evolution of Plant Species*. Cornell University Press, Ithaca, N.Y.

Clausen, J., D. D. Keck, and W. M. Hiesey, 1940. "Experimental Studies on the Nature of Species. I. Effect of Varied Environments on Western North American Plants." Carnegie Inst. Wash. Publ. No. 520, pp. 1–452.

Clausen, J., D. D. Keck, and W. M. Hiesey, 1947. "Heredity of Geographically and Ecologically Isolated Races," *Amer. Nat.* 81:114–133.

Clausen, J., D. D. Keck, and W. M. Hiesey, 1948. "Experimental Studies on the Nature of Species. III. Environmental Responses of Climatic Races of *Achillea.*" Carnegie Inst. Wash. Publ. No. 581, pp. 1–129.

Clausen, J., and W. M. Hiesey, 1958. "Experimental Studies on the Nature of Species. IV. Genetic Structure of Ecological Races." Carnegie Inst. Wash. Publ. No. 615, pp. 1–312.

Cook, L. M., 1971. *Coefficients of Natural Selection. An Account of the Theory of Natural Selection as Used in Ecological Genetics.* Hutchinson, London.

Creed, R., ed., 1971. *Ecological Genetics and Evolution.* Blackwell, Oxford.

Crow, J. F., 1966. "Evolution of Resistance in Hosts and Pests." In *Scientific Aspects of Pest Control.* NAS/NRC, Washington, D.C.

Dice, L. R., 1940. "Ecologic and Genetic Variability Within Species of *Peromyscus,*" *Amer. Nat.* 74:212–221.

Dobzhansky, T., 1970. *Genetics of the Evolutionary Process.* Columbia University Press, New York.

Fenner, F., 1965. "Myxoma Virus and *Oryctolagus cuniculus:* Two Colonizing Species." In *The Genetics of Colonizing Species,*" H. G. Baker and G. L. Stebbins, eds. Academic, New York.

Ford, E. B., 1971. *Ecological Genetics,* 3rd ed. Chapman and Hall, London.

Grant, V., 1963. *The Origin of Adaptations.* Columbia University Press, New York.

Jain, S. K., 1969. "Comparative Ecogenetics of Two *Avena* Species Occurring in Central California," *Evol. Biol.* 3:73–118.

Lerner, I. M., 1963. "Ecological Genetics: Synthesis," *Proc. 11th Int. Cong. Genet.* 2:489–494.

Lerner, I. M., and F. K. Ho, 1961. "Genotype and Competitive Ability of *Tribolium* Species," *Amer. Nat.* 95:329–343.

Levins, R., 1968. *Evolution in Changing Environments.* Princeton University Press, Princeton, N.J.

Mayr, E., 1963. *Animal Species and Evolution.* Harvard University Press, Cambridge, Mass.

Merrell, D. J., 1960. "Heterosis in DDT Resistant and Susceptible Populations of *Drosophila melanogaster,*" *Genetics* 45:573–581.

Merrell, D. J., 1968. "A Comparison of the Estimated Size and 'Effective Size' of Breeding Populations of the Leopard Frog, *Rana pipiens,*" Evol. 22:274–283.

Merrell, D. J., 1970. "Migration and Gene Dispersal in *Rana pipiens,*" *Amer. Zoologist* 10:47–52.

Merrell, D. J., 1973. "Studies in Ecological Genetics with *Rana pipiens.*" In *Evolutionary Biology of Anurans,* J. L. Vial, ed. University of Missouri Press, Columbia.

Moore, J. A., 1949. "Patterns of Evolution in the Genus *Rana.*" In *Genetics, Paleontology, and Evolution,* G. L. Jepsen, E. Mayr, and G. G. Simpson, eds. Princeton University Press, Princeton, N.J.

Pimentel, D., 1961. "Animal Population Regulation by the Genetic Feedback Mechanism," *Amer. Nat.* **95**:65–79.

Pimentel, D., 1968. "Population Regulation and Genetic Feedback," *Science* **159**: 493–504.

Sammeta, K. P. V., and R. Levins, 1970. "Genetics and Ecology," *Ann. Rev. Genet.* **4**:469–488.

Sheppard, P. M., 1959. "The Evolution of Mimicry: A Problem in Ecology and Genetics," *Cold Spring Harbor Symp. Quant. Biol.* **24**:131–140.

Stebbins, G. L., 1950. *Variation and Evolution in Plants.* Columbia University Press, New York.

Stebbins, G. L., 1959. "The Role of Hybridization in Evolution," *Proc. Amer. Philos. Soc.* **103**:231–251.

Streams, F. A., and D. Pimentel, 1961. "Effects of Immigration on the Evolution of Populations," *Amer. Nat.* **94**:201–210.

Turesson, G., 1922. "The Genotypical Response of the Plant Species to the Habitat," *Hereditas* **3**:211–350.

Turesson, G., 1923. "The Scope and Import of Genecology," *Hereditas* **4**:171–176.

Turesson, G., 1925. "The Plant Species in Relation to Habitat and Climate," *Hereditas* **6**:147–236.

Turrill, W. B., 1940. "Experimental and Synthetic Plant Taxonomy." In *The New Systematics,* J. Huxley, ed. Oxford University Press, London.

Underhill, J. C., and D. J. Merrell, 1966. "Fecundity, Fertility, and Longevity of DDT-Resistant and Susceptible Populations of *Drosophila melanogaster,*" *Ecology* **47**:140–142.

Questions

28.1. If organisms growing under different environmental conditions show different phenotypes, how can you determine to what extent these phenotypic differences are influenced by heredity?

28.2. What sorts of practical problems can best be investigated using the methodology of ecological genetics?

28.3. What parameters are necessary to explore the statics and the dynamics of natural populations?

28.4. Define evolution. Does the development of industrial melanism, insecticide resistance, and rabbits resistant to myxoma virus fit your definition of evolution?

28.5. When new species of game or fish are introduced into an area, what precautions are advisable to ensure the success of the introduction and the absence of any undesirable side effects?

28.6. Present-day timber operations often involve reforestation of large areas. What procedures seem advisable in choosing the seeds or seedlings to be used in reforestation projects?

28.7.   What arguments can you offer against the use of antibiotics in cough drops, mouth washes, tooth paste, and so on?

28.8.   Do you think radiation resistant strains of bacteria could evolve in populations of bacteria exposed to X-rays? Do you think this could happen in eukaryotes?

28.9.   The gypsy moth and the Japanese beetle are troublesome pests in the United States, where they have been introduced within the past century, yet they are not a problem in their original area of distribution. What explanation can you suggest for this difference?

28.10. Mallard ducks reared in captivity soon begin to show significant differences in physiological, morphological, and behavioral traits from their wild ancessors of just a few generations ago. Do you think that release of progeny from these semiwild or semidomesticated mallards will have an adverse effect on the gene pool of the wild population? What kinds of information would you need to provide an adequate answer to this question?

CHAPTER TWENTY-NINE

Genetic Counseling and Eugenics

Human population problems fall into two major categories, quantitative and qualitative. The quantitative problems arise from sheer numbers of people. The most pressing biological problem facing mankind is the rapid increase in his own numbers. This population expansion is a major underlying cause for many present-day economic, social, and political problems. Poverty, famine, pollution, the urban crisis, and international tensions are attributable in large measure to a human population expanding so rapidly that its basic needs for food, clothing, housing, education, and so on are not being effectively met. This rapid and accelerating rate of growth is a relatively recent phenomenon: The human population grew very slowly until less than 500 years ago. Now man must be regarded as the ecologically dominant species on earth. His position no longer seems threatened by any species other than his own, but the threat from his own reproductive capacity is a very real one. The population crisis, or the population explosion as it has been called, has been discussed in many books and articles in recent years, and is gradually coming to be widely recognized as a truly serious problem. Solutions, barring the catastrophic, are apt to be slow and difficult. The problem is essentially quantitative rather than qualitative, and hence is not especially germane in a book on genetics. The additional readings at the end of this chapter include some that offer further information on the subject. However, since it is generally agreed that the population explosion would be more humanely contained by lowering the birth rate than by increasing the death rate, research on fertility control has been actively pursued. The information thus acquired has proved useful in dealing with qualitative problems as well. Genetic counseling and eugenics are two aspects of human genetics that have qualitative implications for the human gene pool.

29.1. Genetic Counseling

As understanding of heredity increased, geneticists began to modify, change, and "improve" the genetic characteristics of domesticated species of plants and animals. It is not surprising that eugenics, the improvement of the genetic qualities of mankind, seized the imagination of the early students of heredity. Two kinds of eugenic approach are possible. With positive eugenics, ways are sought to increase the frequency of desirable genes in the population. There can be little doubt that changes could be achieved in human populations comparable to those observed in domesticated species. However, any positive eugenics program is confronted by two great stumbling blocks. The first is the problem of deciding which genes are desirable; the second is the question of who is to make these decisions. The specter of Hitler rising in the background is enough to make most men pause before advocating a program of positive eugenics. The fall from grace of the concept and even of the word eugenics can be traced in large measure to the excesses of Hitler and to the overly optimistic views of the early advocates of positive eugenics programs.

In negative eugenics, means are sought to decrease the frequency of deleterious genes in the population. While general agreement on desirable genes would be difficult to obtain, most people would probably agree that it would be desirable to reduce the frequency of hereditary deafness, or blindness, or mental retardation, or some of the grotesque physical anomalies described earlier. Nevertheless, here too, value judgments are involved, and the need to protect individual rights is clear. At present, relatively few advocates of eugenics can be found, even for negative eugenics programs. The reason, apart from the stigma Hitler gave the subject, is that an increased understanding of human population genetics has led to the realization that it is not as simple a matter to change gene frequencies in human populations as the early eugenicists thought. The approach to these problems today is considerably less optimistic, but also more realistic than before.

Genetic counseling may be considered as a branch of public health medicine. In general, the genetic counselor is not particularly concerned about changing gene frequencies, but rather with helping parents to avoid having a child with a serious genetic defect. Although the prevention of the birth of a single defective child may have little or no effect on overall gene frequencies, the effort is more than worthwhile in terms of the human tragedy averted.

The primary function of the genetic counselor is to provide information and, if necessary, instruction to persons seeking his advice. With this information they can then make a more informed and rational decision about their future course of action. This decision almost has to be theirs, for the counselor is ordinarily in no position to appreciate the imponderables that may also enter the situation, nor does he have to live with the consequences of the decision.

Although practicing physicians generally get most initial inquiries about genetic problems, in many cases they do not have the detailed knowledge or expertise in human genetics needed to provide satisfactory answers to the questions they receive. In recent years centers for genetic counseling have opened in many of the major medical centers in the United States. In these centers specialists in human and medical genetics are available for consultation. Many of the people they serve are referred to them by physicians, but other cases may come from lawyers and from public units such as adoption agencies while many inquiries come directly from the public at large.

The range in the nature of the questions asked covers a surprisingly wide spectrum, from inquiries about simple Mendelian traits such as albinism or Huntington's chorea to questions about the appearance years hence of a baby about to be placed for adoption or the genetic risks entailed in marriage to one's mother's half-sister. Obviously, in some situations it is possible to give much more detailed and specific information than in others, but in nearly all cases the genetic counselor can provide useful information. He seldom deals in certainties, and would be ill-advised to try to do so. Most counseling centers are familiar with instances in which the mother of a defective child has been reassured by her physician that she could never have another such child, but then she does. No matter how small the risk of recurrence, it is always there, and the event may just happen. Hence, even though the doctor's motives may be worthy in that he wishes to spare his patients from worry, nevertheless he runs the risk of jeopardizing his own reputation and destroying his patient's confidence in him. Moreover, a few such cases are enough to raise doubts about the value of genetic counseling itself, and thus to hamper the development of medical genetics.

Rather than certainties, the counselor deals in probabilities, the chances or odds that a particular situation will arise. For simple Mendelian traits, these odds can be figured quite precisely, but even for conditions in which the genetics are obscure, empirical risk figures can be used to provide an estimate of the hazards involved. The empirical risk figures are based on surveys carried out on enough cases to provide a reliable empirical estimate of probabilities. For example, the genetics of spina bifida and related conditions are obscure, but surveys of couples who have had one child with spina bifida have revealed that the chances that a subsequent child will be affected are 1 in 20. If two affected children have been born, the risk for a subsequent child is greater, of the order of 1 in 10. These figures at least give the parents some idea of the odds confronting them.

Since the genetic counselor can nearly always provide useful information, his advice is nearly always helpful. Even when the prognosis is quite bad, it is often helpful to the parents to know the odds they face rather than to suffer from vague and poorly defined fears. Since in many cases the odds are not nearly as bad as the couples expect, the counselor is often in the position of being able to reassure them that the situation is not as serious as they had thought. The psychological aspects of genetic counseling are almost as important as the genetic aspects, for people seeking advice are often beset not only by fears, but by guilt or shame as well. It is to be hoped that these attitudes toward hereditary diseases can be eased in time as people come to realize that hereditary diseases stem from changes in the genetic code and need not be regarded as a family stigma or some form of punishment.

Since most hereditary diseases in man are rare, it may be wondered whether enough cases arise to justify the creation of centers for genetic counseling. However, it is estimated that about 4% of all human infants have tangible defects that are detectable in infancy. Of these about one-half are primarily genetic in origin. Therefore, on the average, about one baby in 50 is born with some type of genetic defect. Clearly, there is a need not only for counseling centers but also for adequate training in human genetics for physicans so that they have the competence to handle a greater proportion of the more routine counseling cases.

The fact that in the population about one pregnancy in 50 results in some sort of genetic anomaly has another type of significance. It furnishes a standard against which to measure the odds for any given counseling case. If, for example, it turns out that the chances are 1 in 25 of an affected child, this is not too much greater than the odds for any random pregnancy. The reaction to this added risk may vary from one couple to another,

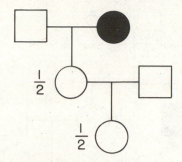

Fig. 29.1. A pedigree of Huntington's chorea. □ = normal male; ○ = normal female; ● = female with Huntington's chorea. **The horizontal lines connect marriage partners, the vertical lines indicate their offspring.**

but the general risk figure helps to give perspective to prospective parents. The reaction to risk figures will also, of course, vary in accordance with the seriousness of the condition and whether it can be treated, and whether it causes inevitable early death or years of invalidism, since the former is actually less of a burden than the latter.

Genetic counseling depends, first, on adequate diagnosis; second, on a thorough family history; and finally, on the interpretation of this information on the basis of what is already known about the trait. The diagnosis is important, especially for traits that have a number of clinically similar forms, so that the proper risk figures will be given. The family history helps to establish the mode of inheritance, and shows whether the clinical data and the family history are consistent with other similar conditions previously described. The interpretation forms the basis for determining the odds. These odds or probabilities are ordinarily the basic information given to persons seeking counsel, on which they can base their decisions.

We shall now consider several types of counseling problems to illustrate some of the principles involved. For simple dominant traits, the probabilities can be readily determined. There is a 50:50 chance that any child of a person affected by a heterozygous dominant gene will be similarly affected. If the age of onset is delayed, as in Huntington's chorea, questions may arise about the risks for grandchildren as well as children. For instance, the chances that a grandchild of a 45-year-old woman just diagnosed as having Huntington's chorea will also eventually become choreic can be diagrammed as shown in Fig. 29.1. The chance that the daughter will become choreic is ½. The chance that the granddaughter will become choreic if the daughter is choreic is also ½. Since these are two independent events, in a statistical sense, the probability that the granddaughter will become choreic is $½ \times ½ = ¼$. If the daughter later becomes choreic, the grand-

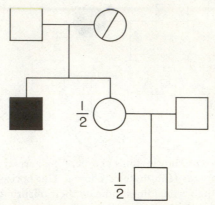

Fig. 29.2. A pedigree of hemophilia. □ = normal male; ○ = normal female; ■ = hemophilic male; ⊘ = heterozygous normal female, carrier of the gene for hemophilia.

daughter's chances then increase to 1 in 2, but if the daughter lives beyond the age at risk, the granddaughter's chances of becoming choreic drop to zero, except for the possibility of a rare new mutation.

Similar reasoning can be applied to sex-linked genes. For example, if a hemophilic son is born to normal parents, the most likely assumption is that his mother is a heterozygous carrier of the gene. Therefore, the chances are ½ that any particular sister is a carrier or that any brother born subsequently will also be hemophilic. The chances that a son of the sister of a hemophilic male will be similarly affected are 1 in 4, as shown in Fig. 29.2. However, if the problem is stated with a slight difference, the probabilities change. The chances that the sister of a hemophilic male will have a hemophilic son are 1 in 8 rather than 1 in 4. The independent probabilities are

½ (that the sister is a carrier) ×
½ (that her child will be a son) ×
½ (that the son will be hemophilic)
= ⅛

For autosomal recessive traits, if two normal but heterozygous parents produce an affected child, the chances that a subsequent child will be similarly affected are 1 in 4. Of course, such parents may be lucky and produce only several normal children. On the other hand, they may be unlucky and produce two or even more affected children. The odds for such eventualities can be calculated. For example, if prospective parents both known to be heterozygous for an autosomal recessive, say because each has a parent homozygous for albinism, wished to know their chances of producing three normally pigmented children, the probability would be $(\frac{3}{4})^3 = \frac{27}{64}$, or a little less than ½. Similarly, their chances of producing a family of three albinos is $(¼)^3 = ¹⁄₆₄$. Other possibilities can be obtained from the binomial equation $(p + q)^n$, where $p = ¾$, $q = ¼$, and $n = 3$. Hence, the chances of one albino and two normal children are $3(¾)^2(¼) = ²⁷⁄₆₄$, while the chances of one normal and two albino children are $3(¾)(¼)^2 = ⁹⁄₆₄$. Thus, for such a couple the chance that they would have at least one albino child if they have a family of three children is the sum of $²⁷⁄₆₄ + ⁹⁄₆₄ + ¹⁄₆₄ = ³⁷⁄₆₄$, or better than 50%. The probabilities will be further complicated if it is asked, for example, what the chances are of having two normal sons. In this case, the probabilities are

$$½ \text{ (the chance of having a son)} \times$$
$$¾ \text{ (the chance of normal pigment)} =$$
$$⅜ \text{ (the chance of a normal son)}$$

The chance of two normal sons in succession is, therefore, $(⅜)^2 = ⁹⁄₆₄$. Other, similar questions could be posed, but the principles are essentially the same. Care is necessary, however, to ensure that all the possibilities are taken into account. A further cautionary word is that the type of problem just considered must be clearly distinguished from the situation where an affected child has already been born. In such cases, the probabilities are usually stated as the chances that the next child of the couple will also be affected. In the problems just considered dealing with families, the probabilities are calculated for entire families before any children have been born at all. As soon as a child is born, if the trait is expressed at birth it is no longer a probability but a certainty as to whether or not he is affected. Therefore, unless in some way his birth changes the odds relating to subsequent births, he no longer enters the calculations.

It is evident that the counselor will be greatly aided in his work if the heterozygous carriers of autosomal or sex-linked recessive genes can be detected or if traits with a delayed age of onset can be recognized in their preclinical stages. Fortunately, for an increasing number of traits, detection of such carriers is becoming feasible.

At this point, perhaps it should be pointed out that the probabilities calculated above were based on genes known to be present in the family. The remote possibility that a new mutation may enter the picture is always present, as is the chance that an allele of independent origin may enter the family through one of the marriage partners. These more or less remote possibilities, depending on the mutation rate and on the frequency of heterozygous carriers, introduce an added risk into the situation and should make the counselor cautious about being too arbitrary in his statements of probabilities.

Even for relatively rare recessive traits like albinism, the frequency of heterozygotes is surprisingly high. Albinos have a frequency of about 1 in 20,000. If the Castle-Hardy-Weinberg equilibrium is assumed, then

$$q^2 = \frac{1}{20,000}$$

$$q = \sqrt{\frac{1}{20,000}} \cong 0.007$$

$$p = 0.993$$

and

$$2pq = 0.0139 \cong \tfrac{1}{70}$$

In other words, although only 1 in 20,000 persons is an albino, 1 in 70 can be expected to be a heterozygous carrier of the gene for albinism. Hence, if an albino individual marries an unrelated person from the general population, the chances are 1 in 70 that that person will be a carrier of the albinism gene, and the chances are 1 in 140 that their first child (or any one child) will be albino:

aa (albino individual—given)
$\tfrac{1}{70}$ (chance that normal marriage partner is heterozygous Aa) ×
$\tfrac{1}{2}$ (chance that a child of $aa \times Aa$ mating is aa) =
$\tfrac{1}{140}$

The chances that a normal sib of an albino individual with normal parents is heterozygous are $\tfrac{2}{3}$. This probability may be surprising, since most probabilities in Mendelian genetics have denominators divisible by 2. However, the progeny of heterozygous parents ($Aa \times Aa$) will be, on the average, $\tfrac{3}{4}$ normal and $\tfrac{1}{4}$ albino. Of the $\tfrac{3}{4}$ who are normal, the expectations are that $\tfrac{1}{4}$ will be homozygous and $\tfrac{2}{3}$ will be heterozygous; in other words, two of the three normals will be heterozygous. A frequent question is the chance that the normal sib of an affected person will have a similarly affected child. In this particular case, these chances can be calculated as follows:

$\tfrac{2}{3}$ (chance that the normal sib is heterozygous)
$\tfrac{1}{70}$ (chance that normal marriage partner from population at large is heterozygous)
$\tfrac{1}{4}$ (chance that a child of this mating will be albino) =
$\tfrac{1}{420}$

Finally, it may be of interest to calculate the probability that two normal persons from the general population will have an albino child, which is as follows:

$$\frac{1}{70} \times \frac{1}{70} \times \frac{1}{4} = \frac{1}{19,600} \cong \frac{1}{20,000}$$

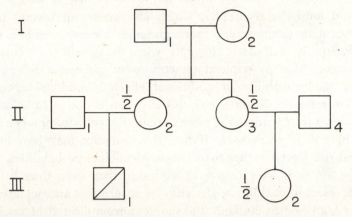

Fig. 29.3. Diagram to aid in the calculation of the probability that two first cousins will carry the same recessive gene. Roman numerals are used to indicate generations; Arabic numbers indicate individuals within each generation.

This figure, of course, agrees well with the incidence figure given above, and reflects the fact that nearly all albinos born are the children of normal but heterozygous parents.

A somewhat different type of problem relates to the question of the added risks involved in marrying a relative. For example, it may be asked, what is the probability that first cousins both carry the same deleterious autosomal recessive gene. In this case, and in most others, it is helpful to diagram the relationships and probabilities involved, as we have done in Fig. 29.3. If the boy in Fig. 29.3 (III-1) carries a particular recessive, the chance is ½ that he received it from his mother (II-2). If his mother has the gene, the chances are ½ that her sister (II-3) also carries it. If she has it, the chances are ½ that she transmitted it to her daughter. Therefore, the chances are 1 in 8 that these first cousins carry the same deleterious recessive gene. The chances are, then, 1 in 32 that a given child of this couple will be homozygous recessive for this gene. In this sort of problem, no affected persons appear in the pedigree, and a most important question is how common deleterious recessive genes are in the human population. The best estimates at present are that nearly everyone carries at least one seriously harmful recessive gene in his genotype and that the average is of the order of four or five. Therefore, the various taboos and restrictions against the marriage of close relatives seem to have a sound genetic basis.

The probabilities discussed so far have been related to clear-cut Mendelian traits, for which the probabilities are easily determined. Difficulties may emerge, however, if clinically similar but genetically distinct entities are involved, as for example, retinitis pigmentosa, for which several dominant, recessive, and sex-linked forms are known as well as sporadic cases, and muscular dystrophy, which also occurs in several hereditary forms. Adequate diagnosis and thorough family histories will help to resolve the difficulties in such cases. However, when the genetics of a condition are more obscure due to reduced penetrance or polygenic inheritance, for instance, the probability of additional affected children becomes more difficult to estimate. Nevertheless, this situation is not as troublesome as it might seem, for in general, the more obscure the genetics of a condition, the lower the risk of recurrence. Hence, the counselor may have to rely on empirical risk figures rather than simple Mendelian probabilities, but he at least can say with some degree of assurance that, even though he cannot calculate exact probabilities, the risks of recurrence are not great, rarely being as high as 1 in 20. Only the simple, uncomplicated Mendelian traits have high risks of recurrence. Therefore, the counselor is in the advantageous position of being able to make the most exact calculations for those cases in which they are most necessary.

29.2. Eugenics

The essence of the eugenic concept is the genetic betterment of man through the study of human genetics and the application of the results of these studies to human reproduction. This idea can be traced at least as far back as the Greeks—it emerged in Plato's *Republic*—and has appeared in most of the Utopias written since. However, its modern origins stem from Darwin's *Origin of Species,* which established that man, as well as other species, is subject to evolutionary change, and from the development of genetics, which for the first time provided insight into the mechanism of evolution. Darwin's cousin, Sir Francis Galton, is generally recognized as the founder of the eugenics movement; he coined the word eugenics in 1883 from the Greek word meaning "well-born." Galton's wide-ranging scientific interests made him a pioneer and leading figure in the development of such fields as statistics, psychological testing, dermatoglyphics (the study of skin patterns, as in fingerprints and footprints), the study of twins, and human genetics. Stimulated by Darwin's findings, Galton made the study of heredity and variation in man the focus of his life's work. Moreover, he did not hesitate to explore the implications and the possible applications of his results to the betterment of mankind. However, he tended to overestimate the importance of heredity and to underestimate the role of the environment in the development of human traits, and to be overly optimistic about the ease with which genetic improvement could be effected in man. Galton's successor, Karl Pearson, was even more hereditarian in his views. Galton emphasized human betterment by encouraging reproduction by superior human stocks, but Pearson and most other early advocates of eugenics were more interested in controlling the reproduction of the "unfit." One of the fields contributing to the developing concepts of eugenics was criminal anthropology, in which it was generally agreed that heredity rather than the environment was responsible for most of the criminals, the insane, the paupers, the mentally retarded, and the alcoholic. Today the views of the early eugenicists seem naive, misguided, and in some cases strongly racist in tone.

The eugenics movement spread throughout the world in the early 1900s following the advent of Mendelism. In the United States, The Eugenics Record Office, established at Cold Spring Harbor on Long Island in 1910 with C. B. Davenport as director, became the major center for eugenics research and propaganda. The movement gained acceptance because by the late 1800s the idea had already taken hold that defective heredity was a major cause for people being put in asylums or prisons or on relief, and this in turn led to the wish that the "unfit" be prevented from reproducing. These attitudes were a reflection of an outlook that has come to be known as Social Darwinism, in which social attitudes were supposedly based on Darwinian principles of natural selection, survival of the fittest, competition, and evolution. Along with the basic concepts of the eugenics movement, which grew out of advances in the study of evolution and genetics, other movements of the day were intertwined with eugenics, among them the movements for peace, prohibition, sex education, and birth control. Another group of adherents was interested in restricting the wave of immigration to the United States that marked the turn of the century. They seized upon eugenics in support of their essentially racist position. Those whose primary interest lay in the scientific study of human genetics were never in the forefront as leaders of the eugenics movement although they were well aware of the implications of their research. Thus, the eugenics movement was supported by a strange conglomerate of adherents, many with their own particular axes to grind. While many were sincere, well-meaning people, others were extremists and reactionaries. The scientific underpinnings for the eugenics movement were never too secure in the early days of the movement although it gained fairly wide support prior to World War I. The movement continued to grow into the 1920s, but as the more extreme elements came to the fore and advances in genetics made some of the basic assumptions dubious, the eugenics movement collapsed as a scientific enterprise during the 1930s. The concepts were perverted by Hitler to meet his political purposes and by racists in the United States to support their views of white supremacy; the extremists held the field. The reaction of serious students of human genetics was to dissociate themselves from the eugenics concept as far as possible, even to the point of avoiding the use of the word. *Annals of Eugenics,* a journal with a long and honorable history, became *Annals of Human Genetics,* and more recently *Eugenics Quarterly* became *Social Biology.*

Human genetics has come into its own as a separate discipline in the years since World War II. As a result there has been a vast increase in the knowledge and understanding of the genetics of man, especially in the area of medical genetics. Inevitably, there has been some revival of interest in eugenic ideas, but for the most part, the reformer's zeal, so notable in the early part of the century, has been conspicuously absent, and the human geneticists have tended to play down the eugenic aspects of their work and to stress the importance of pushing ahead on basic research in human genetics. The explosion of the atomic bomb caused an exception to be made, for geneticists were among the first to point out the genetic hazards from radioactive fallout, and have been among the leaders in efforts to limit the use of such weapons and thus to reduce the danger of mutagenic effects. In fact, all forms of radiation hazard, medical, dental, industrial, and commercial, have been publicized, particularly by H. J. Muller, in an effort to reduce the exposure of the public, which might otherwise be unaware of the inherent risks involved. More recently, this concern has extended to nuclear power installations, and a society has been formed to fight against all forms of environmental mutagens, chemical as well as radioactive. In this respect at least, there has been no reluctance to take up the cudgels in behalf of eugenics.

However, even though no formal eugenics program as such has been established anywhere, several factors can be identified that are already affecting the human gene pool. For this reason, it seems desirable to review briefly what some of these factors are and some of the possibilities that may arise in the future. In some cases, such as genetic counseling, the primary purpose is eugenic, but in others there are eugenic implications even though other considerations are primary. Thus, much that is happening today with eugenic overtones is being done unwittingly.

The main thrust of genetic counseling in a eugenic sense is negative—the prevention of the birth of defective children. To the extent that this is averted, the counseling can be regarded as successful. However, the effect of successful counseling on gene frequencies is more difficult to assess. For example, if heterozygotes for a deleterious recessive gene were persuaded not to marry one another, the incidence of the harmful trait would be reduced, but not that of the gene. On balance, though, it seems probable that genetic counseling will help to reduce the frequency of deleterious genes.

Over the years, mankind has manipulated the birth rate in a variety of ways. One general way is the restriction of mating, a second is the prevention of conception, and a third is the elimination of unwanted zygotes. Today, some of these practices are generally accepted, others are not, and still others are the subject of debate between different segments of society. At the present time attitudes toward many of these practices seem to be in a state of flux.

The commitment of persons to prisons, mental hospitals, and other similar institutions is not primarily for eugenic reasons. However, the effect of institutionalization is to remove these individuals from the mating population. To the extent that the gene pool of the institutionalized population differs from that of the population at large, there will be a genetic effect. Since we have already seen that there appears to be a genetic component to many types of mental illness and mental deficiency, the net effect would seem to be a reduction in the frequency of some types of genes.

In some societies, such as Ireland, the economic situation has led to delays in marriage until relatively late in life. Voluntary celibacy, as in the Catholic clergy, also removes a group of individuals from the mating population. The genetic impact of such restrictions on marriage is unknown, but it seems unwarranted to assume that it is nil without further information.

Some of the ways of preventing conception were known in Biblical times and probably before, but more recently new and more reliable contraceptive methods have been discovered. In the past castrations were performed to produce eunuchs to work in harems or to sing in church choirs, but this particular method of contraception fortunately seems to have waned. However, vasectomy, the severing of the vas deferens or sperm duct in the male, is being used today. In the female, salpingectomy or tubal ligation, the tying off of the Fallopian tubes, and hysterectomy, the removal of the uterus, are also methods in use at present. The operation in women is more serious than vasectomy since it requires opening the abdomen. The passage of laws permitting the sterilization of the insane and the feeble-minded was a major part of the program of the eugenics movement in the early part of this century. By 1931, 30 states had passed sterilization laws, and 27 states still had them in effect. However, they were not widely enforced, and California, which performed nearly half of all the sterilizations under these laws, had carried out only 7,548 operations between 1909 and 1931. By 1958, only 60,926 such sterilizations had been carried out in the entire country, a very trivial proportion of the total number of insane and feeble-minded living in the United States since 1907 when the first law was passed. Thus, even though the movement was relatively successful in getting laws passed, the laws were not widely accepted or enforced. Furthermore, in many minds eugenics came to be associated primarily with involuntary sterilization, an idea that was repugnant because of the possibilities of abuse, and hence the eugenics movement itself suffered. Sterilization is, however, an extremely effective permanent method of birth control, and when requested voluntarily by patients confronted with a serious genetic problem, may represent the best solution to their problem.

Some other methods of preventing conception now widely used are the "rhythm method," the intrauterine device or IUD, and the contraceptive pill. The rhythm method, condoned by the Catholic church, is not particularly reliable. The IUD is not tolerated or retained by a significant proportion of women. The "pill," while reliable, has undesirable side effects in some cases. Thus, the existing methods of contraception suffer to varying degrees from a lack of reliability or acceptability or both, and research for more effective methods continues. Nevertheless, their use is spreading, but the net genetic effect is in doubt. However, couples who wish to limit their families for eugenic reasons as a result of genetic counseling now have several safe, relatively simple, and effective methods to choose from.

In some ancient or primitive societies, unwanted or defective infants were simply exposed to the elements, a practice not in general use today. Abortion is another matter. Although abortion is banned in many countries, illegal abortions with their attendant risks continue to be performed. However, the present trend seems to be toward liberalization of the laws with respect to abortion. A technique known as amniocentesis has recently been devised that enables the physician to check for chromosome anomalies and genetic defects in the developing embryo while it is still within the mother's uterus. If the indications are that the fetus is seriously defective, a therapeutic abortion is then possible. Amniocentesis is a useful new weapon in the struggle against hereditary defects.

Most of the measures discussed thus far fall within the realm of negative eugenics. Certain factors at work in our society, however, seem to be dysgenic. Since there is no threshold to the mutagenic effect of ionizing radiation, the increased amounts of such radiation and of other environmental mutagens will inevitably be harmful. The big question seems to be how much they will add to our load of mutations. Moreover, to the extent that modern medicine succeeds in counteracting the effects of deleterious genes so that affected individuals are able to survive and reproduce, the effect is dysgenic. Since physical and mental standards are set for acceptance into military service, with those not meeting them excused, the overall effect would seem to be dysgenic.

On the other hand, a few factors can be recognized as having, potentially, a positive eugenic effect, for example, loans, scholarships, dependency allowances, and the like for able college and university students. Such measures may encourage them to have children they might not otherwise have, but the eugenic impact of such measures is probably not great.

Another activity of potential positive eugenic impact is the growing use of artificial insemination for women whose husbands are sterile. Understandably, statistics on this sort of semiadoption and data on the sperm donors are not readily available, but on the assumption that the donors are selected with reasonable care, a positive eugenic effect can be predicted. Since it is estimated that thousands of children conceived in this way are born in the United States each year, the potential effect is of some significance. The primary concern of the clinician, however, is often directed toward the production of a child showing some resemblance to the husband so that gossip can be averted, and the identity of the donor or donors is kept secret from the parents.

H. J. Muller has argued that a far more effective positive eugenics program is possible based on "germinal choice." The essence of his plan is the creation of sperm banks with frozen sperm from outstanding individuals of diverse types with unusual intelligence, talent, character, or physical fitness. The sperm can be maintained indefinitely in a medium containing glycerine at the temperature of liquid nitrogen. Prospective parents can then, after studying the records on the donors and the donors' relatives, select the particular donor whose characteristics they most admire and wish to perpetuate. Muller's argument for this proposal is based on the fact that the techniques required are already available, and that we will be remiss if we do not start to use them at once to improve man's genetic constitution. Many of the other suggested eugenic measures are not yet technically feasible and may not be for some time to come. Even though germinal choice is technically within reach, and is already widely used in cattle breeding, for example, one cannot help but wonder whether human society is yet ready to accept the idea. We know so little about the selective factors operating in human reproduction that it is not possible to say whether a woman, given the choice of having a child sired by Beethoven, Einstein, George Washington, or her own husband, would opt for one of them rather than her own beloved Joe Doakes, or whether her husband would accept the situation if she did. And the old problem of positive eugenics again rears its head: Who is to judge which sperm are to go into the sperm bank? Questions such as these, bearing on the psychological and social aspects of germinal choice, make it seem unlikely that such a program will be adopted soon. Nevertheless, since artificial insemination is responsible for thousands of babies born each year, it seems highly desirable that the physicians carrying on the practice give considerable attention to the genetic qualities of the donors they recruit.

If we start to speculate about other techniques that might be used in the future to modify or improve the human genotype, a number of possibilities come to mind. Ova transplants or the transfer of fertilized eggs from the reproductive tract of one female to the uterus of another have already been successfully performed in mammals. Such a procedure might be useful to a busy professional woman or actress who desired to have children of her own but felt she could not afford the time required for pregnancy. The eugenic benefit in such a case is a moot question, but the technique seems feasible. On the other hand, methods for preserving human eggs in an "egg bank" comparable to what can be done with sperm have not yet been perfected.

Recent advances in cell and tissue culture of various types of human somatic cells raise the interesting possibility that entire individuals might be generated by mitotic cell division. The test tube babies of science fiction would become reality, and twin research would boom if it could involve dozens or hundreds of individuals with identical genotypes rather than just two. On the other hand, the prospect of a hundred or a thousand Napoleons or Stalins or even Albert Schweitzers might be more than the world is willing to face. In this case the process would be uncomplicated by genetic recombination; a known genotype could be perpetuated. However, the problem of choosing the appropriate genotypes would probably be even more troublesome.

The recent developments in molecular biology and the breaking of the genetic code have led to speculations about the possibility of direct manipulation of the human genotype by "genetic surgery" or "genetic engineering." The discovery of chemical mutagens with some degree of specificity holds promise that eventually it will be possible to induce specific directed mutations at will. In this event it could then become feasible to attempt to induce defective genes to mutate to their "normal" alleles. Transformation and transduction in bacteria hold similar promise. In transformation an alien DNA segment is incorporated into the genome; in transduction the transfer is mediated by a virus. However, these processes were discovered in microorganisms, and the more complex mammalian cell poses more difficult problems. Hence, the perfection of such techniques to the point where they can be reliably used to "cure" genetic defects in man requires further research. Other possibilities along similar lines are substitution of whole chromosomes or of a chromosome segment. Whole-chromosome substitutions by micromanipulation are conceivable but have not yet been performed. Substitution of one chromosome segment for another bearing a deleterious gene could be carried out by a process analogous to crossing over. In both cases it may be essential that the substituted chromosome or chromosome segment be eliminated from the cell. Moreover, unless the total genetic content of the substituted chromosome or fragment is known, the substituted material may be no improvement on the existing genotype.

Therefore, at present it seems that such techniques as directed mutation or the substitution of favorable genes for harmful ones do not offer immediate hope for eugenic improvement in man. Since these and other suggested eugenics programs do not seem destined to have much immediate impact, it has been urged that programs in euphenics and euthenics be pushed vigorously. Euphenics is the improvement of the human phenotype by biological means. Many activities of this sort are already being practiced, ranging from insulin therapy for diabetics to heart transplants. In this way the harmful effects of many genetic defects are being counteracted so that the affected individuals can live normal productive lives. However, if they also live a normal reproductive life, the effect of euphenics is dysgenic. Euthenics seeks human betterment through improving the environment. Improved nutrition, better education, and the like will help each genotype to realize its full potential.

If we were to attempt to evaluate the potential for human betterment in the near future through euphenics, euthenics, and eugenics, eugenics would probably come in a poor third. Many measures are presently available for alleviating or correcting human ailments, and public health measures, public education, and adequate nutrition can provide at least some of the basic requirements favorable to human development. For one reason or another the existing proposals for eugenics programs seem unlikely to be adopted or to be very effective in producing planned changes in human gene frequencies. This is true whether we are thinking of something as radical as Muller's "germinal choice" plan or as modest as Osborn's suggestion that the most favorably endowed individuals be encouraged to have large families. Genetic counseling may in time have some eugenic impact, but presently it is not great.

However, the human species has evolved rather rapidly over the past 2 million years, which is the present extent of our knowledge of the fossil record of the Hominidae. Therefore, natural selection must have been operating with considerable effect on man's evolving pool of genes. There is no reason to suppose that human evolution has ceased or that natural selection has been suspended. The simple fact is that we have little idea as to what selection pressures in the past caused man's evolution to his present state, nor do we have any very clear idea as to what selective forces are presently acting on the human gene pool. In the early days of the eugenics movement, the histories of the Jukes and the Kallikaks led to dire predictions that the high reproductive rate of the feeble-minded would lead to a gradual decline in the average intelligence of the general population. A major impetus in the eugenics movement was the effort to stem this trend and all the ills assumed to be associated with it. Only recently has it been realized that the high reproductive rate assumed to exist among the mentally retarded was based on faulty data that failed to include the many retarded individuals who never marry. Thus this particular fear, a cornerstone of belief in the early eugenics movement, is now known to be false. Selection is not acting to lower the IQ; the trend, if any, is in the opposite direction. But again, it bears repeating that we really know very little about the selective pressures now operating in man. We do know that positive assortative mating occurs, suggesting some form of sexual selection. We also know that close to 15% of the population in the United States fails to reproduce, either because they never marry or, if married, do not have children. Furthermore, 50% of the women produce 85% of all children born in the United States.

It is obvious that, because of medical progress in the past century in controlling infectious diseases, the selection pressures in man must have changed at least in this respect. Nonetheless, we have very little information about the possible effects of assortative mating or the control of infectious diseases on changes in human gene frequencies. Neither do we know whether there are significant genetic differences between the individuals having many children and those with few or none. If there were, strong selection pressures could be operative of which we are unaware. It seems desirable that before we go overboard again on any sort of eugenics program, we gain at least some notion of the evolutionary forces presently at work in man. Man has evolved to his present condition without the benefit of a formal eugenics program. Negative eugenic measures to reduce the frequency of hereditary blindness, deafness, mental deficiency, and the like seem reasonable and necessary. However, before we tamper too much more with man's evolutionary machinery, it behooves us to learn more about how the machinery actually is working at the present time.

Additional Reading

Davis, B. D., 1970. "Prospects for Genetic Intervention in Man," *Science* **170**: 1279–1283.

Dubos, R., 1965. *Man Adapting.* Yale University Press, New Haven, Conn.

Dunn, L. C., 1962. "Cross Currents in the History of Human Genetics," *Amer. J. Human Genet.* **14**:1–13.

Ehrlich, P. R., 1968. *The Population Bomb.* Ballantine, New York.

Ehrlich, P. R., and A. H. Ehrlich, 1970. *Population, Resources, Environment.* Freeman, San Francisco.

Haldane, J. B. S., 1924. *Daedalus; or Science and the Future.* Dutton, New York.

Haller, M. H., 1963. *Eugenics: Hereditarian Attitudes in American Thought.* Rutgers University Press, New Brunswick, N.J.

Hammons, H. G., ed., 1959. *Heredity Counseling.* Medical Department of Harper & Row, New York.

Hardin, G., ed., 1969. *Population, Evolution and Birth Control,* 2nd ed. Freeman, San Francisco.

Hauser, P. M., 1963. *The Population Dilemma.* Prentice-Hall, Englewood Cliffs, N.J.

Human Genetics in Public Health, 1964. Minnesota Dept. of Health.

Lynch, H. T., 1969. *Dynamic Genetic Counseling for Clinicians.* Charles C Thomas, Springfield, Ill.

Malthus, T. R., 1807. *An Essay on the Principle of Population,* 4th ed. Johnson, London.

Medawar, P. B., 1959. *The Future of Man.* Mentor, New York.

Muller, H. J., 1960. "The Guidance of Human Evolution." In *Evolution after Darwin.* Vol. 2. *The Evolution of Man,* S. Tax, ed. University of Chicago Press, Chicago.

Muller, H. J., 1965. "Means and Aims in Human Genetic Betterment." In *The Control of Human Heredity and Evolution,* T. M. Sonneborn, ed. Macmillan, New York.

Osborn, Fairfield, 1948. *Our Plundered Planet.* Little, Brown, Boston.

Osborn, Fairfield, ed., 1962. *Our Crowded Planet.* Doubleday, Garden City, N.Y.

Osborn, Frederick, 1951. *Preface to Eugenics,* rev. ed. Harper & Row, New York.

Osborn, Frederick, 1968. *The Future of Human Heredity.* Weybright and Talley, New York.

Petersen, W., 1969. *Population,* 2nd ed. Macmillan, New York.

Ramsey, P., 1970. *Fabricated Man: The Ethics of Genetic Control.* Yale University Press, New Haven, Conn.

Reed, S. C., 1963. *Counseling in Medical Genetics,* 2nd ed. Saunders, Philadelphia. (Also in paperback as *Parenthood and Heredity,* Wiley, New York.)

Sax, K., 1955. *Standing Room Only.* Beacon, Boston.

Sonneborn, T. M., ed., 1965. *The Control of Human Heredity and Evolution.* Macmillan, New York.

Thompson, W. S., 1948. *Plenty of People.* Ronald, New York.

Thompson, W. D., and D. T. Lewis, 1965. *Population Problems,* 5th ed. McGraw-Hill, New York.

Vogt, W., 1948. *Road to Survival.* Sloane, New York.

WHO Expert Committee on Human Genetics, 1969. "Genetic Counselling." World Health Org. Tech. Rept. Ser. No. 416, pp. 1–23.

Wolstenholme, G., ed., 1963. *Man and His Future.* Little, Brown, Boston.

Questions

29.1. Discuss the relative merits of programs of positive eugenics as compared with programs of negative eugenics.

29.2. What is the most useful service a genetic counselor can provide?

29.3.   What proportion of infants born have tangible defects detectable in infancy that are primarily genetic in origin?

29.4.   If you are heterozygous for a rare autosomal recessive, what are the chances that a particular one of your first cousins is also heterozygous for this same recessive?

29.5.   If you married your first cousin, what would be the probability that your first child would be affected by this autosomal recessive trait?

29.6.   If 20% of all the children of matings of the type $Aa \times Aa$ show the recessive phenotype, what is the penetrance?

29.7.   The frequency of red hair, an autosomal recessive trait, is about 4% in a population.

    a. What is the estimated frequency of heterozygotes in this population?

    b. What is the probability that two heterozygotes for red hair will marry?

    c. What is the probability that the first child of a non-red-haired couple picked at random from this population will have red hair?

29.8.   What is the probability that the daughter of the sister of a hemophilic male will be a carrier of the gene for hemophilia?

29.9.   What factors are at work in human populations that have a negative eugenic effect?

29.10.   What factors are at work in human populations that have a positive eugenic effect?

29.11.   What dysgenic factors may be affecting present-day human populations?

29.12.   If you were asked to make recommendations for the genetic improvement of the human gene pool, what policies would you recommend?

Statistics

A.1. Biological Variation

Variability is one of the great facts of life for the biologist. No matter what organisms or what trait he studies, he finds that individuals tend to differ from one another. If significant results are to be obtained, some method must be found to deal with this variability so that it becomes both manageable and meaningful. Statistics is the tool by which the data are reduced to manageable form and tested for significance.

Two major types of variation can be recognized, discontinuous and continuous. Discontinuous variables are exemplified by a series of integers. In such a discrete series it is possible to find two integers between which no other integer exists; for instance, there is no integer between 99 and 100. All sorts of enumeration data fall in this category: number of scales in the lateral line of a fish, number of tomatoes on a plant or eggs in a nest, or numbers of bristles on a fly. This type of variation is known as *meristic* variation. In genetics another sort of discontinuous variation is often encountered, which is inherently different. For example, in crosses it is often possible to categorize flower color as red or white, or reaction to disease as susceptible or resistant, or leaf shape as simple or compound. In these cases the variables are discrete, but they do not fall into some sort of logical numerical order. The average number of eggs in a clutch is a useful statistic, but an average leaf shape would have little meaning. It is possible, however, to assign arbitrary values to such variables if such assignments will facilitate analysis of the data and then use the binomial (see Chapter 6). In some crosses, of course, flower color or disease resistance may form an apparent continuum rather than discrete classes.

Continuous variables can take any value over a given interval; such traits as weight or length are continuous. No matter how close two points on the continuum may be, there is still some point, in fact an infinity of points, between them. While variability itself may be either continuous or discrete, in practice, because of the limitations in the precision of measurement, all measurements, even of continuous variation, are discrete variables. Thus, length can be measured to the nearest centimeter, or millimeter, or micron, or millimicron, depending on the equipment used, but even between 3 and 4 millimicrons there remains an infinity of intervals that cannot be measured. Therefore, similar statistical methods can be used for both continuous and discontinuous variation. Discontinuous data can be treated as continuous if enough different values are available. On the other hand, any type of data, continuous or discontinuous, can be treated with binomial methods by groupings such as "less than 5" versus "5 or more."

A.2. Populations and Samples

A fundamental distinction in statistics is that between a population and a sample. A population consists of all the members of a group with certain specified attributes, for example, all the citizens of the United States. A sample is a group on which observations have actually been made. Since it is generally impractical or impossible to study an entire population, the reason for taking a sample is to obtain an unbiased estimate of the characteristics of the population from the much smaller number actually studied in the sample. If the estimate is to be unbiased, the sample must be random. For example, if the average age of the citizens of the United States were of interest, a sample taken among the citizens of any one state such as North Dakota is unlikely to be representative or reliable.

The purpose of statistics is to make possible inferences about the characteristics of populations based on samples drawn from these populations. Obviously, the validity of the inferences drawn depends on the reliability of the sampling method, and proper sampling is crucial to the success of any statistical study, as the *Literary Digest* learned to its sorrow when it predicted the defeat of Franklin D. Roosevelt.

The various values calculated from measurements made on samples are referred to as *sample values*, or *estimates*, or *statistics*; the true values for the entire population are called *parameters*. The statistics based on the samples are used to estimate the population parameters in lieu of a complete census. For example, a sample containing one ear from each row in a field of corn could be used to estimate the average weight of the ears in that field. The parameter, the true mean, could be obtained only by weighing all the ears in the field and calculating the mean. Parameters are designated by Greek letters, and the statistics by the corresponding Latin letters. For a given population, the population mean is a parameter and has a single value. Samples drawn from this population will ordinarily have different means because of sampling variation; the smaller the size of the sample, the greater the sampling variation. The sample values thus cannot be expected to equal the true value (the parameter) or each other, but as we shall see, they can be used to estimate the parameter in the population.

A.3. Frequency Distributions

The first step in bringing order from the chaos of biological variation is usually to arrange the observations in the form of a frequency distribution. For this purpose it is customary to group the data in categories and then to indicate the number of observations falling in each category. The categories must be established in such a way that each observation fits unambiguously in only one category and there is a category for every observation. For example, if weight were measured in tenths of grams, and the grouping were to the nearest gram, the categories could not be 20.5–21.5, 21.5–22.5, and so on, because it is not clear whether 21.5 would be classed as 21 or 22. To solve this problem, various arbitrary conventions have been adopted; for example, grouping 20.5–21.4 as 21, 21.5–22.4 as 22, and so on, or else rounding off such numbers to the nearest even integer, 20.5 to 20 and 21.5 to 22. However, even these conventions introduce certain biases, the first because the values do not average out to the value used, the second because an excess of even numbers will appear. Nevertheless, with care it is possible to avoid serious bias in grouping.

Although grouping in tabular form aids in understanding the data, a graphical representation of the frequency distribution is even easier to grasp. The graph takes the form of a plot of the various categories for the observations against the frequency of observations in each category. The frequencies may be in terms of absolute frequency, the number of observations in each category, or relative frequency, the number of observations in each category divided by the total number of observations. Various kinds

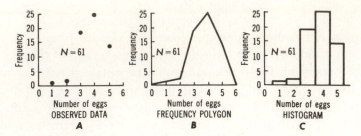

Fig. A.1. Three different types of graphical representation of a discontinuous frequency distribution, number of eggs per nest in the song sparrow. A. Dot diagram. B. Frequency polygon. C. Histogram. (Simpson, G. G., A. Roe, and R. C. Lewontin, 1960. *Quantitative Zoology*, 2nd ed. Harcourt Brace Jovanovich, p. 52, Fig. 4.)

of graphs are used, but in all cases the values of the variate are found on the horizontal X axis or abscissa, and the frequencies on the vertical Y axis or ordinate. One type of graph is a simple dot diagram, another is a frequency polygon, and a third is a histogram (Fig. A.1.).

It is conceivable that all observations of biological variation for a given trait would be equally frequent, the graph being a straight horizontal line, or else that the graph would have two or three or more well-scattered peaks of high frequency. Nevertheless, the types of frequency distributions actually observed are relatively few, and they rather closely resemble well-known theoretical distributions. Occasionally, J-shaped, U-shaped, or S-shaped curves may be observed, which require special statistical methods. However, the three most common and most useful distributions, all rather closely related, are the binomial, the Poisson, and the normal distributions.

Probability Distribution

Probability can be defined empirically as the relative frequency of an event in an infinite number of trials. In a finite series of trials, the observed frequency of the event will seldom equal the true probability, but the larger the number of trials, the closer the observed frequency will come to the true probability. Since probability is defined in terms of the relative frequency of events in an infinite series of trials, and a relative frequency distribution gives the relative frequency of different "events" (different values in the variable under study) in a population, the concept emerges that a probability distribution can be looked upon as simply a relative frequency distribution of an infinitely large number of observations. The probability distribution is then the limit that an observed relative frequency distribution approaches as the number of observations increases. Just as the sum of all the probabilities for a given event equals one, the sum of all the relative frequencies in a frequency distribution also equals one. Furthermore, the area under the curve in a probability distribution can be regarded as equal to unity, and the area under the curve between any two points on the X axis is then a measure of the probability that a given observation of this variable will fall within the limits set by these two points.

The Binomial Distribution

The binomial distribution has already been considered in Chapter 6. It takes the form $(p + q)^n = 1$, where p can be regarded as the probability of occurrence of an event, q the probability that it will not occur, and n the number of trials. The exact probability for a given combination of successes and failures can be obtained from the expression

$$\cdots \frac{n!}{r! \, (n-r)!} p^r q^{n-r}$$

where r, the exponent of p in the desired term, is the number of successes. The binomial distribution can be used only when the values for both p and n are known, and is used in the analysis of data dealing with discontinuous or discrete variation. The binomial expansion deals with just two alternative traits. It can, however, be extended to deal with more than two possibilities, in which case it takes the form of a multinomial distribution:

$$(p + q + r + s + \cdots + x)^n = 1$$

The Normal Distribution

When the binomial distribution is plotted with low values for n, a rather jagged frequency polygon is obtained. As n increases, however, the curve becomes smoother and approximates another frequency distribution known as the normal distribution (also Laplace's normal curve or the Gaussian curve or the normal curve of errors). The normal curve is the limit that the binomial distribution approaches as n increases, provided that neither p nor q is of the order of $1/n$ or less.

The equation for the normal curve is

$$Y = \frac{1}{\sigma \sqrt{2\pi}} e^{-(X-\mu)^2/2\sigma^2}$$

where $X =$ the value of the variate, $Y =$ the height of the ordinate for a

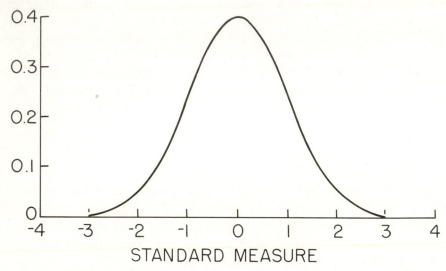

Fig. A.2. The normal curve. The abscissas are standard measure, that is, deviations from the mean expressed in terms of the population standard deviation as a unit. The ordinates are relative frequencies. (Reprinted by permission from *Statistical Methods* by G. W. Snedecor, fourth edition, © 1946 by Iowa State University Press, Ames, Iowa.)

given value of X, π = the constant 3.1416, e = the constant 2.7183, μ = the mean, and σ = the standard deviation.

In this equation two new terms appear, the parameters μ and σ. The values for p and n necessary for the use of the binomial distribution do not appear in this formula. These values are ordinarily established from a detailed knowledge of an experiment, which permits a theoretical determination for the value of the probability p, and an exact determination of the value of n. In many experiments, however, numerical values for p and n cannot be established. For instance, in the case of the chances of albinism among the children of heterozygous normally pigmented parents, p is readily seen to be 0.25. However, the height of these children is also variable, due to both genetic and environmental influences. It is impossible in this case to determine the number of factors involved or the magnitude of their effects, so there is no way to arrive at a theoretical determination of p on which to base a binomial distribution curve. In this sort of situation the normal curve is of great use, because the parameters μ and σ, the mean and the standard deviation, are seldom fixed by hypothesis, but are instead fairly easily estimated from the actual experimental data.

The normal curve is a symmetrical bell-shaped frequency distribution curve (Fig. A.2). The mean μ is the average value for the entire theoretical population of infinite size. The standard deviation σ is a measure of the dis-

persion of probabilities around the mean. In other words, σ gives an estimate of the variability in the population. If σ is small, the variability is low
and the probabilities will be tightly clustered around the mean. If σ is large,
the variability is great, and the frequency distribution curve will be wider
and flatter. A few other observations may be made about the normal curve:
Small deviations from the mean are more frequent than large, plus and
minus deviations from the mean are equally frequent, and extremely large
deviations from the mean have a very low probability of being due solely
to chance.

When a curve is symmetrical but not normal in that it is either flatter or
more pointed than the corresponding normal curve, it is said to show
kurtosis. The flatter curves, indicating high variability, are termed **platy-
kurtic**. The sharper peaked curves are called **leptokurtic**.

A curve is said to be *skewed* when the frequency distribution is asymmetrical as compared to the normal curve. A positively skewed curve has a
long tail to the right; a negatively skewed curve has a long tail to the left.
Although it might be thought that the deviations from the normal curve
due to skewness or kurtosis might introduce serious error into calculations
based on the assumption of a normal distribution, it has in fact been shown
that such errors are small enough to be negligible.

The Poisson Distribution

As just shown above, the binomial distribution approaches a normal curve
as n increases, as long as p and q are both larger than $1/n$. However,
when either p or q is of the same magnitude as $1/n$, the limiting curve
approached by the binomial expansion is not the normal distribution but
the Poisson distribution. A Poisson distribution is observed when the probability of an event happening in any one trial is very small, but the number
of trials is so large that the event will nevertheless be observed with some
frequency. A classic example cited in the statistical literature deals with the
chances of soldiers in 10 Prussian army corps being kicked to death by a
mule. The chance of any one soldier being killed this way in any one year
was very small, but over a 20-year period in 10 army corps, 122 soldiers
were killed by mule kicks.

The calculation of the exact binomial probabilities for such situations
is extremely laborious even if shortcuts are attempted. For example, if n
were, say 60,000 and p were 0.0017, the calculation of an exact probability
from the formula

$$\frac{n!}{r!(n-r)!} p^r q^{(n-r)}$$

would be quite a chore. It is at this point that the Poisson distribution comes to the rescue, for the binomial probability is closely approximated by the Poisson probability

$$\frac{e^{-np}(np)^X}{X!} = E\{X\}$$

where e = the constant 2.7183, n = the number of trials, p = the probability of "success," X = the value of the variate, and $E\{X\}$ = the expected value of X.

The terms n and p are not found separately but only as the product np. This product, np, equals the mean for both the Poisson distribution and the binomial distribution. For a binomial distribution, both n and p must be known separately. However, a theoretical Poisson distribution can be determined from the sample mean of an observed distribution even if the values of n and p are unknown. Therefore, the normal distribution involves two parameters, the mean and the standard deviation, and the binomial distribution is determined by the values of both p and n, but the shape of the Poisson distribution is determined by just a single parameter, the mean.

A.4. Measures of Central Tendency

One of the characteristics of biological variation is that in a frequency distribution some classes are more frequent than others, and the frequencies of the other classes diminish or tail off in either direction the more removed they are from the most common types. One important attribute of a distribution is some measure of the central tendency, of the point around which the observations tend to cluster. The most familiar measure of the central tendency is the *arithmetic mean*, which is defined as the sum of all of the observations divided by the number of observations

$$\bar{X} = \frac{\Sigma X}{N}$$

where $\bar{X}$ = the arithmetic mean, X = the value of a single observation of the variate, Σ = the Greek letter sigma, indicating summation of all the values represented by the symbol or symbols following it, and N = the total number of observations.

The arithmetic mean is the most frequently used measure of central tendency; it is readily understood and easy to calculate, with only the number and sum of the observations necessary for computation. (When the number of measurements is large, the calculation of the mean or other statistics may be expedited by special calculating formulas. Reference should be made to statistics books for such simplifying formulas.) The major drawback to the arithmetic mean is that its value may be distorted by a few extreme observations and therefore in such cases it is not a particularly good indicator of central tendency. For example, if the Duke of Shushire had an income equivalent to a million dollars a year, and the nine peasant families living in the same village had incomes equivalent to $4000 a year each, they might consider the statement that the average annual income in the village of Shushire was $103,600 to be a misrepresentation of their circumstances. So might the Duke.

Another common measure of central tendency is the *median*, which is the value of the middle item when the observed values have been arranged according to size. This definition holds strictly only for an odd number of items, but for an even number of items, the midpoint is taken to be the arithmetic mean of the two central items. The important fact about the median is that half the observations lie below it and half lie above it. The median is easier to determine than the mean and is less distorted by extreme values than the mean. (In Shushire the median income is $4000 per year.) In most other respects, however, the median is a much less useful type of average than the arithmetic mean.

A third measure of central tendency is the *mode*, which is the most frequent value in a group of observations. For discrete variables the mode can be determined by inspection; for continuous variables, the data must be grouped before the mode can be determined. As a measure of central tendency, the mean is affected most by extreme observations, the median somewhat less than the mean, while the mode is not affected at all by extreme values. The mode, therefore, is particularly useful as a measure of central tendency in skewed frequency distributions. A sample mode, however, is less satisfactory than the median or the mean because it is so sensitive to sampling fluctuations. Grouping the observed values so that a smoother curve is obtained will minimize the effects of sampling fluctuations on the mode.

Since the mode is the most frequent value observed, it can be regarded as "typical" of the population. It is easily determined by inspection and is insensitive to extreme values. Nevertheless, like the median, it is less useful than the arithmetic mean because it is of limited value for further calculations or in making comparisons of sets of data. It should be pointed out that for a perfectly symmetrical frequency distribution, the mean, the median, and the mode are identical. For skewed distributions the mode remains at the highest point of the distribution, the median is shifted in the direction of skewness to the point where it divides the area under the curve in half, and the mean, affected as it is by extreme values, is shifted further in the direction of skewness than the median.

Still other measures of central tendency have been used in special situations, but they are ordinarily of limited value in the study of biological variation. Among the better known are the *geometric mean*,

$$\bar{X}_G = \sqrt[N]{X_1 X_2 X_3 \cdots X_N}$$

the *quadratic mean*,

$$\bar{X}_Q = \sqrt{\frac{\Sigma X^2}{N}}$$

the *harmonic mean*,

$$\frac{1}{\bar{X}_H} = \frac{1}{N} \sum \frac{1}{X}$$

and the *range midpoint*,

$$\bar{X}_R = \frac{X_{Lo} + X_{Hi}}{2}$$

A.5. Measures of Dispersion

The various kinds of averages given above measure one of the fundamental characteristics of frequency distributions, their tendency to cluster around some central point. This tendency makes possible the use of a single value, the measure of central tendency or average, to represent the accumulated data. The average, however, gives no indication of the range of variability in the frequency distribution. For this purpose some measure of the dispersion of the observations about the mean is necessary.

Range

The simplest measure of dispersion is the range, which is the difference between the lowest and highest values observed. It is ordinarily expressed in terms of the minimum and maximum values observed rather than as the difference between them. Despite its simplicity and ease of determination, the range is a rather poor estimate of dispersion because it is dependent on just two of the observed values and if, by any chance, these are aberrant, the given range may be quite misleading.

Mean Deviation

Another simple measure of dispersion is the mean deviation, which measures the average distance of the observed values from a given point, usually the arithmetic mean. Since deviations from the arithmetic means are both positive and negative and their sum is zero, their signs must be ignored in calculating the mean deviation, for which the formula is

$$MD = \frac{\Sigma |d|}{N}$$

where $|d|$ = the absolute deviation of an observation from the arithmetic mean and N = the number of observations.

The Standard Deviation and the Variance

The most widely used measure of dispersion is the standard deviation and the closely related variable, the variance. The standard deviation measures the average deviation from the mean in a way different from the mean deviation. The essence of the method is to square the difference between each observation and the mean, thus making all the deviations positive. The squared deviations are then summed, divided by the number of observations (or by $N-1$), and the square root of this value is then taken. The standard deviation is therefore the quadratic mean (see above) of the deviations from the arithmetic mean, and is sometimes referred to as the "root-mean-square deviation about the mean." The variance is simply the square of the standard deviation, or the "mean-square deviation about the mean."

At this point we must return to our earlier discussion of populations and samples and of parameters and statistics. The population variance, a parameter, equals

$$\sigma^2 = \frac{\Sigma(X-\mu)^2}{N}$$

and the population standard deviation is therefore

$$\sigma = \sqrt{\frac{\Sigma(X-\mu)^2}{N}}$$

where σ^2 = the variance, σ = the standard deviation, μ = the population mean, a parameter, X = the value of a single observation of the variate, and N =

(here) the total population size. However, the estimates of the variance and the standard deviation based on samples are equal to

$$s^2 = \frac{\Sigma(X - \bar{X})^2}{N - 1}$$

and

$$s = \sqrt{\frac{\Sigma(X - \bar{X})^2}{N - 1}}$$

where s^2 = the variance of the sample, s = the standard deviation of the sample, $\bar{X}$ = the sample mean, and N = the sample size or the number of observations.

The difference between these two sets of formulas, obviously, is that the denominator in the first case is N and in the second, $(N - 1)$. The reason for the use of $(N - 1)$ in the sample estimates is that in the estimation of the mean one degree of freedom is lost so that the appropriate denominator is $N - 1$ rather than N.

The calculation of the variance in the form given above can be quite laborious, and various equivalent but simplifying formulas are available in statistics texts, one of which is

$$s^2 = \frac{\Sigma X^2 - (\Sigma X)^2/N}{N - 1}$$

The advantages of s^2 and of s as measures of dispersion are not immediately obvious. They lie in the usefulness of these statistics in conducting tests of statistical hypotheses. In these estimates of dispersion, every observation enters into the estimate, but the more extreme observations have greater weight because the deviations are squared. Although s^2 is also a measure of dispersion, s is easier to grasp because the standard deviation is given in the same units as the deviations themselves.

When a completely normal distribution curve exists, the standard deviation, σ, is the distance from the peak of the curve (the mean) to the points of greatest slope in the curve (Fig. A.2). There are two such points of inflection in the normal curve, one at $+\sigma$ and one at $-\sigma$. For a normal curve, 68.26% of the area lying under the curve lies between $\pm\sigma$; 95.46% of the area under the curve lies between $\pm2\sigma$; 99.73% of the area lies between $\pm3\sigma$, and 99.994% between $\pm4\sigma$ (Table A.1). Therefore, practically all the values lie within $\pm3\sigma$ of the mean, and one standard deviation covers about one-sixth of the range.

For the Poisson distribution, it turns out that the variance is equal to the mean; that is, $\mu = Np = \sigma^2$. Thus, if the mean is known, the variance is also

Table A.1 Areas and Ordinates of a Standardized Normal Curve

σ	Ordinate	Area
0	0.399	0.0000
±0.50	0.352	0.3829
±1.00	0.242	0.6826
±1.50	0.130	0.8664
±2.00	0.054	0.9546
±2.50	0.018	0.9876
±3.00	0.0044	0.9973
±4.00	0.0001	0.99994

known. For the binomial distribution, $(p + q)^N$, again the mean, $\mu = Np$, but the variance, $\sigma^2 = Npq$. These formulas give theoretical values for the parameters directly rather than estimates for these parameters.

The above measures of dispersion plus such measures of range as quartiles and percentiles are absolute measures of dispersion. However, in many cases it is desirable to compare the degree of variability in traits that cannot be directly compared, either because of differences in the units of measurement or simply because the means of the populations being compared are so different. In such cases, relative measures of dispersion may be used. The most useful relative measure of dispersion is the **coefficient of variation,**

$$V = \frac{s}{\bar{X}} \cdot 100$$

where V = the coefficient of variation, s = the standard deviation, and $\bar{X}$ = the mean. Since both mean and standard deviation are expressed in the same units, whatever they may be, the units cancel out, leaving a pure number which is expressed as a percentage due to the multiplication by 100. Thus, if one wished to compare the variability in wing length and antenna length in a group of insects or the variability in wing length in an insect population with that in a population of birds, the coefficient of variation is the obvious means for making such comparisons. For example, corolla length can be seen to be more variable in the F_2 than in the F_1 from the cross *Nicotiana langsdorfii* × *N. alata* from the following data and from Fig. A.3.

	$\bar{X}$ (mm)	s (mm)	V (%)
F_1	40.78	2.20	5.39
F_2	38.30	5.99	15.64

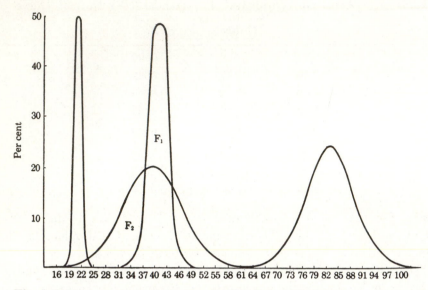

Fig. A.3. Frequency curves for corolla length in *Nicotiana*. Left, *N. Langs-dorffii;* right, *N. Alata;* center, the F_1 and F_2 means are about the same, but the F_2 is more variable than the F_1. (From: H. P. Riley, 1948, *Intro-duction to Genetics and Cytogenetics.* After East.)

A.6. Estimation of Parameters

Thus far, we have been concerned with the description of various frequency distributions and with measures of central tendency and of dispersion for such distributions. Now we shall deal with the estimation of the parameters of a population based on samples drawn from that population.

A parameter has a single fixed value, which can be known only if the entire population can be measured. If this is possible, there is just one value, the true value, for the mean and the standard deviation. Ordinarily, an entire population cannot be studied, either because it is a hypothetical or infinitely large population or because it is so large that only a sample can be obtained. Obviously, the reliability of any estimates based on samples is dependent on the validity of the samples, which must be representative of the population from which they are drawn. The technique of getting adequate, unbiased samples requires careful thought and study. Before undertaking any sampling, you would do well to learn how to avoid possible pitfalls by consulting a statistical text dealing with sampling techniques.

Although a parameter has a single fixed value, this is not true of the estimates of parameters based on samples. Different samples will ordinarily give somewhat different estimates of the mean or the standard deviation or whatever parameter is being estimated. Just as a series of observations has a frequency distribution about the true mean of a population, the means of a series of samples from a population will also have a frequency distribution about the true mean of the population. This distribution of the sample means will be different from the distribution of the individual observations, and is of considerable importance because the precision of the estimate of the true mean can be calculated from it. The variance of a series of observations with a normal distribution has been given as s^2; the variance of the distribution of means is equal to s^2/N. The standard deviation of the distribution of the sample mean, $\bar{X}$, equals $s/\sqrt{N}$, and is usually called the **standard error of the mean**, with the symbol $s_{\bar{x}}$. When a mean is given, it is usually accompanied by its standard error, thus $\bar{X} \pm s_{\bar{x}}$, rather than the standard deviation. If N is given, however, it is a simple matter of calculate the standard deviation from the relation $s_{\bar{x}} = s/\sqrt{N}$.

Thus, we have the following symbols:

μ = true mean of a population, a parameter
$\bar{X}$ = a sample mean, an estimate of μ
σ^2 = the population variance
s^2 = the variance of a sample, an estimate of σ^2
σ = the true standard deviation of a population
s = a sample estimate of σ
$s_{\bar{x}}$ = standard error of the mean, equal to $s/\sqrt{N}$.

The importance of the estimates based on samples is the extent to which they tell us something about the parameters for the population. Even though it may be impossible ever to know the true values of the population parameters, it is possible from the sample estimates to determine the limits within which the true values lie. These limits are called **confidence limits**, and the range or interval within which the true value is thought to lie is called the **confidence interval**. In statistics, it is just as important to set confidence limits on an estimate, that is, to make some sort of probability statement about its range, as it is to make the estimate itself. A mean without a standard error conveys no information about how far from the true mean the estimated mean may be.

The standard error provides a method for setting confidence limits for the mean. An important point to note about it is that the larger N becomes, the smaller $s_{\bar{x}}$ becomes. Thus, the larger the sample, the better the estimate of the true mean. The confidence interval is really a probability statement. If, for example, the 95% confidence interval is given, this means that if we assume that the true mean lies within this interval, we will be right 95% of the time. Obviously, if we make the interval wider by using 99 or 99.9% confidence limits, we shall improve our chances of being right, but the information gained is less useful. Setting confidence limits is a matter of judgment, and the appropriate limits may differ under different circumstances. By convention, and apparently for no other good reason, 95% and 99% confidence limits are the ones ordinarily chosen.

The standard error can be used to set confidence limits if the estimated value to which it is attached is approximately normal in distribution. However, the concept of confidence limits can be used with a number of other types of distribution as well as the normal distribution. A point to note is that even if the distribution of the observed values in the samples is not normal, the distribution of the means based on these observations tends to be normal, especially if there are many observations. Therefore, even though the frequency distribution of the original data is other than normal, it may still be possible to use the standard error to interpret the data.

A.7. Statistical Inference

The use of statistics to estimate parameters leads us to the concept of statistical inference. Just as it is possible to set a confidence interval with a certain probability that the true value lies within this interval, it is also possible to test a particular hypothesis and make a probability statement about whether the hypothesis should be accepted or rejected. When a hypothesis is formulated about a population, it then becomes possible to compare the observed values with the expected values based on the hypothesis. The observations may differ from the expected values either because of chance sampling errors in the finite sample or because the hypothesis itself is not true. Statistical tests of significance are designed to distinguish between these two possibilities. The procedure for all tests of significance is essentially the same. The first step in testing a hypothesis is the choice of the proper hypothesis to test, again a decision dependent on the judgment of the experimenter. Next, based on this hypothesis, the probability distribution for various possible observed values must be determined. Then the actual observed values are compared with this probability distribution to find the probability of getting a fit with the hypothesis as bad or worse due solely to chance sampling error. If this probability is below an acceptable level of significance, it is assumed that the hypothesis is inadequate to account for the data, this is, that the observed deviations are too large to be due solely to chance, and that there is a significant difference between the observations and the hypothesis.

In tests of hypotheses, there are two possible types of error, one that we reject a hypothesis when in fact it is true, the other that we accept a hypothesis when in fact it is false. We have already discussed (Chapter 5), in relation to χ^2, the choice of levels of significance (a) that lead us to reject hypotheses, noting that the more serious the consequences of wrongly rejecting the hypothesis, the smaller a should be. For instance, if $a = 0.05$, there is one chance in 20 that an observed deviation from theory will be greater than expected and yet due solely to chance. Therefore, there is one chance in 20 that a correct hypothesis will be wrongly rejected. If $a = 0.01$, this probability is reduced to one in 100.

The second type of error, known as β, is the chance of accepting as true a false hypothesis. Obviously, a and β are related, because if a is made smaller, β will increase. For a fixed sample size, we can choose the level of significance (a) as we wish, but β is then fixed. The only way we can reduce both a and β is to increase the size of the sample by making more observations. In most statistical work, the β type of error is not considered, but one should be aware that this possibility exists.

The form of most tests of significance is essentially a test of what is called the *null hypothesis*: a test to show that the observations do not differ significantly from expectations based on theory, but are due to chance variations. If the test is significant, say that $p = 0.001$ or less (i.e., that the probability of getting this result due to chance alone is one in 1000 or less), then either the null hypothesis is wrong or else the null hypothesis is true and an extremely improbable event has occurred. If p were 0.001, the null hypothesis would undoubtedly be rejected and a better explanation sought for the observations. On the other hand, if $p = 0.4$, such results could be expected due to chance alone 40% of the time, so that they are not uncommon or unexpected under the given hypothesis. However, they cannot be considered to supply definitive proof to the hypothesis in the sense that a mathematical proof can be given. Instead, they merely show that a particular set of observations do not differ significantly from expectations based on the hypothesis.

A.8. Tests of Significance

A variety of tests of significance is used in statistics. We shall now consider a few of the more useful and representative types to gain some insight into their use.

Comparison of an Individual with a Population

The Normal Deviate. Let us suppose that we have discovered a single individual and wish to know whether he is a member of a particular population. Let us further suppose that the population is normally distributed and that the true mean (μ) and the standard deviation (σ) are known. It is then possible to express the deviation of this individual from the population mean, not in terms of the units of measurement used, but in terms of standard deviation units as follows:

$$d = \frac{X - \mu}{\sigma}$$

where X = the observed value for the individual, μ = the population mean, σ = the standard deviation of the population, and d = the normal deviate in standard units. We have already seen that all normal curves are dependent on two parameters, μ and σ, and since these values are not fixed, there is an infinite possible number of normal curves. The standardized normal deviate given above reduces all these normal distributions to a single normal distribution with $\mu = 0$ and $\sigma = 1$. Thus, for any normal curve, a given deviation has a probability dependent on d, the normal deviate, and tables have been prepared relating this probability to the value of d (Table A.2). We have already seen that 95.46% of the area under the normal curve lies between $\pm 2\sigma$; in other words, 4.5% (or close to 5%) of the area lies beyond $\pm 2\sigma$ (Table A.1). At the 5% level of significance customarily used, the value of the normal deviate is 1.960 rather than 2. It should be noted that this test of significance is a ***two-tailed*** (or "two-sided") test of significance, in that 2.5% of the area under the curve will lie beyond $+1.96d$ and 2.5% of the area will lie below $-1.96d$ (Table A.3). Therefore, the probability given in Table A.2 indicates the chances of departure from the mean in either direction. If only the probability of a positive deviation were of interest, its value would be half as great as that given in Table A.2 and is more readily obtained from Table A.3. Such a test is called a ***one-tailed*** (or "one-sided") test of significance.

The t Test. The normal deviate has been given as the ratio of the deviation from the mean to the standard deviation, σ. However, the value of σ is often unknown. In such cases, the estimated standard deviation, s, can be used in place of σ. When s is used, a new distribution is obtained, known as "Student's t" distribution. In this case

$$t = \frac{X - \bar{X}}{s}$$

Table A.2   Table of Normal Deviates

d	Probability [a]
0.13	0.90
0.39	0.70
0.67	0.50
1.04	0.30
1.28	0.20
1.64	0.10
1.96	0.05
2.33	0.02
2.58	0.01
3.29	0.001

[a] In this table the probabilities represent the areas of the normal curve lying beyond the given value of d in both plus and minus directions.

Table A.2 is taken from Table II of Fisher and Yates: *Statistical Tables for Biological, Agricultural and Medical Research*, published by Longman Group Ltd., London (previously published by Oliver & Boyd, Edinburgh), and by permission of the authors and publishers.

As the sample size increases, the sampling error of s decreases so that the distribution of t approaches the distribution of the normal deviate as a limit for infinitely large samples. For practical purposes, as long as the number of degrees of freedom is 30 or more (in this case, the degrees of freedom equal $N-1$, with N the sample size), s can be substituted for σ and the table of normal deviates can be consulted with no serious inaccuracy resulting. However, if s is based on a small sample, its use in the calculation of a

Table A.3   Cumulative Normal Frequency Distribution of a Standardized Normal Curve

σ	Area [a]	σ	Area [a]
−3.090	0.0010	+3.090	0.9990
−3.000	0.0013	+3.000	0.9987
−2.326	0.0100	+2.326	0.9900
−2.000	0.0228	+2.000	0.9772
−1.960	0.0250	+1.960	0.9750
−1.645	0.0500	+1.645	0.9500
−1.000	0.1587	+1.000	0.8413
0.000	0.5000	0.000	0.5000

[a] In this table, the area given is the area under the normal frequency curve from minus infinity to the given value of σ.

normal deviate could lead to serious error. In such cases, the t distribution must be used. This table is similar to the table of normal deviates except that it is two-dimensional and includes the number of degrees of freedom. The table must be entered not only at the appropriate probability, but also at the proper number of degrees of freedom (Table A.4).

Table A.4 Table of t

N	0.90	0.80	0.70	0.50	0.30	Probability 0.20	0.10	0.05	0.02	0.01	0.001
1	0.16	0.33	0.51	1.00	1.96	3.08	6.31	12.71	31.82	63.66	636.62
2	0.14	0.29	0.45	0.82	1.39	1.89	2.92	4.30	6.97	9.93	31.60
3	0.14	0.28	0.42	0.77	1.25	1.64	2.35	3.18	4.54	5.84	12.94
4	0.13	0.27	0.41	0.74	1.19	1.53	2.13	2.78	3.75	4.60	8.61
5	0.13	0.27	0.41	0.73	1.16	1.48	2.02	2.57	3.37	4.03	6.86
6	0.13	0.27	0.40	0.72	1.13	1.44	1.94	2.45	3.14	3.71	5.96
7	0.13	0.26	0.40	0.71	1.12	1.42	1.90	2.37	3.00	3.50	5.41
8	0.13	0.26	0.40	0.71	1.11	1.40	1.86	2.31	2.90	3.36	5.04
9	0.13	0.26	0.40	0.70	1.10	1.38	1.83	2.26	2.82	3.25	4.78
10	0.13	0.26	0.40	0.70	1.09	1.37	1.81	2.23	2.76	3.17	4.59
11	0.13	0.26	0.40	0.70	1.09	1.36	1.80	2.20	2.72	3.11	4.44
12	0.13	0.26	0.40	0.70	1.08	1.36	1.78	2.18	2.68	3.06	4.32
13	0.13	0.26	0.39	0.69	1.08	1.35	1.77	2.16	2.65	3.01	4.22
14	0.13	0.26	0.39	0.69	1.08	1.35	1.76	2.15	2.62	2.98	4.14
15	0.13	0.26	0.39	0.69	1.07	1.34	1.75	2.13	2.60	2.95	4.07
16	0.13	0.26	0.39	0.69	1.07	1.34	1.75	2.12	2.58	2.92	4.02
17	0.13	0.26	0.39	0.69	1.07	1.33	1.74	2.11	2.57	2.90	3.97
18	0.13	0.26	0.39	0.69	1.07	1.33	1.73	2.10	2.55	2.88	3.92
19	0.13	0.26	0.39	0.69	1.07	1.33	1.73	2.09	2.54	2.86	3.88
20	0.13	0.26	0.39	0.69	1.06	1.33	1.73	2.09	2.53	2.85	3.85
22	0.13	0.26	0.39	0.69	1.06	1.32	1.72	2.07	2.51	2.82	3.79
24	0.13	0.26	0.39	0.69	1.06	1.32	1.71	2.06	2.49	2.80	3.75
26	0.13	0.26	0.39	0.68	1.06	1.32	1.71	2.06	2.48	2.78	3.71
28	0.13	0.26	0.39	0.68	1.06	1.31	1.70	2.05	2.47	2.76	3.67
30 [a]	0.13	0.26	0.39	0.68	1.06	1.31	1.70	2.04	2.46	2.75	3.65

[a] When N is greater than 30, t may be treated as a normal deviate without serious inaccuracy resulting.

Table A.4 is taken from Table III of Fisher and Yates: *Statistical Tables for Biological, Agricultural and Medical Research*, published by Longman Group Ltd., London (previously published by Oliver & Boyd, Edinburgh), and by permission of the authors and publishers.

It should be noted that the t test is an exact test of significance based only on the observational data and involves no assumptions about any unknown parameter. However, it does assume that the observations are normally distributed. No actual population ever has a perfectly normal distribution, but the t test has been shown to be relatively insensitive to departures from normality.

Nevertheless, it may be advisable to test whether the variances of the two populations differ significantly. For this purpose, the variance-ratio or F test is made. F is simply the ratio of the larger to the smaller of the two variance estimates being compared, and thus can never be less than one. Therefore, if s_1^2 is greater than s_2^2,

$$F = \frac{s_1^2}{s_2^2}$$

The significance of the value of F must then be determined from tables of the distribution of F values (Table A.5). These tables are three-dimensional in the sense that there are three variables used to enter the table. One is the degrees of freedom for s_1^2, the second the degrees of freedom for s_2^2, and the third the p value chosen as the level of significance. Unlike many other F tests, which are one-sided, this F test is two-tailed or two-sided, and p values must be selected accordingly.

Comparison of a Sample with a Population

Just as we have compared an individual with a population to determine whether the individual can properly be considered a member of the population, we can also compare a sample of individuals with a population to determine whether this sample belongs to the population. The procedure is similar to that used above, except that the standard error of the mean is used rather than the standard deviation to estimate the normal deviate or alternatively to calculate t. If we knew the true mean and its standard error for the population, the formula would take the form

$$d = \frac{\bar{X} - \mu}{\sigma_{\bar{X}}}$$

and d would again be treated as a normal deviate with zero mean and unit standard deviation. The assumption of zero means signifies that the average difference between $\bar{X}$ and μ is expected to equal zero, in other words, that the two means are the same. Hence, the null hypothesis that is being tested is $\bar{X} = \mu$.

However, frequently the true mean and its standard error will not be

Table A.5(a)   Table of the Variance Ratio, F

0.10 Probability (two-sided test)
0.05 Probability (one-sided test)

N_2 \ N_1	1	2	3	4	5	6	12	24	∞
1	161.4	199.5	215.7	224.6	230.2	234.0	243.9	249.0	254.3
2	18.5	19.0	19.2	19.3	19.3	19.3	19.4	19.5	19.5
3	10.1	9.6	9.3	9.1	9.0	8.9	8.7	8.6	8.5
4	7.7	6.9	6.6	6.4	6.3	6.2	5.9	5.8	5.6
5	6.6	5.8	5.4	5.2	5.1	5.0	4.7	4.5	4.4
6	6.0	5.1	4.8	4.5	4.4	4.3	4.0	3.8	3.7
7	5.6	4.7	4.4	4.1	4.0	3.9	3.6	3.4	3.2
8	5.3	4.5	4.1	3.8	3.7	3.6	3.3	3.1	2.9
9	5.1	4.3	3.9	3.6	3.5	3.4	3.1	2.9	2.7
10	5.0	4.1	3.7	3.5	3.3	3.2	2.9	2.7	2.5
11	4.8	4.0	3.6	3.4	3.2	3.1	2.8	2.6	2.4
12	4.8	3.9	3.5	3.3	3.1	3.0	2.7	2.5	2.3
13	4.7	3.8	3.4	3.2	3.0	2.9	2.6	2.4	2.2
14	4.6	3.7	3.3	3.1	3.0	2.9	2.5	2.3	2.1
15	4.5	3.7	3.3	3.1	2.9	2.8	2.5	2.3	2.1
16	4.5	3.6	3.2	3.0	2.9	2.7	2.4	2.2	2.0
17	4.5	3.6	3.2	3.0	2.8	2.7	2.4	2.2	2.0
18	4.4	3.6	3.2	2.9	2.8	2.7	2.3	2.1	1.9
19	4.4	3.5	3.1	2.9	2.7	2.6	2.3	2.1	1.9
20	4.4	3.5	3.1	2.9	2.7	2.6	2.3	2.1	1.8
22	4.3	3.4	3.1	2.8	2.7	2.6	2.2	2.0	1.8
24	4.3	3.4	3.0	2.8	2.6	2.5	2.2	2.0	1.7
26	4.2	3.4	3.0	2.7	2.6	2.5	2.2	2.0	1.7
28	4.2	3.3	3.0	2.7	2.6	2.4	2.1	1.9	1.7
30	4.2	3.3	2.9	2.7	2.5	2.4	2.1	1.9	1.6
60	4.0	3.2	2.8	2.5	2.4	2.3	1.9	1.7	1.4
120	3.9	3.1	2.7	2.5	2.3	2.2	1.8	1.6	1.3
∞	3.8	3.0	2.6	2.4	2.2	2.1	1.8	1.5	1.0

Table A.5(b)   Table of the Variance Ratio, F

0.02 Probability (two-sided test)
0.01 Probability (one-sided test)

N_2 \ N_1	1	2	3	4	5	6	12	24	∞
1	4,052	4,999	5,403	5,625	5,764	5,859	6,106	6,234	6,366
2	98.5	99.0	99.2	99.3	99.3	99.3	99.4	99.5	99.5
3	34.1	30.8	29.5	28.7	28.2	27.9	27.1	26.6	26.1
4	21.2	18.0	16.7	16.0	15.5	15.2	14.4	13.9	13.5
5	16.3	13.3	12.1	11.4	11.0	10.7	9.9	9.5	9.0
6	13.7	10.9	9.8	9.2	8.8	8.5	7.7	7.3	6.9
7	12.3	9.6	8.5	7.9	7.5	7.2	6.5	6.1	5.7
8	11.3	8.7	7.6	7.0	6.6	6.4	5.7	5.3	4.9
9	10.6	8.0	7.0	6.4	6.1	5.8	5.1	4.7	4.3
10	10.0	7.6	6.6	6.0	5.6	5.4	4.7	4.3	3.9
11	9.7	7.2	6.2	5.7	5.3	5.1	4.4	4.0	3.6
12	9.3	6.9	6.0	5.4	5.1	4.8	4.2	3.8	3.4
13	9.1	6.7	5.7	5.2	4.9	4.6	4.0	3.6	3.2
14	8.9	6.5	5.6	5.0	4.7	4.5	3.8	3.4	3.0
15	8.7	6.4	5.4	4.9	4.6	4.3	3.7	3.3	2.9
16	8.5	6.2	5.3	4.8	4.4	4.2	3.6	3.2	2.8
17	8.4	6.1	5.2	4.7	4.3	4.1	3.5	3.1	2.7
18	8.3	6.0	5.1	4.6	4.3	4.0	3.4	3.0	2.6
19	8.2	5.9	5.0	4.5	4.2	3.9	3.3	2.9	2.5
20	8.1	5.9	4.9	4.4	4.1	3.9	3.2	2.9	2.4
22	7.9	5.7	4.8	4.3	4.0	3.8	3.1	2.8	2.3
24	7.8	5.6	4.7	4.2	3.9	3.7	3.0	2.7	2.2
26	7.7	5.5	4.6	4.1	3.8	3.6	3.0	2.6	2.1
28	7.6	5.5	4.6	4.1	3.8	3.5	2.9	2.5	2.1
30	7.6	5.4	4.5	4.0	3.7	3.5	2.8	2.5	2.0
60	7.1	5.0	4.1	3.7	3.3	3.1	2.5	2.1	1.6
120	6.9	4.8	4.0	3.5	3.2	3.0	2.3	2.0	1.4
∞	6.6	4.6	3.8	3.3	3.0	2.8	2.2	1.8	1.0

Table A.5 (a & b) is taken from Table V of Fisher and Yates: *Statistical Tables for Biological, Agricultural and Medical Research,* published by Longman Group Ltd., London (previously published by Oliver & Boyd, Edinburgh), and by permission of the authors and publishers.

known, and the comparison then must really be between two samples. The test of the null hypothesis for the difference between means for large samples (degrees of freedom both larger than 30) is then based on the formula

$$d = \frac{\bar{X}_1 - \bar{X}_2}{\sqrt{(s_1^2/N_1) + (s_2^2/N_2)}} = \frac{\bar{X}_1 - \bar{X}_2}{\sqrt{s_{\bar{x}_1}^2 + s_{\bar{x}_2}^2}}$$

The more exact t test for the significance of the difference between means when sample sizes are small is given by the equation.

$$t = \frac{\bar{X}_1 - \bar{X}_2}{s\sqrt{(1/N_1) + (1/N_2)}}$$

In the two equations above, the symbols have the same meaning as before. The subscripts 1 and 2 indicate which sample gave rise to the estimated values. In the formula for t, the s is an estimate of the standard deviation based on both samples. However, it is not obtained simply by pooling all of the data. Instead the sums of squares of deviations from the means must be calculated separately and then added together and divided by the total number of degrees of freedom (i.e., $N_1 - 1 + N_2 - 1 = N_1 + N_2 - 2$). When the t test is used to compare means, the variances of the two populations being compared are assumed to be equal. Even though this may not be true, the test has been found to be little affected by differences in the variances. In this case

$$s^2 = \frac{\Sigma(X_1 - \bar{X}_1)^2 + \Sigma(X_2 - \bar{X}_2)^2}{N_1 + N_2 - 2}$$

You may have noticed in these equations dealing with differences between means that we are adding rather than subtracting the variances. Let us see why this should be done.

Very briefly, suppose that we have equal numbers of normally distributed independent measurements from two different populations. If the measurements are added $(X_1 + X_2)$, the mean $\bar{X}_{1+2}$ will equal $\bar{X}_1 + \bar{X}_2$ and the variance s^2_{1+2} will equal $s_1^2 + s_2^2$. If the measurements are subtracted $(X_1 - X_2)$, the mean $\bar{X}_{1-2}$ will equal $\bar{X}_1 - \bar{X}_2$, but the variance s^2_{1-2} will still equal $s_1^2 + s_2^2$. The reason for this is that the variance is based on random deviations from the mean in both negative and positive directions. Subtracting X_2 instead of adding it to X_1 may change the sign of these deviations, but not their overall magnitude. Thus the amount of variability about the mean remains the same, whether this mean is $\bar{X}_{1+2}$ or $\bar{X}_{1-2}$.

If X_1 and X_2 are independent and normally distributed, then $X_1 + X_2$ and $X_1 - X_2$ will also be normally distributed and so will the difference between the means, $\bar{X}_1 - \bar{X}_2$. Just as a given mean has a variance and a standard error of the mean, so does a difference between means. We have seen that the variance for the difference between means equals $s_1^2 + s_2^2$. From this it is easy to see that the standard error for the difference between means should take the form

$$s^2 \bar{x}_1 - \bar{x}_2 = \frac{s_1^2}{N_1} + \frac{s_2^2}{N_2}$$

which was used as the denominator for the calculation of the normal deviate d for the difference between means in large samples.

We were careful to postulate that the observations X_1 and X_2 were independent of one another. If, for some reason, they are related—for example, if when a given X_1 is large, the corresponding X_2 also tends to be large—then the observations would not be independent. In such cases, studies of covariance, correlation, or regression may be necessary.

Confidence Limits and Tests of Significance for Binomial and Poisson Distributions

It was pointed out earlier that the distribution of means may be approximately normal even when the measurements on which the means are based have some other type of frequency distribution. In such cases, it is usually necessary that the samples be large, but when they are, tests of significance based on the normal distribution can be used for data with a binomial or Poisson distribution.

For example, if N is greater than 30, confidence limits for binomial distributions can be determined from the estimated mean plus its standard error presented as

$$p_0 \pm \sqrt{\frac{p_0 q_0}{N}}$$

where in this case $p_0 = $ the observed ratio m/N, $m = $ the observed number of successes, $N = $ the number of trials or sample size, and $q_0 = 1 - p_0$.

From the table of normal deviates, the appropriate value of d for obtaining 95% confidence limits is 1.96. Thus the confidence limits range from $p - 1.96 \sqrt{pq/N}$ to $p + 1.96 \sqrt{pq/N}$.

Furthermore, if we wished, for example, to compare some observed ratio with the theoretical ratio expected in a cross, the value of d in relation to the difference between observed and expected can be calculated as follows:

$$d = \frac{p_0 - p}{\sqrt{pq/N}} = \frac{m - Np}{\sqrt{Npq}}$$

where p = the expected proportion of successes, q = the expected proportion of failures, and p_0 = the observed proportion of successes.

Again, if N is greater than 30, the significance of d can be determined from a table of normal deviates. In the second, equivalent form of the equation given above, numerical values rather than proportions are used.

If, however, we simply wished to test for a significant difference between ratios in two large samples, then the equation for d would take the form

$$d = \frac{p_1 - p_2}{\sqrt{pq[(1/N_1) + (1/N_2)]}}$$

In this case $p_1 = m_1/N_1$, $p_2 = m_2/N_2$, and $p = (m_1 + m_2)/(N_1 + N_2)$.

As always, d is a normally distributed variable with zero mean and unit standard deviation. For this test to be appropriate neither p_1 nor p_2 should be too close to zero or one.

For the Poisson distribution, the variance equals the mean. Therefore, the mean and its standard error are estimated as

$$\bar{X} \pm \sqrt{\frac{\bar{X}}{N}}$$

If the product $N\bar{X}$ is larger than 30, the distribution of $\bar{X}$ is approximately normal, and confidence limits can be set by using the appropriate value of d from the table.

If we wish to compare two Poisson distributions, we can use the relation

$$d = \frac{\bar{X}_1 - \bar{X}_2}{\sqrt{\bar{X}_1/N_1 + \bar{X}_2/N_2}}$$

and again look up d in the table of normal deviates to find if the difference is significant. This test is appropriate if both $N_1\bar{X}_1$ and $N_2\bar{X}_2$ are greater than 30.

We have already discussed the use of the normal deviate and the t test to determine the significance of a single deviation or difference. We have also had occasion to mention the chi-square (χ^2) test and the variance ratio, F. χ^2 and the variance ratio can be regarded as more general tests of the joint significance of a number of deviations. Thus, d and t are special cases of the more general χ^2 and F tests.

In the t test there is a single difference in the numerator, while the denominator has a number of degrees of freedom dependent on the estimate of s used. The F test or variance ratio is a generalization of the t test in that estimated variances (or, in some instances, standard deviations) are used in both the numerator and the denominator, and the degrees of freedom are determined by these two estimates. It should be noted that it is the magnitude of the variances, not of the degrees of freedom, that determines which variance is placed in the numerator of a variance ratio. The values of F have a three-dimensional probability distribution, the dimensions being the probability and the two different degrees of freedom. Tables of F are usually presented in two dimensions, each table with a different fixed probability. Since the value of F is dependent on two observed variances, the F test, like the t test, gives an exact test of significance based on observed data, and is independent of the parameters for these variances.

χ^2 may be regarded as a special case of the variance ratio, in which the denominator is fixed by hypothesis. Thus, χ^2 gives the probability distribution of the ratio of an observed variance (actually a sum of squares) to a variance fixed by hypothesis. χ^2 can be thought of as a generalized form of the normal deviate, d, for which the denominator is also fixed by hypothesis. (In reality, χ^2 is a generalized d^2, and F is a generalized t^2.) The degrees of freedom for χ^2 are derived from the numerator, in contrast to t where the degrees of freedom refer to the denominator.

Since all deviations are squared to estimate the variance, the lowest possible value for s^2 is 0, and the probability distribution of χ^2 does not go below zero. The distribution of χ^2 is asymmetrical and quite different from a normal distribution, which is symmetrical in both positive and negative directions. However, this asymmetrical χ^2 distribution approaches a normal curve as the number of degrees of freedom increases. χ^2 tables give degrees of freedom only as high as 30. Above 30, a normal approximation for large samples can be calculated from the formula $\sqrt{2\chi^2} - \sqrt{2f - 1}$, where f is the number of degrees of freedom. The value so obtained can be treated as a normal deviate, d, with zero mean and unit standard deviation, and can be looked up in the appropriate table, using the rules for a one-sided test of significance.

Therefore, the tests of significance, d, t, χ^2, and F all turn out to be interrelated. Although not all in the same form, they can all be regarded as essentially the ratio of two variances, and d, t, and χ^2 can be considered as special cases of F. The choice as to which of these probability distributions to use in a particular case is dependent on the nature of the data.

A.9. The Chi-Square Distribution

We have already seen how χ^2 is used to test the goodness-of-fit of observed data to expected Mendelian ratios (Chapter 6 and Table 6.1). χ^2 can be used in several other ways. For example, it can be used to set confidence limits on a variance. In fact, the χ^2 distribution is based on the distribution of s^2 when the observed values have a normal distribution.

However, despite the fact that the distribution of χ^2 is continuous, it is very often used in dealing with discontinuous data. This is acceptable because, as long as the expected numbers in each class are greater than 5, the χ^2 test gives a fairly good approximation of the probabilities. If the expected numbers are small, the exact probabilities can be calculated from a binomial or multinomial distribution, but this procedure is so cumbersome that it is customary to use Yates' correction for continuity. This correction is made before χ^2 is calculated by changing the observed numbers in each class in such a way that the observed deviation from expectation is reduced by 0.5. Thus some values will be increased and others decreased by 0.5. Yates' correction obviously reduces the size of χ^2, and is used because in small samples there is a consistent tendency to get a large χ^2 and a low p value, and hence to reject the null hypothesis even when it probably should be accepted. As a general rule, if every class has more than 10 observations, an uncorrected χ^2 is adequate, but if any class has an expected value of 5 or less, χ^2 should not be used at all. Some statisticians make Yates' correction routinely, but it obviously has little effect when numbers in each class are large.

A.10. Contingency Tables

Data from observations or experiments are essentially of two types, measurements or frequencies. Measurements are obtained for continuous traits; frequencies are obtained for discontinuous traits when the frequency of individuals in different classes is counted. Most of our discussion thus far in the appendix has dealt with statistical methods for handling measurements. However, the χ^2 test for the goodness-of-fit of observed data to Mendelian ratios used frequencies rather than measurements despite the fact that the χ^2 distribution itself is continuous. The χ^2 test can be used to handle many other types of frequency data. It is particularly useful in relation to contingency or association tables.

In the simplest contingency table, the 2×2 or fourfold table, the data are classified with respect to two variables, J and K. Each individual can then be classed as J or not J, K or not K, and will fall into one of four categories:

	K	not K	Totals
J	a	b	$a+b$
not J	c	d	$c+d$
Totals	$a+c$	$b+d$	$N = a+b+c+d$

The purpose of the χ^2 test in this case is to find whether there is a significant association between the variables or whether they are in fact independent. If there is an association, observations will tend to be concentrated in one or more categories more often than would be expected by chance alone. χ^2 is used to measure the magnitude of the deviations from expectations based on chance.

In this case the expected values for each class may be obtained as follows:

$$\frac{a+b}{N} = \text{observed frequency of } J$$

$$\frac{a+c}{N} = \text{observed frequency of } K$$

Then on the null hypothesis that there is no association, the expected frequency of category a is the product of the two independent frequencies,

$$\left(\frac{a+b}{N}\right)\left(\frac{a+c}{N}\right)$$

and the number of individuals expected in class a, which we shall call A, will equal

$$A = N\left(\frac{a+b}{N}\right)\left(\frac{a+c}{N}\right) = \frac{(a+b)(a+c)}{N}$$

Similarly,

$$B = \frac{(a+b)(b+d)}{N}$$

$$C = \frac{(c+d)(a+c)}{N}$$

$$D = \frac{(c+d)(b+d)}{N}$$

After the expected values are obtained, χ^2 can be calculated from the usual equation

$$\chi^2 = \sum \frac{(O-E)^2}{E}$$

The above method outlines the steps in arriving at a χ^2 value and gives some insight into the rationale behind the use of χ^2. However, short-cut, completely equivalent equations are available to simplify the calculation of χ^2. One such formula is

$$\chi^2 = \frac{N(ad-bc)^2}{(a+b)(c+d)(a+c)(b+d)}$$

x^2 should not be used if any expected number is 5 or less, but Yates' correction can be used if any numbers seem small. In essence, 0.5 is subtracted from each of the two classes giving the larger cross product (i.e., a and d or b and c) and added to each of the two classes giving the smaller cross product. The correction is made automatically with the equation

$$\chi^2 = \frac{N(|ad - bc| - \tfrac{1}{2}N)^2}{(a+b)(c+d)(a+c)(b+d)}$$

Here the vertical lines indicate that the absolute (positive) value of the difference between cross products is to be used, and its size reduced by subtracting the correction, $\tfrac{1}{2}N$.

An even simpler uncorrected formula for calculating x^2 for a 2×2 contingency table is the following: Let

$$E = \frac{a}{a+b} - \frac{c}{c+d}$$

$$F = \frac{a}{a+c} - \frac{b}{b+d}$$

Then

$$\chi^2 = EFN$$

The degree of freedom for the x^2 from a 2×2 contingency table is always just one.

A.11. Degrees of Freedom

We have referred to degrees of freedom previously, but it is now desirable to examine the determination of appropriate degrees of freedom more carefully. In the earlier discussion of x^2 in tests for goodness-of-fit of Mendelian ratios, we gave the number of degrees of freedom as $N - 1$, where N is the number of classes of data. Just now, for the 2×2 table, it was stated that there is just one degree of freedom. Since there are four classes of data in a 2×2 table, this statement seems to contradict the previous rule. However, in this case, the number of degrees of freedom equals the number of classes to which values may be assigned arbitrarily. In a fourfold table the total number of observations and the subtotals for the rows and columns are fixed so that as soon as a number is arbitrarily assigned to one class, the numbers in all the other classes are fixed since they must add up to the proper subtotals. Hence, there is but one degree of freedom.

In a more general sense, if a fixed number of observations is classified into N different groups, then there are $N - 1$ degrees of freedom. The reason is that $N - 1$ groups can be assigned numbers of observations arbitrarily, but the number in the last group is then fixed by the requirement that the total number of all observations is fixed. For every added independent restriction or constraint, another degree of freedom is lost.

The fourfold table is a special case of the more generalized contingency table, which may have any number of rows (r) and columns (c). In such a table the number of groups equals $r \times c$. Since the marginal subtotals are fixed, this introduces $r + c$ restrictions. However, these restrictions are not all independent, because the total for the rows must equal the total for the columns so that one degree of freedom is regained. Therefore the independent restrictions equal $r + c - 1$ and the degrees of freedom are

$$df = rc - (r + c - 1) = (r - 1)(c - 1)$$

Hence, for the 2×2 table, $df = (2 - 1)(2 - 1) = 1 \times 1 = 1$; for a 5×3 table, $df = (5 - 1)(3 - 1) = 4 \times 2 = 8$.

Degrees of freedom have entered into our discussion of other statistical tests in addition to χ^2. In general, for these other cases, a degree of freedom is lost every time a parameter is estimated from the data. For example, if some observed distribution is to be fitted to a theoretical distribution (Poisson, normal, etc.) for which the parameters are fixed by hypothesis rather than estimated, then the number of degrees of freedom is one less than the number of classes. In this case, the only constant used is the sample size, which is needed to convert theoretical frequencies into expected numbers. If, in addition, the true mean and standard deviation were estimated from the data, two more degrees of freedom would be lost. Obviously, care must be taken to determine the proper number of degrees of freedom because the correct probability for a given value of χ^2 (or t or F) can be obtained only if the tables are entered with the appropriate number of degrees of freedom.

A.12. Tests of Homogeneity

In many cases two or more sets of data are obtained that are thought to be replicates of the same experiment, or samples from the same population. If the data are essentially homogeneous, they can be pooled before further analysis, but if they are heterogeneous, they will have to be treated separately. Such tests of homogeneity are no more than a special case of the association or contingency table with one of the variables the source of the data (which may be different experiments, different geographic areas, etc.) If χ^2 turns out not to be significant, this result indicates that the samples are drawn from the same population and that the data can be pooled.

Two final points should be noted about the χ^2 distribution. The first is that separately calculated values of χ^2 can be added, and the sum of these values has a χ^2 distribution with the number of degrees of freedom equal to the sum of the separate degrees of freedom. The other point is that we have concentrated on large values of χ^2 as indicating significant deviations from expectations. You should also be aware that extremely small values of χ^2 also have a low probability, and if you get too many small χ^2 values, you should begin to look for some sort of bias that is giving you too good a fit of your data to your hypothesis. Such considerations have led to the suggestion that Mendel's data were a little too good to be true and may represent selected data to illustrate the 3:1 and 9:3:3:1 ratios.

A.13. Correlation and Regression

The association or contingency tables test for a relationship between two or more kinds of observations. They determine whether observations in one category of one set of data occur more frequently in association with observations in some category of the other set of data than can be expected by chance alone. They are most useful for situations where the data fall into just a few categories, either because they are not numerical (e.g., male or female, geographical location) or else the numerical categories are few in number (e.g., more than 10, less than 10; 6, 7, or 8 fin rays).When numerical data with a number of different values are available for two continuous variables, the most appropriate methods of analysis are those of correlation and regression.

A variety of kinds of data in biology and many other fields suggest that a relationship exists between two variables. One may wish, for example, to test for the relationship between height and weight, or height and age, or IQ score and academic performance, or milk production of mother and daughter in cattle. A simple way to test for such a relation is to make a scatter diagram, plotting one of the variables on the X axis and the other on the Y axis. If there is a perfect relationship, the points will fall on a straight line. If there is no relationship, the points will be randomly scattered on the graph. If the two variables are imperfectly associated, the points will tend to cluster along some line or curve, but will deviate or scatter from this line to a greater or lesser degree depending on the strength of the relationship (Fig. A.4). Although curvilinear correlation and regression occur, we shall only consider linear forms of association.

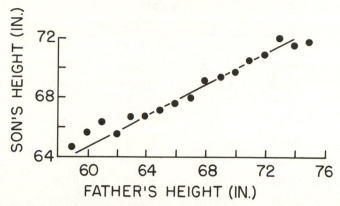

Fig. A.4. The regression of the height of sons on the height of their fathers. 1078 families gave the equation $Y = 0.516X + 33.73$. (Reprinted by permission from *Statistical Methods* by G. W. Snedecor, fourth edition © 1946, by Iowa State University Press, Ames, Iowa.)

The Correlation Coefficient

If the correlation coefficient is to be used, both variables should have approximately normal distributions. Since each of the variables X and Y forms a normal curve, with mean μ and standard deviation σ, the two variables together can be thought of as forming a bivariate normal surface. In the equation for this surface a new parameter, the correlation coefficient, ρ (rho), appears, which is an index of the association between -1 and $+1$. When ρ is zero, there is no association or correlation; when $\rho = +1$, the correlation is perfect; when $\rho = -1$, there is a complete negative correlation. Just as other parameters have been estimated from observations, r is the estimate of the parameter ρ based on sampling. Like other estimates, r will have a sampling distribution, and hence significance tests are needed to find whether an observed r is significantly different from zero. Two equations for estimating the correlation coefficient are

$$r = \frac{\Sigma(X - \bar{X})(Y - \bar{Y})}{\sqrt{\Sigma(X - \bar{X})^2 \, \Sigma(Y - \bar{Y})^2}} = \frac{C}{s_X s_Y}$$

where s_X and s_Y are the standard deviations of X and Y, and C is a new quantity called the **covariance**. The second equation is obtained from the first by dividing both numerator and denominator by $(n-1)$, the number of degrees of freedom. The covariance then equals

$$C = \frac{\Sigma(X - \bar{X})(Y - \bar{Y})}{n-1}$$

which is the sum of cross products of X and Y, $\Sigma(X - \bar{X})(Y - \bar{Y})$, divided

Table A.6   The Correlation Coefficient [a]

Degrees of freedom	Value of P				
	0.10	0.05	0.02	0.01	0.001
1	0.9877	0.99692	0.99951	0.99988	0.9999988
2	0.9000	0.9500	0.9800	0.9900	0.9990
3	0.805	0.878	0.9343	0.9587	0.9911
4	0.729	0.811	0.882	0.9172	0.9741
5	0.669	0.754	0.833	0.875	0.9509
6	0.621	0.707	0.789	0.834	0.9249
7	0.582	0.666	0.750	0.798	0.898
8	0.549	0.632	0.715	0.765	0.872
9	0.521	0.602	0.685	0.735	0.847
10	0.497	0.576	0.658	0.708	0.823
11	0.476	0.553	0.634	0.684	0.801
12	0.457	0.532	0.612	0.661	0.780
13	0.441	0.514	0.592	0.641	0.760
14	0.426	0.497	0.574	0.623	0.742
15	0.412	0.482	0.558	0.606	0.725
16	0.400	0.468	0.543	0.590	0.708
17	0.389	0.456	0.529	0.575	0.693
18	0.378	0.444	0.516	0.561	0.679
19	0.369	0.433	0.503	0.549	0.665
20	0.360	0.423	0.492	0.537	0.652
25	0.323	0.381	0.445	0.487	0.597
30	0.296	0.349	0.409	0.449	0.554
35	0.275	0.325	0.381	0.418	0.519
40	0.257	0.304	0.358	0.393	0.490
45	0.243	0.288	0.338	0.372	0.465
50	0.231	0.273	0.322	0.354	0.443
60	0.211	0.250	0.295	0.325	0.408
70	0.195	0.232	0.274	0.302	0.380
80	0.183	0.217	0.257	0.283	0.357
90	0.173	0.205	0.242	0.267	0.338
100	0.164	0.195	0.230	0.254	0.321

[a] The table gives percentage points for the distribution of the estimated correlation coefficient r, when the true value ρ is zero. Thus when there are 10 degrees of freedom (i.e., in samples of 12), the probability of observing an r greater in *absolute value* than 0.576 (i.e., < -0.576 or $> +0.576$) is 0.05 or 5%.
Table A.6 is taken from Table VII of Fisher and Yates: *Statistical Tables for Biological, Agricultural and Medical Research,* published by Longman Group Ltd., London (previously published by Oliver & Boyd, Edinburgh), and by permission of the authors and publishers.

by the number of degrees of freedom. It should be intuitively clear that if there is little or no association between X and Y, the sum of cross products of X and Y will be very close to zero and so will r.

For very large samples (N at least 500), the estimate and its standard error take the form

$$r \pm \frac{1-r^2}{\sqrt{N}}$$

To test whether $\rho = 0$ when the sample size is small, r can be used to get a value of t for a t test as follows:

$$t = \frac{r\sqrt{N-2}}{\sqrt{1-r^2}}$$

However, tables of significance are available for the correlation coefficient that obviate the need to calculate this t (Table A.6). For r values not equal to zero or for comparisons of r values, a transformation is made for which statistics texts should be consulted.

The correlation coefficient is a measure of the strength of the association between two continuous variables. It is worth noting that a correlation coefficient without an adequate test of significance has little meaning and that interpretation of the meaning of a correlation coefficient requires some caution. Just because the variables X and Y are correlated does not necessarily tell us anything about the cause of this relationship, merely that they are in some way associated. It does not imply that X is a cause of Y or Y of X. Having found a correlation, we must then search further for the cause.

Regression

The regression coefficient also measures the relationship between the variables X and Y but not in the same way as r, which measures the intensity of the association. Instead, the regression of one factor on the other is a measure of how much of a change in one factor may be expected from a unit change in the other. Because of its greater predictive function and quantitative nature, the use of regression analysis is more widespread than the use of the correlation coefficient. Furthermore, regression analysis is not restricted by the requirement that both variables have normal distributions, as is the case for correlation.

The basic equation for linear regression takes the form

$$Y = a + \beta X$$

This is the equation for the true regression line for the regression of Y on X. The slope of the line is determined by β, the regression coefficient, which measures the amount of change in Y for a unit increase in X. The constant a indicates the Y intercept when X equals zero, and hence fixes the position of the regression line. Some assumptions, departure from which is not too serious, are necessary in the use of linear regression: one that the true Y means lie on a straight line when plotted against the various values of X, that the distribution of Y values for any value of X is normal, and that the variances of Y for different values of X must be equal.

The estimates, a and b, of the constant a and the parameter β are made as follows:

$$b = \frac{\Sigma(X-\bar{X})(Y-\bar{Y})}{\Sigma(X-\bar{X})^2} = \frac{C}{s_X^2}$$

Again C is the covariance and the second equation is obtained from the first by dividing both numerator and denominator by $(n-1)$.

The value of a is estimated from the equation

$$a = \bar{y} - b\bar{x}$$

and the fitted regression line is

$$Y = a + bX$$

In this case Y is called the dependent variable and X the independent variable, and the equation predicts values of Y for given values of X. It is possible to reverse the situation, having X as the dependent and Y the independent variable, and then to study the regression of X on Y.

From the equations already given, it can easily be verified that the correlation coefficient has a simple relationship to the two regression coefficients of Y on X and X on Y, which is

$$r = \sqrt{b_{XY}b_{YX}}$$

A point to note is that the sign of the correlation coefficient calculated thus will be the same as the sign of the regression coefficients. From the above equation, it can be seen that r is the geometric mean of the regression coefficients. This relation is valid if the restrictions mentioned above are met, and the estimates, r, b_{XY}, and b_{YX}, all can be calculated from the following information alone: $\Sigma(X-\bar{X})^2$, $\Sigma(Y-\bar{Y})^2$, and $\Sigma(X-\bar{X})(Y-\bar{Y})$.

Furthermore, it is desirable to learn something of the variation about the regression line. If the sample values of Y have a normal distribution for a given X, and all these distributions have the same variance, σ^2, this variance can be estimated from the equation,

$$s^2 = \frac{1}{N-2}\left\{ \Sigma(Y-\bar{Y})^2 - \frac{[\Sigma(X-\bar{X})(Y-\bar{Y})]^2}{\Sigma(X-\bar{X})^2} \right\}$$

From this equation the standard deviation can be obtained and hence some estimate of the variation about the regression line of values of Y.

For a significance test of the regression coefficient when N is greater than 30 to test whether the estimate is significantly different from zero, the estimate with its standard error can be written as

$$b \pm \frac{s}{\sqrt{\Sigma(X-\bar{X})^2}}$$

A comparison of the estimate, b, with some predetermined coefficient, β, can be made by testing the difference, $b-\beta$, with the same standard error, and using the table of normal deviates.

If N is below 30, the t test is used in the form

$$t = \frac{b-\beta}{s/\sqrt{\Sigma(X-\bar{X})^2}}$$

with $N-2$ degrees of freedom (one lost with the estimate of the mean and one for the estimate of b). For other types of tests of significance for regression, statistics texts should be consulted. Our purpose in this section on correlation and regression has been primarily to introduce the concepts involved. Far more sophisticated and complex problems can be handled than have been indicated here.

A.14. The Analysis of Variance

Thus far we have dealt with frequency distributions, measures of central tendency, measures of dispersion, methods of estimating different parameters for a population, the determination of confidence intervals for these estimates, and various tests of the significance of the findings. In addition, we have discussed ways to find whether two sets of data are independent or associated in some way, and also methods to determine whether two populations are essentially the same or differ in some significant manner. This material serves as an introduction to some of the basic statistical methods available. However, for the sake of simplicity we have confined ourselves to the comparison of just two populations with the assumption that these populations, if different at all, differ with respect to only a single factor. In actual practice, especially in biology, it is often desirable to compare three or more different populations, and it turns out to be impractical or impossible to hold all other factors constant except the one under study. For example, if one wished to compare two different methods of teaching reading to children, it would hardly be a fair test to study just two classes exposed to the two different approaches. Many other factors might well enter into the success of the two groups; the competence of the teachers, the age, sex, and IQ of the children, the time of day when reading is taught, and so on. Each of these factors may play a role in the reading performance of the children. Therefore, if a valid test of the two methods of teaching is to be made, the relative importance of all of the possible factors affecting reading success should be assessed. The statistical technique known as the analysis of variance is a method for testing for differences among several different populations at once and for measuring the relative importance of the various factors affecting the trait under study. The concepts used in the analysis of variance have already been introduced; their application in this context permits a more general approach to a wider range of problems. The analysis of variance provides information not only about the effect of each of the factors individually, but also about the effect of possible interactions among them.

The essence of the analysis of variance is a partitioning of the variance in such a way that the variances associated with the different factors can be compared by the variance ratio or F test. If the factor is not an important cause of variation in the trait, its variance should not be significantly greater than that due to sampling error. On the other hand, if it is important, a significant value of F will be found.

To carry this description of the analysis of variance further does not seem warranted, since you will not have immediate occasion to use it. However, it is desirable that you should be aware of one of the most powerful tools in the toolbox of the statistician.

Mark Twain once classified falsehoods into three groups: lies, damn lies, and statistics. In this section we have tried to familiarize you with some of the concepts and methods in statistics that are so important in genetics, especially in the genetics of quantitative characters. It is hoped that you have come to appreciate that, while Twain's complaint may have been justified, the moral deficiencies he cites were those of the statistician and not the statistics.

Additional Reading

Bailey, N. T. J., 1959. *Statistical Methods in Biology*. Wiley, New York.
Bishop, O. N., 1966. *Statistics for Biology*. Houghton Mifflin, Boston.
Bliss, C. A., 1967. *Statistics in Biology*, Vol. I. McGraw-Hill, New York.

Croxton, F. E., 1963. *Elementary Statistics with Applications in Medicine and the Biological Sciences*. Dover, New York.

Dixon, W. J., and F. J. Massey, 1969. *Introduction to Statistical Analysis,* 3rd ed. McGraw-Hill, New York.

Fisher, R. A., 1958. *Statistical Methods for Research Workers,* 13th ed. rev. Hafner, New York.

Goldstein, A., 1964. *Biostatistics*. Macmillan, New York.

Hoel, B. G., 1960. *Elementary Statistics*. Wiley, New York.

Kempthorne, O., T. A. Bancroft, J. W. Gowen, and J. L. Lush, 1954. *Statistics and Mathematics in Biology*. Iowa State University Press, Ames.

Li, C. C., 1959. *Numbers from Experiments*. Boxwood, Pittsburgh.

Li, C. C., 1964. *Introduction to Experimental Statistics*. McGraw-Hill, New York.

Moroney, M. J., 1956. *Facts from Figures*. Penguin, Baltimore.

Mather, K., 1946. *Statistical Analysis in Biology,* 2nd ed. Wiley-Interscience, New York.

Simpson, G. G., A. Roe, and R. C. Lewontin, 1960. *Quantitative Zoology,* rev. ed. Harcourt Brace Jovanovich, New York.

Snedecor, G. W., and W. G. Cochran, 1967. *Statistical Methods,* 6th ed. Iowa State University Press, Ames.

Sokal, R. R., and F. J. Rohlf, 1969. *Biometry*. Freeman, San Francisco.

Weaver, W., 1963. *Lady Luck. The Theory of Probability.* Doubleday, Garden City, N.Y.

Wolf, F. L., 1962. *Elements of Probability and Statistics*. McGraw-Hill, New York.

Glossary

acentric. Lacking a centromere.

acrocentric. Having the centromere very near one end of the chromosome so that the chromosome has one short arm and one long arm.

allele. One of an array of alternative forms of a gene, occupying the same locus in homologous chromosomes. (Originally allelomorph.)

allopatric. Individuals or populations spatially isolated from one another.

allopolyploid. An organism with more than two sets of chromosomes derived from two or more species by hybridization. At meiosis synapsis is primarily between homologous chromosomes of like origin.

allosteric. Refers to an enzyme with two distinct receptor sites. One site binds the substrate to the enzyme and is responsible for the catalytic activity of the enzyme. The other site, the allosteric site, binds reversibly to a molecule of low molecular weight in such a way that the conformation of the enzyme, and hence its activity, are modified.

allotetraploid. An organism diploid for two different genomes, each from a different species.

amorph. An allele that is inactive compared to the wild type.

amphidiploid. Allotetraploid.

analogous. Similar in function but different in structure and origin.

aneuploid. Having a chromosome number that is not an exact multiple of the basic haploid number.

antibody. A specific protein produced by an organism in response to the introduction of an antigen. The antibodies tend to combine with the antigens and neutralize their harmful effects.

785

antigen. A foreign substance, usually a protein, that usually induces antibody formation in a living organism.

antimorph. Allele with action opposite to that of the wild type.

antiserum. A serum containing antibodies.

apomixis. Asexual reproduction in which the outward appearance of sexual reproduction is retained but no fertilization occurs. Usually refers to plants.

ascospore. One of the haploid spores of the ascomycetes (fungi). Normally, eight are formed in an ascus as the result of two meiotic divisions followed by a single mitosis.

ascus (pl. asci). The spore-bearing sac of the ascomycetes.

asexual. Any mode of reproduction not involving fertilization, conjugation, or genetic recombination. Progeny have the same genotype as the parent.

assortative mating. Nonrandom mating in which like tends to pair with like (positive) or unlike types pair (negative).

autogamy. A process of self-fertilization in a single *Paramecium* leading to the fusion of a pair of identical haploid nuclei to form a homozygous diploid nucleus.

autopolyploid. An organism with more than two homologous sets of chromosomes in its somatic cells, derived from a single parent species.

autoradiograph. A photograph showing the location of radioactive substances in cells or tissues, obtained by exposing a photographic emulsion in the dark to radioactive emissions from the preparation and then developing the latent image.

autosome. Chromosomes other than the sex chromosomes, ordinarily found in equal numbers in both males and females.

autotroph. A microorganism able to synthesize its own macromolecules and to obtain energy from simple inorganic compounds such as ammonia and carbon dioxide.

auxotroph. A mutant microorganism requiring supplements to the basic food medium required by the wild type (autotroph or prototroph) for growth.

back mutation. A mutation of a mutant gene causing it to revert to its original state.

backcross. The mating of a hybrid to one of the parental types used to produce the hybrid.

bacteriophage. A bacterial virus; a virus whose host is a bacterial cell.

balanced lethals. Lethal genes so closely linked that crossing over is rare, the genes remain in repulsion, both homozygotes die, and only the heterozygote survives.

balanced polymorphism. Two or more genetically distinct types of individuals coexisting in the same breeding population, actively maintained by selection.

bivalent. A pairing configuration during first meiotic prophase consisting of two synapsed homologous chromosomes. The number of bivalents equals the haploid number of chromosomes.

carcinogenic. Cancer inducing.

Castle-Hardy-Weinberg law. In a large, random mating population in the absence of migration, mutation, and selection, gene frequencies remain constant.

centric. Having a centromere.

centromere. The region of the chromosome with which the spindle fibers become associated during meiosis and mitosis. Essential for orderly chromosome movement during nuclear division. Sometimes diffuse. (Also called kinetochore.)

chiasma (pl. **chiasmata**). A visible change in pairing affecting two nonsister chromatids out of the four chromatids in a tetrad or bivalent in the first meiotic prophase. The point of exchange of partners is the chiasma. Cytological manifestation of crossing over.

chromatids. Half-chromosomes joined by a single centromere resulting from longitudinal duplication of a chromosome, observable during prophase and metaphase, and becoming daughter chromosomes at anaphase.

chromosome. Nucleoprotein bodies in the nucleus, usually constant in number for any given species, and bearing the genes in linear order.

cis configuration. Both mutant alleles of two linked genes or heteroalleles of a cistron are on one homologue and both wild-type alleles are on the other. In the trans configuration, the heterozygote has one wild type and one mutant on each homologue. Cis corresponds to the older term coupling used in linkage studies; trans corresponds to repulsion.

cistron. A section of the DNA molecule that specifies the amino acid sequence for a particular polypeptide chain.

cline. A geographical gradient in genotypic or phenotypic frequencies in a continuous population.

clone. All the individuals descended from a single individual by asexual reproduction.

codominant. Refers to alleles both of which are expressed independently in the heterozygote, as, for example, in the AB blood type in man.

codon. A set of three adjacent nucleotides in a polynucleotide chain that codes for an amino acid at a specific position in a polypeptide chain.

coincidence. The ratio of observed double crossovers to expected double crossovers calculated on the basis of independent occurrence. This ratio is used as a measure of interference in crossing over.

colinearity. The concept that the order of the amino acids along a polypeptide chain corresponds to the order of the codons in the cistron specifying that polypeptide.

complementation test. A genetic test to determine whether two mutant genes occur in the same functional unit or cistron.

conjugation. May refer to the pairing or union of chromosomes, nuclei, cells, or individuals.

constitutive mutant. A mutant that leads to enzyme production in fixed amounts independent of the presence or absence of the substrate, in contrast to inducible and repressible enzymes.

covalent bond. The chemical bond formed by shared electrons between the atoms in a molecule.

crossing over. The reciprocal exchange of corresponding segments between two nonsister chromatids of homologous chromosomes.

cytokinesis. Division of the cytoplasm, as distinct from nuclear division or karyokinesis.

cytoplasm. The protoplasm of the cell except that found within the nucleus, which is called nucleoplasm.

dauermodification. An induced phenotypic change that persists through several to many cell generations in the absence of the inducing stimulus, but eventually diminishes and disappears.

deficiency (deletion). The absence of a segment from a chromosome, which may vary in size from a single nucleotide to a number of genes.

deme. A local interbreeding population.

dicentric. A chromosome or chromatid with two centromeres.

dioecious. With male and female gametes produced by different unisexual individuals.

diploid. Having two sets of chromosomes ($2n$). Somatic cells of higher plants and animals are ordinarily diploid in contrast to the haploid gametes ($1n$).

dominant. Used with reference to both genes and characters. An allele is dominant if its phenotypic effect is the same in both the heterozygous and the homozygous condition. (See recessive.)

duplication. The occurrence of a chromosome segment more than once in the same chromosome or haploid genome.

dysgenic. Tending to be harmful to the hereditary qualities of a species.

ecotype. An ecological race with genotypes adapted to a particular restricted habitat as the result of natural selection.

effective population size. The number of individuals in a local breeding population that actually contribute genes to the next generation. If the population deviates from an "ideal" population, for example, because of unequal numbers of males and females or periodic fluctuations in size, the actual breeding size is converted to a number equivalent to the number of individuals in the "ideal" population in order to estimate random genetic drift.

enzyme. Protein catalyst in living organisms, typically formed from a protein part (apoenzyme) conferring specificity and a nonprotein part (co-enzyme) necessary for activity.

enzyme induction. The synthesis of an enzyme in response to the addition of the metabolite (the inducer) on which it acts. Thus an inducible enzyme is synthesized only in the presence of its substrate.

episome. A genetic element that may exist in an autonomous state in the cell, replicating independently of the chromosomes, or in an integrated state associated with a chromosome and replicating in synchrony with it.

epistasis. The suppression of the expression of a gene or genes by other genes not allelic to the genes suppressed. Similar to dominance but involving the interaction of nonallelic genes. Sometimes used to refer to all nonallelic interactions.

erythroblasts. Cells giving rise to red blood cells or erythrocytes.

ethology. The study of animal behavior.

eukaryotic. Refers to those plants and animals having true nuclei with nuclear envelopes and chromosomes and with nuclear divisions by mitosis or meiosis.

euploid. An exact multiple of the haploid chromosome number.

expressivity. The degree of phenotypic expression of a gene or genotype.

feedback inhibition. Inhibition of an enzyme in a biosynthetic pathway by the end product of that pathway.

fertilization. The fusion of gametes to form a zygote.

fitness. The number of offspring left by an individual of a given genotype as compared with the average of the population of which it is a member or compared to individuals of different genotypes.

fluctuation test. A statistical test used to prove that selected variants such as phage-resistant bacteria arose as spontaneous mutants rather than as the result of treatment with the selective agent.

gamete. A haploid reproductive cell.

gametophyte. The gamete-forming haploid generation in higher plants.

gene conversion. The phenomenon in which the products of meiosis in a heterozygote are not recovered in a 1:1 ratio. For example, if the gametes produced by Aa are $1A$ and $3a$ rather than $2A$ and $2a$, it appears that an A has been "converted" to a.

gene flow. The spread of genes from one breeding population to others as the result of migration.

gene frequency. The ratio of the number of alleles of one type to the total of all alleles of all types at a given gene locus in a breeding population.

gene pool. The sum total of the genes in a given breeding population at a particular time.

genetic drift. Changes in gene frequency from generation to generation in small breeding populations due to chance fluctuations.

genome. The chromosome complement of a gamete or a complete haploid set of genes or chromosomes.

genotype. The entire genetic constitution of an organism.

gynandroid. In the wasp *Habrobracon,* haploid individuals sometimes arise from unfertilized binucleate eggs in which the two nuclei contain different sex alleles. Female characteristics appear along the zone of contact between the two types of haploid tissues, while the remainder of the individual is male.

gynandromorph. An individual that is a sexual mosaic or chimera with some portions of the body typically male and others typically female.

haplodiploidy. A system of sex determination in animals such as bees and wasps (Hymenoptera) in which males develop from unfertilized eggs and are haploid, and females develop from fertilized eggs and are diploid.

haploid. Having only a single set of chromosomes or a single genome.

hemizygous. A gene present only once in the genotype. Used most often with reference to sex-linked genes in the heterogametic sex, but may also be used with respect to haploids, or diploids lacking a chromosome or a chromosome segment.

heritability. The proportion of the total phenotypic variance that is genetic in origin. Usually defined as the ratio of the additive genetic variance to the phenotypic variance.

hermaphrodite. An individual with functional ovaries and testes.

het. A partially heterozygous bacteriophage.

heterocatalytic. Any non-self-replicating enzymatic reaction. In contrast to autocatalytic reactions.

heterochromatin. Chromosome regions that stain differentially compared to the rest of the genome, the euchromatin. Typically, heterochromatic segments are maximally stained during late telophase, interphase, and early prophase.

heteroduplex. A nucleic acid molecule consisting of polynucleotide strands derived from two different parental molecules.

heterogametic. Producing unlike gametes, especially with regard to the sex chromosomes. Where the male is XY, he is heterogametic.

heterogenote. Partially diploid bacteria carrying a genome segment (the exogenote) from a donor cell in addition to their own genome (the endogenote) and heterozygous for pairs of marker genes in the segments concerned.

heterokaryon. A fungal cell that contains genetically different unfused nuclei in a common cytoplasm.

heteromorphic. Having more than one form. Used with reference to homologous chromosomes that differ morphologically.

heteropycnotic. Of chromosomes or chromosome regions that are out of phase with the rest of the genome in their coiling cycle and staining properties. A characteristic of heterochromatin.

heterosis. Hybrid vigor. The superiority of heterozygous genotypes with respect to one or more traits in comparison with the corresponding homozygotes.

heterothallic. Sexual fusion of genetically different nuclei from different thalli in fungi.

heterozygous. An individual having two different alleles at one or more gene loci.

holandric. Appearing only in males. Used, for example, for traits governed by Y-linked genes (XY male sex determination) and passed directly from father to all sons.

hologynic. Appearing only in females. Used, for example, for traits governed by W-linked genes (ZW female sex determination) and passed directly from mother to all daughters.

homogametic. Producing gametes of only one type, especially with regard to the sex chromosomes. An XX female is the homogametic sex.

homologous. 1. Chromosomes in which the same gene loci occur in the same sequences. 2. Similarity of structure due to similar hereditary and developmental origin.

homothallic. Sexual fusion of genetically similar nuclei normally derived from the same thallus in fungi.

homozygous. Having identical alleles at corresponding loci on homologous chromosomes so that the organism breeds true with respect to these loci.

hybrid. Heterozygous at one or more gene loci.

hybridization. The crossing of two genetically different individuals.

hypermorph. A mutant allele whose effect is similar to, but greater than, that of the wild-type allele.

hypomorph. A mutant allele whose effect is similar to, but less than, that of the wild-type allele.

in vitro. Refers to biological processes occurring in isolation from the whole organism. (Literally, in glass.)

in vivo. Refers to biological processes occurring within the living organism.

inbred. Resulting from matings between relatives.

independent assortment. Segregation of one factor pair occurring independently of the segregation of other factor pairs located on different chromosomes.

insertion. Transposition of a chromosome fragment to a nonterminal location within a chromosome. Three breaks are required.

intercalary. Refers to internal chromosomal changes such as deletions or insertions in contrast to terminal changes.

interference. The effect of one crossover in either reducing (positive interference) or increasing (negative interference) the probability of another crossover occurring in its vicinity.

intersex. An individual from a dioecious species with sexual characteristics intermediate between those of males and females. Different from both hermaphrodites and gynandromorphs.

introgressive hybridization. The addition of genes from one species to the gene pool of another species through hybridization and backcrossing.

inversion. Rotation of a chromosome segment through 180° so that the linear order of the genes in the segment is reversed relative to the rest of the chromosome.

iso-alleles. Alleles so similar in their effects that special techniques are needed to distinguish between them.

isolating mechanism. Any genetically controlled factor that prevents or reduces interbreeding and gene flow between different populations.

kappa. A DNA-containing, self-replicating, complex particle found as a symbiont in the cytoplasm of some strains of *Paramecium aurelia*. Maintenance of kappa requires presence of nuclear gene *K* in killer strains.

karyotype. The somatic chromosomal complement of an individual or a related group of individuals.

Lamarckism. Usually, the erroneous theory that evolution may occur through the inheritance of acquired characteristics.

lambda. A temperate phage of *Escherichia coli*.

lethal. A gene or genotype which, when expressed, is fatal to its bearer.

linkage. The association of genes in inheritance due to their being on the same chromosome. Genes borne on homologous chromosomes belong to the same linkage group.

locus (pl. **loci**). The position of a gene on a chromosome.

lysis. The destruction of a cell by the rupture of its cell membrane.

lysogeny. The integration of the genome of a temperate bacteriophage in the form of a prophage into the genome of its bacterial host. Occasionally these lysogenic bacteria will undergo lysis and release mature phage particles.

mean. The sum of a group of quantities divided by the number in the group. The arithmetic average.

median. The middle value in a group of numbers arranged according to size. Thus, half of the numbers will be less than the median and half will be greater.

meiosis. The reduction divisions during which the chromosome number is reduced from diploid to haploid. Two nuclear divisions during which the chromosomes divide only once.

meristic. Of variation in traits that can be counted, such as fin rays in fish, bristles in *Drosophila*, or caudal vertebrae in mice.

merogenote. A fragment of a bacterial chromosome.

merozygote. A partially diploid bacterial cell with the complete genome of a recipient cell and a partial genome, the merogenote, from a donor cell. May be hetero- or homogenotic.

metabolism. The sum total of the chemical processes in living cells by which energy is provided, new materials assimilated or synthesized, and wastes removed.

metacentric. Of chromosomes with the centromere in a medial position.

missense mutation. A mutation that converts a codon coding for one amino acid into a codon coding for another amino acid.

mitochondrion. A self-replicating, DNA-containing organelle in the cytoplasm of all eukaryotic cells. Mitochondria are the principle energy source of the cell and the site of oxidative phosphorylation leading to ATP formation.

mitosis. The process by which the nucleus is divided into two daughter nuclei, each with a complement of genes and chromosomes identical to that of the original nucleus.

mode. The most frequent value in a statistical frequency distribution.

modifying factor. A gene that affects the expression of genes at other loci. Often without other known effects.

monoecious. Producing male and female gametes in a single individual.

monomer. A small molecule serving as the basic subunit from which, by repetition of a single reaction, a polymer is made.

monosomic. Having one chromosome missing from an otherwise diploid chromosome complement ($2n - 1$).

morgan. A genetic map unit. One morgan equals 1% crossing over. (Sometimes also called centimorgan.)

mu. A particle in the cytoplasm of certain strains of *Paramecium aurelia* (mate-killers) that causes the death of sensitive paramecia only as the result of direct contact during conjugation.

multigenic. Influenced by genes at a number of loci.

multiple alleles. A series of more than two alternative forms of a gene at a single locus.

multiple factors. Arrays of genes at a number of different loci assumed to have similar or complementary cumulative effects on a quantitative trait, each locus having only a small effect.

mutagenic. Capable of inducing mutations.

mutation. In the broad sense, any sudden heritable change in the hereditary material not due to segregation or to genetic recombination. Thus, mutations could be genic, chromosomal, or extrachromosomal. In the narrow sense, gene mutations only.

mutation pressure. The continued recurrent production of an allele by mutation, tending to increase its frequency in a population.

mutation rate. The frequency with which a particular mutation occurs per locus per generation. Mutation rates are sometimes estimated for entire chromosomes or genomes.

mutator gene. A gene that influences the mutation rate of genes at other loci.

muton. The smallest unit of DNA capable of mutation (a single nucleotide.)

neomorph. A mutant allele producing an effect qualitatively different from that of the wild-type allele.

nondisjunction. The failure of a homologous pair of chromosomes to segregate at metaphase so that one daughter nucleus receives both chromosomes of the pair and the other daughter nucleus receives neither. Nondisjunction may be mitotic or meiotic.

nonsense mutation. A mutation that converts a codon coding for a particular amino acid into a chain-terminating codon.

normal curve. A symmetrical, bell-shaped curve often approximated when frequency distributions are plotted from observations on biological materials.

nucleoside. A purine or pyrimidine base attached to a ribose or a deoxyribose sugar.

nucleotide. A nucleoside-phosphoric acid complex. Composed of a purine or pyrimidine base, a sugar (ribose or deoxyribose), and a phosphate group (PO_4).

operon. A unit consisting of a number of structural genes or cistrons that control a series of biochemical reactions in a biosynthetic pathway, an operator gene controlling the activity of the structural genes, and a promoter region lying between the operator and the first structural gene.

overdominance. Originally defined with reference to a single locus where $A^1A^1 < A^1A^2 > A^2A^2$. Now sometimes used to refer to superiority of any heterozygote over its corresponding homozygotes. This superiority may be due to pseudooverdominance, a property of blocks of genes rather than any single locus.

paracentric. An inversion that does not include the centromere, but is entirely within one arm of the chromosome.

parameter. In statistics, the true value of a quantity for an entire population, as distinguished from estimates based on samples taken from the population.

parthenogenesis. The development of a new individual from a germ cell (usually an egg) without fertilization. May be either haploid or diploid.

penetrance. The frequency in percent with which a gene or gene combination is expressed phenotypically among individuals all having the same genotype.

pericentric. An inversion that includes the centromere; hence both chromosome arms are involved.

phenocopy. Environmentally induced, nonhereditary phenotypic imitations of the effects of mutant genes.

phenotype. The sum total of the observable or measurable characteristics of an organism produced by its genotype interacting with the environment.

phylogeny. The evolutionary history of a taxonomic group.

pleiotropy. The apparently unrelated multiple phenotypic effects of a single gene.

polarity. In some cases a mutant in an operon results not only in loss of the corresponding enzymatic activity at its own locus but also reduces the rate of synthesis of enzymes coded by structural genes distal to it from the operator gene. Such genes are called polarity mutants, and indicate the polarity of the operon, that is, that the DNA code is read in one direction from a fixed starting point.

polaron. Small chromosome region within which gene conversion and recombination occur with an increasing probability in a fixed direction on the genetic map.

polycistronic. Of a large messenger RNA molecule produced by adjacent cistrons in the same operon that codes for more than one polypeptide chain.

polygene. A gene of small individual effect on the phenotype that in conjunction with several or many similar genes controls a quantitative character. Synonym for multiple factor. Also used in the restricted sense of Mather's theory of quantitative inheritance in contrast to genes of major effect or oligogenes.

polymer. A large molecule composed of a covalently bonded array of repeating subunits or monomers.

polymorphism. The existence of two or more discontinuous variants within a single breeding population. Usually used to refer to genetic polymorphism, which may be balanced or transient. To qualify as polymorphism, frequencies of the morphs must be greater than could be accounted for by recurrent mutation.

polyploid. Having more than two haploid sets of chromosomes.

polysome. A linear array of a group of ribosomes held together by a molecule of messenger RNA. A contraction of polyribosome. The active site of cellular protein synthesis.

polytypic. A species composed of two or more geographical races or subspecies.

position effect. A change in the phenotypic effect of a gene due to a change in its position with respect to other genes in the genome. May be stable (cis-trans type) or variegated, stemming from chromosomal rearrangements.

prototroph. A nutritionally independent, wild-type strain of a microorganism.

pseudoalleles. Very closely linked genes, usually affecting the same trait, that show a mutant phenotype in trans heterozygotes (thus resembling alleles) but the wild-type phenotype in cis heterozygotes.

race. A subspecies or a geographical subdivision of a species. Partially isolated populations of the same species that differ in the frequencies of certain of the genes in their gene pools.

random mating. Any individual of one sex has an equal probability of mating with any individual of the opposite sex.

recessive. Used with reference to both genes and characters. An allele is recessive if it is expressed only in the homozygous condition. Thus a recessive allele is not expressed in the heterozygote. (See dominant.)

reciprocal cross. A second cross similar to the first but with the sexes of the parents interchanged. A ♀ × B ♂ and B ♀ × A ♂ are reciprocal crosses.

recon. The smallest unit of DNA (a single nucleotide) capable of recombination.

rem. Roentgen equivalent man. The dosage of any ionizing radiation that will cause the same amount of biological damage to human tissue as one roentgen of X-rays.

rep. Roentgen equivalent physical. The amount of ionizing radiation that will result in the absorption in tissue of 93 ergs per gram.

replicon. A unit of replication of the hereditary material.

repression. The inhibition of the synthesis of all the enzymes in a metabolic pathway by the end product of that pathway. In other words, the metabolite blocks transcription of the operon responsible for the enzymes controlling the steps in its own synthesis.

reproductive isolation. Genetically controlled barriers to interbreeding between members of different breeding populations.

ribosome. A cellular particle, 10–20 nm in diameter, that is composed of two unequal subunits, each made up of roughly equal parts of ribosomal RNA and protein. Ribosomes bound together by messenger RNA to form polysomes are the site of protein synthesis.

roentgen (r). The unit of measurement of dosage for ionizing radiation. Equal to the amount of radiation that in air at STP will produce 2.1×10^9 ion pairs per cubic centimeter or in tissue approximately two ionizations per cubic micron. A gram of tissue exposed to 1 r of gamma rays absorbs about 93 ergs.

segmental allopolyploid. An allopolyploid in which some chromosome segments from the parent species are homologous and others are not.

segregation. The separation of maternal from paternal chromosomes at meiosis and hence the basis for Mendel's first law.

sex chromosomes. Homologous chromosomes that differ in the heterogametic sex and exert genetic control over sex determination.

sex reversal. A change in the sexual character of an individual from male to female or vice versa.

sex-duction. The transfer of a group of bacterial genes from one bacterium to another when a segment of the bacterial chromosome becomes attached to the sex factor F and is transferred with F during conjugation.

sex-influenced. Refers to traits whose expression is modified by the sex of the individual manifesting the trait.

sex-limited. Refers to traits expressed in one sex but not the other.

sex-linked. Refers to the mode of inheritance of genes located on the sex chromosomes.

sexual. A mode of reproduction normally involving the fusion of gametes and genetic recombination.

sexual isolation. Reproductive isolation due to a tendency toward homogamic mating.

sigma. A virus that produces CO_2 sensitivity in *Drosophila melanogaster*.

somatic. Refers to the vegetative or body cells and tissues, as contrasted with the germinal cells and tissues giving rise to the germ cells or gametes.

sporophyte. The spore-forming diploid generation in higher plants.

standard deviation. The square root of the sum of the squared deviations from the mean divided by one less than the number of observations. A measure of the variability of a population of individuals.

standard error. The standard deviation divided by the square root of the number of observations. A measure of the variation of a population of means.

sympatric. Coexisting in the same area, with the implication that interbreeding is at least possible.

synapsis. The pairing of homologous chromosomes of maternal and paternal origin during the first meiotic prophase. Also observed occasionally in somatic cells, e.g., salivary gland chromosomes in *Drosophila*.

tautomerism. The existence of two reversible isomeric forms of a molecule with somewhat different chemical properties.

template. A macromolecular mold for the synthesis of another macromolecule. The unique sequence of subunits in one molecule specifies a unique sequence in the other.

test cross. Cross of a heterozygote to the corresponding homozygous recessive type. Especially useful in the determination of linkage.

tetrad. The quartet of chromatids in the bivalent during the first meiotic division.

trans configuration. See cis configuration.

transcription. The formation of messenger RNA on a DNA template.

transduction. Transfer of bacterial genetic material from one bacterium to another via a bacteriophage.

transformation. Transfer of genetic information in bacteria when DNA from one strain is incorporated into the genome of another strain.

transgressive. Appearance in a segregating generation of genotypes falling outside the limits of variation in the phenotypes observed in the parents or the F_1.

transient polymorphism. Temporary polymorphism observed while one adaptive type is replacing another.

transition. A mutation caused by the substitution of a purine for a purine or a pyrimidine for a pyrimidine.

translation. The conversion of the DNA-RNA code into the amino acid language during protein synthesis. The formation of a polypeptide directed by a messenger RNA molecule.

translocation. Change in position of a chromosome segment to another part of the same chromosome or to a different chromosome. Reciprocal translocation is the exchange of segments between two chromosomes.

transversion. A mutation caused by the substitution of a purine for a pyrimidine or vice versa.

trisomic. An organism, otherwise diploid, that has three chromosomes of one type.

variance. The mean squared deviation from the mean. The square of the standard deviation.

wild type. The customary phenotype. Also the most frequent allele in wild populations.

zygote. The diploid cell produced at fertilization by the union of the haploid gametes. Also the individual derived from this cell.

Answers
to Selected
Questions

Chapter 2

2.1. d.1:2:1
2.2. ¼
2.3. Autosomal dominant
2.4. ½
2.5. Black is dominant to brown. Female A is heterozygous, while female B is homozygous for black.
2.6. Cross red (RR) × white (rr).
2.7. 50% short-tailed; 50% long-tailed
2.8. Yellow is a dominant allele of black and is lethal when homozygous.
2.9. Nine different phenotypes in the F_2, with the ratio 4:2:2:2:2:1:1:1:1
2.10. 25% heterozygous at both loci; 25% homozygous at both loci
2.11. $^{27}/_{64}$
2.12. $^{27}/_{64}$
2.13. $^{9}/_{64}$
2.14. The father was probably a plain red bird homozygous for both the recessive plain allele and the dominant red allele.

2.15.   *RrBb* × *rrbb*
2.16.   *Ffpp* × *FfPp*
2.17.   ¹⁄₁₆
2.18.   Parents' genotypes: *ee Vg vg* × *Ee vg vg*
        Progeny: ¼ wild type *Ee Vg vg*
                 ¼ vestigial *Ee vg vg*
                 ¼ ebony *ee Vg vg*
                 ¼ ebony vestigial *ee vg vg*
2.19.   a. Phenotypic ratio: 3 round tall:3 round dwarf:1 wrinkled tall:1 wrinkled dwarf
        b. Genotypic ratio: 2 *WwDd*:1 *WWDd*:2 *Wwdd*:1 *WWdd*:1 *wwDd*:1 *wwdd*
2.20.   Cross him to cows known to carry these deleterious recessives.
2.21.   *a/a C/c R/R*
2.22.   50%

Chapter 3

3.2.   a. $2^3 = 8$
        b. $2^2 = 4$
        c. $2^5 = 32$
3.3.   ¹⁄₁₆
3.4.   a. 20
        b. 40
3.5.   43
3.6.   C, D, E, G
3.7.   One chromosome pair lacks marker genes.

Chapter 4

4.3.   The wing color difference controlled by this locus is a sex-limited trait, expressed only in females.
4.4.   A sex-influenced trait expressed in heterozygous males but not heterozygous females.
4.5.   The father was agouti, with genotype *CCAa*.
4.6.   The parents were *CcAa* × *ccAa*.
4.7.   The parents were *ccAa* × *Ccaa*.
4.8.   ⅝
4.9.   ¼ purple-centered; ¾ yellow-centered
4.10.   *Rrpp* × *RrPp*
4.11.   a. ¼ colored; ¾ white
        b. 9 colored:7 white
4.12.   3 colored:13 white
4.13.   Parents: *Hw Hw su su* × *hw hw Su Su*
              F₁: *Hw hw Su su*
              F₂: 9 Hairy wing:7 wild type
4.14.   The albino parent must have been homozygous for the recessive spotting gene and the black parent heterozygous at this locus, that is, *c c a a sp sp* × *C C a a Sp sp*.

Chapter 5

5.1. 10
5.2. 2
5.3. Joe
5.4. 100%
5.5. 100% if parents are AA and A^BA^B; 50% if parents are Aa and A^BA^B or AA and A^Ba; 25% if parents are Aa and A^Ba.
5.6. ¼
5.7. Aa; A^Ba; aa
5.8. ¼
5.9. $\frac{1}{16}$
5.10. ½
5.11. 18

Chapter 6

6.1. $(\frac{1}{2})^6$
6.2. $\frac{1}{6}$
6.3. ¼
6.4. 3%
6.5. $(0.13)^2$
6.6. a. ¼
b. ¼
6.7. $\frac{1}{8}$
6.8. 7
6.9. a. $(\frac{1}{8})\ (\frac{3}{8})\ (\frac{1}{8}) = \frac{3}{8^3}$
b. $6\ (3/8^3)$
6.10. $\frac{1}{7}$
6.11. $\frac{3}{8}$
6.12. $\frac{3}{16}$
6.13. $\frac{1}{8}$ if she and her husband are fertile
6.14. ¼
6.15. $\frac{13}{16}$
6.16. $\frac{1}{32}$
6.17. a. $\frac{1}{8}$
b. $\frac{3}{8}$
6.18. $\frac{1}{16}$
6.19. $\chi^2 = 8.64$; $p_{1\text{df}} < 0.01$
6.20. $\chi^2 = 2.54$; $p_{1\text{df}} > 0.05$
6.21. $\chi^2 = 3.61$; $p_{2\text{df}} > 0.05$
6.22. $\chi^2 = 2.66$; $p_{3\text{df}} > 0.05$

Chapter 7

7.1. a. Males
b. Females
7.2. a. Females
b. Males
7.3. bb ♂♂ × BW ♀♀

7.4.   a. Tortise-shell females and black males

b. 50% of the sons of a tortise-shell female will be black, 50% will be yellow.

7.5.   ⅛

7.6.   Sons: ¼ normal (*AaHY*); ¼ hemophilic (*AahY*); ¼ albino (aaHY); ¼ albino hemophilic (*aahY*)

Daughters: ½ normal (¼ *AaHH*; ¼ *AaHh*); ½ albino (¼ *aaHH*; ¼ *aaHh*)

7.7.   Green males and cinnamon females

7.8.   A sex-linked recessive lethal is responsible.

Chapter 8

8.2.   23

8.3.   Equal to the haploid number of chromosomes

8.4.   43.7%

8.5.   None

8.6.   2.5%

8.7.   All Beaded

8.8.   *B A C*

8.9.   Two

8.10.   12%

8.11.   a. 20% crossing over

b. Tall pear and dwarf sphere

8.12.   20%

8.13.   Linkage with 32% crossing over

8.14.   White 1.5; cut 20.0

8.15.
bm_1 *pr* v_2

0   22.3   65.7

More complete mapping shows bm_1 at 7, *pr* at 32, and v_2 at 73.

8.16.   15.5

Chapter 9

9.11.   Aves, the birds

9.13.   Attached X chromosomes in the female

9.14.   Hexaploidy; in fact, allohexaploidy

9.15.   15

9.16.   a. 60

b. 30

9.18.   52

9.19. $2n = 26$; 13 small pairs of chromosomes

Chapter 10

10.6.   a. 34 cm

b. 10^8 turns of the double helix

10.8.   a. ... TCGAATCCGCTAG ...

b. ... UCGAAUCCGCUAG ...

Chapter 11

11.6. a. 160,000 base pairs per minute; 2667 base pairs per second
b. 16,000 revolutions per minute; 267 revolutions per second
11.7. 18% uracil; 18% adenine; 32% guanine; 32% cytosine

Chapter 12

12.9. a. 4
b. 16
c. 64
12.10. 20
12.13. The deletion should be 81 nucleotides long.

Chapter 13

13.7. The number of mutations expected would be the same.
13.8. The number of terminal deletions expected would be the same.
13.9. The greatest number of two-hit rearrangements would be expected in 800 cells exposed to 100 r in just 5 min.
13.10. 2×10^{-15} cells
13.13. a. 2%
b. 20%
13.15. Methionine
13.20. A 2:1 ratio of transversions to transitions

Chapter 16

16.12. Three males to one female

Chapter 17

17.7. All dextral
17.8. All sinistral
17.12. a. Pale, green, or variegated
b. All pale

Chapter 18

18.1. 18% heterozygotes; the Castle-Hardy-Weinberg equilibrium and its underlying assumptions are assumed.
18.2. a. 0.8
b. 32% heterozygotes
c. Take a sample of black moths from the population and cross them to mottled white recessive moths to determine the proportions of homozygous and heterozygous black moths.
18.3. 0.36% color-blind women; 11.28% of the women will be heterozygous carriers of the gene.

18.4.   1% albino mice
18.5.   36%
18.7.   9.75%
18.8.   No
18.9.   92.16% type O, 7.84% type A
18.12.   1%
18.13.   5
18.14.   392 per 10,000 births
18.15. $u = 4/10,000$ gametes
18.16. $W \cong 0.98$
18.17.   Zero, barring the incidence of new mutations
18.18.   0.0000487 as compared to 0.0000490
18.19. $\hat{p} = 0.32$; $\hat{q} = 0.68$
18.20. $m = 0.01$
18.21.   Three loci

Chapter 19

19.7.   Four pairs of genes
19.8.   The range of variation will decrease as the number of gene pairs increases.
19.9.   Five gene pairs
19.13.   C.V. egg production = 13.3%. C.V. body weight = 8.9%. Egg production is more variable.
19.16.   Height: mean = 69.8 in., variance = 3.73, standard deviation = 1.93
    Length of forearm: mean = 18.85 in., variance = 0.56, standard devia-
            tion = 0.75
    Weight: mean = 156.5 lb, variance = 200.28, standard deviation = 14.15
19.17.   Height: C.V. = 2.8%
    Forearm: C.V. = 4.0%
    Weight: C.V. = 9.0%
19.18.   Height: standard error of mean = 0.61
    Forearm: standard error of mean = 0.24
    Weight: standard error of mean = 4.48
19.19.   Height-forearm length: $r = 0.75$; $b = 0.29$
    Height-weight: $r = 0.70$, $b = 5.15$
19.21. $t_{20df} = 2.602$; $p < 0.02$. Therefore these two samples of students differ significantly in height at the 2% level.
19.22.   From the fourfold table, $\chi^2_{1df} = 56.26$; $p < 0.01$. Therefore inoculation does help to protect against typhoid fever.

Chapter 20

20.2.   Self-fertilization
20.16.   87 cm
20.17.   Selection differential is 5 g; heritability is 0.
20.18.   Heritability is 0.5.

Chapter 23

23.10.   36% monozygotic

Chapter 24

24.6.   The 1970 death rate was 54.6% that of 1900. The life span increased. It is not possible to calculate life span.

Chapter 25

25.5.   a. 0.25% color-blind females
        b. 0.17% color-blind females
        c. no
        d. If they are allelic, mothers with both protan and deutan sons should be color blind.

Chapter 29

29.4.   1 in 8
29.5.   1 in 32
29.6.   80% penetrance
29.7.   a. 32%
        b. 10.24%
        c. 2.56%
29.8   1 in 4

Index

Italicized page numbers indicate illustrations.